Modern Systems Research
for the Behavioral Scientist

Modern Systems Research for the Behavioral Scientist

A SOURCEBOOK

edited by
WALTER BUCKLEY
UNIVERSITY OF CALIFORNIA, SANTA BARBARA

foreword by
ANATOL RAPOPORT
UNIVERSITY OF MICHIGAN

ALDINE PUBLISHING COMPANY / *Chicago*

First published 1968 by

ALDINE Publishing Company

320 West Adams Street

Chicago, Illinois 60606

Library of Congress Catalog Card Number 68–19888

Designed by Bernard Schleifer

Printed in the United States of America

For Mark, Helen, and Cicely

Contents

PART V. *Cybernetics: Purpose, Self-Regulation, and Self-Direction*

A. Cybernetics and Purpose

B. Homeostasis and Evolution

PART VI. *Self-Regulation and Self-Direction in Psychological Systems*

Preface

ALTHOUGH IT IS HOPED that this sourcebook will be of interest to those of any discipline concerned with new developments in science of the last quarter century, it is addressed principally to the student of human behavior as that study is approached from the social side. This is so partly because that is the editor's main competence and interest, but primarily because the study of human behavior is the general area of science that has least responded to the exciting challenge of the modern systems outlook. Yet it is precisely this general area that stands to gain the most from insights into the workings of the more complex types of systems.

The particular selections that follow (and their organization) have evolved over a three-year period as a by-product of the editor's research into the potential of the newer systems approach as the basis for a more adequate model of society to replace the overworked equilibrium and organismic models in vogue for some time. (*Sociology and Modern Systems Theory*, Englewood Cliffs, N.J.: Prentice-Hall, Inc., 1967.) Materials in modern systems research (taken here to refer primarily to fields more officially known as general systems research, cybernetics, and information or communication theory) were found to be widely scattered and often not easily accessible, especially the earlier pioneering statements of basic principles. Hence, the publication of this collection may serve to relieve the social scientist of one previously valid excuse for not looking more closely.

My organization of selections can aim only at the moderate level of success of most sets of readings, particularly in view of the extensive range of substantive foci and theoretical perspectives of the burgeoning materials in this area. An account of the intentions underlying the overall organization of Parts and Sections is given in the General Introduction. As for general criteria of inclusion, I have attempted to present not only a fair selection of recent systems research in behavioral science, but also to provide an extensive selection of important statements of general principles, including several already considered classics though they are only fifteen or twenty years old. Hence, this sourcebook may function in part as a principles text, exposing the initiate to original pioneering statements as well as later work inspired by them, and—I hope—alerting the sizeable number of underexposed scholars who are over-familiar with the few terms such as *feedback*, *boundary*, *input*, and *output*, that there are much greater depths to plumb than meet the eye in semipopular accounts of cybernetics.

On the negative side, there is little or no representation of fields often closely identified with systems research, such as operations research, game theory, computerology, or automatic control systems engineering. All of these require a greater degree of mathematical treatment for meaningful exposition than the average social scientist is trained for or of a sentiment to tolerate. Since it is particularly these scholars whom I wish to reach, papers on any subject that involve extensive mathematics have been avoided wherever possible. Other areas of study have also been unavoidably neglected. For example, the many empirical studies in areas of biology demonstrating beyond doubt the fruitfulness of modern systems analysis—including research on animal behavior highly pertinent to our focus—could not be sampled for lack of space. Such deficiencies are remedied in a

small way by inclusion of wide-ranging Selected
References at the end of the book.

Over a period of years one easily forgets many
of the sources of intellectual stimulation and aid
that have contributed nevertheless to the finished
product. Such is the case with many discussions
and debates enjoyed with colleagues and students,
to whom I acknowledge indebtedness as an
anonymous group. Especially welcomed encouragement, as well as helpful suggestions in the organization and selection of material, is gratefully
acknowledged from Ludwig von Bertalanffy,
Kenneth Boulding, Anatol Rapoport, and Donald
M. MacKay. A number of other contributors to
the book also gave aid in selecting or editing their
material, including Magoroh Maruyama, Richard
Held, and Russell Ackoff. All of the authors are
due my thanks for permission to include their
work, as are the many publishers who made
materials easily and graciously available. (In only
a very few cases were deletions or substitutions
necessary due to publisher permission refusals or
stipulation of exorbitant fees.) Finally, I wish to
thank the several research assistants, particularly
Gloria Nakagawa and Robert Cates, who con-
scientiously aided in the library search, manu-
script preparation, and permission requests, as
well as the Research Committee of the University
of California, Santa Barbara, for the support that
made such assistance possible.

Foreword

THE PARADIGM OF A scientific assertion is "If so . . . then so." Given a state of affairs, certain deductions can be drawn. These deductions may involve a state of affairs concomitant with a given one, or one that preceded the given one, or one that will follow it. A proposition of geometry is an example of a simultaneous consequence: "If two straight lines intersect, the vertical angles are equal." An example of a past event inferred from the present is: "This female is pregnant; therefore she must have copulated with a male." An example of a future event predicted from present circumstances is: "In view of the present positions of the sun and the moon, a total solar eclipse will be visible from [place] at [time]."

If we accept the fundamental positivist postulate, namely the verifiability criterion of truth, then all three types of "If so . . . then so" statements can be interpreted as predictions of future events on the basis of observing presently occurring ones. Thus the inference of a past event can be established as true only with reference to future observations of a *record* of that past event. For instance, the truth of the assertion—"Napoleon died on St. Helena"—can be verified only if the statement is translated into a prediction, such as "You *will* find statements to that effect in such-and-such archives, publications, memoirs, etc." The same holds for geometry, viewed as a physical science (of space) instead of as a mathematical science (of pure inference). "Measure the angles, and you *will* find them equal (within a margin of error)."

The criterion of verifiability is so firmly established in science that it is usually included by inference in the very definitions of scientific truth. Consequently, the criterion of predictability is also held to be an indispensable component of scientific truth. Mathematical physics is accordingly offered as the paradigm of the most mature science; for no branch of inquiry has been as successful as mathematical physics in making predictions of future events from known conditions, predictions which are both precise and often quite unexpected a priori.

The "If so . . . then so" statement can be interpreted as an assertion of causality. The prediction paradigm of science interprets all scientific assertions as assertions of valid causality relations. In the language of mathematical physics, however, assertions of causality do not play an explicit role (although they can be deduced). The assertions of mathematical physics are formulated as *equations*, typically differential equations. A solution of such an equation is usually a family of *trajectories*, that is, a set of time courses of a variable. The particular trajectory of the family is determined when initial or boundary conditions of the differential equation are fixed. For example, the propagation of heat is described in mathematical physics by a certain partial differential equation, which relates the rate of change of temperature at an arbitrary point in space to the temperature gradients along the spatial axes at that point and to the characteristics of the conducting medium. If a sphere at a given temperature is suddenly immersed in a heat bath at another temperature, the boundary conditions of the differential equation are fixed; namely, the temperature at every point on the boundary of the sphere is given at the time of immersion. Then the solution of the differential equation with the imposed boundary conditions determines the temperature of every point within the sphere for every moment of time thereafter. This trajectory can be read as an infinite number of causality statements, such as "If the temperature on the boundary of the sphere at time zero was T_1 and at every point within the sphere T_2, then the temperature of a point at such and such a distance from the center of the sphere will be such and such at time."

The formal language of mathematical physics is literally infinitely richer than the "vulgate" language of causality, because the equation which embodies a physical law (such as that of the propagation of heat or of electro-magnetic waves, or the law of gravity) contains within it literally an infinity of "If so . . . then so" statements, one for each choice of values substituted for the variables of the equation.

Nor is this all. When we speak of causal relations in ordinary language, we tend to establish them pair-wise: a cause linked with an effect. "Why is the water in the kettle boiling?" "Because the kettle was placed over a fire." On further reflection, we realize that such answers are far from adequate. It is not enough to place the kettle over a fire to cause water to boil. The fire must be large enough and hot enough for a given volume of water to be brought to a boil; the kettle must stand over the fire long enough; and the length of time required depends on the conductivity of the material from which the kettle is made, the temperature of the fire, the quantity of water, and the atmospheric pressure. If we tried to combine all of these factors into a single causality statement (in the "vulgate"), the statement would be incomprehensible. Ordinary language is too clumsy to deal with the intricate web of all these relations. Mathematical language, however, has been designed or, better, has evolved in just the way required to deal with situations where not only several "causes" converge on a single "effect" but also where the "causes" and "effects" all interact with each other.

The connection between the language of equations and the "vulgate" language of causality is established by holding constant all of the variables except one pair. This enables us to say; "Other things being equal, the thinner the bottom of the kettle, the sooner the water will boil"; or, "Other things being equal, the greater the atmospheric pressure, the longer it will take to bring the water to a boil," etc. Thus "common-sense" causal relations are included in the equation and are deduced by holding constant all the variables except those of interest.

In assuming that the equation in which all the causal factors were combined was given, we have, of course, assumed that all of the relations were known at once. In actuality they are often determined one by one. These separate determinations are made possible by the method of controlled experiment. In order to bring out some causal relation free of disturbances by other factors, we deliberately try to hold constant all those factors suspected of having some influence. Thus the basic assumption underlying the empirical study of physical phenomena is that we can eliminate all disturbing phenomena and study the relation of interest alone. Next, by establishing several pairs of such relations, we can (we assume) combine them into a more general causality law, that is, an equation in which all the contributing factors appear as variables. This is called the *analytic* method. It has been phenomenally successful in the physical sciences.

If we ask why it is that the analytic method has been so successful in the physical sciences, we might ponder on the possible answer that the physical sciences are just those where the method could be successful. We call the systematic study of life phenomena "biological science" and of human behavior and institutions "social science." However, there is no evidence that the determining events of life and of what we call social events are *not* physical events. In fact, philosophers of the materialist persuasion are convinced that they *are*, in the last analysis, physical events; that is, exchanges of matter and energy, propagations of electrical fields, etc. There is only one good reason why we do not subsume the study of life phenomena under "physical science": such phenomena do not yield easily to the analytic methods of physical science. At least the applicability of these methods to the study of life processes is limited.

We know that living beings are physical objects and that, when these objects are involved in physical events, they are subject to physical laws. Occasionally we can apply our knowledge of physical laws to explain some manifestations of the life process. For instance, in answer to the question, "Why does an elephant have thick legs," we can say that thin legs could not support his bulky body. That is to say, from the fact that an elephant exists at all, we can infer that his legs (being made of the material elephant legs are made of) must have a certain thickness. Similarly, we can explain the flight of birds by physical principles, many physiological processes by physico-chemical principles, events in the nervous system by electrochemical principles, etc. But of course, such explanations have only a very limited scope. We can explain why a bird is able to fly by invoking the principles of physics, but not why a bird takes off in the first place. We can prove that the ability of the elephant to stand is consistent with the laws of mechanics, but the laws of mechanics do not tell us how the elephant got that way. We can estimate how the tide of a battle depends on the relative positions and fire powers of two opposing armies (that is, on physical variables), but not why large numbers of people are shooting at each other. In short, the analytic method of physical science seems to reach just so far. But we can also say with equal justification that we have subsumed under physical science precisely those phenomena which yield to the analytic method.

We must keep in mind that until the middle of the last century physics encompassed several branches of inquiry dealing with apparently unrelated phenomena: mechanics, light, sound, heat, electricity, magnetism. The physical links among these, for example, between mechanics and heat, between light and electromagnetism, were established only later. However, even before the common physical bases of these phenomena were understood, they were subsumed under "physics" or "natural philosophy," as it was once called. The unifying factor was the *method* of physics, which was essentially the analytic method; that is, an examination of the relationship between pairs of variables at a time and construction of a mathematical theory in which the causally interrelated factors were combined into single equations. The *physical* unification of all these phenomena has not been completed to our day. For example, a common principle underlying both gravity and electromagnetism is suspected by some physicists (who believe in a general field theory) but not yet known.

The unifying principle of the analytic method has been the language of mathematics. Phenomena of widely different content, for example, the propagation of heat and the distribution of a gravitational field, can be described by laws which have similar mathematical forms, namely partial differential equations of second order. In this way a common theoretical framework is provided for apparently unrelated phenomena, and consequently the principle of "economizing thought" is served.

As we have said, attempts to extend the analytic method to the study of living processes was only partially successful. Views on why this is so differed. According to one view, called vitalism, the extension is in principle impossible, because living processes are not governed by the same laws as nonliving processes. According to the opposing views, variously called mechanism, physicalism, or reductionism, the difficulty lies not in an irreducible difference between physical and biological laws but only in the tremendous complexity of living processes. The reductionists assumed implicitly that if we knew enough about how living beings were put together, we could write down the equations that govern their behavior; and if we were clever enough in mathematics, we could solve the equations and so determine the "trajectories" of behavior.

Both views were supported by reasonable arguments. The vitalists could always exhibit aspects of the living process which eluded the sort of explanation sought by the physical scientist, specifically causal relations derived from a mathematical model constructed on the basis of known physical laws. Their opponents could always point out that no aspect of the living process has ever been *proved* to violate known physical laws. In fact, many such processes, which were once thought to violate physical laws, were later shown to obey them. It was once thought, for example, that the so-called organic compounds could be produced only within a living organism (which is how these compounds got their name). However, in 1828 an organic compound, urea, was synthesized in the laboratory. Thereafter thousands of "organic" compounds were synthesized, and the vitalists' argument was shown to be without foundation.

At one time the argument was advanced that living organisms did not obey the laws of physics, because they were capable of "autonomous movement." This argument, too, was proved unfounded when it was shown that the law of conservation of energy was quite as valid for living processes as for all physical processes. Indeed the history of the vitalist view is one of repeated retreats. Many times processes which were declared to be impossible except in the context of life were shown to be possible outside this context. Nevertheless the vitalists never gave up, since there were always "unexplained" aspects of life to point out. For the reductionists, however, this was only a challenge to extend the range of the analytic method.

One position firmly defended by the vitalists (by some to the present day) was that living beings are governed by goals and purposes, while inanimate objects, obeying the laws of physics, were not. Hans Driesch (1867–1941), a prominent biologist of the vitalist persuasion, thought that he demonstrated this difference by an ingenious experiment. He took an embryo of a sea urchin at a very early state of development, when it consisted of just a few cells, and separated the two halves. Thereupon each half developed into a full animal. Driesch argued that this demonstrated the operation of a Principle of Equifinality, which governs the behavior of living things but not of nonliving ones. If the two portions of the embryo were nothing but "mechanisms," Driesch argued, they would have developed into two separate halves, as if neither "knew" that the other had been taken away. The fact that each half developed into a whole animal shows that the development was guided by a "goal," namely that of becoming a sea urchin. Thus, teleological causality was presumably operating in the biological process, a principle which the physicists have excluded from physics.

It is instructive to see why this argument is fallacious. It rests on a purely verbal deduction, not on the sort of analysis constructed by a physicist. It is true that "teleology has been excluded from physics," but this was not done by a philosophical edict. Teleological causality was dropped because it became unnecessary in explaining physical events. Dependence on teleological explanations was a heritage of Aristotelian philosophy in European thought, in which motions of material bodies, both living and nonliving, appeared as a consequence of displacing bodies from their "natural" positions. Thus the ability of birds to fly was attributed to the fact that the "natural" position of the bird was in its nest, which was in a tree. The tendency of stones to fall to the ground was attributed to the fact that the "natural" position of a stone was on the earth because of the predominance of the element Earth in the stone, while the tendency of smoke to rise was attributed to the predominance of the element Fire in smoke, which made its "natural" position somewhere beyond the stars in the region of eternal fire.

The maturation of mechanics put an end to these speculations. Mechanical explanations of motion were deduced from the operation of forces acting on the moving object at each point of its trajectory at the moment when it is at that point, without reference to its possible future states. In Aristotelian terms, this amounts to ignoring the "final causes which beckon from ahead" in favor of "efficient causes which push from behind," as it were. The success of the mechanical view in subsuming a vast variety of motions under a single theory contributed to the demise of teleology in physics.

Now, a mechanical explanation of an event is a mathematical one, not a verbal one. It is not an assertion to the effect that efficient causes are operating and teleological ones are not. As we have said, the mechanical explanation is typically a differential equation whose solutions are trajectories. The particular trajectory is determined by the initial or boundary conditions. According to a mechanical explanation, the course of development of an embryo is conceived as a trajectory or, rather, a great many trajectories, each involving some measurable quantity, it being assumed that the totality of the values of these quantities at a given moment is a complete description of the state in which the embryo finds itself at the moment. In separating the two halves of the embryo, Driesch imposed *a different initial condition* on the process, since a half embryo without the other half present is certainly in a different situation than it is with the other half present. It is, therefore, unreasonable to expect the two trajectories to be the same. Starting from the initial condition with the other half present, the half embryo follows a trajectory toward "half an organism"; with the other half absent, it follows the trajectory toward a "whole organism." Driesch's error was in assuming mechanical motion to be determined entirely by an initial impulse. If this were the case—if, for example, the fertilization of the egg were the initial impulse which determined the subsequent fate of each particle of matter in the embryo—then it would continue to develop "blindly" without regard for the environment. But this interpretation of "mechanical" stems from the connotations of the word, not from its meaning in physics. A "mechanism" brings to mind a clockwork which, once set in motion, continues on a prescribed course without "regard for a purpose" of the motion. We often call a rigid performance devoid of nuances a "mechanical" one, etc. These uses of the word "mechanical" have little or nothing to do with mechanics.

However, it is important to recognize that in demolishing Driesch's argument for vitalism, we have merely shown that the development of the embryo is not inconsistent with the principles of mechanics (or of physics), as Driesch thought to have shown. We have not actually derived the trajectories of development from those laws, which would have been a definitive coup de grace to vitalism in this context. Therefore our conviction (which here coincides with the conviction of the physicalist) that living processes can be explained in terms of physical laws remains a belief and no more. We are justified in holding on to it because we have not arrived at any contradiction with it; but we have not shown that this belief provides us with sufficient leverage to advance theoretical biology on the basis of physics and chemistry alone.

So far this has not been possible. Biology would be crippled if it did not depend on concepts outside the scope of physical concepts: organism, life, birth, death, sex, viability, adaptation, behavior, cell, organ, evolution, species, genus, class, phenotype, genotype, mutation, selection, clone, embryo, etc., etc. To be sure, links have been established between some of these concepts and those of physics and chemistry, and, indeed, inroads of physical science methods into biology are being constantly broadened. Still an irreducible residue remains. Biological processes are simply too complex to yield to the analytic method. For example, although the physicalist can *conceive* of representing the development of an embryo or some act performed by an animal in terms of trajectories of variables, he cannot actually do it, and it is futile to attempt it. The analytic method is powerless to deal with such complexities.

When we turn to attempts to subject human behavior to scientific analysis, the problem becomes even more severe. We perceive human behavior (and for that matter the behavior of other animals) as sequences of "acts." Frequently these acts become "understandable" in terms of assigned goals and purposes consciously or unconsciously pursued. However, the criteria of "understanding" in these contexts are not those of a physicist, who demands that all processes be expressed as trajectories derived from equations and, moreover, that the variables in which the trajectories are expressed be physically measurable quantities (positions, masses, electric charges, magnetic moments, etc.). But this is clearly an impossible task in the context of human behavior. To take an example, suppose a man quits his job. Examining the circumstances, we find that he had quarreled with his supervisor, had no hope of promotion, and was offered another job at twice the pay. By any common sense criteria of "understanding" we understand why the man quit his job. Nor is this sort of understanding devoid of predictive power. We can state with some confidence that such an event will occur in most cases under similar circumstances. Nevertheless, convinced as we may be that the whole situation is "ultimately" describable in terms of specific impulses impinging on the nervous system of our man and on the transformation of these impulses into others leading to the activation of effectors, this outlook is all but useless for analyzing the event into its constituents, which is what the analytic method requires. We understand the event *directly* by perceiving wholes rather than parts: the man, his circumstances, his preferences, etc.

It follows that understanding cannot be extended beyond the scope of physical science without introducing concepts which embody irreducible wholes in place of physically measurable variables. The concept of organism is indispensable in biology; the concept of the individual in psychology; the concepts of the institution and social class in sociology; the concept of a nation in contemporary political science; the concept of a culture in anthropology. Each of these wholes presents itself naturally, because we perceive it as such. We *recognize* an organism, an individual, a nation; and we assume that in proper circumstances it *acts* as a whole. Still, if we confined our attention exclusively to the grossly observable patterns of these wholes, we would not make much progress toward understanding this behavior. We gain a deeper understanding of how an organism performs an act if we understand how the components of the act are integrated by its nervous system. We may gain a deeper understanding of why a nation reacts as it does to the "acts" of other nations if we understand how decisions are made within the body politic and how they are enforced, that is: how the actions of perhaps millions of individuals are meshed with each other to result in an act attributable to the nation as a whole.

A whole which functions as a whole by virtue of the interdependence of its parts is called a *system*, and the method which aims at discovering how this is brought about in the widest variety of systems has been called general system theory. General system theory seeks to classify systems by the way their components are *organized* (interrelated) and to derive the "laws," or typical patterns of behavior, for the different classes of systems singled out by the taxonomy.

The system-theoretical point of view is present also in physics. There, however, the systems approach derives directly from the analytic, so that there is complete continuity between the two methods. This is because physical systems are usually simple enough to be understood entirely in terms of the interrelation of their parts. To take an example, consider one of the simplest physical systems, a gas in a closed container isolated from the environment. It is known that such a gas will soon be in a condition of equilibrium, that is, the temperature and the pressure will be the same throughout the container. The system will, accordingly, be in a certain "state" which is entirely characterized by the volume enclosed, the pressure, and the temperature throughout the volume. These three variables define the system, and the system behavior is defined in terms of their interrelation. This interrelation is expressed by an equation, whose form is characteristic for different gases. In the simplest case of the so-called "perfect gas," the equation reads

$$PV = RT,$$

where P is the pressure, V the volume, T the temperature (measured from absolute zero), and R is a constant for all such gases. The behavior of the system is completely determined by this equation if by "behavior" we mean the equilibrium state attained. That is to say, if we impose certain values on two of the three variables, P, V, or T, the third will "adjust" in order to satisfy the equation. Note that this behavior already embodies "equifinality" of sorts: if we disturb the equilibrium of the system, in whatever direction, equilibrium will be restored. However, this is not the sort of equifinality which impresses the biologist. For the state of equilibrium of a gas in a closed container is the equilibrium of maximum chaos. If we associate with the idea of organization (hence of an organism) a differentiation of its parts, then a gas in equilibrium is completely disorganized, because the conditions throughout the enclosed volume (pressure and temperature) are the same. Within such a system no work could be done, since work requires the existence of gradients (for example, gradients of pressure which create fields of force). The biologist is impressed by the existence of just these nonequilibrium gradients in living organisms, and he wants to know how they are maintained against the tendency of physical systems toward equilibrium states.

The answer to this question resides in the fundamental difference between so-called closed and open systems. The former are isolated from their environment and therefore are subject to the action of the Second Law of Thermodynamics, which states that the entropy (roughly a measure of disorder) of a closed system will always increase toward a maximum, attained in equilibrium. At one time the vitalists invoked this law to argue that living systems "disobeyed" the Second Law, since at least in their developmental stage they became more organized instead of disorganized. However, this argument is invalidated in view of the fact that no living system is a closed system, and so the Second Law does not apply to it. Still the refutation is only a negative argument. A constructive argument would have to show how nonequilibrium states can actually be maintained in open systems.

This can be shown in the simplest case of an open system. Consider a system of interrelated chemical equations in which the rate of concentration of each substance depends on a sum of terms each proportional to the concentrations of other substances. Positive constants of proportionality

indicate that the corresponding substances facilitate the formation of the substance in question; negative ones indicate that the corresponding substances inhibit its formation or contribute to its rate of destruction. The system is closed if no substances are permitted to enter the system from the outside nor to leave it; in other words, if there are no sources or sinks. An open system is one which contains sources and sinks. It can be shown mathematically that a closed system will approach an equilibrium state in which the proportions of the various substances depend on the initial concentrations. Consequently, if the initial concentrations are changed by adding or removing substances, a different equilibrium state will be approached. Such a system will not exhibit "equifinality." If, however, the system is open, it can be shown that the final state will *not* depend on the initial concentrations. It will be determined entirely by the properties of the system itself, that is, by the *constants* of proportionality which are independent of the conditions *imposed* on the system. Such a system will appear to exhibit "equifinality," or, metaphorically speaking, to have a "goal of its own." Nor will the final state be an equilibrium state. Depending on the geometric and physical properties of the boundary within which the reactions take place, gradients will be established which could not be maintained if the system were closed. If the boundaries of the system should suddenly become impermeable, the gradients would disappear.

Here we see how quasi-purposeful behavior can be manifested by an open physical system which is not necessarily "alive." Since all living systems are open (if a living system is isolated from its environment, it must die), we have a conceptual link between living and nonliving systems. Nonliving systems acquire *some* of the properties of living ones by virtue of being open.

This finding is not merely a philosophical conclusion. It has become the underpinning of the whole new complex technology associated with the so-called Second Industrial Revolution. The significance of this development is best seen against the background of the First Industrial Revolution.

Underlying the First Industrial Revolution was the principle of transformation of energy from one form to another. The heat engine is based on such transformation: the energy locked in fuels is released in controlled combustion in the form of heat, which is used to establish a temperature difference between two systems (for example, between the boiler and the cooler of a steam engine). In flowing from the source at higher temperature to the sink at lower temperature, heat can be transformed into work. Now, considered as a system which does work, the living organism is also an engine, in which the energy contained in food is transformed into work. The First Industrial Revolution resulted in the replacement of the "natural engines" (human beings and animals) as sources of mechanical work by artificially constructed engines. As we now know, this replacement not only increased tremendously the amount of energy available to human beings but had far-reaching consequences in transforming the organization of human societies. Heat engines became the "slaves" of those societies. They remained, however, stupid slaves, and their work still had to be directed by human beings. Released from physical drudgery, human beings became "mental appendages" to the machines, tending them and "directing" their work through routine control operations. As the complexity of the industrial enterprises and their auxiliary social and political organizations increased, the number of white collar workers (performing routine control operations like bookkeeping and other clerical work) also increased and eventually caught up with the number of industrial workers in technically advanced societies.

The Second Industrial Revolution came into being with the appearance of machines designed to process not energy but information. Business machines, engaged as automatic bookkeepers, clerks, etc., are well-known examples of these devices. The origins of this technology are rooted in the military problems of World War II. With the advent of bomber aircraft, the antiaircraft gun became a major weapon of defense. To be effective, this gun must be properly aimed at a rapidly moving target. This means that the future position of the target must be anticipated on the basis of readings taken of its successively observed positions. Moreover, the calculations must be performed extremely rapidly if the information is to be of any use. Accordingly the needs of the new military technology stimulated the development of extremely rapid automatic computing machines, that is, devices able to process enormous amounts of information quickly and to arrive at appropriate "decisions."

Devices were also developed which operated on the principle of *correcting* their performance on the basis of taking stock of what is happening. Such a device compares its current state with some preset "goal" state and adjusts its performance on the basis of the "observed" difference. Here, then, is a technological device which exhibits the "principle of equifinality" or "purposefulness" in a way which strikingly resembles the behavior of a living system. The postwar burgeoning growth of automation is a result of ever widening applications of this principle, and the science in which this principle is central has been called by Norbert Wiener *cybernetics* from the Greek κυβερνήτης = steersman (*Cybernetics*, New York: John Wiley & Sons, 1948).

Cybernetics is the science of communication and control. As such, it does not examine transformations of energy. It examines patterns of signals by means of which information is transmitted within a system and from one system to another. Transmission of information is essential in control, and the capacity of a system to exercise control depends on how much information it can process and store. In fact, the concept "quantity of information" is central in cybernetics. In this context, "quantity of information" is unrelated to the meaning of the information, its significance, or its truth. Quantity of information is related simply to the number of "decisions" which must be made in order to reduce the range of possible answers to the question one asks; to put it in another way, to reduce uncertainty.

Note that an "If so . . . then so" statement, expanded to include several levels of conditionality, amounts to increasing the number of decisions required to establish the conclusion. A typical item on an "intelligence test" may read "If 3 is greater than 5 or if January is a winter month, cross out the fourth word in this sentence, unless New York is in France, in which case cross out the word following the eleventh word, but only if it is a noun." Giving the correct response depends here on the ability to make conditional sequential decisions. This is essentially what goes on in an automatic information-processing device like a computer or a servo-mechanism.

The appearance of complex information-processing machines suggested a new concept of the living organism, namely one which, in addition to being an engine (a device for transforming energy from one form to another) and a chemical laboratory (a device for transforming matter from one form to another), was also a decision-making system (a device for processing, storing, and retrieving information). The concept was not new, of course. The apparent "purposefulness" of living processes, especially of behavior, has always suggested that organisms "make decisions." What was new was a set of concepts susceptible to logical (or mathematical) operations, from which the "purposeful" or "intelligent" aspects of living systems could be derived. Besides suggesting nonvitalistic explanations of these aspects of life, the concept of information-processing clarified the role of "organization" in a living organism. Information can be processed only if the decisions (that is, events within the organism conditional upon other events) are integrated and meshed in certain ways. Moreover, the development of devices within the organism which effect such integration and processing of signals (for example, a nervous system) itself depends on the processing of information. The discovery that information is actually coded on the genes of a living organism in the form of "instructions" on how the organism is to be put together during gestation provided further evidence that the understanding of a living process is intimately bound up with an understanding of information processing.

If so, then the difference between "living" and "nonliving" becomes obscure. Systems that are "living," in the common sense or biological sense of the word, share many features with systems that are not; and these common features derive from the way systems are *organized*. This suggests a generalization of the concept of "organism" to the concept of "organized system." Organized systems include organisms. Certain methods of studying behavior apply to all organized "living" or "nonliving" systems, namely inquiries concerning the *structure* of the system, in particular the means by which it is enabled to receive, to store, to process, and to recall information; inquiries concerning the *functioning* of the system, in particular the way in which, by means of processed and stored information, the system responds by "behavior outputs" to "sensory inputs" from the environment; and finally, inquiries concerning the *evolution* of a system type.

Any organized system, living or nonliving, can be seen from these three perspectives which, for this reason, encompass the broadest scope of a general system theory.

Let us draw a parallel between a living organized system and a nonliving one, say, between a cat and a computer.

The study of the cat's structure is called *anatomy*. In particular, neural anatomy is concerned with the arrangement of parts in the cat's nervous system. The computer, too, has a structure, most of it being its "nervous system," but some parts also fulfilling the functions of support (analogous to a skeleton) and functions of maintenance (utilizing energy supplied from outside sources to activate the parts).

The study of the cat's functioning is called *physiology* and *psychology*. The two join in the problem of tracing the way physico-chemical events in the cat's nervous system determine its responses to the environment (physiological pyschology). The "psychology" of the computer is described in the way a *program* determines the transformation of inputs into outputs.

The study of the cat's evolution is called *phylogeny* and is concerned with the long-term changes in the line of descent, of which the cat is the present specimen. These changes are known to be adaptations of species to their environment, effected by natural selection. Man has been an important component in the cat's environment. In particular, man has taken over some of the natural selection functions with respect to cats (even more with respect to other domesticated animals) by conscious or unconscious selective breeding. The computer, too, undergoes an evolution, in which man plays a decisive role. Man introduces modifications in the design of computers to suit his purposes and decides which types will continue to "reproduce" and which are allowed to become "extinct." (For the time being the computer is dependent for its reproductive function entirely on man, but this need not be. It has been shown that an artificial information-processing system of a certain minimal complexity can reproduce itself,[1] which adds still another feature, formerly thought to be exclusively characteristic of living things, to the repertoire of general organized systems!)

The evolution of artifacts can be traced as clearly as the evolution of organisms. It is amusing, for example, to note the gradual disappearance of running boards on automobiles, some transitional models showing "vestigial" (that is, rudimentary and nonfunctional) remnants. More serious evidence is found in the evolution of termite nests (also artifacts), where "vestigial" galleries (once functional, but now deadend) are found. These are quite analogous to vestigial anatomical parts, such as the vermiform appendix in human beings.

Once it is recognized that structure, function, and evolution (or *being*, *acting*, and *becoming*) are fundamental aspects of all organized systems, the concept of organism can be broadened still further to include, for example, whole complexes of living organisms plus the inanimate artifacts functionally related to their structure, behavior, and development. Such are *societies*, conceived in the broadest sense.

Sometimes a society reveals a striking similarity

to an organism. In a bee hive, the bees are meaningful analogues to the cells of an individual, even to the extent of being differentiated according to their function. Some are reproductive (the queen and the drones); some perform only maintenance functions (the workers). In ants, the differentiation is carried still further (for example, soldier ants). A bee hive "reproduces" in a way similar to the mitosis of a cell. When a second queen bee appears (the division of the "nucleus"), the hive splits in two parts, each around a queen, and the hive becomes two hives.

Human social aggregates (families, institutions, communities, nations) exhibit all the features of organized systems. The degree of organization varies, of course, as does the robustness and the "viability" of these systems. It therefore makes sense to speak of the "pathology" of such systems, a concept which has a profound intuitive meaning to many social philosophers.

The analogical way of thinking is as old as thinking. It has traditionally been riddled with naïve anthropomorphic ideas and inspiring but scientifically sterile metaphors. With the advent of the analytic method in the physical sciences, the system point of view (with its heavy reliance on directly perceived analogies and teleological explanations) receded into the background. Sophisticated biology eschewed teleological explanations. Empirically oriented social scientists abandoned social philosophy in favor of strictly circumscribed investigations in the best traditions of controlled experiment, where one variable is "pitted" against another without reference to how the relationships obtained can be fitted into a "picture." Scientific psychology turned away from its original subject matter, the psyche, toward piecemeal investigations in thoroughly "sterilized" settings.

These developments were both inevitable and indispensable for the advancement of a science of man, for it was necessary to learn the *discipline* of scientific inquiry in studying man, something very difficult to achieve because of our personal involvement with the object of study. However, there was a price to pay for increasing the resolving power of our observations: the narrowing of the field of vision. The proliferation of disciplines, subdisciplines, and specialities threatened to fractionate the scientific community into mutually isolated enclaves unable to communicate with each other. Science threatened to become an avalanche of "findings" which in their totality no more add up to knowledge, let alone wisdom, than a pile of bricks adds up to a cathedral.

The modern system point of view is a response to this threat. However, the new system approach must not be viewed as a return to the older ways. It is built on recent developments which show promise of reestablishing holistic approaches to knowledge without abandoning scientific rigor. The analogies established or conjectured in system theory are not "mere" metaphors. They are rooted in actual isomorphisms or homomorphisms between systems or theories of systems.

Although the backgrounds of individual scientists vary widely, it can perhaps be assumed that those behavioral scientists who have developed an interest in the systems approach to research have been sufficiently exposed to the underlying ideas to read this volume with profit. This is particularly true with regard to mathematics: a certain minimal level of mathematical literacy (to be distinguished from technical knowledge) can be assumed to be possessed by any one doing research in psychology, linguistics, sociology, or economics. Mathematical ways of thinking and reasoning have also penetrated to some extent into political science and even into anthropology.

This volume therefore is offered to the behavioral scientist as an overview of thinking in the last two or three decades which reflects a trend toward the system point of view. Some of the chapters are philosophical: they discuss the significance of the trend as a development in the contemporary philosophy of science. Some are inevitably detailed and technical, since a formulation of principles governing the behavior of systems rigorously defined is not possible without the use of some symbolic machinery of deduction, for example, symbolic logic and mathematics. Still other chapters discuss the relevance of some particular concepts that are central in the system approach (for example, information theory, feedback, organization as a structural concept), to particular fields of research. The picture that emerges is far from that of a unified theory, and it is an open question whether much progress can be made by attempts to construct a "unified theory of systems" on some rigorous axiomatic base. The materials in this book should be seen rather as a cross-section of contemporary thought by people who have put their hope in synthesizing *directions* rather than in a dramatic synthesis of the sort achieved by Aristotle, Newton, and Darwin.

Perhaps the time for such clearly seen milestones in the history of science is past. Science may have become too vast an enterprise to be organized by ideas springing up in single minds. Perhaps the variety of currents in the system approach portends the emergence of a "collective scientific mind." Vast technological complexes come into being through collective efforts without any individual being able to grasp the complexes in their totality. The situation may be analogous in science, as the meaning of the term is expanded to include the study of man. An acquaintance with the system approach may give the behavioral scientist at least a glimpse of the whole, even though a total understanding (in the sense of understanding the solar system through celestial mechanics) is out of the question.

There is also an ethical justification for disseminating the system approach among behavioral scientists. Behavioral scientists can be roughly divided in two groups: those who aspire to the scientific status of physical scientists and, in consequence, tend to select research problems that yield to the analytic method; and those who are moved by a need to "understand man." The former stand in danger of trivializing the study of man and, what is often worse, of placing their expertise at the service of groups having power to manipulate man for their own purposes. The latter stand in danger of obscuring the study of man in free-wheeling speculations without sufficient anchorage in facts or testable hypotheses.

Both these dangers have been aggravated by the seductive potential of the new system concepts. The conceptualization of a living system, in particular a social system, as an "information-processing entity" plus the formulation of quantitative measures of information and the discovery of the relation between information, entropy, and "open systems" has triggered an avalanche of speculations which, upon closer examination, reveal only a lack of understanding of what is involved in quantifying "amount of information" or "amount of organization." Similar illusions arose here and there around the rigorous formulations of conflict theory on the basis of game-theoretical analysis. Not the least of the dangers associated with these sanguine extrapolations is that the intellectually responsible behavioral scientist, upon discovering the sterility of such speculations, will dismiss the new ideas as altogether irrelevant. This, I think, would be a mistake. The ideas of the system approach can be a rich source of inspiration in the advancement of behavioral science, but only if the actual content and scope of these ideas are clearly understood. They can be understood only if the behavioral scientist temporarily turns away from his central problems to examine the ideas in the light of their own inner logic.

Once this logic is grasped, the system approach to the study of man can be appreciated as an effort to restore meaning (in terms of intuitively grasped understanding of wholes) while adhering to the principles of *disciplined* generalizations and rigorous deduction. It is, in short, an attempt to make the study of man both scientific and meaningful.

ANATOL RAPOPORT

Note

1. John von Neumann, *Theory of Self-Reproducing Automata*, ed. and completed by Arthur W. Burks (Urbana: University of Illinois Press, 1966).

General Introduction

By way of introduction to a book that is itself largely introductory, the following is concerned primarily to offer a rationale for the general organization of material. The particular choice of selections can of course only be taken as one man's sampling and guide to the rapidly proliferating and widely scattered research that may be seen as part of, or especially pertinent to, the modern systems perspective. But the overall organization is intended to suggest a chapter in the history and philosophy of science, at least as seen from the point of view of a student of society and human behavior. To be told adequately, this recently crystallizing story would have to cover, at a minimum: (1) the renewed attention being given to the continuity of nature, particularly that between the physical, the biological, and the psycho-socio-cultural; the fully interactive nature of these various aspects (which have become so compartmentalized in the academy; and the consequent newly insistent demand for interdisciplinary integration; (2) the more acute and constructive debate over problems of reductionism and emergence or holism, and the specific ways in which each, when tempered, may be accepted as necessary and fruitful; (3) the recognition of the many subtle questions of epistemology raised by modern research, and the signs that empirical research and theory or model building are adapting —if not too self-consciously—to the challenge; and (4) the settling of differences between, or even the merging of, well-worn dualities such as that between mechanism and organism, behaviorism and subjectivism, or psychologism and sociologism.

It must be left to the reader to take note of these themes, and others, that run through the selections—often explicitly, sometimes implicitly. They all underlie, however, the organization of Parts, which may be summarized in the following manner.

The development and contagion of the modern systems perspective (Part I provides three overviews) can be traced in part to the concern of several disciplines to treat their subject matter— whether the organism, the species, or the social group—as a whole, an entity in its own right, with unique properties understandable only in terms of the whole, especially in the face of a more traditional reductionistic or mechanistic focus on the separate parts and a simplistic notion of how these parts fit together. Part II, "Parts, Wholes, and Levels of Integration," suggests that we are thus beginning to take seriously, and as not incompatible, the principles of the continuity of nature on the one hand, and the emergence of qualitatively different wholes on the other. Each of the basic scientific disciplines can be seen to treat as its unit of discourse some kind of whole whose parts constitute the wholes studied by the discipline on the next lower level of integration, and which in turn becomes only a component of the whole treated on the next higher level. This viewpoint seems not so problematic at the lower levels of the atomic nucleus, the atom and the molecule, but encounters increasing resistance as we reach the higher levels of the complex organism, the species, the ecosystem, and especially the human society. Nevertheless, it becomes increasingly difficult for any discipline to claim that it is dealing with a "real entity" or "substance" while another's subject matter is an abstraction or mental construct. And should the practitioner of any discipline claim that the unit of focus on the next higher level must be explained basically in terms of *his* unit of analysis, then he must be prepared to give up his own autonomy in the face of the similar claims of the discipline just below him. The end point of such argument, of course, is the not very helpful evaporation of everyone's unit of analysis into a swirl of electromagnetic fields and nuclear forces.

The dethronement of material substance as the only reality, the bedrock, has shifted the focus to the fact of *organization* per se as the more fundamental problem for study. This in turn means a shift from statics, structure, summativity and one-way causality to dynamics, process, emergence, and complex mutual interactions and feedback cycles. It is not the nature of the parts alone that are basic to any whole, but the way they are interrelated that gives them their characteristic properties—that makes the difference between one chemical compound and another built from the same components, between a living glob of protoplasm and an inert solution of organic compounds, or between an authoritarian social group and a democratic one. Further than that, in the more complex systems, the parts come to take on properties that they owe specifically to being components of a larger whole. From this point of view, the notion of *system* may be seen as simply a more self-conscious and generic term for the dynamic interrelatedness of components which may stabilize in any of a number of different organizations of varying significance for the system itself and its surroundings. Finally, in focusing on organization or relationships per se rather than immutable substances, we are less inhibited from engaging in a comparative study of systems of all kinds regardless of their "substantive" nature, with an eye to their structural and processual similarities, isomorphisms, and differences. And a focus on relations per se begs for a tie-in with that most abstract and fundamental science of relations, mathematical logic, and its consequent use in developing a theory of "automata" or the "Machine"—that is, the complex system in general. Such are the themes that we attempt to sample in Part III, "Systems, Organization, and the Logic of Relations."

Since we are particularly concerned with the higher-level systems of the complex, adaptive type, we focus in Part IV, "Information, Communication, and Meaning," on a crucial difference between this type and the typical lower-level, non-living, natural system. In the latter, the interrelations of the parts with one another, and of the whole with its surroundings, are mediated by physical contacts and/or direct energy exchanges, and they are thereby entropic. In the former type of system, on the other hand, the relations among parts and between the whole and its environment are more and more mediated by *information* exchange taken in the modern broad sense, thus making possible higher levels of flexibility, complexity, and dynamic interchanges with the environment. Such systems thus become locally negentropic, maintaining certain higher levels of available energy and perpetuating or elaborating their own organization seen as a continual mapping or matching of the relevant environment. This latter isomorphism between system organization and environmental structure is made possible by the fact that information—itself a generalized form of, or template for, organization—can thus mediate between the mapping of one structural arrangement by another. In more familiar terms, such systems are open, living, and adaptive. And at the highest level of integration, that of the human sociocultural, the combination of the advanced cortex and the manipulation of purely arbitrary symbols has made possible a literally self-conscious system whose components are interrelated into a more or less viable whole almost entirely by way of the communication of not merely mapped information of the physical world, but *meanings* that are largely creations of the social system itself. These may run the gamut from highly direct and perfected mappings of the physical and social world to rather indirect and grossly inaccurate matchings. Such a system, because of its symbolic communicative linkages, may be highly adaptive and viable but unstable in its constant elaboration and change of structure; but its huge successes are built on the same foundations as its colossal failures.

With Part V, "Cybernetics: Purpose, Self-Regulation and Self-Direction," we continue to narrow in on the peculiar characteristics of the higher behavioral systems and the recent attempts to define the mechanisms underlying these characteristics in a scientifically respectable way. The features of concern center around the concepts of self-regulation and self-direction, or homeostasis, purpose and evolution. The scientific field of direct relevance is cybernetics, defined by Norbert Wiener as the study of control and communication in the animal and the machine, and its key explanatory mechanism is the feedback loop carrying a continual flow of information between the system, its parts, and the environment. It is in these terms that a good many scientists have come to feel that at least a number of the essential features of goal-seeking or purpose, as well as of the blinder processes of self-regulation and adaptation, are to be understood in principle. And, once again, we note the benefit deriving from a focus on the detailed nature of the inter-relations or organization *per se* of components, regardless of whether the focus is considered to be physiological, psychological, sociocultural, or even electromechanical.

As suggested above, the complex adaptive system consists of components which are them-selves complex and which, in particular, take on certain of their important properties by virtue of being parts of the larger whole. This is especially true of the sociocultural system, whose com-ponent individuals are homeostatic biological organisms with purposive, adaptive psychological properties—both of these facets being to a very high degree a function of their participation in the larger sociocultural milieu. This means, among other things, the necessity for the joint study of the components and the larger whole, a full appreciation of their reciprocal determination, and a consequent resolution of the several philo-sophical problems associated with the "mind-body" and "individual-group" controversies. In Part VI, focusing on psychological systems, it is seen how modern systems theoretic concepts—in conjunction with earlier ones of closely kindred spirit—illuminate the fully active, open, trans-actional nature of the psychological component as an analytical unit that cannot be adequately specified apart from the physical and social situation that it, in turn, helps to construct. Of special note in this specification is the strongly felt need to get inside the "black box," despite a behaviorist stance, in order to trace out the complex information flow, the symbol decoding and encoding processes, and the feedback control mechanisms that alone seem to approach an adequate explanation of the complexities of behavior.

Finally, Part VII, dealing with sociocultural systems, comes close to a recapitulation of the main themes of the book as a whole, starting as it does with the part–whole problem and utilizing concepts and principles from each of the other Parts. By the time the reader has reached this last Part it is hoped that he will be appreciative of the sense in which we can speak of society as a system (in some degree) which, like any of the other levels, comes to take on certain holistic properties depending on the particular ways the components are bonded or interrelated in rela-tively stable ways. For it is at this point that it becomes "all too human" (or perhaps "all too Western") to balk at a supra-individual level that appears to cut into the principle of the "self-determination" of the individual. Nevertheless, and partly for this very reason, it becomes impor-tant to recognize that the "self" in the "self-regulation" and "self-direction" of this level of system does not refer to individuals—for example, group leaders—but to the larger sociocultural system of which they are interrelated, not com-pletely autonomous, components. This means that the often conflicting plans and purposes of individuals and subgroups are operating factors to which the larger system may, under certain conditions, adjust or adapt in ways not anticipated or desired by those components. It is to a fuller appreciation and deeper understanding of that rather crucial principle that this sourcebook in modern systems research is largely dedicated.

I
General Systems Research: Overview

ALTHOUGH none of the three authors contributing to this first Part can rightly be confined within the bounds of any one specific discipline, to point out that their fields of major concentration run the interdisciplinary gamut from social science to biology to mathematics and its physical applications is to strike a keynote of the modern systems perspective. The two areas they introduce here—General Systems Research, overviewed by economist Kenneth Boulding and critically reviewed several years later by biologist Ludwig von Bertalanffy, and Cybernetics, introduced by its founder—are only two of the several in the modern spirit that might conceivably have been included, given unlimited space. Thus, we might also have introduced information theory, game theory, and operations research. The first, however, we reserve for fairly extensive treatment in Part IV; the second is given very little space (in Part VII), and the third is given none at all, primarily because they both lean heavily on mathematical exposition which, as stated in the Preface, we have elected to avoid.

1.

General Systems Theory—The Skeleton of Science

KENNETH E. BOULDING

GENERAL SYSTEMS THEORY is a name which has come into use to describe a level of theoretical model-building which lies somewhere between the highly generalized constructions of pure mathematics and the specific theories of the specialized disciplines. Mathematics attempts to organize highly general relationships into a coherent system, a system however which does not have any necessary connections with the "real" world around us. It studies all thinkable relationships abstracted from any concrete situation or body of empirical knowledge. It is not even confined to "quantitative" relationships narrowly defined— indeed, the developments of a mathematics of quality and structure is already on the way, even though it is not as far advanced as the "classical" mathematics of quantity and number. Nevertheless because in a sense mathematics contains all theories it contains none; it is the language of theory, but it does not give us the content. At the other extreme we have the separate disciplines and sciences, with their separate bodies of theory. Each discipline corresponds to a certain segment of the empirical world, and each develops theories which have particular applicability to its own empirical segment. Physics, Chemistry, Biology, Psychology, Sociology, Economics and so on all carve out for themselves certain elements of the experience of man and develop theories and patterns of activity (research) which yield satisfaction in understanding, and which are appropriate to their special segments.

From Kenneth Boulding, "General Systems Theory —the Skeleton of Science," *Management Science*, 2 (1956), 197–208. Reprinted with the permission of *Management Science* and the author.

In recent years increasing need has been felt for a body of systematic theoretical constructs which will discuss the general relationships of the empirical world. This is the quest of General Systems Theory. It does not seek, of course, to establish a single, self-contained "general theory of practically everything" which will replace all the special theories of particular disciplines. Such a theory would be almost without content, for we always pay for generality by sacrificing content, and all we can say about practically everything is almost nothing. Somewhere however between the specific that has no meaning and the general that has no content there must be, for each purpose and at each level of abstraction, an optimum degree of generality. It is the contention of the General Systems Theorists that this optimum degree of generality in theory is not always reached by the particular sciences. The objectives of General Systems Theory then can be set out with varying degrees of ambition and confidence. At a low level of ambition but with a high degree of confidence it aims to point out similarities in the theoretical constructions of different disciplines, where these exist, and to develop theoretical models having applicability to at least two different fields of study. At a higher level of ambition, but with perhaps a lower degree of confidence it hopes to develop something like a "spectrum" of theories— a system of systems which may perform the function of a "gestalt" in theoretical construction. Such "gestalts" in special fields have been of great value in directing research towards the gaps which they reveal. Thus the periodic table of elements in chemistry directed research for many decades towards the discovery of unknown elements to fill gaps in the table until the table was completely

3

filled. Similarly a "system of systems" might be of value in directing the attention of theorists towards gaps in theoretical models, and might even be of value in pointing towards methods of filling them.

The need for general systems theory is accentuated by the present sociological situation in science. Knowledge is not something which exists and grows in the abstract. It is a function of human organisms and of social organization. Knowledge, that is to say, is always what somebody knows: the most perfect transcript of knowledge in writing is not knowledge if nobody knows it. Knowledge however grows by the receipt of meaningful information—that is, by the intake of messages by a knower which are capable of reorganizing his knowledge. We will quietly duck the question as to what reorganizations constitute "growth" of knowledge by defining "semantic growth" of knowledge as those reorganizations which can profitably be talked about, in writing or speech, by the Right People. Science, that is to say, is what can be talked about profitably by scientists in their role as scientists. The crisis of science today arises because of the increasing difficulty of such profitable talk among scientists as a whole. Specialization has outrun Trade, communication between the disciples becomes increasingly difficult, and the Republic of Learning is breaking up into isolated subcultures with only tenuous lines of communication between them—a situation which threatens intellectual civil war. The reason for this breakup in the body of knowledge is that in the course of specialization the receptors of information themselves become specialized. Hence physicists only talk to physicists, economists to economists—worse still, nuclear physicists only talk to nuclear physicists and econometricians to econometricians. One wonders sometimes if science will not grind to a stop in an assemblage of walled-in hermits, each mumbling to himself words in a private language that only he can understand. In these days the arts may have beaten the sciences to this desert of mutual unintelligibility, but that may be merely because the swift intuitions of art reach the future faster than the plodding leg work of the scientists. The more science breaks into sub-groups, and the less communication is possible among the disciplines, however, the greater chance there is that the total growth of knowledge is being slowed down by the loss of relevant communications. The spread of specialized deafness means that someone who ought to know something that someone else knows isn't able to find it out for lack of generalized ears.

It is one of the main objectives of General Systems Theory to develop these generalized ears, and by developing a framework of general theory to enable one specialist to catch relevant communications from others. Thus the economist who realizes the strong formal similarity between utility theory in economics and field theory in physics is probably in a better position to learn from the physicists than one who does not. Similarly a specialist who works with the growth concept—whether the crystallographer, the virologist, the cytologist, the physiologist, the psychologist, the sociologist or the economist—will be more sensitive to the contributions of other fields if he is aware of the many similarities of the growth process in widely different empirical fields.

There is not much doubt about the demand for general systems theory under one brand name or another. It is a little more embarrassing to inquire into the supply. Does any of it exist, and if so where? What is the chance of getting more of it, and if so, how? The situation might be described as promising and in ferment, though it is not wholly clear what is being promised or brewed. Something which might be called an "interdisciplinary movement" has been abroad for some time. The first signs of this are usually the development of hybrid disciplines. Thus physical chemistry emerged in the third quarter of the nineteenth century, social psychology in the second quarter of the twentieth. In the physical and biological sciences the list of hybrid disciplines is now quite long—biophysics, biochemistry, astrophysics are all well established. In the social sciences social anthropology is fairly well established, economic psychology and economic sociology are just beginning. There are signs, even, that Political Economy, which died in infancy some hundred years ago, may have a re-birth.

In recent years there has been an additional development of great interest in the form of "multisexual" interdisciplines. The hybrid disciplines, as their hyphenated names indicate, come from two respectable and honest academic parents. The newer interdisciplines have a much more varied and occasionally even obscure ancestry, and result from the reorganization of material from many different fields of study. Cybernetics, for instance, comes out of electrical engineering, neurophysiology, physics, biology, with even a dash of economics. Information theory, which originated in communications engineering, has important applications in many fields stretching from biology to the social sciences. Organization theory comes out of economics, sociology, engineering, physiology, and Management Science itself is an equally multidisciplinary product.

On the more empirical and practical side the interdisciplinary movement is reflected in the development of interdepartmental institutes of many kinds, Some of these find their basis of unity in the empirical field which they study, such as institutes of industrial relations, of public administration, of international affairs, and so on. Others are organized around the application of a common methodology to many different fields and problems, such as the Survey Research Center and the Group Dynamics Center at the University of Michigan. Even more important than these visible developments, perhaps, though harder to perceive and identify, is a growing dissatisfaction in many departments, especially at the level of graduate study, with the existing traditional theoretical backgrounds for the empirical studies which form the major part of the output of Ph.D. theses. To take but a single example from the field with which I am most familiar. It is traditional for studies of labor relations, money and banking, and foreign investment to come out of departments of economics. Many of the needed theoretical models and frameworks in these fields, however, do not come out of "economic theory" as this is usually taught, but from sociology, social psychology, and cultural anthropology. Students in the department of economics however rarely get a chance to become acquainted with these theoretical models, which may be relevant to their studies, and they become impatient with economic theory, much of which may not be relevant.

It is clear that there is a good deal of interdisciplinary excitement abroad. If this excitement is to be productive, however, it must operate within a certain framework of coherence. It is all too easy for the interdisciplinary to degenerate into the undisciplined. If the interdisciplinary movement, therefore, is not to lose that sense of form and structure which is the "discipline" involved in the various separate disciplines, it should develop a structure of its own. This I conceive to be the great task of general systems theory. For the rest of this paper, therefore, I propose to look at some possible ways in which general systems theory might be structured.

Two possible approaches to the organization of general systems theory suggest themselves, which are to be thought of as complementary rather than competitive, or at least as two roads each of which is worth exploring. The first approach is to look over the empirical universe and to pick out certain general *phenomena* which are found in many different disciplines, and to seek to build up general theoretical models relevant to these phenomena. The second approach is to arrange the empirical fields in a hierarchy of complexity of organization of their basic "individual" or unit of behavior, and to try to develop a level of abstraction appropriate to each.

Some examples of the first approach will serve to clarify it, without pretending to be exhaustive. In almost all disciplines, for instance, we find examples of populations—aggregates of individuals conforming to a common definition, to which individuals are added (born) and subtracted (die) and in which the age of the individual is a relevant and identifiable variable. These populations exhibit dynamic movements of their own, which can frequently be described by fairly simple systems of difference equations. The populations of different species also exhibit dynamic interactions among themselves, as in the theory of Volterra. Models of population change and interaction cut across a great many different fields —ecological systems in biology, capital theory in economics which deals with populations of "goods," social ecology, and even certain problems of statistical mechanics. In all these fields population change, both in absolute numbers and in structure, can be discussed in terms of birth and survival functions relating numbers of births and of deaths in specific age groups to various aspects of the system. In all these fields the interaction of population can be discussed in terms of competitive, complementary, or parasitic relationships among populations of different species, whether the species consist of animals, commodities, social classes or molecules.

Another phenomenon of almost universal significance for all disciplines is that of the interaction of an "individual" of some kind with its environment. Every discipline studies some kind of "individual"—electron, atom, molecule, crystal, virus, cell, plant, animal, man, family, tribe, state, church, firm, corporation, university, and so on. Each of these individuals exhibits "behavior," action, or change, and this behavior is considered to be related in some way to the environment of the individual—that is, with other individuals with which it comes into contact or into some relationship. Each individual is thought of as consisting of a structure or complex of individuals of the order immediately below it—atoms are an arrangement of protons and electrons, molecules of atoms, cells of molecules, plants, animals and men of cells, social organizations of men. The "behavior" of each individual is "explained" by the structure and arrangement of the lower individuals of which it is composed, or by certain principles of equilibrium or homeostasis according to which certain "states" of the individual are "preferred." Behavior is described in terms of the restoration of these preferred states when they are disturbed by changes in the environment.

Another phenomenon of universal significance is growth. Growth theory is in a sense a subdivision of the theory of individual "behavior," growth being one important aspect of behavior. Nevertheless there are important differences between equilibrium theory and growth theory, which perhaps warrant giving growth theory a special category. There is hardly a science in which the growth phenomenon does not have some importance, and though there is a great difference in complexity between the growth of crystals, embryos, and societies, many of the principles and concepts which are important at the lower levels are also illuminating at higher levels. Some growth phenomena can be dealt with in terms of relatively simple population models, the solution of which yields growth curves of single variables. At the more complex levels structural problems become dominant and the complex interrelationships between growth and form are the focus of interest. All growth phenomena are sufficiently alike however to suggest that a general theory of growth is by no means an impossibility.

Another aspect of the theory of the individual and also of interrelationships among individuals which might be singled out for special treatment is the theory of information and communication. The information concept as developed by Shannon has had interesting applications outside its original field of electrical engineering. It is not adequate, of course, to deal with problems involving the semantic level of communication. At the biological level however the information concept may serve to develop general notions of structuredness and abstract measures of organization which give us, as it were, a third basic dimension beyond mass and energy. Communication and information processes are found in a wide variety of empirical situations, and are unquestionably essential in the development of organization, both in the biological and the social world.

These various approaches to general systems through various aspects of the empirical world may lead ultimately to something like a general field theory of the dynamics of action and interaction. This, however, is a long way ahead.

II   A second possible approach to general systems theory is through the arrangement of theoretical systems and constructs in a hierarchy of complexity, roughly corresponding to the complexity of the "individuals" of the various empirical fields. This approach is more systematic than the first, leading towards a "system of systems." It may not replace the first entirely, however, as there may always be important theoretical concepts and constructs lying outside the systematic framework. I suggest below a possible arrangement of "levels" of theoretical discourse.

(i) The first level is that of the static structure. It might be called the level of *frameworks*. This is the geography and anatomy of the universe—the patterns of electrons around a nucleus, the pattern of atoms in a molecular formula, the arrangement of atoms in a crystal, the anatomy of the gene, the cell, the plant, the animal, the mapping of the earth, the solar system, the astronomical universe. The accurate description of these frameworks is the beginning of organized theoretical knowledge in almost any field, for without accuracy in this description of static relationships no accurate functional or dynamic theory is possible. Thus the Copernican revolution was really the discovery of a new static framework for the solar system which permitted a simpler description of its dynamics.

(ii) The next level of systematic analysis is that of the simple dynamic system with predetermined, necessary motions. This might be called the level of *clockworks*. The solar system itself is of course the great clock of the universe from man's point of view, and the deliciously exact predictions of the astronomers are a testimony to the excellence of the clock which they study. Simple machines such as the lever and the pulley, even quite complicated machines like steam engines and dynamos fall mostly under this category. The greater part of the theoretical structure of physics, chemistry, and even of economics falls into this category. Two special cases might be noted. Simple equilibrium systems really fall into the dynamic category, as every equilibrium system must be considered as a limiting case of a dynamic system, and its stability cannot be determined except from the properties of its parent dynamic system. Stochastic dynamic systems leading to equilibria, for all their complexity, also fall into this group of systems; such is the modern view of the atom and even of the molecule, each position or part of the system being given with a certain degree of probability, the whole nevertheless exhibiting a determinate structure. Two types of analytical method are important here, which we may call, with the usage of the economists, comparative statics and true dynamics. In comparative statics we compare two equilibrium positions of the system under different values for the basic parameters. These equilibrium positions are usually expressed as the solution of a set of simultaneous equations. The method of comparative statics is to compare the solutions when the parameters of the equations are changed. Most simple mechanical problems are solved in this way. In true dynamics on the other hand we exhibit the system as a set of difference or differential equations, which are then solved in the form of an explicit function of each variable with time. Such a system may reach a position of stationary equilibrium, or it may not—there are plenty of examples of explosive dynamic systems, a very simple one being the growth of a sum at compound interest! Most physical and chemical reactions and most social systems do in fact exhibit a tendency to equilibrium—otherwise the world would have exploded or imploded long ago.

(iii) The next level is that of the control mechanism or cybernetic system, which might be nicknamed the level of the *thermostat*. This differs from the simple stable equilibrium system mainly in the fact that the transmission and interpretation of information is an essential part of the system. As a result of this the equilibrium position is not merely determined by the equations of the system, but the system will move to the maintenance of any *given* equilibrium, within limits. Thus the thermostat will maintain *any* temperature at which it can be set; the equilibrium temperature of the system is not determined solely by its equations. The trick here of course is that the essential variable of the dynamic system is the *difference* between an "observed" or "recorded" value of the maintained variable and its "ideal" value. If this difference is not zero the system moves so as to diminish it; thus the furnace sends up heat when the temperature as recorded is "too cold" and is turned off when the recorded temperature is "too hot." The homeostasis model, which is of such importance in physiology, is an example of a cybernetic mechanism, and such mechanisms exist through the whole empirical world of the biologist and the social scientist.

(iv) The fourth level is that of the "open system," or self-maintaining structure. This is the level at which life begins to differentiate itself from not-life: it might be called the level of the *cell*. Something like an open system exists, of course, even in physico-chemical equilibrium systems; atomic structures maintain themselves in the midst of a throughput of atoms. Flames and rivers likewise are essentially open systems of a very simple kind. As we pass up the scale of complexity of organization towards living systems, however, the property of self-maintenance of structure in the midst of a throughput of material becomes of dominant importance. An atom or a molecule can presumably exist without throughput: the existence of even the simplest living organism is inconceivable without ingestion, excretion and metabolic exchange. Closely connected with the property of self-maintenance is the property of self-reproduction. It may be, indeed, that self-reproduction is a more primitive or "lower level" system than the open system, and that the gene and the virus, for instance, may be able to reproduce themselves without being open systems. It is not perhaps an important question at what point in the scale of increasing complexity "life" begins. What is clear, however, is that by the time we have got to systems which both reproduce themselves and maintain themselves in the midst of a throughput of material and energy, we have something to which it would be hard to deny the title of "life."

(v) The fifth level might be called the genetic-societal level; it is typified by the *plant*, and it dominates the empirical world of the botanist. The outstanding characteristics of these systems are first, a division of labor among cells to form a cell-society with differentiated and mutually dependent parts (roots, leaves, seeds, etc.), and second, a sharp differentiation between the genotype and the phenotype, associated with the phenomenon of equifinal or "blueprinted" growth. At this level there are no highly specialized sense organs and information receptors are diffuse and incapable of much throughput of information—it is doubtful whether a tree can distinguish much more than light from dark, long days from short days, cold from hot.

(vi) As we move upward from the plant world towards the animal kingdom we gradually pass over into a new level, the "animal" level, characterized by increased mobility, teleological behavior, and self-awareness. Here we have the development of specialized information-receptors (eyes, ears, etc.) leading to an enormous increase in the intake of information; we have also a great development of nervous systems, leading ultimately to the brain, as an organizer of the information intake into a knowledge structure or "image." Increasingly as we ascend the scale of animal life, behavior is response not to a specific stimulus but to an "image" or knowledge structure or view of

8

the environment as a whole. This image is of course determined ultimately by information received into the organism; the relation between the receipt of information and the building up of an image however is exceedingly complex. It is not a simple piling up or accumulation of information received, although this frequently happens, but a structuring of information into something essentially different from the information itself. After the image structure is well established most information received produces very little change in the image—it goes through the loose structure, as it were, without hitting it, much as a sub-atomic particle might go through an atom without hitting anything. Sometimes however the information is "captured" by the image and added to it, and sometimes the information hits some kind of a "nucleus" of the image and a reorganization takes place, with far reaching and radical changes in behavior in apparent response to what seems like a very small stimulus. The difficulties in the prediction of the behavior of these systems arises largely because of this intervention of the image between the stimulus and the response.

(vii) The next level is the "human" level, that is of the individual human being considered as a system. In addition to all, or nearly all, of the characteristics of animal systems man possesses self consciousness, which is something different from mere awareness. His image, besides being much more complex than that even of the higher animals, has a self-reflexive quality—he not only knows, but knows that he knows. This property is probably bound up with the phenomenon of language and symbolism. It is the capacity for speech—the ability to produce, absorb, and interpret *symbols*, as opposed to mere signs like the warning cry of an animal—which most clearly marks man off from his humbler brethren. Man is distinguished from the animals also by a much more elaborate image of time and relationship; man is probably the only organization that knows that it dies, that contemplates in its behavior a whole life span, and more than a life span. Man exists not only in time and space but in history, and his behavior is profoundly affected by his view of the time process in which he stands.

(viii) Because of the vital importance for the individual man of symbolic images and behavior based on them it is not easy to separate clearly the level of the individual human organism from the next level, that of social organizations. In spite of the occasional stories of feral children raised by animals, man isolated from his fellows is practically unknown. So essential is the symbolic

image in human behavior that one suspects that a truly isolated man would not be "human" in the usually accepted sense, though he would be potentially human. Nevertheless it is convenient for some purposes to distinguish the individual human as a system from the social systems which surround him, and in this sense social organizations may be said to constitute another level of organization. The unit of such systems is not perhaps the person—the individual human as such—but the "role"—that part of the person which is concerned with the organization or situation in question, and it is tempting to define social organizations, or almost any social system, as a set of roles tied together with channels of communication. The interrelations of the role and the person however can never be completely neglected—a square person in a round role may become a little rounder, but he also makes the role squarer, and the perception of a role is affected by the personalities of those who have occupied it in the past. At this level we must concern ourselves with the content and meaning of messages, the nature and dimensions of value systems, the transcription of images into a historical record, the subtle symbolizations of art, music, and poetry, and the complex gamut of human emotion. The empirical universe here is human life and society in all its complexity and richness.

(ix) To complete the structure of systems we should add a final turret for transcendental systems, even if we may be accused at this point of having built Babel to the clouds. There are however the ultimates and absolutes and the inescapable unknowables, and they also exhibit systematic structure and relationship. It will be a sad day for man when nobody is allowed to ask questions that do not have any answers.

One advantage of exhibiting a hierarchy of systems in this way is that it gives us some idea of the present gaps in both theoretical and empirical knowledge. Adequate theoretical models extend up to about the fourth level, and not much beyond. Empirical knowledge is deficient at practically all levels. Thus at the level of the static structure, fairly adequate descriptive models are available for geography, chemistry, geology, anatomy, and descriptive social science. Even at this simplest level, however, the problem of the adequate description of complex structures is still far from solved. The theory of indexing and cataloguing, for instance, is only in its infancy. Librarians are fairly good at cataloguing books, chemists have begun to catalogue structural formulae, and anthropologists have begun to catalogue culture trails. The cataloguing of events, ideas, theories, statistics, and empirical data has hardly begun. The very multiplication of records however as time goes on will force us into much more adequate cataloguing and reference systems than we now have. This is perhaps the major unsolved theoretical problem at the level of the static structure. In the empirical field there are still great areas where static structures are very imperfectly known, although knowledge is advancing rapidly, thanks to new probing devices such as the electron microscope. The anatomy of that part of the empirical world which lies between the large molecule and the cell however, is still obscure at many points. It is precisely this area however— which includes, for instance, the gene and the virus— that holds the secret of life, and until its anatomy is made clear the nature of the functional systems which are involved will inevitably be obscure.

The level of the "clockwork" is the level of "classical" natural science, especially physics and astronomy, and is probably the most completely developed level in the present state of knowledge, especially if we extend the concept to include the field theory and stochastic models of modern physics. Even here however there are important gaps, especially at the higher empirical levels. There is much yet to be known about the sheer mechanics of cells and nervous systems, of brains and of societies.

Beyond the second level adequate theoretical models get scarcer. The last few years have seen great developments at the third and fourth levels. The theory of control mechanisms ("thermostats") has established itself as the new discipline or cybernetics, and the theory of self-maintaining systems or "open systems" likewise has made rapid strides. We could hardly maintain however that much more than a beginning had been made in these fields. We know very little about the cybernetics of genes and genetic systems, for instance, and still less about the control mechanisms involved in the mental and social world. Similarly the processes of self-maintenance remain essentially mysterious at many points, and although the theoretical possibility of constructing a self-maintaining machine which would be a true open system has been suggested, we seem to be a long way from the actual construction of such a mechanical similitude of life.

Beyond the fourth level it may be doubted whether we have as yet even the rudiments of theoretical systems. The intricate machinery of growth by which the genetic complex organizes the matter around it is almost a complete mystery. Up to now, whatever the future may hold, only God can make a tree. In the face of living systems we are almost helpless; we can occasionally co-operate with systems which we do not understand: we cannot even begin to reproduce them. The ambiguous status of medicine, hovering as it does uneasily between magic and science, is a testimony to the state of systematic knowledge in this area. As we move up the scale the absence of the appropriate theoretical systems becomes ever more noticeable. We can hardly conceive ourselves constructing a system which would be in any recognizable sense "aware," much less self conscious. Nevertheless as we move towards the human and societal level a curious thing happens: the fact that we have, as it were, an inside track, and that we ourselves *are* the systems which we are studying, enables us to utilize systems which we do not really understand. It is almost inconceivable that we should make a machine that would make a poem: nevertheless, poems *are* made by fools like us by processes which are largely hidden from us. The kind of knowledge and skill that we have at the symbolic level is very different from that which we have at lower levels—it is like, shall we say, the "knowhow" of the gene as compared with the knowhow of the biologist. Nevertheless it is a real kind of knowledge and it is the source of the creative achievements of man as artist, writer, architect, and composer.

Perhaps one of the most valuable uses of the above scheme is to prevent us from accepting as final a level of theoretical analysis which is below the level of the empirical world which we are investigating. Because, in a sense, each level incorporates all those below it, much valuable information and insights can be obtained by applying low-level systems to high-level subject matter. Thus most of the theoretical schemes of the social sciences are still at level (ii), just rising now to (iii), although the subject matter clearly involves level (viii), Economics, for instance, is still largely a "mechanics of utility and self interest," in Jevons' masterly phrase. Its theoretical and mathematical base is drawn largely from the level of simple equilibrium theory and dynamic mechanisms. It has hardly begun to use concepts such as information which are appropriate at level (iii), and makes no use of higher level systems. Furthermore, with this crude apparatus it has achieved a modicum of success, in the sense that anybody trying to manipulate an economic system is almost certain to be better off if he knows some economics than if he doesn't. Nevertheless at some point progress in economics is going to depend on its ability to break out of these low-level systems, useful as they are as first approximations, and utilize systems which are more directly appropriate to its universe—when, of course, these systems are discovered. Many other examples could be given—the wholly inappropriate use in psychoanalytic theory, for instance, of the concept of energy, and the long inability of psychology to break loose from a sterile stimulus-response model.

Finally, the above scheme might serve as a mild word of warning even to Management Science. This new discipline represents an important breakaway from overly simple mechanical models in the theory of organization and control. Its emphasis on communication systems and organizational structure, on principles of homeostasis and growth, on decision processes under uncertainty, is carrying us far beyond the simple models of maximizing behavior of even ten years ago. This advance in the level of theoretical analysis is bound to lead to more powerful and fruitful systems. Nevertheless we must never quite forget that even these advances do not carry us much beyond the third and fourth levels, and that in dealing with human personalities and organizations we are dealing with systems in the empirical world far beyond our ability to formulate. We should not be wholly surprised, therefore, if our simpler systems, for all their importance and validity, occasionally let us down.

I chose the subtitle of my paper with some eye to its possible overtones of meaning. General Systems Theory is the skeleton of science in the sense that it aims to provide a framework or structure of systems on which to hang the flesh and blood of particular disciplines and particular subject matters in an orderly and coherent corpus of knowledge. It is also, however, something of a skeleton in a cupboard—the cupboard in this case being the unwillingness of science to admit the very low level of its successes in systematization, and its tendency to shut the door on problems and subject matters which do not fit easily into simple mechanical schemes. Science, for all its successes, still has a very long way to go. General Systems Theory may at times be an embarrassment in pointing out how very far we still have to go, and in deflating excessive philosophical claims for overly simple systems. It also may be helpful however in pointing out to some extent *where* we have to go. The skeleton must come out of the cupboard before its dry bones can live.

Since creative thought is the most important thing that makes people different from monkeys, it should be treated as a commodity more precious than gold and preserved with great care.—A. D. HALL, *A Methodology for Systems Engineering.*

2.
General System Theory—A Critical Review

LUDWIG VON BERTALANFFY

IT IS MORE THAN 15 YEARS since the writer has first presented, to a larger public, the proposal of a General System Theory. Since then, this conception has been widely discussed and was applied in numerous fields of science. When an early reviewer found himself "hushed into awed silence" by the idea of a General System Theory, now in spite of obvious limitations, different approaches and legitimate criticism, few would deny the legitimacy and fertility of the interdisciplinary systems approach.

Even more: The systems concept has not remained in the theoretical sphere, but became central in certain fields of applied science. When first proposed, it appeared to be a particularly abstract and daring, theoretical idea. Nowadays "systems engineering," "research," "analysis" and similar titles have become job denominations. Major industrial enterprises and government agencies have departments, committees or at least specialists to the purpose; and many universities offer curricula and courses for training.

Thus the present writer was vindicated when he was among the first to predict that the concept of "system" is to become a fulcrum in modern scientific thought. In the words of a practitioner of the science [R. L. Ackoff]:

In the last two decades we have witnessed the emergence of the "system" as a key concept in scientific research. Systems, of course, have been studied for centuries, but something new has been added. . . . The tendency to study systems as an entity rather than as a conglomeration of parts is consistent with the tendency in con-temporary science no longer to isolate phenomena in narrowly confined contexts, but rather to open interactions for examination and to examine larger and larger slices of nature. Under the banner of *systems research* (and its many synonyms) we have also witnessed a convergence of many more specialized contemporary scientific developments. . . . These research pursuits and many others are being interwoven into a cooperative research effort involving an ever-widening spectrum of scientific and engineering disciplines. We are participating in what is probably the most comprehensive effort to attain a synthesis of scientific knowledge yet made.

This, however, does not preclude but rather implies that obstacles and difficulties are by no means overcome as is only to be expected in a major scientific reorientation. A reassessment of General Systems Theory, its foundations, achievements, criticisms and prospects therefore appears in place. The present study aims at this purpose.

According to the Preface to the VIth volume of *General Systems* by Meyer, the greatest number of enquiries made asks for "new statements describing the method and significance of the idea." Another central theme is "the organismic viewpoint." As one of the original proponents of the *Society for General Systems Research* and founders of the organismic viewpoint in biology, the author feels obliged to answer this challenge as well as readily admitted limitations of his knowledge and techniques permit.

1. The Rise of Interdisciplinary Theories

The motives leading to the postulate of a general theory of systems can be summarized under a few headings.

From Ludwig von Bertalanffy, "General System Theory—A Critical Review," *General Systems*, VII (1962), 1–20. Reprinted with permission of the author and the Society for General Systems Research.

1. Up to recent times the field of science as a nomothetic endeavor, i.e., trying to establish an explanatory and predictive system of laws, was practically identical with theoretical physics. Few attempts at a system of laws in non-physical fields gained general recognition; the biologist would first think of genetics. However, in recent times the biological, behavioral and social sciences have come into their own, and so the problem became urgent whether an expansion of conceptual schemes is possible to deal with fields and problems where application of physics is not sufficient or feasible.

2. In the biological, behavioral and sociological fields, there exist predominant problems which were neglected in classical science or rather which did not enter into its considerations. If we look at a living organism, we observe an amazing order, organization, maintenance in continuous change, regulation and apparent teleology. Similarly, in human behavior goal-seeking and purposiveness cannot be overlooked, even if we accept a strictly behavioristic standpoint. However, concepts like organization, directiveness, teleology, etc., just do not appear in the classic system of science. As a matter of fact, in the so-called mechanistic world view based upon classical physics, they were considered as illusory or metaphysical. This means, to the biologist for example, that just the specific problems of living nature appeared to lie beyond the legitimate field of science.

3. This in turn was closely connected with the structure of classical science. The latter was essentially concerned with two-variable problems, linear causal trains, one cause and one effect, or with few variables at the most. The classical example is mechanics. It gives perfect solutions for the attraction between two celestial bodies, a sun and a planet, and hence permits to exactly predict future constellations and even the existence of still undetected planets. However, already the three-body problem of mechanics is unsolvable in principle and can only be approached by approximations. A similar situation exists in the more modern field of atomic physics. Here also two-body problems such as that of one proton and electron are solvable, but trouble arises with the many-body problem. One-way causality, the relation between "cause" and "effect" or of a pair or few variables cover a wide field. Nevertheless, many problems particularly in biology and the behavioral and social sciences, essentially are multivariable problems for which new conceptual tools are needed. Warren Weaver, cofounder of information theory, had expressed this in an often-quoted statement. Classical science, he

stated, was concerned either with linear causal trains, that is, two-variable problems; or else with unorganized complexity. The latter can be handled with statistical methods and ultimately stems from the second principle of thermodynamics. However, in modern physics and biology, problems of organized complexity, that is, interaction of a large but not infinite number of variables, are popping up everywhere and demand new conceptual tools.

4. What has been said are not metaphysical or philosophic contentions. We are not erecting a barrier between inorganic and living nature which obviously would be inappropriate in view of intermediates such as viruses, nucleoproteins and self-duplicating units in general which in some way bridge the gap. Nor do we protest that biology is in principle "irreducible to physics" which also would be out of place in view of the tremendous advances of physical and chemical explanation of life processes. Similarly, no barrier between biology and the behavioral and social sciences is intended. This, however, does not obviate the fact that in the fields mentioned we do not have appropriate conceptual tools serving for explanation and prediction as we have in physics and its various fields of application.

5. It therefore appears that an expansion of science is required to deal with those aspects which are left out in physics and happen to concern just the specific characteristics of biological, behavioral, and social phenomena. This amounts to new conceptual models to be introduced. Every science is a model in the broad sense of the word, that is a conceptual structure intended to reflect certain aspects of reality. One such model is the system of physics—and it is an incredibly successful one. However, physics is but *one* model dealing with certain aspects of reality. It needs not to have monopoly, nor is it *the* reality as mechanistic methodology and metaphysics presupposed. It apparently does not cover all aspects and represents, as many specific problems in biology and behavioral science show, a limited aspect. Perhaps it is possible to introduce other models dealing with aspects outside of physics.

These considerations are of a rather abstract nature. So perhaps some personal interest may be introduced by telling how the present author was led into this sort of problem.

When, some 40 years ago, I started my life as a scientist, biology was involved in the mechanism–vitalism controversy. The mechanistic procedure essentially was to resolve the living organism into parts and partial processes: the organism was an aggregate of cells, the cell one of colloids and organic molecules, behavior a sum of unconditional and conditioned reflexes, and so forth. The problems of organization of these parts in the service of maintenance of the organism, of regulation after disturbances and the like were either by-passed or, according to the theory known as vitalism, explainable only by the action of soul-like factors, little hobgoblins as it were, hovering in the cell or the organism—which obviously was nothing less than a declaration of backruptcy of science. In this situation, I was led to advocate the so-called organismic viewpoint. In one brief sentence, it means that organisms are organized things and, as biologists, we have to find out about it. I tried to implement this organismic program in various studies on metabolism, growth, and biophysics of the organism. One way in this respect was the so-called theory of open systems and steady states which essentially is an expansion of conventional physical chemistry, kinetics and thermodynamics. It appeared, however, that I could not stop on the way once taken and so I was led to a still further generalization which I called "General System Theory." The idea goes back for some considerable time—I presented it first in 1937 in Charles Morris' philosophy seminar at the University of Chicago. However, at this time theory was in bad reputation in biology, and I was afraid of what Gauss, the mathematician, called the "clamor of the Boeotians." So I left my drafts in the drawer, and it was only after the war that my first publications in this respect appeared.

Then, however, something interesting and surprising happened. It turned out that a change in intellectual climate had taken place, making model building and abstract generalizations fashionable. Even more: quite a number of scientists had followed similar lines of thought. So General System Theory, after all, was not isolated or a personal idiosyncrasy as I have believed, but rather was one within a group of parallel developments.

Naturally, the maxims enumerated above can be formulated in different ways and using somewhat different terms. In principle, however, they express the viewpoint of the more advanced thinkers of our time and the common ground of system theorists. The reader may, for example, compare the presentation given by Rapoport and Horvath which is an excellent and independent statement and therefore shows even better the general agreement.

There is quite a number of novel developments intended to meet the goals indicated above. We may enumerate them in a brief survey:

(1) Cybernetics, based upon the principle of feedback or circular causal trains providing mechanisms for goal-seeking and self-controlling behavior.

(2) Information theory, introducing the concept of information as a quantity measurable by an expression isomorphic to negative entropy in physics, and developing the principles of its transmission.

(3) Game theory, analyzing in a novel mathematical framework, rational competition between two or more antagonists for maximum gain and minimum loss.

(4) Decision theory, similarly analyzing rational choices, within human organizations, based upon examination of a given situation and its possible outcomes.

(5) Topology or relational mathematics, including non-metrical fields such as network and graph theory.

(6) Factor analysis, i.e., isolation by way of mathematical analysis, of factors in multivariable phenomena in psychology and other fields.

(7) General system theory in the narrower sense (G.S.T.), trying to derive from a general definition of "system" as complex of interacting components, concepts characteristic of organized wholes such as interaction, sum, mechanization, centralization, competition, finality, etc., and to apply them to concrete phenomena.

While systems theory in the broad sense has the character of a basic science, it has its correlate in applied science, sometimes subsumed under the general name of Systems Science. This development is closely connected with modern automation. Broadly speaking, the following fields can be distinguished:

Systems Engineering, i.e., scientific planning, design, evaluation, and construction of man–machine systems;

Operations research, i.e., scientific control of existing systems of men, machines, materials, money, etc.

Human Engineering, i.e., scientific adaptation of systems and especially machines in order to obtain maximum efficiency with minimum cost in money and other expenses.

A very simple example for the necessity of study of "man–machine systems" is air travel. Anybody crossing continents by jet with incredible speed and having to spend endless hours waiting, queuing, being herded in airports can easily realize that the physical techniques in air travel are at their best, while "organizational" techniques still are on a most primitive level.

Although there is considerable overlapping, different conceptual tools are predominant in the individual fields. In systems engineering, cybernetics and information theory, also general system theory are used. Operations research uses tools such as linear programming and game theory. Human engineering, concerned with the abilities, physiological limitations and variabilities of human beings, includes biomechanics, engineering psychology, human factors, etc., among its tools.

The present survey is not concerned with applied systems science; the reader is referred to Hall's book as an excellent textbook of systems engineering. However it is well to keep in mind that the systems approach as a novel concept in science has a close parallel in technology. The systems viewpoint in recent science stands in a similar relation to the so-called "mechanistic" viewpoint, as stands systems engineering to physical technology.

All these theories have certain features in common. *Firstly*, they agree in the emphasis that something should be done about the problems characteristic of the behavioral and biological sciences, but not dealt with in conventional physical theory. *Secondly*, these theories introduce concepts and models novel in comparison to physics: for example, a generalized system concept, the concept of information compared to energy in physics. *Thirdly*, these theories are particularly concerned with multivariable problems, as mentioned before. *Fourthly*, these models are interdisciplinary and transcend the conventional fields of science. If, for example, you scan the *Yearbooks* of the *Society for General Systems Research*, you notice the breadth of application: Considerations similar or even identical in structure are applied to phenomena of different kinds and levels, from networks of chemical reactions in a cell to populations of animals, from electrical engineering to the social sciences. Similarly, the basic concepts of cybernetics stem from certain special fields in modern technology. However, starting with the simplest case of a thermostat which by way of feedback maintains a certain temperature and advancing to servomechanisms and automation in modern technology, it turns out that similar schemes are applicable to many biological phenomena of regulation or behavior. Even more, in many instances there is a formal correspondence or isomorphism of general principles or even of special laws. Similar mathematical formulations may apply to quite different phenomena. This entails that general theories of systems, among other things, are labor-saving devices: A set of

principles may be transferred from one field to another, without the need to duplicate the effort as has often happened in science of the past. *Fifthly* and perhaps most important: Concepts like wholeness, organization, teleology and directiveness appeared in mechanistic science to be unscientific or metaphysical. Today they are taken seriously and as amenable to scientific analysis. We have conceptual and in some cases even material models which can represent those basic characteristics of life and behavioral phenomena.

An important consideration is that the various approaches enumerated are not, and should not be considered to be monopolistic. One of the important aspects of the modern changes in scientific thought is that there is no unique and all-embracing "world system." All scientific constructs are models representing certain aspects or perspectives of reality. This even applies to theoretical physics: far from being a metaphysical presentation of ultimate reality (as the materialism of the past proclaimed and modern positivism still implies) it is but one of these models and, as recent developments show, neither exhaustive nor unique. The various "systems theories" also are models that mirror different aspects. They are not mutually exclusive and often combined in application. For example, certain phenomena may be amenable to scientific exploration by way of cybernetics, others by way of general system theory; or even in the same phenomenon, certain aspects may be describable in the one or the other way. Cybernetics combine the information and feedback models, models of the nervous system net and information theory, etc. This, of course, does not preclude but rather implies the hope for further synthesis in which the various approaches of the present toward a theory of "wholeness" and "organization" may be integrated and unified. Actually, such further syntheses, e.g., between irreversible thermodynamics and information theory, are slowly developing.

The differences of these theories are in the particular model conceptions and mathematical methods applied. We therefore come to the question in what ways the program of systems research can be implemented.

2. Methods of General Systems Research

Ashby has admirably outlined two possible ways or general methods in systems study:

Two main lines are readily distinguished. One, already well developed in the hands of von Bertalanffy and his co-workers, takes the world as we find it, examines the various systems that occur in it—zoological, physiological, and so on—and then draws up statements about the regularities that have been observed to hold. This method is essentially empirical. The second method is to start at the other end. Instead of studying first one system, then a second, then a third, and so on, it goes to the other extreme, considers the set of all conceivable systems and then reduces the set to a more reasonable size. This is the method I have recently followed.

It will easily be seen that all systems studies follow one or the other of these methods or a combination of both. Each of these approaches has its advantages as well as shortcomings.

(1) The first method is empirico-intuitive; it has the advantage that it remains rather close to reality and can easily be illustrated and even verified by examples taken from the individual fields of science. On the other hand, the approach lacks mathematical elegance and deductive strength and, to the mathematically minded, will appear naïve and unsystematic.

Nevertheless, the merits of this empirico-intuitive procedure should not be minimized.

The present writer has stated a number of "system principles," partly in the context of biological theory and without explicit reference to G.S.T., partly in what emphatically was entitled an "Outline" of this theory. This was meant in the literal sense: It was intended to call attention to the desirability of such field, and the presentation was in the way of a sketch or blueprint, illustrating the approach by simple examples.

However, it turned out that this intuitive survey appears to be remarkably complete. The main principles offered such as wholeness, sum, centralization, differentiation, leading part, closed and open system, finality, equifinality, growth in time, relative growth, competition, have been used in manifold ways (e.g., general definition of system; types of growth; systems engineering; social work). Excepting minor variations in terminology intended for clarification or due to the subject matter, no principles of similar significance were added—even though this would be highly desirable. It is perhaps even more significant that this also applies to considerations which do not refer to the present writer's work and hence cannot be said to be unduly influenced by it. Perusal of studies such as those by Beer and Kremyanskiy on principles, Bradley and Calvin on the network of chemical reactions, Haire on growth or organizations, etc., will easily show that they are also using the "Bertalanffy principles."

(2) The way of deductive systems theory was

followed by Ashby. A more informal presentation which summarizes Ashby's reasoning lends itself particularly well to analysis.

Ashby asks about the "fundamental concept of machine" and answers the question by stating "that its internal state, and the state of its surroundings, defines uniquely the next state it will go to." If the variables are continuous, this definition corresponds to the description of a dynamic system by a set of ordinary differential equations with time as the independent variable. However, such representation by differential equations is too restricted for a theory to include biological systems and calculating machines where discontinuities are ubiquitous. Therefore the modern definition is the "machine with input": It is defined by a set S of internal states, a set I of input and a mapping f of the product set $I \times S$ into S. "Organization," then, is defined by specifying the machine's states S and its conditions I. If S is a product set $S = \pi_i T_i$, with i as the parts and T is specified by the mapping f. A "self-organizing" system, according to Ashby, can have two meanings, namely: (1) The system starts with its parts separate, and these parts then change toward forming connections (example: cells of the embryo, first having little or no effect on one another, join by formation of dendrites and synapses to form the highly interdependent nervous system). This first meaning is "changing from unorganized to organized." (2) The second meaning is "changing from a bad organization to a good one" (examples: a child whose brain organization makes it fire-seeking at first, while a new brain organization makes him fire-avoiding; an automatic pilot and plane coupled first by deleterious positive feedback and then improved). "There the organization is bad. The system would be 'self-organizing' if a change were automatically made" (changing positive into negative feedback). But "*no machine can be self-organizing in this sense*" (author's emphasis). For adaptation (e.g., of the homeostat or in a self-programming computer) means that we start with a set S of states, and that f changes into g, so that organization is a variable, e.g., a function of time $\alpha(t)$ which has first the value f and later the value g. However, this change "cannot be ascribed to any cause in the set S; *so it must come from some outside agent, acting on the system* S *as input*" (our emphasis). In other terms, to be "self-organizing" the machine S must be coupled to another machine.

This concise statement permits observation of the limitations of this approach. We completely agree that description by differential equations is not only a clumsy but, in principle, inadequate way to deal with many problems of organization. The author was well aware of this emphasizing that a system of simultaneous differential equations is by no means the most general formulation and is chosen only for illustrative purposes.

However, in overcoming this limitation, Ashby introduced another one. His "modern definition" of system as a "machine with input" as reproduced above, supplants the general system model by another rather special one: the cybernetic model, i.e., a system open to information but closed with respect to entropy transfer. This becomes apparent when the definition is applied to "self-organizing systems." Characteristically, the most important kind of these has no place in Ashby's model, namely, systems organizing themselves by way of progressive differentiation, evolving from states of lower to states of higher complexity. This is, of course, the most obvious form of "self-organization," apparent in ontogenesis, probable in phylogenesis, and certainly also valid in many social organizations. We have here not a question of "good" (i.e., useful, adaptive) or "bad" organization which, as Ashby correctly emphasizes, is relative on circumstances; increase in differentiation and complexity—whether useful or not—is a criterion that is objective and at least on principle amenable to measurement (e.g., in terms of decreasing entropy, of information). Ashby's contention that "no machine can be self-organizing," more explicitly, that the "change cannot be ascribed to any cause in the set S" but "must come from some outside agent, an input" amounts to exclusion of self-differentiating systems. The reason that such systems are not permitted as "Ashby machines" is patent. Self-differentiating systems that evolve toward higher complexity (decreasing entropy) are, for thermodynamic reasons, possible only as open systems, i.e., systems importing matter containing free energy to an amount over-compensating the increase in entropy due to irreversible processes within the system ("import of negative entropy"). However, we cannot say that "this change comes from some outside agent, an input"; the differentiation

within a developing embryo and organism is due to its internal laws of organization, and the input (e.g., oxygen supply which may vary quantitatively, or nutrition which can vary qualitatively within a broad spectrum) makes it only possible energetically.

The above is further illustrated by additional examples given by Ashby. Suppose a digital computer is carrying through multiplications at random; then the machine will "evolve" toward showing even numbers (because products even × even as well as even × odd give numbers even), and eventually only zeros will be "surviving." In still another version Ashby quotes Shannon's Tenth Theorem, stating that if a correction channel has capacity H, equivocation of the amount H can be removed, but no more. Both examples illustrate the working of closed systems: The "evolution" of the computer is one toward disappearance of differentiation and establishment of maximum homogeneity (analog to the Second Principle in closed systems); Shannon's Theorem similarly concerns closed systems where no negative entropy is fed in. Compared to the information content (organization) of a living system, the imported matter (nutrition, etc.) carries not information but "noise." Nevertheless, its negative entropy is used to maintain or even to increase the information content of the system. This is a state of affairs apparently not provided for in Shannon's Tenth Theorem, and understandably so as he is not treating information transfer in open systems with transformation of matter.

In both respects, the living organism (and other behavioral and social systems) is not an Ashby machine because it evolves toward increasing differentiation and inhomogeneity, and can correct "noise" to a higher degree than an inanimate communication channel. Both, however, are consequences of the organism's character as an open system.

Incidentally, it is for similar reasons that we cannot replace the concept of "system" by the generalized "machine" concept of Ashby. Even though the latter is more liberal compared to the classic one (machines defined as systems with fixed arrangement of parts and processes), the

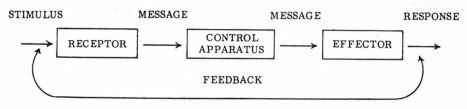

Figure 1.—Simple feedback model.

objections against a "machine theory" of life remain valid.

These remarks are not intended as adverse criticism of Ashby's or the deductive approach in general; they only emphasize that there is no royal road to General Systems Theory. As every other scientific field, it will have to develop by an interplay of empirical, intuitive and deductive procedures. If the intuitive approach leaves much to be desired in logical rigor and completeness, the deductive approach faces the difficulty of whether the fundamental terms are correctly chosen. This is not a particular fault of the theory or of the workers concerned but a rather common phenomenon in the history of science; one may, for example, remember the long debate as to what magnitude—force or energy—is to be considered as constant in physical transformations until the issue was decided in favor of $mv^2/2$.

In the present writer's mind, G.S.T. was conceived as a working hypothesis; being a practicing scientist, he sees the main function of theoretical models in the explanation, prediction and control of hitherto unexplored phenomena. Others may, with equal right, emphasize the importance of axiomatic approach and quote to this effect examples like the theory of probability, non-Euclidean geometries, more recently information and game theory, which were first developed as deductive mathematical fields, and later applied in physics or other sciences. There should be no quarrel about this point. The danger, in both approaches, is to consider too early the theoretical model as being closed and definitive—a danger particularly important in a field like general systems which is still groping to find its correct foundations.

3. Homeostasis and Open Systems

Among the models mentioned, cybernetics in its application as homeostasis, and G.S.T. in its application to open systems lend themselves most readily for interpretation of many empirical phenomena. The relation of both theories is not always well understood, and hence a brief discussion is in place.

The simplest feedback scheme can be represented as follows (Fig. 1). Modern servomechanisms and automation, as well as many phenomena in the organism, are based upon feedback arrangements far more complicated than the simple scheme (Fig. 1) but the latter is the elementary prototype.

In application to the living organism, the feedback scheme is represented by the concept of homeostasis.

Homeostasis, according to Cannon, is the ensemble of organic regulations which act to maintain the steady states of the organism and are effectuated by regulating mechanisms in such a way that they do not occur necessarily in the same, and often in opposite, direction to what a corresponding external change would cause according to physical laws. The simplest example is homeothermy. According to Van't Hoff's rule in physical chemistry, a decrease in temperature leads to slowing down of the rate of chemical reactions, as it does in ordinary physico-chemical systems and also in poikilothermic animals. In warm-blooded animals, however, it leads to the opposite effect, namely, to an increase of metabolic rate, with the result that the temperature of the body is maintained constant at approximately 37°C. This is effectuated by a feedback mechanism. Cooling stimulates thermogenic centers in the brain thalamus which "turn on" heat-producing mechanisms in the body. A similar feedback pattern is found in a great variety of physiological regulations. Regulation of posture and the control of actions in animals and man toward a goal are similarly controled by feedback mechanisms.

In contradistinction to cybernetics concerned with feedback arrangements, G.S.T. is interested in dynamic interaction within multivariable systems. The case particularly important for the living organism is that of open systems. It amounts to saying that there is a system into which matter is introduced from outside. Within the system, the material undergoes reactions which partly may

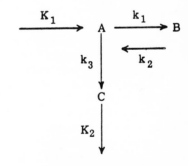

Figure 2.—Model of a simple open system. The component A is introduced into the system and transformed, in a reversible reaction, into B; it is catabolized in an irreversible reaction, into C which eventually is excreted. K_1, K_2 are constants of import and export, respectively; k_1, k_2, k_3 are reaction constants. The model approximately corresponds, for example, to protein turnover in an animal organism, A representing amino acids, B proteins, and C products of excretion.

yield components of a higher complexity. This is what we call anabolism. On the other hand, the material is catabolized and the end products of catabolism eventually leave the system. A simple model of an open system is indicated in Figure 2.

A few main characteristics of open as compared to closed systems are in the fact that, appropriate system conditions presupposed, an open system will attain a steady state in which its composition remains constant, but in contrast to conventional equilibria, this constancy is maintained in a continuous exchange and flow of component material. The steady state of open systems is characterized by the principle of equifinality; that is, in contrast to equilibrium states in closed systems which are determined by initial conditions, the open system may attain a time-independent state independent of initial conditions and determined only by the system parameters. Furthermore, open systems show thermodynamic characteristics which are apparently paradoxical and contradictory to the second principle. According to the latter, the general course of physical events (in closed systems) is toward increasing entropy, leveling down of differences and states of maximum disorder. In open systems, however, with transfer of matter import of "negative entropy" is possible. Hence, such systems can maintain themselves at a high level, and even evolve toward an increase of order and complexity—as is indeed one of the most important characteristics of life processes.

The open-system model also has a wide application. According to its character, it is particularly applicable to phenomena showing non-structural, dynamic interaction of processes, such as those of metabolism, growth, metabolic aspects of excitation, etc.

Speaking more generally, living systems can be defined as hierarchically organized open systems, maintaining themselves, or developing toward a steady state. Disease is the life process regulating toward normalcy after disturbance, owing to the equifinality of biological systems and with the assistance of the physician. In this way, the *vis medicatrix naturae* of old is divested of its metaphysical paraphernalia; it is not a vitalistic agent but an expression of the dynamics of living systems, maintaining and reestablishing, so far as possible, the steady state.

In this way, the theory of open systems accounts for basic characteristics of the living organism which have baffled physicists, biologists, and philosophers, and appeared to be violations of the laws of physics, explainable only by vitalistic factors beyond the competence of science and scientific explanation.

Thus "feedback" and "open systems" are two models for biological and possibly behavioral phenomena. It should be made clear that the term "homeostasis" can be used in two ways. It is either taken in the original sense as proferred by Cannon and illustrated by examples like maintenance of body temperature and other physiological variables by feedback mechanisms—or else the term is often used as a synonym for organic regulation and adaptation in general. This is a question of semantics. However, it is a wise rule in the natural sciences to use terms in the sense originally attached to them by their authors. So I propose to use the word homeostasis in its narrower but well-defined sense, and this has important consequences, as it reveals certain limitations which are often forgotten.

As was already emphasized, regulations of the homeostasis or feedback type are abundant in the mature higher organism. However, it is clear from the scheme (Fig. 1) or any other flow diagram that feedback represents a machine-like arrangement, that is, an order of processes based upon fixed arrangements and representing linear, though circular, causal trains. The primary phenomena of organic regulation, e.g., the regulations in early embryonic development, in regeneration, etc., appear to be of a different nature. It seems that the primary regulations in the organism result from dynamic interaction within a unitary open system that reestablishes its steady state. Superimposed upon this by way of progressive mechanization are secondary regulatory mechanisms governed by fixed structures especially of the feedback type.

Although the homeostasis model transcends older mechanistic models by acknowledging directiveness in self-regulating circular processes, it still adheres to the machine theory of the organism. This also applies to a second aspect. An essential component of the mechanistic view is a utilitarian conception which is deeply connected with the economic outlook of the 19th and early 20th centuries. This is well-known, for example, in the history of Darwinism: Struggle for existence and survival of the fittest are a biological version of the economic model of free competition. This utilitarian or economic viewpoint also prevails in the concept of homeostasis: The organism is essentially envisaged as an aggregate mechanism for maintenance of minimum costs. However, there seem to be plenty of non-utilitarian structures and functions in the living world.

The concept of homeostasis also retains a third aspect of the mechanistic view. The organism is essentially considered to be a reactive system. Outside stimuli are answered by proper responses in such a way as to maintain the system. The feedback model (Fig. 1) is essentially the classical stimulus-response scheme, only the feedback loop being added. However, an overwhelming amount of facts shows that primary organic behavior as, for example, the first movements of the fetus, are not reflex responses to external stimuli, but rather spontaneous mass activities of the whole embryo or larger areas. Reflex reactions answering external stimuli and following a structured path appear to be superimposed upon primitive automatisms, ontogenetically and phylogenetically, as secondary regulatory mechanisms. These considerations win particular importance in the theory of behavior, as we shall see later on.

In this sense, it appears that in development and evolution dynamic interaction (open system) precedes mechanization (structured arrangements particularly of a feedback nature). In a similar way, G.S.T. can logically be considered the more general theory; it includes systems with feedback constraints as a special case, but this assertion would not be true vice versa. It need not be emphasized that this statement is a program for future systematization and integration of G.S.T. rather than a theory presently achieved.

4. Criticism of General System Theory

A discussion of G.S.T. must take account of the objections raised, both to clarify misunderstanding and to utilize criticism for improvement.

A "devastating" criticism of "General Behavior Systems Theory" by Buck would hardly deserve discussion were it not for the fact that it appeared in the widely read *Minnesota Studies in the Philosophy of Science*, a leading publication of modern positivism. In passing, it should be noted that the lack of interest in, or even hostility of logical positivists against, G.S.T. is a rather remarkable phenomenon. One would expect that a group whose program is "Unified Science" should be concerned with a novel approach to this problem, however immature it may still be. The opposite is the case; no contribution or even pertinent criticism came forward from these quarters. The reason is not difficult to see. Abandoning the debatable but challenging position of Logical Positivism and replacing it by a rather tame "Empirical Realism," modern positivists have come back to what is generally agreed among modern scientists, avoiding commitments which trespass and would imply an adventure of thought. It needs to be said that modern positivism has been a singularly sterile movement. It is paradoxical that the declared "philosophers of science" have neither contributed any empirical research nor new idea to modern science—while professional or half-time philosophers who were justly censored for their "mysticism," "metaphysics," or "vitalism," indubitably did. Eddington and Jeans in physics, Driesch in biology, Spengler in history are but a few examples.

Buck's critique is not directed against the present author but against J. G. Miller and his Chicago group. Its essence is in the "So what?" argument: Supposing we find an analogy or formal identity in two "systems," it means nothing. Compare, for example, a chessboard and a mixed dinner party; a general statement expressing the alternation of black and white squares on the one hand, and of men and women on the other can be made. "If one is tempted to say 'All right, so they're structurally analogous, so what?' my answer is 'So, nothing.'" In the same vein, Buck pokes fun at some of Miller's more hazardous comparisons, such as of the behavior of slime molds and Londoners during the *blitz*. He asks, "What are we to conclude from all this? That Londoners are a form of slime mold? That myxamoebae are a sort of city dweller?" Or "if no conclusion, why bother with the analogy at all?"

As proof of the emptiness of analogies Buck offers the example of a scientist, A, who finds a formula for the rate of formation of frost in a refrigerator; of another, B, formulating the rate of carbon deposit in an automobile motor; and a "general systems theorist," C, who notices that both formulas are the same. The similarity of mathematical expressions and models is, according to Buck, "sheer coincidence"—it does not prove that a refrigerator is an automobile or vice versa, but only that both are "systems" of some sort. This, however, is a meaningless statement; for

One is unable to think of anything, or of any combination of things, which could not be regarded as a system. And, of course, a concept that applies to everything is logically empty.

Regardless of the question whether Miller's is a particularly felicitous presentation, Buck has simply missed the issue of a general theory of systems. Its aim is not more or less hazy analogies; it is to establish principles applicable to entities not covered in conventional science. Buck's criticism is, in principle, the same as if one would criticize Newton's law because it draws a loose "analogy" between apples, planets, ebb and tide and many other entities; or if one would declare the theory of probability meaningless because it is concerned with the "analogy" of games of dice, mortality statistics, molecules in a gas, the distribution of hereditary characteristics, and a host of other phenomena.

The basic role of "analogy"—or rather of isomorphisms and models in science—has been lucidly discussed by Ashby. Hence a few remarks in answer to Buck will suffice.

The "So what?" question mistakes a method which is fundamental in science although—like every method—it can be misused. Even Buck's first example is not a meaningless pseudoproblem; in the "analogy" of chessboard and dinner party topology may find a common structural principle that is well worth stating. Generally speaking, the use of "analogy" (isomorphism, logical homology) —or, what amounts to nearly the same, the use of conceptual and material models—is not a half-poetical play but a potent tool in science. Where would physics be without the analogy or model of "wave," applicable to such dissimilar phenomena as water waves, sound waves, light and electromagnetic waves, "waves" (in a rather Pickwickian sense) in atomic physics? "Analogies" may pose fundamental problems, as for example, the analogy (logically not dissimilar from that of chessboard and dinner party) of Newton's and Coulomb's law which raises the question (one of the most basic for "Unified Science") of a general field theory unifying mechanics and electrodynamics. It is commonplace in cybernetics that systems which are different materially; e.g., a mechanical and an electrical system, may be formally identical; far from considering this as a meaningless So what? the researcher has to work out the common structure (flow diagram), and this may be of incomparable value for practical technology.

A similar lack of understanding is manifest in the criticism of the system concept. By the same token ("One is unable to think of anything" which would not show the properties in question) mechanics would have to be refused as "logically empty" because every material body shows mass, acceleration, energy, etc. In the following paragraphs of his paper, Buck has some glimpse of this truism, but he soon comes back to ridiculing Miller's use of "analogies."

Although Buck justly criticizes certain un-

fortunate formulations, his misunderstanding of the basic problems involved makes one wonder how his essay found its way into a treatise on "Philosophy of Science."

At an incomparably higher level stands the criticism by the Soviet authors, Lektorsky and Sadovsky. The writers give a sympathetic and fair presentation of Bertalanffy's G.S.T. sketching diligently its gradual evolution from "organismic biology" and the theory of open systems. In view of the above criticism by Buck, the following quotation is of interest:

Bertalanffy emphasizes the idea that a general system theory is not an investigation of hazy and superficial analogies. ... Analogies as such have little value, since differences can always be found among phenomena as well as similarities. Bertalanffy declares that the kind of isomorphism with which general system theory is concerned is a consequence of the fact that in some respects corresponding abstractions and conceptual models can be applied to different phenomena.

"We can only welcome [the] goal [of G.S.T.]," write Lektorsky and Sadovsky, "i.e., the attempt to give a general definition of the concept of 'organized system,' to classify logically various types of systems and to work out mathematical models for describing them . . . Bertalanffy's theory of organization and of organized complexes is a special scientific discipline. At the same time it certainly fulfills a definite methodological function" (i.e., avoiding duplication of effort in various disciplines by a single formal apparatus). "Its mathematical apparatus can be utilized for analyzing a comparatively large class of systemed objects of interest to biologists, chemists, biochemists, biophysicists, psychologists and others."

The criticism of the Russian authors is directed against imperfections of G.S.T. which, unfortunately, cannot be denied: "Bertalanffy's definition is rather a description [not pretending to precision] of the class of events which we may call systems than a strictly logical definition." "The description contains no trace of logical elegance." "Elementary methods of analysis and synthesis are insufficient for the analysis of systems." Fairly enough the authors concede that "The flaws we have noted speak only for the fact that general system theory, like any scientific theory, should develop further and in the process of development should strive for more adequate reflection of the objects of investigation."

The "main flaws of the theory," according to Lektorsky and Sadovsky, are in the lack of "methodology" (i.e., presumably of rules to establish and to apply system principles) and in considering G.S.T. "a philosophy of modern science." With respect to the first item, the present study is devoted to just this problem. The second point is a misunderstanding. G.S.T. in its present form is one—and still very imperfect—model among others. Were it completely developed, it would indeed incorporate the "organismic" world view of our time, with its emphasis on problems of wholeness, organization, directiveness, etc., in a similar way as when previous philosophies have presented a mathematical world view (philosophies *more geometrico*), a physicalistic one (the mechanistic philosophy based upon classical physics), etc., corresponding to scientific development. Even then, this "organismic" picture would not claim to be a "Nothing-but" philosophy: It would remain conscious that it only presents certain aspects of reality (richer and more comprehensive than previous ones, as corresponds to the advance of science), but never exhaustive, exclusive or final.

According to the authors, Marxist–Leninist philosophy "formulates a series of most important methodological principles of analysis of complex systems"; Soviet scientists "attempt to give a general definition of the notion of systems and to obtain a classification." Difficulties in international communication make it unfortunately impossible to the present writer to evaluate these claims.

Another criticism backed by the same *weltanschauung* is that of Kamarýt. The main arguments are:

(1) Underestimation of the structural and morphologic aspects of organization of the theory of open systems (and implicitly in G.S.T.). The theory of open system does not "solve" the problem of life, its origin and evolution which is successfully attacked in modern biochemistry, submicroscopic morphology, physiological genetics, etc. The reply to this is that the functional and processual aspect has been emphasized in the theory, particularly in contradistinction to structural, homeostatic mechanisms. But neither the importance of the latter is denied, nor of course the specificity of the material basis of life. "Morphology and physiology are different and complementary ways of studying the same integrated object." If one wishes, this may be called a "dialectic unity of structure and function."

(2) Neglect of "qualitative specifity" of biological open system and of the specific "chemodynamics" of the first. The reply is: Thermodynamic considerations (of machines, chemical reactions, organisms, etc.) permit balance statements regarding the system as a whole, without entering into, or even knowing partial reactions, components, organization, etc., in detail. Hence part of the "theory of open systems" is concerned with such over-all balances of the system as a whole. If, however, the theory is applied to individual processes such as formation of proteins, behavior of tracers in the organism, ionic steady states, etc., the "specifity" of the respective components enters as a matter of course.

5. Advances of General System Theory

The decisive question is that of the explanatory and predictive value of the "new theories" attacking the host of problems around wholeness, teleology, etc. Of course, the change in intellectual climate which allows us to see new problems which were overlooked previously, or to see problems in a new light, is in a way more important than any single and special application. The "Copernican Revolution" was more than the possibility somewhat better to calculate the movement of the planets; general relativity more than an explanation of a very small number of recalcitrant phenomena in physics; Darwinism more than a hypothetical answer to zoological problems; it was the changes in the general frame of reference that mattered. Nevertheless, the justification of such change ultimately is in specific achievements which would not have obtained without the new theory.

There is no question that new horizons have been opened up but the relations to empirical facts often remain tenuous. Thus, information theory has been hailed as a "major breakthrough" but outside the original technological field contributions have remained relatively scarce. In psychology, they are so far limited to rather trivial applications such as rote learning, etc. When, in biology, DNA is spoken of as "coded information" and of "breaking the code" when the structure of nucleic acids is elucidated, this is more a *façon de parler* than added insight into the control of protein synthesis. "Information theory, although useful for computer design and network analysis, has so far not found a significant place in biology." Game theory, too, is a novel mathematical development which was considered to be comparable in scope to Newtonian mechanics and the introduction of calculus; again, "the applications are meager and faltering" (the reader is urgently referred to Rapoport's discussions on information and game theory which admirably analyze the problems here mentioned). The same is seen in decision theory from which considerable gain in applied systems science was expected; but as regards the much-advertised military and business games, "there has been no controlled evaluation of their performance in training, personnel selection, and demonstration."

A danger in recent developments should not remain unmentioned. Science of the past (and partly still the present) was dominated by one-sided empiricism. Only collection of data and experiments were considered as being "scientific" in biology (and psychology); "theory" was equated with "speculation" or "philosophy," forgetting that a mere accumulation of data, although steadily piling up, does not make a "science." Lack of recognition and support for development of the necessary theoretical framework and unfavorable influence on experimental research itself (which largely became an at-random, hit-or-miss endeavor) was the consequence. This has, in certain fields, changed to the contrary in recent years. Enthusiasm for the new mathematical and logical tools available has led to feverish "model building" as a purpose in itself and often without regard to empirical fact. However, conceptual experimentation at random has no greater chances of success than at-random experimentation with biological, psychological, or clinical material. In the words of Ackoff, there is the fundamental misconception in game (and other) theory to mistake for a "problem" what actually is only a mathematical "exercise." One would do well to remember the old Kantian maxim that experience without theory is blind, but theory without experience a mere intellectual play.

The case is somewhat different with cybernetics. The model here applied is not new; although the enormous development in the field dates from the introduction of the name, Cybernetics, application of the feedback principle to physiological processes goes back to R. Wagner's work nearly 40 years ago. The feedback and homeostasis model has since been applied to innumerable biological phenomena and—somewhat less persuasively—in psychology and the social sciences. The reason for the latter fact is, in Rapoport's words that

usually, there is a well-marked negative correlation between the scope and the soundness of the writings. . . . The sound work is confined either to engineering or to rather trivial applications; ambitious formulations remain vague.

This, of course, is an ever-present danger in all approaches to general systems theory: doubtless, there is a new compass of thought but it is difficult to steer between the scylla of the trivial and the charybdis of mistaking neologisms for explanation.

The following survey is limited to "classical" general system theory—"classical" not in the sense that it claims any priority or excellence, but that the models used remain in the framework of "classical" mathematics in contradistinction to the "new" mathematics in game, network, information theory, etc. This does not imply that the theory is merely application of conventional mathematics. On the contrary, the system concept poses problems which are partly far from being answered. In the past, system problems have led to important mathematical developments such as Volterra's theory of integro-differential equations, of systems with "memory" whose behavior depends not only on actual conditions but also on previous history. Presently important problems are waiting for further developments, e.g., a general theory of non-linear differential equations, of steady states and rhythmic phenomena, a generalized principle of least action, the thermodynamic definition of steady states, etc.

It is, of course, irrelevant whether or not research was explicitly labeled as "general system theory." No complete or exhaustive review is intended. The aim of this unpretentious survey will be fulfilled if it can serve as a sort of guide to research done in the field, and to areas that are promising for future work.

OPEN SYSTEMS

The theory of open systems is an important generalization of physical theory, kinetics and thermodynamics. It has led to new principles and insight, such as the principle of equifinality, the generalization of the second thermodynamic principle, the possible increase of order in open systems, the occurrence of periodic phenomena of overshoot and false start, etc. The possibility of measuring organization in terms of entropy ("chain entropy" of high molecular compounds showing a certain order of component molecules) deserves further attention.

The extensive work done cannot be reviewed here. . . . It should be briefly mentioned, however, that apart from theoretical developments, the field has two major applications, i.e., in industrial chemistry and in biophysics.

The applications of "open systems" in biochemistry, biophysics, physiology, etc., are too numerous to permit more than brief mentioning in the present study. The impact of the theory follows from the fact that the living organism, the cell as well as other biological entities essentially are steady states (or evolving toward such states). This implies the fundamental nature of the theory in the biological realm, and a basic reorientation in many of its specialties. Among others, the theory was developed and applied in such fields as, e.g., the network of reactions in photosynthesis, calculation of turnover rates in isotope experiments, energy requirements for the maintenance of body proteins, transport processes, maintenance of ion concentrations in the blood, radiation biology, excitation and propagation of nerve impulses, and others. The organism is in a steady state not only with respect to its chemical components, but also to its cells; hence the numerous modern investigations on cell turnover and renewal have also to be included here. Beside the work already cited, results and impending problems in biophysics and related fields may be found in Netter (1959).

There are certainly relations between irreversible thermodynamics of open systems, cybernetics, and information theory, but they are still unexplored. First approaches to these problems are those by Foster, Rapoport and Trucco, and by Tribus. Another interesting approach to metabolizing systems was made by Rosen (1960) who instead of conventional reaction equations, applied "relational theory" using mapping by way of block diagrams.

Beyond the individual organism, systems principles are also used in population dynamics and ecologic theory. Dynamic ecology, i.e., the succession and climax of plant populations, is a much-cultivated field which, however, shows a tendency to slide into verbalism and terminological debate. The systems approach seems to offer a new viewpoint. Whittacker has described the sequence of plant communities toward a climax formation in terms of open systems and equifinality. According to this author, the fact that similar climax formations may develop from different initial vegetations is a striking example of equifinality, and one where the degree of independence of starting conditions and of the course development has taken appears even greater than in the individual organism. A quantitative analysis on the basis of open systems in terms of production of biomass, with climax as steady state attained, was given by Patten.

The open-system concept has also found application in the earth sciences, geomorphology and meteorology, drawing a detailed comparison of modern meteorological concepts and Bertalanffy's organismic concept in biology. It may be remembered that already Prigogine in his classic mentioned meteorology as one possible field of application of open systems.

GROWTH-IN-TIME

The simplest forms of growth which, for this reason, are particularly apt to show the isomorphism of law in different fields, are the exponential and the logistic. Examples are, among many others, the increase of knowledge of number of animal species, publications on drosophila, of manufacturing companies. Boulding and Keiter have emphasized a general theory of growth.

The theory of animal growth after Bertalanffy (and others)—which, in virtue of using overall physiological parameters ("anabolism," "catabolism") may be subsumed under the heading of G.S.T. as well as under that of biophysics—has been surveyed in its various applications.

RELATIVE GROWTH

A principle which is also of great simplicity and generality concerns the relative growth of components within a system. The simple relationship of allometric increase applies to many growth phenomena in biology (morphology, biochemistry, physiology, evolution).

A similar relationship obtains in social phenomena: Social differentiation and division of labor in primitive societies as well as the process of urbanization (i.e., growth of cities in comparison to rural population) follow the allometric equation. Application of the latter offers a quantitative measure of social organization and development, apt to replace the usual, intuitive judgments. The same principle apparently applies to the growth of staff compared to total number of employees in manufacturing companies.

COMPETITION AND RELATED PHENOMENA

The work in population dynamics by Volterra, Lotka, Gause and others belongs to the classics of G.S.T., having first shown that it is possible to develop conceptual models for phenomena such as the "struggle for existence" that can be submitted to empirical test. Population dynamics and related population genetics have since become important fields in biological research.

It is important to note that investigation of this kind belongs not only to basic but also to applied biology. This is true of fishery biology where theoretical models are used to establish optimum conditions for the exploitation of the sea (survey of the more important models: Watt). The most elaborate dynamic model is by Beverton and Holt developed for fish populations exploited in commercial fishery but certainly of wider application. This model takes into account recruitment (i.e., entering of individuals into the population), growth (assumed to follow the growth equations after Bertalanffy), capture (by exploitation), and natural mortality. The practical value of this model is illustrated by the fact that it has been adopted for routine purposes by the Food and Agriculture Organization of the United Nations, the British Ministry of Agriculture and Fisheries and other official agencies.

Richardson's studies on armaments races, notwithstanding their shortcomings, dramatically show the possible impact of the systems concept upon the most vital concerns of our time. If rational and scientific considerations matter at all, this is one way to refute such catch words as *Si vis pacem para bellum*.

The expressions used in population dynamics and the biological "struggle for existence," in econometrics, in the study of armament races (and others) all belong to the same family of equations. A systematic comparison and study of these parallelisms would be highly interesting and rewarding. One may, for example, suspect that the laws governing business cycles and those of population fluctuations according to Volterra stem from similar conditions of competition and interaction in the system.

In a non-mathematical way, Boulding has discussed what he calls the "Iron Laws" of social organizations: the Malthusian law, the law of optimum size of organizations, existence of cycles, the law of oligopoly, etc.

SYSTEMS ENGINEERING

The theoretical interest of systems engineering and operations research is in the fact that entities whose components are most heterogeneous—men, machines, buildings, monetary and other values, inflow of raw material, outflow of products and many other items—can successfully be submitted to systems analysis.

As already mentioned, systems engineering employs the methodology of cybernetics, information theory, network analysis, flow and block diagrams, etc. Considerations of G.S.T. also enter. The first approaches are concerned with structured, machine-like aspects (yes-or-no decisions in the case of information theory); one would suspect that G.S.T. aspects will win increased importance with dynamic aspects, flexible organizations, etc.

PERSONALITY THEORY

Although there is an enormous amount of theorizing on neural and psychological function in the cybernetic line based upon the brain–computer comparison, few attempts have been made to apply G.S.T. in the narrower sense to the theory of human behavior. For the present purposes, the latter may be nearly equated with personality theory.

We have to realize at the start that personality theory is at present a battlefield of contrasting and controversial theories. Hall and Lindzey have justly stated: "All theories of behavior are pretty poor theories and all of them leave much to be desired in the way of scientific proof"—this being said in a textbook of nearly 600 pages on "Theories of Personality."

We can therefore not well expect that G.S.T. can present solutions where personality theorists from Freud and Jung to a host of modern writers were unable to do so. The theory will have shown its value if it opens new perspectives and viewpoints capable of experimental and practical application. This appears to be the case. There is quite a group of psychologists who are committed to an organismic theory of personality, Goldstein and Maslow being well-known representatives. Biological considerations may therefore be expected to advance the matter.

There is, of course, the fundamental question whether, first, G.S.T. is not essentially a physicalistic simile, inapplicable to psychic phenomena; and secondly whether such model has explanatory value when the pertinent variables cannot be defined quantitatively as is in general the case with psychological phenomena.

(1) The answer to the first question appears to be that the systems concept is abstract and general enough to permit application to entities of whatever denomination. The notions of "equilibrium," "homeostasis," "feedback," "stress," etc., are no less of technologic or physiological origin but more or less successfully applied to psychological phenomena. System theorists agree that the concept of "system" is not limited to material entities but can be applied to any "whole" consisting of interacting "components." . . . Systems engineering is an example where components are partly not physical and metric.

(2) If quantitation is impossible, and even if the components of a system are ill-defined, it can

at least be expected that certain principles will qualitatively apply to the whole *qua* system. At least "explanation on principle" (see below) may be possible.

Bearing in mind these limitations, one concept which may prove to be of a key nature is the organismic notion of the organism as a spontaneously active system. In the present author's words,

Even under constant external conditions and in the absence of external stimuli the organism is not a passive but a basically active system. This applies in particular to the function of the nervous system and to behavior. It appears that internal activity rather than reaction to stimuli is fundamental. This can be shown with respect both to evolution in lower animals and to development, for example, in the first movements of embryos and fetuses.

This agrees with what von Holst has called the "new conception" of the nervous system, based upon the fact that primitive locomotor activities are caused by central automatisms that do not need external stimuli. Therefore, such movements persist, for example, even after the connection of motoric to sensory nerves had been severed. Hence the reflex in the classic sense is not the basic unit of behavior but rather a regulatory mechanism superimposed upon primitive, automatic activities. A similar concept is basic in the theory on instinct. According to Lorenz, innate releasing mechanisms (I.R.M.) play a dominant role, which sometimes go off without an external stimulus (in-vacuo or running idle reactions): A bird which has no material to build a nest may perform the movements of nest building in the air. These considerations are in the framework of what Hebb called the "conceptual C.N.S. of 1930–1950." The more recent insight into activating systems of the brain emphasizes differently, and with a wealth of experimental evidence, the same basic concept of the autonomous activity of the C.N.S.

The significance of these concepts becomes apparent when we consider that they are in fundamental contrast to the conventional stimulus-response scheme which assumes that the organism is an essentially reactive system answering, like an automation, to external stimuli. The dominance of the S-R scheme in contemporary psychology needs no emphasis, and is obviously connected with the *zeitgeist* of a highly mechanized society. This principle is basic in psychological theories which in all other respects are opposite, for example, in behavioristic psychology as well as in psychoanalysis. According to Freud it is the supreme tendency of the organism to get rid of

tensions and drives and come to rest in a state of equilibrium governed by the "principle of stability" which Freud borrowed from the German philosopher, Fechner. Neurotic and psychotic behavior, then, is a more of less effective or abortive defense mechanism tending to restore some sort of equilibrium (according to D. Rappaport's analysis of the structure of psychoanalytic theory: "economic" and "adaptive points of view").

Charlotte Buhler, the well-known child psychologist, has aptly epitomized the theoretical situation:

In the fundamental psychoanalytic model, there is only one basic tendency, that is toward *need gratification* or *tension reduction*. . . . Present-day biologic theories emphasize the "spontaneity" of the organism's activity which is due to its built-in energy. The organism's autonomous functioning, its "drive to perform certain movements" is emphasized by Bertalanffy. . . . These concepts represent *a complete revision of the original homeostasis principle* which emphasized exclusively the tendency toward equilibrium. It is the original homeostasis principle with which psychoanalysis identified its theory of discharge of tensions as the only primary tendency (Emphasis partly ours).

In brief, we may define our viewpoint as "Beyond the Homeostasis Principle":

(1) The S-R scheme misses the realms of play, exploratory activities, creativity, self-realization, etc.;

(2) The economic scheme misses just specific, human achievements—the most of what loosely is termed "human culture";

(3) The equilibrium principle misses the fact that psychological and behavioral activities are more than relaxation of tensions; far from establishing an optimal state, the latter may entail psychosis-like disturbances as, e.g., in sensory-deprivation experiments.

It appears that the S-R and psychoanalytic model is a highly unrealistic picture of human nature and, in its consequences, a rather dangerous one. Just what we consider to be specific human achievements can hardly be brought under the utilitarian, homeostasis, and stimulus-response scheme. One may call mountain climbing, composing of sonatas or lyrical poems "psychological homeostasis"—as has been done—but at the risk that this physiologically well-defined concept loses all meaning. Furthermore, if the principle of homeostatic maintenance is taken as a golden rule of behavior, the so-called well-adjusted individual will be the ultimate goal, that is a well-oiled robot maintaining itself in optimal biological, psychological and social homeostasis. This is a *Brave*

New World—not, for some at least, the ideal state of humanity. Furthermore, that precarious mental equilibrium must not be disturbed. Hence in what somewhat ironically is called progressive education, the anxiety not to overload the child, not to impose constraints and to minimize all directing influences—with the result of a previously unheard-of crop of illiterates and juvenile delinquents.

In contrast to conventional theory, it can safely be maintained that not only stresses and tensions but equally complete release from stimuli and the consequent mental void may be neurosogenic or even psychosogenic. Experimentally this is verified by the experiments with sensory deprivation when subjects, insulated from all incoming stimuli, after a few hours develop a so-called model psychosis with hallucinations, unbearable anxiety, etc. Clinically it amounts to the same when insulation leads to prisoners' psychosis and to exacerbation of mental disease by isolation of patients in the ward. In contrast, maximal stress need not necessarily produce mental disturbance. If conventional theory were correct, Europe during and after the war, with extreme physiological as well as psychological stresses, should have been a gigantic lunatic asylum. As a matter of fact, there was statistically no increase either in neurotic or psychotic disturbances, apart from easily explained acute disturbances such as combat neurosis.

We so arrive at the conception that a great deal of biological and human behavior is beyond the principles of utility, homeostasis and stimulus-response, and that it is just this which is characteristic of human and cultural activities. Such new look opens new perspectives not only in theory, but in practical implications with respect to mental hygiene, education, and society in general.

What has been said can also be couched in philosophical terms. If existentialists speak of the emptiness and meaninglessness of life, if they see in it a source not only of anxiety but of actual mental illness, it is essentially the same viewpoint: that behavior is not merely a matter of satisfaction of biological drives and of maintenance in psychological and social equilibrium but that something more is involved. If life becomes unbearably empty in an industrialized society, what can a person do but develop a neurosis? The principle which may loosely be called spontaneous activity of the psychophysical organism, is a more realistic formulation of what the existentialists want to say in their often obscure language. And if personality theorists like Maslow or Gardner Murphy speak of self-realization as human goal,

it is again a somewhat pompous expression of the same.

Theoretical History

We eventually come to those highest and ill-defined entities that are called human cultures and civilizations. It is the field often called "philosophy of history." We may perhaps better speak of "theoretical history," admittedly in its very first beginnings. This name expresses the goal to form a connecting link between "science" and the "humanities"; more in particular, between the "social sciences" and "history."

It is understood, of course, that the techniques in sociology and history are entirely different (polls, statistical analysis against archival studies, internal evidence of historic relics, etc.). However, the object of study is essentially the same. Sociology is essentially concerned with a temporal cross-section as human societies *are*; history with the "longitudinal" study how societies *become* and develop. The object and techniques of study certainly justify practical differentiation; it is less clear, however, that they justify fundamentally different philosophies.

The last statement already implies the question of constructs in history, as they were presented, in grand form, from Vico to Hegel, Marx, Spengler, and Toynbee. Professional historians regard them at best as poetry, at worst as fantasies pressing, with paranoic obsession, the facts of history into a theoretical bed of Procrustes. It seems history can learn from the system theorists, not ultimate solutions but a sounder methodological outlook. Problems hitherto considered to be philosophical or metaphysical can well be defined in their scientific meaning, with some interesting outlook at recent developments (e.g., game theory) thrown into the bargain.

Empirical criticism is outside the scope of the present study. For example, Geyl and many others have analyzed obvious misrepresentations of historical events in Toynbee's work, and even the non-specialist reader can easily draw a list of fallacies especially in the later, Holy-Ghost inspired volumes of Toynbee's *magnum opus*. The problem, however, is larger than errors in fact or interpretation or even the question of the merits of Marx's, Spengler's or Toynbee's theories; it is whether, in principle, models and laws are admissible in history.

A widely held contention says that they are not. This is the concept of "nomothetic" method in science and "idiographic" method in history. While science to a greater or less extent can establish "laws" for natural events, history, concerned with human events of enormous complexity in causes and outcome and possibly determined by free decisions of individuals can only describe, more or less satisfactorily, what has happened in the past.

Here the methodologist has his first comment. In the attitude just outlined, academic history condemns constructs of history as "intuitive," "contrary to fact," "arbitrary," etc. And, no doubt, the criticism is pungent enough vis-à-vis Spengler or Toynbee. It is, however, somewhat less convincing if we look at the work of conventional historiography. For example, the Dutch historian, Peter Geyl, who made a strong argument against Toynbee from such methodological considerations, also wrote a brilliant book about Napoleon, amounting to the result that there are a dozen or so different interpretations—we may safely say, *models*—of Napoleon's character and career within academic history, all based upon "fact" (the Napoleonic period happens to be one of the best documented) and all flatly contradicting each other. Roughly speaking, they range from Napoleon as the brutal tyrant and egotistic enemy of human freedom to Napoleon the wise planner of a unified Europe; and if one is a Napoleonic student (as the present writer happens to be in a small way), one can easily produce some original documents refuting misconceptions occurring even in generally accepted, standard histories. You cannot have it both ways. If even a figure like Napoleon, not very remote in time and with the best of historical documentation, can be interpreted contrarily, you cannot well blame the "philosophers of history" for their intuitive procedure, subjective bias, etc., when they deal with the enormous phenomenon of universal history. What you have in both cases is a conceptual model which always will represent certain aspects only, and for this reason will be one-sided or even lopsided. Hence the construction of conceptual models in history is not only permissible but, as a matter of fact, is at the basis of any historical interpretation as distinguished from mere enumeration of data, i.e., chronicle or annals.

If this is granted, the antithesis between idiographic and nomothetic procedure reduces to what psychologists are wont to call the "molecular" and "molar" approach. One can analyze events within a complex whole—individual chemical reactions in an organism, perceptions in the psyche, for example; or one can look for over-all laws covering the whole such as growth and development in the first or personality in the second instance. In terms of history, this means detailed study of individuals, treaties, works of art, singular causes and effects, etc., or else over-all phenomena with the hope of detecting grand laws. There are, of course, all transitions between the first and second considerations; the extremes may be illustrated by Carlyle and his hero worship at one pole and Tolstoy (a far greater "theoretical historian" than commonly admitted) at the other.

The question of a "theoretical history" therefore is essentially that of "molar" models in the field; and this is what the great constructs of history amount to when divested of their philosophical embroidery.

The evaluation of such models must follow the general rules for verification or falsification. First, there is the consideration of empirical bases. In this particular instance it amounts to the question whether or not a limited number of civilizations—some twenty at the best—provide a sufficient and representative sample to establish justified generalizations. This question and that of the value of proposed models will be answered by the general criterion: whether or not the model has explanatory and predictive value, i.e., throws new light upon known facts and correctly foretells facts of the past or future not previously known.

Although elementary, these considerations nevertheless are apt to remove much misunderstanding and philosophical fog which has clouded the issue.

(1) As had been emphasized, the evaluation of models should be simply pragmatic in terms of their explanatory and predictive merits (or lack thereof); *a priori* considerations as to their desirability or moral consequences do not enter.

Here we encounter a somewhat unique situation. There is little objection against so-called "synchronic" laws, i.e., supposed regularities governing societies at a certain point in time; as a matter of fact, beside empirical study this is the aim of sociology. Also certain "diachronic" laws, i.e., regularities of development in time, are undisputed such as, e.g., Grimm's law stating rules for the changes of consonants in the evolution of Indo-Germanic languages. It is commonplace that there is a sort of "life-cycle"—stages of primitivity, maturity, baroque dissolution of form and eventual decay for which no particular external causes can be indicated—in individual fields of culture, such as Greek sculpture, Renaissance painting or German music. Indeed, this even has its counterpart in certain phenomena of biological evolution showing, as in ammonites or dinosaurs, a first explosive phase of formation of new types, followed by a phase of speciation and eventually of decadence.

Violent criticism comes in when this model is applied to civilization as a whole. It is a legitimate question—Why often rather unrealistic models in the social sciences remain matters of academic discussion, while models of history encounter passionate resistance? Granting all factual criticism raised against Spengler or Toynbee, it seems rather obvious that emotional factors are involved. The highway of science is strewn with corpses of deceased theories which just decay or are preserved as mummies in the museum of history or science. In contrast, historical constructs and especially theories of historical cycles appear to touch a raw nerve, and so opposition is much more than usual criticism of a scientific theory.

(2) This emotional involvement is connected with the question of "Historical Inevitability" and a supposed degradation of human "freedom." Before turning to it, discussion of mathematical and non-mathematical models is in place.

Advantages and shortcomings of mathematical models in the social sciences are well known. Every mathematical model is an oversimplification, and it remains questionable whether it strips actual events to the bones or cuts away vital parts of their anatomy. On the other hand, so far as it goes, it permits necessary deduction with often unexpected results which would not be obtained by ordinary "common sense."

In particular, Rashevsky has shown in several studies how mathematical models of historical processes can be constructed.

On the other hand, the value of purely qualitative models should not be underestimated. For example, the concept of "ecologic equilibrium" was developed long before Volterra and others introduced mathematical models; the theory of selection belongs to the stock-in-trade of biology, but the mathematical theory of the "struggle for existence" is comparatively recent, and far from being verified under wildlife conditions.

In complex phenomena, "explanation on principle" by qualitative models is preferable to no explanation at all. This is by no means limited to the social sciences and history; it applies alike to fields like meteorology or evolution.

(3) "Historical inevitability"—subject of a well-known study by Sir Isaiah Berlin—dreaded as a consequence of "theoretical history," supposedly contradicting our direct experience of having free choices and eliminating all moral

judgment and values—is a phantasmagoria based upon a world view which does not exist any more. As, in fact, Berlin emphasizes, it is founded upon the concept of the Laplacean spirit who is able completely to predict the future from the past by means of deterministic laws. This has no resemblance with the modern concept of "laws of nature." All "laws of nature" have a statistical character. They do not predict an inexorably determined future but probabilities which, depending on the nature of events and on the laws available, may approach certainty or else remain far below it. It is nonsensical to ask or fear more "inevitability" in historical theory than is found in sciences with relatively high sophistication like meteorology or economics.

Paradoxically, while the cause of free will rests with the testimony of intuition or rather immediate experience and can never be proved objectively ("Was it Napoleon's free will that led him to the Russian Campaign?"), determinism (in the statistical sense) can be proved, at least in small-scale models. Certainly business depends on personal "initiative," the individual "decision" and "responsibility" of the entrepreneur; the manager's choice whether or not to expand business by employing new appointees, is "free" in precisely the sense as Napoleon's choice whether or not to accept battle at Austerlitz. However, when the growth curve of industrial companies is analyzed, it is found that "arbitrary" deviations are followed by speedy return to the normal curve, as if invisible forces were active. Haire states that "the return to the pattern predicted by earlier growth suggests the operation of *inexorable forces* operating on the social organism" (our emphasis).

It is characteristic that one of Berlin's points is "the fallacy of historical determinism (appearing) from its utter inconsistency with the common sense and everyday life of looking at human affairs." This characteristic argument is of the same nature as the advice not to adopt the Copernican system because everybody can see that the sun moves from morning to evening.

(4) Recent developments in mathematics even allow to submit "free will"—apparently the philosophical problem most recalcitrant against scientific analysis—to mathematical examination.

In the light of modern systems theory, the alternative between molar and molecular, nomothetic and idiographic approach can be given a precise meaning. For mass behavior, system laws would apply which, if they can be mathematized, would take the form of differential equations of the sort of those used by Richardson mentioned above. Free choice of the individual would be described by formulations of the nature of game and decision theory.

Axiomatically, game and decision theory are concerned with "rational" choice. This means a choice which "maximizes the individual's utility or satisfaction," that "the individual is free to choose among several possible courses of action and decides among them at the basis of their consequences," that he "selects, being informed of all conceivable consequences of his actions, what stands highest on his list," he "prefers more of a commodity to less, other things being equal," etc. Instead of economical gain, any higher value may be inserted without changing the mathematical formalism.

The above definition of "rational choice" includes everything that can be meant by "free will." If we do not wish to equate "free will" with complete arbitrariness, lack of any value judgment and therefore completely inconsequential actions (like the philosopher's favorite example: It is my free will whether or not to wiggle my left little finger) it is a fair definition of those actions with which the moralist, priest or historian is concerned: free decision between alternatives based upon insight into the situation and its consequences and guided by values.

The difficulty to apply the theory even to simple, actual situations is of course enormous; so is the difficulty in establishing over-all laws. However, without explicit formulation, both approaches can be evaluated in principle—leading to an unexpected paradox.

The "principle of rationality" fits—not the majority of human actions but rather the "un-reasoning" behavior of animals. Animals and organisms in general do function in a "ratio-morphic" way, maximizing such values as maintenance, satisfaction, survival, etc.; they select, in general, what is biologically good for them, and prefer more of a commodity (e.g., food) to less.

Human behavior, on the other hand, falls far short of the principle of rationality. It is not even necessary to quote Freud to show how small is the compass of rational behavior in man. Women in a supermarket, in general, do not maximize utility but are susceptible to the tricks of the advertiser and packer; they do not make a rational choice surveying all possibilities and consequences; and do not even prefer more of the commodity packed in an inconspicuous way to less when packed in a big red box with attrac-

tive design. In our society, it is the job of an influential specialty—advertisers, motivation researchers, etc.—to *make* choices irrational which essentially is done by coupling biological factors —conditioned reflex, unconscious drives—with symbolic values.

And there is no refuge by saying that this irrationality of human behavior concerns only trivial actions of daily life; the same principle applies to "historical" decisions. That wise old mind, Oxenstierna, Sweden's Chancellor during the Thirty Years' War, has perfectly expressed this by saying: *Nescis, mi fili, quantilla ratione mundus regatur*—you don't know, my dear boy, with what little reason the world is governed. Reading newspapers or listening to the radio readily shows that this applies perhaps even more to the 20th than the 17th century.

Methodologically, this leads to a remarkable conclusion. If one of the two models is to be applied, and if the "actuality principle" basic in historical fields like geology and evolution is adopted (i.e., the hypothesis that no other principles of explanation should be used than can be observed as operative in the present)—then it is the statistical or mass model which is backed by empirical evidence. The business of the motivation and opinion researcher, statistical psychologist, etc., is based upon the premise that statistical laws obtain in human behavior; and that, for this reason a small but well-chosen sample allows for extrapolation to the total population under consideration. The generally good working of a Gallup poll and prediction verifies the premise— with some incidental failure like the well-known example of the Truman election thrown in, as is to be expected with statistical predictions. The opposite contention—that history is governed by "free will" in the philosophical sense (i.e., rational decision for the better, the higher moral value or even enlightened self-interest) is hardly supported by fact. That here and there the statistical law is broken by "rugged individualists" is in its character. Nor does the role played in history by "great men" contradict the systems concept in history; they can be conceived as acting like "leading parts," "triggers" or "catalyzers" in the historical process—a phenomenon well accounted for in the general theory of systems.

(5) A further question is the "organismic analogy" unanimously condemned by historians. They combat untiringly the "metaphysical," "poetical," "mythical" and thoroughly unscientific nature of Spengler's assertion that civilizations are a sort of "organisms," being born, developing

according to their internal laws and eventually dying. Toynbee takes great pains to emphasize that he did not fall into Spengler's trap—even though it is somewhat difficult to see that his civilizations, connected by the biological relations of "affiliation" and "apparentation," even (according to the latest version of his system) with a rather strict time span of development, are not conceived organismically.

Nobody should know better than the biologist that civilizations are no "organism." It is trivial to the extreme that a biological organism, a material entity and unity in space and time, is something different from a social group consisting of distinct individuals, and even more from a civilization consisting of generations of human beings, of material products, institutions, ideas, values, and what not. It implies a serious underestimate of Vico's, Spengler's (or any normal individual's) intelligence to suppose that they did not realize the obvious.

Nevertheless, it is interesting to note that, in contrast to the historians' scruples, sociologists do not abhor the "organismic analogy" but rather take it for granted. For example, in the words of Rapoport and Horvath:

There is some sense in considering a real organization as an organism, that is, there is reason to believe that this comparison need not be a sterile metaphorical analogy, such as was common in scholastic speculation about the body politic. Quasibiological functions are demonstrable in organizations. They maintain themselves; they sometimes reproduce or metastasize; they respond to stresses; they age, and they die. Organizations have discernible anatomies and those at least which transform material inputs (like industries) have physiologies.

Or Sir Geoffrey Vickers:

Institutions grow, repair themselves, reproduce themselves, decay, dissolve. In their external relations they show many characteristics of organic life. Some think that in their internal relations also human institutions are destined to become increasingly organic, that human cooperation will approach ever more closely to the integration of cells in a body. I find this prospect unconvincing (and) unpleasant. (N.B. so does the present author.)

And Haire:

The biological model for social organizations—and here, particularly for industrial organizations—means taking as a model the living organism and the processes and principles that regulate its growth and develop-

ment. It means looking for lawful processes in organizational growth.

The fact that simple growth laws apply to social entities such as manufacturing companies, to urbanization, division of labor, etc., proves that in these respects the "organismic analogy" is correct. In spite of the historians' protests, the application of theoretical models, in particular, the model of dynamic, open and adaptive systems to the historical process certainly makes sense. This does not imply "biologism," i.e., reduction of social to biological concepts, but indicates system principles applying in both fields.

(6) Taking all objections for granted—poor method, errors in fact, the enormous complexity of the historical process—we have nevertheless reluctantly to admit that the cyclic models of history pass the most important test of scientific theory. The predictions made by Spengler in the *Decline of the West*, by Toynbee when forecasting a time of trouble and contending states, by Ortega y Gasset in the *Uprise of the Masses*—we may as well add *Brave New World* and *1984*—have been verified to a disquieting extent and considerably better than many respectable models of the social scientists.

Does this imply "historic inevitability" and inexorable dissolution? Again, the simple answer was missed by moralizing and philosophizing historians. By extrapolation from the life cycles of previous civilizations nobody could have predicted the Industrial Revolution, the Population Explosion, the development of atomic energy, the emergence of underdeveloped nations, and the expansion of Western civilization over the whole globe. Does this refute the alleged model and "law" of history? No—it only says that this model—as every one in science—mirrors only certain aspects or facets of reality. Every model becomes dangerous only when it commits the "Nothing-but" fallacy which mars not only theoretical history, but the models of the mechanistic world picture, of psychoanalysis and many others as well.

We have hoped to show in this survey that General System Theory has contributed toward the expansion of scientific theory; has led to new insights and principles; and has opened up new problems that are "researchable," i.e., are amenable to further study, experimental or mathematical. The limitations of the theory and its applications in their present status are obvious; but the principles appear to be essentially sound as shown by their application in different fields.

3.
Cybernetics in History

NORBERT WIENER

SINCE THE END of World War II, I have been working on the many ramifications of the theory of messages. Besides the electrical engineering theory of the transmission of messages, there is a larger field which includes not only the study of language but the study of messages as a means of controlling machinery and society, the development of computing machines and other such automata, certain reflections upon psychology and the nervous system, and a tentative new theory of scientific method. This larger theory of messages is a probabilistic theory, an intrinsic part of the movement that owes its origin to Willard Gibbs and which I have described in the introduction.

Until recently, there was no existing word for this complex of ideas, and in order to embrace the whole field by a single term, I felt constrained to invent one. Hence "Cybernetics," which I derived from the Greek word *kubernētēs*, or "steersman," the same Greek word from which we eventually derive our word "governor." Incidentally, I found later that the word had already been used by Ampère with reference to political science, and had been introduced in another context by a Polish scientist, both uses dating from the earlier part of the nineteenth century.

I wrote a more or less technical book entitled *Cybernetics* which was published in 1948. In response to a certain demand for me to make its ideas acceptable to the lay public, I published the first edition of *The Human Use of Human Beings*

in 1950. Since then the subject has grown from a few ideas shared by Drs. Claude Shannon, Warren Weaver, and myself, into an established region of research. Therefore, I take this opportunity occasioned by the reprinting of my book to bring it up to date, and to remove certain defects and inconsequentialities in its original structure.

In giving the definition of Cybernetics in the original book, I classed communication and control together. Why did I do this? When I communicate with another person, I impart a message to him, and when he communicates back with me he returns a related message which contains information primarily accessible to him and not to me. When I control the actions of another person, I communicate a message to him, and although this message is in the imperative mood, the technique of communication does not differ from that of a message of fact. Furthermore, if my control is to be effective I must take cognizance of any messages from him which may indicate that the order is understood and has been obeyed.

It is the thesis of this book that society can only be understood through a study of the messages and the communication facilities which belong to it; and that in the future development of these messages and communication facilities, messages between man and machines, between machines and man, and between machine and machine, are destined to play an ever-increasing part.

When I give an order to a machine, the situation is not essentially different from that which arises when I give an order to a person. In other words, as far as my consciousness goes I am aware of the order that has gone out and of the signal of compliance that has come back. To me, personally, the fact that the signal in its intermediate stages

From Norbert Wiener, "Cybernetics in History," *The Human Use of Human Beings: Cybernetics and Society* (Garden City, N.Y.: Doubleday Anchor, 1954), Chapter I. Reprinted with permission of Houghton Mifflin Company.

has gone through a machine rather than through a person is irrelevant and does not in any case greatly change my relation to the signal. Thus the theory of control in engineering, whether human or animal or mechanical, is a chapter in the theory of messages.

Naturally there are detailed differences in messages and in problems of control, not only between a living organism and a machine, but within each narrower class of beings. It is the purpose of Cybernetics to develop a language and techniques that will enable us indeed to attack the problem of control and communication in general, but also to find the proper repertory of ideas and techniques to classify their particular manifestations under certain concepts.

The commands through which we exercise our control over our environment are a kind of information which we impart to it. Like any form of information, these commands are subject to disorganization in transit. They generally come through in less coherent fashion and certainly not more coherently than they were sent. In control and communication we are always fighting nature's tendency to degrade the organized and to destroy the meaningful; the tendency, as Gibbs has shown us, for entropy to increase.

Much of this book [*The Human Use of Human Beings*] concerns the limits of communication within and among individuals. Man is immersed in a world which he perceives through his sense organs. Information that he receives is co-ordinated through his brain and nervous system until, after the proper process of storage, collation, and selection, it emerges through effector organs, generally his muscles. These in turn act on the external world, and also react on the central nervous system through receptor organs such as the end organs of kinaesthesia; and the information received by the kinaesthetic organs is combined with his already accumulated store of information to influence future action.

Information is a name for the content of what is exchanged with the outer world as we adjust to it, and make our adjustment felt upon it. The process of receiving and of using information is the process of our adjusting to the contingencies of the outer environment, and of our living effectively within that environment. The needs and the complexity of modern life make greater demands on this process of information than ever before, and our press, our museums, our scientific laboratories, our universities, our libraries and textbooks, are obliged to meet the needs of this process or fail in their purpose. To live effectively is to live with adequate information. Thus, communication and control belong to the essence of man's inner life, even as they belong to his life in society.

The place of the study of communication in the history of science is neither trivial, fortuitous, nor new. Even before Newton such problems were current in physics, especially in the work of Fermat, Huygens, and Leibnitz, each of whom shared an interest in physics whose focus was not mechanics but optics, the communication of visual images.

Fermat furthered the study of optics with his principle of minimization which says that over any sufficiently short part of its course, light follows the path which it takes the least time to traverse. Huygens developed the primitive form of what is now known as "Huygens' Principle" by saying that light spreads from a source by forming around that source something like a small sphere consisting of secondary sources which in turn propagate light just as the primary sources do. Leibnitz, in the meantime, saw the whole world as a collection of beings called "monads" whose activity consisted in the perception of one another on the basis of a pre-established harmony laid down by God, and it is fairly clear that he thought of this interaction largely in optical terms. Apart from this perception, the monads had no "windows," so that in his view all mechanical interaction really becomes nothing more than a subtle consequence of optical interaction.

A preoccupation with optics and with message, which is apparent in this part of Leibnitz's philosophy, runs through its whole texture. It plays a large part in two of his most original ideas: that of the *Characteristica Universalis*, or universal scientific language, and that of the *Calculus Ratiocinator*, or calculus of logic. This Calculus Ratiocinator, imperfect as it was, was the direct ancestor of modern mathematical logic.

Leibnitz, dominated by ideas of communication, is, in more than one way, the intellectual ancestor of the ideas of this book, for he was also interested in machine computation and in automata. My views in this book are very far from being Leibnitzian, but the problems with which I am concerned are most certainly Leibnitzian. Leibnitz's computing machines were only an offshoot of his interest in a computing language, a reasoning calculus which again was in his mind, merely an extention of his idea of a complete artificial language. Thus, even in his computing machine, Leibnitz's preoccupations were mostly linguistic and communicational.

Toward the middle of the last century, the work of Clerk Maxwell and of his precursor, Faraday, had attracted the attention of physicists once more to optics, the science of light, which was now regarded as a form of electricity that could be reduced to the mechanics of a curious, rigid, but invisible medium known as the ether, which, at the time, was supposed to permeate the atmosphere, interstellar space and all transparent materials. Clerk Maxwell's work on optics consisted in the mathematical development of ideas which had been previously expressed in a cogent but non-mathematical form by Faraday. The study of ether raised certain questions whose answers were obscure, as, for example, that of the motion of matter through the ether. The famous experiment of Michelson and Morley, in the nineties, was undertaken to resolve this problem, and it gave the entirely unexpected answer that there simply was no way to determine the motion of matter through the ether.

The first satisfactory solution to the problems aroused by this experiment was that of Lorentz, who pointed out that if the forces holding matter together were conceived as being themselves electrical or optical in nature, we should expect a negative result from the Michelson–Morley experiment. However, Einstein in 1905 translated these ideas of Lorentz into a form in which the unobservability of absolute motion was rather a postulate of physics than the result of any particular structure of matter. For our purposes, the important thing is that in Einstein's work, light and matter are on an equal basis, as they had been in the writings before Newton, without the Newtonian subordination of everything else to matter and mechanics.

In explaining his views, Einstein makes abundant use of the observer who may be at rest or may be moving. In his theory of relativity it is impossible to introduce the observer without also introducing the idea of message, and without, in fact, returning the emphasis of physics to a quasi-Leibnitzian state, whose tendency is once again optical. Einstein's theory of relativity and Gibbs' statistical mechanics are in sharp contrast, in that Einstein, like Newton, is still talking primarily in terms of an absolutely rigid dynamics not introducing the idea of probability. Gibbs' work, on the other hand, is probabilistic from the very start, yet both directions of work represent a shift in the point of view of physics in which the world as it actually exists is replaced in some sense or other by the world as it happens to be observed, and the old naïve realism of physics gives way to something on which Bishop Berkeley might have smiled with pleasure.

At this point it is appropriate for us to review certain notions pertaining to entropy which have already been presented in the introduction. As we have said, the idea of entropy represents several of the most important departures of Gibbsian mechanics from Newtonian mechanics. In Gibbs' view we have a physical quantity which belongs not to the outside world as such, but to certain sets of possible outside worlds, and therefore to the answer to certain specific questions which we can ask concerning the outside world. Physics now becomes not the discussion of an outside universe which may be regarded as the total answer to all the questions concerning it, but an account of the answers to much more limited questions. In fact, we are now no longer concerned with the study of all possible outgoing and incoming messages which we may send and receive, but with the theory of much more specific outgoing and incoming messages; and it involves a measurement of the no-longer infinite amount of information that they yield us.

Messages are themselves a form of pattern and organization. Indeed, it is possible to treat sets of messages as having an entropy like sets of states of the external world. Just as entropy is a measure of disorganization, the information carried by a set of messages is a measure of organization. In fact, it is possible to interpret the information carried by a message as essentially the negative of its entropy, and the negative logarithm of its probability. That is, the more probable the message, the less information it gives. Clichés, for example, are less illuminating than great poems.

I have already referred to Leibnitz's interest in automata, an interest incidentally shared by his contemporary, Pascal, who made real contributions to the development of what we now know as the desk adding-machine. Leibnitz saw in the concordance of the time given by clocks set at the same time, the model for the pre-established harmony of his monads. For the technique embodied in the automata of his time was that of the clockmaker. Let us consider the activity of the little figures which dance on the top of a music box. They move in accordance with a pattern, but it is a pattern which is set in advance, and in which the past activity of the figures has practically nothing to do with the pattern of their future activity. The probability that they will diverge from this pattern is nil. There is a message, indeed; but it goes from the machinery of the music box to the figures, and stops there. The figures themselves have no trace of communication with the outer world, except this one-way stage of communication with the pre-established mechanism of the music box. They are blind, deaf, and dumb, and cannot vary their activity in the least from the conventionalized pattern.

Contrast with them the behavior of man, or indeed of any moderately intelligent animal such as a kitten. I call to the kitten and it looks up. I have sent it a message which it has received by its sensory organs, and which it registers in action. The kitten is hungry and lets out a pitiful wail. This time it is the sender of a message. The kitten bats at a swinging spool. The spool swings to its left, and the kitten catches it with its left paw. This time messages of a very complicated nature are both sent and received within the kitten's own nervous system through certain nerve end-bodies in its joints, muscles, and tendons; and by means of nervous messages sent by these organs, the animal is aware of the actual position and tensions of its tissues. It is only through these organs that anything like a manual skill is possible.

I have contrasted the prearranged behavior of the little figures on the music box on the one hand, and the contingent behavior of human beings and animals on the other. But we must not suppose that the music box is typical of all machine behavior.

The older machines, and in particular the older attempts to produce automata, did in fact function on a closed clockwork basis. But modern automatic machines such as the controlled missile, the proximity fuse, the automatic door opener, the control apparatus for a chemical factory, and the rest of the modern armory of automatic machines which perform military or industrial functions, possess sense organs; that is, receptors for messages coming from the outside. These may be as simple as photoelectric cells which change electrically when a light falls on them, and which can tell light from dark, or as complicated as a television set. They may measure a tension by the change it produces in the conductivity of a wire exposed to it, or they may measure temperature by means of a thermocouple, which is an instrument consisting of two distinct metals in contact with one another through which a current flows when one of the points of contact is heated. Every instrument in the repertory of the scientific-instrument maker is a possible sense organ, and may be made to record its reading remotely through the intervention of appropriate electrical apparatus. Thus the machine which is conditioned by its relation to the external world, and by the things happening in the external world, is with us and has been with us for some time.

The machine which acts on the external world by means of messages is also familiar. The automatic photoelectric door opener is known to every person who has passed through the Pennsylvania Station in New York, and is used in many other buildings as well. When a message consisting of the interception of a beam of light is sent to the apparatus, this message actuates the door, and opens it so that the passenger may go through.

The steps between the actuation of a machine of this type by sense organs and its performance of a task may be as simple as in the case of the electric door; or it may be in fact of any desired degree of complexity within the limits of our engineering techniques. A complex action is one in which the data introduced, which we call the *input*, to obtain an effect on the outer world, which we call the *output*, may involve a large number of combinations. These are combinations, both of the data put in at the moment and of the records taken from the past stored data which we call the *memory*. These are recorded in the machine. The most complicated machines yet made which transform input data into output data are the high-speed electrical computing machines, of which I shall speak later in more detail. The determination of the mode of conduct of these machines is given through a special sort of input, which frequently consists of punched cards or tapes or of magnetized wires, and which determines the way in which the machine is going to act in one operation, as distinct from the way in which it might have acted in another. Because of the frequent use of punched or magnetic tape in the control, the data which are fed in, and which indicate the mode of operation of one of these machines for combining information, are called the *taping*.

I have said that man and the animal have a kinaesthetic sense, by which they keep a record of the position and tensions of their muscles. For any machine subject to a varied external environment to act effectively it is necessary that information concerning the results of its own action be furnished to it as part of the information on which it must continue to act. For example, if we are running an elevator, it is not enough to open the outside door because the orders we have given should make the elevator be at that door at the time we open it. It is important that the release for opening the door be dependent on the fact that the elevator is actually at the door; otherwise something might have detained it, and the passenger might step into the empty shaft. This control of a machine on the basis of its *actual* performance rather than its *expected* performance is known as *feedback*, and involves sensory members which are actuated by motor members and perform the function of *tell-tales* or *monitors*—that is, of elements which indicate a performance. It is the function of these mechanisms to control the mechanical tendency toward disorganization; in other words, to produce a temporary and local reversal of the normal direction of entropy.

I have just mentioned the elevator as an example of feedback. There are other cases where the importance of feedback is even more apparent. For example, a gun-pointer takes information from his instruments of observation, and conveys it to the gun, so that the latter will point in such a direction that the missile will pass through the moving target at a certain time. Now, the gun itself must be used under all conditions of weather. In some of these the grease is warm, and the gun swings easily and rapidly. Under other conditions the grease is frozen or mixed with sand, and the gun is slow to answer the orders given to it. If these orders are reinforced by an extra push given when the gun fails to respond easily to the orders and lags behind them, then the error of the gun-pointer will be decreased. To obtain a performance as uniform as possible, it is customary to put into the gun a control feedback element which reads the lag of the gun behind the position it should have according to the orders given it, and which uses this difference to give the gun an extra push.

It is true that precautions must be taken so that the push is not too hard, for if it is, the gun will swing past its proper position, and will have to be pulled back in a series of oscillations, which may well become wider and wider, and lead to a disastrous instability. If the feedback system is itself controlled—if, in other words, its own entropic tendencies are checked by still other controlling mechanisms—and kept within limits sufficiently stringent, this will not occur, and the existence of the feedback will increase the stability of performance of the gun. In other words, the performance will become less dependent on the frictional load; or what is the same thing, on the drag created by the stiffness of the grease.

Something very similar to this occurs in human action. If I pick up my cigar, I do not will to move any specific muscles. Indeed in many cases, I do not know what those muscles are. What I do is to turn into action a certain feedback mechanism;

namely, a reflex in which the amount by which I have yet failed to pick up the cigar is turned into a new and increased order to the lagging muscles, whichever they may be. In this way, a fairly uniform voluntary command will enable the same task to be performed from widely varying initial positions, and irrespective of the decrease of contraction due to fatigue of the muscles. Similarly, when I drive a car, I do not follow out a series of commands dependent simply on a mental image of the road and the task I am doing. If I find the car swerving too much to the right, that causes me to pull it to the left. This depends on the actual performance of the car, and not simply on the road; and it allows me to drive with nearly equal efficiency a light Austin or a heavy truck, without having formed separate habits for the driving of the two. I shall have more to say about this in the chapter in this book on special machines, where we shall discuss the service that can be done to neuropathology by the study of machines with defects in performance similar to those occurring in the human mechanism.

It is my thesis that the physical functioning of the living individual and the operation of some of the newer communication machines are precisely parallel in their analogous attempts to control entropy through feedback. Both of them have sensory receptors as one stage in their cycle of operation: that is, in both of them there exists a special apparatus for collecting information from the outer world at low energy levels, and for making it available in the operation of the individual or of the machine. In both cases these external messages are not taken *neat*, but through the internal transforming powers of the apparatus, whether it be alive or dead. The information is then turned into a new form available for the further stages of performance. In both the animal and the machine this performance is made to be effective on the outer world. In both of them, their *performed* action on the outer world, and not merely their *intended* action, is reported back to the central regulatory apparatus. This complex of behavior is ignored by the average man, and in particular does not play the role that it should in our habitual analysis of society; for just as individual physical responses may be seen from this point of view, so may the organic responses of society itself. I do not mean that the sociologist is unaware of the existence and complex nature of communications in society, but until recently he has tended to overlook the extent to which they are the cement which binds its fabric together.

We have seen [here] the fundamental unity of

a complex of ideas which until recently had not been sufficiently associated with one another, namely, the contingent view of physics that Gibbs introduced as a modification of the traditional, Newtonian conventions, the Augustinian attitude toward order and conduct which is demanded by this view, and the theory of the message among men, machines, and in society as a sequence of events in time which, though it itself has a certain contingency, strives to hold back nature's tendency toward disorder by adjusting its parts to various purposive ends.

II

Parts,
Wholes,
and Levels of
Integration

THE general plan of Part II may perhaps be inferred from the titles of the selections included. Moving from the physical through the biological, social and cultural levels of organization, we find that each demonstrates the principle that at certain stages of complexity in the interrelations of components, a "discontinuity" (as the mathematician might call it), or emergent level of organization with novel features, may develop.

A long, hard scientific struggle was required to recognize that the difference between inert matter and living material does not lie in any inherent qualitative differences in the substance as such, but in the way it is *organized*. Thus, the crucial difference between the living organism and the corpse is the systemic organization of parts and processes—and nothing else.

But this "nothing else," while implying the amenability of organization or component interactions to exact scientific study, does not destroy its wondrous nature. As Purcell's article shows, even at the molecular level the emergence of differentiated structure (for example, clusterings of molecules) out of a homogeneous mass poses no mean problem of explanation, that is, is not easily reduced to a simple function of component properties and their interactions. How much more true this becomes at the biological level of organization is brought out in the next two papers by Khailov and Gerard, both of which argue forcefully for the modern direct attack on systemic organization per se as the only perspective that promises to carry us another step forward.

Though written a quarter century ago, Robert Redfield's unifying summary of a symposium on parts, wholes and levels of integration admirably brings out both the continuity and the great variety and complexity of the transitional events bridging the biological and sociocultural levels. Perhaps it is a question of hindsight rather than insight, but a reading of this material sees it as begging the need for the general systems movement that was then just gathering momentum.

4.
Parts and Wholes in Physics

EDWARD PURCELL

PHYSICS IS SURELY THE PLACE where one expects few surprises as one goes from part to whole. Most exploration in physics is directed the other way, and it has taken us into the domain of the elementary particles, where, at the moment, it seems the deepest questions lie. But, there are interesting and difficult problems in physics which are presented as going from part to whole. It is this family that I would like to review. Indeed, I would like to pose, in the context of physics, the question which I assume this title evokes in all the subjects represented in this series. The question is: Is the behavior of the whole system richer in any essential way than the behavior of its parts?

In biology, perhaps, and certainly in sociology, one may suspect that even when one has unraveled the elementary interactions one will still be a long way from comprehending the life of the system as a whole, and the barrier that remains may or may not be described adequately as one of mathematical complexity only. That is to say, there may be deeper reasons to mistrust analogies between the physics of a many-particle system and a social or biological organism. But I need not face this question for my aim is to make a much weaker point. I want to explore with you one branch of physics where the problem is to understand the behavior of the aggregate in terms of the elementary laws governing its individual parts and where there can be no doubt at all that we are in possession of all the relevant facts about the parts and also about their elementary interactions.

From Edward Purcell, "Parts and Wholes in Physics." Reprinted with permission of The Free Press from *Parts and Wholes*, edited by Daniel S. Lerner. Copyright © 1963 by Massachusetts Institute of Technology.

Now the surprising thing is that—and this is the only point I want to make this evening—even at this very primitive level, where there is no question of a *new* organizing principle coming in as we proceed from the individual to the system, we find a subtle problem and something very like qualitatively new phenomena.

The general problem is one which we call in physics the problem of cooperative phenomena—or, more specially, order–disorder transitions. Some complex systems—complex in the physicist's sense, which merely implies a very large number of particles—can change from a disordered state to one that is orderly, or vice versa. Both of these conditions are properties of the ensemble as a whole, not of the individual. That is, you cannot look at an individual and tell whether it is orderly or not; orderliness implies the *pattern* of the whole arrangement of parts. Still, the behavior of the whole must, in some way, come about from the primitive laws which govern the interaction of the parts. This is a problem best explained by an example. My particular example will be a model in physics that has had a celebrated history. It is simple enough to understand even if you haven't studied any physics for a long time. One can describe it in many ways and it is always, in essence, the same thing.

My example has its origin in the problem of magnetism. When you have a piece of iron which appears magnetic, it does so because all the elementary magnets in the iron, or most of them, point the same way. If they are not almost all pointed the same way, the outside effect is small, and we say the iron is not magnetized. The early theories of magnetism attempted to formulate some picture of what goes on inside iron when the little magnets line up. One of the difficulties was that, at that time, one did not know exactly what the elementary magnets were. But that was not really essential to this problem then, nor is it to our problem now. We know that a magnet consists of a large number of elementary magnets (actually we know nowadays that these are electrons in the iron atoms), which, if they all line up the same way, will cause the whole block of iron to exhibit a magnetic effect. So the question of magnetism is a question of order. The elementary magnets, whether ordered or disordered, are all identical. In the disordered state they are simply pointing in different directions. One of the early and very limited attempts to make a theory of this generated what has become a classic problem in physics, and also in physical chemistry. This problem is named after the man who introduced it in his doctor's thesis in 1924, one Ernest Ising, who published the substance of his thesis in a short paper in the *Zeitschrift für Physik* in 1925. The Ising problem is concerned with a model that might imitate the "cooperative" behavior of the elementary magnets in iron.

If we take a material such as iron, which is capable of being a magnet, and measure the strength of its magnetization at different temperatures, what happens typically is the following. If the iron is very hot, it is not magnetic at all. As we cool it off, there comes a particular temperature (in the case of iron, somewhere below a red heat) where it changes abruptly into a magnetic material. Below this critical temperature (called the "Curie point" after Pierre Curie who performed the early experiments) the elementary magnets in any neighborhood within the metal spontaneously line up with one another so that a large majority point in the same direction. In different regions or "domains" different directions may be chosen so that the specimen as a whole may appear unmagnetized, if it consists of many domains. But this complication need not concern us, for it is the spontaneous ordering of the millions of elementary magnets within one domain that is the nub of the problem. The degree of order is found to increase very rapidly as this material is cooled below its Curie point. At absolute zero, of course, the order becomes complete.

That is the actual behavior of matter. The next thing is strictly a mental construct—a model, the analysis of which is the so-called Ising problem. In the simplest case we have a row of objects each of which has two possible conditions, either pointing up or pointing down (Figure 1).[1] The nature of the two conditions is actually not

involved in the puzzle, but we shall identify them as "pointing up" and "pointing down" because we have elementary magnets in mind. The state of this whole chain of elements is described by saying how many are pointing up and how many are pointing down. Now, let us assume that whether a given magnet is pointing down or up, whether it wants to point down or up, depends *only* on what its neighbors are doing. In fact, let me assume that if two neighbors become parallel energy is given off. That is, we shall associate negative energy with a parallel arrangement of two neighbors, positive energy with an opposite or "anti-parallel" arrangement. Thus, pairs that try to get into the lowest energy state will tend to be parallel. In other words, this is a society in which everyone wants to do what everyone else does, but in which each man has a view only of his nearest neighbor on either side. That's all there is to the problem, except for the basic assumptions that temperature has a meaning for this system and that thermodynamics applies to it.

"Long-range" Order

"Short-range" Order

Figure 1.

We ask now what, at some given temperature —that is, at some given average energy per particle—is the likely condition of this system. It is clear that if we go to absolute zero, the individuals have to seek the lowest state, and they will all be parallel. But the question is, as we cool the system, but before we get that low, do we reach a point where suddenly there is a fad or a movement that sweeps the whole system, in which a single direction becomes universally popular? Do the individuals get together and say, "Let's all point north; it looks like those fellows over there are pointing north?" The system might thus snap into partial order. Were this to happen, one would have a model that had some relation to the problem of magnetism. This question presents a simple mathematical problem which Ising solved in his thesis. He found that, unfortunately, nothing very interesting happens. When you cool this system down, it gradually acquires more order until finally, at absolute zero, it is fully in order. But nowhere in between does anything sudden happen to the whole community. There is no discontinuity in its behavior.

There are different kinds of order that one can anticipate. If we reach a condition where nearly everyone is pointing up and only a few individuals happen to point down, then we would say we have *long-range* order through the whole chain—if the arrows are up at one end they will also be up at the other end. On the other hand, we can have an order manifested only on a short-range basis; you might find four individuals together pointing down and five together pointing up. This is a highly ordered state compared to a random assembly, but it has no external manifestation because in the physics problem that gave rise to this discussion, these things are only one atomic distance apart, and so there are millions and millions of them in a row. If they are ordered only in this way, we call it *short-range* order. In the case of the one-dimensional Ising chain, it turns out that you can get long-range order only at absolute zero, and magnetism is associated only with long-range order. The kind of order you get before that is just a degree of short-range order.

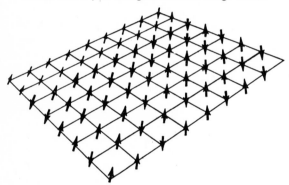

Figure 2.—The Two-dimensional Ising Lattice.

But this is obviously a far cry from a physical system that has its elements distributed in three dimensions, and the next step is to look at a two-dimensional array of things. A two-dimensional Ising lattice has the same elements, but they are distributed on the corners of squares (Figure 2). Imagine the lattice stretching off to infinity in all directions, a sheet of magnets whose individuals can point up or down. The difference is that now each magnet has four neighbors whom it sees and seeks to emulate, rather than two. But it is not the increase in the number of nearest neighbors so much as the new topology of the interconnections that has changed the problem. Assuming that a magnet's preference for up or down depends on its four nearest neighbors only, the question is: does this two-dimensional array behave like a real magnet?

Ising, in his paper, offered a simple and plausible proof that it comes no closer to such behavior than the one-dimensional array. He noted that any time we strengthen the interaction between two magnets it increases their tendency to be parallel. That surely should *promote* the effect we are looking for. Therefore, let us take *all* the interactions crosswise and strengthen them so much that the members of a transverse row all will have to be parallel under any conditions. But now the question of the most likely orientation of the rows as units is simply the one-dimensional problem, already solved, that showed no Curie transition.

Ising remarked that this result unfortunately does not seem to show any of the aspects of ferromagnetism, and on this rather wistful note he took his leave of the problem and the literature. This apparently is the only paper he ever published, and its last conclusion is wrong. Plausible as his argument is and simple as his model is, the conclusion is wrong. Although the one-dimensional chain does not show spontaneous magnetism, the two-dimensional array must do so. This was first proved by the physicist R. E. Peierls in 1936, by a very ingenious argument.

This was the beginning of an assault upon the Ising problem by a long list of very distinguished theoretical physicists and chemists. The history of the problem from then on contains names like that of Kramers, Onsager, and finally, most recently, Yang and Lee. In 1941 Kramers and Wannier found a way to get an appropriate answer in the two-dimensional case and finally in 1944 Professor Onsager of Yale found an exact solution for the thermodynamic properties of the two-dimensional Ising array. It took intellectual troops of this power to crack this problem because the fact is that even the two-dimensional array poses an exceedingly subtle and difficult puzzle. I cannot do justice to the strategy of Onsager's attack but we may look at the result, the predicted behavior of the system as the temperature is changed. Instead of plotting the magnetization, I have plotted the "specific heat," the amount of energy you have to put in to change the temperature by one degree (Figure 3). This reflects the changes that are going on in the system. Now we can see what happens in the one-dimensional Ising chain—the row of people who can only see their immediate neighbors on either side—and what happens with the people who can see four neighbors instead of two. The two-dimensional system has a singularity in its specific heat curve. The curve goes to infinity logarithmically, and it is at this point that one has the Curie temperature where order sets in. Indeed, in 1952, Yang produced a formula for the magnetization of this system as it is cooled off, which looks very much like the graph for a real magnetic material (Figure 3).

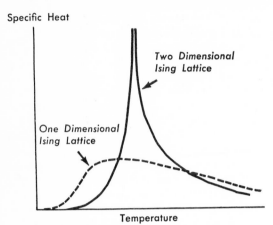

Specific Heat

Two Dimensional
Ising Lattice

One Dimensional
Ising Lattice

Temperature

Figure 3.

The main point I want to make here is that in spite of the conceptual simplicity of the model, the behavior of the system is astonishingly subtle. It is not hard to calculate, but also, remarkably, this simple two-dimensional system, as it cools down, abruptly enters a cooperative phase in which long-range order appears. The only physics involved is the assumption that nearest neighbors interact so as to prefer being parallel.

Actually, this model has applications to other things and fits some of them better than it does magnetism. It is a very good model—or rather, the three-dimensional version of it would be a good model—for certain behavior that metal alloys exhibit. If you have an alloy of copper and gold, the question is whether the copper atoms and the gold atoms are arranged in an orderly way in the crystal lattice, or whether they are scrambled. There one finds the same kind of behavior as in the example of magnetism; below a certain temperature, an ordered arrangement prevails. The Ising model can also imitate certain properties of a gas, and a remarkable paper by Yang and Lee in 1952 showed that the condensation of a gas from its vapor, with a very simple model of the gas, can be put into exact correspondence with the Ising model. I think this is the first model of a gas in which it has been possible to calculate the transition between liquid and vapor. It is totally a mental construct. Quantitatively, this is not like any gas, but it is the first model that had the feature with which we are all

familiar—when you cool the vapor, you get the liquid. Previous models of a gas, from kinetic theory on, give only a hint that a thing like condensation can happen. This, too, is, in a sense, a transition from a disordered state to an ordered state.

Let me show you quite another approach to problems of molecular order and disorder, a rather direct approach to the behavior of a system of any identical particles. The idea is to look in and find out what the particles are doing—except, again, one uses a model. Suppose we want to know what happens to a dense liquid or gas when molecules are running around bumping one another. Is there anything in this behavior which suggests the onset of long-range order? Well, we know there is in nature. There is *freezing*, in which a liquid changes from a more or less disorderly collection of molecules to an orderly one of almost the same density. Somehow, the molecules decide that the time has come for them to arrange themselves in neat, orderly layers and stop knocking one another about in a disorderly way. Yet the forces between the molecules are the same in one case as in the other. No organizing principle enters, other than that expressed in just those forces.

The approach I shall describe has been made possible by the fast computing machines. There are really two different approaches. Each deals with a model that consists of so-called hard spheres. There is a box and the spheres in the box are to be imagined as molecules. The molecules may be pictured as "billiard balls" in the sense that they cannot penetrate one another and each molecule exerts no force on the other unless the two actually collide. This is a very primitive model of a gas, but it has done yeoman service in kinetic theory for a century and is worth trying here. One shortcoming is that there aren't very many molecules in the box. In the example I am going to describe there are only thirty-two.

The idea of Adler and Wainwright, who developed this approach at the Lawrence Laboratory at Livermore, was to follow the actual history of the individual spheres as they fly about hitting one another. The imaginary spheres are started off with some arbitrary directions and speeds. The computing machine follows every sphere, keeping track of it, finding where it hits the next one, where the two go after that, where the first hits another one, and so on. You can see right away why you cannot do this for very many objects. If so few as thirty-two molecules were closely confined by a box, one would expect the walls of the box to influence strongly the configuration of the system. To suppress this influence the walls are "eliminated" by a trick. The computer is instructed to replace any molecule about to bump the north wall by a new molecule *entering* as if through a hole in the south wall. It is fascinating to see the random distribution developing. For example, if you start the system out with all molecules moving at exactly the same speed, after about two collisions per molecule they all have acquired different speeds, distributed in very close agreement with the theoretical distribution curve for an infinite assembly thoroughly randomized.

But this is not yet a cooperative phenomenon; it is just the working out of the statistics of a molecular assembly. Now suppose I squeeze the box down so the molecules do not have room to move. If I squeeze it down far enough, they have to be packed in there, like cannon balls, in some neat way. If I give them a little room, they will presumably jostle around and get a little irregular. What actually happens under *those* conditions? The results of Adler and Wainwright are interesting and surprising. A picture that they published last year shows these molecules as seen from the top of the box (Figure 4). A point of light traces

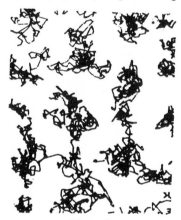

Figure 4.

the path of the center of a molecule, showing how it moved in a time during which the whole assembly enjoyed three thousand collisions. Since there are thirty-two molecules, that's roughly one hundred collisions per molecule. One can see that the system is not *completely* disorderly. But although the molecules favor certain positions, they straggle around a good deal.

One can also find a different state of affairs (Figure 5) in which the molecules are well localized; each is vibrating about its home base. So the molecules settle into what looks pretty much

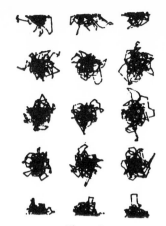

Figure 5.

like our notion of a crystal—namely, an orderly array. Now the remarkable thing is that these two pictures are taken at exactly the same conditions in respect to relative volume of molecules and free space between them. This is another three thousand collisions worth, merely shown at a different time in the computing cycle. The system changes abruptly from one structure to the other once in a while. For three thousand collisions it will run along looking like one structure, and then it will change to a different average pressure, very much like a liquid changing to a crystal or vice versa. Where the statistics are so difficult to come by, even with an IBM 704 computer, it is a little hard to follow the changeover. Some people at Los Alamos are working with what appears superficially to be the same method, but is actually a different one—the Monte Carlo method—in which they do not follow the molecules in detail, but really do statistical mechanics. They find a similar behavior: at this density, the system will go toward either structure, and it will stay at one solution for a long time; then it will jump to the other solution. Now there is nothing here showing this cooperative behavior but hard spheres of a certain radius.

These hypothetical systems, the Ising model and the hard-sphere gas, teach us that the most elementary interactions can generate, in a large assembly, cooperative behavior the prediction of which challenges our most powerful methods of analysis.

I am not proposing the Ising model as a very deep system. Indeed, I am not particularly interested in it . . . as a model of physics. I am interested in it as a model of a model because I feel that perhaps it has a bearing on the problem of going from parts to wholes in other situations, indeed, other disciplines. I cannot feel, for example, that the elementary laws that govern an organism, be it a cell or a society, are less subtle than those that govern the Ising model; I am quite sure that anything else one has to deal with is more difficult. Still, we have seen how even this simple system works itself into great complexity when one tries to understand the essential problem of cooperative behavior. I suggest that the astonishing stubbornness of the Ising problem stands as a sober warning to anyone who attempts to carve a path of rigorous deduction from the part to the whole.

Notes

1. The figures used in this selection are from the article by T. Wainwright and B. J. Adler that appeared in the *Supplemento* to *Il Nuovo Cimento*, No. 1, Vol. 9, 1958.

5.
The Problem of Systemic Organization in Theoretical Biology

K. M. KHAILOV

I. Introduction

The last decade has been characterized by an unmistakable decline of the level of theoretical activity in biology, both in the USSR and abroad. This is sometimes explained by the supposed impasse in classical genetics which has been the essential source of nurture of evolutionary biology. However, to agree with this explanation would be to simplify the matter drastically. The fact is that in the past twenty or thirty years biology has made large strides forward and has transcended the organism with which its theoretical constructs had been bound up. Probably it was primarily for this reason that evolutionary theory, which had served for a long time as the basis of biological synthesis, has ceased to function in this complex role. Classical evolutionary theory could not incorporate many of the new data and concepts which arose in biology itself and in adjoining disciplines, such as cybernetics and information theory. Consequently, the development of the classical theory was slowed down. At the same time, however, shoots of an essentially new kind of integrative thinking appeared; and, what is especially important, this new method was able to master the modern material which had been escaping the classical theory.

This new synthesis arose on the basis of observations and generalizations embracing not only biology but many other fields of knowledge. Curiously, the content of the synthesis was not the idea of evolution; it was the idea of the organization of living matter. For some time the idea may have been open to doubts, but now its definitive success in science leads us to believe that the new theoretical movement has captured a property of living matter as universal as that of potential for development. Indeed there is no unorganized living matter in nature as there is no non-evolving life. This compels us to view the new movement with great attention. Although it began abroad, it has deep roots in the U.S.S.R., and evidently its future successes will in large measure depend on the developments already carried out by Soviet theoretical biologists.

Although the theory of organization has already captured secure positions in theoretical biology, strangely the question has not yet been raised about the meaning of this new conception and of its relation to classical evolutionary biology. Can it claim only a modest role or a position on an equal footing with the evolutionary conception? Finally, what is the origin of the new ideas and to what extent has their appearance been inevitable?

One of the most elementary of scientific principles is the circumstance that an individual,[1] whether an organism or a community, may be viewed in two aspects, distinct but equally fundamental: (1) the origin of its structure and function (genesis) and (2) its organization, which insures the existence and the functioning of the individual. However, as often happens in science, some one aspect of a phenomenon, more accessible than others, first becomes the object of extensive investigations; and the initial theory is constructed

From K. M. Khailov, "The Problem of Systemic Organization in Theoretical Biology," translated by Anatol Rapoport from "Problema sistemnoi organizovannosti v teoreticheskoi biologii," *Zhurnal Obshchei Biologii*, 24 (1963), 324–332, in *General Systems*, IX (1964), 151–157. Reprinted by permission of the translator.

upon this aspect. Consequently, the initial theory encompasses more than it ought. In the world of biological phenomena, the genesis of the individual is, as a rule, the more accessible. From the time the variability of the living world was connected with its changing character, the idea of evolution and later the developed theory of evolution has assumed a central position in biology. It was supposed that the theory, in revealing the mechanism of evolution, would at the same time explain the phenomenon of organization. In the framework of this theory, the model of the organic world as a whole emerged. Naturally this was a model of phylogenetic unity and phylogenetic wholeness of life. But it was supposed that the organized nature of life was likewise reflected in the model. This last illusion was nurtured by the identification of the organism's organization with its structure.

Certainly the necessity of investigating the phylogenetic aspect of life cannot be doubted. But it is easy to assume that the phylogenetic scheme, having arisen before a serious study of organization had been developed, will not be complete and will produce a tendency to view the organizational aspect as being both simpler and less essential than it really is, that is, to believe it to have been already sufficiently clarified in the present conception.

Indeed the wholeness, the unity, and the organized nature of life, side by side with the complex problems of phylogenetic evolution have appeared as a truth too simple and obvious to deserve special attention. Paradoxical as this may be, until most recently these principles were illustrated in terms of the same general categories as in the middle of the last century—by the unity of the world, the unity of the substrate of life, the generality of the laws of development, and by the obvious fact of increasing complexity in the course of evolution.

In this way, the other potential half of the theory of life—the concept of organization—has remained undeveloped in classical biology and did not play any essential role in the history of ideas, although the actual fact of organization did not elicit any doubts.

The shift of biology in the direction of the ideas of organization was made in two main directions. First, in the framework of the organism, attention shifted from separate processes to their interactions. Inasmuch as the essence of mutual interaction consists of establishing the couplings in the transfer of information along channels, the first major success in the analysis

of interactions were achieved only very recently with the appearance of communication theory and cybernetics. This group of modern multidisciplines, having encompassed in themselves the principles of physics, biology, and technology, has advanced entirely new scientific principles applicable both to physical and biological entities as well as to systems. The concept of system, so commonplace in the physics and suitable in technological automata, has turned out to be indispensable in biology as well, since biological individuals, considered in the aspect of their internal interactions, are typical systems. It has turned out that fundamental complex phenomena of life, which had long remained unexplained on the basis of pre-systemic principles, have to do with these very interactions, i.e., with the nature of systems.

In the first instance, this encompasses the phenomena of self-regulation and regeneration, of genetic and physiological homeostasis, and of higher nervous activity. A specific feature of living systems is the fact that they are "open" in the sense of thermodynamics and that their stability depends on their flexibility. Finally, another achievement of extraordinary importance, having to do with a new approach to life phenomena was the discovery of a precise quantitative relation between the fundamental characteristics of living and non-living matter and correspondingly between the principles of physics and biology—the relation between order and organization on the one hand and entropy on the other.

Another channel, along which the ideas of organization have penetrated into biology is related to the broadening of the sphere of investigation beyond the limits of the individual and to the development of bio-phyto- and zoo-cenology, the theory of the biosphere, of ecosystems, and of biogeocenoses. These developments have placed alongside the individual several new supraorganismic entities. In place of a single "unit" of nature—an individual (or species)—there appeared a whole sequence of progressively more complex units, and these have been awarded, along with the organism, the right of being included in the picture of the organic world and in the theory of its evolution. However, it is of significance that, although this necessity is apparent, the supra-organismic individuals could not be included in the phylogenetic picture, for the simple reason that they were not studied in that scheme. Their study was developed simultaneously with the organization scheme. Thus, in the process of being applied to different kinds of living individuals, namely, to organisms and to communities, two distinct but equally important approaches met: the approach through genesis and the approach through organization. This is why the theory of the organism based on evolutionary phylogenetic principles could not be immediately merged with the theory of communities based on the organizational principle, in spite of the fact that in the framework of general theoretical biology such a unification is not only natural but also necessary.

This split in theoretical biology appeared to be in the process of being overcome by the fact that the study of communities has led to a new specialty (sometimes called symbiology) or else was referred to the competence of ecology. However, this only led to the broadening of ecology into another general-biological conceptualization, parallel to the evolutionary one, which has repeatedly elicited reservations and has been a subject of discussion. It is becoming increasingly evident that we are dealing essentially with two aspects of life and that a unified theory is needed. The resulting situation strongly reminds one of the well known situation in physics where the wave and the particle theories of light, having been merged into a single theory, expressed two aspects of the same phenomenon. We may expect that as a consequence of further development in biology, the idea of organization, which for the time being is the stronger idea in symbiology, will be developed also with reference to the organism. The success of cybernetic ideas indicate that the process has already begun. On the other hand, the classical evolutionary phylogenetic idea must, apparently, pass into symbiology. We are witnessing also the latter process, and unquestionably priority in this belongs to Soviet phytocenology and biogeocenology.

In this way, essential changes in biology are seen to be in the circumstance that the fact of organization is becoming the subject of profound attention in the same measure as the fact of evolution had been two centuries ago. We are witnesses to ever more evidently illusory character of the old faith that organization is simpler and secondary to evolution, that it can be understood as a consequence derived from evolution. Modern biology and its neighboring disciplines reveal in ever greater measure the fact that this property of living matter is as fundamental and as complex as its development potential. Observing organic evolution wherever organized life exists (there is no other) we could as well consider the potential

for evolution as a consequence of organization as vice versa.

The interchangeability of these two theses compels us to suspect that both are erroneous: neither evolution nor organization precedes the other nor explains the other, because each is a fundamental, equally important property of matter in general. In this connection, it is appropriate to recall that physics is inclined more than formerly to view "disorganization" as an abstraction, in relation to which different degrees of organization are referred, rather than a real state of matter.

At the present time, it is hardly possible to define the degree to which the great expectations expressed herein are justified. However, the following has been long apparent: the state of the "open mind" to which Wiener appeals corresponds most closely to the modern position in biological science.

II. General System Theory in Biology

In any field of knowledge, the study of interactions leads logically to the concept of system organization. Developing along parallel lines almost everywhere, also quite intensively beyond the scope of biology, such concepts naturally had to elicit attempts to juxtapose them and to be revealed as the carriers of essentially the same idea. The first serious attempt to realize a program of this sort was made by L. von Bertalanffy and has since become the General System Theory described by him and a group of other investigators.

The central concept of the general theory, namely "system," is used by these authors evidently in a very broad sense, inasmuch as the theory unites the fundamental principles of organization and encompasses most diverse natural, social, and intellectual phenomena. A system is defined in this context as "the totality of elements in interaction with each other," "unity consisting in mutually interacting parts," "a type of structure which functions in the form of definite sequence of operations," "the totality of objects together with their mutual interactions," etc. The essence of all these definitions is the coupling among the components and the system organization resulting from such couplings.

The biological component of general system theory is still very feebly developed. Practically speaking, at the present time, it can be viewed in the form of the principal problems comprising this component. However, these problems are sufficiently defined to reveal definite outlines of a theory and the immediate opportunities contained in it.

Problem number one is to define the notion of a "living system" inasmuch as the above mentioned system characteristics do not reflect biological specificity. This specificity is contained in the fact that living systems are (1) natural, discrete individuals; (2) functionally complex; and (3) developing. The first criterion distinguishes living systems from arbitrary artificially created ones; the second from systems of only structural complexity, whose parts are not necessarily functionally differentiated; the third is a uniquely biological feature and needs no explanation. Therefore, in the first approximation, a living system may be defined as a natural, discrete totality of functionally differentiated elements, which have a potential for development and are in mutually complementary and interacting relations vis-à-vis each other.

Problem number two is the enumeration of objects which can be studied in their system aspect. This problem also offers many difficulties. These stem not from a lack of objects, but rather from their abundance. It is natural to begin a listing of living systems with some cellular or pre-cellular proto-organisms, although S. Wright begins with the gene. Above the organism come the supra-organismic systems, the phyto- and zoo-cenoses, the biocenoses, including phyto- and zoo-components which are singled out by most ecologists. Then come systems comprising the entire living population of a given territory and the related inorganic environment. These latter are known under different names, e.g., microcosm, holocene, ecosystem, biosystem, biogeocenosis. All of these reflect approximately the same sort of complex. The complex of the ecosystem and biocenosis of the entire planet comprises an even more general system—the biosphere. Besides, some biologists add to this list the population and species. Sometimes systems like flora and fauna are specifically singled out.

It is quite evident that a simple enumeration of all the complexes mentioned by the various authors by no means solves the problem posed, since the list includes biological entities, which to begin with, satisfy our definition of living system to an unequal degree and secondly are integrated in different degrees by different kinds of mechanisms. For example, an organism, a biogeocenosis and, to an extent, a biocenosis are integrated on the basis of interactions in the course of metabolic exchanges among functionally distinct compon-

ents; in the organism this is an internal metabolic cycle; in the biocenosis it is an exchange among producers, consumers, and reducers; in the biogeocenosis it is a biological vortex including living matter in any of its forms of organization as well as biogenic elements and minerals. Systems distinguished by formation type, such as fauna and flora, and also phyto- and zoo-cenoses do not comprise complete metabolic cycles. However, the functional differentiation and integration of phyto- and zoo-cenoses is realized in the ecological plan, although it is manifested in different degrees. As for organizations like the population and species, and a forteriori the gene, the mechanism of their integration which could be called genetic is quite a specific phenomenon, distinct from the metabolic, physiological, and ecological forms of interaction.

Doubtless the question is considerably more complex than it can be pictured here, but our task here is not the discussion of the extremely problematic status of cenoses and other units of nature but rather in posing the question as part of the theory of biological systems. Evidently not all biological individuals satisfy the definition of system equally, and only a deeper study will show which of them are members of a hierarchical series.

Finally, problem number three concerns the establishment of a hierarchy of living systems. The obvious increase of complexity or organization which puts the community historically above the organism by no means solves the whole problem since the position of biogeocenoses (or ecosystems) and of the biosphere as a whole remains controversial: do these stand in the beginning of a historical sequence of living systems or should macro-systems of this type be placed on the upper rungs of the organization ladder? On the other hand, the question about the hierarchy of systems has not only a chronological significance but is related also to the mechanism of interaction. For example, organisms, biocenoses and biogeocenoses constructed essentially upon inner-metabolic cycles are connected by inclusion, that is, each of them is included in another system and is open to the latter. The organism is open to the biocenosis, the biocenosis is open to the biogeocenosis, and in the structure of a single system—the biosphere—all three cycles are coupled by the same type of interaction.

The problem of subordinating and supraordinating systems probably has great theoretical significance since it leads directly to the formulation of a single integrating scheme of living nature in the world of non-living matter.

III. The Principle of System Organization and Some Controversial Questions of Evolutionary Biology

The overly bold claims of system theory undoubtedly make one wary. However, in evaluating these claims one must first of all realize that this new approach by no means bids to become a substitute for the evolutionary approach. Nevertheless, it is natural to expect that evolutionary theory will derive some sort of benefits from its neighbor. And this may become the acid test. A survey of such benefits demands a detailed examination, which will be done elsewhere. Here we shall only compare the effectiveness of evolutionary phylogenetic theory before and after it was combined with system theory in the solution of certain debatable questions.

A conspicuous feature and at the same time a limitation of evolutionary phylogenetic approach is in the fact that this approach permits us to classify only phylogenetic connections among organisms; while for the most part the inter-relations established among organisms and between the latter and their environment do not yield to the phylogenetic scheme of classification. For this reason, the problem of connections and of life factors has remained outside the competence of the fundamental theoretical discipline—the evolution doctrine where it has primary significance as a problem related to the driving force [of evolution]. The problem is seriously attacked primarily in the realm of ecology, a discipline which has not concerned itself much with evolutionary problems. Accordingly, most progress in the classification of non-genetic interactions and factors was made in recent times in ecology, while in evolutionary theory this problem and with it the problem of driving factors of evolution have remained least developed and most confused. In evolutionary theory there appeared a characteristic tendency of viewing living nature as consisting only of organisms (and species). The specific characteristic of recent conceptualizations in our country has been also an assertion that the essence of organic evolution and of life in general is unity—the unity of organism and its environment, hence a single decisive integration—that with the environment and accordingly a single driving force—the "assimilation of external conditions." In ecology, on the contrary, a tendency prevailed to view living nature as a "segmented" whole and to sub-divide factors among the units ("segments") of life, i.e., among systems. Indeed, the system principle appears to be the natural basis for classifying non-genetic interrelations, and the interactions of factors, since living systems—organisms, communities, biogeocenoses—appear from the point of view developed here the natural bundles of interactions and factors connected by inter-system bonds in the structure of the biosphere.

It is not surprising that many of our workers in evolution theory, having been torn from their soil, have in recent years lost the concrete point of view and were basing their views on generalities which sometimes reflected an older level of scientific knowledge. The often repeated descriptions of life and evolution in terms of "metabolism," "assimilation and dissimilation," "emergence and extinction," "the struggle between the old and the new," etc., while contributing nothing new, not only devalue theoretical investigations but also conceal the specificity of events, that characteristic "segmentation" of life which becomes ever clearer in concrete investigations. The "segmentation" of life is revealed in the discrete nature of systems. This is why biological system theory, its generality notwithstanding, points precisely to a concretization of the scheme of life and thus indicates a way out of the impasse of general discussions about all-encompassing metabolism, assimilation, dissimilation, etc. On the basis of system concepts it is possible (for the time being only approximately) to carry out an objective classification of interconnections, in particular to attempt to orient oneself in the labyrinthine discussions concerning the primacy of life factors. One can assert that metabolic exchanges between living and non-living matter throughout the whole history of life from proto-organisms to human society have developed only in the system of the biosphere and of biogeocenoses and that therefore this type of coupling is important only on that level; while all the other kinds of coupling related to other systems (within and among organisms) are not fundamental for the system in question. Similarly, the couplings and interactions among organisms are fundamental only in systems representing communities, symbiotic and parasitic groupings, and correspondingly the couplings of organisms with the abiotic environment cannot be viewed as fundamental in communities. Similarly in the organism one cannot but view as fundamental the biochemical and physiological types of couplings and interactions, relegating all the others to secondary status. Thus as we pass from system to system, the principal and secondary couplings and interactions keep being interchanged and so there is no type of coupling which is principal *in general* without reference to some concrete system. On this basis, we cannot agree with those who turn the coupling (unity) of organism and environment into an absolute as something fundamental for life in general.

In our opinion, it is, in principle, incorrect to put the question about leading factors of life and evolution, because each system has its own factor which can be called leading and which corresponds to the intra-system interaction which has created the system. It is natural to suppose that the evolution of biogeocenoses and of the biosphere, for example, is determined principally by the interaction within the boundaries of these systems; the evolution of communities by the interactions of organisms, while the development of specific system characteristics of the organism is determined by interactions within the organism. Therefore we cannot share the conviction of those who suppose that the prime mover, the leading factor in the evolution of organisms and of living nature as a whole is the interaction with the "assimilated" material factors of the environment. In equal measure it is difficult to agree to view the interrelations among organisms as the leading factor of organic evolution.

An analogous discussion has been carried on for a long time concerning "external" and "internal" factors of evolution. The recognition of factors as external is sometimes viewed as a devaluation of their role in evolution. Consequently, a tendency arises to promote them to the status of internal factors and so to insure for them a respectable position of leading factors in evolutionary theory. Thus it has been proposed that a part of the external environment of an organism ("conditions of life") should be considered as an internal factor relative to the organism and even to include it in the definition of organism, something to which the majority of biologists could not, of course, subscribe. Evidently every living system from the organism to the biosphere has its own internal and external factors and there is no complex of factors, abiotic or biotic, valid for all systems such that interaction with them should play a unique leading role in biological evolution on all of its levels of organization simultaneously.

IV. Summary

We have examined quite cursorily the logical basis of the idea of system organization, have noted that this idea has been formulated as an interdisciplinary "general system theory" and

have found that the biological part of this theory has not yet been sufficiently developed, but that it already comprises three posed problems leading eventually to a formulation of a theory of system organization applied to living nature.

Claiming to reflect a property of living matter as fundamental as the potential for development, the concept of system organization has penetrated theoretical biology and strives to occupy in it a position as central as that of evolutionary phylogenetic conception. It is supposed that a synthesized image of life should embrace the two aspects, evolutionary and organizational. The new mode of synthetic thinking, by introducing into biology the notion that the organic world is organizationally discrete, amounts to an abandonment of a unitary view of living nature pictured as some monolith consisting of organisms in a single environment, one or two "leading" factors, and a single decisive unity (with the "condition of life"). At the same time a renewed and differentiated conception of fundamental material factors of evolution is re-introduced into biology. The essence of the innovation is that different evolutionary factors are assumed to be acting on the different levels of organization of living matter. Naturally, this does not in any manner contradict the classical theory of evolution on the level of the organism.

Besides, evolutionary theory is endowed with an opportunity of widely utilizing the ideas of cybernetics and of information theory which had been essentially separated from general biology. In this way, one of the modern branches of science, reflecting the behavior and regulation of complex systems (which living systems are on various levels of organization), is adjoined to a unified general theory of life. Finally the prospect appears of putting in modern terms the question of the thermodynamic foundations of evolution, primarily of the connection between organization and entropy. Undoubtedly, the concept of organization deserves a great measure of attention that it has received recently in philosophical literature.[2]

Notes

1. In the author's usage "individual" means a biological entity on any level, what R. W. Gerard sometimes calls an "org." What general system theorists call "individual," the author calls "organism."— Translator.

2. See, for example, the selection from V. I. Kremyanskiy in Part III of this volume.

6.
Units and Concepts of Biology

R. W. GERARD

IN THIS CONFERENCE,[1] which is concerned with the concepts and units of all science, my assignment is the sector of biology. The goal is thus set to consider the life sciences in the context of all science—to compare and contrast, with attention to both similarities and dissimilarities. An approach that is too general will lead into the problems of philosophy; one that is too particular, to the separate subdisciplines. Attention to the sector boundaries, or junctions, thus seems the most efficient, and I shall therefore emphasize the boundaries between physical and biological science and between biological and social science. Since the former boundary has had far more attention than the latter, and since differences have been noted more vividly than have similarities, my emphasis is on biosocial comparisons, a topic that has occupied a portion of my effort in recent years.

Entities

It is not chance that the cleavage between natural and social science is greater than that between the sectors of natural science; it is a cleavage between substance and action, body and soul, the objective and the subjective. Inquiring man scans the universe with his sensory end organs, orders and classifies the information thus obtained, and so imposes a structure on the world he recognizes. Here is William James' "blooming, buzzing confusion"; here, in Henry

From R. W. Gerard, "Units and Concepts of Biology," *Science*, 125 (1957), 429–33. Reprinted by permission of the author and *Science*.

Adams' figure, is man "on a sensuous raft adrift in a supersensuous sea"; and here is the impact of Kronecker's dictum on mathematics, "God made the integers, man did all the rest." Some inhomogeneity must exist for man even to be, and emphatically for him to divide his world into classes. And, since man depends mostly on visual information (two-thirds of all the nerve fibers that enter the human central nervous system come from the eyes), and since the eye detects primarily patterns of spatial extension, man first sees his universe as a collection of material objects. An entity is distinguished from its ground and, given appropriate duration, an individual is born to the perceiver. This is the basic event.

The individuals that people man's primitive world are necessarily commensurate with his own dimensions, his sensory range, and his time span; indeed, they even seem to conform to his status as a living being, for they are strongly personified. As technology offers instruments that reveal the lesser and the greater, as man's senses are extended (and mainly, again, his vision), new entities engage his attention; but this is a later development. More immediately, various observed individuals are recognized as having some common attributes and so as being amenable to grouping or classifying. This is the taxonomic stage of knowledge, and it follows the stage of simple observation and description just because differences are more likely to command attention than are likenesses. Man thus types his observed concrete entities into sets and, as the second abstraction (the first being entity from ground), draws sharp boundaries about them.

But, as Whitehead well said, "Nature doesn't come as clean as you can think it"; and, with growing sophistication, man replaces his plateau-like typology with the graded slopes of a probability distribution in a population of non-identical individuals. A mere collection of seemingly unrelated entities is first given meaning or pattern in terms of perceived similarities; only later is it possible to look more closely at the individuals and to reintroduce differences, but now ordered differences with significance to the larger whole. Moreover, once the initial integration (or induction) has been achieved, progressive differentiations (and deductions) can be meaningful, and subclasses can be conceived and identified, later to become graded subpopulations. This is a sign of growing familiarity with the entities of attention, indicating more interest and ordering and leading to subdivision of effort, or to fragmenting of science. Attention to a subject matter reveals finer differences and new attributes, first of whole individuals and then of their parts and structure, calls for new words to characterize these, and adds new digits to a decimal number as subclasses and sub-subclasses become significant. Here knowledge is in the morphologic stage.

So far, we have considered primarily material entities and their grouping on the basis of sensible, mainly visual, attributes. Clearly, animate and inanimate objects are more alike than are objects and the behaviors of objects; so the physical and life sciences, concerned (as a first approximation) with these two types of object, are closer to each other than they are to social science, which is concerned (still as an approximation) with the behavior of one variety of animate object. But the real shift here is from a focus on organization to a focus on action, from being to behaving, from form to function, from pattern to process, from the timeless to the temporal. "Being" is the cross section of an entity in time, and those aspects of the organization which appear relatively unchanged in a series of such instants constitute the essential structure of the entity or organism. Invariance in time helps to identify the significant units of a mature system. Conversely, along a longitudinal section in time appear the transient and reversible changes, often repetitive, that constitute "behaving" or functioning, and the enduring and irreversible changes, often progressive, that constitute "becoming" or developing. And with this shift in orientation to time there occurs a shift in the entity of concern—from an object, a pattern of matter in space, to a behavior, a pattern of events in time.

Let me briefly recapitulate. Man's attention is first drawn to particular objects in his experienced world. These objects are grouped into classes and subdivided into components, at first with Procrustean rigidity and later with more freedom of variance, and then interest shifts to particular processes. But, as a group is more removed from concrete prehension than are its component entities—a species seems more abstract to us than does an individual organism—and as its component entities become more or less interchangeable, so is a process more abstract and universal than are the acting objects. Permeability, to some thing—ion, gene, idea, person—and in some degree, is a property of all boundaries (indeed, a boundary may be characterized as a zone of lowered permeability), as irritability, quantified as a threshold to some environmental change, is a property of all responsive systems. This shift in approach is like that from phenotype to genotype or that from observation to model building.

For study of permeability or irritability, the particular system examined is not initially so important as is what is done with it, and the common denominator shifts from the object to the variable. Some attribute of the object is selected for attention and measurement as the object is manipulated. New methods arise, experiment dominates observation, and changes of the observable in time are almost universally examined. Not the nerve but the nerve impulse is the entity of concern, and this can be studied on any particular nerve that comes conveniently to hand. True, important differences appear between impulses in invertebrate and vertebrate nerves, in nerves of the frog and man, and even in separate fibers of a single nerve; but these differences are mainly quantitative ones in the same parameters. Indeed, the behavioral similarities are so significant that eventually a functional criterion may help to define a material set: a nerve cell is one that conducts an impulse, and a smooth muscle cell is sometimes best distinguished from a connective tissue cell by its ability to contract. This is the essential shift from the morphologic to the physiologic mode. It must not be forgotten, however, that manipulation remains limited to the material object: the influence of temperature on conduction rate is determined by warming a nerve, not a nerve impulse; the answer to the question, Did you ever see a dream walking? is *No*.

The pure morphologist, then, is concerned with the structure of particular objects and attempts to make his description ever more complete. Here is the gross anatomist and naturalist of the past as well as the old organic chemist describing a substance or the visiting anthropologist describing a village. (The electron microscopist or cytochemist or ethologist of the present, as well as the modern macromolecule chemist or the factor analyst seeking primary abilities or the sociometrist noting contacts or quantifying opinion, is often busy with the specific case but is usually concerned really with the class.) He observes what is; and he seeks ever more powerful tools to identify a system and fix it at an instant of time, to reveal its finer detail, to discriminate its more subtle differences, and to do this more precisely on more limited samples. His concern is primarily with the individual instance, like the clinician's with his patient, the humanistic historian's with his character or period, the artist's with his poem or painting or other unique creation of man. When a class property becomes the focus of interest, comparative studies replace those of the individual, and descriptive morphology gives way to comparative morphology or systematics or physiology or genetics or some other discipline concerned with relation or function or development. As the class or property replaces the individual—as the actuarial approach replaces the clinical approach—there is greater distance between the operator and his material, the material becomes more objectified—not Tabby, but a cat; not John and Mary Smith, but a family—and analysis is added to description.

The entities or units which are significant, which are invariable (organization) or repetitive (function) or progressive (development) in time, similarly shift from concrete material entities to abstracted properties of classes of material entities.[2] It is an interest in such attributes as personality traits or social roles and connections as the units of structure (aside from personal behavior and role-playing as actions), rather than in the individual or groups that exhibit them, that makes some aspects of social science seem different from life science.

Levels

An entity of interest, then, may be an object or some property of it or a class of objects or some property of the class. But the class is, of course, a kind of individual; and the more the members of the class interact—even to the extent of developing into differentiated subclasses—rather than coexist, the more does the superordinate group become a true individual rather than a collection of ordinate individuals. The shift from separated cells to a reproducing body, shown by the slime mold, remains an excellent example of such individuation; a species with interbreeding individuals determining the properties of its gene pool is a continuing superordinate individual. This is the now widely accepted relation of hierarchy and levels. The atom, an individual or unit, is built of subordinate differentiated and interacting units, the various nucleons and other ultimate (for the moment) particles, and is built into a superordinate molecule. The individual molecule, in turn, with like or unlike fellows, becomes a crystal or a colloid or some other material aggregate; the colloids form particulates; these, cells; cells, tissues and organs; these, organisms; and organisms, species and larger taxonomic categories, or, in another way, groups, communities and larger ecosystems. I have found the word *org* convenient for those material systems or entities which are individuals at a given level but are composed of subordinate units, lower level orgs, and which serve as units in superordinate individuals, higher level orgs. The important levels are those whose orgs (entities) are relatively enduring and self-contained. Thus, a cell is more likely to continue as an individually maintained entity than is a given colloidal micelle or a cell particulate, and an organism will vary less in time than will its parts or organs.

In the course of cosmic evolution, assuming a start in chaos, orgs at a given level become more highly integrated—with a clearer boundary, with more differentiated and interdependent units, and so with a more complex structure and more powerful regulating mechanisms—and new levels are superposed. But each added level permits the combination of old level orgs, now subunits of the new org, in various ways. It is because of the explosive increase in richness of pattern with rise in level that there appears to be an emergence of unpredictable novelty. The particular org that forms is indeed understandable in terms of the units, and their relationships, of which it is built; in this sense the situation is reductionist. But the particular org and its properties are rarely predictable a priori, because of the great number of possible outcomes, with either known or unknown probabilities and with that strong dependence on unspecified values of unidentified conditions which we call chance.

Thus, higher level orgs are likely to have a greater variety than lower order ones, and they are likely to depend more on their particular past; they are more individual. But they are also less plentiful, since several subordinate units contribute to each. A handful of ultimate particles form a hundred species of atoms, or perhaps a thousand, noting isotopes, and a hundred kinds of atoms form millions of molecular species. But the total of molecules is less than the total of atoms, and the total number of members of an average species of molecule is far less than of the average species of atom. The kinds of living organisms compoundable from the subpopulations of atoms and molecules (and combined or macromolecules, as nucleoproteins from amino acid and nucleotide building blocks) must be infinite in any meaningful sense; and the groups compoundable from organisms must be even more so. Yet the number of members of each kind of organism is vastly less than of each kind of molecule; and of groups, again less. In fact, whether from insufficiency of material or of time, the actually realized species or organisms are probably fewer than those of molecules (certainly they will be with the creative meddling of synthetic chemists); and the realized groups or ecosystems or societies, far less again.

Certainly all realizable molecules or cells or organisms or groups have not been realized, and the null members are not distributed at random. Particular combinations, orgs, are more stable in a given environment than are others, and these will be "successful"—that is, will occur in larger numbers—while that environment lasts. Furthermore, individual orgs form sets, as species form genera (and elements thus fit into columns of the Mendeleev table); molecules divide as to ionization or polymerization or what not into inorganic, organic, and macromolecules; organisms classify into kingdoms and phyla and down, because of a kind of periodicity in the patterns of formation. And, the grandest dichotomies of all, the hierarchy of levels has branches. The physics of atoms is unitary. The chemistry of molecules is strained between inorganic and organic and bio but still retains a unity.

At the next level, however, is an unquestioned split into the complex inanimate orgs of geology and astronomy (and meteorology and oceanography, perhaps even of architecture and engineering) and the complex animate orgs of biology. Physics and chemistry thus are subordinate to biology, the earth sciences coordinate with it; and in many ways the earth sciences are more comparable to biology than are the former. The entities of concern in biology and the earth sciences, as compared with physics and chemistry, are more individual and there are more species of them to deal with; and as unique orgs at the supermolecular level they are closer to ourselves, are more a subject (*thou*) than an object (*it*).

In the domain of biology between the cell and the multicellular individual, the tissue is an org with cell units, and so also is an organ. The first is a population, with cells related by origin or history, and a loose org; the second is a tight well-integrated org, with cells related by function. The same threads extend beyond the individual— to species and larger population categories, based on descent, and to ecogroups of various sizes and levels, based on fundamental interrelation. Moreover, above the level of the individual occurs another major branching, with societies—especially but not exclusively of man, or even of any single species—diverging from the population axis. Here population genetics and systematics leave ethology and ecology, and here social science separates from biology. The ecosystem of a lake or forest, or of an ant hill or flock of starlings, is thus coordinate with human spatial and functional groups, the village or the tradesmen, as is the clone with the clan as a lineage group. Again there is a jump in individuality and a diminution in kinds, a greater nearness to man and a consequent shift from object relationship toward subject relationship, at the social level.

Becoming; History

History, or becoming, I said in a preceding section, is a regular change, normally progressive, in a system along the time axis; function, or behaving, is a repetitive perturbation along this secular trend; and structure, or being, is the instantaneous status. The units and subunits of an org are nodes of stability, relatively constant in time. These are the structural residues of past action, the molecules or organs or institutions that have become fixed, yet which also carry the cumulative changes of becoming. It is critical, however, that, whatever role process initially played, traces of the past can be carried forward only by concrete material entities, not by abstracted units. The neurone can evolve or develop by changes in its components, and so the nerve impulse can also change, in speed, intensity, and what not; but the impulse, per se, does not develop; it is a single action and has no history. So also, the individual person—or generations or groups of persons—carries the history of a society, even though the role or status is the unit of interest. And, of course, the gene—or generations and arrays of genes—carries the heredity of the organism and the population. It is well to note, to prevent confusion, that secondary material products of the primary entities may also be carriers of the information and ordering, the amount and kind of matter, which the past imposes on the present—wooden vascular tubes, seed cases and egg albumen, chitin or bony skeletons, elastic fibers and plasma antibodies, nests and burrows, buildings and machines, books and recordings, are examples—but these separate material carriers only reemphasize the point.

In the becoming of a given entity, there may be a shift in emphasis from one carrier of the past to another, and the shift is normally up the hierarchy levels. A gene, if it is a nucleoprotein molecule, is the product of vast chemical adventures, from the formation of atoms and simple molecules in the distant past of its ancestral lineage, to relatively minor shifts in kind or arrangement of atom or radical, the mutations of its macromolecular maturity. The cell is directed in its development, first by the information stored in its genes, later by the structures and substances that have been formed partly under their influence —reduplicating particulates, somatic mutations, adaptive enzymes, and the like. The organism, in turn, develops by virtue of the various cell types that are differentiated early in its individual existence and their later patterning and other modification as tissues and organs. And the group, finally, changes as its component individuals learn differential roles and skills. It is hardly surprising, then, that higher level orgs are more individualized than lower level ones, that they are more determined by their particular experience, and that they carry a richer and more characterizing past. A society becomes what it is through learning by its individuals, morphogenetic development by their cells, reduplication with mutation by their genes, and so, by regress, into the domain of chemistry.

Attention was focused, in the preceding paragraph, on the units as carriers of the past. Equally essential in shaping each present from its immediate past is the environment acting on the unit. Indeed, at each stage of development of an org, the heredity is fixed in the units entering that stage; and the environment, interacting with these units, leads to new fixations—irreversible changes —in these units or in superordinate ones. Thus, of course, arises the progressive specification and differentiation of orgs, an amorphous totipotentiality yielding to a concrete realization. And the magnificent inventions of gene reduplication and recombination, of heredity and sex, insure stability with variation; as the environment, operating through mutation and selection, insures guided

change. At other levels, the mechanisms of becoming are less understood; but there is little doubt that they are similar in broad principle, dissimilar in all else.

Since at each stage and level future development could be along any one of a number of branching paths, depending on the vicissitudes at the moments of decision, the difficult problem is not that of diversity but that of uniformity. More than 40 cell generations lie between human egg and baby, and at each division a slight difference in cell properties or arrangement could magnify through subsequent ones; yet billions of babies have been born within the fantastically narrow range of "normality," only a negligible scattering of monsters outside of it. Of course, too great an abnormality cannot continue its development and is cut off by death; but, aside from such selection, there are self-regulating or homeostatic mechanisms in operation at all times and levels. If enzyme molecules are too active, a fall in substrate concentration and an increase in end-products will slow the reaction. If cells multiply over-rapidly, they become too far removed from a source of nourishment and are retarded. If a liver is lagging in its many functions or a nerve trunk is failing to innervate its peripheral field, the structures will grow or regenerate—controlled by still unknown mechanisms—until performance is adequate. Populations of predators and prey regulate each other in quantitatively predictable ways. And if a man deviates far from the norm of his culture, social pressures and sanctions—by better understood mechanisms than the morphogenetic ones—bring him into line or exclude him. Homeostatic processes nudge orgs toward a uniform state. The interaction of units to form a superordinate org is regulated, as is their action to maintain it.

Behaving; Regulation

This viewpoint has been little applied to the secular changes of long-range becoming, but it is the daily bread of moment-to-moment behaving. The vast bulk of the functioning of any enduring system is as displacement-correcting responses. Here is the negative feedback of engineering or the adaptive or self-regulating or homeostatic response of physiology. All orgs maintain themselves in a dynamic or flux equilibrium by mobilizing internal reserves to oppose environmentally imposed change; or, more rigorously, each unit responds to loads imposed on it by its environment (which may be the superordinate org of which it is part) with responses of its subordinate units that tend to eliminate the stress on the whole, even at the expense of a greater temporary displacement of the part. It is in the particular mechanisms and sequences that different orgs, and especially different level orgs, differ from one another; and each case must be examined individually, as for its structure. Yet here also important commonalities exist.

The organization of an org, its function–structure complex, is investigated by imposing displacements on it. Ordinarily an input is presented, and the output is observed, the stimulus-response situation. But the thruput in the system can also be manipulated—as in plucking a piano string, stimulating a neurone pool or cutting a nerve tract, or blocking an artery or a highway—and the spread of, or adjustment to, the disturbance tells much about the system. The quantitative relation between magnitude of displacement and strength of restoring influence—linear, concave, convex, sigmoid, or more complex—as also the existence of different or like mechanisms for restoring displacements from opposite directions, and whether the return is oscillating or damped, might serve to group widely different orgs into classes. There is a limit of homeostatic tolerance, an amount of displacement of a system beyond which it will not return. Change is then irreversible, and process leaves behind it structure—behaving shifts to becoming and alters being, sometimes leading to pathology and dissolution.

General questions can also be asked at this level about the degree of displacement tolerated in relation to kind, repetition, frequency, direction, and other aspects of the load; about the safety factor; about the speed of change of physiological zero, or adaptation; and about many other matters. Moreover, since structure is a product of history, or irreversible process, the character of the material change can serve as an index to properties of the action. A highly regular structure, as a honeycomb, indicates a highly determined process, even though this is a behavior of a group of organisms. Striated muscle fibers are highly ordered longitudinally and more variable in section; presumably the micelles are arranged very powerfully, once formed, but the number in a fiber is determined more by chance.

Being; Organization

An organism has organization, an ordering of material in space and of events in time. Any random arrangement is an order; the essence of ordering is that some particular order, out of all possible ones, will be produced. The particular one can be defined in relation to the observer, as near and far; or to some polarity he chooses, as large and small; or to a functionally related object, as key to lock; or to a generatively related object, as parent to offspring; or to an unrelated object, as a photograph or model to the original. Of these, the ability to reproduce itself, along with any fixed aberration, is the most demanding and is especially characteristic of biology; and life has been defined as "the repetitive production of ordered heterogeneity." The guiding information is carried, and the given arrangement is imposed or reproduced by various means, from electric fields around linked pyrimidines in nucleic acids (four of which can "code" the building of the 20 amino acids of proteins), through the protein antigens of cellular immunity, the metabolic and allied gradients of morphogenesis, the engrams of racial or individual experience, to the coded tapes of calculators and the culture traits, especially language, of civilizations.

Communication of information across org boundaries, between entities at the same or different levels, is not only the means of fixing the past; it is also the means of responding to the present. Nerve impulses and hormones, like talk and books, are transient or more enduring signals (or symbols). Perhaps hearing is more important than vision to social man, as is often claimed, because speech is the vehicle for the immediate communication of information in ongoing interacting behavior. Indeed, a major difference between physical, biological, and social orgs may be in the relative importance for them of the energy, substance, information, and meaning that cross their boundaries. The higher the level, the more do individuality and specificity enter and the more is the system coded to, or discriminating of, differential environmental stimuli or information.

The more also does the study of higher level orgs involve the use of experimental methods and mental tools dealing with patterns of relatedness. The forking paths of a nerve impulse through a brain, or an infection in a population, or a rumor in a community reveal connectivity patterns; and for the analysis of these relations of "organized complexity" are coming to hand the new techniques of set, game, and probability theory, of topology and stochastics, and of other nascent branches of mathematics or logic.

History produces structure, and structure determines function—becoming gives being, and this is capable of behaving; order is produced and maintained—but the relations are so intimate and seemingly reciprocal that the distinctions sometimes seem artificial. Further consideration shows that this is not the case. For the function of an org at its level depends on the structure of its subordinate units, and the structure of these subordinate-level orgs depends, in turn, on the history of their sub-subordinate units. Contraction of a muscle fiber is possible by virtue of the fibrillar and membrane structures, and these are produced by processes involving macromolecules, enzymes, and other submicroscopic units. It would lead too far afield to develop the notion, but it deserves thought, that history, structure, and function stand in relation to one another as do cause, org, and purpose. In both triads, time runs, say, from left to right through past, present, and future; and levels rise from left to right through subordinate, ordinate, and superordinate. Incidentally function (the noun), with an overtone of duty, relates an ordinate unit to a superordinate org; at its own level, functioning has an overtone of pleasure.

Conclusion

In the remaining space, I can merely suggest the concrete application of these considerations. Again a brief recapitulation. The units of man's attention are first concrete objects, directly sensible. These are classified, then seen as populations with variance; dissected or combined, to sub- and superordinate units forming hierarchical levels; compared and analyzed so that functional units replace or add to structural ones; and considered in relation to time, both as to irreversible development and to maintenance or restoration of equilibrium; and in relation to order and the information carriers that reproduce it. The world of organized experience thus plots on a map, with orgs at different hierarchical levels—molecule, cell (or crystal), organism, group, population (or society)—along the ordinate; and with their properties—becoming, being, behaving—along the abscissa. The entities, the disciplines concerned with them, the manipulative and rational methods for studying them, and the resulting concepts about them, can be classified into appropriate squares of such a table.

The hierarchy has two major branchings: (i) above the molecule level, into more organized entities with or without the collective properties that describe the living; and (ii) above the organ-ism level, into entities based on human or non-human components. Biology is thus superordinate to physics and chemistry and, at its lower levels, coordinate to the earth sciences; it is subordinate to and, at its higher levels, coordinate to the social sciences. The boundaries are reasonably sharp; yet the biochemist or biophysicist or electron microscopist, concerned with molecular traffic and macromolecular edifices, is much closer in attitudes and operations to the physical scientist, while the systematist or population geneticist or ecologist, concerned with organism traffic and population edifices, is much closer to the social scientist, than these different-level biologists are to one another. And perhaps the biologists operating between cell and organism levels are most akin to, say, meteorologists and might find rich mental nutrition by learning how they handle such problems as storms by the study of individual hurricanes, from Alice to Zelda.

The attributes that help define living orgs are (i) highly ordered and clearly bounded heterogeneity, spread over many levels and with many differentiated units at each; (ii) dependable mechanisms for reproducing units and patterns, by reduplication of the information carriers, and for altering them, by recombination of carriers and by the innovating (mutagenic, imprinting) and selective action of environment on the carriers; (iii) powerful homeostatic mechanisms for maintaining and regaining equilibrium, including especially the use of transported material, transmitted activation by energy or signs or signals, and stationary dominance-subordination gradients. The gene, materialized as a macromolecule, and the idea, materialized as an engram, chiefly among the transmitters of enduring order, are the bearers of structural and behavioral heredity; they carry the past of the entity and account for its individuality. The hormone and the nerve impulse and the sound or gesture, chiefly among the transmitters of transient order, are the bearers of information and instruction to and from orgs or their subsystems and evoke adaptive or innovative behavior that maintains the entity or modifies it in conformity with environmental pressures.

The student of the living stands between the students of the material and of the human on an ordinal scale. He deals with entities or orgs or systems that are less when compared to the latter, but more when compared to the former; more various and more individualized: more highly ordered and capable of more varied behaviors; more dependent on a particularized past and a discriminated present and so on fixed or transient information; more devoted to self-maintenance and self-duplication over the short run—by stability, supplied by feedback and inheritance—and more devoted to change and adaptation over the long run—by modifiability, supplied by learning and gene shift and guided by environmental selection; more sensitive to more environmental variables and more able to dissociate the response from the stimulus in magnitude, kind, and time interval; more personified and closer to the observer and harder to dimensionalize and quantify; more "free" to achieve their "purposes" and reach their "values," including survival and "progress," and to be "aware" of the attendant experience of inner "private" and outer public "reality"—if these words add anything to what has been said.

Such an exercise in analysis and integration is more than an exciting mental adventure; it can have important and useful consequences. The attention of an investigator may be directed to other disciplines from which ideas or skills or information can be plucked ready to apply to his own. Social scientists have been slow in exploiting biology in this way, but they could profit much from its approach and content. Acculturation as a stress, culture as a self-regulated internal environment, institutions as organs, ideas as heritable social mutations and subject to the same factors or pressures as operate for organic evolution, ideologies as polar or balanced views of man as a whole entity and man as a unit in a larger unity—such viewpoints can demonstrably aid understanding of the social epiorganism or the body politic.

The interrelations of subdisciplines in an investigator's own field may be exhibited, so that the great unities are not lost in the small particulars. This may spark the seminal insight that leads to a new structuring of the universe of interest and, failing this, must reveal areas of research emptiness or duplication. And this also should favor presentation of biology as a dynamic whole, with a few penetrating concepts replacing a legion of detailed facts and words, to our students and to our public. Life science is a great entity, and part of a greater one; biology, all science, will attract more and better members and more generous and enthusiastic support when, in all senses, the forest is added to the trees.

Notes

1. This article, prepared explicitly for the AAAS symposium on "Fundamental units and concepts of science" that was held on December 27–28, 1956, nonetheless rests heavily on the conference on Concepts of Biology (and on work with colleagues at the Mental Health Research Institute) and can serve to help integrate the rich discussion of the conference.

2. Much confusion has arisen from the use of a given word as a noun, a structural connotation, and as a verb, a functional one. Thus, *function* in physiology, *adaptation* in systematics, and *role* in sociology, as examples, when used as nouns, refer to an existing state; as verbs, to an action. The state, in each case, carries an implication of purpose and value, of the org at one level of the superordinate system; and the action, similarly, implies a behavior of the unit that has utility in the larger setting. Adaptation of the individual, in adaptive amplification, is different from the adaptation it has acquired in an environmental situation; the adaptive radiating of a population is different from the adaptive radiations of a phylum.

7.
Levels of Integration in Biological and Social Systems

ROBERT REDFIELD

THESE PAPERS formed a symposium held in September, 1941, in connection with the celebration at the University of Chicago of the fiftieth anniversary of the University. This symposium had a double origin. Representatives of the Division of the Social Sciences planned a program of papers having to do with some of the more comprehensive and underlying aspects of society. The program was to emphasize three borderland fields of recent research interest—borderland from the point of view of the student of human society. In the first place there was the disposition in recent years for students of primitive society on the one hand and of modern society on the other to study their subjects in common terms: the significant event here was the rapprochement of anthropology and sociology. In the second place recent investigations of the social behavior of monkeys and apes had made a fresh contribution to the understanding of the origins of human society. In the third place the rapidly developing work of students of mammalian and bird societies had aroused the interests of sociologists and anthropologists. So the social scientists arranged for papers to be read respectively by an animal ecologist, a student of the social behavior of monkeys and apes, an anthropologist, and a sociologist known for his understanding of modern urban society. The essential idea was to present human society as an example within a class, societies, and to have a look at some of the resemblances and differences among examples of the class.

From Robert Redfield, "Introduction," in Robert Redfield (Ed.), *Levels of Integration in Biological and Social Systems* (Lancaster, Pa.: Jacques Catell Press, 1942), pp. 5–26. Reprinted with permission of Jacques Catell Press.

In the meantime biologists at the University were making ready a program of papers concerned with the ways in which parts are organized into wholes in life forms. Here again there was a wish to represent new frontiers of research, and to consider special problems in wider contexts. The problem of the appearance of multicellularity, questions as to the mechanisms of integration of parts of an organism into the whole organism, problems presented by the study of populations and of the organization of populations into entities greater than and different from the component individuals—all these problems had had recent attention with significant results. There was, furthermore, a disposition to recognize that the integration of parts into wholes within an organism, and the integration of parts into wholes within a population or social aggregation, were not entirely separate problems, but that they could be considered in relation to each other, and together. So there resulted a program of seven papers beginning with a consideration of the evolution of the metazoa and concluding with a report on the social relations within certain avian and mammalian aggregations.

.   .   .   .   .

The order to which the placing of contributions has been chiefly obedient is that implied by the conception of a developmental sequence of levels of integration. This, the basic assumption of the symposium, is expressed by Gerard in terms offered by him to include all the part–whole systems viewed by the symposium: "The panorama of evolution shows a strong tendency for animorgs, indeed for orgs in general, to evolve towards increasing integration at any one level and to superpose additional levels of complexity."

The papers direct attention to successive levels of integration. The first two are concerned with the integration of parts into an individual organism. Hyman deals with evolutionary events: the transition from unicellular to multicellular organisms. Buchanan is concerned with the mechanisms whereby the parts of an organism are subordinated to the whole. The third paper, that of Gerard, might have come first, or perhaps last, for it offers a set of terms by means of which the results of successive levels of integration may be considered together. . . . In T. Park's paper and in part of Jennings' paper the larger entity is not an organism, but a population; the mechanisms of integration are still simple, and have largely to do with common responses of the individuals of the population to the environment; the relations of individuals to each other are unreported, or elementary. In Allee's paper on birds and mice the inter-individual relationships reported are clearly social, and the integration depends on more complex stimuli-response connections and on the effects of previous such connections. Emerson's paper advances the line of thought in the direction of comprehensiveness, for it offers a statement of general factors which have brought about the development of societies at two different organic levels, those of insects and of men. This paper explicitly recognizes the adaptive character of increasing integration. Carpenter's paper on sub-human primates reports a society in which sign mechanisms play a large role in bringing about integration; in Kroeber's paper on primitive societies the reader encounters the enormously developed mechanism of conventionalized symbolic behavior which distinguishes our own species; and R. E. Park's paper points to the significance of important differences in levels of integration within the human and therefore culturally organized societies.

What these papers seem to be saying, in most general terms, is this: The organism and the society are not merely analogues; they are varieties of something more general: the disposition, in many places in the history of life, for entities to undergo such modification of function and such adjustment to other similar entities as result in the development and persistence of larger entities inclusive of the smaller. "Fitness may mean cooperation for mutual benefit both between species and within integrated intraspecific populations as well as between parts of the organism." Departing from the language of science, one might say that the individual metazoan, the infusorian population, the ant colony, the flock of fowl, the tribe,

and the world-economy, are all exemplifications of nature's grand strategy.

.   .   .   .   .

Hyman's paper on "The Transition from the Unicellular to the Multicellular Individual" offers a hypothesis, in several parts, to account for the evolutionary event identified in the title. This paper does not represent the lowest level of integration in the animal world, for subcellular integration, on a molecular level, comes to mind, as well as the increasing complexity of single-celled organisms; but the first paper is certainly concerned with an early development of new integration of major importance, and one on the biological history of which a good deal can be said. The author considers alternative possibilities along the series of steps in the evolutionary sequence, and, after weighing the evidence, makes a choice. The choice leads to consideration of possibilities to account for the next evolutionary event, and another choice is made; in this manner a compound hypothesis, or most tenable explanation, is developed.

In summary the sequence proposed is the following. It is probable that the first metazoan came about, not through formation of walls in a multinucleate protozoan, but through consolidation of a protozoan colony. For certain reasons, it is probable that the protozoa were colored flagellates, which lost chlorophyll and adopted holozoic habits. But here a difficulty is encountered: the asexual cell division in the cells of Volvox (a holophytic colonial flagellate) bears little resemblance to cleavage in a metazoan egg, especially in that cleavage in Volvox is meridional, while in metazoan eggs it is partly equatorial. The appearance of equatorial cleavage must be accounted for. The occurrence of equatorial cleavage, following meridional cleavage, in certain sponges, suggests that equatorial cleavage appeared in some hollow flagellated colony. It was equatorial cleavage, together with differential relations of members of the colony to the outside world, which probably established polarity, upon which metazoan organization depends. The next development to be accounted for is the establishment of two layers in the flagellate colony. Among possible explanations, Hyman chooses that of Metschnikoff: certain cells wandered into the interior of the colony. Probably the inversion seen in sponges did not characterize these flagellate colonies; with inversion eliminated, and equatorial cleavage established, the anterior part of the colony became the more active part. There resulted a blastula-like organism in which the upper cells had epithelial-mesenchyme tendencies, and in which the lower cells had food-storing tendencies. The latter tended to pass into the interior.

Hyman is thus showing us how a spherical colony of protozoans, with no polarity, in which each cell acts independently and can perform all functions, becomes a metazoan individual, in which the colony is polarized and the cells act in coordination. While Hyman deals with this phylogenetic problem, Buchanan ("Intermediate Levels of Organismic Integration") deals with a physiological problem. While Hyman is (in part) concerned with the way in which polarity became established in evolutionary event, for Buchanan the central question is, What are the mechanisms whereby regulation of the whole through dominance of a part—an aspect or derivative of polarity—is maintained within the single organism? After establishing the fact of combination of local independence of parts with considerable centralization of control in flatworms and annelids, and the fact that there are great differences among closely related species of these animals as to lability and functional centralization, Buchanan takes up his main question. A review of experimental data leads to cautious conclusions. Dominant regions of the organism are controlled by more active physiological systems; these regional differences may be experimentally demonstrated in terms of many characteristics; but it is not clear which of these characteristics (more rapid oxidative metabolism, higher electrical potential with respect to subordinate regions, greater susceptibility to toxic agents) are causal and which concomitant. Nor can we say with assurance whether transportation of materials or transmission of impulses is the more effective mechanism of integration. It is at least clear, Buchanan concludes, that the process of establishing the regulative center within an organism appears to involve the setting up of a differential in the rate of physiological processes. A conspicuous feature of the differential is the relatively high rate of oxidative metabolism in the regulative center. This higher rate depends, at least, on the presence of oxidizing enzymes, the concentration of the enzymes, the approach to optimum for that enzyme of hydrogen ion concentration, the amount of water, the permeability of the system and the age of the tissue. Once established, a dominating center operates through electric impulses, or by the secretion and diffusion of substances, or through both.

The scope of Gerard's paper ("Higher Levels of Integration") is almost as wide as the entire symposium. Gerard's topics may be reduced to two headings: (1) the nature and evolution of transmissive mechanisms; and (2) the part–whole relationship, especially the parallel between the organism and the human society. The first topic connects with the main theme of Buchanan's paper; Buchanan declares that mechanisms, part mechanical, part electrical, are common to all life forms in making dominance and subordination possible. Gerard reviews the evolutionary improvement of propagation of impulse within an organism, and shows that with evolutionary development the speed of propagation of impulses within an organism increases, as does also the degree of specialization of transmissive organs.

His second subject represents the transition from the first major subject of the symposium to its second: from the question, How does an organism come about out of its component parts? to the question, How does a society come about out of its component individuals?

Gerard's answer takes chiefly the form of an expansion of the organismic analogy. Human society, he says, has developed as a new "org" by routes similar to those followed by the sequence of organisms in the improvement of their integration. The social group is like the single biological unit in that both are composed of units made of the same substances, similarly organized; in that both manifest the same developmental changes; in that both exhibit the same activities and the same mechanisms. The growth curve is the same for organisms and for societies. Both maintain adaptive amplification. Communication parallels transmission of impulses. And finally he argues that evolution into more integrated orgs continues, and that this goes along with increase in altruism.

Interesting is his reference to the slime mold as an instance of an organism the component cells of which behave as separate individuals during part of their existence, coming together only at other periods to form a single many-celled organism. The slime mold is a part-time multicellular organism, so to speak. In the one phase the slime mold suggests to one following Gerard's analogy, the "mass" of the human sociologist; in the other it suggests the organized social group. From another point of view the slime mold has an interest for the subject of this symposium. Its two phases also suggest that it is not always easy to distinguish between integration of cells into a many-celled organism and integration of spatially separate and mobile individuals into a society. The slime mold is at times multicellular, but it does not require great effort to think of it as a simple society with highly impacted constituent members.

As is apt to be the case when an analogy is consistently developed, questions are aroused as to the limitations of the analogy, and not all such questions are answered in Gerard's paper. One such may be mentioned here, in order that reference may be made to it in a later connection. "A more integrated org," says this writer, is one which "exhibits a greater degree of control of the units by the org as compared to control of the whole by the parts." He also declares that "societies are evolving into more integrating orgs" and cites the ever-widening authority of government. Might it not be asked if the widening authority of government shows merely an integration with scope over more entities or units, but not necessarily a greater degree of control of the whole over the parts? As appears from papers later in the series, there is reason to suppose that the degree of integration—the extent to which the whole, small though it is, exerts control over the units of the whole—is in some respects greater in a primitive society than in a modern society. Is this a limitation upon the applicability of the analogy to the course of human evolution?

Burrows' paper ("Synergistic Aspects of Bacterial Populations") turns the reader's attention to a very general and early developed form of association among organisms. This writer reviews evidence that populations of two specifically different bacteria in association produce effects which neither one alone can produce: (1) changes in the rate of growth of the population; (2) quantitative alteration in biochemical changes produced; (3) the bringing about of biochemical effects which neither population can produce alone (synergism).

If the main axis upon which these eleven papers are hung is the series of levels of integration, it might be argued that this paper logically should come last, for it does not deal with integration among units within an organism, but within a society composed of specifically different populations. The conjunct bacterial populations may suggest comparison with symbiotic societies, animal and human.

On the other hand, because of the contribution made by this paper to understanding problems as to the mechanisms of interaction (enzyme systems) between units of a bacterial system that bring about certain of the effects reported, effects by the whole not characterizing the parts separately, the paper belongs early in the series. In this aspect the paper relates to Buchanan's exposition

of the mechanisms of transmission of impulse within an organism, though, of course, in the bacterial populations the concepts of dominance or polarity are hardly applicable.

More fundamentally considered, the paper stands for that widespread and important form of integration which has been called non-conscious cooperation or automatic mutualism. From this point of view it is of no great importance that the interacting bacterial populations represent different species. If adequate evidence were available for the existence of functional mechanisms of integration within what are known as homogeneous bacterial populations, such populations, rather than mixed populations, might have served as the basis for the discussion here.

The essential point is that there are, even at a very low level of the organization and differentiation of the individual organism, functionally integrating mechanisms. The different bacterial populations interact so as to cooperate—if the word may be used to refer to joint activity without consciousness—and thereby show us association, if not society, among organisms. In other papers, notably in those of Gerard, Jennings and Emerson, there is reference to or exemplification of this recurrent disposition among life forms for organisms to act together so as to accomplish effects, often favorable to themselves, which they cannot accomplish separately. In this sense Burrows' paper indicates a disposition of students to examine the cooperative aspects of developing integration and to regard the competitive aspects as hitherto overstressed.

Unless Burrows' paper be so considered, the fifth paper, by Jennings ("The Transition from the Individual to the Social Level") is the first concerned directly with the question, What is the nature of societies among the simplest organisms? This contributor indeed distinguishes two levels of the social, and provides corresponding definitions: There is social action (or social relations?) where individuals influence each other to their negative or positive social value. There is social organization where there is functional differentiation among the individuals.

The two levels are represented in one or another of the protozoan aggregations studied by Jennings. In certain infusorian aggregations the individuals react to each other simply as physical bodies, clinging to each other. In certain species individuals produce chemicals of such a nature that other individuals of the same species react negatively when they get outside the area where the chemicals are present. In neither case does it appear that the interaction results to the value of the individuals. On the other hand it appears that individuals of some species reproduce more rapidly in the presence of individuals of the same species; in other cases the opposite effect occurs. Here interaction is beneficial or the opposite. A form of inter-individual relationship among protozoa that is more easily to be called social appears in the temporary paired swimming of certain ciliate infusoria. The motions of the individuals of the pair are specialized and coordinated, and the behavior plays a role in the processes of mating. Apparent cases of social organization, in Jennings' sense, are represented in the partial fusion and zygotic reproduction of paired infusoria. In some species there are several reproduction types some of which may mate with all the other types and some with only certain of the other reproductive types. (Some species have varieties each with reproductive types.) In these cases the individuals of the aggregates react to each other selectively, as individuals, and are functionally differentiated.

Attention also attaches to Jennings' account of the clone: an entity composed of all the individuals of successive generations, vegetatively produced, from a single parent. The infusorian clone is of course recognizable only as maintained as a segregated persisting aggregation in the laboratory. In certain species the clone has a life cycle. A period of youth, when conjugate reproduction does not occur is followed by a period of maturity when conjugate reproduction is frequent with vitality of offspring, and this in turn gives way to a period of old age when conjugate reproduction, and even reproduction by fission, decrease, with lowered vitality of offspring, until death of the clone occurs.

The clone thus bears comparison to the individual organism. Both are composed of many cells; the cells in each kind of entity ("org") are of the same sex type; both show similar life cycles. But in the infusorian clone the individuals are physically discrete, and there is no differentiation of function except that having to do with reproduction.

Emphasis in Jennings' paper is placed on the differentiation of function in connection with reproduction and the incidental social interactions of individuals as the beginnings of societies in certain unicellular organisms. The system within which integration occurs is the aggregation, in many cases dense and easily recognized, of infusorian individuals. In the next paper in the series, by Thomas Park ("Integration in Infra-Social Insect Populations") the system considered

by the author is not alone the individuals making up the aggregation. The system is the individuals of the (insect) population plus the influences, or "pressures," arising from either genetic or environmental sources, which interact to form the system which is integrated. The pressures are interrelated with one another. "These pressures are integrated in the sense that, as in an organism, change in one affects another and always results in some compensatory regulation in the system. There is nothing mystical in this statement."

Park considers one character of the insect population: its size. By reason of the integration of the system, this character is very comprehensive, and a discussion of it leads into consideration of the major factors operating on this level. The result of Park's analysis is a new formulation of these major factors and of the relations among them. Current formulations of the problem make use of a single variable, environmental resistance, regarding reproduction as a constant. Park here asserts that "both natality and mortality can fluctuate from a high to a low value and that certain environmental (and genetic) influences affect these rates in either a positive or a negative direction." The discussion distinguishes density-independent factors from those which are density-dependent, and exemplifies two kinds of each of these, those favoring population increase, and those favoring population decrease.

In what sense is the non-social insect population with which Park is concerned an integrated system? The answer probably is: the individuals influence each other (and are influenced by factors outside them) to the result that the population appears as a new entity with properties which cannot be predicated of the component individual organisms. One may be able to say that the population is dying, for example, when some of the component organisms are not yet senescent, and one may characterize the birth-rate of the population while the individual organisms are without this property. Thus it is the sense of this paper that an aggregation of individual organisms is integrated if (a) one may describe the aggregation as defined in space–time (a population), (b) with properties additional to or different from those characterizing the component individuals, (c) which properties are the result of interaction among the individuals—and of the individuals with their environment. The terms "society" or "social relations" might then be reserved for more special forms of integration in which there is some adjustment of the respective interests of the individuals of the aggregation by convergence of interest or limitation of conflicts; or Jennings' more inclusive category might be adopted which recognizes as social any relations between individuals resulting in either positive or negative value to the individuals. (If the latter choice is made, relations between predator and prey are social.) Paramecia engaged in paired swimming probably illustrate integration amounting to social relations. And colonies characterized by division of labor as well as specialization of function among the organs of an organism are cases of integration on a higher level than that represented in the populations described by Thomas Park. The infra-social insect population bears comparison with the infusorian clone. Apart from differences as to genetic origin and as to observation of the entity in nature or in the laboratory, the clone is described by Jennings with little said as to the character of the inter-relations (other than genetic) among the individuals. Park, dealing with more complex organisms and a more complexly differentiated environment, presents in systematic fashion some of the relations among the individuals, and the relations they have with the environment, which bring about certain properties of the population.

Except, probably, certain colonial protozoa described in earlier papers, the fowl and mice described by Allee ("Social Dominance and Subordination Among Vertebrates") are the first instance presented in this symposium of a population integrated in the form of a society. In the rank-order of the individuals composing the population we have a case of limitation of conflict among the individuals resulting in relatively stable forms of relationship, which are social. Not merely is it true that the individuals affect each other so as to bring about properties in the population which cannot be declared to characterize the component individuals, but the interests of the individuals are limited in terms of the interests of the others. Moreover, the aggregation has structure analogous to that occurring among the organs constituting an organism. Indeed in at least some of the cases reported, the social aggregation has, as a result of order established chiefly in terms of dominance and subordination, what amounts to division of labor, or specialization of function. The master cocks and the sub-cocks of the sage grouse are the mating "organs" of the strutting-ground society, while the guard cocks have the special auxiliary function suggested by the name the observer has given them. Furthermore, the existence of a territorial basis in the formation of society is demonstrated in these aggregations.

From the point of view of evolution, the establishment of rank-order, with its demonstrated influence over reproduction, is effective in two ways, as Allee indicates: (1) the most assertive of combative individuals have greatest opportunity to produce offspring; (2) the organization of the strutting-ground group makes it effective in competing with other such groups.

Allee's paper is chiefly concerned with the mechanisms by which rank-order is brought about in birds and mammals. He lists twelve factors contributing to (relative) dominance in one individual as compared with another. Special interest attaches to what he calls "psychological factors" for it appears, whatever may be true in invertebrate societies, that the experience within social relations occurring among fowl or mammals leaves a residue in each individual which affects its response in subsequent social relations. Among these organisms society is in part a result of the inter-individual experiences, and individuals assume, with reference to other individuals, special relations which give each individual a corresponding special acquired nature that anticipates the enormously more developed phenomenon called personality in human societies.

In Emerson's paper ("Basic Comparisons of Human and Insect Societies") the comparison between subhuman and human societies which was largely implicit in the paper by Allee becomes explicit and the central subject to be examined. Emerson does more than state the many analogies between animal and human societies; he tells us that these resemblances constitute another instance of convergent evolution. The mechanisms which have produced the two kinds of societies are, of course, different: biological heredity in the one case and learned symbolic behavior (or culture) in the other.

His important contribution lies in the exploration of the general characteristics of evolutionary change in order therein to discover common causative factors for societies at both insect and human levels. The effort requires a conception of evolution enlarged to consider evolutionary factors such as natural selection as operating on groups or populations as well as on individuals and to regard variation as occurring in both organism and environment with cause and effect relation between.

Adaptive radiation, Emerson says, has most often occurred through an internal control over an external fluctuation. This control has resulted not merely through genetic modification of the individual, but through coordination among organisms. Just as greater control over inner environment resulted when the multicellular organisms came about, so greater control over inner environment was achieved by coordination among organisms. As reproductive adaptation tended to control the environment of the individual during early stages of development, so adaptations on the social level (family, etc.) tended to control environment during later stages. Coordination resulting in greater control of the environment is not only intraspecific, but also interspecific as in symbiosis, and in ecological communities. "The demarkation between the social system and the ecological community is not sharp."

In that this paper deals with the evolutionary changes which have achieved ever higher levels of integration in life forms, it may be broadly seen as resuming a theme begun in Hyman's paper. While Hyman dealt with the particular evolutionary steps whereby coordination of single cells resulted in establishment of many-celled organisms, Emerson deals with the general basic evolutionary factors which have resulted in the establishment of societies. "Society," he writes, "is surely a manifestation of fundamental life attributes which are shared with the biological systems and the division between the social and the non-social is not sharp." Remembering Hyman's paper and that of Gerard, one is tempted to add that the division between the multicellular organism and society is not absolute either. Integration has come about by coordination of various kinds of biological entities into systems of varying complexity and varying distinctions.

In Carpenter's paper ("Societies of Monkeys and Apes") there are presented in summary form data, much of it only recently reported, describing what is beyond doubt an animal society. The populations of fowl or of mice described by Allee are societies, so far as reported, chiefly by reason of the rank-order among the individuals in the society. It may be supposed that rank-order is not the only aspect of social relations among individuals which give mice or fowl populations a social structure. The relations between parent, or at least mother, and offspring, if not yet other relations, must help to make up those societies. Allee was concerned with the one aspect: rank-order, and with the mechanisms whereby it is established or changed.

Carpenter reports on sub-human primate societies with interest directed toward aspects of social relations within them, of which rank-order is only one. He deals with the relations between sex- and age-groups, with intergroup relations, with relations of solitary individuals to organized groups, with inter-specific relations, and with the relations of the group to its environment.

He establishes that each sub-human primate genus, or species, has its own characteristic form of grouping, with respect to number, sex-distribution, and other features, and that these patterns are highly persistent. He tells something about the ways in which new groups are formed. He shows that the social structure can be investigated in part in terms of the prevailing nearness or farness in space of individuals to one another. He establishes the territorial basis of group-life among these primates, and shows that the territorial limits expand or contract with the relatively great or small dominance of the leading male of the group. It appears that intergroup dominance exists also, and that the prevailing hostility between local groups of the same (but not of different) species prevents federation of groups. In the discussion of dominance special interest attaches to the fact that there is a "prestige effect," which suggests the "psychological carry-over" of Allee's mice.

Carpenter distinguishes social coordination from integration. He does not define the latter, but by the term he apparently means articulation of individuals, as parts, into a whole, society. But "the coordinated actions of two or more primates involve reciprocal exchange of stimuli and responses, acting on a background of previous social integration." This concept of social coordination involves the establishment of "sets," dispositions to behave, established by experience, in part, and communication between individuals which cause them to act in accordance with these "sets." He shows that these primates have gesture-symbols and vocalized symbols. They are symbols (signs or indicators) in that the act suggests to the performer and the communicant the performance of coordinated responses. It is not clear what part is played by learning in the using of these indicators. They refer only to immediate situations; the range of indicators which may be employed is small and is narrowly fixed by hereditary determinants; and the power to characterize, represent, or abstract situations is limited.

It is obvious that the comparison of importance raised by this paper is that between these sub-human primate societies and human societies. "Monkeys and apes," writes Carpenter, "have only limited capacity to characterize, represent or abstract a perceived situation." Animal societies are in fact constituted in great part of genetically related individuals, but this situation they cannot abstractly recognize, any more than they can any other. The animal mother knows her child, but she does not know children. Animals are kin to one another, but kinship, being a category, plays no part in their societies.

Kroeber's paper ("The Societies of Primitive Man") deals with kinship, as a body of abstractions, in the societies of primitive men. In primitive societies men characteristically give abstract recognition to the fact that their groups are composed in considerable part of genetically related individuals, and express this recognition in systematically organized categories. Animals distinguish their own kind from others, and members of the group with which they ordinarily associate from members of other groups. They accomplish this by mechanisms, whether genetically determined or affected by learning, that allow the individual animal to respond to each other animal so as to maintain these distinctions. But they do not distinguish cross-cousins from parallel cousins, or even parents as a class from children as a class, for the power of making categories is an aspect of or is dependent upon the capacity to establish conventional symbols, and this capacity is absent, wholly or almost wholly, in animals.

Kroeber's paper looks two ways. It takes up this important point of difference between animal and human societies which Carpenter mentioned more briefly, and reminds us that at the human level a new mechanism of integration is introduced which does not wait upon modification of genes. At this level it is not enough to say that the individuals composing aggregations are integrated by specialization of function enabling the aggregation to persist as such. It must also be said that the specialization of function is conventional, that is, depends upon recognized abstractions which are communicated through symbols in which meaning has been stored by communicated experience; and it may further be said that these abstractions, or categories, have systematic interrelation among themselves. It is these systems, removed by several degrees of abstraction from the elements of the overt behavior of individual men which the Carpenters of human sociology may set down in their notebooks, which we call, with varying shades of emphasis, "customs and institutions," "folkways and mores," or "culture."

Kroeber presents one of the major conventional categories upon which human societies depend: kinship. This is selected because it is universal for all such societies, and because it assumes such a great variety of special forms. It is the great

variety of combinations of elements, with evidence that the same kinship form may be derived, in two different groups, from different beginnings, and the general impression of recency and instability of certain of these institutions, that lead Kroeber to make his point that many of these kinship institutions seem to be "true luxury products." It is far from clear in every case that the form of marriage rule is more adaptive, in that society, than some other marriage rule would be. It is probably a good working hypothesis that institutions tend to take forms and to assume relations among one another which make the society an entity integrated more effectively to survive as such in its total environment, but it is not at all clear just how each institution is thus adaptive. "They serve some function, but it may be a minor one among major possible ends which are left formally unprovided for."

In his paper, written with special reference to insect societies, Emerson pleaded for search for common principles of societal development which will explain both animal and human societies. "Most often," he writes, "adaptive radiation occurred through an internal control over an external fluctuation." The many and varied instances of kinship institutions which Kroeber reports suggest the great variation of organic form within a general type which is common in the animal world. Perhaps these kinship institutions represent another form of radiation. Because changes in human societies can occur so much more rapidly than can changes in animal societies, the significant variation is on the level of convention or culture. The adaptiveness of each form is a matter to be investigated, but we may perhaps assume that in the long run the more stable forms are adaptive. They contribute to "internal control over an external fluctuation." The establishment of ethical systems, with the associated equally human phenomenon of morale, represents another level of integration, yet another way in which the aggregation, now a society with "culture," is enabled to resist fluctuations in the external environment, whether they be food shortages or wars.

The second connection of Kroeber's paper is of course with that which follows it and which deals with the societies of modern men. Here the significant point is the "luxuriance" of kinship structure in primitive society without important political institutions, as compared with the association, in other societies, of complex political institutions along with economic and technological complexity. This simple and important observation alone provides justification for recognition of a major dichotomy among human societies: primitive *versus* modern or urban. Kroeber's paper concludes with identification of two major levels of integration within the human society-section of the great total panorama of integration: an earlier level based upon personal relations and expressed chiefly in kinship categories and institutions, and a later and superimposed level in which technological, economic and political categories and institutions, somehow related, come to characterize social aggregations.

The concluding paper, by R. E. Park ("Modern Society"), is concerned in large part with the point raised at the end of Kroeber's paper: the interrelation of technological, economic, and political fact in modern societies. It is true that R. E. Park declares that these aspects of modern society are in relative independence of each other, but he immediately stresses the fact that the independence is only relative, and that the "economic, political, or the moral and religious aspects of society" can be fully comprehended only if we view them "as anthropologists do, as integral parts of a single organism." "What is fundamentally unique about modern society, as perhaps of every other, is probably just the relation which these different levels of integration—the economic, political and moral—bear to one another and to the societal total of which they are parts."

The interrelations which characterize modern society, as R. E. Park gives them to us, are approximately these: the increase of the size of the aggregate within which social relations exist is accompanied by a complexity which is chiefly dependent upon the great increase in the division of labor; along with these characteristics goes greatly increased mobility; the development of technology which is closely related to the characteristics already mentioned is related to the prevailing mechanistic, secular and rational character of modern life. The personal and intimate relationships which in primitive societies get one conspicuous expression in the kinship institutions described by Kroeber, give way in part to impersonal and formal relations. The total aspect of modern societies is so different from that prevailing in primitive societies as to justify recognition, even against the broad background of all life forms, of modern societies as constituting yet another kind of integration.

In the last section of R. E. Park's paper, under the heading "Freedom," we are confronted with comparisons between primitive and modern society which raise another question: Is this a case where the development of a new level of integration is accompanied by loss of integration on a lower level? The general conclusion to which the reader is drawn from a reading of the entire series of papers is that on the whole, in the development of life forms, the addition of new levels of integration has not involved the abandonment of integration at lower levels. The appearance of multicellularity did not involve reduction in the contribution of cellular complexity to the total picture of integration, nor did the development of societies, or other "epiorganisms," carry with it the disappearance of or decline in the multicellular organisms.

R. E. Park points out that in modern societies men are much freer, as compared with men in primitive societies, to move about, to compete for a job, to pick a wife, and to act inconsistently with tradition and previously established moral rule. Indeed, it is only in modern society that the concept of individual freedom has reached so high a development. But the exercise of these freedoms has been "profoundly disturbing to all the (socially) inherited forms of society." As Park points out, social distinctions are, or seem to be, abolished, and modern man has in large part ceased to believe in the old gods and the old standards. But it was these social distinctions, as systematized in the social structure of primitive societies, and the common beliefs in old gods and long-established standards, that largely constituted the integration of primitive societies. Modern cities "where men live together in relations that are symbiotic rather than social (cultural) have not yet, it seems, developed a tradition, a body of mores, or a moral solidarity sufficient to insure either their perpetuation of existing social institutions, or the orderly succession of these economic, political and cultural changes which embody the aspirations of this modern world."

These considerations suggest a review of Gerard's conclusion that "societies are evolving into more integrated orgs . . . this means a greater influence of the whole on its parts, a greater subordination of the individual man to the larger group." It is true that the course of human development has been such as to increase the number and complexity of social relations so that the individual man is a component unit in "orgs" of greater and greater size, and it is true that on certain of the levels characterizing human society, the economic and the political, the "orgs" are more and more integrated. On the other hand, R. E. Park's paper requires us to recognize that on the level of common understandings, especially with reference to moral values, there has been, in recent times, disintegration rather than increasing integration. If, as R. E. Park says, the relations of modern men are symbiotic rather than cultural, it must appear that there has been, recently, a relative decline in the importance of a level of integration, that of organized common moral and religious values, which appeared, in the whole series of developing levels of integration, after symbiosis and not before. And how is R. E. Park's emphasis on modern freedom of the individual to be reconciled with Gerard's conclusion that the evolution of societies means a greater subordination of the individual to the larger group? Again the answer seems to lie in recognition of two different levels of integration in human societies: the ecological–economic–symbiotic, and the moral and religious. With respect to the former, individual men in modern societies are greatly subordinated to the larger group; with respect to the latter, they are less subordinated than is true in primitive societies.

Whether this condition represents a turning point in the great history of integration among life-forms, or is merely a temporary condition, is a question not answered in this symposium, nor perhaps can it be answered. The disposition of the contributors to this volume is more favorable to the latter view than to the former. Gerard pro-phesies that mankind as a whole will become an integrated cooperating unit. R. E. Park declares that the fact that men will still make war for a way of life indicates that the development of integration is for human societies, unfinished, and implies that integration on the moral level is still of great importance, and may become greater again. The common conscience and trust of scientists and scholars, of which this symposium is one small expression, is, in a wide sense, moral, and it is no small part of the way of life which many men will make war to preserve. The world of science and scholarship rests on qualities of common understanding that tempt one to look upon it as the beginning of an integration wider than has existed before. About this R. E. Park writes "what gives this vast and more or less unconscious cooperation of scholars its real importance is the fact that it is based on a mutual understanding and solidarity which makes communication and collaboration possible, in spite of racial, religious and political differences. And this in turn makes men of science participants in a common enterprise designed to create and preserve a fund of knowledge which is the common heritage of all nations and all peoples." It was to celebrate this mutual understanding and solidarity, as much as the fiftieth birthday of the University of Chicago, that this symposium and the program of September, 1941, were instituted.

III

Systems, Organization, and the Logic of Relations

PART II argued for an emergent, holistic view of the various levels of integration, seen as complexes of systemically organized components. Part III focuses directly on this problem of organization and includes a number of pioneering attempts to study its basic nature and particularly the various mechanisms (that is, the peculiar organizations of components) that can account for those features characteristic of the higher complex levels. Thus, Rapoport and Horvath suggest the foundations of a theory of organized complexity which will embrace in a scientifically respectable way the "teleological" nature of such systems, that is, a theory that will get at the goal-seeking feature as a function of the organized complexity itself (the earlier story of the specification of the mechanisms underlying this feature is reviewed in Part V). This capability, they argue, can be analyzed in terms of the modern mathematical equivalent of taxonomy, namely, branches of topology such as graph theory or network theory, which study the basic logic of the relations of parts making up the "organized complexity" of concern. Kremyanskiy further illuminates the nature of organized complexity by comparing it to "chaotic aggregates," or what Rapoport has called "chaotic complexity," as well as intermediate levels of organization. Of particular interest also is Kremyanskiy's attempt to specify objective criteria for "progress" based on the differential features of the various levels of integration or complexity.

Although the sport of attempting a precise definition of the concept of "system" is often not a very helpful one, the specification of Hall and Fagen has been one of the more useful, going, as it does, beyond mere definition to a delineation of characteristics distinguishing different types of systems and their relations to an environment.

But even more abstract and fundamental to the analysis of organization have been theories of the archetype of the Machine, or theories of automata. These operate at the level of the basic logic of relations and attempt to determine by rigid logical proof what kinds of constructions or systems and their behaviors are possible. Thus we have the "Turing Machine," named after the British logician A. M. Turing, the general theory of automata of the late John von Neumann, McCulloch and Pitts's network model of the neurophysiological system as a finite automaton, and W. Ross Ashby's theory of "The Machine." Each of these now classic or near-classic positions can be seen as a highly generalized theory of a dynamic system of interrelated

69

components whose possible sequences of transformations make predictable or under-standable the complex states of the whole. Thus McCulloch and Pitts, in the sections of their paper reproduced here, develop a network model for the neurophysiology of the central nervous system, showing that the binary action of the neurons interrelated in certain ways makes for a logical calculus underlying the generation of propositional ideas characteristic of human thought. John von Neumann's equally famous theory of automata builds on the "Turing Machine" to prove logically that a complex system may not only be self-reproducing, but may be constructed so that it can produce other machines even more complex than itself. And finally, Ashby draws conclusions from his earlier work on a cybernetic theory of "The Machine" about the nature and possibilities of the self-organizing system.

Although such work may seem to some to be "mere" abstract speculation, to many others it has provided a solid logical foundation for, and thus taken the mystery from, many of the emergent characteristics of the complex systems of special interest to the behavioral scientist.

8.
Thoughts on Organization Theory

ANATOL RAPOPORT AND WILLIAM J. HORVATH

THE CHARGE brought against his age (some 30 years ago) by Alfred North Whitehead was that the age was living on the intellectual capital accumulated in the seventeenth century. The implication was that if no new sources of "intellectual income" are discovered, the resources will be squandered. Elsewhere Whitehead said that the culture that cannot burst through the framework of its own concepts is doomed.

These remarks are metaphors, and if they remained unelucidated by a more concrete analysis, they would have to be viewed like so many other examples of poetically expressed discontent characteristic of pessimistic philosophers from Heraclitus to Schopenhauer. Fortunately Whitehead had a profound understanding of science and was able to spell out just what the philosophical limitations were which hemmed in *scientific* thought and just where the threat of impending sterility lay.

The constraining framework of thought, according to Whitehead, was the overwhelming predominance of analytic thinking in science, stemming from the hegemony of physics as *the* model of exact science. Analysis is an attempt to understand a complexity by examining its constituent parts. The parts being simpler, they are supposedly more amenable to understanding. The idea of analysis, then, is to understand the working of the parts. Note how in presenting the analytical view, it is natural to fall into the language appropriate for describing a mechanism —the "working of the parts." The implied hope is that it is possible to "build up" the understanding

From Anatol Rapoport and William J. Horvath, "Thoughts on Organization Theory," *General Systems*, 4 (1959), 87–91. Reprinted by permission of the authors and the Society for General Systems Research.

of a complexity by "superimposing" the workings of the various parts.

There is no denying that this approach is effective in specific areas of investigation. To understand how a complex machine works, it is often sufficient to understand the separate links in its construction: this lever transmits a force to effect a torque on this wheel; when the wheel has turned through a certain angle, this relay is closed, which allows a current to flow through this wire; the rotation of the wire, in turn, gives rise to a magnetic field, which . . ., etc. One has broken up the whole operation into a temporal chain of events, all connected by determinate "causal" relations.

Another good example is the explanation of the solar system (i.e., the motions of its components) in terms of classical mechanics. Having isolated certain measurable quantities (distance, time, mass) and having postulated certain laws of motion and a universal law of gravitation, the mathematical physicist deduces first the behavior in empty space of two point masses, subject to mutual gravitational attraction. Upon his solution he *super-imposes* perturbations due to the presence of other point masses. The scheme works because of the *additivity* of the fundamental effects if the perturbing influences are small compared with the interactions between the original pair. And of course underlying the whole scheme is the additivity of the fundamental quantities. Distances and forces are added according to the vector laws of addition; times and masses are added like scalars. In electro-magnetic theory too, additivity, stemming from the linearity of the basic partial differential equations, is the fundamental *tenable* assumption, which allows the super-position of vector fields (physical additivity) and the super-position of the many particular solutions of partial differential equations, multiplied by appropriate constants and added, to obtain the solution satisfying specific boundary conditions (mathematical additivity). Thus the classical picture "works" where it is applicable. And because of its success it has a terrible seductive power over the human mind. One is tempted to ask whether all complexities cannot be explained in this way. Is it not only a matter of super-posing more and more parts, each of which is understandable in itself, until the entire complexity is accounted for? Or, consequently, is not the understanding of any complexity to be gained by analyzing it further and further into its constituent parts? The essence of Whitehead's critique is to answer this question in the negative.

Whitehead's critique has been echoed during recent decades in many quarters. Most of the echoes have been solemn but vague. We are told that scientific thought needs to be spiked with "organismic" and "holistic" ideas. The holists even have a motto inscribed on their banner: "The whole is greater than the sum of its parts." In the writings of the proponents of a neo-organismic philosophy, we often read that Goethe had the right idea all along, since he thought "holistically" and "organismically." Some point to biology as the science to serve as a model for a new organismic orientation. Others proclaim the dawn of the new era as being ushered in by a preoccupation with "organized complexity," the region between the extremes of "organized simplicity" and "chaotic complexity," with which the exact scienitst has been hitherto exclusively concerned.

In spite of the vagueness of the language, much of this has a ring of truth. Still the ideas remain elusive, and they stand in constant danger of becoming clichés through repetition. Occasionally the arguments are supported by concrete examples, which can serve as points of anchorage for further exploration.

The notion of the organism, for example, is taken to be central in most holistic expositions of method. Well, one might ask, what about the organism? To begin with, what is it? This is a legitimate question, but it should not be pushed too hard if the "holist" is to be given a fair chance to explain his position. In order to avoid futile hair-splitting right at the start, we should not insist on rigor in the definition of "organism." An extensional definition will do. Organisms are exemplified by living things, most of which are sufficiently individuated to be recognized as separate entities. Next, what about the organism? How does placing the organism at the center of one's conceptual scheme help one get away from "limited mechanistic concepts"? The only answer that suggests itself is related to the well-known methodological differences between biological and the physical sciences.

First, the biological sciences depend to a far greater extent than the physical on categorization, and classification. Second, the biological sciences have often resorted to teleological explanations, which are taboo in physics (except for pseudo-teleological explanations to be dealt with below).

To illustrate the first distinction, we need only notice the importance of taxomony in biology: a classification of living things is its natural point of departure; important insights reflect themselves in reclassifications. In medicine, too, an art nurtured by biology, classification of diseases is a central problem. In short, the problem of *recognition* is an important one for the biologist. Before he begins to say anything about *events*, he must have a proper language to talk about *entities*, of which some can be directly observed and others must be specifically invented for the purpose.

Physics, as we understand it today, does *not* start with a classification of "entities." It starts with what have become to us intuitively evident abstractions from experience, which are *immediately* quantified. The development, then, proceeds by elaborate mathematical deductions from "first principles." Observations are not the wide-eyed explorations of the naturalist; the observations of physics are pre-arranged and carefully delineated: they are the observations of controlled experiment. They are recorded not as meticulous and intricate descriptions of a talented observer (the naturalist's art) but rather routinely, according to preconceived rules, practically always as measurements, often mechanically. Thus the physicist and the (old-time) biologist differ fundamentally in their approach to observing nature and describing what they observe, and this fundamental difference is reflected in the great importance of classification as a tool of cognition in biology and its relative unimportance in physics.

To illustrate the second difference, the early explanations given by biologists for observed events relevant to that discipline were in terms of the needs of the organism or, on the grand scale, in terms of the "strivings of the race" (as in Lamarckian evolutionary theory). Such explanations were also dominant in pre-Galilean physics. But with the mathematization of mechanics, such teleological causation was expelled from physics except when such "causation" could be shown logically equivalent to a mechanistic one. For example, the Hamiltonian Principle of Least Action *sounds* like a teleological explanation (trajectories are "chosen" so as to minimize certain integrals taken over the traversed path), but the Principle is mathematically derivable from Newtonian laws, in which no reference is made to future events as "causes" of present ones. The particles of Newtonian mechanics have no goals. The forces acting upon them act *here* and *now*.

It will be obvious to any contemporary biologist that in comparing the methods of biology and physics, we have not drawn a picture of modern biology at all. Natural history, taxonomy, collections, etc. are now only a small portion of the biologist's world. He is now quite as at home in the laboratory and even sometimes with mathematics as is the physicist. He certainly performs controlled experiments and does not confine himself to mere descriptions of what he has seen. Also the biologist is relying less and less on teleological explanations. He seeks to explain physiological processes in terms of physics and chemistry, which means that he is no longer satisfied with relating events to the needs of the organism but relies on the same laws of "here and now causality" as the physicist. Teleological ideas are no more welcome in evolution: genetic theories, based on random mutations and the theory of natural selection have replaced the Lamarckian outlook.

In short, the methods and outlook of biology approach more and more those of the physical sciences. What then can a philosopher mean when he suggests that the method of biology may have more to offer than physics to the science of the future, the science of organized complexity? Can he mean that the modern scientist should place more reliance on those aspects of biology which are carry-overs from a previous age? Such a position would be difficult to defend. Yet a defense of the "holist's" position can be made, provided we can spell out just how the older schemes of cognition (the "static" logic of classification and the teleological explanation) can fit into the new modes of analysis.

The key concept is "organized complexity" as exemplified by, say, a living organism or any "organized" collection of entities, that is, a collection interconnected by a complex net of relations, as distinguished from (1) organized simplicity and (2) chaotic complexity. We will try to clarify the difference by giving examples of (1) and (2), by

showing how these have been treated by classical methods, and by assuming what is needed to extend those methods to the study of "organized complexity"—i.e., what is needed for a theory of organization.

The organization of a system is simple if the system is a serial or an additive complex of components, each of which is understood. We have already given examples of such systems. One is a machine whose working is a time-linear chain of events, each a determinate consequence of the preceding one; in other words, a system without closed loops in the causal chain. Another is, say, a propagated wave form which can be analyzed into additively super-imposed sinusoidal components.

As soon as strict sequential sequence or linear additivity is transcended, an "organized system" becomes rapidly more complex, usually too complex for detailed analysis into super-posable parts or effects. Already the three-body problem (three mass points of the same order of magnitude in space interacting gravitationally) becomes enormously complex for determinate analysis.

At the other extreme is "chaotic complexity" where the number of entities involved is so vast that the interactions can be described in terms of continuously distributed quantities or gradients, i.e., do not need to be specifically identified with regard to the individual entities. Such systems can be described by the methods of statistical mechanics which merge with those of classical mechanics when the collections of entities are treated as continuous. Thus it can be said that the class of events amenable to exact treatment has been extended from "organized simplicity" to "chaotic complexity" when a new conceptual scheme was added to the physicist's kit, namely the mathematics of probability (which does not enter classical mechanics at all). Hence it was not altogether true that in the first decades of this century physics still remained imprisoned in the conceptual framework of the seventeenth century as Whitehead implied. The new metaphysics of probability had been added to that of determinate causality and had been digested.

The question before us is whether we can foresee what conceptual tools must be further added in order to extend systematic rigorous theoretical methods to "organized complexity," with which the holists are presumably concerned. There are at least two such classes of concepts discernible, and both are reminiscent of the ways of thinking of the biologist. It is this resemblance that gives the proposal to give more attention to the ways of thinking derived from biology an aura of plausibility. Of course, in the process of being incorporated into modern scientific thought, these concepts lost their resemblance to their origins, but the origins remain recognizable, namely teleology and taxonomy.

When we say teleology, we mean teleology in its modern garb, made respectable by cybernetics. For cybernetics has concerned itself with a new phylum of machines, machines which behave as if they had goals. The mathematics of cybernetics has shown how, on the basis of physical laws alone *plus* certain principles of constructing networks of causal relations (networks including closed loops, which, recall, we have excluded from our simple-chain linkages of relations), "goal-seeking" behavior of mechanisms emerges.

Such machines can simulate goal-seeking behavior through the feedback principle, according to which the output of the system is modified by the error between its output and some pre-set goal. It is not unlikely that simple goal-seeking actions of living organisms are effected by similar organizational principles. If so, then the apparently teleologically determined behavior can be reduced to non-teleological principles (plus organizational principles hitherto not considered in the study of non-living systems). Once this possibility of reduction is established *in principle*, the biologist can rely on "teleological" (now pseudo-teleological) theories with an eased conscience. He can say with some justification that it is the cybernetician's job, not his, to show how a "teleological" theory is to be reduced to a non-teleological one; he assumes it can be done, and this frees him to construct theories "on a higher level" so to speak.

In discussing the rehabilitation of teleological principles, we have already touched on taxonomic ones. For example, we have distinguished between machines with and without feedback loops, only the first of which can exhibit "goal-seeking" behavior. The distinction is a topological one. If "causal relations" are indicated schematically by directed segments, the systems so described appear as directed graphs. It appears, then, that systems described by graphs containing cycles can exhibit forms of behavior which systems described by cycle-free graphs ("trees") cannot. This was recognized already in the early formulation of the isomorphism between symbolic logic and network theory by McCulloch and Pitts in 1943. The development of that theory begins with the explicit assumption of cycle-free networks. Many theorems do not hold if cycles are allowed, and even the conceptual repertoire becomes insufficient. For

example, the introduction of cycles into the topology necessitates the introduction of an existential operator into the logic.

Come to think of it, topology can be viewed as a "taxonomic" branch of mathematics. It is qualitative rather than quantitative in spirit. Its findings resemble classification more than computation (witness the concept "genus" as applied to surfaces or to more general topological manifolds). The theorems of topology have an "all or none" rather than a "continuous" flavor. They frequently are assertions that something does or does not exist; that something is or is not possible, rather than expressions of functional relations among variables taking on a continuum of values.

Thus, we hope it is clear now what was meant by the statement that the "older" ways of thinking characteristic of biology, namely classification and teleological explanation, appear in modern development in a new garb. Cybernetics has shown that a teleological way of thinking is not incompatible with the metaphysics of physical determinism; topology has shown that a taxonomic way of thinking is not inconsistent with mathematical sophistication. The two disciplines are related: cybernetics is a dynamics superimposed on a topology.

It seems likely that these two disciplines will be at the foundation of that branch of science which deals with "organized complexity," i.e., organization theory.

There has also been a third line of development, quite unknown at the time Whitehead wrote his critique of the analytic method, but which has since come to be viewed as the third corner stone of organization theory. This is the theory of decisions, comprising both "games against nature" (including statistical inference) and game theory proper. Obviously this orientation is much more relevant to the "theory of organizations" than to organization theory. The distinction between the two we see as follows: We see organization theory as dealing with general and abstract organizational principles; it applies to any system exhibiting organized complexity. As such, organization theory is seen as an extension of mathematical physics or, even more generally, of mathematics designed to deal with organized systems. The theory of organizations, on the other hand, purports to be a social science. It puts real human organizations at the center of interest. It may study the social structure of organizations and so can be viewed as a branch of sociology; it can study the behavior of individuals or groups as members of organizations and so can be viewed as a part of social psychology; it can study power relations and principles of control in organizations and so fits into political science. The "decision," i.e., a choice based on examining a state of affairs *and* the range of possible outcomes is the fundamental event of decision theory. Such events are frankly based on teleological principles (I choose this action, because I hope to attain that goal). But we have already seen that what appear as teleological determinants can *in principle* be reduced to non-teleological ones; therefore it is not necessary to assume that a permanent gulf will persist between the methods of physical and social science. The decision-maker may one day be described as a complex goal-seeking system. Until then, the psychologist and the social scientist can simply take his goal-seeking behavior for granted.

To summarize, three lines of rigorous theoretical development have been intensively pursued in the past fifteen years or so, none of which had played an important part in classical physics and all of which show promise as the theoretical underpinnings of organization theory in general and perhaps in the theory of organizations specifically, namely:

(1) cybernetics—the theory of complex interlocking "chains of causation," from which goal-seeking and self-controlling forms of behavior emerge;

(2) topology or relational mathematics;

(3) decision theory, whose specific application potential is to the theory of human organizations.

The subdivisions of these disciplines of particular interest are:

(1) information theory, related to cybernetics;

(2) network theory, related to topology;

(3) game theory, i.e., that branch of decision theory in which decisions must be made not only in the face of uncertainty but also taking into account the presence of other decision makers, some or all of whose interests may be opposed to those of ego.

We cannot expect, however, that these developments will be rigorously applied in every serious study of real organizations. Indeed, we expect that premature application will be largely sterile. Information theory has been extensively developed only in connection with the problems of telecommunication, where only syntactic charac-teristics of language are relevant. Attempts to extend information-theoretical methods to semantic and pragmatic aspects of communication often degenerate into metaphorical speculations and contribute nothing to rigorous analysis. Attempts at rigorous application of game theory have fared no better. Network terminology has been useful in providing a descriptive language for communication patterns but the deductive powers of topology (specifically of graph theory) have been tapped only to a limited extent in application.

"Scientism" is still a predominant childhood disease of organization theory as it is applied in social science. Application of rigorous methods of analysis to real organizations is still largely a pious hope rather than a genuine opportunity. Therefore the impact of the new ideas can be expected to be at the time being only indirect. We expect that the study of human organizations will continue to depend on the insights and methods of the social scientist, i.e., the psychologist, the sociologist, the economist, and the political scientist. With the growth of interdisciplinary research we may expect the biologist to make additional contributions. There is some sense in considering a real organization as an organism, that is, there is reason to believe that this comparison need not be a sterile metaphorical analogy, such as was common in scholastic speculations about the body politic. Quasi-biological functions are demonstrable in organizations. They maintain themselves; they grow; they sometimes reproduce or metastasize; they respond to stresses; they age, and they die. Organizations have discernible anatomies and those at least which transform material inputs (like industries) have physiologies. All have "neural physiologies," since organization without internal communication, integration, and control is unthinkable.

In totality, then, we have today a variety of approaches to the study of organization (as an abstract principle) and a variety of approaches to the study of organizations (i.e., human aggregates with certain specified relations of interdependence among the members). The two developments are destined to travel along two separate roads for a while. Occasionally, a connecting path will be discerned, along which ideas can trickle from one stream to the other. Eventually, it is hoped, the two streams of ideas will actually merge.

9.
Certain Peculiarities of Organisms as a "System" from the Point of View of Physics, Cybernetics, and Biology

V. I. KREMYANSKIY

IN THE HEATED DISCUSSION going on in our country about the significance of similarities and differences between automatic electronic devices and living organisms, the cyberneticists have usually been reproached for their extreme universalism, and also, on occasion, for their inadequate knowledge of biology. The opposition has, on the other hand, often pointed to the indistinctiveness of organisms or to such distinctive qualities of theirs which have already or could become properties of cybernetic machines.

The present article represents a biologist's attempt to introduce greater accuracy into the posing of certain questions and into the defining of the distinctions of living entities regarded as material systems of a special type.

Which Organisms Are Involved?

In the discussion of these problems most of the attention has been given to those plans of automatic machines of modulating systems which are designed to increase the effectiveness of the brain or to study it. As a rule, reference is made only to the nervous system of humans and the higher animals. It is not difficult to see that this limitation is broadened at the discussion in question. The reasons for it are clear—these devices

From V. I. Kremyanskiy, "Certain Peculiarities of Organisms as a 'System' from the Point of View of Physics, Cybernetics, and Biology," a translation of a Russian article prepared by U.S. Joint Publications Research Service. Original publication in *Voprosy Filosofii* (*Problems of Philosophy*), August, 1958, pp. 97–107. Translated from the Russian by Anatol Rapoport, in *General Systems*, 5 (1960), 221–24. Reprinted by permission of the translator and publisher.

are, after all, designed to raise the productivity of mental labor.

Nevertheless, the question of the significance of the similarities is often posed in a most general form. For example: Can cybernetic automatic devices be acknowledged as "living?" Biologists are often disinclined to admit these machines into their domain. Is it possible, however, to solve the problem of understanding organic life at all if as evidence we use facts relating to society, i.e., another form of material existence, or only to the highest forms of activity of the highest animals? In that case the concept "living entity" does not include the lower animals and those organisms which have no nervous system—the plants and most of the microbes.

Within the organic world itself, there evidently exist very profound qualitative differences between different organisms. It is a universally known fact that life is possible only in highly organized bodies; but natural scientists cannot, of course, be satisfied with general ideas concerning the higher or lower organization of living bodies. There exists a vast literature dealing with matters concerning the basic levels or degrees of organization which have developed in the organic world and beyond its limits, the peculiarities characteristic of each of these degrees, and the reason that one level is indeed "higher" or "lower" than another. However, there does not yet exist a strictly objective criterion of progress. Without dwelling upon the history of these problems or indulging in unfavorable criticism of the respective theories and nomenclatures, we attempt primarily to give a very brief definition of the most important differences between the fundamental types of material systems.

The Fundamental Types of Material Systems

In mathematical logic there is a theory of numbers which concerns the grouping of objects, phenomena and concepts, including those whose combination might be hypothetical or abstract. Contrary to this theory, the theories of material systems pertain solely to actually existing associations. One of the simplest types of association includes such things, for example, as undifferentiated gas and dust mist or small amounts of gaseous substances at "moderate" temperature and pressure, which are not in eddy or stream movement. Associations of this type might properly be called unorganized systems or *chaotic aggregates.*

It would be wrong to think that such systems contain no unity or integrity, or that the properties of a chaotic aggregate are no different from the simple sum or simple physical resultant of the properties of its elements. Contrary to the view widely held among foreign scientists, no material association is absolutely additive in this sense. The existence of any *relationship* between two objects or two changes creates a situation which could not exist in either of them separately. It is true, however, that from a number of important aspects, or from a single aspect, the properties of the chaotic aggregate may not actually differ from the simple sum of the components' properties.

In chaotic aggregates, the interconnections between the elements are comparatively uniform, but they are particularly simple when a relative organizational simplicity typifies the elements themselves. The nature of the elements is not changed by entering or leaving the aggregate. Where there is a large number of elements, changes in the chaotic whole depend more on changes in many elements than on solitary or small groups of elements. The total of its internal interconnections, and hence the internal causal conditioning of the changes, bears a predominantly statistical or "probability" character. Not only in the inorganic world, but also in the organic world (and even in society), do relationships of this type exist. But it is only in chaotic aggregates that they are the leading relationships and the main basis for the most fundamental changes in the aggregate. (The idea of a main basis or leading moment should not be confused with the idea of the historically primary basis, to which we will come back later.)

Numbers can belong to classes of different degree. The class of each successive degree includes a certain number of classes from the preceding degree as their members. This is also true of

material systems. It should be noted that when studying material associations in the various degrees one need not carry the analysis down to "last elements" (whose existence we will leave aside), but only to a given number of preceding degrees of complexity.

Thus number of degrees is not always the same. When studying chaotic aggregates it is usually sufficient to carry the analysis to the elements representing the objects and changes of one preceding degree. Thus, when studying the regularity of changes in a small amount of unsuperheated gas, it is usually sufficient to carry the analysis to the molecules, viewing them as "simple" particles. It is a different matter when studying material associations of other types. The more developed their internal and external connections, and the more complex the partial systems (subsystems) forming the material association, the more the whole is dependent upon the individual components; or, in other words, the deeper the analysis must penetrate in order to reveal the details which are needed to understand the whole.

Having taken cognizance of these incomplete definitions of the peculiarities of chaotic aggregates, let us proceed further. In contrast to chaotic aggregates, organized systems are typified by basically regulated, varied and deep-seated internal connections among the elements. Since the latter now exist in different relationships and can alter fundamentally, it would be better to find another term for "elements." In the "general theory of systems," which is now popular in the United States of America through the initiative of L. von Bertalanffy, the word "systems" is used only for organized associations, the members of the systems being called "objects." But for a number of reasons we think it preferable to use the term components, an expression which has long been used in describing physical systems and automatic devices.

The more varied and complex the interconnections between components or subsystems (groups of components), the deeper the changes in the components (usually in only the first or second immediately preceding degrees). But these components can change only to the extent of their inherent capacity for change. For example, atoms change in molecules, and inorganic molecules change in crystals, solutions and cells; but there is far greater change in large polymerized molecules (macromolecules) in cells, and cells in multicellular organisms. The most profound changes occur in multicellular animals in the higher-degree systems.

Furthermore, the essential features of the components can do more than change. They can be newly created through the creative capacities of the system (or subsystem). For example, the overwhelming majority of complex organic substances are synthesized only in the cell organelles, i.e., in the cell "subsystems," under the influence of enzymes which are disposed and which act in a definite order. Analogous examples are numerous in nonliving nature as well, but here there are important differences which are usually either totally ignored or are given inadequate attention by the authors of existing systems theories.

Very serious attention is given in many branches of natural science, in cybernetics and in the general theory of systems, to the relationship between the system and the environment. Below, we will trace out several very fundamental differences in this relationship, which radically change its nature at the various levels of development, but for the moment we will point up only the most general aspects. The environment includes the objects and changes which exert considerable influence on the material system without being part of it. Account must also be taken of influences which become considerable in their totality rather than separately, as, for example, the effects of extremely distant cosmic bodies. The environment furthermore includes all objects and phenomena which feel the strong and direct (or not too remotely mediated) effects of the system.

Material systems are called, according to their type of relationship with the environment, isolated, closed or open. Absolutely isolated systems are, of course, purely abstract and hypothetical. In the changes of closed systems, the exchange of elements and energy with the environment does not play a very important role for a very long time. The entropy of the closed system as a rule only grows, whereas the system as a whole, being subservient to the environment and incapable of renewing itself, is inevitably destroyed, without, moreover, leaving a successor. After systems of this type, no matter how sturdy their structure, degenerate and disintegrate, other systems of this type must emerge as systems, anew. However, the succession of "embryos" can also have considerable significance in nonliving matter, such as the "primer" in crystallization.

In open systems, on the other hand, a periodic or continuous exchange of elements and energy with the environment is typical. This exchange can serve as the basis for the perpetuation of this form of existence and as the basis for the decrease or relative constancy of entropy only when the system possesses certain features of internal organization and interaction with the environment. There is no need to explain that these are then living bodies, and that they contain far more regulated internal interconnections which associate the components so intimately that the system becomes, as is said, an "organic whole." This does not necessarily signify that there are organic substances containing carbon or silicon in its make-up. To avoid confusion it would be better to call material systems of this type differently, e.g., *organic* systems, with this notion also including those open systems which cannot definitely be classified either as living or as consisting of organic substances.

The Relationships between Different Degrees of Organizational Development of Systems

Following the rather widespread custom, we will call these degrees the organizational *orders* of matter. The relationship between the organizational orders of matter consists in that the organic systems of each succeeding order contains the systems of the preceding order as its basic components, not directly, however, but mainly as part of the subsystem, e.g., as part of the organelles of cells, or organs of multicellular organisms.

Here the whole contains the parts "under itself," as Hegel said. The whole is bigger than the sum but not bigger than the organized system of its parts, in all their connections and intermediaries. In the organic world these latter are extremely complex. We know that in all complex combinations of interconnections we must be able to isolate the leading link. But to do this we must make a careful study of all the details having essential significance for the entire system. Through how many degrees should biological analysis be carried? We can immediately say through more than one, since this would be sufficient only in the case of chaotic aggregates or only in the narrow examination, e.g., when studying the birth rate in certain rodents, insects, etc. But these are not the major preoccupations of biology. When studying the most important phenomena of organic life it is mandatory to delve into the "smaller-scale" phenomena whose significance increases as the level of material organization heightens. Hence, such sciences as cytology, biochemistry and biophysics are not only vitally pertinent in studies of microbes, but also in studies of multicellular organisms. It is not enough simply to "recognize" these sciences; their achievements and methods must be put to as wide use as possible. Indeed, the under-estimation of *details* has caused our cytological and biophysical research to fall behind perceptibly, a situation which should be rectified as quickly as possible.

Those subsystems (organs and organ systems) which form the leading link in the particular living system as a rule also serve as the main link for its interconnections with the environment. This is due, of course, to the fact that plants and animals dominate only over secondary spheres in their environment (which are ever-expanding, however) and are obliged for the most part to adapt themselves to the environment. In special conditions of embryonic development this rule is broken, and the leading role can temporarily be played by the subsystems which in these stages do not perform the functions of interconnection with the environment; but this is only possible insofar as this embryonic development is ensured by the previous generation's struggle with the environment, by means of its accumulated energy and materials.

Not only the properties and changes of the organic system as a whole, but also the specific properties of its major subsystems, are completely *absent* in the independently existing bodies of the type of the system's individual components. It is a different matter when bodies of this type unite and develop into a system of the following order. In the course of development, as we mentioned above, the nature of the components may take on the basic "imprint" of the whole. Thus, to compare them with systems of subsequent orders, we must not select *their* components, but the separately existing bodies of the type of these components. When comparing organic systems of the same order, the highest significance may be had by the distinctions acquired by these systems as a result of their being included or newly formed in systems of higher orders.

It follows from this, incidentally, that the adherence of an animal and automatic device to the same type of organic system would not signify a truly fundamental similarity between them. A computing machine can possess many of the basic features of an animal, but the former represents a component of society which was created newly by man (or by other automatic devices, the possibility of which was pointed out by Neiman).

The developed components of an organic whole always contain something more than they would alone, but never more than the given whole and its environment (and also its history). This is not taken into consideration by those foreign cyber- neticists, and particularly philosophers, who arrive at idealist conclusions by supposing that the information involved in the operation of, say, memory devices or communication channels is something that can be reduced neither to matter nor energy. The idea that this information is hence something "non-material" is very much like the "overcoat point of view," in which only the trappings are considered matter, while society with its internal and external relationships is not considered matter. The word *matter* should not be reduced to the word substance in its physico-chemical sense, as is generally the trend outside the sphere of gnoseology.

Let us now see how the problems raised at the beginning of this article could be handled on the basis of the aforesaid. Let us first frame an objective criterion for progress, and then outline the relative importance of the similarities and differences between living systems of the organizational orders, representing the most important stages of development in the organic world.

But we first must prove in a more complete exposition that the *leading* properties of organic systems are actually superior in certain stages of development to the subordinate properties or subordinate variables (where domination does not always mean superiority) and that the development of ways and means of variability offers the organism more, and not fewer, advantages than the development of dynamic stability. We note that cyberneticists emphasize in theory the significance of ultrastability and multistability (phenomena discovered by Ashby), but in practice they also concern themselves with variability. It is known that the latter serves as a means for attaining stability (stabilizing variability); but, in addition (and this is far more important for biological systems), it serves as a means for developing new forms both of the species' stability and also of the variability itself (form-creating variability). It is easily shown that an increased number of components (the limit to which is specific for each order depending upon the available integration means and upon the environment conditions) increases variability and has several other advantages, of which we will speak in time. There apparently exists a phenomenon of "ultravariability" (the repeated and qualitatively heightened capacity for change).

In the light of these preliminary schematic observations and that which was said earlier, we can objectively, if in "qualitative" form only, draw the following conclusions with regard to the questions posed.

1. One organic system is truly superior to another organic system if:

(a) the first possesses every essential property of the second, but in addition to this relationship of essential similarity or kinship (not to be confused with the relationship of identity) with the latter, it possesses properties which are essentially lacking in the second, whereupon (b) the total of these original properties of the first, with the pre-eminence of one of them, acts as the leading factor in its internal and external interconnections, that is, relates to the properties of the second system as major to secondary (and, in the end, better ensures preservation and development).

It is easily seen that it is just this relationship that does indeed exist between the organic systems of each succeeding order and the systems of each preceding order. Consequently, the organic system of each succeeding order is indeed superior to the system of each preceding order (when certain conditions exist, which should also be defined in a more complete way).

Organic life has passed through at least three major stages of development in types of transition to higher orders of organizations (cells, multicellular organisms, and family–herd groups). Since the relationship of organizational orders is the same as the relationship of progress, it becomes impossible to deny the progressive nature of development in the organic world. (There are still biologists and philosophers who deny the existence of progressive development). All progress is relative, and bound up with regression (Engels), but progress can relate to the leading properties or *major* moments in a form of existence, while regression can relate to the secondary moments and subordinate properties. Then, with all the relativity of progress, the development as a whole proves to be progressive. True progress enriches the forms of existence.

We are of the opinion that the elaboration of a system of concepts connected with this or some other objective criterion of superiority, might open the way to *calculating progress*, i.e., to a technique for the strictly objective and quantitative determination of the degree of true progressiveness in new or planned changes.

2. When comparing organic systems of different orders, the exclusion of everything similar results in the isolation of everything which does not belong to the systems of the *lower* order. These distinctions are precisely what we must know about the living systems of the higher order, for it is in these distinctions that the leading moments of their life forms lie. Thus, if we are comparing organic systems of different orders of organization, the differences between them are of greater importance than the similarities.

Corresponding to these profound differences between the most important degrees of progressive development of organic life are the highly essential differences in the types of *relationship of living bodies to their own existence.*

10.
Definition of System

A. D. HALL AND R. E. FAGEN

1. Introduction

The plan of the present paper is to discuss properties of systems more or less abstractly; that is, to define *system* and to describe the properties that are common to many systems and which serve to characterize them.

2. Definition of "System"

Unfortunately, the word "system" has many colloquial meanings, some of which have no place in a scientific discussion. In order to exclude such meanings, and at the same time provide a starting point for exposition we state the following definition:

> *A system is a set of objects together with relationships between the objects and between their attributes.*

Our definition does imply of course that a system has properties, functions or purposes distinct from its constituent objects, relationships and attributes.

The "definition" above is certainly terse and vague enough to merit further comments, the first of which should, in all fairness, be a note of caution. The "definition" is in no sense intended or pretended to be a definition in the mathematical or philosophical sense. Definitions of the mathe-

matical or philosophical type are precise and self contained, and settle completely and unambiguously the question of the meaning of a given term. The definition given above certainly does not meet these requirements; indeed, one would be hard pressed to supply a definition of system that does. This difficulty arises from the concept we are trying to define; it simply is not amenable to complete and sharp description.

In order to reduce the vagueness inherent in our definition, we now elaborate on the terms *objects*, *relationships*, and *attributes*.

2.1 OBJECTS

Objects are simply the parts or components of a system, and these parts are unlimited in variety. Most systems in which we will be interested consist of physical parts: atoms, stars, switches, masses, springs, wires, bones, neurons, genes, muscles, gases, etc. We also admit as objects abstract objects such as mathematical variables, equations, rules and laws, processes, etc.

2.2 ATTRIBUTES

Attributes are properties of objects. For example, in the preceding cases, the objects listed have, among others, the following attributes:

atoms—the number of planetary electrons, the energy states of the atoms, the number of atomic particles in the nucleus, the atomic weight.

stars—temperature, distances from other stars, relative velocity.

From A. D. Hall and R. E. Fagen, "Definition of System," revised introductory chapter of *Systems Engineering* (New York: Bell Telephone Laboratories), reprinted from *General Systems*, I (1956), 18–28. Reprinted by permission of the authors and Bell Telephone Laboratories.

switches—speed of operation, state.

masses—displacement, moments of inertia, momentum, velocity, kinetic energy, mass.

springs—spring tension, displacement.

wires—tensile strength, electrical resistance, diameter, length.

2.3 RELATIONSHIPS

The relationships to which we refer are those that "tie the system together." It is, in fact, these relationships that make the notion of "system" useful.

For any given set of objects it is impossible to say that no inter-relationships exist since, for example, one could always consider as relationships the distances between pairs of the objects. It would take us too far afield to try to be precise and exclude certain "trivial" relationships or to introduce a philosophical notion such as causality as a criterion. Instead we will take the attitude that the relationships to be considered in the context of a given set of objects depend on the problem at hand, important or interesting relationships being included, trivial or unessential relationships excluded. The decision as to which relationships are important and which trivial is up to the person dealing with the problem; i.e. the question of triviality turns out to be relative to one's interest. To make the idea explicit, let us consider a few simple examples.

3. Examples of Physical Systems

First, suppose the parts are a spring, a mass, and a solid ceiling. Without the obvious connections, these components are unrelated (except for some logical relationships that might be thought of, such as being in the same room, etc.). But hang the spring from the ceiling and attach the mass to it and the relationships (of physical connectedness) thus introduced give rise to a more interesting system. In particular, new relationships are introduced between certain attributes of the parts as well. The length of the spring, the distance of the mass from the ceiling, the spring tension and the size of the mass are all related. The system so determined is *static*; that is, the attributes do not change with time. Given an initial displacement from its rest position however, the mass will have a certain velocity depending on the size of the mass and the spring tension; its position changes with time, and in this case the system is *dynamic*.

A more complex example is given by a high-fidelity sound system. The parts of this system are more numerous, but for simplicity we could consider only the turntable and arm of the record player, the amplifier, the speaker and the cabinet. Again, without connections, these parts in themselves would not behave as a sound reproducing system. With connections, in this case electrical coupling of input to output, these parts and their attributes are related in that the performance in each stage is dependent on performance in the other stages; mechanical vibrations in the speaker are related to currents and voltages in the amplifier, etc.

4. Examples of Abstract or Conceptual Systems

An example of a nonphysical nature is given by a set of real variables. The most obvious property of a real variable is its numerical size; in other words in this example *object* and *attribute* are closely related (in fact, in any example an object is ultimately specified by its attributes). Familiar relationships between variables take the form of equations. For concreteness, consider two variables x_1 and x_2 satisfying the two linear equations.

$$a_1 x_1 + a_2 x_2 = c_1$$
$$\tag{1}$$
$$b_1 x_1 + b_2 x_2 = c_2$$

The equations provide constraints on the variables; together the two equations constitute a system of linear equations; the parts of the system are the variables x_1 and x_2, the relationships being determined by the constants and the simultaneous restrictions on the given quantities. The system of equations (1) might be termed *static*, by way of analogy with the static spring and mass system. The analogy is determined by the fact that the numbers which satisfy the equations are fixed, just as the length of the spring is fixed in the mechanical analogue.

On the other hand, introduction of a time parameter t gives rise, for example, to equations of the form

$$\frac{dx}{dt} = a_1 x_1 + a_2 x_2$$
$$\tag{2}$$
$$\frac{dx}{dt} = b_1 x_1 + b_2 x_2$$

The system (2) might, by further analogy with

the spring and mass example, be termed *dynamic*. Here the solutions are functions of time just as the length of the spring in the dynamic system is a function of time.

The terms "static" and "dynamic" are always in reference to the system of which the equations are an abstract model. Abstract mathematical and/or logical relationships are themselves always timeless.

5. Abstract Systems as Models

The two examples of the preceding section provide more than incidental illustrations of the idea of system; they suggest one of the most fruitful ways of analyzing physical systems, a way that will be immediately recognized as a fundamental method of science: the method of abstraction.

A return to the simple example of the coupled mass and spring provides a direct illustration of the idea. In the static case, the attributes of interest are the spring constant K, displacement x, and weight W. These are related (within elastic limits by Hooke's law) by the linear equation

$$Kx = W \qquad (3)$$

which is of the form (1) for one variable. This further suggests the intimate relationship between an abstract system such as (1) and its *physical realization*. To study the physical system, we substitute for it an abstract system with analogous relationships and the problem becomes a mathematical one. In the dynamic case as well, it is not hard to show that the same sort of analogy obtains, the system being replaced in this case by a differential equation instead of a linear algebraic equation.

This practice is certainly a familiar one to physicists, chemists and engineers; usually it is spoken of as the creation of a mathematical *model*. The extent to which a model agrees with the actual behavior of a system is a measure of the applicability of the particular model to the situation in question. On the other hand, the ease with which a given system can be represented accurately by a mathematical model is a measure of the ease of analyzing the given system.

In order to be completely amenable to mathematical analysis, a system must possess rather special properties. First, the relationships must be known explicitly; secondly, the attributes of importance must be quantifiable and not so numerous as to defy listing, and finally the mode of behavior (as would be given by a physical law

such as Hooke's law), under the given set of relationships must be known. Unfortunately, it is a rare system indeed that has all these properties; more exactly, systems possess these qualities in degrees, the more interesting systems such as living organisms exhibiting less of a conformance than simpler systems such as mechanical systems of which the spring and mass is a special case.

6. Definition of Environment

At this point it seems worthwhile to introduce the notion of *environment* of systems. Environment for our purposes can best be defined in a manner quite similar to that used to define system, as follows:

> *For a given system, the environment is the set of all objects a change in whose attributes affect the system and also those objects whose attributes are changed by the behavior of the system.*

The statement above invites the natural question of when an object belongs to a system and when it belongs to the environment; for if an object reacts with a system in the way described above should it not be considered a part of the system? The answer is by no means definite. In a sense, a system together with its environment makes up the universe of all things of interest in a given context. Subdivision of this universe into two sets, system and environment, can be done in many ways which are in fact quite arbitrary. Ultimately it depends on the intentions of the one who is studying the particular universe as to which of the possible configurations of objects is to be taken as the system. A few examples may serve to illustrate this idea.

7. Systems and Their Environments

First, let us return to one of our original examples, the high-fidelity sound system. Suppose the whole system is situated in a living room, and that a record is being played over the system. The environment of the system could consist of the record being played, the room in which it is situated, and the listener. It is easily seen that each of these objects bears some relationship to the behavior of the system; the record determines the succession of electrical impulses and mechanical vibrations in the various stages of the system. The output of the system, in turn, affects the pattern of sound waves in the room as well as the mental state of the listener (which for a high-fi "bug" might range from sheer ecstasy to nervous apprehension depending on the excellence of the output). Any or all of these environmental objects could be considered to be part of the system instead of the environment. For certain purposes this might be an artificial designation. Each time a different record is played, one would be considering a different system in this case, whereas actually the system of interest to a sound engineer would not include any specific record, and so would not change in nature from record to record. On the other hand, if one is interested in a system to reproduce one specific announcement, it would make more sense to consider the record as part of the system.

The example above is cited only to make clear what is meant by system and environment and why the dichotomy of sets of related objects into system and environment depends essentially on the point of view at hand. However, the general problem of specifying the environment of a given system is far from trivial. To specify completely an environment one needs to know all the factors that affect or are affected by a system; this problem is in general as difficult as the complete specification of the system itself. As in any scientific activity, one includes in the universe of system and environment all those objects which he feels are the most important, describes the inter-relationships as thoroughly as possible and pays closest attention to those attributes of most interest, neglecting those attributes which do not play essential roles. One "gets away" with this method of idealization rather well in physics and chemistry; mass-less strings, frictionless air, perfect gases, etc. are commonplace assumptions and simplify greatly the description and analysis of mechanical and thermodynamical universes. Biologists, sociologists, economists, psychologists, and other scientists interested in animate systems and their behavior are not so fortunate. In these fields it is no mean task to pick out the essential variables from the nonessential; that is, specification of the universe and subsequent dichotomization into system and environment is in itself, apart from analysis of the inter-relationships, a problem of fundamental complexity.

8. Subsystems

It is clear from the definition of system and environment that any given system can be further subdivided into subsystems. Objects belonging to one subsystem may well be considered as part of the environment of another subsystem. Consideration of a subsystem, of course, entails a new set of relationships in general. The behavior of the subsystem might not be completely analogous with that of the original system. Some authors refer to the property *hierarchical order* of systems; this is simply the idea expressed above regarding the partition of systems into subsystems. Alternatively, we may say that the elements of a system may themselves be systems of lower order.

In passing it may be worthwhile to note that this idea of examining subsystems and their behavior has a rather widespread significance in mathematics, particularly in set theory and modern algebra. Just to mention an example, the study of groups (collections of mathematical objects having certain algebraic properties) includes considerations of the properties of subgroups; moreover, subgroups do not necessarily "behave" (behavior here is in the algebraic sense) the same as their parent groups in all respects.

Returning to our example of the high-fidelity system, we see that the idea of division into subsystem is clearly illustrated. The amplifier itself is a system of considerable complexity; the pick-up arm and speaker, themselves systems of a different character can be quite naturally considered as parts of the environment of amplifier. In turn, the amplifier could be further divided into its stages, and each circuit considered as a separate subsystem.

9. Macroscopic vs. Microscopic Views of Systematic Behavior

One technique for studying systems which are exceedingly complex is to consider in detail the behavior of certain of its subsystems. Another method is to neglect the minute structure and observe only the macroscopic behavior of the system as a whole. Both of the methods above are common and familiar in many fields, and are of fundamental importance. Before discussing these ideas further, we cite a familiar example.

The difference between these two approaches can be seen by considering the roles of the physiologist and psychologist in the study of the human system. The physiologist is interested in the internal properties and characteristics of the body; he isolates and studies separately the functions of the various internal organs in relationship to bodily activity. When studying the heart, for example, the blood stream, lungs, kidneys, etc. might well be considered as parts of the environment. On the other hand, the psychologist, while not completely neglecting visceral conditions, is primarily concerned with patterns of behavior of the system under various external conditions. It may well be that the psychologist could theoretically improve his knowledge by a complete physiological approach. From the practical standpoint this may be virtually impossible. The variables and their relationships are still beyond description and comprehension; the psychologist is left with the realization that his investigation of behavior is more fruitful from a macroscopic point of view.

10. Some Macroscopic Properties of Systems

So far we have been talking in detail about systems as though by implication there were in the background some sort of unified theory of systems. Actually, there is as yet no such theory, although attempts have been made at one. It is always a good idea when considering such general theories to be sure the types of system under discussion are clearly understood and, where generalizations to systems of other types are claimed, to see if all the analogies and correspondences used are valid.

Nevertheless, there are some properties that belong to certain classes of systems, and are worth mentioning briefly. Also, there are some valid and useful analogies concerning the behavior and properties of certain types of systems that often aid in analysis, at least conceptually, of particular systems. As a notable example, the concept of entropy, useful in thermodynamic systems, has an interesting and valuable analogue in the concept of entropy as defined for message sources in information theory. Other familiar examples are found in the close analogies between electrical, mechanical and acoustical systems, a simple instance being an R-L-C circuit and its mechanical analogue, the coupled mass, spring and resistive dashpot.

Properties that are frequently mentioned by various authors in discussing systems are:

Wholeness and Independence. In our definition of system we noted that all systems have relationships between objects and between their attributes. If every part of the system is so related to every other part that a change in a particular part causes a change in all the other parts and in the total system, the system is said to behave as a *whole* or *coherently*. At the other extreme is a set of parts that are completely unrelated: that is, a change in each part depends only on that part alone. The variation in the set is the physical sum of the variations of the parts. Such behavior is called *independence* or *physical summativity*.

Wholeness or coherence and independence or summativity are evidently not two properties, but extremes of the same property. We may speak of 100% wholeness being at the same end of a scale with 0% independence, but such use of these terms would be merely a matter of verbal convenience. While wholeness and independence may be matters of degree, no sensible method of measuring them yet exists. Nevertheless, the property provides a useful qualitative notion. In fact, since all systems have some degree of wholeness, this property is used by some writers to define "system."

Since all systems have wholeness in some degree, we have no difficulty illustrating the property. Near the 100% end of the scale we have such systems as passive electrical networks and their mechanical analogues. At the other end of the scale we have difficulty finding examples. In fact, most of the literature uses the term "heap" or "complex" to describe a set of parts which are mutually independent and the term "system" is used only when some degree of wholeness exists. We prefer to call sets of parts with complete independence "degenerate systems" because, as we noted before, it is impossible to deny systematic relationships in a heap of sand or odds and ends, or for mechanical forces acting according to the parallelogram of forces.

Progressive Segregation. The concepts of wholeness and summativity can be used to define another qualitative property often observed in physical systems. Most non-abstract systems change with time. If these changes lead to a gradual transition from wholeness to summativity, the system is said to undergo *progressive segregation*. We can illustrate this very simply with equations (2) by letting the "mutual" or "transfer" terms a_2 and b_1 become functions of time. If these terms decrease to zero as a limit we will have two independent systems represented by the equations, or we can say that the larger system, consisting of two simultaneous equations, becomes a "degenerate system."

We can distinguish two kinds of progressive segregation. The first, and simplest kind, illustrated above, corresponds to decay. It is as though, through much handling, the parts of a jigsaw puzzle become so rounded that a given piece no longer fits the other pieces better than another. Or suppose an open-wire carrier telephone system were suddenly deprived of maintenance. Vacuum tubes would wear out, poles would rot, and so on, and eventually there would be a group of parts that no longer behaved as a system.

The second kind of progressive segregation corresponds to growth. The system changes in the direction of increasing division into subsystems and sub-subsystems or differentiation of functions. This kind of segregation seems to appear in systems involving some creative process or in evolutionary and developmental processes. An example is embryonic development, in which the germ passes from wholeness to a state where it behaves like a sum of regions which develop independently into specialized organs. Another example, often observed in the creation and development of a new communication system, occurs when an idea appears, or a need is defined, and the original conception of a system segregates through planning effort into subsystems whose design and development eventually proceed almost independently.

Progressive Systematization. This is simply the opposite of progressive segregation, a process in which there is change toward wholeness. It may consist of strengthening of pre-existing relations among the parts, the development of relations among parts previously unrelated, the gradual addition of parts and relations to a system, or some combination of these changes. As an example, consider the development of the long distance telephone network. First, local telephone exchanges sprang up about the country. Then exchanges were joined with trunk lines. As transmission techniques improved, more exchanges were added at greater distances. Later, toll dialing was added, placing the network at the command of operators and eventually at the command of customers. The record has been one of increasing unification of the whole system.

It is possible for progressive segregation and systematization to occur in the same system. These two processes can occur simultaneously, and go on indefinitely so that the system can exist in some kind of steady state as with the processes of anabolism and catabolism in the human body. These processes can also occur sequentially. Consider the early history of America during which groups of people colonized various parts of the country. These groups became more and more independent of their parent countries. Gradually, the new country became more coherent as more interchanges occurred between the groups, a new government was formed, etc.

Centralization. A *centralized* system is one in which one element or subsystem plays a major or dominant role in the operation of the system. We may call this the *leading part* or say that the system is *centered* around this part. A small change in the leading part will then be reflected throughout the system, causing considerable change. It is like a trigger with a small change being amplified in the total system. An example from politics might be a totalitarian regime, decisions of an autocrat affecting behavior of the entire system.

Either progressive segregation of progressive systematization may be accompanied by *progressive centralization*; as the system evolves one part emerges as a central and controlling agency. In the case of embryonic development previously noted, segregation does not proceed to the limit for several reasons, the most important perhaps is that the brain emerges as the controlling and unifying part.

11. Natural and Man-Made Systems

To enhance the meaning of "system" we distinguish natural systems and man-made systems. Engineers are directly interested in man-made systems; however in the environment of these man-made systems are natural systems which also require investigation since their properties interact with the system under study. Furthermore, there are certain properties that both types of systems have in common; man-made systems are often copies of natural systems or at least are constructed to perform analogous functions.

11.1 NATURAL SYSTEMS

The description of these is the task of the astronomer, physicist, chemist, biologist, physiologist, etc., and again the amount one can say about a given natural system depends on the number of essential variables involved.

Open and Closed Systems. Most organic systems are *open*, meaning they exchange materials, energies, or information with their environments. A system is *closed* if there is no import or export of energies in any of its forms such as information, heat, physical materials, etc., and therefore no change of components, an example being a chemical reaction taking place in a sealed insulated container. An open system becomes closed if ingress or egress of energies is cut off.

Whether a given system is open or closed depends on how much of the universe is included in the system and how much in the environment. By adjoining to the system that part of the environment with which an exchange takes place, the system becomes closed. For instance, in thermodynamics, the second law is universally applicable to closed systems; it seems to be violated for organic processes. For the organic system and its environment, however, the second law still holds.

Adaptive Systems. Many natural systems, especially living ones, show a quality usually called *adaptation*. That is, they possess the ability to react to their environments in a way that is favorable, in some sense, to the continued operation of the system. It is as though systems of this type have some prearranged "end" and the behavior of the system is such that it is led to this end despite unfavorable environmental conditions. The "end" might be mere survival; evolutionary theory is based heavily on the notion of adaptation to environment.

There are many examples of adaptive behavior in the body. Many of these are mechanisms that tend to keep within certain physiological limits various bodily conditions such as body temperature, physical balance, etc. Mechanisms of this sort are sometimes called "homeostatic mechanisms." One example is the inborn reaction to cold by shivering, tending to resist a drop in body temperature by a compensating movement producing warmth. Closely related to the concept of adaptation, learning and evolution is the notion of *stability*.

Stable Systems. A system is stable with respect to certain of its variables if these variables tend to remain within defined limits. The man-made thermostat is an example of a device to insure stability in the temperature of a heating system; the notion of stability is familiar also in mechanics and especially in the communications field. Note that a system may be stable in some respects and unstable in others. An adaptive system maintains stability for all those variables which must, for favorable operation, remain with limits. In physiology, "motor co-ordination" is intimately connected with stability; clumsiness, tremor, and ataxia are examples of deficient or impaired motor coordination and instability.

Systems with Feedback. Certain systems have the property that a portion of their outputs or behavior is fed back to the input to affect succeeding outputs. Such systems are familiar enough to the communications engineer; servomechanisms in general are man-made systems utilizing the principle of feedback. Systems with feedback occur quite frequently in nature as well; posture control in the human body is an example. It is a well known fact that the nature, polarity, and degree of feedback in a system have a decisive effect on the stability or instability of the system.

11.2 MAN-MADE SYSTEMS

Man-made systems exhibit many of the properties possessed by natural systems; simple notions such as wholeness, segregation and summativity have meaning for both types of system. On the other hand, it has not been until recently that man-made machines have shown what might be termed adaptive behavior even on a modest scale. Other kinds of man-made systems, such as language and systems of social organization, have always shown adaptive behavior.

Adaptation for man-made systems is not strictly analogous to that for natural systems; in fact, what might be considered mystical behavior on the part of a natural system is perfectly explainable for the man-made system. Any seemingly purposeful or intelligent behavior on the part of a machine has been built into it by its designer. Also, adaptive behavior on the part of a machine is not to ensure the survival of the machine necessarily, but instead to insure a specified performance in some respect.

There are, in addition to the differences above, some additional considerations in connection with man-made systems that seem to have less bearing on natural systems.

Compatibility (or Harmony). Often the problem arises of constructing a system to match a given environment, or what amounts to virtually the same thing, of adding new parts to already existing systems, or of connecting two systems to operate in tandem. There is no guarantee that a system constructed for a given purpose will function properly if its environment is changed (not all fountain pens write under water). Similarly, two systems independently might be quite satisfactory in certain respects, but in tandem could have completely different and not necessarily favorable characteristics.

Systems might be compatible in some respects and incompatible in others; it depends on the purpose for which the systems are introduced as well as the environmental factors. Also, systems may be compared as to the degree of compatibility with a given system. In terms of the high-fidelity system, we might consider as an example the problem of matching a speaker to the rest of the outfit. Different speakers would function with varying degrees of success; some of the environmental factors might be the size of the room, the amount of money available to spend on the speaker, etc. A speaker with perfectly matched impedance and excellent mechanical construction might produce beautiful results in the given setting, but if it cost a few thousand dollars it could easily be called incompatible with respect to at least one environmental factor.

Optimization. Compatibility considerations lead naturally to the problem of optimization. As the term implies, it means adapting the system to its environment to secure the best possible performance in some respect. Optimum performance in one respect does not necessarily mean optimum performance in another; again it is a question of intent on the part of the system planner. Often, the factor of interest in an optimization problem is economic: how much bandwidth to allocate to a telephone channel, how many interoffice trunks to provide, etc. Note that the optimum bandwidth for transmitting all the subtle voice characteristics is not the same as the optimum from an economic standpoint.

11.3 SYSTEMS WITH RANDOMNESS

In either natural or man-made systems it is sometimes necessary to take into account random behavior. What randomness means and when to introduce it in analysis of a system are questions that can be hotly debated by philosophers. In practice it is usually introduced as a factor when the variables that may affect a given attribute are so great in number or so inaccessible that there is no choice but to consider behavior as subject to chance. One example is the noise in a vacuum tube due to random emission of electrons from the cathode.

Random variables enter in at both the microscopic and macroscopic levels. Statistical mechanics and modern physics are both dependent on assumptions of microscopic randomness. Economic conditions, numbers of potential customers, etc. are macroscopic factors also subject to chance fluctuation.

The operation of some systems with randomness can best be described in terms of stochastic processes (also called random processes or time series). Familiar examples in the field of communications are random message sources and disturbing noise in information theory, and the theory of waiting lines in telephone traffic.

12. Isomorphism

As has been suggested before, there are instances in many sciences where the techniques and general structure bears an intimate resemblance to similar techniques and structures in other fields. A one-to-one correspondence between objects which preserves the relationships between the objects is called an *isomorphism*. For instance in the electrical–mechanical duality, an R-L-C circuit is isomorphic to its mechanical dual since each circuit element has its corresponding mechanical interpretation and the relationships are formally the same.

Isomorphisms of this type are rather numerous; in fact, their prevalence has led to several attempts at unifying various fields of science using the idea of "system" as a fundamental concept, but these attempts are as yet incomplete. There are, however, several disciplines with more modest aims that have achieved notable success. To quote a well-known mathematician [W. Feller]:

As for practical usefulness, it should be borne in mind that for a mathematical theory to be applicable it is by no means necessary that it be able to provide accurate models of observed phenomena. Very often in applications the constructive role of mathematical theories is less important than the economy of thought and experimentation resulting from the ease with which qualitatively reasonable working hypotheses can be eliminated by mathematical arguments. For example, in geology we are confronted with random processes which have been going on for millions of years, some of them covering the surface of the earth. We observe that certain species go through a period of prosperity and steady increase, only to die out suddenly and without apparent reason. Is it really necessary to assume cataclysms working one-sidedly against certain species, or to find other explanations? The Volterra–Lotka theory of struggle for existence teaches us that even under constant conditions situations are bound to arise which would appear to the naive observer exactly like many of the cataclysms of geology. Similarly, although it is impossible to give an accurate mathematical theory of evolution, even the simplest mathematical model of a stochastic process, together with observations of age, geographical distribution, and sizes of various genera and species, makes it possible to deduce valuable information concerning the influence on evolution of various factors such as selection, mutation and the like. In this way undecisive qualitative arguments are supplemented by a more convincing quantitative analysis.

In addition to the Volterra–Lotka theory mentioned in the quotation above, there are other theories of the same nature unifying several subdomains of science. Mathematical biology, for instance, has had considerable success in this direction. There have been attempts at proposing a mathematical theory of history, cybernetics is widely quoted (and seldom understood) as unifying the communication field with the study of the behavior of living organisms, demography is a study of the growth and spread of populations, etc., but these attempts, while offering hope that certain areas will be unified eventually, are yet incomplete.

That there are isomorphisms, either total or partial, is neither accidental nor mystical. It just amounts to the fact that many systems are structurally similar when considered in the abstract. For example telephone calls, radioactive disintegrations and impacts of particles, all considered as random events in time have the same abstract nature and can be studied by exactly the same mathematical model. It is not surprising then that properties shown by systems of gases with diffusion are useful in analyzing waiting lines of telephone calls and vice versa.

13. The State-Determined System

As an example of the notion of isomorphism, to illustrate some of the macroscopic properties discussed, and to enhance further in a more concrete way the meaning of "system," we will examine the so-called state-determined system. Known to mathematicians as the time-invariant system, it has simple properties and widespread interpretations.

13.1 DEFINITION OF STATE-DETERMINED SYSTEM

Suppose that a system is completely specified by n variables x_1, x_2, ... x_n. Then the state of the system is uniquely describable by a set of n numbers. To borrow terminology from physics, the set of all points in n-dimensions describing possible states of a system is called *phase space*.

To describe the behavior of a system of this type, it is sufficient to specify the possible paths in phase space or in other words the succession of states through which the system passes. For simplicity let us assume that two variables determine the system. Then phase space is the ordinary Euclidean plane, and possible paths are curves in the plane.

If a system has the property that, given an initial state, the path is uniquely determined regardless of how the system arrived at the initial state, the system is called state-determined.

Such systems have the following important mathematical property which we shall state but not prove:

For a system to be state-determined it is necessary and sufficient that its variables satisfy a system of equations of the form

$$\frac{dx_1}{dt} = f_1(x_1, \ldots, x_n)$$
$$\vdots \qquad \vdots$$
$$\frac{dx_n}{dt} = f_n(x_1, \ldots, x_n)$$

(4)

where the functions f_1, ..., f_n are single-valued functions of their arguments. For instance, the system described in (2) is by this theorem state-determined.

The absence of t in equations (4) is what Margenau regards as the essence of causality. The same set is used by Ashby to define "absolute" systems, and by von Bertalanffy to demonstrate the possibility of a "General Systems Theory." When the constants of the set become functions of time, as in progressive segregation or systematization, the definition is no longer satisfied.

As examples, first of a system which is state-determined and then of one which is not to illustrate the theorem above, consider the following system with lines of behavior given by the equations

$$x_1 = a + bt + t^2$$
$$x_2 = b + 2t.$$

(5)

This system is state-determined. For if the curves so defined are plotted in the (x_1, x_2) plane it is easily seen that they are all parabolas with vertices on the x_1 axis and opening to the right; thus they do not intersect and exactly one of the parabolas passes through each point of the plane (see Figure 1a).

Also, the curves determined by (5) satisfy the differential equations

$$\frac{dx_1}{dt} = x_2$$
$$\frac{dx_2}{dt} = 2$$

(6)

which are of the form (4) with $f_1(x_1, x_2) = x_2$ and $f_2(x_1, x_2) = 2$: a verification of the theorem.

On the other hand, a system with lines of behavior given by

$$x_1 = a + bt + t^2$$

$$x_2 = b + t \tag{7}$$

is not state-determined. The curves in this case are again parabolas, but this time the vertices do not all lie on the x_1 axis, so that a given parabola will intersect other parabolas of the family (see Figure 1b). By differentiating (7) and substituting

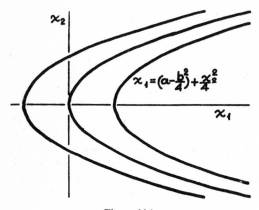

Figure 1(a).

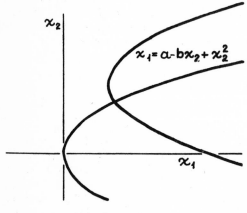

Figure 1(b).

in (7) it can be seen that the curves so defined satisfy the differential equations.

$$\frac{dx_1}{dt} = x_2 + t$$

$$\frac{dx_2}{dt} = 1 \tag{8}$$

which is not of the form (4) since $\dfrac{dx_1}{dt}$ depends not only on x_2 but on t as well.

Systems of equations of the form (4) arise in many fields. The most familiar are in mechanics, circuit theory, etc. That they should frequently occur is borne out by the underlying structure; whenever a system is *state-determined* (that is, from a given state the future progress, or line of behavior, is completely determined), it can be described by equations of this type.

13.2 PROPERTIES OF STATE-DETERMINED SYSTEMS

We can use the system (4) to illustrate some of the concepts we have spoken of previously.

In case $f_1, ., f_n$ are all zero, the system is static. That is the variables too do change with time. If this is not the case, the system is dynamic; at least one of the variables does change with time. The degree of "wholeness" of the system is determined by the nature of the functions $f_1, ., f_n$. If each of these functions depends strongly on each of the variables, the system shows a high degree of wholeness; a change in any variable then affects appreciable changes in all the rest. On the other hand if each of the functions depends on only one of the variables, the system does not have strong connections. In particular, if the equations degenerate to the form

$$\frac{dx_1}{dt} = f_1(x_1)$$

$$\frac{dx_n}{dt} = f_n(x_n), \tag{9}$$

the parts function independently, a change in any variable depending only on the condition of that variable.

On the other hand, the pair of equations below illustrate the type of relationships that might characterize a system with wholeness.

$$\frac{dx_1}{dt} = x_1 x_2$$

$$\frac{dx_2}{dt} = x_1{}^2 + x_2{}^2 \tag{10}$$

In (10) each variable affects the other in a completely symmetric manner.

Finally, a centralized system might have the form

$$\frac{dx_1}{dt} = x_1$$

$$\frac{dx_2}{dt} = x_1{}^3 + x_2, \ x_1 > x_2 > 1 \tag{11}$$

the part represented by x_1 playing the central role in determining changes in the system.

One simple form in which equations of the type (4) frequently appear is

$$\frac{dx}{dt} = ax; \quad x(0) = x_0 \qquad (12)$$

This applies, for instance, whenever the rate of change of the number of elements in a system is proportional to the number already there; for example to unlimited growth of populations when a is positive, or to the decay of radium when a is negative. The solution (line of behavior) is of course the exponential

$$x = x_0 e^{at} \qquad (13)$$

A second simple example of this type is

$$\frac{dx}{dt} = 2x + bx^2 \quad x(0) = x_0 \qquad (14)$$
$$b < 0$$

the solution of which has the form

$$x = \frac{ax_0}{(a + bx_0)e^{-at} - bx_0} \qquad (15)$$

The curve described in (15) in contrast to that in (13) approaches the limiting value $-\frac{a}{b}$ as t gets large. This curve is usually called a logistic curve, and arises in applications concerned with growth processes occurring in situations where the growth is limited by factors analogous to saturation. Specific instances in which growth processes are described fairly well by a logistic curve include the growth of human populations in a limited living space (from demography), the growth of a railway or communications network in a given area, and the law of autocatalytic reaction (from physical chemistry) describing how a compound formed in a closed reaction vessel catalyzes its own formation until all the molecules are transformed.

The examples above are given because the solutions are simple and have many familiar interpretations. The general case, in which there are n variables and hence n equations is more complicated; examples with more than one variable would require more mathematical notation than it seems worthwhile to include, and so will not be considered.

Finally, the notion of stability can be illustrated for state-determined systems by considering the lines of behavior in phase space. Consider again, for simplicity, the two-variable case where lines of behavior are a family of curves in the plane. For a given region in the plane, a line of behavior from a point in the region is stable if it never leaves the region. This corresponds with our earlier definition of stability; that is, if the line of behavior is confined to a certain prescribed region when starting from a point in the region, it follows that the variables involved are constrained to operate within given limits. As a simple example suppose a vehicle is operating on a fuel supply, and that the velocity of the vehicle depends on the rate of flow of fuel which is in turn controlled by a governor sensitive to velocity. The governor, for instance, might be arranged to allow an increasing rate of flow until a critical velocity is attained and then cut down the rate of flow until the velocity drops below another critical level, and then increase the flow again, etc. Plotting a typical line of behavior (rate of flow against velocity), one sees that a closed loop (of a shape depending on the details of the system) is obtained. The system is stable because both rate of fuel flow and velocity are constrained to vary within prescribed limits.

14. Summary and Additional Remarks

The preceding sections discuss the notion of system and introduce some related ideas frequently encountered in the literature pertaining to systems in general. We admit that the ideas so introduced and the examples illustrating them are for the most part simple and familiar ones, and that the level of sophistication involved is considerably below that required for solution of actual technical problems.

The role of the scientist or engineer is complex and important. His work involves among other things analyses of systems, synthesis of systems, and evaluation of systems operations.

To analyze systems, a scientist must be aware of models available as aids to analysis as well as their limitations. Knowledge of when to use a mathematical model, and which model to use is vital; appreciation of the inter-play between a theoretical and an empirical approach to systems analysis is equally essential. There are models other than mathematical ones; at times a physical model, whether a scale model of the actual system or an analogy to it is far more effective and accurate for analysis of a given system than is an abstract model which fits poorly and is overly complex. On the other hand, as mentioned in the section on isomorphism, there are occasions when a mathematical model, even though simple and relatively inaccurate, can introduce surprising clarifications and simplifications. The section on state-determined systems was included to emphasize this idea; to be state-determined a system must have very special properties, and it would be naive to suspect that complex systems of interest are so simple. Yet, with the proper amount of care and understanding, one can use the framework of state-determined systems with good results in surprisingly diverse situations. Appreciation of similarities or isomorphisms often leads to discovering new and unsuspected connections and unifications.

Synthesis of systems is much more difficult. Here science and engineering begin to take on aspects of art. A systems designer or planner not only must construct systems that work harmoniously individually and in tandem, he must also know a lot about the environment that the system is intended to match. Consideration of environmental factors requires foresight and experience; no one can ever foresee all the variables of importance and a choice of which to include is often a difficult one to make.

Finally, in evaluating system performance, the scientist is confronted with a problem somewhat different in nature from analysis or synthesis of systems. Often one is concerned with the evaluation of large scale operations which must be studied without interrupting the process; a good example is a traffic study in an operating telephone central office. Deciding what the level of performance is requires certain criteria where often no quantitative criteria exist. For instance, in evaluating a traffic system, one must decide what effect delays have on the quality of service. If the traffic service is providing dial tone, an average delay of a few minutes is probably unacceptable, whereas a similar average delay for overseas service is commonplace. It is often necessary to adopt arbitrary levels of performance as standard; again this requires a combination of sound judgment and a knowledge of environment.

In summary, a scientist in his analysis, evaluation and synthesis of systems is not concerned primarily with the pieces of hardware that make up a system, but with the concept of system as a whole; its internal relations, and its behavior in the given environment. In this paper we have given explicitly a few of the notions concerning system and environment that enter implicitly or tacitly into any piece of scientific work.

Because of the "all-or-none" character of nervous activity, neural events and the relations among them can be treated by means of propositional logic. It is found that the behavior of every net can be described in these terms, with the addition of more complicated logical means for nets containing circles; and that for any logical expression satisfying certain conditions, one can find a net behaving in the fashion it describes. It is shown that many particular choices among possible neurophysiological assumptions are equivalent, in the sense that for every net behaving under one assumption, there exists another net which behaves under the other and gives the same results, although perhaps not in the same time. Various applications of the calculus are discussed.

11.
A Logical Calculus of the Ideas Immanent in Nervous Activity

WARREN S. MCCULLOCH AND WALTER H. PITTS

THEORETICAL NEUROPHYSIOLOGY rests on certain cardinal assumptions. The nervous system is a net of neurons, each having a soma and an axon. Their adjunctions, or synapses, are always between the axon of one neuron and the soma of another. At any instant a neuron has some threshold, which excitation must exceed to initiate an impulse. This, except for the fact and the time of its occurrence, is determined by the neuron, not by the excitation. From the point of excitation the impulse is propagated to all parts of the neuron. The velocity along the axon varies directly with its diameter, from less than one meter per second in thin axons, which are usually short, to more than 150 meters per second in thick axons, which are usually long. The time for axonal conduction is consequently of little importance in determining the time of arrival of impulses at points unequally remote from the same source. Excitation across synapses occurs predominantly from axonal terminations to somata. It is still a moot point whether this depends upon irreciprocity of individual synapses or merely upon prevalent anatomical configurations. To suppose the latter requires no hypothesis *ad hoc* and explains known exceptions, but any assumption as to cause is compatible with the

calculus to come. No case is known in which excitation through a single synapse has elicited a nervous impulse in any neuron, whereas any neuron may be excited by impulses arriving at a sufficient number of neighboring synapses within the period of latent addition, which lasts less than one quarter of a millisecond. Observed temporal summation of impulses at greater intervals is impossible for single neurons and empirically depends upon structural properties of the net. Between the arrival of impulses upon a neuron and its own propagated impulse there is a synaptic delay of more than half a millisecond. During the first part of the nervous impulse the neuron is absolutely refractory to any stimulation. Thereafter its excitability returns rapidly, in some cases reaching a value above normal from which it sinks again to a subnormal value, whence it returns slowly to normal. Frequent activity augments this subnormality. Such specificity as is possessed by nervous impulses depends solely upon their time and place and not on any other specificity of nervous energies. Of late only inhibition has been seriously adduced to contravene this thesis. Inhibition is the termination or prevention of the activity of one group of neurons by concurrent or antecedent activity of a second group. Until recently this could be explained on the supposition that previous activity of neurons of the second group might so raise the thresholds of internuncial neurons that they could no longer be excited by neurons of the first group, whereas the impulses of the first group must sum with the impulses of these internuncials to excite the now inhibited neurons. Today, some inhibitions have been shown to consume less than one millisecond. This excludes internuncials and requires synapses

Reprinted from *The Bulletin of Mathematical Biophysics*, 5 (1943), 115–33, with permission of the authors and editor. To conserve space, the tentative mathematical sections II and III have been omitted. For more recent and precise work in this area, see S. C. Kleene, "Representation of Events in Nerve Nets and Finite Automata," in C. E. Shannon and J. McCarthy (Eds.), *Automata Studies* (Princeton, N.J.: Princeton University Press, 1956); and I. M. Copi, C. C. Elgot, and J. B. Wright, "Realization of Events by Logical Nets," *J. Assn. Computing Machinery*, 5 (1958), 181–96.

through which impulses inhibit that neuron which is being stimulated by impulses through other synapses. As yet experiment has not shown whether the refractoriness is relative or absolute. We will assume the latter and demonstrate that the difference is immaterial to our argument. Either variety of refractoriness can be accounted for in either of two ways. The "inhibitory synapse" may be of such a kind as to produce a substance which raises the threshold of the neuron, or it may be so placed that the local disturbance produced by its excitation opposes the alteration induced by the otherwise excitatory synapses. Inasmuch as position is already known to have such effects in the case of electrical stimulation, the first hypothesis is to be excluded unless and until it be substantiated, for the second involves no new hypothesis. We have, then, two explanations of inhibition based on the same general premises, differing only in the assumed nervous nets and, consequently, in the time required for inhibition. Hereafter we shall refer to such nervous nets as *equivalent in the extended sense*. Since we are concerned with properties of nets which are invariant under equivalence, we may make the physical assumptions which are most convenient for the calculus.

Many years ago one of us, by considerations impertinent to this argument, was led to conceive of the response of any neuron as factually equivalent to a proposition which proposed its adequate stimulus. He therefore attempted to record the behavior of complicated nets in the notation of the symbolic logic of propositions. The "all-or-none" law of nervous activity is sufficient to insure that the activity of any neuron may be represented as a proposition. Physiological relations existing among nervous activities correspond, of course, to relations among the propositions; and the utility of the representation depends upon the identity of these relations with those of the logic of propositions. To each reaction of any neuron there is a corresponding assertion of a simple proposition. This, in turn, implies either some other simple proposition or the disjunction or the conjunction, with or without negation, of similar propositions, according to the configuration of the synapses upon and the threshold of the neuron in question. Two difficulties appeared. The first concerns facilitation and extinction, in which antecedent activity temporarily alters responsiveness to subsequent stimulation of one and the same part of the net. The second concerns learning, in which activities concurrent at some previous time have altered the net permanently, so that a stimulus which would previously have been in-

adequate is now adequate. But for nets undergoing both alterations, we can substitute equivalent fictitious nets composed of neurons whose connections and thresholds are unaltered. But one point must be made clear: neither of us conceives the formal equivalence to be a factual explanation. *Per contra!*—we regard facilitation and extinction as dependent upon continuous changes in threshold related to electrical and chemical variables, such as after-potentials and ionic concentrations; and learning as an enduring change which can survive sleep, anaesthesia, convulsions and coma. The importance of the formal equivalence lies in this: that the alterations actually underlying facilitation, extinction and learning in no way affect the conclusions which follow from the formal treatment of the activity of nervous nets, and the relations of the corresponding propositions remain those of the logic of propositions.

The nervous system contains many circular paths, whose activity so regenerates the excitation of any participant neuron that reference to time past becomes indefinite, although it still implies that afferent activity has realized one of a certain class of configurations over time. Precise specification of these implications by means of recursive functions, and determination of those that can be embodied in the activity of nervous nets, completes the theory.

We shall make the following physical assumptions for our calculus.

1. The activity of the neuron is an "all-or-none" process.

2. A certain fixed number of synapses must be excited within the period of latent addition in order to excite a neuron at any time, and this number is independent of previous activity and position on the neuron.

3. The only significant delay within the nervous system is synaptic delay.

4. The activity of any inhibitory synapse absolutely prevents excitation of the neuron at that time.

5. The structure of the net does not change with time.

.   .   .   .   .

Consequences

Causality, which requires description of states and a law of necessary connection relating them, has appeared in several forms in several sciences, but never, except in statistics, has it been as

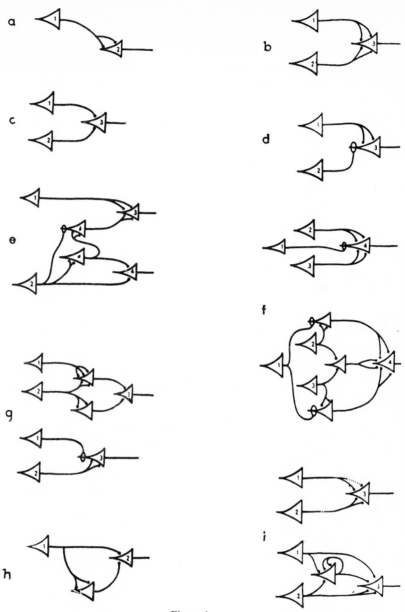

Figure 1.

EXPRESSION FOR THE FIGURES

In the figure the neuron c_i is always marked with the numeral i upon the body of the cell, and the corresponding action is denoted by 'N' with i as subscript, as in the text.

Figure 1a. $N_2(t) . \equiv . N_1(t-1)$

Figure 1b. $N_3(t) . \equiv . N_1(t-1) \lor N_2(t-1)$

Figure 1c. $N_3(t) . \equiv . N_1(t-1) . N_2(t-1)$

Figure 1d. $N_3(t) . \equiv . N_1(t-1) . \sim N_2(t-1)$

Figure 1e. $N_3(t) : \equiv : N_1(t-1) . \lor . N_2(t-3) . \sim N_2(t-2)$
$\qquad\qquad N_4(t) . \equiv . N_2(t-2) . N_2(t-1)$

Figure 1f. $N_4(t) : \equiv : \sim N_1(t-1) . N_2(t-1) \lor N_3(t-1) . \lor . N_1(t-1) . N_2(t-1) . N_3(t-1)$
$\qquad\qquad N_4(t) : \equiv : \sim N_1(t-2) . N_2(t-2) \lor N_3(t-2) . \lor . N_1(t-2) . N_2(t-2) . N_3(t-2)$

Figure 1g. $N_3(t) . \equiv . N_2(t-2) . \sim N_1(t-3)$

Figure 1h. $N_2(t) . \equiv . N_1(t-1) ; . N_1(t-2)$

Figure 1i. $N_3(t) : \equiv : N_2(t-1) . \lor . N_1(t-1) . (Ex)i-1 . N_1(x) . N_2(x)$

irreciprocal as in this theory. Specification for any one time of afferent stimulation and of the activity of all constituent neurons, each an "all-or-none" affair, determines the state. Specification of the nervous net provides the law of necessary connection whereby one can compute from the description of any state that of the succeeding state, but the inclusion of disjunctive relations prevents complete determination of the one before. Moreover, the regenerative activity of constituent circles renders reference indefinite as to time past. Thus our knowledge of the world, including ourselves, is incomplete as to space and indefinite as to time. This ignorance, implicit in all our brains, is the counterpart of the abstraction which renders our knowledge useful. The role of brains in determining the epistemic relations of our theories to our observations and of these to the facts is all too clear, for it is apparent that every idea and every sensation is realized by activity within that net, and by no such activity are the actual afferents fully determined.

There is no theory we may hold and no observation we can make that will retain so much as its old defective reference to the facts if the net be altered. Tinnitus, paraesthesias, hallucinations, delusions, confusions and disorientations intervene. Thus empiry confirms that if our nets are undefined, our facts are undefined, and to the "real" we can attribute not so much as one quality or "form." With determination of the net, the unknowable object of knowledge, the "thing in itself," ceases to be unknowable.

To psychology, however defined, specification of the net would contribute all that could be achieved in that field—even if the analysis were pushed to ultimate psychic units or "psychons," for a psychon can be no less than the activity of a single neuron. Since that activity is inherently propositional, all psychic events have an intentional, or "semiotic," character. The "all-or-none" law of these activities, and the conformity of their relations to those of the logic of propositions, insure that the relations of psychons are those of the two-valued logic of propositions. Thus in psychology, introspective, behavioristic or physiological, the fundamental relations are those of two-valued logic.

Hence arise constructional solutions of holistic problems involving the differentiated continuum of sense awareness and the normative, perfective and resolvent properties of perception and execution. From the irreciprocity of causality it follows that even if the net be known, though we may predict future from present activities, we can deduce neither afferent from central, nor central from efferent, nor past from present activities—conclusions which are reinforced by the contradictory testimony of eye-witnesses, by the difficulty of diagnosing differentially the organically diseased, the hysteric and the malingerer, and by comparing one's own memories or recollections with his contemporaneous records. Moreover, systems which so respond to the difference between afferents to a regenerative net and certain activity within that net, as to reduce the difference, exhibit purposive behavior; and organisms are known to possess many such systems, subserving homeostasis, appetition and attention. Thus both the formal and the final aspects of that activity which we are wont to call *mental* are rigorously deducible from present neurophysiology. The psychiatrist may take comfort from the obvious conclusion concerning causality—that, for prognosis, history is never necessary. He can take little from the equally valid conclusion that his observables are explicable only in terms of nervous activities which, until recently, have been beyond his ken. The crux of this ignorance is that inference from any sample of overt behavior to nervous nets is not unique, whereas, of imaginable nets, only one in fact exists, and may, at any moment, exhibit some unpredictable activity. Certainly for the psychiatrist it is more to the point that in such systems "Mind" no longer "goes more ghostly than a ghost." Instead, diseased mentality can be understood without loss of scope or rigor, in the scientific terms of neurophysiology. For neurology, the theory sharpens the distinction between nets necessary or merely sufficient for given activities, and so clarifies the relations of disturbed structure to disturbed function. In its own domain the difference between equivalent nets and nets equivalent in the narrow sense indicates the appropriate use and importance of temporal studies of nervous activity: and to mathematical biophysics the theory contributes a tool for rigorous symbolic treatment of known nets and an easy method of constructing hypothetical nets of required properties.

12.
The General and Logical Theory of Automata

JOHN VON NEUMANN

I HAVE TO ASK your forbearance for appearing here, since I am an outsider to most of the fields which form the subject of this conference. Even in the area in which I have some experience, that of the logics and structure of automata, my connections are almost entirely on one side, the mathematical side. The usefulness of what I am going to say, if any, will therefore be limited to this: I may be able to give you a picture of the mathematical approach to these problems, and to prepare you for the experiences that you will have when you come into closer contact with mathematicians. This should orient you as to the ideas and the attitudes which you may then expect to encounter. I hope to get your judgment of the modus procedendi and the distribution of emphases that I am going to use. I feel that I need instruction even in the limiting area between our fields more than you do, and I hope that I shall receive it from your criticisms.

Automata have been playing a continuously increasing, and have by now attained a very considerable, role in the natural sciences. This is a process that has been going on for several decades. During the last part of this period automata have begun to invade certain parts of mathematics too—particularly, but not exclusively, mathematical physics or applied mathematics. Their role in mathematics presents an interesting counterpart to certain functional aspects of organization in nature. Natural organisms are, as a rule, much more complicated and subtle, and therefore much less well understood in detail, than are artificial automata. Nevertheless, some regularities which we observe in the organization of the former may be quite instructive in our thinking and planning of the latter; and conversely, a good deal of our experiences and difficulties with our artificial automata can be to some extent projected on our interpretations of natural organisms.

Preliminary Considerations

Dichotomy of the Problem: Nature of the Elements, Axiomatic Discussion of Their Synthesis. In comparing living organisms, and, in particular, that most complicated organism, the human central nervous system, with artificial automata, the following limitation should be kept in mind. The natural systems are of enormous complexity, and it is clearly necessary to subdivide the problem that they represent into several parts. One method

From John von Neumann, "The General and Logical Theory of Automata," in Lloyd A. Jeffress (Ed.), *Cerebral Mechanisms in Behavior: The Hixon Symposium* (New York: John Wiley and Sons, 1951), pp. 1–2, 15–31. Reprinted by permission of the author and publisher. To conserve space, two sections unessential to von Neumann's theory of automata have been omitted; they are entitled "Discussion of Certain Relevant Traits of Computing Machines" and "Comparisons between Computing Machines and Living Organisms." To the complete paper from which this selection is excerpted, the author attached this note: "This paper is an only slightly edited version of one that was read at the Hixon Symposium on September 20, 1948, in Pasadena, California. Since it was delivered as a single lecture, it was not feasible to go into as much detail on every point as would have been desirable for a final publication. In the present write-up it seemed appropriate to follow the dispositions of the talk; therefore this paper, too, is in many places more sketchy than desirable. It is to be taken only as a general outline of ideas and of tendencies."

of subdivision, which is particularly significant in the present context, is this: The organisms can be viewed as made up of parts which to a certain extent are independent, elementary units. We may, therefore, to this extent, view as the first part of the problem the structure and functioning of such elementary units individually. The second part of the problem consists of understanding how these elements are organized into a whole, and how the functioning of the whole is expressed in terms of these elements.

The first part of the problem is at present the dominant one in physiology. It is closely connected with the most difficult chapters of organic chemistry and of physical chemistry, and may in due course be greatly helped by quantum mechanics. I have little qualification to talk about it, and it is not this part with which I shall concern myself here.

The second part, on the other hand, is the one which is likely to attract those of us who have the background and the tastes of a mathematician or a logician. With this attitude, we will be inclined to remove the first part of the problem by the process of axiomatization, and concentrate on the second one.

The Axiomatic Procedure. Axiomatizing the behavior of the elements means this: We assume that the elements have certain well-defined, outside, functional characteristics; that is, they are to be treated as "black boxes." They are viewed as automatisms, the inner structure of which need not be disclosed, but which are assumed to react to certain unambiguously defined stimuli, by certain unambiguously defined responses.

This being understood, we may then investigate the larger organisms that can be built up from these elements, their structure, their functioning, the connections between the elements, and the general theoretical regularities that may be detectable in the complex syntheses of the organisms in question.

I need not emphasize the limitations of this procedure. Investigations of this type may furnish evidence that the system of axioms used is convenient and, at least in its effects, similar to reality. They are, however, not the ideal method, and possibly not even a very effective method, to determine the validity of the axioms. Such determinations of validity belong primarily to the first part of the problem. Indeed they are essentially covered by the properly physiological (or chemical or physical–chemical) determinations of the nature and properties of the elements.

The Significant Orders of Magnitude. In spite of these limitations, however, the "second part" as circumscribed above is important and difficult. With any reasonable definition of what constitutes an element, the natural organisms are very highly complex aggregations of these elements. The number of cells in the human body is somewhere of the general order of 10^{15} or 10^{16}. The number of neurons in the central nervous system is somewhere of the order of 10^{10}. We have absolutely no past experience with systems of this degree of complexity. All artificial automata made by man have numbers of parts which by any comparably schematic count are of the order 10^3 to 10^6. In addition, those artificial systems which function with that type of logical flexibility and autonomy that we find in the natural organisms do not lie at the peak of this scale. The prototypes for these systems are the modern computing machines, and here a reasonable definition of what constitutes an element will lead to counts of a few times 10^3 or 10^4 elements.

.   .   .   .   .

The Future Logical Theory of Automata

Further Discussion of the Factors That Limit the Present Size of Artificial Automata. We have emphasized how the complication is limited in artificial automata, that is, the complication which can be handled without extreme difficulties and for which automata can still be expected to function reliably. Two reasons that put a limit on complication in this sense have already been given. They are the large size and the limited reliability of the componentry that we must use, both of them due to the fact that we are employing materials which seem to be quite satisfactory in simpler applications, but marginal and inferior to the natural ones in this highly complex application. There is, however, a third important limiting factor, and we should now turn our attention to it. This factor is of an intellectual, and not physical, character.

The Limitation Which Is Due to the Lack of a Logical Theory of Automata. We are very far from possessing a theory of automata which deserves that name, that is, a properly mathematical–logical theory. There exists today a very elaborate system of formal logic, and, specifically, of logic as applied to mathematics. This is a discipline with many good sides, but also with certain serious weaknesses. This is not the occasion to enlarge upon the good sides, which I have certainly no intention to belittle. About the inadequacies, however, this may be said: Everybody who has worked in formal logic will confirm that it is one of the technically most refractory parts of mathematics. The reason for this is that it deals with rigid, all-or-none concepts, and has very little contact with the continuous concept of the real or of the complex number, that is, with mathematical analysis. Yet analysis is the technically most successful and best-elaborated part of mathematics. Thus formal logic is, by the nature of its approach, cut off from the best cultivated portions of mathematics, and forced onto the most difficult part of the mathematical terrain, into combinatorics.

The theory of automata, of the digital, all-or-none type, as discussed up to now, is certainly a chapter in formal logic. It would, therefore, seem that it will have to share this unattractive property of formal logic. It will have to be, from the mathematical point of view, combinatorial rather than analytical.

Probable Characteristics of Such a Theory. Now it seems to me that this will in fact not be the case. In studying the functioning of automata, it is clearly necessary to pay attention to a circumstance which has never before made its appearance in formal logic.

Throughout all modern logic, the only thing that is important is whether a result can be achieved in a finite number of elementary steps or not. The size of the number of steps which are required, on the other hand, is hardly ever a concern of formal logic. Any finite sequence of correct steps is, as a matter of principle, as good as any other. It is a matter of no consequence whether the number is small or large, or even so large that it couldn't possibly be carried out in a lifetime, or in the presumptive lifetime of the stellar universe as we know it. In dealing with automata, this statement must be significantly modified. In the case of an automaton the thing which matters is not only whether it can reach a certain result in a finite number of steps at all but also how many such steps are needed. There are two reasons. First, automata are constructed in order to reach certain results in certain pre-assigned durations, or at least in pre-assigned orders of magnitude of duration. Second, the componentry employed has on every individual operation a small but nevertheless non-zero probability of failing. In a sufficiently long chain of operations the cumulative effect of these individual probabilities of failure may (if unchecked) reach the order of magnitude of unity—at which point it produces, in effect, complete unreliability. The

probability levels which are involved here are very low, but still not too far removed from the domain of ordinary technological experience. It is not difficult to estimate that a high-speed computing machine, dealing with a typical problem, may have to perform as much as 10^{12} individual operations. The probability of error on an individual operation which can be tolerated must, therefore, be small compared to 10^{-12}. I might mention that an electromechanical relay (a telephone relay) is at present considered acceptable if its probability of failure on an individual operation is of the order 10^{-8}. It is considered excellent if this order of probability is 10^{-9}. Thus the reliabilities required in a high-speed computing machine are higher, but not prohibitively higher, than those that constitute sound practice in certain existing industrial fields. The actually obtainable reliabilities are, however, not likely to leave a very wide margin against the minimum requirements just mentioned. An exhaustive study and a non-trivial theory will, therefore, certainly be called for.

Thus the logic of automata will differ from the present system of formal logic in two relevant respects.

1. The actual length of "chains of reasoning," that is, of the chains of operations, will have to be considered.

2. The operations of logic (syllogisms, conjunctions, disjunctions, negations, etc., that is, in the terminology that is customary for automata, various forms of gating, coincidence, anti-coincidence, blocking, etc., actions) will all have to be treated by procedures which allow exceptions (malfunctions) with low but non-zero probabilities. All of this will lead to theories which are much less rigidly of an all-or-none nature than past and present formal logic. They will be of a much less combinatorial, and much more analytical, character. In fact, there are numerous indications to make us believe that this new system of formal logic will move closer to another disicpline which has been little linked in the past with logic. This is thermodynamics, primarily in the form it was received from Boltzmann, and is that part of theoretical physics which comes nearest in some of its aspects to manipulating and measuring information. Its techniques are indeed much more analytical than combinatorial, which again illustrates the point that I have been trying to make above. It would, however, take me too far to go into this subject more thoroughly on this occasion.

All of this re-emphasizes the conclusion that was indicated earlier, that a detailed, highly mathematical, and more specifically analytical, theory of automata and of information is needed. We possess only the first indications of such a theory at present. In assessing artificial automata, which are, as I discussed earlier, of only moderate size, it has been possible to get along in a rough, empirical manner without such a theory. There is every reason to believe that this will not be possible with more elaborate automata.

Effects of the Lack of a Logical Theory of Automata on the Procedures in Dealing with Errors. This, then, is the last, and very important, limiting factor. It is unlikely that we could construct automata of a much higher complexity than the ones we now have, without possessing a very advanced and subtle theory of automata and information. A fortiori, this is inconceivable for automata of such enormous complexity as is possessed by the human central nervous system.

This intellectual inadequacy certainly prevents us from getting much farther than we are now.

A simple manifestation of this factor is our present relation to error checking. In living organisms malfunctions of components occur. The organism obviously has a way to detect them and render them harmless. It is easy to estimate that the number of nerve actuations which occur in a normal lifetime must be of the order of 10^{20}. Obviously, during this chain of events there never occurs a malfunction which cannot be corrected by the organism itself, without any significant outside intervention. The system must, therefore, contain the necessary arrangements to diagnose errors as they occur, to readjust the organism so as to minimize the effects of the errors, and finally to correct or to block permanently the faulty components. Our modus procedendi with respect to malfunctions in our artificial automata is entirely different. Here the actual practice, which has the consensus of all experts of the field, is somewhat like this: Every effort is made to detect (by mathematical or by automatical checks) every error as soon as it occurs. Then an attempt is made to isolate the component that caused the error as rapidly as feasible. This may be done partly automatically, but in any case a significant part of this diagnosis must be effected by intervention from the outside. Once the faulty component has been identified, it is immediately corrected or replaced.

Note the difference in these two attitudes. The basic principle of dealing with malfunctions in nature is to make their effect as unimportant as possible and to apply correctives, if they are necessary at all, at leisure. In our dealings with artificial automata, on the other hand, we require an immediate diagnosis. Therefore, we are trying to arrange the automata in such a manner that errors will become as conspicuous as possible, and intervention and correction follow immediately. In other words, natural organisms are constructed to make errors as inconspicuous, as harmless, as possible. Artificial automata are designed to make errors as conspicuous, as disastrous, as possible. The rationale of this difference is not far to seek. Natural organisms are sufficiently well conceived to be able to operate even when malfunctions have set in. They can operate in spite of malfunctions, and their subsequent tendency is to remove these malfunctions. An artificial automaton could certainly be designed so as to be able to operate normally in spite of a limited number of malfunctions in certain limited areas. Any malfunction, however, represents a considerable risk that some generally degenerating process has already set in within the machine. It is, therefore, necessary to intervene immediately, because a machine which has begun to malfunction has only rarely a tendency to restore itself, and will more probably go from bad to worse. All of this comes back to one thing. With our artificial automata we are moving much more in the dark than nature appears to be with its organisms. We are and apparently, have to be, at least at present, much more "scared" by the occurrence of an isolated error and by the malfunction which must be behind it. Our behavior is clearly that of overcaution, generated by ignorance.

The Single-Error Principle. A minor side light to this is that almost all our error-diagnosing techniques are based on the assumption that the machine contains only one faulty component. In this case, iterative subdivisions of the machine into parts permit us to determine which portion contains the fault. As soon as the possibility exists that the machine may contain several faults, these, rather powerful, dichotomic methods of diagnosis are lost. Error diagnosing then becomes an increasingly hopeless proposition. The high premium on keeping the number of errors to be diagnosed down to one, or at any rate as low as possible, again illustrates our ignorance in this field, and is one of the main reasons why errors must be made as conspicuous as possible, in order to be recognized and apprehended as soon after their occurrence as feasible, that is, before further errors have had time to develop.

Principles of Digitalization

Digitalization of Continuous Quantities: the Digital Expansion Method and the Counting Method. Consider the digital part of a natural organism; specifically, consider the nervous system. It seems that we are indeed justified in assuming that this is a digital mechanism, that it transmits messages which are made up of signals possessing the all-or-none character. . . . In other words, each elementary signal, each impulse, simply either is or is not there, with no further shadings A particularly relevant illustration of this fact is furnished by those cases where the underlying problem has the opposite character, that is, where the nervous system is actually called upon to transmit a continuous quantity. Thus the case of a nerve which has to report on the value of a pressure is characteristic.

Assume, for example, that a pressure (clearly a continuous quantity) is to be transmitted. It is well known how this trick is done. The nerve which does it still transmits nothing but individual all-or-none impulses. How does it then express the continuously numerical value of pressure in terms of these impulses, that is, of digits? In other words, how does it encode a continuous number into a digital notation? It does certainly not do it by expanding the number in question into decimal (or binary, or any other base) digits in the conventional sense. What appears to happen is that it transmits pulses at a frequency which varies and which is within certain limits proportional to the continuous quantity in question, and generally a monotone function of it. The mechanism which achieves this "encoding" is, therefore, essentially a frequency modulation system.

The details are known. The nerve has a finite recovery time. In other words, after it has been pulsed once, the time that has to lapse before another stimulation is possible is finite and dependent upon the strength of the ensuing (attempted) stimulation. Thus, if the nerve is under the influence of a continuing stimulus (one which is uniformly present at all times, like the pressure that is being considered here), then the nerve will respond periodically, and the length of the period between two successive stimulations is the recovery time referred to earlier, that is, a function of the strength of the constant stimulus (the pressure in the present case). Thus, under a high pressure, the nerve may be able to respond every 8 milliseconds, that is, transmit at the rate of 125 impulses per second; while under the influence of a smaller pressure it may be able to repeat only every 14 milliseconds, that is, transmit at the rate of 71 times per second. This is very clearly the behavior of a genuinely yes-or-no organ, of a digital organ. It is very instructive, however, that it uses a "count" rather than a "decimal expansion" (or "binary expansion," etc.) method.

Comparison of the Two Methods. The Preference of Living Organisms for the Counting Method. Compare the merits and demerits of these two methods. The counting method is certainly less efficient than the expansion method. In order to express a number of about a million (that is, a physical quantity of a million distinguishable resolution-steps) by counting, a million pulses have to be transmitted. In order to express a number of the same size by expansion, 6 or 7 decimal digits are needed, that is, about 20 binary digits. Hence, in this case only 20 pulses are needed. Thus our expansion method is much more economical in notation than the counting methods which are resorted to by nature. On the other hand, the counting method has a high stability and safety from error. If you express a number of the order of a million by counting and miss a count, the result is only irrelevantly changed. If you express it by (decimal or binary) expansion, a single error in a single digit may vitiate the entire result. Thus the undesirable trait of our computing machines reappears in our digital expansion system; in fact, the former is clearly deeply connected with, and partly a consequence of, the latter. The high stability and nearly error-proof character of natural organisms, on the other hand, is reflected in the counting method that they seem to use in this case. All of this reflects a general rule. You can increase the safety from error by a reduction of the efficiency of the notation, or, to say it positively, by allowing redundancy of notation. Obviously, the simplest form of achieving safety by redundancy is to use the, per se, quite unsafe digital expansion notation, but to repeat every such message several times. In the case under discussion, nature has obviously resorted to an even more redundant and even safer system.

There are, of course, probably other reasons why the nervous system uses the counting rather than the digital expansion. The encoding–decoding facilities required by the former are much simpler than those required by the latter. It is true, however, that nature seems to be willing and able to go much further in the direction of complication than we are, or rather than we can afford to go.

One may, therefore, suspect that if the only demerit of the digital expansion system were its greater logical complexity, nature would not, for this reason alone, have rejected it. It is, nevertheless, true that we have nowhere an indication of its use in natural organisms. It is difficult to tell how much "final" validity one should attach to this observation. The point deserves at any rate attention, and should receive it in future investigations of the functioning of the nervous system.

Formal Neural Networks

The McCulloch–Pitts Theory of Formal Neural Networks. A great deal more could be said about these things from the logical and the organizational point of view, but I shall not attempt to say it here. I shall instead go on to discuss what is probably the most significant result obtained with the axiomatic method up to now. I mean the remarkable theorems of McCulloch and Pitts on the relationship of logics and neural networks.

In this discussion I shall, as I have said, take the strictly axiomatic point of view. I shall, therefore, view a neuron as a "black box" with a certain number of inputs that receive stimuli and an output that emits stimuli. To be specific, I shall assume that the input connections of each one of these can be of two types, excitatory and inhibitory. The boxes themselves are also of two types, threshold 1 and threshold 2. These concepts are linked and circumscribed by the following definitions. In order to stimulate such an organ it is necessary that it should receive simultaneously at least as many stimuli on its excitatory inputs as correspond to its threshold, and not a single stimulus on any one of its inhibitory inputs. If it has been thus stimulated, it will after a definite time delay (which is assumed to be always the same, and may be used to define the unit of time) emit an output pulse. This pulse can be taken by appropriate connections to any number of inputs of other neurons (also to any of its own inputs) and will produce at each of these the same type of input stimulus as the ones described above.

It is, of course, understood that this is an oversimplification of the actual functioning of a neuron. I have already discussed the character, the limitations, and the advantages of the axiomatic method. . . . They all apply here, and the discussion which follows is to be taken in this sense.

McCulloch and Pitts have used these units to build up complicated networks which may be called "formal neural networks." Such a system is built up of any number of these units, with their inputs and outputs suitably interconnected with arbitrary complexity. The "functioning" of such a network may be defined by singling out some of the inputs of the entire system and some of its outputs, and then describing what original stimuli on the former are to cause what ultimate stimuli on the latter.

The Main Result of the McCulloch–Pitts Theory. McCulloch and Pitts' important result is that any functioning in this sense which can be defined at all logically, strictly, and unambiguously in a finite number of words can also be realized by such a formal neural network.

It is well to pause at this point and to consider what the implications are. It has often been claimed that the activities and functions of the human nervous system are so complicated that no ordinary mechanism could possibly perform them. It has also been attempted to name specific functions which by their nature exhibit this limitation. It has been attempted to show that such specific functions, logically, completely described, are per se unable of mechanical, neural realization. The McCulloch–Pitts result puts an end to this. It proves that anything that can be exhaustively and unambiguously described, anything that can be completely and unambiguously put into words, is ipso facto realizable by a suitable finite neural network. Since the converse statement is obvious, we can therefore say that there is no difference between the possibility of describing a real or imagined mode of behavior completely and unambiguously in words, and the possibility of realizing it by a finite formal neural network. The two concepts are co-extensive. A difficulty of principle embodying any mode of behavior in such a network can exist only if we are also unable to describe that behavior completely.

Thus the remaining problems are these two. First, if a certain mode of behavior can be effected by a finite neural network, the question still remains whether that network can be realized within a practical size, specifically, whether it will fit into the physical limitations of the organism in question. Second, the question arises whether every existing mode of behavior can really be put completely and unambiguously into words.

The first problem is, of course, the ultimate problem of nerve physiology, and I shall not attempt to go into it any further here. The second question is of a different character, and it has interesting logical connotations.

Interpretations of This Result. There is no doubt that any special phase of any conceivable form of behavior can be described "completely and unambiguously" in words. This description may be lengthy, but it is always possible. To deny it would amount to adhering to a form of logical mysticism which is surely far from most of us. It is, however, an important limitation, that this applies only to every element separately, and it is far from clear how it will apply to the entire syndrome of behavior. To be more specific, there is no difficulty in describing how an organism might be able to identify any two rectilinear triangles, which appear on the retina, as belonging to the same category "triangle." There is also no difficulty in adding to this, that numerous other objects, besides regularly drawn rectilinear triangles, will also be classified and identified as triangles—triangles whose sides are curved, triangles whose sides are not fully drawn, triangles that are indicated merely by a more or less homogeneous shading of their interior, etc. The more completely we attempt to describe everything that may conceivably fall under this heading, the longer the description becomes. We may have a vague and uncomfortable feeling that a complete catalogue along such lines would not only be exceedingly long, but also unavoidably indefinite at its boundaries. Nevertheless, this may be a possible operation.

All of this, however, constitutes only a small fragment of the more general concept of identification of analogous geometrical entities. This, in turn, is only a microscopic piece of the general concept of analogy. Nobody would attempt to describe and define within any practical amount of space the general concept of analogy which dominates our interpretation of vision. There is no basis for saying whether such an enterprise would require thousands or millions or altogether impractical numbers of volumes. Now it is perfectly possible that the simplest and only practical way actually to say what constitutes a visual analogy consists in giving a description of the connections of the visual brain. We are dealing here with parts of logics with which we have practically no past experience. The order of complexity is out of all proportion to anything we have ever known. We have no right to assume that the logical notations and procedures used in the past are suited to this part of the subject. It is not at all certain that in this domain a real object might not constitute the simplest description of itself, that is, any attempt to describe it by the usual literary or formal-logical method may lead to something less

manageable and more involved. In fact, some results in modern logic would tend to indicate that phenomena like this have to be expected when we come to really complicated entities. It is, therefore, not at all unlikely that it is futile to look for a precise logical concept, that is, for a precise verbal description, of "visual analogy." It is possible that the connection pattern of the visual brain itself is the simplest logical expression or definition of this principle.

Obviously, there is on this level no more profit in the McCulloch–Pitts result. At this point it only furnishes another illustration of the situation outlined earlier. There is an equivalence between logical principles and their embodiment in a neural network, and while in the simpler cases the principles might furnish a simplified expression of the network, it is quite possible that in cases of extreme complexity the reverse is true.

All of this does not alter my belief that a new, essentially logical, theory is called for in order to understand high-complication automata and, in particular, the central nervous system. It may be, however, that in this process logic will have to undergo a pseudomorphosis to neurology to a much greater extent than the reverse. The foregoing analysis shows that one of the relevant things we can do at this moment with respect to the theory of the central nervous system is to point out the directions in which the real problem does not lie.

The Concept of Complication; Self-Reproduction

The Concept of Complication. The discussions so far have shown that high complexity plays an important role in any theoretical effort relating to automata, and that this concept, in spite of its prima facie quantitative character, may in fact stand for something qualitative—for a matter of principle. For the remainder of my discussion I will consider a remoter implication of this concept, one which makes one of the qualitative aspects of its nature even more explicit.

There is a very obvious trait, of the "vicious circle" type, in nature, the simplest expression of which is the fact that very complicated organisms can reproduce themselves.

We are all inclined to suspect in a vague way the existence of a concept of "complication." This concept and its putative properties have never been clearly formulated. We are, however, always tempted to assume that they will work in this way.

When an automaton performs certain operations, they must be expected to be of a lower degree of complication than the automaton itself. In particular, if an automaton has the ability to construct another one, there must be a decrease in complication as we go from the parent to the construct. That is, if A can produce B, then A in some way must have contained a complete description of B. In order to make it effective, there must be, furthermore, various arrangements in A that see to it that this description is interpreted and that the constructive operations that it calls for are carried out. In this sense, it would therefore seem that a certain degenerating tendency must be expected, some decrease in complexity as one automaton makes another automaton.

Although this has some indefinite plausibility to it, it is in clear contradiction with the most obvious things that go on in nature. Organisms reproduce themselves, that is, they produce new organisms with no decrease in complexity. In addition, there are long periods of evolution during which the complexity is even increasing. Organisms are indirectly derived from others which had lower complexity.

Thus there exists an apparent conflict of plausibility and evidence, if nothing worse. In view of this, it seems worth while to try to see whether there is anything involved here which can be formulated rigorously.

So far I have been rather vague and confusing, and not unintentionally at that. It seems to me that it is otherwise impossible to give a fair impression of the situation that exists here. Let me now try to become specific.

Turing's Theory of Computing Automata. The English logician, Turing, about twelve years ago attacked the following problem.

He wanted to give a general definition of what is meant by a computing automaton. The formal definition came out as follows:

An automaton is a "black box," which will not be described in detail but is expected to have the following attributes. It possesses a finite number of states, which need be prima facie characterized only by stating their number, say n, and by enumerating them accordingly: 1, 2, ... n. The essential operating characteristic of the automaton consists of describing how it is caused to change its state, that is, to go over from a state i into a state j. This change requires some interaction with the outside world, which will be standardized in the following manner. As far as the machine is concerned, let the whole outside world consist of a long paper tape. Let this tape

be, say, 1 inch wide, and let it be subdivided into fields (squares) 1 inch long. On each field of this strip we may or may not put a sign, say, a dot, and it is assumed that it is possible to erase as well as to write in such a dot. A field marked with a dot will be called a "1," a field unmarked with a dot will be called a "0." (We might permit more ways of marking, but Turing showed that this is irrelevant and does not lead to any essential gain in generality.) In describing the position of the tape relative to the automaton it is assumed that one particular field of the tape is under direct inspection by the automaton, and that the automaton has the ability to move the tape forward and backward, say, by one field at a time. In specifying this, let the automaton be in the state i ($= 1 \ldots, n$), and let it see on the tape an e ($= 0$, 1). It will then go over into the state j ($= 0, 1 \ldots, n$), move the tape by p fields ($p = 0$, $+1$, -1; $+1$ is a move forward, -1 is a move backward), and inscribe into the new field that it sees f ($= 0$, 1; inscribing 0 means erasing; inscribing 1 means putting in a dot). Specifying j, p, f as functions of i, e is then the complete definition of the functioning of such an automaton.

Turing carried out a careful analysis of what mathematical processes can be effected by automata of this type. In this connection he proved various theorems concerning the classical "decision problem" of logic, but I shall not go into these matters here. He did, however, also introduce and analyze the concept of a "universal automaton," and this is part of the subject that is relevant in the present context.

An infinite sequence of digits e ($= 0$, 1) is one of the basic entities in mathematics. Viewed as a binary expansion, it is essentially equivalent to the concept of a real number. Turing, therefore, based his consideration on these sequences.

He investigated the question as to which automata were able to construct which sequences. That is, given a definite law for the formation of such a sequence, he inquired as to which automata can be used to form the sequence based on that law. The process of "forming" a sequence is interpreted in this manner. An automaton is able to "form" a certain sequence if it is possible to specify a finite length of tape, appropriately marked, so that, if this tape is fed to the automaton in question, the automaton will thereupon write the sequence on the remaining (infinite) free portion of the tape. This process of writing the infinite sequence is, of course, an indefinitely continuing one. What is meant is that the automaton will keep running indefinitely and, given a sufficiently long time, will have inscribed any desired (but of course finite) part of the (infinite) sequence. The finite, premarked, piece of tape constitutes the "instruction" of the automaton for this problem.

An automaton is "universal" if any sequence that can be produced by any automaton at all can also be solved by this particular automaton. It will, of course, require in general a different instruction for this purpose.

The Main Result of the Turing Theory. We might expect a priori that this is impossible. How can there be an automaton which is at least as effective as any conceivable automaton, including, for example, one of twice its size and complexity?

Turing, nevertheless, proved that this is possible. While his construction is rather involved, the underlying principle is nevertheless quite simple. Turing observed that a completely general description of any conceivable automaton can be (in the sense of the foregoing definition) given in a finite number of words. This description will contain certain empty passages—those referring to the functions mentioned earlier (j, p, f in terms of i, e), which specify the actual functioning of the automaton. When these empty passages are filled in, we deal with a specific automaton. As long as they are left empty, this schema represents the general definition of the general automaton. Now it becomes possible to describe an automaton which has the ability to interpret such a definition. In other words, which, when fed the functions that in the sense described above define a specific automaton, will thereupon function like the object described. The ability to do this is no more mysterious than the ability to read a dictionary and a grammar and to follow their instructions about the uses and principles of combinations of words. This automaton, which is constructed to read a description and to imitate the object described, is then the universal automaton in the sense of Turing. To make it duplicate any operation that any other automaton can perform, it suffices to furnish it with a description of the automaton in question and, in addition, with the instructions which that device would have required for the operation under consideration.

Broadening of the Program to Deal with Automata That Produce Automata. For the question which concerns me here, that of "self-reproduction" of automata, Turing's procedure is too narrow in one respect only. His automata are purely computing machines. Their output is a piece of tape with zeros and ones on it. What is needed for the construction to which I referred is an automaton whose output is other automata. There is, however, no difficulty in principle in dealing with this broader concept and in deriving from it the equivalent of Turing's result.

The Basic Definitions. As in the previous instance, it is again of primary importance to give a rigorous definition of what constitutes an automaton for the purpose of the investigation. First of all, we have to draw up a complete list of the elementary parts to be used. This list must contain not only a complete enumeration but also a complete operational definition of each elementary part. It is relatively easy to draw up such a list, that is, to write a catalogue of "machine parts" which is sufficiently inclusive to permit the construction of the wide variety of mechanisms here required, and which has the axiomatic rigor that is needed for this kind of consideration. The list need not be very long either. It can, of course, be made either arbitrarily long or arbitrarily short. It may be lengthened by including in it, as elementary parts, things which could be achieved by combinations of others. It can be made short—in fact, it can be made to consist of a single unit—by endowing each elementary part with a multiplicity of attributes and functions. Any statement on the number of elementary parts required will therefore represent a common-sense compromise, in which nothing too complicated is expected from any one elementary part, and no elementary part is made to perform several, obviously separate, functions. In this sense, it can be shown that about a dozen elementary parts suffice. The problem of self-reproduction can then be stated like this: Can one build an aggregate out of such elements in such a manner that if it is put into a reservoir, in which there float all these elements in large numbers, it will then begin to construct other aggregates, each of which will at the end turn out to be another automaton exactly like the original one? This is feasible, and the principle on which it can be based is closely related to Turing's principle outlined earlier.

Outline of the Derivation of the Theorem Regarding Self-reproduction. First of all, it is possible to give a complete description of everything that is an automaton in the sense considered here. This description is to be conceived as a general one, that is, it will again contain empty spaces. These empty spaces have to be filled in with the functions which describe the actual structure of an automaton. As before, the difference between these spaces filled and unfilled is the difference between the description of a specific automaton and the general description of a general automaton. There is no difficulty of principle in describing the following automata.

(*a*) Automaton *A*, which when furnished the description of any other automaton in terms of appropriate functions, will construct that entity. The description should in this case not be given in the form of a marked tape, as in Turing's case, because we will not normally choose a tape as a structural element. It is quite easy, however, to describe combinations of structural elements which have all the notational properties of a tape with fields that can be marked. A description in this sense will be called an instruction and denoted by a letter *I*.

"Constructing" is to be understood in the same sense as before. The constructing automaton is supposed to be placed in a reservoir in which all elementary components in large numbers are floating, and it will effect its construction in that milieu. One need not worry about how a fixed automaton of this sort can produce others which are larger and more complex than itself. In this case the greater size and the higher complexity of the object to be constructed will be reflected in a presumably still greater size of the instructions *I* that have to be furnished. These instructions, as pointed out, will have to be aggregates of elementary parts. In this sense, certainly, an entity will enter the process whose size and complexity is determined by the size and complexity of the object to be constructed.

In what follows, all automata for whose construction the facility *A* will be used are going to share with *A* this property. All of them will have a place for an instruction *I*, that is, a place where such an instruction can be inserted. When such an automaton is being described (as, for example, by an appropriate instruction), the specification of the location for the insertion of an instruction *I* in the foregoing sense is understood to form a part of the description. We may, therefore, talk of "inserting a given instruction *I* into a given automaton," without any further explanation.

(*b*) Automaton *B*, which can make a copy of any instruction *I* that is furnished to it. *I* is an aggregate of elementary parts in the sense outlined in (*a*), replacing a tape. This facility will be used when *I* furnishes a description of another automaton. In other words, this automaton is nothing more subtle than a "reproducer"—the machine which can read a punched tape and produce a second punched tape that is identical with the first. Note that this automaton, too, can produce objects which are larger and more complicated than itself. Note again that there is nothing surprising about it. Since it can only copy, an object of the exact size and complexity of the output will have to be furnished to it as input.

After these preliminaries, we can proceed to the decisive step.

(*c*) Combine the automata *A* and *B* with each other, and with a control mechanism *C* which does the following. Let *A* be furnished with an instruction *I* (again in the sense of [*a*] and [*b*]). Then *C* will first cause *A* to construct the automaton which is described by this instruction *I*. Next *C* will cause *B* to copy the instruction *I* referred to above, and insert the copy into the automaton referred to above, which has just been constructed by *A*. Finally, *C* will separate this construction from the system *A* + *B* + *C* and "turn it loose" as an independent entity.

(*d*) Denote the total aggregate *A* + *B* + *C* by *D*.

(*e*) In order to function, the aggregate *D* = *A* + *B* + *C* must be furnished with an instruction *I*, as described above. This instruction, as pointed out above, has to be inserted into *A*. Now form an instruction I_D, which describes this automaton *D*, and insert I_D into *A* within *D*. Call the aggregate which now results *E*.

E is clearly self-reproductive. Note that no vicious circle is involved. The decisive step occurs in *E*, when the instruction I_D, describing *D*, is constructed and attached to *D*. When the construction (the copying) of I_D is called for, *D* exists already, and it is in no wise modified by the construction of I_D. I_D is simply added to form *E*. Thus there is a definite chronological and logical order in which *D* and I_D have to be formed, and the process is legitimate and proper according to the rules of logic.

Interpretations of This Result and of Its Immediate Extensions. The description of this automaton *E* has some further attractive sides, into which I shall not go at this time at any length. For instance, it is quite clear that the instruction I_D is roughly effecting the functions of a gene. It is also clear that the copying mechanism *B* performs the fundamental act of reproduction, the duplication of the genetic material, which is clearly the fundamental operation in the multiplication of living cells. It is also easy to see how arbitrary alterations of the system *E*, and in particular of I_D, can exhibit certain typical traits which appear in connection with mutation, lethally as a rule, but with a possibility of continuing reproduction with a modification of traits. It is, of course, equally clear at which point the analogy ceases to be valid. The natural gene does probably not contain a complete description of the object whose construction its presence stimulates. It probably contains only general pointers, general cues. In the generality in which the foregoing consideration is moving, this simplification is not attempted. It is, nevertheless, clear that this simplification, and others similar to it, are in themselves of great and qualitative importance. We are very far from any real understanding of the natural processes if we do not attempt to penetrate such simplifying principles.

Small variations of the foregoing scheme also permit us to construct automata which can reproduce themselves and, in addition, construct others. (Such an automaton performs more specifically what is probably a—if not the—typical gene function, self-reproduction plus production—or stimulation of production—of certain specific enzymes.) Indeed, it suffices to replace the I_D by an instruction I_{D+F}, which describes the automaton D plus another given automaton F. Let D, with I_{D+F} inserted into A within it, be designated by E_F. This E_F clearly has the property already described. It will reproduce itself, and, besides, construct F.

Note that a "mutation" of E_F, which takes place within the F-part of I_{D+F} in E_F, is not lethal. If it replaces F by F', it changes E_F into $E_{F'}$, that is, the "mutant" is still self-reproductive; but its by-product is changed—F' instead of F. This is, of course, the typical non-lethal mutant.

All these are very crude steps in the direction of a systematic theory of automata. They represent, in addition, only one particular direction. This is, as I indicated before, the direction towards forming a rigorous concept of what constitutes "complication." They illustrate that "complication" on its lower levels is probably degenerative, that is, that every automaton that can produce other automata will only be able to produce less complicated ones. There is, however, a certain minimum level where this degenerative characteristic ceases to be universal. At this point automata which can reproduce themselves, or even construct higher entities, become possible. This fact, that complication, as well as organization, below a certain minimum level is degenerative, and beyond that level can become self-supporting and even increasing, will clearly play an important role in any future theory of the subject.

13.
Principles of the Self-Organizing System

W. ROSS ASHBY

QUESTIONS OF PRINCIPLE are sometimes regarded as too unpractical to be important, but I suggest that that is certainly not the case in *our* subject. The range of phenomena that we have to deal with is so broad that, were it to be dealt with wholly at the technological or practical level, we would be defeated by the sheer quantity and complexity of it. The total range can be handled only piecemeal; among the pieces are those homomorphisms of the complex whole that we call "abstract theory" or "general principles". They alone give the bird's-eye view that enables us to move about in this vast field without losing our bearings. I propose, then, to attempt such a bird's-eye survey.

What is "Organization"?

At the heart of our work lies the fundamental concept of "organization". What do we mean by it? As it is used in biology it is a somewhat complex concept, built up from several more primitive concepts. Because of this richness it is not readily defined, and it is interesting to notice that while March and Simon use the word "Organizations" as title for their book, they do not give a formal definition. Here I think they are right, for the word covers a multiplicity of meanings. I think that in future we shall hear the *word* less frequently, though the *operations* to which it corresponds, in

the world of computers and brain-like mechanisms, will become of increasing daily importance.

The hard core of the concept is, in my opinion, that of "conditionality". As soon as the relation between two entities *A* and *B* becomes conditional on *C*'s value or state then a necessary component of "organization" is present. Thus *the theory of organization is partly co-extensive with the theory of functions of more than one variable.*

We can get another angle on the question by asking "what is its converse?" The converse of "conditional on" is "not conditional on", so the converse of "organization" must therefore be, as the mathematical theory shows as clearly, the concept of "reducibility". (It is also called "separability".) This occurs, in mathematical forms, when what looks like a function of several variables (perhaps very many) proves on closer examination to have parts whose actions are *not* conditional on the values of the other parts. It occurs in mechanical forms, in hardware, when what looks like one machine proves to be composed of two (or more) sub-machines, each of which is acting independently of the others.

Questions of "conditionality", and of its converse "reducibility", can, of course, be treated by a number of mathematical and logical methods. I shall say something of such methods later. Here, however, I would like to express the opinion that the method of Uncertainty Analysis, introduced by Garner and McGill, gives us a method for the treatment of conditionality that is not only completely rigorous but is also of extreme generality. Its great generality and suitability for application to complex behavior, lies in the fact that it is applicable to any arbitrarily defined set of states. Its application requires neither linearity,

From W. Ross Ashby, "Principles of the Self-Organizing System," in Heinz von Foerster and George W. Zopf (Eds.), *Principles of Self-Organization* (New York: Pergamon Press, 1962), pp. 255–78. Reprinted by permission of the author and publisher.

nor continuity, nor a metric, nor even an ordering relation. By this calculus, the *degree* of conditionality can be measured, and analyzed, and apportioned to factors and interactions in a manner exactly parallel to Fisher's method of the analysis of variance; yet it requires no metric in the variables, only the frequencies with which the various combinations of states occur. It seems to me that, just as Fisher's conception of the analysis of variance threw a flood of light on to the complex relations that may exist between variations on a metric, so McGill and Garner's conception of uncertainty analysis may give us an altogether better understanding of how to treat complexities of relation when the variables are non-metric. In psychology and biology such variables occur with great commonness; doubtless they will also occur commonly in the brain-like processes developing in computers. I look forward to the time when the methods of McGill and Garner will become the accepted language in which such matters are to be thought about and treated quantitatively.

The treatment of "conditionality" (whether by functions of many variables, by correlation analysis, by uncertainty analysis, or by other ways) makes us realize that the essential idea is that there is first a product space—that of the *possibilities*—within which some sub-set of points indicates the actualities. This way of looking at "conditionality" makes us realize that it is related to that of "communication"; and it is, of course, quite plausible that we should define parts as being "organized" when "communication" (in some generalized sense) occurs between them. (Again the natural converse is that of independence, which represents non-communication.)

Now "communication" from A to B necessarily implies some constraint, some correlation between what happens at A and what at B. If, for given event at A, all possible events may occur at B, then there is no communication from A to B and no constraint over the possible (A, B)-couples that can occur. Thus the presence of "organization" between variables is equivalent to the existence of a *constraint* in the product-space of the possibilities. I stress this point because while, in the past, biologists have tended to think of organization as something extra, something *added* to the elementary variables, the modern theory, based on the logic of communication, regards organization as a restriction or constraint. The two points of view are thus diametrically opposed; there is no question of either being exclusively right, for each can be appropriate in its context.

But with this opposition in existence we must clearly go carefully, especially when we discuss with others, lest we should fall into complete confusion.

This excursion may seem somewhat complex but it is, I am sure, advisable, for we have to recognize that the discussion of organization theory has a peculiarity not found in the more objective sciences of physics and chemistry. The peculiarity comes in with the product space that I have just referred to. Whence comes this product space? Its chief peculiarity is that *it contains more than actually exists in the real physical world*, for it is the latter that gives us the actual, constrained *subset*.

The real world gives the subset of what *is*; the product space represents the uncertainty of the *observer*. The product space may therefore change if the observer changes; and two observers may legitimately use different product spaces within which to record the same subset of actual events in some actual thing. The "constraint" is thus a *relation* between observer and thing; the properties of any particular constraint will depend on both the real thing and on *the observer*. It follows that a substantial part of the theory of organization will be concerned with *properties that are not intrinsic to the thing but are relational between observer and thing*. We shall see some striking examples of this fact later.

Whole and Parts

"If conditionality" is an essential component in the concept of organization, so also is the assumption that we are speaking of a whole composed of parts. This assumption is worth a moment's scrutiny, for research is developing a theory of dynamics that does *not* observe parts and their interactions, but treats the system as an unanalysed whole. In physics, of course, we usually start the description of a system by saying "Let the variables be x_1, x_2, ..., x_n" and thus start by treating the whole as made of n functional parts. The other method, however, deals with unanalysed states, S_1, S_2, ... of the whole, without explicit mention of any parts that may be contributing to these states. The dynamics of such a system can then be defined and handled mathematically; I have shown elsewhere how such an approach can be useful. What I wish to point out here is that we can have a sophisticated *dynamics*, of a whole as complex and cross-connected as you please, that makes no reference to any parts and that therefore does *not* use the concept of organization. Thus the concepts of dynamics and of organization are essentially independent, in that all four combinations, of their presence and absence, are possible.

This fact exemplifies what I said, that "organization" is partly in the eye of the beholder. Two observers studying the same real material system, a hive of bees say, may find that one of them, thinking of the hive as an interaction of fifty thousand bee-parts, finds the bees "organized", while the other, observing whole states such as activity, dormancy, swarming, etc., may see *no* organization, only trajectories of these (unanalysed) states.

Another example of the independence of "organization" and "dynamics" is given by the fact that whether or not a real system is organized or reducible depends partly on the point of view taken by the observer. It is well known, for instance, that an organized (i.e. interacting) linear system of n parts, such as a network of pendulums and springs, can be seen from another point of view (that of the so-called "normal" coordinates) in which all the (newly identified) parts are completely separate, so that the whole is reducible. There is therefore nothing perverse about my insistence on the relativity of organization, for advantage of the fact is routinely taken in the study of quite ordinary dynamic systems.

Finally, in order to emphasize how dependent is the organization seen in a system on the observer who sees it, I will state the proposition that: given a whole with arbitrarily given behavior, a great variety of arbitrary "parts" can be seen in it; for all that is necessary, when the arbitrary part is proposed, is that we assume the given part to be coupled to another suitably related part, so that the two together form a whole isomorphic with the whole that was given. For instance, suppose the given whole, W of 10 states, behaves in accordance with the transformation:

$$W \downarrow \quad \begin{array}{c} p\,q\,r\,s\,t\,u\,v\,w\,x\,y \\[4pt] q\,r\,s\,q\,s\,t\,t\,x\,y\,y \end{array}$$

Its kinematic graph is

$$
\begin{array}{c}
u \\
\searrow \\
t \rightarrow s \longrightarrow q \leftarrow p \\
\nearrow \quad \nwarrow \swarrow \\
v \qquad r \\
w \rightarrow x \rightarrow y \circlearrowright
\end{array}
$$

and suppose we wish to "see" it as containing the part P, with internal states E and input states A:

$$
\begin{array}{cc|cc}
 & & \multicolumn{2}{c}{E} \\
\downarrow & & 1 & 2 \\
\hline
 & 1 & 2 & 1 \\
A & 2 & 1 & 1
\end{array} \Big\} P
$$

with a little ingenuity we find that if part P is coupled to part Q (with states (F, G) and input B) with transformation Q:

$$
(F, G)
$$

$$
\begin{array}{c|cccccc}
\downarrow & 1,1 & 1,2 & 1,3 & 2,1 & 2,2 & 2,3 \\
\hline
1 & 2,1 & 1,2 & 1,2 & 2,1 & 1,2 & 1,2 \\
B & & & & & & \\
2 & . & 2,3 & . & 2,1 & 2,2 & 2,2
\end{array} \Big\} Q
$$

by putting $A = F$ and $B = E$, then the new whole W' has transformation

$$
W': \quad \downarrow \quad
\begin{array}{llll}
1,1,1 & 1,1,2 & 1,1,3 & 1,2,1, \text{etc.} \\[4pt]
2,2,1 & 2,1,2 & 2,1,2 & 1,2,1, \text{etc.}
\end{array}
$$

which is *isomorphic with W* under the one–one correspondence

$$
\begin{array}{llll}
1,1,1 & 1,1,2 & 1,1,3 & 1,2,1, \text{etc.} \\
\downarrow & & & \\
w & s & p & y \quad , \text{etc.}
\end{array}
$$

Thus, subject only to certain requirements (e.g. that equilibria map into equilibria) *any dynamic system can be made to display a variety of arbitrarily assigned "parts"*, simply by a change in the *observer's* view point.

Machines in General

I have just used a way of representing two "parts", "coupled" to form a "whole", that anticipates the question: what do we mean by a "machine" in general?

Here we are obviously encroaching on what has been called "general system theory", but this last discipline always seemed to me to be uncertain whether it was dealing with *physical* systems, and therefore tied to whatever the real world provides, or with mathematical systems, in which the sole demand is that the work shall be free from internal contradictions. It is, I think, one of the substantial advances of the last decade that we have at last identified the *essentials* of the "machine in general".

Before the essentials could be seen, we had to realize that two factors must be *excluded as irrelevant*. The first is "materiality"—the idea that a machine must be made of actual matter, of the hundred or so existent elements. This is wrong, for examples can readily be given showing that what is essential is whether the system, of angels and ectoplasm if you please, *behaves* in a law-abiding and machine-like way. Also to be excluded as irrelevant is any reference to energy, for any calculating machine shows that what matters is the *regularity* of the behavior—whether energy is gained or lost, or even created, is simply irrelevant.

The fundamental concept of "machine" proves to have a form that was formulated at least a century ago, but this concept has not, so far as I am aware, ever been used and exploited vigorously. A "machine" is that which behaves in a machine-like way, namely, that its internal state, and the state of its surroundings, defines uniquely the next state it will go to.

This definition, formally proposed fifteen years ago has withstood the passage of time and is now becoming generally accepted. It appears in many forms. When the variables are continuous it corresponds to the description of a dynamic system by giving a set of ordinary differential equations with time as the independent variable. The *fundamental* nature of such a representation (as contrasted with a merely convenient one) has been recognized by many earlier workers such as Poincaré, Lotka, and von Bertalanffy.

Such a representation by differential equations is, however, too restricted for the needs of a science that includes biological systems and calculating machines, in which discontinuity is ubiquitous. So arises the modern definition, able to include both the continuous and the discontinuous and even the discrete, without the slightest loss of rigor. The "machine with input" or the "finite automaton" is today defined by a set S of internal states, a set I of input or surrounding states, and a mapping, f say, of the product set $I \times S$ into S. Here, in my opinion, we have the very essence of the "machine"; all known types of machine are to be found here; and all interesting deviations from the concept are to be found by the corresponding deviation from the definition.

We are now in a position to say without ambiguity or evasion what we mean by a machine's "organization". First we specify which system we are talking about by specifying its states S and its conditions I. If S is a product set, so that $S = \Pi_i T_i$ say, then the parts i are each specified by its set of states T_i. The "*organization*" between these parts is then specified by the mapping f. Change f and the organization changes. In other words, the possible organizations between the parts can be set into one–one correspondence with the set of possible mappings of $I \times S$ into S. Thus "organization" and "mapping" are two ways of looking at the same thing—the organization being noticed by the observer of the actual system, and the mapping being recorded by the person who represents the behavior in mathematical or other symbolism.

"Good" Organization

At this point some of you, especially the biologists, may be feeling uneasy; for this definition of organization makes no reference to any *usefulness* of the organization. It demands only that there be conditionality between the parts and regularity in behavior. In this I believe the definition to be right, for the question whether a given organization is "good" or "bad" is quite independent of the prior test of whether it is or is not an organization.

I feel inclined to stress this point, for here the engineers and the biologists are likely to think along widely differing lines. The engineer, having put together some electronic hardware and having found the assembled network to be roaring with parasitic oscillations, is quite accustomed to the idea of a "bad" organization; and he knows that the "good" organization has to be searched for. The biologist, however, studies mostly animal species that have survived the long process of natural selection; so almost all the organizations he sees have already been selected to be good ones, and he is apt to think of "organizations" as *necessarily* good. This point of view may often be true in the biological world but it is most emphatically not true in the world in which we people here are working. We *must* accept that

(1) most organizations are bad ones;

(2) the good ones have to be sought for; and

(3) what is meant by "good" must be clearly defined, explicitly if necessary, *in every case*.

What then is meant by "good", in our context of brain-like mechanisms and computers? We must proceed cautiously, for the word suggests some evaluation whose origin has not yet been considered.

In some cases the distinction between the "good" organization and the "bad" is obvious, in the sense that as everyone in these cases would tend to use the same criterion, it would not need explicit mention. The brain of a living organism, for instance, is usually judged as having a "good" organization if the organization (whether inborn or learned) acts so as to further the organism's survival. This consideration readily generalizes to all those cases in which the organization (whether of a cat or an automatic pilot or an oil refinery) is judged "good" if and only if it acts so as to keep an assigned set of variables, the "essential" variables, within assigned limits. Here are all the mechanisms for homeostasis, both in the original sense of Cannon and in the generalized sense. From this criterion comes the related one that an organization is "good" if it makes the system stable around an assigned equilibrium. Sommerhoff in particular has given a wealth of examples, drawn from a great range of biological and mechanical phenomena, showing how in all cases the idea of a "good organization" has as its essence the idea of a number of parts so interacting as to achieve some given "focal condition". I would like to say here that I do not consider that Sommerhoff's contribution to our subject has yet been adequately recognized. His identification of *exactly* what is meant by coordination and integration is, in my opinion, on a par with Cauchy's identification of exactly what was meant by convergence. Cauchy's discovery was a real discovery, and was an enormous help to later workers by providing them with a concept, rigorously defined, that could be used again and again, in a vast range of contexts, and always with exactly the same meaning. Sommerhoff's discovery of how to represent *exactly* what is meant by coordination and integration and good organization will, I am sure, eventually play a similarly fundamental part in our work. [See selection from Sommerhoff in Part VI.]

His work illustrates, and emphasizes, what I want to say here—*there is no such thing as "good organization" in any absolute sense*. Always it is relative; and an organization that is good in one context or under one criterion may be bad under another.

Sometimes this statement is so obvious as to arouse no opposition. If we have half a dozen lenses, for instance, that can be assembled this way to make a telescope or that way to make a microscope, the goodness of an assembly obviously depends on whether one wants to look at the moon or a cheese mite.

But the subject is more contentious than that! The thesis implies that there is no such thing as a brain (natural or artificial) that is good in any absolute sense—it all depends on the circumstances and on what is wanted. Every faculty that

a brain can show is "good" only conditionally, for there exists at least one environment against which the brain is handicapped by the possession of this faculty. Sommerhoff's formulation enables us to show this at once: whatever the faculty or organization achieves, let that be *not* in the "focal conditions".

We know, of course, lots of examples where the thesis is true in a somewhat trivial way. Curiosity tends to be good, but many an antelope has lost its life by stopping to see what the hunter's hat is. Whether the organization of the antelope's brain should be of the type that does, or does not, lead to temporary immobility clearly depends on whether hunters with rifles are or are not plentiful in its world.

From a different angle we can notice Pribram's results, who found that brain-operated monkeys scored higher in a certain test than the normals. (The operated were plodding and patient while the normals were restless and distractible.) Be that as it may, one cannot say which brain (normal or operated) had the "good" organization until one has decided which sort of temperament is wanted.

Do you still find this non-contentious? Then I am prepared to assert that there is not a single mental faculty ascribed to Man that is good in the absolute sense. If any particular faculty is *usually* good, this is solely because our terrestrial environment is so lacking in variety that its usual form makes that faculty usually good. But change the environment, go to really different conditions, and possession of that faculty may be harmful. And "bad", by implication, is the brain organization that produces it.

I believe that there is not a single faculty or property of the brain, usually regarded as desirable, that does not become *un*desirable in some type of environment. Here are some examples in illustration.

The first is Memory. Is it not good that a brain should have memory? Not at all, I reply—only when the environment is of a type in which the future often *copies* the past; should the future often be the *inverse* of the past, memory is actually disadvantageous. A well known example is given when the sewer rat faces the environmental system known as "pre-baiting". The naïve rat is very suspicious, and takes strange food only in small quantities. If, however, wholesome food appears at some place for three days in succession, the sewer rat will learn, and on the fourth day will eat to repletion, and die. The rat without memory, however, is as suspicious on the fourth day as on the first, and lives. Thus, in *this* environment, memory is positively disadvantageous. Prolonged contact with this environment will lead, other things being equal, to evolution in the direction of diminished memory-capacity.

As a second example, consider organization itself in the sense of connectedness. Is it not good that a brain should have its parts in rich functional connection? I say, No—not *in general*; only when the environment is itself richly connected. When the environment's parts are *not* richly connected (when it is highly reducible, in other words), adaptation will go on faster if the brain is also highly reducible, i.e. if its connectivity is small. Thus the *degree* of organization can be too high as well as too low; the degree we humans possess is probably adjusted to be somewhere near the optimum for the usual terrestrial environment. It does not in any way follow that this degree will be optimal or good if the brain is a mechanical one, working against some grossly non-terrestrial environment—one existing only inside a big computer, say.

As another example, what of the "organization" that the biologist always points to with pride—the development in evolution of specialized organs such as brain, intestines, heart and blood vessels. Is not this good? Good or not, it is certainly a specialization made possible only because the earth has an atmosphere; without it, we would be incessantly bombarded by tiny meteorites, any one of which, passing through our chest, might strike a large blood vessel and kill us. Under such conditions a better form for survival would be the slime mould, which specializes in being able to flow through a tangle of twigs without loss of function. Thus the development of organs is not good unconditionally, but is a specialization to a world free from flying particles.

After these actual instances, we can return to theory. It is here that Sommerhoff's formulation gives such helpful clarification. He shows that in all cases there must be given, and specified, first a *set of disturbances* (values of his "coenetic variable") and secondly a goal (his "focal condition"); the disturbances threaten to drive the outcome outside the focal condition. The "good" organization is then of the nature of a *relation* between the set of disturbances and the goal. Change the set of disturbances, and the organization, without itself changing, is evaluated "bad" instead of "good". As I said, there is no property of an organization that is good in any absolute sense; all are relative to some given environment, or to

some given set of threats and disturbances, or to some given set of problems.

Self-Organizing Systems

I hope I have not wearied you by belaboring this relativity too much, but it is fundamental, and is only too readily forgotten when one comes to deal with organizations that are either biological in origin or are in imitation of such systems. With this in mind, we can now start to consider the so-called "self-organizing" system. We must proceed with some caution here if we are not to land in confusion, for the adjective is, if used loosely, ambiguous, and, if used precisely, self-contradictory.

To say a system is "self-organizing" leaves open two quite different meanings.

There is a first meaning that is simple and unobjectionable. This refers to the system that starts with its parts separate (so that the behavior of each is independent of the others' states) and whose parts then act so that they change towards forming connections of some type. Such a system is "self-organizing" in the sense that it changes from "parts separated" to "parts joined". An example is the embryo nervous system, which starts with cells having little or no effect on one another, and changes, by the growth of dendrites and formation of synapses, to one in which each part's behavior is very much affected by the other parts. Another example is Pask's system of electrolytic centers, in which the growth of a filament from one electrode is at first little affected by growths at the other electrodes; then the growths become more and more affected by one another as filaments approach the other electrodes. In general such systems can be more simply characterized as "self-*connecting*", for the change from independence between the parts to conditionality can always be seen as some form of "connection", even if it is as purely functional as that from a radio transmitter to a receiver.

Here, then, is a perfectly straightforward form of self-organizing system; but I must emphasize that there can be no assumption at this point that the organization developed will be a good one. If we wish it to be a "good" one, we must first provide a criterion for distinguishing between the bad and the good, and then we must ensure that the appropriate selection is made.

We are here approaching the second meaning of "self-organizing". "Organizing" may have the first meaning, just discussed, of "changing from unorganized to organized". But it may also mean "changing from a bad organization to a good one", and this is the case I wish to discuss now, and more fully. This is the case of peculiar interest to *us*, for this is the case of the system that changes itself from a bad way of behaving to a good. A well known example is the child that starts with a brain organization that makes it fire-seeking; then a change occurs, and a new brain organization appears that makes the child fire-avoiding. Another example would occur if an automatic pilot and a plane were so coupled, by mistake, that positive feedback made the whole error-aggravating rather than error-correcting. Here the organization is bad. The system would be "self-organizing" if a change were *automatically* made to the feedback, changing it from positive to negative; then the whole would have changed from a bad organization to a good. Clearly, *this* type of "self-organization" is of peculiar interest to us. What is implied by it?

Before the question is answered we must notice, if we are not to be in perpetual danger of confusion, that *no machine can be self-organizing in this sense*. The reasoning is simple. Define the set S of states so as to specify which machine we are talking about. The "organization" must then, as I said above, be identified with f, the mapping of S into S that the basic drive of the machine (whatever force it may be) imposes. Now the logical relation here is that f determines the changes of S:—f is *defined* as the set of couples (s_i, s_j) such that the internal drive of the system will force state s_i to change to s_j. To allow f to be a function of the state is to make nonsense of the whole concept.

Since the argument is fundamental in the theory of self-organizing systems, I may help explanation by a parallel example. Newton's law of gravitation says that $F = M_1 M_2/d^2$, in particular, that the force varies inversely as the distance to power 2. To power 3 would be a different law. But suppose it were suggested that, not the force F but the *law* changed with the distance, so that the power was not 2 but some function of the distance, $\varphi(d)$. This suggestion is illogical; for we now have that $F = M_1 M_2/d^{\varphi(d)}$, and this represents not a law that varies with the distance but *one* law covering all distances; that is, were this the case we would *re-define* the law. Analogously, were f in the machine to be some function of the state S, we would have to re-define our machine. Let me be quite explicit with an example. Suppose S had three states: a, b, c. If f depended on S there would be three f's: f_a, f_b, f_c say. Then if they are

↓	a	b	c
f_a	**b**	a	b
f_b	c	**a**	a
f_c	b	b	**a**

then the transform of a must be under f_a, and is therefore b, so the whole set of f's would amount to the *single* transformation:

$$\begin{matrix} & a & b & c \\ \downarrow & & & \\ & b & a & a \end{matrix}$$

It is clearly illogical to talk of f as being a function of S, for such talk would refer to operations, such as $f_a(b)$, which cannot in fact occur.

If, then, no machine can properly be said to be self-organizing, how do we regard, say, the Homeostat, that rearranges its own wiring; or the computer that writes out its own program?

The new logic of mechanism enables us to treat the question rigorously. We start with the set S of states, and assume that f changes, to g say. So we really have a *variable*, $\alpha(t)$ say, a function of the time that had at first the value f and later the value g. This change, as we have just seen, cannot be ascribed to any cause in the set S; so it must have come from some outside agent, acting on the system S as input. If the system is to be in some sense "*self*-organizing", the "self" must be enlarged to include this variable α, and, to keep the whole bounded, the cause of α's change must be in S (or α).

Thus the appearance of being "self-organizing" can be given only by the machine S being coupled to another machine (of one part):

$$\boxed{S} \begin{matrix} \rightarrow \\ \leftarrow \end{matrix} \boxed{\alpha}$$

Then the part S can be "self-organizing" within the whole $S + \alpha$.

Only in this partial and strictly qualified sense can we understand that a system is "*self*-organizing" without being self-contradictory.

Since no system can correctly be said to be self-organizing, and since use of the phrase "self-organizing" tends to perpetuate a fundamentally confused and inconsistent way of looking at the subject, the phrase is probably better allowed to die out.

The Spontaneous Generation of Organization

When I say that no system can properly be said to be self-organizing, the listener may not be satisfied. What, he may ask, of those changes that occurred a billion years ago, that led lots of carbon atoms, scattered in little molecules of carbon dioxide, methane, carbonate, etc., to get together until they formed proteins, and then went on to form those large active lumps that today we call "animals"? Was not this process, on an isolated planet, one of "self-organization"? And if it occurred on a planetary surface can it not be made to occur in a computer? I am, of course, now discussing the origin of life. Has modern system theory anything to say on this topic?

It has a great deal to say, and some of it flatly contradictory to what has been said ever since the idea of evolution was first considered. In the past, when a writer discussed the topic, he usually assumed that the generation of life was rare and peculiar, and he then tried to display some way that would enable this rare and peculiar event to occur. So he tried to display that there is *some* route from, say, carbon dioxide to the amino acid, and thence to the protein, and so, through natural selection and evolution, to intelligent beings. I say that this looking for special conditions is quite wrong. The truth is the opposite— *every* dynamic system generates its own form of intelligent life, is self-organizing in this sense. (I will demonstrate the fact in a moment.) Why we have failed to recognize this fact is that until recently we have had no experience of systems of medium complexity; either they have been like the watch and the pendulum, and we have found their properties few and trivial, or they have been like the dog and the human being, and we have found their properties so rich and remarkable that we have thought them supernatural. Only in the last few years has the general-purpose computer given us a system rich enough to be interesting yet still simple enough to be understandable. With this machine as tutor we can now begin to think about systems that are simple enough to be comprehensible in detail yet also rich enough to be suggestive. With their aid we can see the truth of the statement that *every isolated determinate dynamic system obeying unchanging laws will develop "organisms" that are adapted to their "environments"*.

The argument is simple enough in principle. We start with the fact that systems in general go to equilibrium. Now most of a system's states are non-equilibrial (if we exclude the extreme case of the system in neutral equilibrium). So in going from *any* state to one of the equilibria, the system is going from a larger number of states to a smaller. In this way it is performing a selection, in the purely objective sense that it rejects some states, by leaving them, and retains some other state, by sticking to it. Thus, as every determinate system goes to equilibrium, so does it select. We have heard *ad nauseam* the dictum that a machine cannot select; the truth is just the opposite: every machine, as it goes to equilibrium, performs the corresponding act of selection.

Now, equilibrium in simple systems is usually trivial and uninteresting; it is the pendulum hanging vertically; it is the watch with its main-spring run down; the cube resting flat on one face. Today, however, we know that when the system is more complex and dynamic, equilibrium, and the stability around it, can be much more interesting. Here we have the automatic pilot success-fully combating an eddy; the person redistributing his blood flow after a severe haemorrhage; the business firm restocking after a sudden increase in consumption; the economic system restoring a distribution of supplies after a sudden destruction of a food crop; and it is a man successfully getting at least one meal a day during a lifetime of hardship and unemployment.

What makes the change, from trivial to interesting, is simply the *scale* of the events. "Going to equilibrium" *is* trivial in the simple pendulum, for the equilibrium is no more than a single point. But when the system is more complex; when, say, a country's economy goes back from wartime to normal methods then the stable region is vast, and much interesting activity can occur within it. The computer is heaven-sent in this context, for it enables us to bridge the enormous conceptual gap from the simple and understandable to the complex and interesting. Thus we can gain a consider-able insight into the so-called spontaneous genera-tion of life by just seeing how a somewhat simpler version will appear in a computer.

Competition

Here is an example of a simpler version. The competition between species is often treated as if it were essentially biological; it is in fact an expression of a process of far greater generality. Suppose we have a computer, for instance, whose stores are filled at random with the digits 0 to 9. Suppose its dynamic law is that the digits are continuously being multiplied in pairs, and the right-hand digit of the product going to replace the first digit taken. Start the machine, and let it "evolve"; what will happen? Now under the laws of this particular world, even times even gives even, and odd times odd gives odd. But even times odd gives even; so after a mixed encounter *the even has the better chance of survival.* So as this system evolves, we shall see the evens favored in the struggle, steadily replacing the odds in the stores and eventually exterminating them.

But the evens are not homogeneous, and among them the zeros are best suited to survive in this particular world; and, as we watch, we shall see the zeros exterminating their fellow-evens, until eventually they inherit this particular earth.

What we have here is an example of a thesis of extreme generality. From one point of view we have simply a well defined operator (the multiplication and replacement law) which drives on towards equilibrium. In doing so it *automatically* selects those operands that are *specially resistant* to its change-making tendency (for the zeros are uniquely resistant to change by multiplication). This process, of progression towards the specially resistant form, is of extreme generality, demanding only that the operator (or the physical laws of any physical system) be determinate and unchanging. This is the general or abstract point of view. The biologist sees a special case of it when he observes the march of evolution, survival of the fittest, and the inevitable emergence of the highest biological functions and intelligence. Thus, when we ask: What was necessary that life and intelligence should appear? the answer is not carbon, or amino acids or any other special feature but only that the dynamic laws of the process should be *unchanging,* i.e. that the system should be *isolated. In any isolated system, life and intelligence inevitably develop* (they may, in degenerate cases, develop to only zero degree).

So the answer to the question: How can we generate intelligence synthetically? is as follows. Take a dynamic system whose laws are unchanging and single-valued, and whose size is so large that after it has gone to an equilibrium that involves only a small fraction of its total states, this small fraction is still large enough to allow room for a good deal of change and behavior. Let it go on for a long enough time to get to such an equilibrium. Then examine the equilibrium in detail. You will find that the states or forms now in being are peculiarly able to survive against the changes induced by the laws. Split the equilibrium in two, call one part "organism" and the other part "environment": you will find that this "organism" is peculiarly able to survive against the disturbances from this "environment". The *degree* of adaptation and complexity that this organism can develop is bounded only by the size of the whole dynamic system and by the time over which it is allowed to progress towards equilibrium. Thus, as I said, every isolated determinate dynamic system will develop organisms that are adapted to their environments. There is thus no difficulty in principle, in developing synthetic organisms as complex or as intelligent as we please.

In *this* sense, then, *every* machine can be thought of as "self-organizing", for it will develop, to such degree as its size and complexity allow, some functional structure homologous with an "adapted organism". But does this give us what we at this Conference are looking for? Only partly; for nothing said so far has any implication about the organization being good or bad; the criterion that would make the distinction has not yet been introduced. It is true, of course, that the developed organism, being stable, will have its own essential variables, and it will show its stability by vigorous reactions that tend to preserve its own existence. To *itself,* its own organization will *always,* by definition, be good. The wasp finds the stinging reflex a good thing, and the leech finds the blood-sucking reflex a good thing. But these criteria come *after* the organization for survival; having seen *what* survives we then see what is "good" for that form. What emerges depends simply on what are the system's laws and from what state it started; there is no implication that the organization developed will be "good" in any absolute sense, or according to the criterion of any outside body such as ourselves.

To summarize briefly: there is no difficulty, in principle, in developing *synthetic organisms as complex, and as intelligent as we please.* But we must notice two fundamental qualifications; first, their intelligence will be an adaptation to, and a specialization towards, their particular environment, with no implication of validity for any other environment such as ours; and secondly, their intelligence will be directed towards keeping their own essential variables within limits. They will be fundamentally selfish. So we now have to ask: In view of these qualifications, can we yet turn these processes to our advantage?

Requisite Variety

In this matter I do not think enough attention has yet been paid to Shannon's Tenth Theorem or to the simpler "law of requisite variety" in which I have expressed the same basic idea. Shannon's theorem says that if a correction-channel has capacity H, then equivocation of amount H can be removed, *but no more*. Shannon stated his theorem in the context of telephone or similar communication, but the formulation is just as true of a biological regulatory channel trying to exert some sort of corrective control. He thought of the case with a lot of message and a little error; the biologist faces the case where the "message" is small but the disturbing errors are many and large. The theorem can then be applied to the brain (or any other regulatory and selective device), when it says that the amount of regulatory or selective action that the brain can achieve is absolutely bounded by its capacity as a channel. Another way of expressing the same idea is to say that any quantity K of appropriate selection demands the transmission or processing of quantity K of information. *There is no getting of selection for nothing.*

I think that here we have a principle that we shall hear much of in the future, for it dominates all work with complex systems. It enters the subject somewhat as the law of conservation of energy enters power engineering. When that law first came in, about a hundred years ago, many engineers thought of it as a disappointment, for it stopped all hopes of perpetual motion. Nevertheless, it did in fact lead to the great practical engineering triumphs of the nineteenth century, because it made power engineering more realistic.

I suggest that when the full implications of Shannon's Tenth Theorem are grasped we shall be, first sobered, and then helped, for we shall then be able to focus our activities on the problems that are properly realistic, and actually solvable.

The Future

Here I have completed this bird's-eye survey of the principles that govern the self-organizing system. I hope I have given justification for my belief that these principles, based on the logic of mechanism and on information theory, are now essentially *complete*, in the sense that there is now no area that is grossly mysterious.

Before I end, however, I would like to indicate, very briefly, the directions in which future research seems to me to be most likely to be profitable.

One direction in which I believe a great deal to be readily discoverable, is in the discovery of new types of dynamic process. Most of the machine-processes that we know today are very specialized, depending on exactly what parts are used and how they are joined together. But there are systems of more net-like construction in which what happens can only be treated statistically. There are processes here like, for instance, the spread of epidemics, the fluctuations of animal populations over a territory, the spread of wave-like phenomena over a nerve-net. These processes are, in themselves, neither good nor bad, but they exist, with all their curious properties, and doubtless the brain will use them should they be of advantage. What I want to emphasize here is that they often show very surprising and peculiar properties; such as the tendency, in epidemics, for the outbreaks to occur in waves. Such peculiar new properties may be just what some machine designer wants, and that he might otherwise not know how to achieve.

The study of such systems must be essentially statistical, but this does not mean that each system must be individually stochastic. On the contrary, it has recently been shown that no system can have greater efficiency than the determinate when acting as a regulator; so, as regulation is the one function that counts biologically, we can expect that natural selection will have made the brain as determinate as possible. It follows that we can confine our interest to the lesser range in which the sample space is over a set of mechanisms each of which is individually determinate.

As a particular case, a type of system that deserves much more thorough investigation is the large system that is built of parts that have many states of equilibrium. Such systems are extremely common in the terrestrial world; they exist all around us, and in fact, intelligence as we know it would be almost impossible otherwise. This is another way of referring to the system whose variables behave largely as part-functions. I have shown elsewhere that such systems tend to show habituation (extinction) and to be able to adapt progressively. There is reason to believe that some of the well-known but obscure biological phenomena such as conditioning, association, and Jennings' law of the resolution of physiological states may be more or less simple and direct expressions of the multiplicity of equilibrial states. At the moment I am investigating the possibility that the transfer of "structure", such as that of three-dimensional space, into a dynamic system—the sort of learning that Piaget has specially considered—may be an *automatic* process when the input comes to a system with many equilibria. Be that as it may, there can be little doubt that the study of such systems is likely to reveal a variety of new dynamic processes, giving us dynamic resources not at present available.

A particular type of system with many equilibria is the system whose parts have a high "threshold"—those that tend to stay at some "basic" state unless some function of the input exceeds some value. The general properties of such systems is still largely unknown, although Beurle has made a most interesting start. They deserve extensive investigation; for, with their basic tendency to develop avalanche-like waves of activity, their dynamic properties are likely to prove exciting and even dramatic. The fact that the mammalian brain uses the property extensively suggests that it may have some peculiar, and useful, property not readily obtainable in any other way.

Reference to the system with many equilibria brings me to the second line of investigation that seems to me to be in the highest degree promising —I refer to the discovery of *the living organism's memory store*: the identification of its physical nature.

At the moment, our knowledge of the living brain is grossly out of balance. With regard to what happens from one millisecond to the next we know a great deal, and many laboratories are working to add yet more detail. But when we ask what happens in the brain from one hour to the next, or from one year to the next, practically nothing is known. Yet it is these longer-term changes that are the really significant ones in human behavior.

It seems to me, therefore, that if there is one thing that is crying out to be investigated it is the physical basis of the brain's memory-stores. There was a time when "memory" was a very vague and metaphysical subject; but those days are gone. "Memory", as a *constraint* holding over events of the past and the present, and a *relation* between them, is today firmly grasped by the logic of mechanism. We know exactly what we mean by it behavioristically and operationally. What we need now is the provision of adequate resources for its investigation. Surely the time has come for the world to be able to find resources for *one* team to go into the matter?

Summary

Today, the principles of the self-organizing system are known with some completeness, in the sense that no major part of the subject is wholly mysterious.

We have a secure base. Today we know *exactly* what we mean by "machine", by "organization", by "integration", and by "self-organization". We understand these concepts as thoroughly and as rigorously as the mathematician understands "continuity" or "convergence".

In these terms we can see today that the artificial generation of dynamic systems with "life" and "intelligence" is not merely simple—it is unavoidable if only the basic requirements are met. These are not carbon, water, or any other material entities but the persistence, over a long time, of the action of any operator that is both unchanging and single-valued. *Every* such operator forces the development of its own form of life and intelligence.

But will the forms developed be of use to *us*? Here the situation is dominated by the basic law of requisite variety (and Shannon's Tenth Theorem), which says that the achieving of appropriate selection (to a degree better than chance) is absolutely dependent on the processing of at least that quantity of information. Future work must respect this law, or be marked as futile even before it has started.

Finally, I commend as a program for research, the *identification of the physical basis of the brain's memory stores*. Our knowledge of the brain's functioning is today grossly out of balance. A vast amount is known about how the brain goes from state to state at about millisecond intervals; but when we consider our knowledge of the basis of the important long-term changes we find it to amount, practically, to nothing. I suggest it is time that we made some definite attempt to attack this problem. Surely it is time that the world had *one* team active in this direction?

IV

Information, Communication, and Meaning

WE have investigated wholes as systems of organized, dynamically interacting parts. Part IV, drawing on modern information theory (or what is perhaps more appropriately called *communication theory* abroad), focuses on the fundamental nature of the mediating linkage underlying these interrelations and interactions in the complex behavior systems of concern to us. The fact that such systems are open, in dynamic interchange with their environment, are self-organizing and adaptive, learn, have memories, are self-aware and goal-seeking, depends on the unique character of "information" and the process of its communication between systems, their components, and their environments.

For our purposes, information theory divides into two aspects, reflected in the two Sections of this Part. The first aspect, taken up in Section A, deals with the abstract logical nature of information, its mathematical measure, and its close affinity with the notions of structure, organization, control, order, and entropy. The second aspect, discussed in Section B, is concerned with that special systemic situation wherein the phenomenon of *meaning* emerges—generated out of the social interaction of human minds capable of manipulating arbitrary symbols as against simply signs or signals.

Part IV is introduced by three articles. First is an overview of the mathematical theory of information as viewed by the behavioral scientist—here, George A. Miller. Next is W. Ross Ashby's complementary discussion, couched in the most abstract terms of "variety" and its "constraints," applied to the nature of system–environment interchanges in the case of adaptive or self-regulating systems. Such interchanges are seen, at bottom, as a mapping of the "constrained variety" (or basic information pool) characterizing the environment into the organization of the machine or organism by way of evolution or learning. It is such mappings that make possible the system's control of its behavior in ways relevant to or coordinate with the nature of its environment such that, in effect, system and environment become interacting components of a larger whole. Such a conceptualization provides a common principle for the otherwise diverse facts of phylogenetic evolution, ontogenetic development, psychological learning, or sociocultural elaboration. Ashby's discussion here lays the groundwork for his later analysis of the basic nature of cybernetic regulation and control, the core of which is presented in Section B of Part V. Part IV's third introductory paper, by Rapoport, provides an excellent rationale for, and introduction to, the ensuing two Sections of this Part, discussing as he does the promises and dangers of attempting to relate information to physical entropy, on the one hand, and to knowledge on the other.

The late Erwin Schrödinger's famous discussion of the nature of "life," culminating in the oft-quoted aphorism, "life feeds on negative entropy," initiates the related string of discussions in the ensuing four papers of Section A (Entropy and Life) by Brillouin, Raymond, and Ostow. Though many will continue to dispute the significance of the close correspondence between the mathematical expression for information and that for the thermodynamic concept of entropy, there is no doubt that we are confronted with one of those rare scientific insights that—until further elucidated—inspires intellectual awe and unrest.

Concluding this section is an especially pertinent bridging article by Heinz von Foerster, which describes in concrete terms an important part of the process by which the organism maps the variety and constraints of its environment into its own structure and is thereby able to adapt. Furthermore, the continuity is shown between the various forms of information and information processing, leading up to the use of the symbol in man. The many abstract ideas discussed in earlier selections are given more direct illustration here.

In the first article of Section B (Behavior and Meaning), which has been edited to remove the mathematical overview of information theory overlapping with Miller's exposition, F. C. Frick does not go all the way to the semantic question as such, but provides a valuable discussion of a number of principles that must underlie any adequate analysis of semantic behavior. He argues that the use of information theory entails a systematic position that views behavior as a statistical process: behavior is not a simple matter of distinct stimulus and response events, but of *sets of alternative possible or potential* stimuli and responses from which *choices* must be made. Thus, recognizing the fundamental principle that all information processes are *relational selection processes*, this viewpoint shifts the focus from the forms of information input and the *structure* of the learning process to the "efficiency" of communication—the analysis of how the person "selects and codes the available stimuli, how effectively he processes his inputs." The social psychologist and sociologist will recognize this as restating the point that the person's *subjective interpretation of the situation of action* must be taken as a central focus of analysis and not swept under the behavioristic rug. The next step to the problem of semantics is made easier by these conceptions.

With Osgood's paper, we turn to a focus on the human cortex and a behavioristic psychologist's interpretation of neurological and informational processes showing distinct convergences with perspectives already developed. But though the mediation theories such as that of Osgood (see also O. H. Mowrer's *Learning Theory and the Symbolic Processes*) have moved significantly in the direction of correcting the perennial weakness of behaviorism in the face of symbolic behavior, they are still seen as inadequate, especially by those who insist on a more holistic and developmental approach, such as the followers of John Dewey and G. H. Mead (who are represented in Parts V and VI). An important facet of this dissatisfaction is brought out by Heinz Werner and Bernard Kaplan in their book *Symbol Formation* (1963) when they contrast Osgood's hypothesis with their own "organismic–developmental" framework:

Osgood's view, rooted in a stimulus-response psychology, is agenetic and does not distinguish fundamentally between mere *reacting* and *knowing*; representational activity is treated as a response essentially no different from other responses. Our view, on the other hand, is genetically oriented and makes a fundamental distinction between *reacting to* and *knowing about*; representation is thus an emergent activity not reducible to the overlap of responses. Finally, for us, but apparently not for Osgood, representational activity goes hand in hand with the *construction* of a world of objects ("knowables"). (p. 24)

It should be noted that Osgood has recently gone beyond the two-stage model presented here to suggest the need for a three-stage model [in his "Psycholinguistics," in Sigmund Koch (Ed.), *Psychology: A Study of a Science*, Vol. 6].

In the course of his discussion in Section A, Mortimer Ostow raised an often-met problem in information theory of particular interest to us here. In asking whether the information value of a code or message is relative to the information state of the receiving unit, it is implied that "information" could conceivably be some kind of independent entity situated somewhere in space, rather than an inherently relational concept. Thus, it is asked if it makes a difference whether the "information" in a system is measured at the transmitting or at the receiving end of a channel; or it is stated or implied that information is "contained" in the head or in a computer memory or on magnetic tape. But here is where the communication engineer's use of "information" to refer mainly to physical signals in a wire, as against the semantic or "meaning" referent of the term, does its mischief. For the electrical events in a wire, the wiggles in a phonograph record groove, or the pencil marks on a piece of paper cannot be presumed to be signals, signs, or symbols without a great deal of prior knowledge of the context within which they appear. Technically, taken by themselves they can be objectively confirmed to be only constrained variety, to use Ashby's terms. Thus, as suggested by Rapoport early in this Part, and to anticipate MacKay's more extensive discussion, information in its fuller sense is first and foremost a relation or mapping between sets of constrained variety contained in behaving systems, such that the transmission of physical signals of a certain sort (a subset drawn from a predetermined variety pool according to set constraints—for example, language) performs a selective function on the systems' repertoire of tendencies to act in certain ways. Hence, being relational, the "meaning" or "information" carried by the physical signals does not reside in these signals but is a function of the larger system made up of the states of the sending and receiving units and their relevant environment. This insight, we believe, becomes a crucial foundation for the *transactional* view of the nature of all open behavioral systems, a view discussed in later selections of this sourcebook.

As one of the few serious attempts to bring over the general principles of information theory to the semantic side, MacKay's work constitutes the beginning of a generalized framework that promises to articulate the psychological, social, and situational forces which, as a dynamic system of ongoing events, must all be embraced in any adequate theory of meaning and its generation and transmission. This is partly, but significantly, brought out in the close similarity between MacKay's definition of meaning in terms of its "selective function" and G. H. Mead's seminal conception developed in his symbolic interactionism (of which MacKay may have been quite unaware). For a quite different approach to a theory of semantic information the reader should be referred to the work of Rudolf Carnap and Yehoshua Bar-Hillel (Y. Bar-Hillel, *Language and Information*, 1964).

The concluding paper of Part IV brings us to the main point: the problem of understanding the behavior of communicatively interacting human beings. Ackoff applies what is very close to MacKay's conception of the selective function of semantic information to a model of the communicative behaviors of two or more individuals, suggesting how the transmitted information selectively links "mind to mind" and makes possible coordinate, mutually influenced behaviors. Thus, it is not simply a "behavioral theory of communication," but also a communicative theory of behavior.

14.
What is Information Measurement?

GEORGE A. MILLER

IN RECENT YEARS a few psychologists, whose business throws them together with communication engineers, have been making considerable fuss over something called "information theory." They drop words like "noise," "redundancy," or "channel capacity" into surprising contexts and act like they had a new slant on some of the oldest problems in experimental psychology. Little wonder that their colleagues are asking, "What is this 'information' you talk about measuring?" and "What does all this have to do with the general body of psychological theory?"

The reason for the fuss is that information theory provides a yardstick for measuring organization. The argument runs like this. A well-organized system is predictable—you know almost what it is going to do before it happens. When a well-organized system does something, you learn little that you didn't already know—you acquire little information. A perfectly organized system is completely predictable and its behavior provides no information at all. The more disorganized and unpredictable a system is, the more information you can get by watching it. Information, organization, and predictability room together in this theoretical house. The key that unlocks the door to predictability is the theory of probability; but once this door is open we have access to information and organization as well.

The implications of this argument are indeed worth making a fuss about. Information, organization, predictability, and their synonyms are not rare concepts in psychology. Each place they occur now seems to be enriched by the possibility of

quantification. One rereads familiar passages with fresh excitement over their experimental possibilities. Well-worn phrases like "perceptual organization," "the disorganizing effects of emotion," "knowledge of results," "stereotyped behavior," "reorganization of the problem materials," etc., being to leap off the pages.

In the first blush of enthusiasm for this new toy it is easy to overstate the case. When Newton's mechanics was flowering, the claim was made that animals are nothing but machines, similar to but more complicated than a good clock. Later, during the development of thermodynamics, it was claimed that animals are nothing but complicated heat engines. With the development of information theory we can expect to hear that animals are nothing but communication systems. If we profit from history, we can mistrust the "nothing but" in this claim. But we will also remember that anatomists learned from mechanics and physiologists profited by thermodynamics. Insofar as living organisms perform the functions of a communication system, they must obey the laws that govern all such systems. How much psychology will profit from this obedience remains for the future to show.

Most of the careless claims for the importance of information theory arise from overly free associations to the word "information." This term occurs in the theory in a careful and particular way. It is not synonymous with "meaning." Only the *amount* of information is measured—the amount does not specify the content, value, truthfulness, exclusiveness, history, or purpose of the information. The definition does not exclude other definitions and certainly does not include all the meanings implied by the colloquial usages

Reprinted from *American Psychologist*, 8 (1963), 3–11, with permission of the author and publisher.

of the word. This garland of "nots" covers most of the objectionable exaggerations. In order to demonstrate some properly constrained associations to the word "information," we need to begin with definitions of some basic concepts.

Basic Concepts

Amount of information. A certain event is going to occur. You know all the different ways this event can happen. You even know how probable each of these different outcomes is. In fact, you know everything about this event that can be learned by watching innumerable similar events in the past. The only thing you don't know is exactly which one of these outcomes will actually happen.

Imagine a child who is told that a piece of candy is under one of 16 boxes. If he lifts the right box, he can have the candy. The event— lifting one of the boxes—has 16 possible outcomes. In order to pick the right box, the child needs information. Anything we tell him that reduces the number of boxes from which he must choose will provide some of the information he needs. If we say, "The candy is not under the red box," we give him just enough information to reduce the number of alternatives from 16 to 15. If we say, "The candy is under one of the four boxes on the left end," we give more information because we reduce 16 to 4 alternatives. If we say, "The candy is under the white box," we give him all the information he needs—we reduce the 16 alternatives to the one he wants.

The amount of information in such statements is a measure of how much they reduce the number of possible outcomes. Nothing is said about whether the information is true, valuable, understood, or believed—we are talking only about *how much* information there is.

Bit. A perfectly good way to measure the amount of information in such statements (but not the way we will adopt) is merely to count the number of possible outcomes that the information eliminates. Then the rule would be that every time one alternative is eliminated, one unit of information is communicated.

The objection to this unit of measurement is intuitive. Most people feel that to reduce 100 alternatives to 99 is less helpful than to reduce two alternatives to one. It is intuitively more attractive to use ratios. The amount of information depends upon the fraction of the alternatives that are eliminated, not the absolute number. In order to convey the same amount of information, the 100 alternatives should be reduced by the same fraction as the two alternatives, that is to say, from 100 to 50.

Every time the number of alternatives is reduced to half, one unit of information is gained. This unit is called one "bit" of information. If one message reduces k to k/x, it contains one bit less information than does a message that reduces k to $k/2x$. Therefore, the amount of information in a message that reduces k to k/x is $\log_2 x$ bits.

For example, if the child's 16 boxes are reduced to two, then x is 8 and $\log_2 8$ is three bits of information. That is to say, 16 has been halved three times: 16 to 8, 8 to 4, and 4 to 2 alternative outcomes.

Source. The communication engineer is seldom concerned with a particular message. He must provide a channel capable of transmitting any message that a source may generate. The source selects a message out of a set of k alternative messages that it might send. Thus each time the source selects a message, the channel must transmit $\log_2 k$ bits of information in order to tell the receiver what choice was made.

If some messages are more probable than the others, a receiver can anticipate them and less information needs to be transmitted. In other words, the frequent messages should be the short ones. In order to take account of differences in probability, we treat a message whose probability is p as if it was selected from a set of $1/p$ alternative messages. The amount of information that must be transmitted for this message is, therefore, $\log_2 1/p$, or $-\log_2 p$. (Note that if all k messages are equally probable, $p = 1/k$ and $-\log_2 p = \log_2 k$, which is the measure given above.) In other words, some messages that the source selects involve more information than others. If the message probabilities are $p_1, p_2, \ldots, p_k$, then the amounts of information associated with each message are $-\log_2 p_1, -\log_2 p_2, \ldots, -\log_2 p_k$.

Average amount of information. Since we want to deal with sources, rather than with particular messages, we need a measure to represent how much information a source generates. If different messages contain different amounts of information, then it is reasonable to talk about the average amount of information per message we can expect to get from the source—the average for all the different messages the source may select. This expected value from source x is denoted $H(x)$:

$$H(x) = \text{the mean value of } (-\log_2 p_i)$$

$$= \sum_{i=1}^{k} p_i(-\log_2 p_i)$$

This is the equation that occurs most often in the psychological applications of information theory. $H(x)$ in bits per message is the mean logarithmic probability for all messages from source x: In all that follows we shall be talking about the average amount of information expected from a source, and not the exact amount in any particular message.

Related sources. Three gentlemen—call them Ecks, Wye, and Zee—are each making binary choices. That is to say, Ecks chooses either heads or tails and simultaneously Wye also makes a choice and so does Zee. They repeat their synchronous choosing over and over again, varying their choices more or less randomly on successive trials. Our job is to predict what the outcome of this triple-choice event will be.

With no more description than this we know that there are eight ways the triple-choice can come out: HHH, HHT, HTH, HTT, THH, THT, TTH, and TTT. Thus our job is to select one out of these eight possible outcomes. If all eight were equally probable, we would need three bits of information to make the decision.

Now suppose that Ecks tells us each time what his next choice is going to be. With Ecks out of the way we are left with only four combinations of double-choices by Wye and Zee, so we can gain one bit of information about the triple-choice from Ecks. Similarly, if Wye tells us what his choice is going to be, that can also be worth one bit of information. Now the question is this: If Ecks and Wye both tell us what they are going to do, how much information do we get?

Case I: Suppose that it turns out that Ecks and Wye are perfectly correlated. In other words, if we know what Ecks will do, we also know what Wye will do, and vice versa. Given the information from either one of them, the other one has no further information to add. Thus the most we can get from both is exactly the same as what we would get from either one alone. Note that if Ecks and Wye always make the same choice, there are actually only four possible outcomes: HHH, HHT, TTH, and TTT, so we need only two bits to select the outcome.

Case II: Next, suppose that Ecks and Wye make their choices with complete independence. Then a knowledge of Ecks' choice tells us absolutely nothing about what Wye is going to do, and vice versa. None of the information from one is duplicated by the other. Thus, if we get one bit from Ecks and one bit from Wye, and if there is no common information at all, we must get two whole bits of information from both of them together.

Case III: Finally, suppose that, as will usually be the case when we apply these ideas, Ecks and Wye are partially but not perfectly correlated. If we know what Ecks will do, we can make a fairly reliable guess what Wye will do, and vice versa. Some but not all of the information we get from Ecks duplicates the information we get from Wye. This case falls in between the first two: the total information is greater than either of its parts, but less than their sum.

The situation in Case III is pictured in Figure 1. The left circle is the information we get from Ecks and the right circle is the information from Wye. The symbols $H(x)$ and $H(y)$ denote the average amounts of information in bits per event expected from sources Ecks and Wye respectively. The overlap of the two circles represents the common information due to the correlation of Ecks and Wye and its average amount in bits per event is symbolized by T. The left half of the left circle is information from Ecks alone, and the right half of the right circle is information from Wye alone. The symbols $H_y(x)$ should be taken to mean the average amount of information per event that remains to be gotten from source Ecks after Wye is already known. The total area enclosed in both circles together represents all the information that both Ecks and Wye can provide. This total amount in bits per event is symbolized by $H(x,y)$.

$H(x)$ is calculated from the probabilities for Ecks' choices according to the equation given above. The same equation is used to calculate $H(y)$ from the probabilities for Wye's choices. And the same equation is used a third time to calculate $H(x,y)$ from the joint probabilities of the double-choices by Ecks and Wye together. Then all the other quantities involved can be calculated by simple arithmetic is just the way Figure 1 would suggest. For example:

$$H_y(x) = H(x,y) - H(y)$$
$$\text{or} \qquad T = H(x) + H(y) - H(x,y).$$

It will be seen that T has the properties of a measure of the correlation (contingency, dependence) between Ecks and Wye. In fact, $1.3863\ nT$ (where n is the number of occurrences of the event that you use to estimate the probabilities involved) is essentially the same as the value of chi square you would compute to test the null hypothesis that Ecks and Wye are independent.

These are the basic ideas behind the general theory. There are many ways to adapt them to specific situations depending on the way the elements of the specific situation are identified

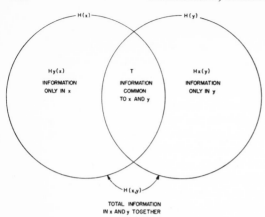

Figure 1.—Schematic representation of the several quantities of information that are involved when messages are received from two related sources.

with the several variables of the theory. In general, however, most applications of the theory seem to fall into one or the other of two types. I shall refer to these two as the *transmission situation* and the *sequential situation*.

The Transmission Situation

When information is communicated from one place to another, it is necessary to have a channel over which it can travel. If you put a message in at one end of the channel, another message comes out the other end. So the communication engineer talks about the "input" to the channel and the "output" from the channel. For a good channel, the input and the output are closely related but usually not identical. The input is changed, more or less, in the process of transmission. If the changes are random, the communication engineer talks about "noise" in the channel. Thus the output depends upon both the input and the noise.

Now we want to identify the variables in this transmission situation with the various quantities of information pictured in Figure 1. In order to do this, we let x be the source that generates the input information and let y be the source that generates the output information. That is to say, y is the channel itself. Since x and y are related sources of information, the overlap or common information is what is transmitted. $H(x)$ is the average amount of input information, $H(y)$ is the average amount of output information, and T is the average amount of transmitted information. (To keep terms uniform, we might refer to T as the average amount of "throughput" information.)

What interpretation can we give to $H_y(x)$ and $H_x(y)$? $H_y(x)$ is information that is put in but not gotten out—it is information *lost* in transmission. $H_y(x)$ is often called "equivocation" because a receiver cannot decide whether or not it was sent. Similarly, $H_x(y)$ is information that comes out without being put in—it is information *added* in transmission. $H_x(y)$ is called "noise" with the idea that the irrelevant parts of the output interfere with good communications.

Finally, $H(x,y)$ is the total amount of information you have when you know both the input and the output. Thus $H(x,y)$ includes the lost, the transmitted, and the added information,

$$H(x,y) = H_y(x) + T + H_x(y),$$

equivocation plus transmission plus noise.

This interpretation of the basic concepts of information theory is ordinarily used with the object of computing T, the amount of information transmitted by the channel. A characteristic of most communication channels is that there is an upper limit to the amount of information they can transmit. This upper limit is called the "channel capacity" and is symbolized by C. As the amount of information in the input is increased, there comes a point at which the amount of transmitted information no longer increases. Thus as $H(x)$ increases, T approaches an upper limit, C. This situation is shown graphically in Figure 2, where T is plotted as a function of $H(x)$.

The obvious psychological analogy to the transmission situation is between the subject in an

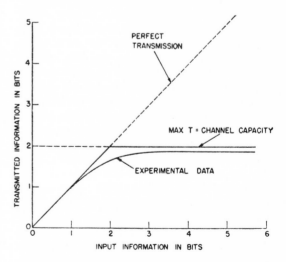

Figure 2.—Illustrative graph showing the amount of transmitted information as a function of the amount of input information for a system with a channel capacity of 2 bits.

experiment and a communication channel, between stimuli and inputs, and between responses and outputs. Then $H(x)$ is the stimulus information, $H(y)$ is the response information, and T measures the degree of dependence of responses upon stimuli. It turns out that T can be considered as a measure of discrimination, and C is the basic capacity of the subject to discriminate among the given stimuli. That is to say, C can be interpreted as a sort of modern version of the traditional Weber-fraction.

In order to explain how T and C measure the discriminative abilities of the subject, a simple example is useful. Imagine a subject can discriminate perfectly among four classes of stimuli. Any two stimuli in the same class are indistinguishable to him, but two stimuli from different classes are never confused. If we pick the stimuli carefully from different classes, therefore, he can distinguish perfectly which one of two, of three, or of four alternative stimuli we present. However, there is no way we can pick five or more stimuli so that he can discriminate them without any mistakes; at least two must be from the same class and so will be confused. If we select k stimuli to test him with, the best he can do is to reduce k to $k/4$ by saying which of his four classes each stimulus belongs in. He can never reduce the range of possible inputs to less than $k/4$. Thus his channel capacity, C, is $\log_2 4$, or 2 bits, and this is the maximum value of T we can get from him. Since 2^C is the maximum number of discriminably different classes of stimuli for this subject, C is a measure of his basic discriminative capacity.

Another psychological analogy to the transmission situation arises in mental testing. A test is a device for discriminating among people, with respect to some psychological dimension. Each person who takes the test has some true value on this dimension. The result of the test is a score that will, with more or less accuracy, tell us what this value is. So we can think of the test as a communication channel. The true values are the input information, and the test scores are the output information. If it is a good test, if T is large and the noise, $H_x(y)$, is small, then the test may discriminate rather accurately among the people who take it. In other words, 2^T would tell us how many classes of people we can distinguish by using this test.

It requires only a slight extension of this analogy to see the similarity between any process of measurement and the transmission situation. Nature provides the input, the process of measurement is the channel, and the measurements themselves are the output. In this context, the information of the communication engineer is quite similar to the information that R. A. Fisher defined many years ago and used as the foundation for his development of a theory of statistical inference. Considered in this sense, the possible applications of information theory are as broad as scientific measurement itself.

For the psychologist interested in the construction of scales of measurement, information theory will be a valuable tool. He will find that most of the things it tells him he could have learned just as well by more traditional statistical procedures, but the analogy to the transmission situation will undoubtedly stimulate insights and suggest new approaches to old problems.

The Sequential Situation

In all that has been said so far it has been implicitly assumed that successive occurrences of the event are independent. When we are dealing with behavioral processes, this assumption is never better than a first approximation. What we are going to do is conditioned by what we have just done, whether we are carrying out the day's work, writing a letter, or producing a random sequence of digits.

Although any behavioral sequence can be analyzed to discover its conditional probabilities, the most interesting example is our own verbal behavior. To take an obvious case, imagine that you are typing a letter and that you have just typed, "I hope we will see you again very." You need at least one more word to complete the sentence. You cannot open the dictionary at random to get this next word. The whole context of the sentence constrains your freedom of choice. The next word depends on the preceding words. Your most probable choice is "soon," although you might choose "often" or "much." You will certainly not choose "bluejay," or "the," or "take," etc. The effect of these constraints built into normal English usage is to reduce the number of alternatives from which successive words are chosen. We have already seen that when the number of possible outcomes of a choice is reduced, some information has been communicated. That is to say, by reducing the range of choice, the context gives us information about what the next item is going to be. Thus when the next word occurs, some of the information it conveys is identical with information we have already received from the context. This repeated information is called "redundancy."

How can the variables in this sequential situation be identified with the various quantities of information pictured in Figure 1? In order to relate them we let x be the source that generates the context and let y be the source that generates the next word. Since x and y are related sources of information, the overlap or common information from x and y is the redundancy. $H(x)$ is the average amount of information in the first $n-1$ words (the context), $H(y)$ is the average amount of information in the nth word, and T is the average amount of redundant information. $H_y(x)$ is the average amount of information in the context that is unrelated to the next word. $H_x(y)$ is the average amount of information in the next word that cannot be obtained from the context. $H(x,y)$ is the total amount of information we have when all n words, the context plus the next word, are known.

When this interpretation of the basic concepts is used, the quantity of major interest is ordinarily $H_x(y)$, the average amount of information per word when the context is known. $H_x(y)$ can be thought of as the additional information we can expect from each new word in the sequence. Thus $H_x(y)$ is closely related to the *rate* at which information is generated by the source; it measures the average number of bits per unit (per word).

If the successive units in a sequence are chosen independently, then the redundancy T is zero and the context tells us nothing about the next unit. If the next unit is completely determined by the context—for example, in English a "q" is always followed by "u"—then the new information $H_x(y)$ is zero and the occurrence of the next unit adds nothing to what we already know.

Sequences of letters in written English have been studied with this model. It has been estimated that a context of 100 letters will, on the average, reduce the effective number of choices for the next letter to less than three possibilities. That is to say, $H_x(y)$ is about 1.4 bits per letter in standard English. We can compare this result with what would happen if successive letters were chosen independently; then each letter would be chosen from 26 alternatives and would carry $\log_2 26$, or about 4.7 bits of information. In other words, we encode about one-fourth as much information per letter as we might if we used our alphabet more efficiently. Our books seem to be about four times as long as necessary.

It is reasonable to ask why we are so redundant. The answer lies in the fact that redundancy is an insurance against mistakes. The only way to catch an error is to repeat. Redundant information is an automatic mistake-catcher built into all natural languages. Of course, if there is no chance of error, then there is no need for redundancy. The large amount of redundancy that we seem to insist on reflects our basic inefficiency as information-handling systems. Compared with the thousands or millions of bits per second that electronic devices can handle, man's performance figures (always less than 50 bits per second and usually much lower if memory is involved) can charitably be called puny. By making our languages redundant we are able to decrease the rate, $H_x(y)$, to a point where we can cope with what is being said.

Knowledge of the redundancy of English is knowledge about our verbal habits. Since so much of man's behavior is conditioned by these verbal habits, any way to measure them should interest a psychologist. For example, a verbal learning experiment might compare the memorization of ten consonant–vowel–consonant nonsense syllables (30 letters in all) with the memorization of a 30-letter sentence from English text. Since the sucessive letters in the nonsense syllables are effectively independent, the learner faces many more possible sequences than he does if he knows that the 30 letters are English text. Since he has already learned the redundancies of English, he is required to assimilate less new information from the sentences than from the nonsense syllables. A knowledge of the information in sequences of letters in English text thus gives us an independent, quantitative estimate of previous learning. In short, the sequential application of information concepts enables us to calibrate our verbal learning materials and so to control in a quantitative way factors that we have always discussed before in qualitative terms.

It is not necessary to confine the sequential interpretation to verbal behavior. It can be applied whenever an organism adopts a reasonably stable "course of action" that can be described probabilistically. If the course of action is coherent in such a way that future conduct depends upon past conduct, we say the behavior is predictable or, to some degree, stereotyped. In such cases, the redundancy T can be used to measure the stereotypy. Arguments about the degree of organization in emotional behavior, for example, might be clarified by such a measure.

Taken together, the sequential and the transmission situations suggest a wide range of possible applications in psychology. The idea of reviewing some of the applications already made by psychologists is tempting, but space prevents it here. . . .

15.
Variety, Constraint, and the Law of Requisite Variety

W. ROSS ASHBY

A SUBJECT in which the concept of a *set* plays an essential part is that of "communication", especially in the theory developed by Shannon and Wiener. At first, when one thinks of, say, a telegram arriving, one notices only the singleness of *one* telegram. Nevertheless, the act of "communication" necessarily implies the existence of a *set* of possibilities, i.e. more than one, as the following example will show.

A prisoner is to be visited by his wife, who is not to be allowed to send him any message however simple. It is understood that they may have agreed, before his capture, on some simple code. At her visit, she asks to be allowed to send him a cup of coffee; assuming the beverage is not forbidden, how is the warder to ensure that no coded message is transmitted by it? He knows that she is anxious to let her husband know whether or not a confederate has yet been caught.

The warder will cogitate with reasonings that will go somewhat as follows: "She might have arranged to let him know by whether the coffee goes in sweetened or not—I can stop that simply by adding lots of sugar and then telling him I have done so. She might have arranged to let him know by whether or not she sends a spoon—I can stop that by taking away any spoon and then telling him that Regulations forbid a spoon anyway. She might do it by sending tea rather than coffee—no, that's stopped because, as they know, the canteen will only supply coffee at this time of day." So his cogitations go on; what is note-

worthy is that at each possibility he intuitively attempts to stop the communication by enforcing a reduction of the possibilities to one—always sweetened, never a spoon, coffee only, and so on. As soon as the possibilities shrink to one, so soon is communication blocked, and the beverage robbed of its power of transmitting information. The transmission (and storage) of information is thus essentially related to the existence of a *set* of possibilities. The example may make this statement plausible; in fact it is also supported by all the work in the modern theory of communication, which has shown abundantly how essential, and how fruitful, is the concept of the set of possibilities.

Communication thus necessarily demands a set of messages. Not only is this so, but the information carried by a particular message depends on the set it comes from. *The information conveyed is not an intrinsic property of the individual message.* That this is so can be seen by considering the following example. Two soldiers are taken prisoner by two enemy countries A and B, one by each; and their two wives later each receive the brief message "I am well". It is known, however, that country A allows the prisoner a choice from

> *I am well,*
> *I am slightly ill,*
> *I am seriously ill,*

while country B allows only the message

> *I am well*

meaning "I am alive". (Also in the set is the possibility of "no message".) The two wives will

From W. Ross Ashby, *An Introduction to Cybernetics* (London: Chapman and Hall, 1956), Chapter 7, pp. 123–134, and Chapter 11, pp. 202–209. Reprinted with permission of the author and publisher.

certainly be aware that though each has received the same phrase, the informations that they have received are by no means identical.

From these considerations it follows that, in this book, we must give up thinking, as we do as individuals, about "this message". We must become scientists, detach ourselves, and think about "people receiving messages". And this means that we must turn our attention from any individual message to the set of all the possibilities

Variety

Throughout this Part we shall be much concerned with the question, given a set, of how many distinguishable elements it contains. Thus, if the order of occurrence is ignored, the set

$$c, b, c, a, c, c, a, b, c, b, b, a$$

which contains twelve elements, contains only three *distinct* elements—*a*, *b* and *c*. Such a set will be said to have a *variety* of three elements. (A qualification is added in the next section.)

Though this counting may seem simple, care is needed. Thus the two-armed semaphore can place each arm, independently of the other, in any of eight positions; so the two arms provide 64 combinations. At a distance, however, the arms have no individuality—"arm *A* up and arm *B* down" cannot be distinguished from "arm *A* down and arm *B* up"—so to the distant observer only 36 positions can be distinguished, and the variety is 36, not 64. It will be noticed that a set's variety is not an intrinsic property of the set: the observer and his powers of discrimination may have to be specified if the variety is well defined.

The word *variety*, in relation to a set of distinguishable elements, will be used to mean either (i) the number of distinct elements, or (ii) the logarithm to the base 2 of the number, the context indicating the sense used. When variety is measured in the logarithmic form its unit is the "bit", a contraction of "BInary digiT". Thus the variety of the sexes is 1 bit, and the variety of the 52 playing cards is 5·7 bits, because $\log_2 52 = 3·322 \log_{10} 52 = 3·322 \times 1·7160 = 5·7$. The chief advantage of this way of reckoning is that multiplicative combinations now combine by simple addition.

To say that a set has "no" variety, that the elements are all of one type, is, of course, to measure the variety logarithmically; for the logarithm of 1 is 0.

Constraint

A most important concept, with which we shall be much concerned later, is that of *constraint*. It is a *relation* between two sets, and occurs when the variety that exists under one condition is less than the variety that exists under another. Thus, the variety in the human sexes is 1 bit; if a certain school takes only boys, the variety in the sexes within the school is zero; so as 0 is less than 1, constraint exists.

Another well-known example is given by the British traffic lights, which have three lamps and which go through the sequence (where "+" means lit and "0" unlit):

	(1)	(2)	(3)	(4)	(1)	
Red:	+	+	0	0	+	...
Yellow:	0	+	0	+	0	...
Green:	0	0	+	0	0	...

Four combinations are thus used. It will be noticed that Red is, at various times, both lit and unlit; so is Yellow; and so is Green. So if the three lights could vary independently, eight combinations could appear. In fact, only four are used; so as four is less than eight, constraint is present.

A constraint may be slight or severe. Suppose, for instance, that a squad of soldiers is to be drawn up in a single rank, and that "independence" means that they may stand in any order they please. Various constraints might be placed on the order of standing, and these constraints may differ in their degree of restriction. Thus, if the order were given that no man may stand next a man whose birthday falls on the same day, the constraint would be slight, for of all the possible arrangements few would be excluded. If, however, the order were given that no man was to stand at the left of a man who was taller than himself, the constraint would be severe; for it would, in fact, allow only one order of standing (unless two men were of exactly the same height). The intensity of the constraint is thus shown by the reduction it causes in the number of possible arrangements.

It seems that constraints cannot be classified in any simple way, for they include all cases in which a set, for any reason, is smaller than it might be. Here I can discuss only certain types of outstanding commonness and importance, leaving the reader to add further types if his special interests should lead him to them.

Constraint in vectors. Sometimes the elements of a set are vectors, and have components. Thus the traffic signal of S.7/8 was a vector of three components, each of which could take two values.

In such cases a common and important constraint occurs if the actual number of vectors that occurs under defined conditions is fewer than the total number of vectors possible without conditions (i.e. when each component takes its full range of values independently of the values taken by the other components). Thus, in the case of the traffic lights, when Red and Yellow are both lit, only Green unlit occurs, the vector with Green lit being absent.

It should be noticed that a set of vectors provides several varieties, which must be identified individually if confusion is not to occur. Consider, for instance, the vector:

(Age of car, Horse-power, Colour).

The first component will have some definite variety, and so will the second component, and the third. The three varieties need not be equal. And the variety in the set of vectors will be different again.

The variety in the set of vectors has, however, one invariable relation to the varieties of the components—it cannot exceed their sum (if we think in logarithms, as is more convenient here). Thus, if a car may have any one of 10 ages, of 8 horse-powers, and of 12 colours, then the variety in the types of car cannot exceed $3·3 + 3·0 + 3·6$ bits, i.e. 9·9 bits.

The components are *independent* when the variety in the whole of some given set of vectors equals the sum of the (logarithmic) varieties in the individual components. If it were found, for instance, that all 960 types of car could be observed within some defined set of cars, then the three components would be said to be "independent", or to "vary independently", within this defined set.

It should be noticed that such a statement refers essentially to what is observed to occur within the set; it need contain no reference to any supposed cause for the independence (or for the constraint).

Degrees of freedom. When a set of vectors does not show the full range of possibilities made available by the components, the range that remains can sometimes usefully be measured by saying how many components with independence would give the same variety. This number of components is called the *degrees of freedom* of the set of vectors. Thus the traffic lights show a variety of four. If the components continued to have two states apiece, *two* components with independence could give the same variety (of four). So the constraint on the lights can be expressed by saying that the three components,

not independent, give the same variety as *two* would if independent; i.e. the three lights have two degrees of freedom.

If all combinations are possible, then the number of degrees of freedom is equal to the number of components. If only one combination is possible, the degrees of freedom are zero.

It will be appreciated that this way of measuring what is left free of constraint is applicable only in certain favourable cases. Thus, were the traffic lights to show three, or five combinations, the equivalence would no longer be representable by a simple, whole, number. The concept is of importance chiefly when the components vary continuously, so that each has available an infinite number of values. A reckoning by degrees of freedom may then still be possible, though the states cannot be counted.

Importance of Constraint

Constraints are of high importance in cybernetics because *when a constraint exists advantage can usually be taken of it.*

Shannon's work displays this thesis clearly. Most of it is directed to estimating the variety that would exist if full independence occurred, showing that constraints (there called "redundancy") exist, and showing how their existence makes possible a more efficient use of the channel.

The next few sections will also show something of the wide applicability and great importance of the concept.

Laws of Nature. First we can notice that the existence of any invariant over a set of phenomena implies a constraint, for its existence implies that the full range of variety does not occur. The general theory of invariants is thus a part of the theory of constraints.

Further, as every law of nature implies the existence of an invariant, it follows that *every law of nature is a constraint.* Thus, the Newtonian law says that, of the vectors of planetary positions and velocities which might occur, e.g. written on paper (the larger set), only a smaller set will actually occur in the heavens; and the law specifies what values the elements will have. From our point of view, what is important is that the law *excludes* many positions and velocities, predicting that they will never be found to occur.

Science looks for laws; it is therefore much concerned with looking for constraints. (Here the larger set is composed of what *might* happen if the behaviour were free and chaotic, and the smaller set is composed of what does actually happen.)

Cybernetics looks at the totality, in all its possible richness, and then asks why the actualities should be restricted to some portion of the total possibilities.

Object as constraint. Constraints are exceedingly common in the world around us, and many of our basic concepts make use of it in an essential way. Consider as example the basic concept of a "thing" or "object", as something handled in daily life. A chair is a thing because it has coherence, because we can put it on this side of a table or that, because we can carry it around or sit on it. The chair is also a collection of parts.

Now any free object in our three dimensional world has six degrees of freedom for movement. Were the parts of the chair unconnected each would have its own six degrees of freedom; and this is in fact the amount of mobility available to the parts in the workshop before they are assembled. Thus the four legs, when separate, have 24 degrees of freedom. After they are joined, however, they have only the six degrees of freedom of the single object. That there *is* a constraint is obvious when one realises that if the positions of three legs of an assembled chair are known, then that of the fourth follows necessarily—it has no freedom.

Thus the change from four separate and free legs to one chair corresponds precisely to the change from the set's having 24 degrees of freedom to its having only 6. Thus the essence of the chair's being a "thing", a unity, rather than a collection of independent parts corresponds to the presence of the constraint.

Seen from this point of view, the world around us is extremely rich in constraints. We are so familiar with them that we take most of them for granted, and are often not even aware that they exist. To see what the world would be like without its usual constraints we have to turn to fairy tales or to a "crazy" film, and even these remove only a fraction of all the constraints.

A world without constraints would be totally chaotic. The turbulent river below Niagara might be such a world (though the physicist would still find some constraint here). Our terrestrial world, to which the living organism is adapted, is far from presenting such a chaos. Later it will be suggested that the organism can adapt just so far as the real world is constrained, and no further.

Prediction and constraint. That something is "predictable" implies that there exists a constraint. If an aircraft, for instance, were able to move,

second by second, from any one point in the sky to any other point, then the best anti-aircraft prediction would be helpless and useless. The latter can give useful information only because an aircraft cannot so move, but must move subject to several constraints. There is that due to continuity—an aircraft cannot suddenly jump, either in position or speed or direction. There is the constraint due to the aircraft's individuality of design, which makes this aircraft behave like an A-10 and that one behave like a Z-20. There is the constraint due to the pilot's individuality; and so on. An aircraft's future position is thus always somewhat constrained, and it is to just this extent that a predictor can be useful.

Machine as constraint. It will now be appreciated that the concept of a "machine", as developed from the inspection of a protocol, comes from recognizing that *the sequence in the protocol shows a particular form of constraint.* Were the protocol to show no constraint, the observer would say it was chaotic or unpredictable, like a roulette-wheel.

When it shows the characteristic form of constraint, the observer can take advantage of the fact. He does this by re-coding the whole protocol into a more compact form, containing only:

(i) a statement of the transformation
and (ii) a statement of the actual input given.

Subsequently, instead of the discussion being conducted in terms of a lengthy protocol, it is conducted compactly in terms of a succinct transformation.

Thus, use of the transformation is one example of how one can turn to advantage the characteristic constraint on behaviour imposed by its being "machine-like".

Within the set of determinate machines further constraints may be applied. Thus the set can be restricted to those that have a certain set of states as operands, or to those that have only one basin, or to those that are not reducible.

A common and very powerful constraint is that of continuity. It is a constraint because whereas the function that changes arbitrarily can undergo *any* change, the continuous function can change, at each step, only to a neighbouring value.

Learning and constraint. For the psychologist, an important example of constraint occurs in learning. Pavlov, for instance, in one experiment gave both thermal and tactile stimuli, as well as reinforcement by meat powder, in the following combinations:

	Thermal	Tactile	Reinforcement
1	+	+	+
2	+	−	−
3	−	+	+
4	−	−	−

(The fourth combination occurred, of course, in the intervals.) Now the total combinations possible are eight; Pavlov presented only four. It was an essential part of the experiment that the full set should not be given, for otherwise there would be nothing particular for the animal to learn. Constraint was an essential feature of the experiment.

The same principle can be seen more simply in learning by association. Suppose one wanted the subject, given a letter, to reply with a number according to the rule

A given: reply with 2
B   „   :   „   „ 5
C   „   :   „   „ 3

The subject might then be given a sequence such as A2, B5, C3, B5, C3, A2, A2, C3, and so on.

Now this sequence, as a sequence of vectors with two components, shows constraint; and if learning is to occur the constraint is necessary; for without constraint A would be followed equally by 2, 3 or 5; and the subject would be unable to form any specific associations. Thus *learning is possible only to the extent that the sequence shows constraint.*

The same is true of learning a maze. For this to occur the maze must retain the same pattern from day to day during the period of learning. Were the maze to show no constraint, the animal would be unable to develop the particular (and appropriate) way of behaving. Thus, *learning is worth while only when the environment shows constraint.*

.   .   .   .   .

Requisite Variety

In [the following] we shall examine the process of regulation itself, with the aim of finding out exactly what is involved and implied. In particular we shall develop ways of *measuring* the amount or degree of regulation achieved, and we shall show that this amount has an upper limit.

The subject of regulation is very wide in its applications, covering as it does most of the activities in physiology, sociology, ecology, economics, and much of the activities in almost every branch of science and life. Further, the types of regulator that exist are almost bewildering in their variety. One way of treating the subject would be to deal seriatim with the various types; . . . [here], however, we shall be attempting to get at the core of the subject—to find what is common to all.

What is common to all regulators, however, is not, at first sight, much like any particular form. We will therefore start anew in the next section, making no explicit reference to what has gone before. Only after the new subject has been sufficiently developed will we begin to consider any relation it may have to regulation.

Play and outcome. Let us therefore forget all about regulation and simply suppose that we are watching two players, R and D, who are engaged in a game. We shall follow the fortunes of R, who is attempting to score an a. The rules are as follows. They have before them Table 11/3/1, which can be seen by both:

Table 11/3/1

		R		
		α	β	γ
	1	b	a	c
D	2	a	c	b
	3	c	b	a

D must play first, by selecting a number, and thus a particular row. R, knowing this number, then selects a Greek letter, and thus a particular column. The italic letter specified by the intersection of the row and column is the *outcome*. If it is an a, R wins; if not, R loses.

Examination of the table soon shows that with this particular table R can win always. Whatever value D selects first, R can always select a Greek letter that will give the desired outcome. Thus if D selects 1, R selects β; if D selects 2, R selects α; and so on. In fact, if R acts according to the transformation

$$\begin{array}{ccc} 1 & 2 & 3 \\ \downarrow & & \\ \beta & \alpha & \gamma \end{array}$$

then he can always force the outcome to be a.

R's position, with this particular table, is peculiarly favourable, for not only can R always force a as the outcome, but he can as readily force, if desired, b and c as the outcome. R has, in fact, complete control of the outcome.

The Table used above is, of course, peculiarly favourable to *R*. Other Tables are, however, possible. Thus, suppose *D* and *R*, playing on the same rules, are now given Table 11/4/1 in which *D* now has a choice of five, and *R* a choice of four moves.

If *a* is the target, *R* can always win. In fact, if *D* selects 3, *R* has several ways of winning. As every row has at least one *a*, *R* can always force the appearance of *a* as the outcome. On the other hand, if the target is *b* he cannot always win. For if *D* selects 3, there is no move by *R* that will give *b* as the outcome. And if the target is *c*, *R* is quite helpless, for *D* wins always.

It will be seen that different arrangements within the table, and different numbers of states available to *D* and *R*, can give rise to a variety of situations from the point of view of *R*.

Table 11/4/1

		R			
		α	β	γ	δ
	1	*b*	*d*	*a*	*a*
	2	*a*	*d*	*a*	*d*
D	3	*d*	*a*	*a*	*a*
	4	*d*	*b*	*a*	*b*
	5	*d*	*a*	*b*	*d*

Can any *general* statement be made about *R*'s modes of play and prospects of success?

If full generality is allowed in the Table, the possibilities are so many, arbitrary and complicated that little can be said. There is one type, however, that allows a precise statement and is at the same time sufficiently general to be of interest. (It is also fundamental in the theory of regulation.)

From all possible tables let us eliminate those that make *R*'s game too easy to be of interest. . . . If a column contains repetitions, *R*'s play need not be discriminating; that is, *R* need not change his move with each change of *D*'s move. Let us consider, then, only those tables in which *no column contains a repeated outcome*. When this is so *R* must select his move on *full* knowledge of *D*'s move; i.e. any change of *D*'s move must require a change on *R*'s part. (Nothing is assumed here about how the outcomes in one column are related to those in another, so these relations are unrestricted.) Such a Table is 11/5/1. Now, some target being given, let *R* specify what his move will be for each move by *D*. What is essential is that, win or lose, he must specify one and only

Table 11/5/1

		R		
		α	β	γ
	1	*f*	*f*	*k*
	2	*k*	*e*	*f*
	3	*m*	*k*	*a*
	4	*b*	*b*	*b*
D	5	*c*	*q*	*c*
	6	*h*	*h*	*m*
	7	*j*	*d*	*d*
	8	*a*	*p*	*j*
	9	*l*	*n*	*h*

one move in response to each possible move of *D*. His specification, or "strategy" as it might be called, might appear:

If *D* selects 1, I shall select γ
„ „ „ 2, „ „ „ α
„ „ „ 3, „ „ „ β
· · ·   · · ·
„ „ „ 9, „ „ „ α

He is, of course, specifying a transformation (which must be single-valued, as *R* may not make two moves simultaneously):

$$1 \quad 2 \quad 3 \ldots 9$$
$$\downarrow$$
$$γ \quad α \quad β \ldots α$$

This transformation uniquely specifies a set of outcomes—those that will actually occur if *D*, over a sequence of plays, includes every possible move at least once. For 1 and γ give the outcome *k*, and so on, leading to the transformation:

$$(1,γ) \quad (2,α) \quad (3,β) \quad \ldots \quad (9,α)$$
$$\downarrow$$
$$k \quad \quad k \quad \quad k \quad \ldots \quad l$$

It can now be stated that the variety in this set of outcomes cannot be less than

$$\frac{D\text{'s variety}}{R\text{'s variety}}$$

i.e., in this case, 9/3.

It is easily proved. Suppose *R* marks one element in each row and concentrates simply on keeping the variety of the marked elements as small as possible (ignoring for the moment any idea of a target). He marks an element in the first row. In the second row he must change to a new column if he is not to increase the variety by adding a new, different, element; for in the initially selected column the elements are all different, by hypothesis. To keep the variety down to one element he must change to a new column at each row. (This is the *best* he can do; it may be that change from column to column is not sufficient to keep the variety down to one element, but this is irrelevant, for we are interested only in what is the least possible variety, assuming that everything falls as favourably as possible.) So if R has n moves available (three in the example), at the n-th row all the columns are used, so one of the columns must be used again for the next row, and a new outcome *must* be allowed into the set of outcomes. Thus in Table 11/5/1, selection of the k's in the first three rows will enable the variety to be kept to one element, but at the fourth row a second element *must* be allowed into the set of outcomes.

In general: If no two elements in the same column are equal, and if a set of outcomes is selected by R, one from each row, and if the table has r rows and c columns, then *the variety in the selected set of outcomes cannot be fewer than r/c.*

The Law of Requisite Variety

We can now look at this game (still with the restriction that no element may be repeated in a column) from a slightly different point of view. If R's move is unvarying, so that he produces the same move, whatever D's move, then *the variety in the outcomes will be as large as the variety in D's moves.* D now is, as it were, exerting full control over the outcomes.

If next R uses, or has available, two moves, then the variety of the outcomes can be reduced to a half (but not lower). If R has three moves, it can be reduced to a third (but not lower); and so on. Thus if the variety in the outcomes is to be reduced to some assigned number, or assigned fraction of D's variety, R's variety *must* be increased to at least the appropriate minimum. *Only variety in R's moves can force down the variety in the outcomes.*

If the varieties are measured logarithmically (as is almost always convenient), and if the same conditions hold, then the theorem takes a very simple form. Let V_D be the variety of D, V_R that of R, and V_O that of the outcome (all measured logarithmically). Then the previous section has proved that V_O cannot be less, numerically, than the value of $V_D - V_R$. Thus V_O's minimum is $V_D - V_R$.

If V_D is given and fixed, $V_D - V_R$ can be lessened only by a corresponding increase in V_R. Thus *the variety in the outcomes,* if minimal, *can be decreased further only by a corresponding increase in that of R. . . .*

This is the law of Requisite Variety. To put it more picturesquely: *only variety in R can force down the variety due to D*; ONLY VARIETY CAN DESTROY VARIETY.

This thesis is so fundamental in the general theory of regulation that I shall give some further illustrations and proofs before turning to consider its actual application.

The law is of very general applicability, and by no means just a trivial outcome of the tabular form. To show that this is so, what is essentially the same theorem will be proved in the case when the variety is spread out in time and the fluctuation incessant—the case specially considered by Shannon. (The notation and concepts in this section are those of Shannon's book.)

Let D, R, and E be three variables, such that each is an information source, though "source" here is not to imply that they are acting independently. Without any regard for how they are related causally, a variety of entropies can be calculated, or measured empirically. There is $H(D,R,E)$, the entropy of the vector that has the three as components; there is $H_D(E)$, the uncertainty in E when D's state is known; there is $H_{ED}(R)$, the uncertainty in R when both E and D are known; and so on.

The condition introduced [above] (that no element shall occur twice in a column) here corresponds to the condition that if R is fixed, or given, the entropy of E (corresponding to that of the outcome) is not to be less than that of D, i.e.

$$H_R(E) \geqslant H_R(D).$$

Now whatever the causal or other relations between D, R and E, algebraic necessity requires that their entropies must be related so that

$$H(D) + H_D(R) = H(R) + H_R(D),$$

for each side of the equation equals $H(R,D)$. Substitute $H_R(E)$ for $H_R(D)$, and we get

$$H(D) + H_D(R) \leqslant H(R) + H_R(E)$$
$$\leqslant H(R,E).$$

But always, by algebraic necessity,

$$H(R,E) \leqslant H(R) + H(E)$$

so $\quad H(D) + H_D(R) \leqslant H(R) + H(E)$

i.e. $\qquad H(E) \geqslant H(D) + H_D(R) - H(R).$

Thus the entropy of the E's has a certain minimum. If this minimum is to be affected by a relation between the D- and R-sources, it can be made least when $H_D(R) = 0$, i.e. *when R is a determinate function of D.* When this is so, then $H(E)$'s minimum is $H(D) - H(R)$, a deduction similar to that of the previous section. It says simply that the minimal value of E's entropy can be forced down below that of D only by an equal *increase* in that of R.

The theorems just established can easily be modified to give a worth-while extension.

Consider the case when, even when R does nothing (i.e. produces the same move whatever D does) the variety of outcome is *less* than that of D. This is the case in Table 11/4/1. Thus if R gives the reply α to all D's moves, then the outcomes are a, b or d—a variety of three, less than D's variety of five. To get a manageable calculation, suppose that within each column each element is now repeated k times (instead of the "once only" of S.11/5). The same argument as before, modified in that kn rows may provide only one outcome, leads to the theorem that

$$V_o \geqslant V_D - \log k - V_R,$$

in which the varieties are measured logarithmically.

An exactly similar modification may be made to the theorem in terms of entropies, by supposing, not that

$$H_R(E) \geqslant H_R(D), \text{ but that}$$

$$H_R(E) \geqslant H_R(D) - K.$$

$H(E)$'s minimum then becomes

$$H(D) - K - H(R),$$

with a similar interpretation.

The law states that certain events are impossible. It is important that we should be clear as to the origin of the impossibility. Thus, what has the statement to fear from experiment?

It has nothing to do with the properties of matter. So if the law is stated in the form "No machine can . . .", it is not to be overthrown by the invention of some new device or some new electronic circuit, or the discovery of some new element. It does not even have anything to do with the properties of the machine in the general sense, for it comes from the *Table*; this Table says simply that certain D–R combinations lead to certain outcomes, but is quite independent of whatever it is that determines the outcome. Experiments can only *provide* such tables.

The theorem is primarily a statement about possible arrangements in a rectangular table. It says that certain types of arrangement cannot be made. It is thus no more dependent on special properties of machines than is, say, the "theorem" that four objects can be arranged to form a square while three can not. The law therefore owes nothing to experiment.

16.
The Promise and Pitfalls of Information Theory

ANATOL RAPOPORT

INFORMATION THEORY, repeatedly hailed as a "major breakthrough" came at a time when mathematical sciences had just become newsworthy—on the heels of nuclear fission. But while the impact of the suddenly publicized atomic physics has been on the entire population through the conventional channels of popular fancy—the awesome possibilities for good and evil of vast amounts of suddenly harnessed energy—the appeal of information theory has been largely confined to academic circles. Yet the nature of this appeal has been not unlike the other. It is felt that a new conceptualization has crystallized which will lead to momentous theoretical reconstructions in biology, psychology, and the social sciences. A vast proliferation of papers devoted to attempts to exploit the new conceptualization reflects this feeling. A 1953 bibliography of information theory [F. L. Stumpers'] contains some 800 entries. Writings on the subject range from sophisticated analysis of radar systems and television circuits to crackpot speculations. Usually, there is a well-marked negative correlation between the scope and the soundness of the writings, so that it is not hard to understand the disappointment with "information theory" as a broad conceptual tool: the sound work is confined either to engineering or to rather trivial applications (rote learning, etc.); ambitious formulations remain vague.

The creators of information theory could, of course, point out that their discipline was conceived in the context of communication engineering, computing machinery, and automation, that they did not coin the phrase, "Life feeds on negative entropy;" that the parallel between entropy and information has been noted merely as a formal identity of mathematical expressions, and that those who wish to speculate about the effects of the Second Law of Thermodynamics on a "closed society" and similar matters should do so at their own risk, etc.

Yet the challenge of extending the concepts of information theory remains, and, as a matter of fact, is traceable to the writings of its founders.[1] It seems useful, therefore, to try to see what the problems of extension and generalization actually are, that is, to try to state the issues as "problèmes bien posés."

Given a set of configurations of any system with associated independent probabilities of occurrence p_i, the *uncertainty* of the set is defined in information theory as

$$H = -\sum p_i \log_2 p_i \qquad [1]$$

Thus if one selects a message from a source of n messages, each selection is a "configuration" characterized by a certain probability. Then H is the uncertainty (per message) associated with the source. The receipt of the message transmitted without error "destroys" the uncertainty of the recipient, with regard to which message will be chosen. Therefore H measures also the amount of information per message.

Now in statistical mechanics, the thermodynamic quantity *entropy* also appears as the same expression, where now the set of configurations comprises all the arrangements of positions and momenta of the particles comprising a

From Anatol Rapoport, "The Promise and Pitfalls of Information Theory," *Behavioral Science*, I (1956), 303–309. Reprinted by permission of the author and publisher.

system that correspond to the thermodynamic *state* of the system. It turns out that states described in thermodynamics as *equilibrium states* have corresponding to them the greatest number of such configurations and so are "most probable." The trend toward equilibrium observed in all systems "left to themselves" is interpreted in statistical mechanics as the predominance of occurrence of the more probable states over the less probable.

A homely example illustrates the principle. In a well-shuffled deck of cards the number of red cards among the first 26 cards will be not too far off from 13. At any rate with continued shuffling, numbers around 13 for reds among the first will appear far more frequently than numbers around 26 or 0. This is so because many more arrangements of the deck are possible where the reds are evenly distributed than when they are all in one half. It is justifiable to say that the "equilibrium condition" (where a red card has as much chance of passing from the first half of the deck to the second as vice versa) is the condition of the greatest "entropy" of the deck.

The condition "if the system is left to itself" is crucial. Nothing is easier than to lower the "entropy" of a deck of cards by deliberate re-arrangements. Once thermodynamic entropy was recognized as a measure of "disorder" in a system it was natural to raise the question of whether the Second Law could be circumvented by the intervention of an "intelligence." The seed of the idea relating information to entropy was sown. The idea first appears in the writings of none other than James Clerk Maxwell himself, one of the founders of statistical mechanics. He writes in *Theory of Heat* ". . . a being whose faculties are so sharpened that he can follow every molecule in his course . . . would be able to do what is at present impossible to us. . . . Let us suppose that a vessel is divided into two portions *A* and *B* by a division in which there is a small hole, and that a being who can see the individual molecules opens and closes this hole, so as to allow only the swifter molecules to pass from *A* to *B*, and only the slower ones to pass from *B* to *A*. He will, thus, without expenditure of work raise the temperature of *B* and lower that of *A*, in contradiction to the second law of thermo-dynamics."

Thus was born Maxwell's Demon, possibly a descendant of the Homunculus to challenge by means of his phenomenal acuity and infinitesimal reaction time the inexorable Second Law.

Maxwell's Demon, or Maxie, as we will call

him for short, now 85 years old, is still a subject of serious theoretical discussion. A paper of historical significance for information theory was devoted entirely to him by L. Szilard. Szilard has shown that in lowering the entropy of a gas by his "decisions" Maxie causes an increase of entropy *elsewhere* which, as calculations show, more than compensates for the decrease and thus vindicates the dictum of the Second Law that the entropy of a *closed* system (which in this case must include Maxie and his apparatus) can only increase. More recently, L. Brillouin emphasized Maxwell's assumption that Maxie *can see* the molecules. Thus he cannot be sitting in the total darkness of the enclosed gas (even if the gas is hot and radiates light, it doesn't help him to distinguish the molecules if the whole system is in equilibrium). To see the molecules Maxie must use a flashlight and thus, as, Brillouin shows, feeds "negative entropy" in the form of a light beam into the system. It is this negative entropy which is more than an ample source to account for the resulting decreased entropy of the gas alone.

Thus Maxie is shown to be cheating like any charlatan exhibiting a perpetual motion machine. These discussions are interesting, because they may lead to insight concerning the way enzymes work. It is the enzymes which are credited with the "local" contraventions of the Second Law which seem to be operating in living systems. On the other hand, it is known that eventually all enzymes become "denatured" and lose their function as "organizers" of living processes. ("For an enzyme to be in equilibrium is to be de-natured."—Wiener.) It has been remarked that besides counting the rise in entropy in the gadgets which Maxie must have for his entropy-lowering activity, one should take into account the rise in entropy within Maxie himself, perhaps the de-naturation of his enzymes reflecting the "entropic costs" of his decisions. This, in turn, leads to speculations concerning similar processes in all "organizing" activity, the perfection of tech-nology, accumulation of scientific knowledge, etc. Are we becoming "denatured" to the extent that we "order" the world to suit our needs? We mention this only to illustrate how easy it is to embark on metaphysical and even theological speculations once one becomes seriously con-cerned with the Second Law.

So much concerning the possible exploitation of the conjectured equivalence of information and physical entropy. Let us turn to the speculations revolving around the concept of information as a tool for quantifying the "amount of knowledge."

Here, even more than in matters dealing with physical entropy the scope of attempted generalizations far outruns available tools of analysis. This is not surprising, since "knowledge" evokes strong intuitively felt similarity to information and since knowledge has been traditionally the subject matter of philosophic speculation. Accordingly, one finds a wealth of discussion about fighting the inexorable Second Law by increasing our knowledge (knowledge = information = negative entropy), about the inevitable collapse of totalitarian societies (totalitarian society = closed society = closed system, in which entropy is bound to increase) and similar matters much removed from the theory originally designed to inquire into conditions which make for efficient transmission of signals over a channel.

We must disavow in passing any suggestion that we consider such speculation as entirely idle. After all, the Copernican revolution had a profound effect much beyond astronomy, and the law of conservation of matter has implications much more far reaching than making for correct bookkeeping in the chemist's laboratory. However, the profound effect of celestial mechanics and the conservation laws on our world view operated, we think it is safe to say, through the resulting success of scientific method, and the latter depends for its operation on precise definitions of situations one deals with. Therefore it seems to us that the search for applications of information theory to communication in a broader sense than it is understood by those who think in terms of wires and tubes (i.e., broadly to problems of human communication), while commendable and attractive, must be tempered by a sober attempt to keep the universe of discourse well defined.

A difficult transition is from the concept of information in the technical (communication engineering) sense to the semantic (theory of meaning) sense. This transition has been undertaken by Carnap, Bar-Hillel, and other workers. The difficulties involved become immediately apparent. Suppose we ask how much information is contained in an average English word. Has the question any meaning? This depends on to whom it is addressed. The telegraph engineer has a most definitive answer. He knows that the redundancy of English reduces the amount of information per letter from about 5 to 2 bits, and since the "average" word has about 5 letters, this makes 10 bits the information content of the average word. The telephone engineer's or the linguist's answer requires a more intricate analysis in terms of phonemes, distinctive features, etc., but his answer can, *in principle*, be equally direct.

Now what does 10 bits per word mean? It should mean that if I think of an English word and you try to guess it, you should be able to do it in about 10 guesses requiring yes or no answers. If you choose your guesses judiciously, you should be able to do it simply guessing letter by letter. But suppose letter guessing is disallowed. You are required to guess, as in the game of Twenty Questions, by narrowing down logical categories. As people who play Twenty Questions know, extremely bizarre notions can be guessed within that limit; so it is possible that an experienced player can guess a single word in ten tries. It is remarkable, however, that while it would be rather simple to program a logical computer so as to guess any word letter by letter, to build a machine that would do so by "logical categories" would involve building into the machine the entire encyclopedia of human knowledge! Moreover, any eight-year-old can be taught to guess any word letter by letter (even unfamiliar words, if the question, "Is that all?" is included). But if the word is not in his repertoire, no amount of advance instruction will enable an eight-year-old to guess it by "logical categories."

This illustrates a cardinal feature of information theory. It is fundamentally a theory of *selection*. Something is selected from a well-defined set. To examine the selective process, it is essential to be able to examine the set.

The modest but, we think, significant applications of information theory to certain psychological experiments owe their success to the fact that in each situation the set in question was strictly defined: a list of syllables to be memorized, associations to be formed, responses selected from, etc. There was, therefore, no difficulty in quantifying the associated "amounts of information" and relating such accounts to certain aspects of performance.

To many these results appear somewhat trivial. We do not agree. If information theory is to break out of its original habitat of bandwidths and modulations, a proper beginning must be made, which usually means a modest beginning. Experimental psychology has made this beginning possible. However, one must admit that the gap between this sort of experimentation and questions concerning the "flow of information" through human communication channels is enormous. So far no theory exists, to our knowledge, which attributes any sort of unambiguous measure to this "flow." It is naive to take simply the flux in signals per second to multiply by bits per signals in the communication engineering sense and call the result "amount of communication" in the sense of transmission of knowledge (labeling everything one does not like "noise"). If there is such a thing as semantic information, it is based on an entirely different kind of "repertoire," which itself may be different for each recipient. To be defined as a quantity of information a signal must be selected from a set or matched with an element of a set. It is misleading in a crucial sense to view "information" as something that can be poured into an empty vessel, like a fluid or even like energy.

Yet some meaning lurks in the expression "to acquire information"! We feel, however vaguely, that as a race we have learned certain things as a result of which we can more effectively "order the universe" to our liking. We have learned to reverse the degradation of energy locally by putting heat to work. We like to think that in increasing our average life span, we are challenging (again only locally) the trend toward chaos (all living things seem to do so, but we think we do it better). At this point we could be accused of the same promiscuous speculation we have implicitly warned against. We can only plead that such speculation is extremely difficult to avoid.

Let us try for a while to walk the tightrope between rigor and speculative abandon. It was probably the now famous remark of Schroedinger, "Life feeds on negative entropy," that has been responsible for speculations which are always tantalizing but which range from profound to irresponsible. A most prosaic interpretation is simply in terms of thermodynamic bookkeeping. The First Law having long been established, it is axiomatic that life feeds on energy: it converts chemical energy of ingested food and of sunlight into work necessary to pursue its activities. The energy equation must balance. However, with the advent of the Second Law in its classical, i.e., purely thermodynamic form, another bookkeeping appears necessary. The entropy inequality must hold. A living organism must ingest at least enough *free* energy, which contains a negative entropy term, to account for any local decrease of entropy concomitant to its *organizational* activity. All this would be perfectly clear if the "increment of organization" inherent in a process were as clearly defined as an increment of energy. One might argue as many do, that organization is simply the negative of entropy. True, entropy is unambiguously defined in classical thermodynamics, but this discipline confines itself only to

processes which take place under continued equilibrium (reversible processes). The classical definition, therefore, is inappropriate for living processes ("For an organism to be in equilibrium is to be dead."—Wiener). The advent of statistical mechanics with its re-definition of entropy as the probability of an aggregate of states does make possible the extension of the entropy concept to non-equilibrium states, indeed in information-theoretical terms. Hence the excitement. One feels it may be possible to calculate the "amount of organization" of configurations arising in the living process and thus establish the bookkeeping demanded by the Second Law to deal with those aspects of life, which, it is felt, are much more crucial than those covered by energetics bookkeeping (First Law).

One such attempt, to give an example, is due to R. C. Raymond. Raymond proposes to define the entropy of a configuration by the sum of two terms, a positive one, equal to the entropy the aggregate would finally attain if left to itself (i.e., its classical entropy) and a negative term equal to the amount of information necessary to reconstruct the original configuration from equilibrium (from complete chaos). This definition is attractive and indeed fits perfectly into the situation previously discussed where Maxie works with 100% efficiency pumping information (= negative entropy) into a system which starts with positive entropy only.

We suspect, however, that there are grave difficulties associated with this definition when one tries to extend it to situations other than the simplest (ideal gas, etc.). The difficulty is the same conceptual one which governs the applications of information theory to semantic problems. What are the elements of a configuration? And what is the set from which messages describing the construction of a configuration from chaos are selected? That is not to say that it is impossible to define these elements, but only that it appears that the definition must be made anew in each instance. How then is one to agree on a definition once for all so as to make the entropy of a configuration unambiguous?

Let us take an example. How much "order" is contained in a particular arrangement of a deck of cards? According to one view, since each arrangement is equiprobable its probability being $(52!)^{-1}$, it follows that $\log_2 (52!)$ yes–no decisions are required to reconstruct a particular order from a thoroughly shuffled deck. Now although $(52!)$ is prodigiously large ($\sim 10^{68}$) its logarithm is not (~ 200), and it is perfectly feasible to guess the arrangement in about 200 guesses, provided one is allowed to guess card by card. But suppose one is to guess by arrangements only? *We simply do not have the language* to classify the card arrangements so as to be able to make judicious dichotomies for yes–no answers. Our knowledge of card arrangements is confined to very few, which we call "ordered" (e.g., arrangements by suits and face values), and we tend to lump all the other arrangements as "meaningless" chaos. Similarly, a dictionary is useful only because its entries are arranged in a *particular* order meaningful to us. It is no consolation to be told that any other arrangement is equally improbable and therefore should be considered as having an equally low "entropy." One suspects that similar considerations apply to biology and make Raymond's definition of order less useful than it appears. Only *certain* arrangements maintain life. It is misleading therefore to measure the amount of order simply by the amount of talking one has to do to give instructions for its reconstruction or by how improbable the configuration in question is. All *particular* configurations are a priori equally improbable. As for the amount of information necessary for reconstruction of order, paradoxically the arrangements which we tend to consider ordered are usually reconstructible with *fewer* instructions. To reconstruct the ordered card arrangement, we need only say, "arrange in descending order of values the spades, clubs, diamonds, and hearts," whereas to reconstruct an arbitrary arrangement from another arbitrary one, we will generally have to name the position of each card separately! So in biological situations one might think certain arrangements have to be first *singled out* as the significant ones. Then one can hope to quantify the degree to which an arrangement approaches the "proper" one. But the discovery of "proper" arrangements is not simply a matter of counting; it is a matter of understanding the life process. Hence much more than understanding the theoretical tool of information theory is required for a proper application of the tool.

With regard to the other proposed extension of the entropy concept (as it appears in information theory), i.e., to the study of human communication (intertwined with semantic problems), the task is at least as difficult. To answer the question, "How much does a man know?" or "How much has he learned?" is possible in the sense of the present state of information theory only if one can specify a set from which selections must be made. Information is measured by the extent to which it makes uncertain selection certain. An attempt to define such repertoires of selection often leads to conclusions which seem, to say the least, unsatisfactory. Suppose, for example, a person does not know whether ghosts exist. Then he reads a book which presents weighty evidence against the existence of ghosts, as a result of which he becomes convinced that ghosts do not exist. How much information has he acquired? One can "reason" as follows. Prior to reading the book he was equally uncertain about which answer is proper, "Yes" or "No." After reading the book, he is certain of the answer "No." Therefore he has acquired one bit of information. We submit, the conclusion is unsatisfactory. (We will refrain from raising the question about the case where the man is convinced by reading a book that ghosts do exist.) Or take another example. A child learns that $2+2 = 4$. How much has he learned? The formal argument could run thus. If the child is acquainted with the names of numbers 1, 2, . . . n and was equally uncertain about which was the right answer, then he has acquired $\log_2 n$ bits of information. This conclusion seems to be equally ludicrous. These examples are given solely to illustrate what pitfalls await those who wish to extend the formal aspects of information theory to applications with more human interest. It is tempting to suggest that the American Revolution was triggered by one bit of information ("One if by land; two if by sea"). In this case the contention is not wholly unjustified: it refers to the simplicity of the arrangement for implementing one of two alternative plans of resistance, depending on the choice of the British commander to move by land or by sea against the insurgents. The simplicity of the arrangement was made possible by previous elaborate organization (of Minutemen, etc.) which it served to trigger. But such reductions of complicated situations to well-specified alternatives are rare.

Lest we seem to pause on a pessimistic note, let us reiterate our faith in the future of information theory. It is sometimes instructive to learn from historical analogy. A precursor of information theory, of which the latter is an elaborate extension, was probability theory, which started from humble beginnings as in intellectual schematization of what happens in games of chance, and which has grown to become one of the fundamental tools of exact science, whose logical foundations are a challenge to most serious philosophers. We note that attempts to extend probability theory to epistemological problems, for example, were also studded with pitfalls. No less a figure than Laplace, starting with a plausible theory of induction, came to rather ridiculous conclusions concerning the probability that the sun would rise on the following day. The axiomatization of probability theory is of recent origin and rests squarely on the definition of a "sample space," i.e., an explicit listing of the range of events over which the probability measure is to be defined. The notion of sample space seemed to have given a *coup de grâce* to the hope of extending exact probabilistic notions to measure a "degree of belief," a subjective concept. (What is the probability that the Democrats will win in 1956?) Yet there were those who refused to give up hope of successful axiomatization of "personal probability," so as to extend the strictly objectivistic view adhered to by the majority. As a result we have, for example, the highly interesting formulation of L. J. Savage in which this axiomatization is achieved, not without the use of rather sophisticated conceptual and mathematical machinery.

We believe that extensions of information theory and of the entropy concept, which is now riding on the coat tails of information theory, are entirely in order. We are afraid, however, that these extensions are no easy matter and cannot be achieved by "semantic suggestion," that is, a tendency to confuse invention of nomenclature with discovery of principles.

Notes

1. Perhaps a historical analogy is found in the proliferation of speculations concerning the broader meanings of the "uncertainty principle" in quantum theory. Actually many of these speculations are found in the writings of Bohr himself, one of the fathers of quantum physics.

A.

ENTROPY AND LIFE

Nec corpus mentem ad cogitandum nec mens corpus ad motum, neque ad quietem nec ad aliquid (si quid est) aliud determinare potest.[1]—SPINOZA, *Ethics*, P. III, Prop. 2.

17.
Order, Disorder, and Entropy

ERWIN SCHRÖDINGER

A Remarkable General Conclusion from the Model

Let me refer to the last phrase in [a preceding section], in which I tried to explain that the molecular picture of the gene made it at least conceivable 'that the miniature code should be in one-to-one correspondence with a highly complicated and specified plan of development and should somehow contain the means of putting it into operation'. Very well then, but how does it do this? How are we going to turn 'conceivability' into true understanding?

Delbrück's molecular model, in its complete generality, seems to contain no hint as to how the hereditary substance works. Indeed, I do not expect that any detailed information on this question is likely to come from physics in the near future. The advance is proceeding and will, I am sure, continue to do so, from biochemistry under the guidance of physiology and genetics.

No detailed information about the functioning of the genetical mechanism can emerge from a description of its structure so general as has been given above. That is obvious. But, strangely enough, there is just one general conclusion to be obtained from it, and that, I confess, was my only motive for writing this book.

From Delbrück's general picture of the hereditary substance it emerges that living matter, while not eluding the 'laws of physics' as established up to date, is likely to involve 'other laws of physics' hitherto unknown, which, however,

From Erwin Schrödinger, *What Is Life?* (Cambridge: Cambridge University Press, 1945), Chapter VI. Reprinted by permission of the publisher.

once they have been revealed, will form just as integral a part of this science as the former.

Order Based on Order

This is a rather subtle line of thought, open to misconception in more than one respect. All the remaining pages are concerned with making it clear. A preliminary insight, rough but not altogether erroneous, may be found in the following considerations:

It has been explained [earlier] that the laws of physics, as we know them, are statistical laws.[2] They have a lot to do with the natural tendency of things to go over into disorder.

But, to reconcile the high durability of the hereditary substance with its minute size, we had to evade the tendency to disorder by 'inventing the molecule', in fact, an unusually large molecule which has to be a masterpiece of highly differentiated order, safeguarded by the conjuring rod of quantum theory. The laws of chance are not invalidated by this 'invention', but their outcome is modified. The physicist is familiar with the fact that the classical laws of physics are modified by quantum theory, especially at low temperature. There are many instances of this. Life seems to be one of them, a particularly striking one. Life seems to be orderly and lawful behaviour of matter, not based exclusively on its tendency to go over from order to disorder, but based partly on existing order that is kept up.

To the physicist—but only to him—I could hope to make my view clearer by saying: The living organism seems to be a macroscopic system which in part of its behaviour approaches to that purely mechanical (as contrasted with thermo-dynamical) conduct to which all systems tend, as the temperature approaches the absolute zero and the molecular disorder is removed.

The non-physicist finds it hard to believe that really the ordinary laws of physics, which he regards as the prototype of inviolable precision, should be based on the statistical tendency of matter to go over into disorder. . . . The general principle involved is the famous Second Law of Thermodynamics (entropy principle) and its equally famous statistical foundation. In [the following] I will try to sketch the bearing of the entropy principle on the large-scale behaviour of a living organism—forgetting at the moment all that is known about chromosomes, inheritance, and so on.

Living Matter Evades the Decay to Equilibrium

What is the characteristic feature of life? When is a piece of matter said to be alive? When it goes on 'doing something', moving, exchanging material with its environment, and so forth, and that for a much longer period than we would expect an inanimate piece of matter to 'keep going' under similar circumstances. When a system that is not alive is isolated or placed in a uniform environment, all motion usually comes to a standstill very soon as a result of various kinds of friction; differences of electric or chemical potential are equalized, substances which tend to form a chemical compound do so, temperature becomes uniform by heat conduction. After that the whole system fades away into a dead, inert lump of matter. A permanent state is reached, in which no observable events occur. The physicist calls this the state of thermodynamical equilibrium, or of 'maximum entropy'.

Practically, a state of this kind is usually reached very rapidly. Theoretically, it is very often not yet an absolute equilibrium, not yet the true maximum of entropy. But then the final approach to equilibrium is very slow. It could take anything between hours, years, centuries, . . . To give an example—one in which the approach is still fairly rapid: if a glass filled with pure water and a second one filled with sugared water are placed together in a hermetically closed case at constant temperature, it appears at first that nothing happens, and the impression of complete equilibrium is created. But after a day or so it is noticed that the pure water, owing to its higher vapour pressure, slowly evaporates and condenses

on the solution. The latter overflows. Only after the pure water has totally evaporated has the sugar reached its aim of being equally distributed among all the liquid water available.

These ultimate slow approaches to equilibrium could never be mistaken for life, and we may disregard them here. I have referred to them in order to clear myself of a charge of inaccuracy.

It Feeds on 'Negative Entropy'

It is by avoiding the rapid decay into the inert state of 'equilibrium', that an organism appears so enigmatic; so much so, that from the earliest times of human thought some special non-physical or supernatural force (vis viva, entelechy) was claimed to be operative in the organism, and in some quarters is still claimed.

How does the living organism avoid decay? The obvious answer is: By eating, drinking, breathing and (in the case of plants) assimilating. The technical term is metabolism. The Greek word (μεταβάλλειν) means change or exchange. Exchange of what? Originally the underlying idea is, no doubt, exchange of material. (E.g. the German for metabolism is Stoffwechsel.) That the exchange of material should be the essential thing is absurd. Any atom of nitrogen, oxygen, sulphur, etc., is as good as any other of its kind; what could be gained by exchanging them? For a while in the past our curiosity was silenced by being told that we feed upon energy. In some very advanced country (I don't remember whether it was Germany or the U.S.A. or both) you could find menu cards in restaurants indicating, in addition to the price, the energy content of every dish. Needless to say, taken literally, this is just as absurd. For an adult organism the energy content is as stationary as the material content. Since, surely, any calorie is worth as much as any other calorie, one cannot see how a mere exchange could help.

What then is that precious something contained in our food which keeps us from death? That is easily answered. Every process, event, happening—call it what you will; in a word, everything that is going on in Nature means an increase of the entropy of the part of the world where it is going on. Thus a living organism continually increases its entropy—or, as you may say, produces positive entropy—and thus tends to approach the dangerous state of maximum entropy, which is death. It can only keep aloof from it, i.e. alive, by continually drawing from its environment negative entropy—which is something very positive as we shall immediately see. What an organism feeds upon is negative entropy. Or, to put it less paradoxically, the essential thing in metabolism is that the organism succeeds in freeing itself from all the entropy it cannot help producing while alive.

What is Entropy?

What is entropy? Let me first emphasize that it is not a hazy concept or idea, but a measurable physical quantity just like the length of a rod, the temperature at any point of a body, the heat of fusion of a given crystal or the specific heat of any given substance. At the absolute zero point of temperature (roughly $-273°$ C.) the entropy of any substance is zero. When you bring the substance into any other state by slow, reversible little steps (even if thereby the substance changes its physical or chemical nature or splits up into two or more parts of different physical or chemical nature) the entropy increases by an amount which is computed by dividing every little portion of heat you had to supply in that procedure by the absolute temperature at which it was supplied—and by summing up all these small contributions. To give an example, when you melt a solid, its entropy increases by the amount of the heat of fusion divided by the temperature at the melting-point. You see from this, that the unit in which entropy is measured is cal./° C. (just as the calorie is the unit of heat or the centimetre the unit of length).

The Statistical Meaning of Entropy

I have mentioned this technical definition simply in order to remove entropy from the atmosphere of hazy mystery that frequently veils it. Much more important for us here is the bearing on the statistical concept of order and disorder, a connection that was revealed by the investigations of Boltzmann and Gibbs in statistical physics. This too is an exact quantitative connection, and is expressed by

$$\text{entropy} = k \log D,$$

where k is the so-called Boltzmann constant ($= 3 \cdot 2983 \, . \, 10^{-24}$ cal./° C.), and D a quantitative measure of the atomistic disorder of the body in question. To give an exact explanation of this quantity D in brief non-technical terms is well-nigh

impossible. The disorder it indicates is partly that of heat motion, partly that which consists in different kinds of atoms or molecules being mixed at random, instead of being neatly separated, e.g. the sugar and water molecules in the example quoted above. Boltzmann's equation is well illustrated by that example. The gradual 'spreading out' of the sugar over all the water available increases the disorder D, and hence (since the logarithm of D increases with D) the entropy. It is also pretty clear that any supply of heat increases the turmoil of heat motion, that is to say increases D and thus increases the entropy; it is particularly clear that this should be so when you melt a crystal, since you thereby destroy the neat and permanent arrangement of the atoms or molecules and turn the crystal lattice into a continually changing random distribution.

An isolated system or a system in a uniform environment (which for the present consideration we do best to include as a part of the system we contemplate) increases its entropy and more or less rapidly approaches the inert state of maximum entropy. We now recognize this fundamental law of physics to be just the natural tendency of things to approach the chaotic state (the same tendency that the books of a library or the piles of papers and manuscripts on a writing desk display) unless we obviate it. (The analogue of irregular heat motion, in this case, is our handling those objects now and again without troubling to put them back in their proper places.)

Organization Maintained by Extracting 'Order' from the Environment

How would we express in terms of the statistical theory the marvellous faculty of a living organism, by which it delays the decay into thermodynamical equilibrium (death)? We said before: 'It feeds upon negative entropy', attracting, as it were, a stream of negative entropy upon itself, to compensate the entropy increase it produces by living and thus to maintain itself on a stationary and fairly low entropy level.

If D is a measure of disorder, its reciprocal, $1/D$, can be regarded as a direct measure of order. Since the logarithm of $1/D$ is just minus the logarithm of D, we can write Boltzmann's equation thus:

$$-(\text{entropy}) = k \log (1/D).$$

Hence the awkward expression 'negative entropy' can be replaced by a better one: entropy, taken with the negative sign, is itself a measure of order. Thus the device by which an organism maintains itself stationary at a fairly high level of orderliness (= fairly low level of entropy) really consists in continually sucking orderliness from its environment. This conclusion is less paradoxical than it appears at first sight. Rather could it be blamed for triviality. Indeed, in the case of higher animals we know the kind of orderliness they feed upon well enough, viz. the extremely well-ordered state of matter in more or less complicated organic compounds, which serve them as foodstuffs. After utilizing it they return it in a very much degraded form—not entirely degraded, however, for plants can still make use of it. (These, of course, have their most powerful supply of 'negative entropy' in the sunlight.)

Notes

1. Neither can the body determine the mind to think, nor the mind the body to move or to rest nor to anything else, if such there be.
2. To state this in complete generality about 'the laws of physics' is perhaps challengeable. . . .

18.
Life, Thermodynamics, and Cybernetics

L. BRILLOUIN

HOW IS IT POSSIBLE to understand life, when the whole world is ruled by such a law as the second principle of thermodynamics, which points toward death and annihilation? This question has been asked by many scientists, and, in particular, by the Swiss physicist, C. E. Guye, in a very interesting book [*L'évolution physico-chimique*]. The problem was discussed at the Collège de France in 1938, when physicists, chemists, and biologists met together and had difficulty in adjusting their different points of view. We could not reach complete agreement, and at the close of the discussions there were three well defined groups of opinion:

(*A*) Our present knowledge of physics and chemistry is practically complete, and these physical and chemical laws will soon enable us to explain life, without the intervention of any special "life principle."

(*B*) We know a great deal about physics and chemistry, but it is presumptuous to pretend that we know all about them. We hope that, among the things yet to be discovered, some new laws and principles will be found that will give us an interpretation of life. We admit that life obeys all the laws of physics and chemistry at present known to us, but we definitely feel that something more is needed before we can understand life. Whether it be called a "life principle" or otherwise is immaterial.

(*C*) Life cannot be understood without reference to a "life principle." The behavior of living organisms is completely different from tha tof

From L. Brillouin, "Life, Thermodynamics, and Cybernetics," *American Scientist*, 37 (October, 1949), 554–68. Reprinted by permission of the author and publisher.

inert matter. Our principles of thermodynamics, and especially the second one, apply only to dead and inert objects; life is an exception to the second principle, and the new principle of life will have to explain conditions contrary to the second law of thermodynamics.

Another discussion of the same problems, held at Harvard in 1946, led to similar conclusions and revealed the same differences of opinion.

In summarizing these three points of view, I have of course introduced some oversimplifications. Recalling the discussions, I am certain that opinions *A* and *B* were very clearly expressed. As for opinion *C*, possibly no one dared to state it as clearly as I have here, but it was surely in the minds of a few scientists, and some of the points introduced in the discussion lead logically to this opinion. For instance, consider a living organism; it has special properties which enable it to resist destruction, to heal its wounds, and to cure occasional sickness. This is very strange behavior, and nothing similar can be observed about inert matter. Is such behavior an exception to the second principle? It appears so, at least superficially, and we must be prepared to accept a "life principle" that would allow for some exceptions to the second principle. When life ceases and death occurs, the "life principle" stops working, and the second principle regains its full power, implying demolition of the living structure. There is no more healing, no more resistance to sickness; the destruction of the former organism goes on unchecked and is completed in a very short time. Thus the conclusion, or question: What about life and the second principle? Is there not, in living organisms, some power that prevents the action of the second principle?

The Attitude of the Scientist

The three groups as defined in the preceding section may be seen to correspond to general attitudes of scientists towards research: (*A*) strictly conservative, biased against any change, and interested only in new development and application of well established methods or principles; (*B*) progressive, open-minded, ready to accept new ideas and discoveries; (*C*) revolutionary, or rather, metaphysical, with a tendency to wishful thinking, or indulging in theories lacking solid experimental basis.

In the discussion just reviewed, most non-specialists rallied into group *B*. This is easy to understand. Physicists of the present century had to acquire a certain feeling for the unknown, and always to be very cautious against over-confidence. Prominent scientists of the previous generation, about 1900, would all be classed in group *A*. Common opinion about that time was that everything was known and that coming generations of scientists could only improve on the accuracy of experiments and measure one or two more decimals on the physical constants. Then some new laws were discovered: quanta, relativity, and radioactivity. To cite more specific examples, the Swiss physicist Ritz was bold enough to write, at the end of the nineteenth century, that the laws of mechanics could not explain optical spectra. Thirty years passed before the first quantum mechanical explanation was achieved. Then, about 1922, after the first brilliant success of quantum mechanics, things came to a standstill while experimental material was accumulating. Some scientists (Class *A*) still believed that it was just a question of solving certain very complicated mathematical problems, and that the explanation would be obtained from principles already known. On the contrary, however, we had to discover wave mechanics, spinning electrons, and the whole structure of the present physical theories. Now, to speak frankly, we seem to have reached another dead-end. Present methods of quantum mechanics appear not to be able to explain the properties of fundamental particles, and attempts at such explanations look decidedly artificial. Many scientists again believe that a new idea is needed, a new type of mathematical correlation, before we can go one step further.

All this serves to prove that every physicist must be prepared for many new discoveries in his own domain. Class *A* corresponds to cautiousness. Before abandoning the safe ground of well established ideas, says the cautious scientist, it must be proved that these ideas do not check with experiments. Such was the case with the Michelson-Morley experiment. Nevertheless, the same group of people were extremely reluctant to adopt relativity.

Attitude *B* seems to be more constructive, and corresponds to the trend of scientific research through past centuries; attitude *C*, despite its exaggeration, is far from being untenable. We have watched many cases of new discoveries leading to limitations of certain previous "laws." After all, a scientific law is not a "decree" from some supernatural power; it simply represents a systematization of a large number of experimental results. As a consequence, the scientific law has only a limited validity. It extends over the whole domain of experimentation, and maybe slightly beyond. But we must be prepared for some strange modifications when our knowledge is expanded much farther than this. Many historical examples could be introduced to support this opinion. Classical mechanics, for instance, was one of the best-established theories, yet it had to be modified to account for the behavior of very fast particles (relativity), atomic structure, or cosmogony.

Far from being foolish, attitude *C* is essentially an exaggeration of *B*; and any scholar taking attitude *B* must be prepared to accept some aspects of group *C*, if he feels it necessary and if these views rest upon a sound foundation.

To return to the specific problem of life and thermodynamics, we find it discussed along a very personal and original line in a small book [*What Is Life?*] published by the famous physicist E. Schrödinger. His discussion is very interesting and there are many points worth quoting. Some of them will be examined later on. In our previous classification, Schrödinger without hesitation joins group *B*:

> We cannot expect [he states] that the "laws of physics" derived from it [from the second principle and its statistical interpretation] suffice straightaway to explain the behavior of living matter. . . . We must be prepared to find a new type of physical law prevailing in it. Or are we to term it a non-physical, not to say a super-physical law?

The reasons for such an attitude are very convincingly explained by Schrödinger, and no attempt will be made to summarize them here. Those who undertake to read the book will find plenty of material for reflection and discussion. Let us simply state at this point that there is a problem about "life and the second principle."

The answer is not obvious, and we shall now attempt to discuss that problem systematically.

The Second Principle of Thermodynamics, Its Successes and Its Shortcomings

Nobody can doubt the validity of the second principle, no more than he can the validity of the fundamental laws of mechanics. However, the question is to specify its domain of applicability and the chapters of science or the type of problems for which it works safely. We shall put special emphasis on all cases where the second principle remains silent and gives no answer. It is a typical feature of this principle that it has to be stated as an inequality. Some quantity, called "entropy," cannot decrease (under certain conditions to be specified later); but we can never state whether "entropy" simply stays constant, or increases, or how fast it will increase. Hence, the answer obtained from the second principle is very often evasive, and keeps a sibyllic character. We do not know of any experiment telling *against* the second principle, but we can easily find many cases where it is useless and remains dumb. Let us therefore try to specify these limitations and shortcomings, since it is on this boundary that life plays.

Both principles of thermodynamics apply only to an isolated system, which is contained in an enclosure through which no heat can be transferred, no work can be done, and no matter nor radiation can be exchanged.[1] The first principle states that the total energy of the system remains constant. The second principle refers to another quantity called "entropy," S, that may only increase, or at least remain constant, but can never decrease. Another way to explain the situation is to say that the total amount of energy is conserved, but not its "quality." Energy may be found in a high-grade quality, which can be transformed into mechanical or electrical work (think of the energy of compressed air in a tank, or of a charged electric battery); but there are also low-grade energies, like heat. The second principle is often referred to as a principle of energy degradation. The increase in entropy means a decrease in quality for the total energy stored in the isolated system.

Consider a certain chemical system (a battery, for instance) and measure its entropy, then seal it and leave it for some time. When you break the seal, you may again measure the entropy, and you will find it increased. If your battery were charged to capacity before sealing, it will have lost some of its charge and will not be able to do the same amount of work after having been stored away for some time. The change may be small, or there may be no change; but certainly the battery cannot increase its charge during storage, unless some additional chemical reaction takes place inside and makes up for the energy and entropy balance. On the other hand, life feeds upon high-grade energy or "negative entropy" [Schrödinger]. A decrease in high-grade energy is tantamount to a loss of food for living organisms. Or we can also say that living organisms automatically destroy first-quality energy, and thus contribute to the different mechanisms of the second principle. If there are some living cells in the enclosure, they will be able for some time to feed upon the reserves available, but sooner or later this will come to an end and death then becomes inevitable.

The second principle means *death by confinement*, and it will be necessary to discuss these terms. Life is constantly menaced by this sentence to death. The only way to avoid it is to prevent confinement. Confinement implies the existence of perfect walls, which are necessary in order to build an ideal enclosure. But there are some very important questions about the problem of the existence of perfect walls. Do we really know any way to build a wall that could not let any radiation in or out? This is theoretically almost impossible; practically, however, it can be done and is easily accomplished in physical or chemical laboratories. There is, it is true, a limitation to the possible application of the second principle, when it comes to highly penetrating radiation, such as ultra-hard rays or cosmic rays; but this does not seem to have any direct connection with the problem of life and need not be discussed here.

Time and the second principle. The second principle is a death sentence, but it contains no time limit, and this is one of the very strange points about it. The principle states that in a closed system, S will increase and high-grade energy must decrease; but it does not say how fast. We have even had to include the possibility that nothing at all might happen, and that S would simply remain constant. The second principle is an arrow pointing to a direction along a one-way road, with no upper or lower speed limit. A chemical reaction may flash in a split second or lag on for thousands of centuries.

Although time is not a factor in the second principle, there is, however, a very definite connection between that principle and the definition of time. One of the most important features about time is its irreversibility. Time flows on, never comes back. When the physicist is confronted with this fact he is greatly disturbed. All the laws of physics, in their elementary form, are reversible; that is, they contain the time but not its sign, and positive or negative times have the same function. All these elementary physical laws might just as well work backward. It is only when phenomena related to the second principle (friction, diffusion, energy transferred) are considered that the irreversibility of time comes in. The second principle, as we have noted above, postulates that time flows always in the same direction and cannot turn back. Turn the time back, and your isolated system, where entropy (S) previously was increasing, would now show a decrease of entropy. This is impossible: evolution follows a one-way street, where travel in the reverse direction is strictly forbidden. This fundamental observation was made by the founders of thermodynamics, as, for instance, by Lord Kelvin in the following paragraphs:

If Nature Could Run Backward

If, then, the motion of every particle of matter in the universe were precisely reversed at any instant, the course of nature would be simply reversed forever after. The bursting bubble of foam at the foot of a waterfall would reunite and descend into the water; the thermal motions would reconcentrate their energy and throw the mass up the fall in drops re-forming into a close column of ascending water. Heat which had been generated by the friction of solids and dissipated by conduction, and radiation with absorption, would come again to the place of contact and throw the moving body back against the force to which it had previously yielded. Boulders would recover from the mud the materials required to rebuild them into their previous jagged forms, and would become reunited to the mountain peak from which they had formerly broken away. And if, also, the materialistic hypothesis of life were true, living creatures would grow backward, with conscious knowledge of the future but with no memory of the past, and would become again, unborn.

But the real phenomena of life infinitely transcend human science, and speculation regarding consequences of their imagined reversal is utterly unprofitable. Far otherwise, however, is it in respect to the reversal of the motions of matter uninfluenced by life, a very elementary consideration of which leads to the full explanation of the theory of dissipation of energy.

This brilliant statement indicates definitely that Lord Kelvin would also be classed in our group *B*, on the basis of his belief that there is in life something that transcends our present knowledge. Various illustrations have been given of the vivid description presented by Lord Kelvin. Moviegoers have had many opportunities to watch a waterfall climbing up the hill or a diver jumping back on the springboard; but, as a rule, cameramen have been afraid of showing life going backward, and such reels would certainly not be authorized by censors!

In any event, it is a very strange coincidence that life and the second principle should represent the two most important examples of the impossibility of time's running backward. This reveals the intimate relation between both problems, a question that will be discussed in a later section.

Statistical interpretation of the second principle. The natural tendency of entropy to increase is interpreted now as corresponding to the evolution from improbable toward most probable structures. The brilliant theory developed by L. Boltzmann, F. W. Gibbs, and J. C. Maxwell explains entropy as a physical substitute for "probability" and throws a great deal of light upon all thermodynamical processes. This side of the question has been very clearly discussed and explained in Schrödinger's book and will not be repeated here.

Let us, however, stress a point of special interest. With the statistical theory, entropy acquires a precise mathematical definition as the logarithm of probability. It can be computed theoretically, when a physical model is given, and the theoretical value compared with experiment. When this has been found to work correctly, the same physical model can be used to investigate problems outside the reach of classical thermodynamics and especially problems involving time. The questions raised in the preceding section can now be answered; the rate of diffusion for gas mixtures, the thermal conductivity of gases, the velocity of chemical reactions can be computed.

In this respect, great progress has been made, and in a number of cases it can be determined *how fast* entropy will actually increase. It is expected that convenient models will eventually be found for all of the most important problems; but this is not yet the case, and we must distinguish between those physical or chemical experiments for which a detailed application of statistical thermodynamics has been worked out, and other problems for which a model has not yet been found and for which we therefore have to rely on classical thermodynamics without the help of statistics. In the first group, a detailed model enables one to answer the most incautious questions; in the second group, questions involving time cannot be discussed.

Distinction between two classes of experiments where entropy remains constant. The entropy of a closed system, as noted, must increase, or at least remain constant. When entropy increases, the system is undergoing an *irreversible* transformation; when the system undergoes a *reversible* transformation, its total entropy remains constant. Such is the case for reversible cycles discussed in textbooks on thermodynamics, for reversible chemical reactions, etc. However, there is *another case* where no entropy change is observed, a case that is usually ignored, about which we do not find a word of discussion in textbooks, simply because scientists are at a loss to explain it properly. This is the case of systems in *unstable equilibrium*. A few examples may serve to clarify the problem much better than any definition.

In a private kitchen, there is a leak in the gas range. A mixture of air and gas develops (unstable equilibrium), but nothing happens until a naughty little boy comes in, strikes a match, and blows up the roof. Instead of gas, you may substitute coal, oil, or any sort of fuel; all our fuel reserves are in a state of unstable equilibrium. A stone hangs along the slope of a mountain and stays there for years, until rains and brooklets carry the soil away, and the rock finally rolls downhill. Substitute waterfalls, water reservoirs, and you have all our reserves of "white fuel." Uranium remained stable and quiet for thousands of centuries; then came some scientists, who built a pile and a bomb and, like the naughty boy in the kitchen, blew up a whole city. Such things would not be permitted if the second principle were an active principle and not a passive one. Such events could not take place in a world where this principle was strictly enforced.

All this makes one thing clear. All our so-called power reserves are due to systems in unstable equilibrium. They are really reserves of negative entropy—structures wherein, by some sort of miracle, the normal and legitimate increase of entropy does not take place, until man, acting like a catalytic agent, comes and starts the reaction.

Very little is known about these systems of unstable equilibrium. No explanation is given. The scientist simply mumbles a few embarrassed words about "obstacles" hindering the reaction, or "potential energy walls" separating systems that should react but do not. There is a hint in these vague attempts at explanation, and, when properly developed, they should constitute a practical theory. Some very interesting attempts at an interpretation of *catalysis*, on the basis of

quantum mechanics, have aroused great interest in scientific circles. But the core of the problem remains. How is it possible for such tremendous negative entropy reserves to stay untouched? What is the mechanism of *negative catalysis*, which maintains and preserves these stores of energy?

That such problems have a stupendous importance for mankind, it is hardly necessary to emphasize. In a world where oil simply waits for prospectors to come, we already watch a wild struggle for fuel. How would it be if oil burned away by itself, unattended, and did not wait passively for the drillers?

Life and Its Relations with the Second Principle

We have raised some definite questions about the significance of the second principle, and in the last section have noted certain aspects of particular importance. Let us now discuss these, point by point, in connection with the problem of life maintenance and the mechanism of life.

Closed systems. Many textbooks, even the best of them, are none too cautious when they describe the increase of entropy. It is customary to find statements like this one: "The entropy of the universe is constantly increasing." This, in my opinion, is very much beyond the limits of human knowledge. Is the universe bounded or infinite? What are the properties of the boundary? Do we know whether it is tight, or may it be leaking? Do entropy and energy leak out or in? Needless to say, none of these questions can be answered. We know that the universe is expanding although we understand very little of how and why. Expansion means a moving boundary (if any), and a moving boundary is a leaking boundary; neither energy nor entropy can remain constant within. Hence, it is better not to speak about the "entropy of the universe." In the last section we emphasized the limitations of physical laws, and the fact that they can be safely applied only within certain limits and for certain orders of magnitude. The whole universe is too big for thermodynamics and certainly exceeds considerably the reasonable order of magnitude for which its principles may apply. This is also proved by the fact that the Theory of Relativity and all the cosmological theories that followed always involve a broad revision and drastic modification of the laws of thermodynamics, before an attempt can be made to apply them to the universe as a whole. The only thing that we can reasonably discuss is the entropy of a conceivable closed structure. Instead of the very mysterious universe, let us speak of our home, the earth. Here we stand on familiar ground. The earth is not a closed system. It is constantly receiving energy and negative entropy from outside—radiant heat from the sun, gravitational energy from sun and moon (provoking sea tides), cosmic radiation from unknown origin, and so on. There is also a certain amount of outward leak, since the earth itself radiates energy and entropy. How does the balance stand? Is it positive or negative? It is very doubtful whether any scientist can answer this question, much less such a question relative to the universe as a whole.

The earth is not a closed system, and life feeds upon energy and negative entropy leaking into the earth system. Sun heat and rain make crops (remember April showers and May flowers), crops provide food, and the cycle reads: first, creation of unstable equilibriums (fuels, food, waterfalls, etc.); then, use of these reserves by all living creatures.

Life acts as a catalytic agent to help destroy unstable equilibrium, but it is a very peculiar kind of catalytic agent, since it profits by the operation. When black platinum provokes a chemical reaction, it does not seem to care, and does not profit by it. Living creatures care about food, and by using it they maintain their own unstable equilibrium. This is a point that will be considered later.

The conclusion of the present section is this: that the sentence to "death by confinement" is avoided by living in a world that is not a confined and closed system.

The role of time. We have already emphasized the silence of the second principle. The direction of any reaction is given, but the velocity of the reaction remains unknown. It may be zero (unstable equilibrium), it may remain small, or it may become very great. Catalytic agents usually increase the velocity of chemical reactions; however, some cases of "anticatalysis" or "negative catalysis" have been discovered, and these involve a slowing down of some important reactions (e.g., oxidation).

Life and living organisms represent a most important type of catalysis. It is suggested that a systematic study of positive and negative catalysts might prove very useful, and would in fact be absolutely necessary before any real understanding of life could be attained.

The *statistical interpretation* of entropy and *quantum mechanics* are undoubtedly the tools with which a theory of catalysis should be built. Some pioneer work has already been done and has proved extremely valuable, but most of it is restricted, for the moment, to the most elementary types of chemical reactions. The work on theoretical chemistry should be pushed ahead with great energy.

Such an investigation will, sooner or later, lead us to a better understanding of the mechanisms of "unstable equilibrium." New negative catalysts may even make it possible to stabilize some systems that otherwise would undergo spontaneous disintegration, and to preserve new types of energies and negative entropies, just as we now know how to preserve food.

We have already emphasized the role of living organisms as catalytic agents, a feature that has long been recognized. Every biochemist now thinks of ferments and yeasts as peculiar living catalysts, which help release some obstacle and start a reaction, in a system in unstable equilibrium. Just as catalysts are working within the limits of the second principle, so living organisms are too. It should be noted, however, that catalytic action in itself is something which is not under the jurisdiction of the second principle. Catalysis involves the velocity of chemical reactions, a feature upon which the second principle remains silent. Hence, in this first respect, life is found to operate along the border of the second principle.

However, there is a second point about life that seems to be much more important. Disregard the very difficult problem of birth and reproduction. Consider an adult specimen, be it a plant or an animal or man. This adult individual is a most extraordinary example of a chemical system in unstable equilibrium. The system is unstable, undoubtedly, since it represents a very elaborate organization, a most improbable structure (hence a system with very low entropy, according to the statistical interpretation of entropy). This instability is further shown when death occurs. Then, suddenly, the whole structure is left to itself, deprived of the mysterious power that held it together; within a very short time the organism falls to pieces, rots, and goes (we have the wording of the scriptures) back to the dust whence it came.

Accordingly, a living organism is a chemical system in unstable equilibrium maintained by some strange "power of life," which manifests itself as a sort of *negative catalyst*. So long as life goes on, the organism maintains its unstable structure and escapes disintegration. It slows down to a considerable extent (exactly, for a lifetime) the normal and usual procedure of decomposition. Hence, a new aspect of life. Biochemists usually look at living beings as possible catalysts. But this same living creature is himself an unstable system, held together by some sort of internal anticatalyst! After all, a poison is nothing but an active catalyst, and a good drug represents an anticatalyst for the final inevitable reaction: death.

N. Wiener, in his *Cybernetics*, takes a similar view when he compares enzymes or living animals to Maxwell demons, and writes: "It may well be that enzymes are metastable Maxwell demons, decreasing entropy. . . . We may well regard living organisms, such as Man himself, in this light. Certainly the enzyme and the living organism are alike metastable: the stable state of an enzyme is to be deconditioned, and the stable state of a living organism is to be dead. All catalysts are ultimately poisoned: they change rates of reaction, but not true equilibrium. Nevertheless, catalysts and Man alike have sufficiently definite states of metastability to deserve the recognition of these states as relatively permanent conditions."

Living Organisms and Dead Structures

In a discussion at Harvard (1946), P. W. Bridgman stated a fundamental difficulty regarding the possibility of applying the laws of thermodynamics to any system containing living organisms. How can we compute or even evaluate the entropy of a living being? In order to compute the entropy of a system, it is necessary to be able to create or to destroy it in a reversible way. We can think of no reversible process by which a living organism can be created or killed: both birth and death are irreversible processes. There is absolutely no way to define the change of entropy that takes place in an organism at the moment of its death. We might think of some procedure by which to measure the entropy of a dead organism, albeit it may be very much beyond our present experimental skill, but this does not tell us anything about the entropy the organism had just before it died.

This difficulty is fundamental; it does not make sense to speak of a quantity for which there is no operational scheme that could be used for its measurement. The entropy content of a living organism is a completely meaningless notion. In the discussion of all experiments involving living organisms, biologists always avoid the difficulty by assuming that the entropy of the living objects remains practically constant during the operation. This assumption is supported by experimental results, but it is a bold hypothesis and impossible to verify.

To a certain extent, a living cell can be compared to a flame: here is matter going in and out, and being burned. The entropy of a flame cannot be defined, since it is not a system in equilibrium. In the case of a living cell, we may know the entropy of its food and measure the entropy of its wastes. If the cell is apparently maintained in good health and not showing any visible change, it may be assumed that its entropy remains practically constant. All experimental measures show that the entropy of the refuse is larger than that of the food. The transformation operated by the living system corresponds to an increase of entropy, and this is presented as a verification of the second principle of thermodynamics. But we may have some day to reckon with the underlying assumption of constant entropy for the living organism.

There are many strange features in the behavior of living organisms, as compared with dead structures. The evolution of species, as well as the evolution of individuals, is an irreversible process. The fact that evolution has been progressing from the simplest to the most complex structures is very difficult to understand, and appears almost as a contradiction to the law of degradation represented by the second principle. The answer is, of course, that degradation applies only to the whole of an isolated system, and not to one isolated constituent of the system. Nevertheless, it is hard to reconcile these two opposite directions of evolution. Many other facts remain very mysterious: reproduction, maintenance of the living individual and of the species, free will, etc.

A most instructive comparison is presented by Schrödinger when he points to similarities and differences between a living organism, such as a cell, and one of the most elaborate structures of inanimate matter, a crystal. Both examples represent highly organized structures containing a very large number of atoms. But the crystal contains only a few types of atoms, whereas the cell may contain a much greater variety of chemical constituents. The crystal is always more stable at very low temperatures, and especially at absolute zero. The cellular organization is stable only within a given range of temperatures. From the point of view of thermodynamics this involves a very different type of organization.

When distorted by some stress, the crystal may to a certain extent repair its own structure and move its atoms to new positions of equilibrium, but this property of self-repair is extremely limited. A similar property, but exalted to stupendous proportions, characterizes living organisms. The living organism heals its own wounds, cures its sicknesses, and may rebuild large portions of its structure when they have been destroyed by some accident. This is the most striking and unexpected behavior. Think of your own car, the day you had a flat tire, and imagine having simply to wait and smoke your cigar while the hole patched itself and the tire pumped itself to the proper pressure, and you could go on. This sounds incredible.[2] It is, however, the way nature works when you "chip off" while shaving in the morning. There is no inert matter possessing a similar property of repair. That is why so many scientists (class *B*) think that our present laws of physics and chemistry do not suffice to explain such strange phenomena, and that something more is needed, some very important law of nature that has escaped our investigations up to now, but may soon be discovered. Schrödinger, after asking whether the new law required to explain the behavior of living matter (see [below]) might not be of a super-physical nature, adds: "No, I do not think that. For the new principle that is involved is a genuinely physical one. It is, in my opinion, nothing else than the principle of quantum theory over again." This is a possibility, but it is far from certain, and Schrödinger's explanations are too clever to be completely convincing.

There are other remarkable properties characterizing the ways of living creatures. For instance, let us recall the paradox of Maxwell's demon, that submicroscopical being, standing by a trapdoor and opening it only for fast molecules, thereby selecting molecules with highest energy and temperature. Such an action is unthinkable, on the submicroscopical scale, as contrary to the second principle.[3] How does it become feasible on a large scale? Man opens the window when the weather is hot and closes it on cold days! Of course, the answer is that the earth's atmosphere is not in equilibrium and not at constant temperature. Here again we come back to the unstable conditions created by sunshine and other similar causes, and the fact that the earth is not a closed isolated system.

The very strange fact remains that conditions forbidden on a small scale are permitted on a large one, that large systems can maintain unstable equilibrium for large time-intervals, and that life is playing upon all these exceptional conditions on the fringe of the second principle.

Entropy and Intelligence

One of the most interesting parts in Wiener's *Cybernetics* is the discussion on "Time series, information, and communication," in which he specifies that a certain "amount of information is the negative of the quantity usually defined as entropy in similar situations."

This is a very remarkable point of view, and it opens the way for some important generalizations of the notion of entropy. Wiener introduces a precise mathematical definition of this new negative entropy for a certain number of problems of communication, and discusses the question of time prediction: when we possess a certain number of data about the behavior of a system in the past, how much can we predict of the behavior of that system in the future?

In addition to these brilliant considerations, Wiener definitely indicates the need for an extension of the notion of entropy. "Information represents negative entropy"; but if we adopt this point of view, how can we avoid its extension to all types of intelligence? We certainly must be prepared to discuss the extension of entropy to scientific knowledge, technical know-how, and all forms of intelligent thinking. Some examples may illustrate this new problem.

Take an issue of the New York *Times*, the book on Cybernetics, and an equal weight of scrap paper. Do they have the same entropy? According to the usual physical definition, the answer is "yes." But for an intelligent reader, the amount of information contained in these three bunches of paper is very different. If "information means negative entropy," as suggested by Wiener, how are we going to measure this new contribution to entropy? Wiener suggests some practical and numerical definitions that may apply to the simplest possible problems of this kind. This represents an entirely new field for investigation and a most revolutionary idea.

Many similar examples can be found. Compare a rocky hill, a pyramid, and a dam with its hydroelectric power station. The amount of "know-how" is completely different, and should also correspond to a difference in "generalized entropy," although the physical entropy of these three structures may be about the same. Take a modern large-scale computing machine, and compare its entropy with that of its constituents before the assembling. Can it reasonably be assumed that they are equal? Instead of the "mechanical brain," think now of the living human brain. Do you imagine its (generalized) entropy to be the same as that for the sum of its chemical constituents?

It seems that a careful investigation of these problems, along the directions initiated by Wiener, may lead to some important contributions to the study of life itself. Intelligence is a product of life, and a better understanding of the power of thinking may result in a new point of discussion concerning this highly significant problem.

Let us try to answer some of the questions stated above, and compare the "value" of equal weights of paper: scrap paper, New York *Times*, *Cybernetics*. To an illiterate person they have the same value. An average English-reading individual will probably prefer the New York *Times*, and a mathematician will certainly value the book on Cybernetics much above anything else. "Value" means "generalized negative entropy," if our present point of view be accepted. The preceding discussion might discourage the reader and lead to the conclusion that such definitions are impossible to obtain. This hasty conclusion, however, does not seem actually to be correct. An example may explain the difficulty and show what is really needed.

Let us try to compare two beams of light, of different colors. The human eye, or an ultraviolet photo-cell, or an infrared receiving cell will give completely different answers. Nevertheless, the entropy of each beam of light can be exactly defined, correctly computed, and measured experimentally. The corresponding definitions took a long time to discover, and retained the attention of most distinguished physicists (e.g., Boltzman, Planck). But this difficult problem was finally settled, and a careful distinction was drawn between the intrinsic properties of radiation and the behavior of the specific receiving set used for experimental measurements. Each receiver is defined by its "absorption spectrum," which characterizes the way it reacts to incident radiations. Similarly, it does not seem impossible to discover some criterion by which a definition of generalized entropy could be applied to "information," and to distinguish it from the special sensitivity of the observer. The problem is certainly harder than in the case of light. Light depends only upon one parameter (wave length), whereas a certain number of independent variables may be required for the definition of the "information value," but the distinction between an absolute intrinsic value of information and the absorption spectrum of the receiver is indispensable. Scientific information represents certainly a sort of negative entropy for Wiener, who knows how to use it for prediction, and may be of no value whatsoever

to a non-scientist. Their respective absorption spectra are completely different.

Similar extensions of the notion of entropy are needed in the field of biology, with new definitions of entropy and of some sort of absorption spectrum. Many important investigations have been conducted by biologists during recent years, and they can be summarized as "new classifications of energies." For inert matter, it suffices to know energy and entropy. For living organisms, we have to introduce the "food value" of products. Calories contained in coal and calories in wheat and meat do not have the same function. Food value must itself be considered separately for different categories of living organisms. Cellulose is a food for some animals, but others cannot use it. When it comes to vitamins or hormones, new properties of chemical compounds are observed, which cannot be reduced to energy or entropy. All these data remain rather vague, but they all seem to point toward the need for a new leading idea (call it principle or law) in addition to current thermodynamics, before these new classifications can be understood and typical properties of living organisms can be logically connected together. Biology is still in the empirical stage and waits for a master idea, before it can enter the constructive stage with a few fundamental laws and a beginning of logical structure.

In addition to the old and classical concept of physical entropy, some bold new extensions and broad generalizations are needed before we can reliably apply similar notions to the fundamental problems of life and of intelligence. Such a discussion should lead to a reasonable answer to the definition of entropy of living organisms and solve the paradox of Bridgman.

A recent example from the physical sciences may explain the situation. During the nineteenth century, physicists were desperately attempting to discover some mechanical models to explain the laws of electromagnetism and the properties of light. Maxwell reversed the discussion and offered an electromagnetic theory of light, which was soon followed by an electromagnetic interpretation of the mechanical properties of matter. We have been looking, up to now, for a physico-chemical interpretation of life. It may well happen that the discovery of new laws and of some new principles in biology could result in a broad redefinition of our present laws of physics and chemistry, and produce a complete change in point of view.

In any event, two problems seem to be of major importance for the moment: a better understanding of catalysis, since life certainly rests upon a certain number of mechanisms of negative catalysis; and a broad extension of the notion of entropy, as suggested by Wiener, until it can apply to living organisms and answer the fundamental question of P. W. Bridgman.

Notes

1. The fundamental definition must always start with an isolated system, whose energy, total mass, and volume remain constant. Then, step by step, other problems may be discussed. A body at "constant temperature" is nothing but a body enclosed in a large thermostat, that is, in a big, closed, and isolated tank, whose energy content is so large that any heat developed in the body under experience cannot possibly change the average temperature of the tank. A similar experimental device, with a closed tank containing a large amount of an ideal gas, leads to the idea of a body maintained at constant pressure and constant temperature. These are secondary concepts derived from the original one.

2. The property of self-repairing has been achieved in some special devices. A self-sealing tank with a regulated pressure control is an example. Such a property, however, is not realized in most physical structures and requires a special control device, which is a product of human ingenuity, not of nature.

3. Wiener discusses very carefully the problem of the Maxwell demon (*Cybernetics*, pp. 71–72). One remark should be added. In order to choose the fast molecules, the demon should be able to see them; but he is in an enclosure in equilibrium at constant temperature, where the radiation must be that of the black body, and it is impossible to see anything in the interior of a black body. The demon simply does not see the particles, unless we equip him with a torchlight, and a torchlight is obviously a source of radiation not at equilibrium. It pours negative entropy into the system. Under these circumstances, the demon can certainly extract some fraction of this negative entropy by using his gate at convenient times. Once we equip the demon with a torchlight, we may also add some photoelectric cells and design an automatic system to do the work, as suggested by Wiener. The demon need not be a living organism, and intelligence is not necessary either. The preceding remarks seem to have been generally ignored, although Wiener says: the demon can only act on information received and this information represents a negative entropy.

19.
Communication, Entropy, and Life

RICHARD C. RAYMOND

BRILLOUIN IN DISCUSSING these subjects has suggested that a generalized definition of entropy, including the commonly defined thermodynamic variety and the information entropy discussed by Wiener, Shannon, and a number of others, would be highly desirable. The work of von Bertalanffy on the theory of open systems, and a demonstration of the writer of the physical equivalence of information entropy and thermodynamic entropy in a particular case, can be generalized to yield the required definition. The consideration of entropy relations in systems in which communication is possible leads to a qualitative explanation of the ability of a living organism to maintain and repair itself.

The thermodynamic definition of entropy is applicable only to equilibrium states in closed systems and is therefore not particularly useful in treating the nonequilibrium states and irreversible processes of life and communication, or in discussing the interchange of energy and entropy among a number of separate open systems. Neither life nor communication can exist in a system at thermodynamic equilibrium. It is the purpose of this paper to suggest that the entropy of a system may be defined quite generally as the sum of the positive thermodynamic entropy which the constituents of the system would have at thermodynamic equilibrium and a negative term proportional to the information necessary to build the actual system from its equilibrium state. This definition may be applied to open systems by closing them momentarily. Since the selection of

boundaries for a thermodynamic system is arbitrary it is possible to consider an open system as a part of a larger closed system and to close it from time to time where necessary for the determination of certain quantities. Communicated information is used in non-equilibrium systems to reduce or maintain at low levels the entropy of selected open systems within them. The storage of information by any physical process involves the selective decrease of the entropy of some physical system. Communicated information is often used in an open system to control the expenditure of energy in amounts far larger than the energies exchanged in the communication process. It may be possible to explain the metastable condition of living organisms through the use of stored and communicated information in open systems to achieve the necessary degree of order in the irreversible physical and chemical processes of metabolism. In a certain sense we may regard the initially stored information in a living individual as the number of choices the individual makes on the basis of his heredity, and the communicated information as the number of choices he makes in response to his environment. One physiological effect of information transfer apparently divorced from energy or material transport is now recognized medically in terms of the psychosomatic diseases.

The Second Law of Thermodynamics Is Obeyed

The above-defined entropy of a non-equilibrium system is consistent with the second law of thermodynamics in all of the cases so far considered. In a

From Richard C. Raymond, "Communication, Entropy, and Life," *American Scientist*, 38 (April, 1950), 273–78. Reprinted by permission of the author and publisher.

closed system which contains a source of information, a communication device, and a recorder, the initial entropy is the thermodynamic entropy which the system constituents would have at equilibrium less the information entropy necessary to construct the system from its equilibrium components. The recorder in particular may be recognized as an open system within the closed system. The initial state of the recorder may possibly be an equilibrium state. After the communication device has transmitted a message to the recorder, both terms have changed. The thermodynamic entropy of the system has increased because of the degradation of energy by the communication device in the transmission of the message, and the information entropy of the recorder has either increased or remained constant. The change in the total entropy of the system through the operation of the communication device has thus a positive term due to the degradation of energy by the communication device, and a negative term due to storage of information in the recorder. The entropy of the open system of the recorder may have increased, decreased, or remained constant. There are three cases of interest:

(a) The transmitted message is one which was previously recorded in the recorder and therefore communicates very little information. The net entropy change is positive.

(b) The transmitted message is not used or stored in the recorder. The net entropy change is positive.

(c) The entire message is used in the recorder to effect physical changes in the system. The net entropy change is the difference between the two terms.

It is not possible, even in case (c), to say whether the entropy of the recorder as an open system is in general increased or decreased through the operation of the communication device, but a consideration of the problem throws some light on the processes which lead to increases or decreases. As will be shown below, the transmission of information to the receiver requires a transfer of energy which must be degraded to heat. If the heat is dissipated by the recorder into the rest of the system as it is produced, the thermodynamic entropy of the open system may be left unchanged by the message. If the heat stays in the recorder, the thermodynamic entropy of the recorder is increased. In either case, the entropy of the closed system containing the open system is increased in the act of communication. It may be argued that there are means of signalling which do not involve

the transfer of any appreciable amount of energy. This is true in systems where such degenerative processes as friction and thermal noise may be ignored; but where noises, statistical fluctuations, or such effects as friction are encountered, no message may be sent without the expenditure and degradation of energy.

The theory of electrical communication has now reached the point where it is possible to make a computation of the increase of thermodynamic entropy and the decrease of information entropy of the closed system under certain assumptions and compare them. If we assume that the communication device in the system under discussion is an electrical system such as a radio or a telegraph, and that the heat capacity of the closed system is so large that the change in temperature with operation of the communication device is negligible, then the rate of increase of thermodynamic entropy during communication is

$$dS'/dt \geq W/T \qquad (1)$$

where S' is thermodynamic entropy, t is time, W is the average power expended in the communication device, and T is the absolute temperature of the system.

The maximum rate of change of information entropy in the recorder is given by

$$dS''/dt \leq -kV \log (1 + W/N) \qquad (2)$$

where k is Boltzmann's constant, V is the bandwidth of the communication device, and N is the average power of a noise uniformly distributed over bandwidth V. Inequality (2) is obtained by adjusting the entropy units used in communication theory by Shannon to agree with thermodynamic entropy units, by the comparison of information storage and entropy in a system involving a perfect gas subject to thermal fluctuations. The right half is the maximum possible rate of information transfer with all of the power of the communication device used in the receiver and with the best possible system of encoding and decoding.

The net rate of change of entropy of the closed system is given by

$$dS/dt \geq W/T - kV \log (1 + W/N) \qquad (3)$$

Violation of the second law requires the right half of inequality (3) to be negative. Since all the individual quantities involved are positive, the sign of the right side depends on the relative magnitudes of the two terms. Because W appears to the first power in the positive term and in a logarithm in the negative term, there is more chance of having the negative term exceed the positive at low values of W. If we assume W/N is small and substitute for N the Johnson noise value

$$N = kVT \qquad (4)$$

we have

$$dS/dt \geqq O \qquad (5)$$

This expression is positive for all values of W. It shows that even the optimum possible operation of the communication device, and the assumption of our new definition for entropy, does not lead us into contradiction of the second law in the case of electrical communication.

If the theories of other communications devices, such as reading and speaking, were as well developed as that of electrical communication it might be possible to develop quantitative treatments of such information storage devices as the Maxwell demon and such communication methods as the printing and reading of books. At present we can merely say that in any known process of sending information from one place to another, and in any method of making a record of transmitted information, there is a kind of work performed which involves the degradation of energy by irreversible processes probably leading to a net increase in the entropy of a closed system including both transmitter and receiver, but that the communicated information may be used to decrease the entropy of an open system within the closed system.

Information Storage and Equilibrium

It is quite easy to misinterpret a steady state or a metastable state of a system as an equilibrium state, particularly if spontaneous changes in the state occur only at very long time intervals. The entropy definition developed here helps to distinguish among these states because it requires that the information storage in a true equilibrium state be as small as possible. The comparison by Brillouin of a copy of *Cybernetics*, the *New York Times*, and a quantity of scrap paper, is not a comparison among systems at thermodynamic equilibrium and can therefore not be made by the ordinary thermodynamic definition of entropy. If we wait a sufficient period of time we may expect that the degenerative processes of nature will eventually bring all three to equilibrium, but when this happens their information contents will have disappeared, and it will not be possible to distinguish among them. Any process which results in

the recording of information produces some degree of order in the record. In magnetic recording, for instance, ferromagnetic materials are taken from an unmagnetized state to a selected pattern of magnetized states. The conversion of raw metal and printed information into a machine results in a lowering of the entropy of the metal through the information stored in its structure. The use of food by an organism for the growth or repair of its body represents a storage of information in the materials used and decreases their entropy.

Brillouin has suggested that we may find the difference between the book, newspaper, and scrap paper of his example by reference to the user or reader. The fact that all three systems are in different metastable states with regard to information entropy may be derived on a basis independent of the value of the information to the observer. The latter value is, however, also an interesting point. In general the response of a reader to printed information corresponds to some level among the alternatives laid out in the first example here. The negative entropy created in the reader during the reading operation is determined by the number of choices the reader makes and records on the basis of the information proffered by the book. This amount depends greatly on the initial condition of the reader with regard to information previously stored and needs for further information. If the reader already knows the material the only information he gains is that someone else knows it also, and the light and other forms of energy degraded in the reading process may have served mainly to increase the entropy of the reader and his vicinity. If the reader does not have the initial information necessary to understand what he reads he does not record it, and the entropy change is still positive. If, however, the information is of value to the reader, and he can understand it and put it to use, the entropy change of the open system including the reader but excluding the reading lamp may easily be negative. Although the net entropy change of the closed system including both reader and lamp is probably positive, the negative entropy change in the reader has an importance that is hard to overemphasize in a world in which order and predictability have their present value.

Catalysis

The definition of entropy presented here suggests that a theoretical explanation of the operation of catalysts may be based on the postulate that the catalyst carries information to direct the specific chemical or physical reaction which is desired. The information serves to order and speed up a process which is thermodynamically possible but either is very slow or is complicated by competing reactions in the absence of a catalyst. The catalyst is able to create negative entropy in the molecules formed in the desired reaction. The use of a catalyst requires knowledge and thus represents an addition of information to the chemical system in which it is employed. A poisoned catalyst is one which has become disorganized and therefore of low information content or high entropy. This view is in accord with the selection by Langmuir of minute silver iodide crystals as condensation nuclei for water vapor, in the hope that the information carried in the crystal structure of the silver iodide would facilitate the formation of water crystals. It suggests also the idea that the simplest living molecule may be one which gathers new atoms to be arranged according to information stored in the structure of the molecule until the essential parts of a similar molecule have been collected. At this point that large twin molecule formed becomes unstable against fission and so reproduces.

Life

Although it is quite true that the operations for the determination of the entropy of living organisms according to current thermodynamic definitions do not exist, it is possible that such operations might be devised in the light of the definition proposed here. Success in this measurement will depend on the operations of synthesis of a living organism. If this synthesis can be achieved, the entropy of the resulting organism may be taken as the equilibrium entropy of the constituents of the organism less the information entropy necessary to the synthesis of the organism from equilibrium components of known entropy. The difficulty of the synthesis indicates that the amount of information entropy is quite large in a living organism.

This view may also be extended to the use of the concept of entropy as an indication of the state of health of an organism. Life processes which are interfered with as a result of poor transfer of matter or energy through the organism may be viewed at times in terms of deficiencies in information for the direction of the required reactions. Death in a complicated organism is not an instantaneous event from the standpoint of the individual parts of the body, and the increase in entropy of the organism which accompanies death proceeds gradually as the structural and behavior information content of the various parts of the body is destroyed. The maintenance of the steady state in the open system as discussed by von Bertalanffy is accomplished in part through the use of fuels of high free-energy content and in part through the use of communicated information to balance out the degenerative processes of nature. In a large organism, communication is carried on through a number of nervous systems, by transmission of pressure through vascular systems, and by the generation and transmission of hormones and enzymes. The control of body growth through a chemical messenger generated in the pituitary gland has been postulated for some time. Relations between the control of sugar metabolism and insulin have reached a point where individuals who would otherwise die from a lack of this information may inject the information into themselves as required. In a very simple case a single-celled organism might exist for some period of time in the absence of added information, if it is provided with food and energy as required for the maintenance of its simplest steady state. Any stimulus applied to the cell, however, even if it is only a change in the food supply, represents a transmission of information to the cell, and is used together with the stored information in the structure of the cell in the determination of the new state of the cell which results from the stimulus. Organisms which contain a sufficient amount of information appropriately keyed to facilitate rapid adjustment to a wide variety of information inputs, are said to be readily adaptable to their environments, and these have historically been the most successful organisms from the point of view of biology and evolution.

Conclusion

The development of a modern theory of communication has led to an interpretation of the processes of information transfer in terms of the creation of negative entropy in places where information is used to direct physical or chemical processes. A consideration of the effects of information storage and information transfer on physical, chemical, biological, psychological, and sociological systems, both open and closed, may help in understanding and predicting many of the aspects of our universe.

20.
Thermodynamics and Information Theory

L. BRILLOUIN

Similarity between Information and Negentropy

In a paper published in this Review, the author discussed the possibilities of extending thermo-dynamical considerations to the theory of information. These problems were investigated more recently on a specific example, where it was shown that every observation in the laboratory requires degradation of energy, and is made at the expense of a certain amount of negative entropy (abbreviation: *negentropy*), taken away from the surroundings. Whether he be a Maxwell's demon or a physicist, the observer can obtain a bit of information only when a certain quantity of negentropy is lost. Information gain means an increase of entropy in the laboratory.

Vice versa, with the help of this information the observer may be in a position to make an appropriate decision, which decreases the entropy of the laboratory. In other words, he may use the information to recuperate part of the loss in negentropy, but the over-all balance is always an increase in entropy (a loss in negentropy).

The cycle under consideration was:

(1) Negentropy → Observation → Information → Decision → Negentropy

This problem offers a new example of the possibilities of changing negentropy into information. It compares with Shannon and Wiener's problems in telecommunications and its cycle:

(2) Information → Telegram transmitted →

From L. Brillouin, "Thermodynamics and Information Theory," *American Scientist*, 38 (October, 1950), 594–99. Reprinted by permission of the author and publisher.

Negentropy on the cable → Telegram received → Information received

Here the process is a different one: signals emitted by the transmitter create a very unstable and improbable distribution of currents on the cable. This represents a physical situation of low probability, hence large negentropy. When these signals are absorbed in the receiver, this negentropy disappears and information is obtained.

In both cases we have examples of trans-formations of negentropy into information, or vice versa. This leads to the conclusion that both quantities are of similar nature. A more detailed discussion, however, will show that caution is required and that precise definitions should be stated before any accurate application of these ideas may become possible.

How to Define Information?

Before we attempt to propose some definitions, let us examine a few points that were raised at different meetings where the author had an opportunity to discuss these problems with various groups of engineers and physicists.

Let us first consider Shannon's cycle (2). During the process of transmission along the cable a certain amount of information will get lost: ohmic resistance and thermal noise will contribute to decrease the quantity of information available. These very same causes result in a decrease of negentropy. At this stage, the similarity is striking, but one point must be noticed: the information available at the transmitter did not get lost!

This is completely different from what we

observe with physical entropy. When a substance A_1 at temperature T_1 loses an amount of heat Q, which is absorbed by a body A_2 at a lower temperature T_2, the process is the following: A_1 loses an entropy $S_1 = Q/T_1$, and A_2 gains $S_2 = Q/T_2$, the total balance being

$$(3) \qquad S_2 - S_1 = Q(1/T_2 - 1/T_1) > 0$$

since $T_2 < T_1$.

In the exchange of information (2), the transmitter emits a certain amount of information (or negentropy $-S_1$), but he can do it without losing anything!

This is a general rule. When an author writes a book, he makes a certain amount of information available to the public. The reader will be able to assimilate and use part of this information; this corresponds to the loss of information during transmission, in cycle (2). If there be a thousand copies sold, and a thousand readers, it seems that the total amount of information has been increased. And again there was no loss of information for the author!

What was said about a writer and his readers applies just as well to a teacher and his pupils.

Let us choose another example. An engineer prepares the blueprints for a new machine. These drawings contain all the information. With their help, we can build one machine or a million of them—and the information contained in the blueprints is still there. Printing, teaching, mass production, broadcasting, television, all these aspects of modern life lead to the same conclusion.

We may have to introduce two different quantities:

Absolute information, which exists as soon as one person has it, and should be counted as the same given amount of information, whether it is known to one man or to millions.

Distributed information, defined as the product of the amount of absolute information and the number of people who share that information. In the process of broadcasting the information, from the point where it originates to all the persons who finally receive it, a certain amount of information may get lost and the total amount of distributed information will be reduced by a certain percentage.

Distribution of information corresponds to the different processes discussed on this page: broadcasting, telecommunications, printing, teaching, advertising, etc. It increases the amount of distributed information without changing the total absolute information available.

Another point of importance is the natural *decay of the information value.* This is another aspect of the similarity between negentropy and information. When an exceptional situation is created somewhere, it corresponds to high negentropy, and its normal evolution is always toward lower negentropy, according to the second principle of thermodynamics. If an observer has been able to record this abnormal situation, he has obtained some absolute information which may be of great value for a short period of time but which will progressively lose its value because of the character of instability of the recorded facts. Weather data, stock exchange quotations, political information belong to this category.

Some other sorts of information seem to possess a relatively stable character: scientific laws and well-established historical facts have a rather permanent value. Nevertheless, there is a law of natural decay to be observed here also: Newton's laws of mechanics were supposed to be absolutely general and universal; but they had to be amended and corrected later on, in the theory of relativity and in quantum mechanics.

Absolute information always loses its value progressively. This is a law of *degradation of absolute information*, very similar to the famous law of degradation of energy stated by Lord Kelvin, and corresponding to the natural decrease of negentropy (second principle of thermodynamics). This process of decay was pointed out by R. C. Raymond in his recent discussion of these problems.

Processes That May Increase Information

There are also other processes by which the amount of absolute information available to mankind is constantly increased, and we have to examine them very carefully in order to discover whether they may constitute exceptions to the second principle of thermodynamics.

First, we may increase information by recording some experimental facts and making observations. This process was discussed by the author, and it was shown that every observation is made at the expense of the negentropy of the surroundings. A methodical discussion of the cycle (1) proved that negentropy was always lost in the over-all process and that there was no chance of escaping the general law of degradation at this stage of the game.

Let us now consider the second class of information discussed in the preceding section: scientific laws. Here the process is much more elaborate. The scientist records the results of a large number of experiments, compares them, and from this discussion draws some general conclusions. He then formulates a hypothetical law, checks the law by new experiments, and finally obtains an "absolute information" of great importance and well-established validity.

When Einstein discovered relativity, or when L. de Broglie invented wave mechanics, an important contribution was certainly added to the general knowledge of mankind. Their work undoubtedly increased the amount of "absolute information."

All this represents another cycle, different from our first cycle (1):

(4) Numerous observations → Comparison → Thinking → Scientific Laws → Practical application of these laws → New apparatus or machines built

The consideration of such a cycle raises certain important questions. The first part involves the use of many experimental observations, hence a large decrease of negentropy of the surroundings. But the next steps are very hard to analyze. They correspond to the work of the human brain, and to the ingenuity of the scientist. It looks as if the final product of this activity, namely "scientific laws," corresponds to really new information, not entirely accounted for in the first step of experimental observation. We may try to introduce at this point the amount of negentropy needed to maintain the scientist alive and to keep his brain working, but this seems to be very much beyond our present knowledge of biology.

The last part of the cycle is no less challenging. With their knowledge of scientific laws, the scientist and the engineer are able to build new machines and obtain all the products of modern industry. They create very improbable structures and manufacture all sorts of apparatus that nature never was able to produce: cars, radios, watches, computing machines, etc. Every one of these machines represents an extremely improbable assembly of parts, and obtains most extraordinary practical results. Our general theory of statistical thermodynamics relates all cases of low probability with examples of high negentropy. . . .

We may add a few remarks to the point. An industrial product is an improbable structure, and it must contain a certain amount of "structural negentropy." This can be proved by the fact that the machine wears out after a while, or rusts, or gets broken, and finally turns into a collection of useless broken pieces and junk. This process of

destruction is just one of the many processes of the second principle. Hence the conclusion: If there is in the machine something that can be destroyed by the second principle, this "something" must be some sort of negentropy.[1]

Here we reach another difficult point: The amount of negentropy should be computed at a rate of so much per machine. If we produce a million machines, we create a million times the original amount of negentropy.

In other words, the negentropy produced is proportional to the "distributed information" defined [above]. On the other hand, the amount of negentropy used in the discovery of a scientific law is proportional to the "absolute information" contained in this law.

Our last cycle (4) raises many serious problems for which we seem to be unable to find any definite answers at present. Altogether, it looks as if the two processes: (a) thinking and discovery of scientific laws, and (b) distribution of information, might lead to contradictions with any extension of the second principle to problems of information. We suspect that such an extension will require a considerable modification of the original principle.

Computation and the Exact Role of Computers

Many authors have emphasized the comparison between mechanical computing devices and the human brain. *Giant Brains* is the title of a recent book on large-scale computing equipment. How far is such a comparison justified? Or is it completely misleading? These are important questions to answer in connection with our present discussion. We wish to show that the *machines are not producing any new "absolute information."* They do not perform any type of active thinking, and the best way to prove it is to analyze methodically the sort of work done by these machines.

At first sight, it looks as if the machine were manufacturing new information. It computes solutions of mathematical problems. It prints numerical tables for Bessel functions, or solves some complicated systems of equations. Nevertheless, we feel that the machine can never be compared to a living organism: a machine-tool is not a worker; a car is different from a horse; a plane is not an eagle. Every machine requires a man to drive it, or better expressed, a machine is a dead structure under the command of the brain of a man. Mathematical machines are no exceptions to this rule: they need a staff of scientists to run them. These scientists think, get the work organized, prepare a program on punched cards or punched tape, and the machine operates blindly according to these orders. An ideal computing machine performs exactly the prepared program, computes without errors, and prints the results of the computation. If the machine is not in perfect condition or when its accuracy is not high enough, it introduces a few errors or some undesirable approximations; the results are inaccurate or may even be completely wrong.

The machine can be compared to the telegraphic cable in Shannon's cycle (2). If everything works fine, the message is correctly transmitted. If there be some errors or misinterpretations, the message is distorted, and part of the information is lost.

This leads to a better representation of the exact role played by the computer. Let us try to investigate the situation more completely. In order to do this, we have to distinguish between two different operations.

Active, creative thinking requires invention, methodical imagination, unexpected comparisons; and it results in a new idea, representing a completely new type of absolute information. We discussed this process when we analyzed the mechanism of scientific discovery and tried to describe the procedure by which a scientist may obtain a new scientific law. An author writing a book, a musician composing a symphony, an artist painting a canvas, are other examples of creative work.

Passive work represents a different sort of occupation. A typical case is found in translation. When a delegate from Iran is speaking at the United Nations, he is presenting his point of view and he has been doing active thinking. Immediately, a staff of interpreters starts translating the talk into English, French, Russian, and other languages. The role of an interpreter is just as passive as the role of a telegraphic cable. The interpreters have to repeat exactly what the Iranian delegate said. They must not omit one sentence, or mistake one word for another. If they do, part of the original information is lost. They are not allowed to add any comments or any remarks of their own, for this too would distort the information. "Translation, treason" is an old saying. It states very correctly the fact that part of the information may be lost in the translation process.

Between these two extreme cases, we discover in human activities a whole field of intermediate situations. A violinist, "interpreting" a Beethoven sonata, is submitted to the limitations of a translator. He is not supposed to add one note of music or to omit one, but his role is not entirely passive. He has to feel and understand the music and to communicate this feeling to the audience.

What is the position of a mathematical machine? It is a typical example of translation, of completely passive work. The machine receives on the punched tape all the information concerning the problem. Similarly, the translator is given a paper to translate, a grammar, a dictionary, and he must know all the rules required to change an Iranian discourse into an English version of it. Machine and translator have to follow the rules exactly, and nothing else.

Let us repeat the example on a slightly different problem. Suppose you receive in your mail a scientific paper written in Japanese. All the information is there, but you cannot read it, you are unable to receive the message. A good translation of the paper enables you to obtain the information. If the translation is inaccurate, some information is lost, the negentropy is decreased.

In a similar way, the machine receives a complete set of information. It contains all the data and all the rules of computation (this means translation). The data logically contain the solution, but we are unable to read it, or too lazy to compute it. Information given to the machine represents a coded message, which requires decoding. The machine blindly applies the rules for decoding and yields a readable message. Machines do not think. Mechanisms have no imagination and no invention. They simply translate the data into a different language which we can understand. That is all, and that represents a great deal.

Let us consider a specific example: the computation of firing tables for a certain gun. We write down on the punched tape all the computations to be performed, according to the laws of motion of the bullet through the atmosphere. We also punch on the tape the initial data: position, direction, and initial velocity. The mathematical problem is thus completely and entirely stated. The data punched on the tape define mathematically the trajectory of the projectile without any uncertainty, but we have not enough time and patience to make the computation ourselves and obtain the point of impact. The information is complete, in its special coded form, but we do not know how to decode it. The machine performs the decoding operation. It does not yield any new information, but simply translates into another language the information it received on punched tape.

We can prove it, by simply operating backward: give the machine the final data and the laws of motion, and instruct it to compute the initial data.

An ideal machine will give you exactly the same initial data from which it started in the first run. An actual machine will yield these initial data with a certain error: some of the information has been lost. No "absolute information" was gained.

We may try a similar experiment with the translator, and let him translate back into Japanese the English text of the scientific paper. Are we going to obtain the original Japanese paper after this double back-and-forth translation? The chances are that the second Japanese paper will only bear a great similarity to the original one. Some details have got lost. Information has decreased. The similarity between the problem of translation and the operation of a mathematical machine is really striking.

In both cases, the absolute information is not modified, but it is presented in a different manner, which makes it understandable to the reader. This is a special example of a general problem that was discussed in our previous paper. We introduced the idea of a sort of "absorption spectrum" for the reader. This comparison can be used here again. We cannot see X rays directly, but we may use a fluorescent screen which changes X rays into visible light, within the range of the "absorption spectrum" of our eyes. The translation acts in a similar way and brings the information within the range of our receptivity. And so does the machine when we are dealing with mathematical problems: it changes information of "invisible" type into visible information.

We may discuss the problem from another angle. Let us look at the "block diagram" of a large-scale computer. We have a system of rectangles, each of which represents some part of the machine. Some of these rectangles are labeled "program," "addition," "multiplication," "memory," and so on, and a system of arrows indicates the flow of information from one part of the machine to another part. There is no block labeled "imagination" nor "unprogrammed comparisons." These essential features of creative thinking are totally missing in the machine. The machine follows a given program and does nothing else.

Active thinking has been done by the designers of the machine and is done by the staff of scientists using the machine. Creative thinking is not to be found in the machinery itself.

Notes

1. In order to avoid any confusion, let us specify that we do *not* suggest that the machine operates contrary to the second principle. The machine works according to physical laws, including the second principle. The only thing it can do (and actually does) is to delay the action of degradation. In a waterfall, energy is definitely degraded. If we build a power station, we transform the energy of the waterfall into electric energy, then into mechanical energy, and finally into heat by friction. The final degradation is inescapable, but human ingenuity manages to delay it.

21.

The Entropy Concept and Psychic Function

MORTIMER OSTOW

IN HIS ARTICLE, "Life, Thermodynamics, and Cybernetics," Dr. L. Brillouin offers an objection to the conception that the recording of information involves a change in entropy as now defined thermodynamically. Such an entropy change would have to take into account whether or not the information was previously known to the observer, whether the observer can understand, and whether he is interested in, the observation. Dr. Brillouin suggests an analogy with the computation of the entropy of light beams which depends upon the absorption spectrum of the recording instrument. A new definition of entropy, he suggests, involving the nature of the receiving instrument, is therefore required for use in biology.

Dr. Richard C. Raymond, in his reply, "Communication, Entropy, and Life," proposes that the entropy of a system be computed as the sum of the entropy of all its components plus a negative term representing the organization of its components. In an open system comprising a reading lamp, recorded information, and the reader, negative entropy may appear within the reader if he is interested in and can understand the information. Dr. Raymond does not state, however, at what point in the open system the entropy increases when it decreases within the reader.

In a second article, "Thermodynamics and Information Theory," Dr. Brillouin concerns himself with the paradox that the amount of information can be multiplied indefinitely by such

processes as printing or photography without really any loss of negative entropy from the original source. That is, it seems that negative entropy can be created indefinitely. To resolve this paradox, he suggests that information be classified in terms of absolute information and distributed information. The absolute information refers to information that exists anywhere in the universe and communication of such information does not contradict the second law. Distributed information, he says, cannot follow this law. However, we are concerned most with distributed information. The fact that one scientist in one laboratory in one corner of the world has arrived at a certain conclusion is of no help to anyone else unless his conclusions are communicated and tested. Most of our thoughts are not original, even with ourselves. If only absolute new information is found to obey the second law, then really the entropy concept cannot be used in much of our work in psychology, sociology, and communications. Brillouin also speaks of "the decay of information value" and states that this corresponds to the second law. However, he fails to demonstrate any connection between the "value" of a piece of information and its negative entropy. Brillouin also attempts to contrast the entropy of "active creative thinking" with the entropy of "passive work." It seems to me that actually in both cases the outcome is implicit in the data and that the difference between the two types of thinking is merely a difference in the technique used to obtain the results. The entropy difference between state A and state B is independent of the pathway taken from state A to state B. If Brillouin believes that the brain can do thinking that machines cannot because of their intrinsic natures, he may be

From Mortimer Ostow, "The Entropy Concept and Psychic Function," *American Scientist*, 39 (1951), 140–44. Reprinted by permission of the author and publisher.

stating a conviction but he has failed to demonstrate the point.

The consideration of the nature of information by physical scientists rather than by biological scientists has led to a concern with the manipulation of data already set down by a thinking being, without regard for the fact that the information arises in a mind and is meant solely for communication to another mind, or even to the same mind at another time. We may think of the mind as a device for the establishment of relationships among the few types of primitive sensory data.

Let us consider a simple visual perception. The light reflected by the object perceived strikes the retina. Those visual elements which are illuminated discharge so that the visual patterns appear a few synapses later at an end station. Now the spatial distribution of the object exists and can be detected at several points: at the object itself, on the retina, and in the brain. Although one may say that the pattern of the object has been communicated to the brain, there has been no actual transfer of material or energy from object to brain and therefore one cannot say that there has been an entropy transfer. Light from an extraneous luminous source has been reflected upon the retina. The negative entropy inherent in the light pattern on the retina derives exclusively from the luminous source with no contribution from the visualized object. Similarly, although it is true that the absorption of light energy is the trigger which initiates the light impulse, the *vis a tergo* for the transmission of the impulse within the central nervous system arises in the metabolic discharge which takes place in the nerve fiber itself. Hence, whatever the nature of the neural alterations by which the spatial pattern of the visualized object is perceived in the brain, the negative entropy inherent in those changes is provided by the intraneural energy sources within the brain. Hence there is no material or energic transfer from object to brain and it is misleading to say that the negative entropy of the space distribution of the object is communicated to the brain unless it is clearly understood that no actual transfer is meant—that this is merely a manner of speaking.

It is more useful to consider that the negative entropy of the brain change is provided by intraneural metabolism and that the retinal image is used merely as a template; similarly that the negative entropy of the retinal image is provided by the luminous source with the object itself used as a template. If we consider auditory or tactile perceptions, in which cases energy is emitted by the object of perception and the energy impinges directly on the sensory organ, the incident energy merely has a trigger function since the energy of centripetal transmission and recording is provided by intraneural metabolism. The mechanical deformation constitutes a temporal template by which the intraneural energy is more efficiently degraded.

From the above discussion, one may infer that organizational negative entropy does not exist *in vacuo* and therefore cannot be independently created or transmitted. The negative entropy of the spatial distribution of the object is inherent in its physical structure and was provided in the act of construction. Although for our purposes we may concern ourselves with abstracting from the total the negative entropy of the pattern alone, it would be fallacious to believe that our abstraction implies the independent existence of this negative entropy. Similarly, the negative entropy of the retinal image was provided by the luminous source, and the negative entropy of the brain representation was provided by the brain metabolism. At no point does organizational negative entropy exist independently. What is called informational negative entropy is provided by the individual impressions in the case of the printed page, by the electric power source in the case of electrical transmission, and by muscular effort in the case of vocalization. The pattern which is reproduced acts merely as a template that contributes neither material nor energy.

Thus the paradox of the endless propagation of negative entropy by circulation of information does not exist. The circulation of information is limited by the physical modes of such circulation. It has not been emphasized in the literature that there is a tremendous disparity between the order of magnitude of the entropy of physical and chemical changes inherent in communication and the order of magnitude of the entropy of the information. Thus, the entropy change inherent in the arrangement of autos in a garage by age or color or price or any other criterion for classification is incommensurate with the entropy change involved in the actual movement of the cars into position—and the classification negative entropy is provided by the combustion of the gasoline. It should be the property of a good device for handling relationships, whether it be a brain, a computing machine, or a communication system, that the ratio of organizational negative entropy to power entropy be as great as possible.

So far we have discussed only perception; that is, the recording within the brain of external objects—their forms, nature, and distribution. The brain has the function of forming inferences with these data, making predictions, and making plans. The details of just how this job is accomplished elude us. However, the solution to any problem the brain has to consider may be stated as: objects A_1, A_2 ... A_k ... A_n bear relationship R_k to each other. To reach this conclusion, the brain must make a proper choice of objects A_k, A_l, A_m, etc., and a proper choice of relationship R_k. It is conceivable that if the total number of objects and possible relationships were known together with their relative probabilities, the probability that the proper conclusion would be reached by a random guess rather than by direct thought could be calculated, and therefrom the negative entropy change of the thinking process. This is true no matter whether the data include one or more of the elements A_k or the relationship R_k, and the problem would be to discover whatever is not given.

Now if the problem is unfamiliar, the conclusion will be a new proposition which will represent a negative entropy change within the brain as indicated above. If the problem is familiar, the cerebration process can be much briefer and may become merely remembering. Nevertheless, the result will be the appearance of the same proposition in consciousness. Clearly, the problem has been solved in each case and the result in negative entropy in consciousness is identical. What differs is merely the amount of cerebration required to obtain the proper result; that is, the costs of the problem solution and the efficiencies differ.

Now that the proposition which constitutes the solution of the problem exists in consciousness and the proposition implies a negative entropy, how is this solution communicated to the outside world? There is no direct communication between the brain and the world. The brain can influence the world only by manipulation of muscles and glands. Any organization and its accompanying negative entropy appearing in the world are accomplished only by muscular effort and energy, and although the external world may be made to conform to a pattern which exists in the brain, the brain pattern is not dissipated thereby. This is essentially the same argument as that pursued above with respect to sensation.

In all these considerations there is no point at which any vitalistic hypothesis may be permitted to enter and no reason to suspect that any modification or limitation of the second law is required. In dealing with information or, more properly,

thought problems, in previous issues of this journal, Dr. Brillouin and Dr. Raymond considered the question of whether the informational negative entropy of a page of type was related to its meaningfulness to a given reader. In a communication to me, Dr. Raymond makes the telling point: Any arrangement of dots or letters, he says, may be used for storage of information by proper choice of code and therefore may be said to contain stored information. The creation of a page of print of given complexity requires the entropy increase no matter what it says. What a given individual can do with it depends upon the individual, except that the complexity (i.e., negative entropy) of the information stored in the central nervous system can never exceed the complexity of the information presented, which, in turn, can never exceed the increase of entropy involved in the recording process.

However, when once we begin to apply the entropy conception to biological problems, we encounter new difficulties. Every change in chemistry and physics has a definite entropy value apparently without regard to the frame of reference. But in chemistry and physics the frame of reference is held constant by a set of conventions based upon what we know of our ability to measure directly or by instruments. That a physical change may have two entropy values depending upon the knowledge of the observer is well illustrated in the paradox cited by Bridgman. Let two gases, A and B, be separated by an impermeable partition within a large container. When the partition is removed the gases will mix and an entropy change of definite and predictable value will occur. Let us now go back to the original situation with the partition still in place and let us assume that A and B are the same gas. Now when the partition is removed exactly the same molecular diffusion process will occur as in the first case, but no increase of entropy will result because there is no way of distinguishing the gas distribution after removal of the partition from that before the removal of the partition. This is not a purely hypothetical situation. The mixing of solutions of a given sugar, one containing sugar of natural origin and one containing synthetic sugar, is accompanied by an increase in entropy only to the individual who has knowledge of polarimetry. Therefore, the value of the entropy change for the process depends upon whether the observer chooses to be aware of or to ignore the technique of polarimetry.

This is the type of frame-of-reference adjustment and definition required in biological work.

With respect to any given system one has the right to contemplate all of its permutations and combinations and to label any group of them B and all the rest A, and to calculate an entropy change value from the probability that from any of the random situations labelled A the system will pass to any of the situations labelled B. In discussing the computation of the entropy change of a diffusion process such as, for example, the diffusion of salt ions through a large body of solvent, it may be stated that the process is one associated with an increase in entropy because it is more likely that the salt ions will be distributed throughout the tank rather than concentrated at one end. The fact is, however, that any particular configuration of ions in which they are homogeneously distributed is in itself no more probable than any configuration in which they are bunched at one end. The essence of the matter lies in the fact that there are more possible configurations which will satisfy measuring instruments as constituting a homogeneous distribution than there are possible configurations which can be appreciated as bunched. Therefore, it is more likely that the salt ions will seem uniformly distributed to our instruments than bunched at one end. Hence, it is not the bunched or concentrated nature of the salt ions that determines the likelihood of the situation, but merely the fact that such a specification limits to a small number those equally probable configurations which will satisfy our requirements. Let us consider any system which contains a large number of small elements in continual independent motion. In the course of changing configurations of the system it will on occasion be noticed that a particular corner (relatively large compared with the size of the elements) of the container is momentarily empty of elements. If we define as state A that state of the system in which this corner is empty, and as state B that state in which the corner is not empty, it is evident that fewer configurations will satisfy the definition of state A than state B. Therefore, the existence of state A is less likely than state B and if the system is found in state A it is more likely to change to state B than not: state A has a lower entropy than state B. If now we define as state A' that state of the system in which two corners are simultaneously empty, it becomes evident that there are even fewer configurations which satisfy the definition of state A' as of state A; therefore state A' will have a lower probability and a lower entropy than state A. In general, the more restricted and limited we make our definition of state A, the lower will be the probability and entropy of that state.

Again, if an animal is cut with a knife, an entropy increase appears in the muscle or instrument responsible for the cutting, and by most physical criteria an increase in entropy appears in the cut structure of the animal. If, however, the cut results in the removal of a tumor or the drainage of an abscess, the effect on the animal can be considered either a decrease in entropy or an increase in entropy, depending on how one defines states A and B. In biological work, one must be prepared to deal with entropy values which vary markedly with subtle differences in definition of initial and final states. However, so long as entropy changes are calculated by probability, the second law holds no matter how the states are defined.

22.
From Stimulus to Symbol: The Economy of Biological Computation

HEINZ VON FOERSTER

MAN'S HERITAGE is of two different kinds. One has been accumulated through perhaps two billion years of evolution and is encoded in the molecular structure of his genetic make-up. The other has been built up during approximately one million years of communication and is encoded in the symbolic structure of his knowledge.

While man evolved as a result of interplay between genetic mutability and environmental selectivity, his self-made symbols evolved as a result of interplay between his flexibility in expressing and his sensitivity in distinguishing. This observation links these two evolutionary processes in a not too obvious way, and gives rise to the formidable problem of demonstrating this link by tracing structure and function of the symbols he uses back to the cellular organization of his body.

It is clear that we are today still far from a solution to this problem. First, we do not yet possess a consistent comprehension of structure and function of our symbols, to wit, the Cyclopean efforts by various linguistic schools to establish a concise language for dealing with language; second, our knowledge of the cellular organization of the body is still meager, despite the incredible amount of knowledge accumulated over the past decades. As a matter of fact, it is indeed doubtful whether with presently available conceptual tools this problem can be solved at all. These tools, however, will permit us to get an insight into the magnitude of this problem.

From Heinz von Foerster, "From Stimulus to Symbol: The Economy of Biological Computation," in Gyorgy Kepes (Ed.), *Sign, Image, Symbol* (New York: George Braziller, 1966). Reprinted with permission from the author and publisher.

An approach which considers symbolization in the framework suggested by the formulation of this problem does have the advantage that it can tie together evidences accumulated in a variety of fields. Moreover, within the framework suggested here it becomes impossible to talk about symbols in a static, ontological way and not consider the dynamic evolution of symbolic presentation. Likewise, it becomes impossible to separate a symbol from its symbolizer, his sensory motor and mental capabilities and constraints. And further, it becomes impossible to separate symbol and symbolizer from his environment which we have to populate with other symbolizers in order that symbolization makes any sense at all.

The following is an attempt to establish clues for the understanding of potentialities and limits of symbolization through the understanding of variety and constraints in the maker and user of symbols and in his environment.

The argument will be presented in three steps. First, the concept of "environment" and the relation "environment–environmentee" will be discussed. The second step will be to briefly sketch some basic principles and some hypotheses of the processes that permit internal representations of environmental features. Third, modes of projecting externally these internal representations will lead to the consideration of possibilities of interaction by symbolization.

Environment: An Analysis

Evolution, like memory, is an irreversible process. The man who once knew a datum, but has forgotten it now, is different from the man who never knew it. Irreversibility in evolution permits one to picture this process in the form of a tree with divergent branch points only. Time runs from bottom to top and the number of different species at any time within each branch is indicated by the width of this branch. A subspecies among mammals called *homo sapiens*, including its entire temporal extension, occupies in this graph but a tiny speak of space in the corner of the mammalian branch.

It is perhaps easy to see that this graph represents paleontological estimates of only those species that were sufficiently stable to leave detectable traces. All instable mutants escape detection, and thus cannot be accounted for. In other words, this graph is essentially a picture of the success story of living forms. This observation permits us to look at this representation in a slightly different way, namely, to consider each point in a branch as being an instant at which a crucial problem is presented to a particular species. If it solves this problem the point will be retained and moves upward an ever so slight amount. If not, the point will be removed, *i.e.*, the species is eliminated. It is clear that the crucial problem referred to here is how to survive, and it is also clear that this crucial problem is posed by the properties of the particular environment which is in interaction with elements of this species or its mutants.

From this viewpoint "environment" is seen in a two-fold way: as a set of properties of the physical world that act upon an organism; and also as an accumulation of successful solutions to the problem of selecting such conditions in the physical world which are at least survivable. In this discussion "environment" will always carry this relative notion as "environment of . . .," where environment and the organism associated with it will be duals to each other in the sense that a particular organism O implies its particular environment E(O), and vice versa, that a particular environment E implies its appropriate organism O(E).

By carving out from the physical universe just that portion E(O) which is "meaningful" for this organism O, one has carved out a portion that is necessarily of compatible complexity with that of the organism. An organism that tolerates a variation of temperature of, say, thirty degrees Fahrenheit around a certain mean, cannot "dare" to move into places where temperatures vary beyond this tolerance.

This statement can be expressed differently. An organism that is matched to its environment possesses in some way or another an internal representation of the order and the regularities of this environment. How this internal representation within the cellular architecture of living systems is achieved will be taken up later in this paper.

At this point the concept of "order" needs further clarification. Intuitively one would associate order with the relation of parts in a whole. But what are parts? Again, intuitively, parts emerge as "separabilia," because the relation among their components is of higher order than that of the parts of the whole. Although this definition is circular, it points in the right direction, for it relates order to the strength of constraints that control the interaction of elements which comprise the whole. These constraints manifest themselves in the structures they produce. The globular star cluster has simple spherical symmetry, because the weak gravitational forces that hold the approximately 100,000 elements of this system in statistical equilibrium have themselves radial symmetry. Of course, much more sophisticated structures are obtained if the constraints are more numerous and stronger. The volume *Structure* of this present series [*Structure in Art and in Science*, Gyorgy Kepes (Ed.)] abounds with beautiful examples from nature and art, where either strong molecular forces (*e.g.*, the paper by Cyril Stanley Smith) or the application of strong principles of construction (*e.g.*, the paper by R. Buckminster Fuller) generate structures of great intricacy and sophistication. Here only one shall be given, the almost inexhaustible variety of hexagonal symmetries in snow crystals. The growth mechanism of these crystals is subjected to a major constraint, namely the triangular shape of the water molecule H_2O which has two hydrogen atoms attached to the big oxygen atom at angles which are close to either 30° to 60°. This slight deviation from the condition that would produce equilateral shapes introduces a certain amount of "freedom" for the molecules to attach themselves to each other, which in turn allows for the large variability within this constraint. Note that in spite of the great difference in the individual shapes of these crystals, no difficulty arises in recognizing these forms at a glance as snow crystals. This suggests that the cognitive apparatus that "figures out"— or computes—the answer to the question "What is this?" is the one thing that is common to all these shapes, and this is the constraint in their growth mechanism. The name we give to this constraint is simply "snow crystal."

In the temporal domain order is again generated by the constraints of the "Laws of Nature" which, on the macroscopic scale of direct observation, control the chain of events. Chaos would permit transitions from any state to any other state, mountains transforming themselves into flying pink elephants, pink elephants turning into yellow goo, etc. Not only are organisms impossible in this world, for by definition, there is no law that holds the organism together, but also this world is indescribable, for description requires names, and names refer to the "invariabilia"—the constraints —in the environment.

One clue of how to compute these constraints from the apparent structure of the environment is suggested by the preceding examples. Structure in space was determined by a law in the growth mechanism that permitted attachment of new neighbor elements only at particular points; structure in time was determined by a law in the transition process that permitted only a particular event to be neighbor to an existing one. In other words, spatiotemporal order is generated by constraints that control spatiotemporal neighborhood relationships. Hence, if these can be "sized up," the constraints can be evaluated.

If chaos permits every event to appear with equal probability, order emerges from chaos when certain transitions of events become more probable than others. Certainty of an event following another creates a perfect, deterministic universe, and the problem of how to survive in such a deterministic universe is reduced to finding the constraints that govern the transitions from one event to the next. Clearly, the simplest of all such deterministic universes is the one where no transitions take place, i.e., where everything is at motionless and uniform tranquility. Hence, the oceans, where temperature variations, changes in the concentration of chemicals, destructive forces, etc., are kept at a minimum, were the cradle for life.

The dual interdependence of organism–environment permits a dual interpretation of the tree of evolution. Instead of interpreting points on this graph as *species of organisms*, one may interpret them as *species of environments*. Thus viewed, this chart represents the evolution of environments which were successively carved out of the physical universe. These environments evolved from simple, almost deterministic ones, to extremely complex ones, where large numbers of constraints regulate the flow of events. An environmental subspecies among mammalian environments, called "E (*homo sapiens*)," occupies in this graph a small speck of space in the upper right corner of branch number 8. Hence, its dual, "*homo sapiens* (E)," sees "his universe" as a result of two billion years of environmental evolution, which step by step carved out from the physical universe an ever increasing number of constraints of all those in this universe that are computable within the limits of the evolving organism.

The diagram shown here below sketches the circular flow of information in the system environment–organism. In the environment constraints generate structure. Structural information is received by the organism which passes this information on to the brain which, in turn, computes the constraints. These are finally tested against the environment by the actions of the organism.

ENVIRONMENT

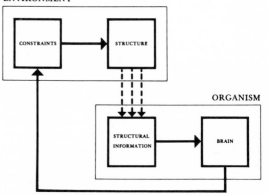

With the emergence of self-reflection and consciousness in higher organisms a peculiar complication arises. A self-reflecting subject may insist that introspection does not permit him to decide whether the world as he sees it is "real," or just a phantasmagory, a dream, an illusion of his fancy. A decision in this dilemma is important in this discussion, since, if the latter alternative should hold true, no problems as to how organisms represent internally the features of their environment would arise, for all environmental features would be just internal affairs in the first place.

In which sense reality indeed exists for a self-reflecting organism will become clear by the argument that defeats the solipsistic hypothesis. This argument proceeds by *reductio ad absurdum* of the thesis: "This world is only in my imagination; the only reality is the imagining 'I.'"

Assume for the moment that [a] gentleman in [a] bowler hat . . . insists that he is the sole reality, while everything else appears only in his imagination.

However, he cannot deny that his imaginary universe is populated with apparitions that are not unlike himself. Hence he has to grant them the privilege, that they themselves may insist that they are the sole reality and everything else is only a concoction of their imaginations. On the other hand, they cannot deny that their fantasies are populated by apparitions that are not unlike themselves, one of which may be *he*, the gentleman with the bowler hat.

With this the circle of contradiction is closed, for if one assumes to be the sole reality, it turns out he is the imagination of someone else who, in turn, insists that *he* is the sole reality.

The resolution of this paradox establishes the reality of environment through evidence of a second observer. Reality is that which can be witnessed; hence, rests on knowledge that can be shared, that is, "together-knowledge," or *con-scientia*.

Internal Representation of Environment: A Physiology

Distributed over the surface of multicellular organisms are highly differentiated cells that establish the interface between the proceedings of the external world and the representations of these proceedings within the organism. To some variables in the physical universe these cells, called sensory receptors, have become specifically sensitive: for example, cells, sensitive to changes in pressure are insensitive to, say, the changes in the concentration of sodium chloride in the water surrounding the organism, etc., etc., and vice versa.

Sensitivity of a receptor cell to a specified perturbation is observed by its response in the form of a short electric discharge, which, after it has been initiated at the surface, travels into the interior of the organism along a thin fiber, the axon, which protrudes from the cell.

The approximate duration of this discharge is several thousandths of a second and its magnitude always about one-tenth of a volt, irrespective of the intensity of the perturbation. A prolonged perturbation produces a sequence of discharges the frequency of which corresponds approximately to the logarithm of the intensity of the perturbation. In engineering language the encoding of an intensity into frequency of a signal is called frequency modulation, or FM, and it may be noted that all sensory information—irrespective of sensory modality—is coded into this common language.

If a perturbation is permanently applied, the interval between pulses slowly increases until the sensor fires at a low frequency—called the resting rate—which is independent of the intensity of the permanent perturbation. This penomenon, "habituation," is one example of computational economy in living organisms, for a property of the universe that does not change in space or time can safely be ignored. Air has no smell. It is the change of things to which an organism must be altered.

A specific perturbation that elicits responses of a sensory receptor is called stimulus. Stimulus and receptor are duals in the same sense as are environment and organism. Consequently, a tree of the evolution of sensory receptors could be drawn which, at the same time, would show the successive acquisition of specified properties of the physical universe that are selectively filtered out from the rest of the universe.

In the higher animals the most intricately developed sensory system is that of their visual organs. Distributed over the human retina are 180 million sensory receptors of essentially two kinds, the rods and the cones. Rods respond to brightness in general and are more concentrated on the periphery, while cones respond to brightness modified by a variety of pigments and are more concentrated in the central part of the retina, the fovea. The fovea, by proper accommodation of the crystalline lens, has the lion's share in transducing the information contained in the inverted image focused on the retina. The concentration of sensors in the fovea is very high indeed. An area on the retina of the magnitude of the small, black, circular spot that indicated termination of the previous sentence contains approximately 20,000 cones and rods. The projected image of this spot, when looked at under normal reading conditions, is "seen" by about 200 cells. Since each cell distinguishes about 60 levels of brightness, the number of images distinguishable by this small ensemble of 200 cells is exactly $(60)^{200}$, or approximately 10^{1556}. This is a meta-astronomical number which, if printed out on this page, spreads over 13 lines.

It is clear that this overwhelming mass of information is neither useful nor desirable, for an organism has to act; and to act requires making a decision on the available information, which in this case is so large that it would take eons of eons to initiate action, even if the evidence were scanned at lightning speed. Moreover, any accidental distortion of the image—may it be ever so slight—caused, say, by light scattering in the vitreous humor, by optical aberrations in the lens,

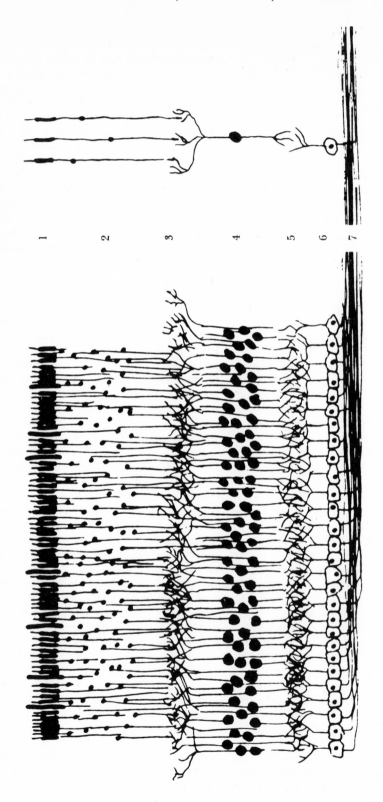

Figure 1.—Semischematic drawing of the postretinal neural network. 1. Rods and cones. 2. Nuclei of cones and rods. 3. Interaction between sensors and bipolars. 4. Bipolar cells. 5. Interaction bipolars and ganglion cells. 6. Ganglion cells. 7. Optic nerve.

such as achromatism, astigmatism, temporary failure of single receptors, etc., etc., would pass as evidence with equal weight and be admitted in the decision-making operation.

What, then, protects the brain from overflow of information?

A first clue was discovered by counting the fibers in the optic tract that is the bundle of nerves which connect the eye with the brain. Here one counts only one million fibers, a reduction by 1/180 compared with the number of sensors. Why this waste on the sensory level, or why this redundancy? Is all this tremendous sensory information just discarded? One has to look and to measure in order to answer these questions.

The anatomy of the postretinal neural structures is known over many decades; the knowledge of its functions emerges only slowly with advances in electronics and the refinement of microelectrodes that permit penetration of single fibers in vivo and thus permit the recording of their activity under controlled conditions of illumination.

Figure 1 shows a semischematic sketch of the multilayered postretinal neural network, that connects sensors with the fibers of the optic tract. Rods, and a few cones, with their associated cell bodies containing the nucleus, comprise layers 1 and 2, the light-sensitive nerve-endings in 1, the nuclei in 2. Their axons descend into layer 3 where contacts are established with fibers emerging from the nuclei of a second layer of cells, the "bipolars," in layer 4. Their axons, in turn, connect in layer 5 with branch-like ramifications, the "dendrites," emerging from cells of a third kind, the first ganglion cells in layer 6, which send their axons into deeper regions of the brain, making up the fibers of the optic tract, layer 7.

Two features of this network should be noted. First, that only a few sensors within a spatial neighborhood contribute to one ganglion cell. Second, that the signal pattern generated at the cones and rods may be modified only in two places, namely in layers 3 and 5 where cells in different layers connect, and thus may act on their successors according to rules of signal transmission from neuron to neuron and according to the local connection scheme.

The mechanisms that determine the response of a successor cell when stimulated by the activity of its predecessor at the place of their junction— the synapse—are still today not clear. Nevertheless, it is clear that two types of interaction can take place, excitation and inhibition. An excitatory synapse will transmit to the successor the oncoming discharge, while an inhibitory synapse will cancel the trigger action of another excitatory synapse.

This observation of the two kinds of signal transmission suffices to see neural interaction in a new light, for it suggests the possibility of seeing the function of a neuron in the form of a logical operation, the affirmative corresponding to excitation, negation corresponding to inhibition. Hence, a network of synapting neurons can be regarded as a system that computes certain logical functions depending upon the type and structure of the connections.

To see clearly the significance of this observation, an idealized two-layer neural network is drawn in Figure 2. The first layer consists of "rods," each of which acts upon precisely three neurons in the second "computing" layers. Two fibers with excitatory synapses connect with the neuron just below, while two other fibers with inhibitory synapses connect with its left- and right-hand neighbor. This we shall call an elementary net. It repeats itself periodically over the entire strip, which is thought to extend far out to both sides of the figure.

What does this net compute? Assume that all sensors are uniformly illuminated. An arbitrary neuron in the computer layer receives from its corresponding sensor immediately above two excitatory stimuli which are, however, cancelled by the two inhibitory stimuli descending from the immediate neighbors of its corresponding sensor. Due to the perfect cancellation of the two "yeses" and the two "noes," the net result is no response at all. Since this is true for all other neurons in the computer layer, the whole net remains silent, independent of the intensity of light projected on the sensors. One property of this scheme is now apparent: the net is insensitive to a uniform light distribution.

What happens if a perturbation is introduced in the light path? Figure 3 illustrates this situation. Again, under regions of uniform darkness or uniform illumination the computer cells do not respond. However, the neuron at the fringe between darkness and light receives no inhibitory signal from the sensor in the shade; double excitation overrides single inhibition and the cell fires. Due to the periodicity of the elementary net, this property, namely, the presence of an edge, will be computed independent of the position of this edge and independent of the level of over-all illumination. Hence, such a network may be called an "edge detector," which when the same principle is extended into two dimensions, may be called a "contour detector."

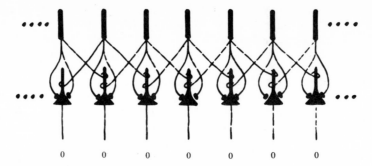

0   0   0   0   0   0   0

Figure 2.—Periodic network of idealized neurons incorporating lateral inhibition.

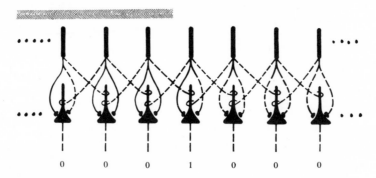

0   0   0   1   0   0   0

Figure 3.—Periodic network with later inhibition computing the property "edge."

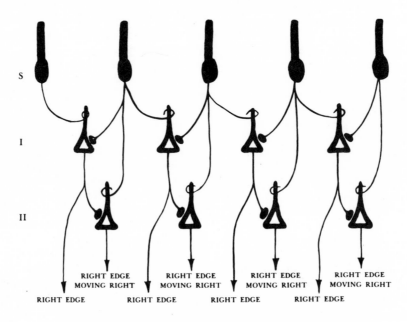

Figure 4.—Anisotropic periodic network computing the property "right edge" and the property "right edge moving right."

Other connections will compute other properties in the visual field. Figure 4 shows a periodic net with two computer layers, 1 and 2, where layer 2 utilizes results computed by layer 1. Inspection of the connection scheme may easily show that layer 1 computes the presence of a right edge (light right, dark left); while layer 2, utilizing the synaptic delay in elements of layer 1, computes a right edge moving right. Of course, no responses are obtained for left edges (light left, dark right), left edges moving left or right, and right edges moving left.

These examples are intended to show that owing to the basic computational properties of the neuron, parallel, periodic arrays of elementary networks are capable of extracting a variety of useful "invariants" in an otherwise complex environment. The theory that connects structure and function of such networks with the invariants they compute is fully developed. Given any universal property to be computed, the appropriate network to carry out this computation can be synthesized.

To establish similar correlations in actual physiological nerve nets is infinitely more difficult. Nevertheless, during the last couple of years in a series of brilliant experiments the computation of invariants by postretinal networks in some vertebrates (frog, pigeon) has been demonstrated. The experimental procedure consists of observing responses of single fibers in the optic tract elicited by the presentation of various visual stimuli to the retina of an anesthetized animal. These observations show indeed that certain fibers respond only if the appropriate invariant is present in their receptor field. Some of these invariants are:

1. Local sharp edges and contrast.
2. The curvature of edge of a dark object.
3. The movement of edges.
4. Local dimmings produced by movement or rapid general darkening.
5. Vertical edges only.
6. Vertical edges moving.
7. Vertical edges moving right (left) only.
   etc.

These abstracts are still on a primitive level, but it is the way in which they are computed that invites further comments. Although only those operations of the perceptive apparatus have been described which are an immediate consequence of the stimulation of sensors, some basic principles are now visible which underlie the translation of environmental features into representations of these features within the cellular architecture of the organism. Perhaps the most fundamental principle involved in this translation is the correspondence between the *neighborhood relationships* that determine environmental structures, and the *neighborhood logics* that are incorporated into neural connectivity which determine the "whether" and "where" of certain environmental properties.

This suggests two levels of computation. First, computation on the grand scale of evolutionary differentiation which incorporates the environmental constraints into the structure of those networks which, on the second level, compute within the limits of their structure spatiotemporal quantities of useful universal parameters. Clearly, the first level refers to the species, the second to the specimen. It is on the first level that the notion of "Platonic Ideas" arises, for they refer to the fabric without which experience cannot be gathered.

The importance of distributed operations that can be carried out on a distributed stimulus is further emphasized by a careful preservation of neighborhood relationships even after the original stimulus has been relayed over many cascades of computational layers into the deeper regions of the brain. [A] topological mapping—that is, a mapping which preserves neighborhoods—of our body with respect to the sensation of touch into the appropriate cortical regions . . . is obtained by registering with microelectrodes those regions in the brain which become active when certain regions of the body are stimulated. Such a "signal representation" must not necessarily conform with original proportions, as seen by the emphasis of organs that convey most of the tactile information. The importance is the preservation of neighborhoods which permit further computation of tactile abstracts.

The reliance on neighborhood relationships can cause peculiar breakdowns of the perceptive apparatus when presented, for instance, with a triple-pronged fork with only two branches. Although in all details (neighborhoods) this figure seems right, as a whole it represents an impossible object.

Similar difficulties arise when the visual system is confronted with unusual projections which do not allow quick reconstruction of the unprojected image. Erhard Schön's anamorphosis seems to picture a somewhat peculiar landscape, but "actually" it portrays the three Emperors, Charles V, Ferdinand I, Francis I, and Pope Paul III. Faces of these personalities, including their names, can easily be recovered by looking at this engraving under a grazing angle from the left.

Since all sensory modalities translate stimuli into the universal language of electric pulse activity, invariants computed by different senses may be compared on higher levels of neural activity. . . . It is on this level where we have to search for the origin of symbolization. . . .

In the light of the preceding discussion it may indeed be argued that in this case the pattern of neural activity, which represents the visual stimulus configuration, is homologous to that generated by configurations of the auditory stimulus. This argument is going in the right direction, but it fails to cope with a strange situation, namely, that earlier experience and learning is not involved in this spontaneous identification process.

Since associations gained from experience are excluded, one must assume that this audio-visual correspondence rests upon the fabric without which experience cannot be gained. The structure of this fabric must permit some cross-talk between the senses, not only in terms of associations, but also in terms of integration. If this structure permits the ear to witness what the eye sees and the eye to witness what the ear hears then there is "together-knowledge," there is *con-scientia*.

Symbolization: A Synthesis

To survive is to anticipate correctly environmental events. The logical canon of anticipation is inductive inference, that is, the method of finding, under given evidence E, the hypothesis H which is highly confirmed by E and is suitable for a certain purpose. This is computation of invariants within the limits of insufficient information, and follows the principles of invariant computations as before, only on a higher level. Knowledge is the sum total of these hypotheses (invariants, laws, regulations) and is accumulated on three levels. First, on the molecular level in the genetic structure which tests the viability of its hypotheses, the mutations, through the vehicle of the developed organism; second, on the level of the individual organism through adaptation and learning; and third, on the social level through symbolic communication which cumulatively passes information on from generation to generation.

Since these are evolutionary processes, and hence irreversible, error would accumulate with knowledge, were it not for a preventative mechanism: death. With death, all registers are cleared and untaught offspring can freshly go on learning. This mechanism works on the first and second levels, but not on the third.

To cumulatively acquire knowledge by passing it on through generations, it must be communicated in symbols and not in signs. This separates man from beast. Communication among social insects is carried out through unalterable signs which are linked to the genetic make-up of the species. While signs refer to objects and percepts, and serve to modify actions and manipulations, symbols refer to concepts and ideas and serve to initiate and facilitate computation.

Since the ultimate relation between symbols and environmental entities is cascaded over the relations symbol/concept and concept/environment, it is in its logical structure very complicated indeed. This gives rise to breakdowns that manifest themselves on various levels of semantic morbidity.

Symbols share with concepts and ideas the property that they do not possess the properties of the entities they represent. The concept of roses "smells" as much, or as little, as the concept of jumping "jumps." The concept of a square is not quadratic. If this point is missed, a number would be just so many fingers and a square with area 2 would have nonexisting sides.

Since symbols refer to concepts and ideas, they too may not have the properties they represent. The symbol of a square may not be quadratic, as can be clearly seen by the string of peculiarly shaped little marks on this paper that have just been used to refer to this geometrical figure. This was, of course, well understood when mystical experience was to be coded into symbols. . . .

What, then, determines the form of a symbol; is it an arbitrary convention, or does it convey its meaning by its shape? Again, ontologically this question cannot be resolved. One has to look into the ontogenesis of symbolic presentations.

We here repeat the diagram seen earlier which represents the information flow between a single organism and its environment:

ENVIRONMENT

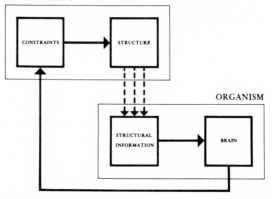

Since symbolization requires at least two interacting subjects who are immersed in an environment that is common to both, we must extend this diagram to admit a second subject. This is done here below:

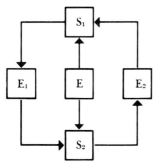

Subjects S_1 and S_2 are coupled to their common environment E. In contrast to the first diagram in which the organism is faced only with an environment with given constraints, now each of these subjects is confronted with the additional complication of seeing his environment populated with at least one other subject that generates events in the environment E. Hence S_2 sees, in addition to the events generated by E, those generated by S_1, and since these take place in E, they shall be labeled E_1; conversely, subject S_1 sees in addition to events generated by E those generated by S_2 which will be called E_2. Thus, in spite of the fact that both S_1 and S_2 are immersed in the same environment E, each of these subjects sees a different environment, namely, S_1 has to cope with (E, E_2), and S_2 with (E, E_1). In other words, this situation is asymmetrical regarding the two subjects, with E being the only symmetrical part.

Assume that E_1 and E_2 are initial attempts by S_1 and S_2 to communicate environmental features to each other. It is clear that these attempts will fail unless—and this is the decisive point—both subjects succeed in eventually converging to like representation for like universal features. This process may be expressed symbolically:

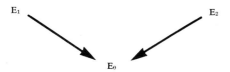

The arrows indicate the convergence process, and E_0 stands for the final universal "language" spoken by both subjects. At this point the initial asymmetry ceases to exist and both subjects see the same environment (E, E_0).

As in all evolutionary systems, the outcome of this process cannot be predicted in the usual sense, because the goal which establishes equilibrium is not directly visible in the final equilibrial state which is a communicable symbol, while the goal is communicability.

Symbols must not necessarily have the shape of the objects they ultimately refer to, yet within that freedom there are constraints working in the evolution of symbolic representation which confine their development within reasonable limits. One of these constraints is dictated by the tools with which these symbols are generated, the other one is their syntactical structure.

An example of the first kind of constraints operating on the development of written symbols is given in the development of highly stylized symbolic forms from initially representational pictograms. This transition is believed to have taken place in the two millennia of Sumerian cultural activity between 4000 and 2000 B.C. It is clearly seen how the constraints imposed by the writing tools—a stylus with triangular cross-section pressed into soft clay—strongly modify the early pictograms. It may be interesting to note that simultaneously with this departure from structural representation goes an increase in the possibility to add modifiers to the original meaning. While the pictogram says "foot," after two thousand years of stylization it may stand for "walking," "running," "delivering a message," or other "foot-connected" actions if associated with appropriate modifiers. Nevertheless, in some instances it seems to be possible to see behind the form of later symbols the shape of earlier pictorial representations.

The other kind of constraint is a structural one and does not show itself in an obvious way, for symbols carry rules of connectivity and not so much rules of entity. Symbols may be compared to atoms which react to particular atoms only to form the molecular compounds, but are inert to all others. Take, for instance, these "molecular" sentences:

"Socrates is identical."
"4 + 4 = purple."

The disturbing thing about these is that they are neither true nor false; they are nonsensical. The connection rules of the symbols have been violated in these examples. "Identical" sets up a relation between two entities. "Socrates is identical with Plato" is a sentence that makes sense although it happens to be a false proposition. The compound "4 + 4 =   " requires a number to follow. Putting "6" at the end is a good guess, but "purple" is an operator with an entirely different structure. This indicates that somehow structure is still preserved in symbolical discourse, but in a syntactical and not in a representational sense. The language of symbols has, so to speak, its own logical grammar. Uniqueness in symbolic expressions is established in a way similar to that of a jigsaw puzzle in which pieces can be put together in one, and only one way. It is the some-time far-extending neighborhood relationship among the pieces—the symbols—that puts them into place.

It is clear that the constraints expressed in the neighborhood relationships of symbols reflect constraints in the environment. For instance, a sentence that refers to two particular persons must employ two proper names. To establish connection rules among symbols of speech is the linguistic problem. One of the most primitive connectivities among words is the probability of their succession. With the following two examples the emergence of order by tightening the constraints of succession will be demonstrated. These examples are random sequences of words generated by a chance device which, however, takes into account the various probabilities by which a particular English word follows a number of precursors. In the first example the number of precursor words is two:

. . . THE HEAD AND IN FRONTAL ATTACK ON AN ENGLISH WRITER THAT THE CHARACTER OF THIS POINT IS THEREFORE ANOTHER METHOD FOR THE LETTERS THAT THE TIME OF WHO EVER TOLD THE PROBLEM FOR AN UNEXPECTED. . . .

In the second example the constraints are tightened by extending the neighborhood relationship up to four words:

. . . HOUSE TO ASK FOR IS TO EARN OUR LIVING BY WORKING TOWARDS A GOAL FOR HIS TEAM IN OLD NEW YORK WAS A WONDERFUL PLACE WASN'T IT EVEN PLEASANT TO TALK ABOUT AND LAUGH HARD WHEN HE TELLS LIES HE SHOULD NOT TELL ME THE REASON WHY YOU ARE IS EVIDENT. . . .

Symbols are no proxy for their objects. There are two morbid states of the mind, magical thinking and schizophrenia, in which this distinction is erased. In both cases symbol and object become indistinguishable. In purpose-oriented Jou Jou and in Voodoo the identity of symbol with object is used to manipulate the world by manipulating the symbol. In schizophrenia symbol and object are freely interchanged to produce peculiar hierarchies of identities. In order to comprehend in depth the modality of this affliction, a short passage of the extensive description of the case of a six-year-old boy by the name of Walter (= water) is given here:

Late in November, 1936, presumably because he had heard a rumor about a child killed in an accident in an elevator there, he became terrified when taken to a department store. He trembled, cried, vomited and remained "hysterical" for two days during which time he made little jerking movements of his body and shoulders and said scarcely a word. The following day, Dr. Hamill was for the first time able to make out that he failed to distinguish between himself (Walter) and water. Walter shifted to water, thence to Deanna Durbin who played in "Rainbow on the River" and so to water again. Being water, he felt he could not be drowned, but might be imprisoned in the radiator. On hearing the knocking of water in the radiator, he said, "elevator just came up and gave the kid a knock" and again, "they are killing the kid," which terrified him because he was the kid. Then followed "the telephone burnt and got water after Suzy burnt."

(Dr.: "Where does water come from?") "I come from the show." (Dr.: "You thought water and Walter were the same thing.") "My father used to take me across the river." (Dr.: "And he called you Walter?") "And got drowned. I do not live on Springfield. Bad boys drink water. They do not drink milk. Good boys live on Springfield. I used to live on Springfield—Mississippi River."

It may be speculated that evolution did not weed out mental diseases that afflict proper use of symbols because the survival value of the ability to symbolize is so enormous that occasional morbid deviations of this ability in individuals and in whole cultures could still be tolerated. The enormous advantage of organisms that are able to manipulate symbols over those who can only react to signs is that all logical operations have not to be acted out, they can be computed. It is obvious that this saves considerable amounts of energy. But the really crucial point here is that errors in reasoning are not necessarily lethal.

The recognition of the fact that information is a precious commodity and can be processed by manipulating symbols gave rise to the quick emergence of the fast and large electronic computer systems. These systems manipulate symbols only and do not know objects. The laws of algebra and logic are incorporated in their structure. Hence, they cannot err by confusing modality as does a schizophrenic, nor can they err in syntax and generate nonsense. The only error they can make is confusing true with false and false with true.

The human retina with its associated, genetically structured networks may be compared to these computer systems from a purely quantitative point of view, namely by the sheer amount of information that is processed. The retina, with its 180 million sensors which operate in parallel at millisecond intervals, performs equivalently to a modern digital computer system that occupies 800 square feet of floor space and uses 4 tons of highly sophisticated electronics. In comparison, the retina's extensions are 2 square inches by 4/1000 of an inch, and it weighs approximately 100 milligrams. This may be taken as an indication of the economy of biological computation.

BEHAVIOR AND MEANING

23.

The Application of Information Theory in Behavioral Studies

F. C. FRICK

Introduction

Information theory is a formal—as opposed to substantive—theory. It could be regarded as simply an extension of correlation theory. It is not a model of behavior, not even communicative behavior, but rather a tool that may be used in the construction of such models.

. . . . .

Unfortunately it is still too early to evaluate the effects on psychology of information theory itself, or the conceptual position which it represents. The present author's biases in this respect should, however, be stated. Essentially they are: (1) the viewpoint represented by the theory is fruitful and important, and (2) the mathematical formalism does not lend itself to uncritical, mechanical application. Even if the data are believed to meet the mathematical requirements, the computation of information transmission, say, does not in itself insure any deeper understanding of the process under study. This is, of course, characteristic of formal theory in the sense that we have been using the term; it is, however, a point that may be disguised by the computational complexities involved.

The insight that led to the development of information theory was the realization that all the processes which might be said to convey information are basically selection processes. Speech, for example, can be regarded as a sequence of selections from a number of possible choices—the

Condensed from F. C. Frick, "Information Theory," in *Psychology: A Study of a Science*, Vol. 2, pp. 611–15, 629–36, edited by Sigmund Koch. Copyright © 1959 by McGraw-Hill, Inc. Used by permission of the author and McGraw-Hill Book Co.

phonemes, letters, or words of the language. Alternatively, we need information only when we are faced with a choice of some sort. If I know the road to Boston, I do not need a route sign at the intersection. In the same sense, a completely determinate process would not be regarded as an information source. For as it moves through its appointed course, I can predict its motions in advance, and I receive no new information when it actually carries out its function.

Thus, information and ignorance, choice, prediction, and uncertainty are all intimately related. On the other hand, complete ignorance or indeterminance also precludes information transmission. A lecture in German is not informative to a listener who does not understand German. There must be some degree of agreement, some sort of common language established between the information source and the receiver. Put somewhat more precisely: information processes are selection processes, but these selections must be made from a *specific* set of alternatives, and if the sequence of selections is to convey information, the possible choices must be known to the receiver. To say "yes" or "no" to someone who has not asked a question is not informative in any usual sense of the word. It is this restriction on the set of alternatives that makes it possible to speak precisely and quantitatively about information.

Within these bounds of complete knowledge and complete ignorance, it seems intuitively reasonable to speak of degrees of uncertainty. The wider the choice, the larger the set of alternatives open to use, the more uncertain we are as to how to proceed—the more information we require in order to make our decision. Thus, color alone will distinguish an apple from an orange, but we will have to know more than just color if we are to distinguish among apples, oranges, grapefruit, and lemons.

In much the same way, the informational value of a choice seems also to depend upon the likelihood of that choice. If we feel certain of what a man is going to say, it is not very informative when he says it. If a process is highly predictable, we need only a few observations to establish its present state quite accurately. Our uncertainties are intimately tied up with probability estimates and if we are to fit our intuitive notions regarding information, we must consider not only the range of choices available but the probabilities associated with each. This reflects the second basic insight behind the development of information theory. We may summarize the position by stating two premises: (1) information is associated with a selection process. (2) Such a process is basically statistical in the sense that it involves probability considerations.

From these premises, it has been possible to develop a measure of amount of information and a set of limit theorems which justifies the particular measure chosen. The ultimate justification of the theory will lie in the extent to which it gives precision to our intuitive notions regarding information and information transmission and opens the way to a clearer understanding of the basic processes involved.

.   .   .   .   .

The Application of Information Theory in Behavioral Studies

Information theory is a *probability* model and suffers from all the conceptual difficulties associated with probability theory. In particular, the definitions and theorems of the theory assume that it is meaningful to speak of infinite samples of past symbol occurrences and the expected value for infinite samples of future messages. The theory assumes that we know the parameter values of a statistical process and is thus subject to the same criticisms in application that have been leveled at Bayes' theorem. In fact the expression for T can be derived from Bayes' result.

This is not a new problem to psychologists, but it is worth pointing out that the passage from probability theory to statistics has not yet been achieved for information measure. In applications of the theory, the experimenter must be prepared to assume that the frequencies with which events occur are reasonable estimates of the true prob-

abilities with which the model deals. In fact, the maximum likelihood estimates, $\hat{H}$ and $\hat{T}$, have been shown to be biased and general corrections for these biases are not available. Nor do we know the sampling distribution of H.

Despite these clear difficulties with the application of information measures to finite sequences and limited data, the large number of experimental studies that have appeared in recent years using information measures is an indication that many researchers are willing to make statements about the probability distributions characterizing their data. Actually, the use of information theory to describe empirical data is not as unreasonable as it might at first appear.

In the first place, as a formal and normative theory, in the sense discussed in the introduction, information theory does not lead to deductions amenable to critical test, where precision of measurement might be vital. Rather, it establishes boundary conditions for maximally efficient coding. In general, it is the observed large deviations from the model and the search for their sources that have proven of most interest to psychologists and engineers who have applied information theory in their work.

In the second place, as Fano has demonstrated, the statistical characteristics assumed a priori can be quite different from the frequencies actually observed without the efficiency of transmission or coding being lowered very much. This arises from the fact that H varies rather slowly with any one of the $p(i)$ unless that $p(i)$ is close to zero or unity. In a similar fashion, the fundamental coding theorem assumes an unlimited delay, or message length, to achieve the optimum representation given by H. Yet we were able, in our illustration of coding procedure, to achieve a better than 99 per cent efficient code by considering messages only out to length three.

On the assumption, then, that applications to empirical data are to some extent tenable, information theory has played a number of roles in psychology. Not the least of these has been that of a general reorientation in thinking about various behavioral problems. It is very easy to substitute psychological terms into the verbal structure of the model, or to extend the technical terms of the formal theory to describe the classic concerns of the psychologist. Thus, the inputs and outputs of a channel are easily and naturally translated into the stimuli and responses of the organism. Proceeding in the other direction, as Crossman suggests, we can say: "Instead of a *stimulus* causing a *reaction* when the *threshold* is exceeded, we now think rather in terms of a *signal* which may be obscured by *noise*, providing the *information* needed to *select* a response."

Verbal recodings of this sort are not simply playful. They influence the choice of variables and the design of experiments. Crossman, for example, continues his discussion by pointing out that, unlike a stimulus, a signal (which should be regarded as the output of a transmitter as we have defined it) implies a set of alternatives and thus emphasizes the effect on behavior of what might have been as well as what is immediately present. Furthermore, a signal in this sense functions purely as the basis for response selection. It can, according to the theory, be coded into a variety of physical forms and embedded in a variety of signal sets, without effect on its selective function.

The respective dependence and invariance suggested by this reformulation of the basic psychophysical problem have stimulated a great deal of research in recent years. This work has been directed to the study of human information processing and considers, in Miller's terminology, the "transmission situation." The observer is regarded as a "channel" and the interest has been in estimating his channel capacity under various conditions and for various input stimuli.

The apparent contradiction in terms—channel capacity is defined in the theory as the maximum transmission for all possible input sources—is avoided by the intuitively reasonable assumption that man is a multi-channel system. Because the information measure is invariant under operations of coding, it becomes possible to compare information transmission among these various channels and to determine the efficiency of various coding schemes. For example, the same amount of information can be presented to an observer by exchanges between the number of symbols per second and the amount of information per symbol. Similar exchanges can be made by distribution of the input simultaneously over different channels, or by permitting different aspects of the stimuli to vary and hence "carry" information.

Studies like these which are directed at optimizing the display of information in the sense of maximizing the amount of information transmitted have been of considerable value to the growing field of engineering psychology. Perhaps, of most importance, the results of these studies indicate the existence of a minimal discriminative difference, proportional to the range of variation and invariant for different discriminable aspects of the stimulus. This is the stuff of which sub-

stantive theory is made, and indeed, it has suggested hypotheses regarding the manner in which people process, or organize, the sensory inputs from their environment. It must, however, be emphasized that such developments are not an extension of information theory. The theory simply provides a conceptual framework and set of measures by means of which we can analyze a number of diverse situations. Any "model building," in the usual sense, must develop from the empirical results obtained.

As we have attempted to show in our development, the theory deals primarily with the coding problem and in particular with the question of optimal coding. Such a theory requires some measure of variation and of contingency, and Shannon has demonstrated that H and T are peculiarly appropriate for the analysis of a communication system, which is his main interest. The abstract structure of such a system is, however, much more general than the terminology associated with the theory might suggest, and mean logarithmic probability measures are applicable to many situations where terms like "information transmission" are awkward or quite inappropriate.

Consider, for instance, conditional probability matrices. These present the possible relations between an input and an output, or, as we have suggested, between stimuli and responses. More generally, we might say that [they] represent the various ways in which any set of events can be mapped into another set of events. Phrased this way, it is apparent that we could also consider mapping the set that we are concerned with into itself. Here x and y represent the same set of responses and the contingency could be sequential or temporal dependence. In this case, we would read for the conditional probabilities $p_x(y)$, "the probability that the response x is followed by the response y." As x and y are representative of the same set, we can adopt our earlier notation and represent the alternative responses and their probabilities by $p(i)$, where $i = 1, 2, \ldots, i, \ldots, k$, the set of alternative responses recorded. As before, $p(ij)$ is the probability of the responses i and j occurring in that order and equals $p(i) \, p_i(j)$. It should then be apparent that, by a somewhat devious route, we have arrived back at the point in our earlier development where we considered the definition of H for nonindependent source selections. In short, H may be used as a measure of the internal dependencies, or degree of patterning, within a sequence of responses.

This result is obtained, much more elegantly,

by Shannon in a theorem which shows that a series of approximations to H can be obtained by considering sequences extending over 1, 2, . . ., n symbols. The estimates of H thus obtained are a monotonic decreasing function of n. The function will equal H in the limit, or when the length of sequence considered has exhausted the statistical dependencies within the sequence.

It is thus apparent that information theory offers a ready-made tool for the description of behavior patterns and the analysis of sequential behavior. In this connection, it appears to be most convenient to consider the relative uncertainty of the sequence, which is the ratio of the obtained value of H to the maximum value that would obtain for the same set of alternatives if they were equally likely and occurred independently. One minus this quantity is the percentage of *redundancy* in the sequence, and in application to response sequences, redundancy may be used as a measure of the degree of stereotypy, or organization, of the observed behavior.

In fact, informational measures have been little used in the description of behavioral strategies and organization except in the study of language. As suggested in the introduction, the effect of the theory may well be indirect. Psychologists have long recognized that successive responses are seldom independent. Information theory, or more properly the probability model on which it is based, explicitly recognizes sequential dependencies and demonstrates that they can be taken into account in a reasonable fashion. Again, the particular formulation of the theory may not be optimal, and indeed, in a particularly simple demonstration of the effects of an anxiety-producing stimulus on the pattern of responding an alternative time series analysis was employed. However, it is exactly these peripheral effects that give power to the formal theory, emphasizing its generality as well as the distinction between information measures and the mathematical theory of communication for which they were originally derived. . . .

24.
A Behavioristic Analysis of Perception and Language as Cognitive Phenomena

CHARLES E. OSGOOD

PSYCHOLOGISTS, when they are behaving like psychologists, limit themselves to observing what goes into the organism (stimuli) and what comes out (responses). Between these two observation points lies a Great Unknown, the nervous system. Nowadays it is fashionable to refer to this region as "a little black box." In any case, psychological theory, as distinct from psychological observation, is made up of hunches about what goes on in this little black box. Theories of hearing and color vision, principles of association, generalization, and reinforcement, notions about cohesive forces between like processes in a visual field—all imply certain conceptions about how the nervous system works. If these conceptions are made explicit, as Hebb has done, for example, one is said to "neurologize," but, explicit or not, psychological theories select from among neurophysiological alternatives.

Behavior theories are often divided into two general classes—the S–S and S–R models. Each of these models is insufficient, an incomplete theory. The S–S model may adequately handle relations among input events and between these and central, "meaningful" events, but it says little or nothing about how they eventuate in behavior. For example, we are not told by Köhler how a pattern of direct currents in the visual brain elicits those responses in vocal muscles which constitute saying "circle" or "square." Similarly, the S–R model may adequately handle rather simple relations

between stimulus and response variables, but it says little or nothing about either the integration of sensory events (perception) or the integration of response events (motor skill). And neither model has had much to contribute to an understanding of symbolic processes.

Language is challenging to the behavior theorist because it includes at once the most complex organizations of perceptual and motor skills and the most abstract, symbolic processes of which the human animal is capable. It is also a necessary first step in the application of psychological principles to social behavior, because it is mainly via language that one nervous system establishes relationship with others. Perception presents equally difficult problems. Phenomena that have been called perceptual range the gamut from projection-system dynamics to meaningful processes, and certainly the integrational character of perception, which Gestalt psychologists have stressed, has been the Waterloo of contemporary behaviorism—I know of no S–R model that gives a convincing interpretation of standard perceptual phenomena. It is my hope that a combined analysis of language and perception may shed some light on both.

In the body of this paper I shall describe a highly speculative conception of behavior, which at least pretends to be a complete theory, in scope although certainly not in detail. It will necessarily imply a conception of how the nervous system operates—how it determines the relations we observe between stimulus inputs and response outputs—but I shall try to phrase the theory itself in psychological terms. It is a model that has gradually developed in the course of my work on language behavior. It envisages two stages

186

and three levels of organization between stimulus and response in the complete behavioral act. The first stage is what I shall call *decoding*, the total process whereby physical energies in the environment are interpreted by an organism. The second stage is what I shall call *encoding*, the total process whereby intentions of an organism are expressed and hence turned again into environmental events. The three levels of organization are assumed to apply to both sides of the behavioral equation, to both decoding and encoding: (1) *a projection level* of organization, which relates both receptor and muscle events to the brain via "wire-in" neural mechanism; (2) an *integration level*, which organizes and sequences both incoming and outgoing neural events; and (3) a *representation* or *cognitive level*, which is at once the termination of decoding operations and the initiation of encoding operations. We have evidence for all three of these levels, but the principles that apply most parsimoniously to one do not apply easily to the others.

Projection

The receptor surface of the organism is rather precisely mapped upon the sensory cortex. Similarly, the voluntary muscle system is rather precisely mapped upon the motor cortex. The most direct evidence for these statements is the predictability of experienced sensations or muscle contractions when the sensory or motor cortex is explored electrically. One general principle of the projection level, then, is *isomorphism*. This does not mean that the projection level is simply an uncomplicated relay system. At the successive synaptic junctures between periphery and cortex transverse connections make possible lateral interactions of limited scope and kind. For example, across any band of impulse-bearing fibers at any synaptic level there seems to be in operation a principle of lateral facilitation and inhibition—more rapidly firing elements in the band are further facilitated by summation with impulses received laterally from more slowly firing elements, and conversely the firing of the slower elements is relatively damped by receiving laterally a more rapid, subthreshold barrage. Something of this sort seems to underlie sharpening of contours and segregation of figure from ground in vision, as well as the phenomenon of masking in audition.

One can also, I think, handle the major characteristics of both color and brightness con-

trast, such phenomena as the apparent solidity of objects viewed binocularly, the continuity of optimum visual movement, and figural after-effects with projection level mechanisms. This argument has been given in more detail in my book [*Method and Theory in Experimental Psychology*] and in the paper by Heyer and myself, which proposed an interpretation of figural after-effects alternative to that offered by Köhler and Wallach. It relies heavily on the work of Marshall and Talbot and others on the functioning of the projection system.

The main point here is that there are many so-called perceptual phenomena that will probably be shown to depend upon projection mechanisms and hence be entirely predictable from knowledge of the stimulus and knowledge of projection dynamics. Such phenomena represent changes in the sensory signal itself, as I have defined it, rather than subsequent utilization of it in interaction with other signals.

Another characteristic of the projection level is that *its functioning is not modifiable by experience*. I know of no evidence showing that "what leads to what" in either sensory or motor projection systems can be modified by learning. The projection system is a perpetual *tabula rasa*—a centrally fixated object produces the same activity in Area 17 at twenty years as it did at twenty months, even though the subsequent utilization of these signals may be quite different. The experiments of Sperry and others, in which segments of either sensory or motor projection systems are transplanted in embryo, also provide impressive evidence for the absence of functional modifications at this level—an animal operated upon in this manner will continue to lift the left limb when the right limb is shocked, for example, with no evidence of learning. Appropriately, the work of Senden, with human adults recovering sight for the first time, and Riesen, with chimpanzees reared in darkness, shows that certain so-called perceptual functions are independent of experience—primitive isolation of figure from ground, fixation of an object in space, contour formation, color and brightness differentiation, and certain others.

The salient point for the behavior theorist is this: because the projection systems do display these two characteristics—isomorphism and inability to modify through experience—we can depend on stimulus-and-response observations as faithful indices of the sensory and motor signals with whose more central interactions I think our science of behavior is concerned.

Integration

Even the crudest observations of behavior reveal that certain patterns and sequences of responses are more readily executed than others and that certain patterns and sequences of stimuli have priority over others. Apparently both motor and sensory signals are capable of becoming structured or organized. I think there is a very simple property of nervous tissues that accounts for such structuring, and D. O. Hebb has already put his finger on it. *Whenever central neural correlates of projection-level signals are simultaneously active and in fibrous contact, either directly or mediately, an increased dependence of one upon the other results.* A few explanatory comments are in order about this statement. First, we must say that it is the more central neural correlates, rather than the sensory or motor signals themselves, which can thus be associated, because the projection systems are not modifiable through experience, as we have seen. There is no requirement that the central correlates of signals be isomorphic with these signals; in fact, existing evidence indicates that strict isomorphism breaks down beyond the sensory projection level. Secondly, strict simultaneity among the signals whose more central correlates are to be associated is not necessary; the work of Lorente de No and others describes reverberatory circuits which would prolong activation and hence make possible integration over time.

In a greatly oversimplified way, Figure 1 attempts to illustrate what I have in mind here. The isomorphic relations between stimuli and sensory signals and between responses and motor signals are shown on lower left and lower right respectively. It is assumed that cells at the termination of the projection system (for example, sensory or motor signals, as I have called them) have ample synaptic contacts with certain more central cells to guarantee exciting them (in the case of sensory decoding) or being excited by them (in the case of motor encoding). These are the cells in the integration level, a, b, and c, which I call "central correlates." This utilizes what I believe is a general principle of central nervous tissue: the probability of an antecedent neurone being a sufficient condition for the firing of a subsequent neurone is some direct function of the density of fibrous contact at their synapse. The control exercised by one cell over the firing of another may be increased, of course, by determining bombardment via mediate, circuitous routes; this is illustrated by cell x in the sensory integration system.

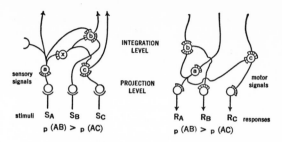

Figure 1.

The conditions given in this diagram are such that *in the stimulus input* to this organism the contingency of events A and B is greater than that between events A and C, and similarly for responses A and B versus A and C. This means that the resultant tendency for central correlate a to activate central correlate b should be greater than its tendency to activate c. I have indicated this in the diagram by a thicker band of contacts on cell b from a than on cell c from a.

Now, as I said before, I wish to outline a psychological theory, not a neurological one. What kind of psychological principle seems to be embodied here? Given isomorphism between observables and signals, which makes it possible for me to deal directly with observable stimuli and responses, I can phrase the following pair of principles: *sensory integration—the greater the frequency with which stimulus events A and B are associated in the input to an organism, the greater will be the tendency for the central correlates of one*, a, *to activate the central correlates of the other*, b. This principle says in effect that patternings, regularities, and orderings of events in the stimulating environment of an organism come to be mirrored in the structuring of its sensory nervous system. *Motor integration—the greater the frequency with which response events A and B are associated in the output of an organism, the greater will be the tendency for the central correlates of one*, a, *to activate the central correlates of the other*, b. This says in effect that patternings, regularities, and orderings of the overt behavior of an organism, no matter how established, will come to be paralleled by equivalent organizations within its motor nervous system.

It should be stressed that what I am calling sensory and motor integrations function as classes of intervening variables, anchored directly to antecedent and subsequent observables, respectively, via a simple frequency-of-co-occurrence function. How would varying this frequency factor be expected to affect what is observed?

I suggest that *with high frequency of stimulus or response pairing the central correlates of one will become a sufficient condition for the excitation of the correlates of the other*. I shall call this an *evocative relation*. Behaviorally it means that the occurrence of some of a set of related stimulus events will produce the central experience of the others as well and be reported or responded to as such, without their external correlates necessarily being given at all; it means that the initiation of some elements of a response pattern will set the whole pattern going. *With a lower frequency of stimulus or response pairing, the central correlates of one will become merely a condition for "tuning up" the correlates of the other*. I shall call this a *predictive relation*. Perhaps here the energies delivered from the antecedent cell to the dependent cell, while not adequate to fire that cell, are available for summation with energies being delivered via the direct path. Behaviorally this would mean an experience-determined increase in the *stability* of both decoding and encoding processes by the organism—perception of certain cues would increase the probability of also perceiving others, in competition with many simultaneous stimuli; initiating certain responses would increase the probability of also initiating others, again in competition with many simultaneous action tendencies. In a sense we would have here a mechanism for reducing the "noise" in both decoding and encoding.

Let us look into some of the behavioral implications of this principle. In ordinary perceiving we seldom receive complete information—the environment is inspected with rapid, flighty samplings, and intensity–duration factors in the projection system certainly imply that these samplings must yield only partial signals—nevertheless, perceptual experiences are usually wholistic. I assume that what I have called *evocative relations*, based on high frequencies of input pairing or redundancy, underlie the well-documented closure and "filling in" phenomena. Directly relevant are some recent papers by Fred Attneave in which he demonstrates that "various Gestalt-factors including symmetry, good continuation, and other forms of regularity may all be considered to constitute redundancy in visual stimulation and be quantified accordingly within a framework of information theory." I have not attempted myself the application of information-theory statistics to the phenomena subsumed under this integration principle, but I suspect they would be quite appropriate. We refer to "closure" when actual stimulus events, as

independently measured, correspond to what is perceived, but the same tendency toward completion of an integrational unit lies at the base of many perceptual illusions, where the actual stimulus events do *not* correspond to what is integrated. I have an electric clock at home which can't be reset after the current has gone off briefly; I have to stop it and wait a day until time catches up. Every once in a while I glance up to find the time, and momentarily I see the sweep-second hand moving! Considering the thousands of times clock-face signals have been followed in my experience by sweeping second-hand signals, this illusion becomes understandable.

Merely *predictive relations* in perceptual decoding are also familiar in everyday experience. It is easier to follow a familiar juke-box tune than an unfamiliar tune against the uproar in a local tavern; the more familiar the camouflaged object in a complex picture, the more readily its contour can be traced. On the experimental side, I would interpret the findings of Bruner and Postman on the perception of incongruity along these lines—the most common response to trick cards, say a *black* six of hearts, presented tachistoscopically, was to complete the integration set in motion by either the color *or* the form, but, as would be expected from the lack of reciprocal "tuning up," exposure times for decoding trick cards were significantly longer than for normal cards. Similarly, Hake and Hyman have found that subjects will come to reflect in their predictions about successive stimuli the sequential dependencies built into the series, even though they may be unaware that these dependencies exist. Hake says, "It appears that the mechanism by which we develop expectancies about the occurrence of probabilistic events operates such that over longer series of trials or choice points we [come to] expect events about as often as they appear."

Turning now to integrations in *ordinary motor encoding*, it may be noted first that S–R behaviorists have always relied upon proprioceptive feedback as the mechanism for organizing motor skills—and this despite the fact that as long ago as 1917 Lashley pointed out that there simply wasn't enough time in rapidly executed skills for impulses to be carried to and from the sequentially activated muscle groups. In a more recent and very stimulating paper given in the Hixon Symposium, he makes this point again: "Sensory control of movement seems to be ruled out in such acts. They require the postulation of some central nervous mechanism which fires with predetermined intensity and duration or activates different muscles in predetermined order. The mechanism might be represented by a chain of effector neurons, linked together by internuncials to produce successive delays in firing." This does not mean that proprioceptive feedback mechanisms are unimportant. On the contrary, I think that three stages in skill formation could be traced: (1) a very slow and uncertain patterning or ordering of responses on the basis of exteroceptive controls, as in imitating the seen movements of another person; this makes possible (2) a transfer gradually to proprioceptive controls (feedback), accompanied by considerably increased speed of execution; and this more rapid and stable organization in turn makes possible (3) a transfer to central programming in the integrational motor system which we are discussing.

Here again, a very high frequency of pairing should result in the formation of *evocative relations* among motor events. I call such tightly integrated patterns "motor skill components." All the complex acts with which we deal as psychologists seem to be compounded of such components— "opening the door," for example, is a complex act involving an arm-extending-and-hand-opening component, a hand-closing component, a wrist-twisting component, and an arm-flexing component. These same elements, just like the syllables of spoken language, enter in various combinations into the myriad activities of everyday life. Motor integrations may lead to errors of completion analogous to perceptual illusions—in typing, my favorite error is regularly to add an "n" to the word *ratio*, presumably because of the tendency to complete the very common "ion" that terminates words like *action*, *fashion*, and, of course, *ration*. Based on lower orders of frequency, many response–response integrations become merely *predictive motor relations*—unbuttoning one's shirt is predictive of peeling it off, lighting one's cigarette is predictive of blowing out the match (much to my occasional embarrassment when someone just then indicates the need of a light!) I do not mean that stimulus controls are absent in such predictive motor sequences; rather, the motor preparation decreases the probability of disturbance through ordinary stimulus changes. In other words, there seems to be a syntax of behavior just as there is a syntax of language, and this provides a stability of customary action that frees it from constant voluntary supervision.

The integrative mechanisms we have been discussing appear even more clearly in *language behavior*, and this is because the units of both decoding and encoding have been more sharply etched by linguists than have the units of nonlanguage behavior by psychologists. Generally speaking, we find evocative integrations in the smallest skill units of both speaking and listening, and predictive integrations in the grammatical mechanisms that interrelate larger message events.

The minimal units in language decoding are called *phonemes*. These are classes of similar sounds having a common significance in the code —for example, the initial sounds in "key" and "cool" are both members of the "k" phoneme, and their differences in auditory quality are entirely predictable from the message environment, in this case the following vowel. There are only some 32 phonemes in the English code; in other words, on the basis of amazingly high frequency of occurrence we have developed about 32 evocative auditory integrations, each one of which is set in motion by a *class* of input signals that varies in the elements actually present in any instance. Testimony to the general validity of our principle is the fact that ordinarily we are incapable of perceiving the differing members of these phoneme classes—*allophones*, as they are called. Only by adopting the analytic attitude of the linguist, which means listening to our language as sounds rather than meanings, can most of us hear the differences between the allophones of, say, the "p" in "pin," "spin," and "nip," in the allophones of "t" in "I bought a bitter bottle," yet these auditory distinctions are in themselves sufficient for speakers of other languages to arrange separate phoneme categories upon them. It is also interesting in this connection that our decoding of phonemes is on an all-or-nothing basis—when I say "he was a *trader*," some of you heard "traitor" (to his country) and others heard "trader" (on the stockmarket), but none of you heard both at once or any compromise between the "t" and "d" phonemes. In other words, evocative integrations function as all-ornothing units.

In language encoding, the smallest functional units are probably *syllables*. This is at least suggested by the work of Grant Fairbanks on delayed auditory feedback, in which he finds the interval of maximum interference to correspond to the rate of syllable production in ordinary speech (about 4/sec.); it is also suggested by the fact that slowing down one's speech is usually accomplished by prolongation of syllabic boundaries, for example, "syll-a-ble pro-duct-ion." Here again we have a limited number—much larger than the number of phonemes, to be sure

—of motor patterns and sequences used with such high frequency that they become evocative integrations. Think of the number of word units in which the syllable "bit" appears—the word "bit" itself, "habit," "arbitrary," "bitter," "prohibit," "bitsy," and so on. These syllables involve both simultaneous and sequential integrations of many motor elements in the vocal system, and they come to function as units in behavior.

The operation of *predictive integrations* in language has a number of excellent experimental demonstrations. Postman, Bruner, and Walk, for example, have shown that imbedding a single reverse-printed letter in a meaningful word lengthens the tachistoscopic exposure time at which that letter can be reported *as* reversed more than imbedding it in a series of unrelated consonants—the normal configuration of the familiar word is thus highly predictive of its components—and this was true despite the fact that the average exposure time for letters in meaningful words was very much shorter than for letters in nonsense sequences, which also follows from the integration hypothesis. An experiment by Miller, Postman, and Bruner shows that varying the sequential probabilities of orthographic materials affects their recognition times in the expected way. And we may add Shannon's finding that the guesses of subjects as to what letters should follow sequences of varying length matched very closely redundancy measurements made on large samples of English texts.

But it is in the *grammar* of a language that one observes the most remarkable predictions over time—a phenomenon which, interestingly enough, Lashley, in the Hixon Symposium mentioned earlier, took as his jumping-off point for an analysis of serial order in behavior. While in the rapidly flowing tide of conversation, both speakers and listeners attend to the lexical units in messages that represent semantic choices, leaving the complex grammatical and syntactical regularities to take care of themselves. It would be safe to say that the lay user of a language is almost never aware of its grammatical structure, couldn't possibly describe its laws, and yet follows them faithfully. When analyzed linguistically, the rules of grammar prove to be elaborate cases of redundancy or predictiveness. One such grammatical redundancy mechanism is *congruence*: in the present tense in English, the occurrence of a singular subject sets up a readiness for a verb ending in *s* (The boy run*s* but the boy*s* run); a time marker sets up a readiness for the appropriate tense tag on the verb (*Yesterday* in the city I *bought* a hat);

a dependent clause marker sets up a readiness for the major clause (*When* I come, *open the door*). In terms of our model, it is the frequency with which such grammatical redundancies have been heard and produced that sets up in the nervous system predictive integrations that match the structure of the language. As would be expected, the longer the interval between congruent elements, the weaker becomes the set and the more likely errors.

Being a relatively uninflected language, English depends heavily upon syntactical *ordering* mechanisms, another grammatical redundancy. "John loves Mary" is quite a different proposition from "Mary loves John," as many a jilted lover has discovered. In Latin these words could be kept in the same order and the difference in implication borne by inflectional endings. If I say "the happy, little ————," all of you feel a strong tendency to fill in *some* noun. If I say "the farmer killed the ————," you have essentially two structural alternatives, a noun or a noun phrase (for example, *duck* or *ugly duckling*). If I say "the old man eats ————," the set of structural alternatives is larger, but still limited (a noun, *dinner*, *meat*; an adverb, *swiftly*, *heartily*; a prepositional phrase, *with his hands*, *on the table*, and so on). At each point in a language message, then, we have a hierarchy of structural alternatives, this hierarchy varying in its probabilistic character with the grammatical restrictions in the language as a whole. The closer the language user's nervous system can come to matching these restrictions with its own predictive integrations, the smoother become both decoding and encoding processes and the fewer decisions have to be handled by the semantic system.

By way of evidence that these grammatical redundancies do facilitate decoding and encoding, we might cite the following: Wilson Taylor, using his "cloz" procedure in which the subject fills in the gaps in mutilated messages, finds that with both sides of the gap given as in his method structural determinism is almost perfect (for example, in filling in "the old man ———— along the road," all subjects will fill in a verb form even though they vary semantically in what verb they choose). Miller and Selfridge and others have demonstrated that ease of learning and retention of meaningful materials varies with the degree of approximation to English structure. Along similar lines, one of my students, Mr. Albert Swanson, compared the ease of learning nonsense sequences that retained the structure of the English sentences from which they were derived, for example,

The maff vlems oothly um the glox nerfs

with matched materials in which the grammatical cues had been eliminated, for example,

maff vlem ooth um glox nerf.

Despite the greater absolute amount of material in the structured forms, they were learned significantly more easily than the matched strings of nonsense items.

Before leaving this integrational level, we should deal at least briefly with a problem raised by the existence of these *hierarchies of alternatives*. In grammatical ordering mechanisms we have seen that at each choice-point in a message the speaker or hearer has available a set of alternative constructions, having different probabilities attached to them; similarly in perceptual decoding a particular subset of signals will be predictive of a hierarchy of potential integrations, each having a different probability associated with it. Take as examples the two sets of letters shown in Figure 2. Based on the absolute frequencies of occurrence in this type-face, integrations associated with O,

C, Q, and G will have varying conditional probabilities upon occurrence of the subset of signals from the lefthand, which are common to all; their *absolute frequencies* of occurrence, however, are all probably sufficient to produce what I have called evocative integrations. Yet, given the arc as a stimulus set in the tachistoscope, one never sees all of these possibilities at once—perception would be a perpetual jumble if such were the case! Rather, we **experience** one alternative or another, that having the highest over-all momentary conditional probability. The larger context may select among **alternatives**—C is the only possibility in QUICK, but C and O are competitive in the ambiguous context of LOCK. The same arguments can be made on the basis of the second set of letters—B, D, P, R—where the composite jumble would be even more unlikely.

These facts require some extension of our notion of how the integrational level operates. In the first place, I think we must distinguish between frequency of co-occurrence of signals and redundancy within hierarchies of alternative integrations. The former seems to be the basis for setting up integrations, while the latter is the basis for selection among alternative integrations containing the same subset of initiating signals. Two integration hierarchies may have identical redundancy characteristics (for example, a .50, .30, .10, .05, .05 probability structure) and yet on the basis of absolute frequencies of occurrence one may be evocative throughout and the other merely predictive throughout; in either case, only that alternative which is momentarily dominant will be effective. And this implies this notion: *selection of the momentarily most probable integration among the hierarchy of alternatives based upon the same subset of signals serves to inhibit all other potential integrations.* I am not going to speculate upon the possible neural basis for such selection among alternatives, except to point out that there are known to be in the cortex "suppressor areas" whose excitation produces generalized spread of inhibition and Ruch tentatively identifies them with "attentional" functions (Area 19 seems to exert such an effect upon the visual system). I might also point out that many thoughtfully introspective people, psychologists among them, have reported the "singleness" of awareness or consciousness—as if at any one moment we are capable of handling only one item of information, the apparent multiplicity of attention being in reality a rapid succession. The integrational system seems to operate in such a way that many are called but only one is chosen.

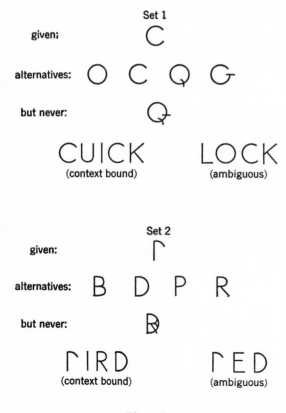

Figure 2.

Representation

Let's take a moment to see what kind of organism we've constructed so far. On the input side it is capable of rather faithfully recording as signals what events take place on its receptor surface; it is also able, on the basis of experience, to integrate these motor signals into evocative and predictive units that reflect redundancies in its own past behavior. But something is obviously missing: this organism does not connect with its own activities events that are happening in the world; it is completely "mindless" in the colloquial sense. In other words, we have so far dealt exclusively with S–S and with R–R relations, and what is missing are S–R relations.

Stimulus events may be related to response events at all levels of organization, and these associations may be either innate or acquired. Some sensory signals have an innate "wired-in" connection with specific responses (unconditioned reflexes) and additional classes of signals may acquire such direct connection with motor signals (conditioned reflexes). Similarly, at the integrational level, associations between complex patterns of sensory and motor signals may be innate—certainly, the complicated organization of instinctual sex behavior falls in this rubric, as does the "freezing" reaction of baby birds to certain complex retinal patterns. It also seems to be a general characteristic of the central nervous system that S–R relations *originally* organized on the "voluntary" level will, if repeated sufficiently often, become autonomous integrations—most sensory-motor skills seem to suffer this fate, reading aloud and typing as well as tying one's shoes and brushing one's teeth.

But the most important mechanism for associating sensory events with motor events—certainly in the human, and I suspect in the higher vertebrates in general—is via a *two-stage mediation process*. The essential notion here is that in the course of associating external stimuli with overt behavior some representation of this overt behavior becomes anticipatory, producing self-stimulation that has a symbolic function. There is nothing highly original about postulating such mediation processes. My own use of the notion stems directly from Hull's conception of the "pure-stimulus-act," which incidentally he suggested would prove to be the basis of abstraction and symbolism in behavior; the same idea is used by Guthrie as "movement-produced-stimuli," and Tolman's basic conception of a "sign–significate–expectation" can, I think, be shown to be functionally identical. But whereas Hull and Guthrie, at least, only called upon this device in dire extremities, when single-stage mechanisms proved insufficient, I consider it to be the usual form of S–R learning. And furthermore, taking Hull's suggestion about symbolism very seriously, I have tried to show that the representational character of the mediation process provides the basis for a theory of sign behavior—or, if you will, of cognition.

We may start with the fact that certain stimulus events have a "wired-in" association with certain response events; for the hungry infant the taste and feel of warm milk in the mouth are reflexly associated with swallowing, salivating, and digestive activities, and the pressure of a yielding object against the lips is reflexly associated with sucking and head-turning. This type of stimulation I call a *significate*. However, since I want this class to include previously learned as well as wired-in relations, I would define a significate as *any stimulus that, in a given situation, reliably elicits a predictable pattern of behavior*. Thus all unconditional stimuli in Pavlov's sense are significates, but the reverse is not true. Now, there is an infinitude of stimuli that are *not* initially capable of eliciting specific patterns of behavior—the sight of the breast or the infant's bottle does not initially produce salivating, for example. Under what conditions will such a pattern of stimulation become a *sign*?

I would state the conditions this way: *whenever a non-significate stimulus is associated with a significate, and this event is accompanied by a reinforcing state of affairs, the non-significate will acquire an increment of association with some fractional portion of the total behavior elicited by the significate.* I call such fractional behavior a representational mediation process. It is representational because although now elicited by another stimulus it is part of the behavior produced by the significate itself—this is why the bottle becomes a sign of milk–food object and not any of a thousand other things. It is meditational because the self-stimulation it produces can become associated with various overt responses appropriate to the object signified—sight of the bottle can thus mediately evoke "yum-yum" noises and reaching out the arms. Just what portions of the total behavior to the significate will appear redintegratively in the mediation process? At least the following determinants should be operating: (1) *energy expenditure*—the less the effortfulness of any component of the total behavior elicited by the significate, the more likely is this component to appear in the portion elicited by the sign; (2)

interference—the less any component interferes with on-going instrumental (goal-directed) behavior, the more likely it is to be included; (3) *discrimination*—the more discriminable any component from those elicited by other signs, the more likely it is to be included. There is considerable evidence in the conditioning literature that certain components of UR, particularly "light-weight" and autonomic components, appear earlier in the CR than other components.

Figure 3 diagrams the theoretical development of *perceptual and linguistic decoding*. We may take as illustration the object, BALL. The large S at the top of the diagram refers to those stimulus characteristics of this object (its resilience, its shape, its weight, and so on) which reliably produce certain total behavior (rotary eye-movements, grasping, bouncing, squeezing, and even the pleasurable autonomic reactions associated with play-behavior), all of which are symbolized by RT. Now, according to the mediation hypothesis, the sight of this ball as a visual sensory integration, *initially meaningless*, will come to elicit some distinctive portion of the total behavior to the object as a representational mediation process $(r_m - s_m')$. To the extent that this process occurs, the visual pattern becomes a *perceptual sign* (Š) signifying BALL object, e.g., this is *a unit in perceptual decoding*. In other words, here at the ground floor in the development of meaning is the development of perceptual significance. Long before the child begins to use language, most of the sensory signals from its familiar environment have been lifted from their original Jamesian

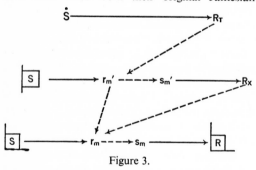

Figure 3.

Now, whereas perceptual signs bear a necessary physical relation to the objects they represent, linguistic signs bear an arbitrary relation. It is characteristic of human societies that adults, when interacting with children, often vocalize those lexical items in their language code which refer to the objects being used and the activities underway. Thus Johnny is likely to hear the noise "ball," a linguistic sign (Š), in frequent and close continuity with the visual sign of this object. As shown on the lower portion of this figure, the linguistic sign must acquire, as its own mediation process $(r_m - s_m)$, some part of the total behavior to the perceptual sign and/or object—presumably the mediation process already established in perceptual learning includes the most readily short-circuited components of the total behavior and hence should tend to be transferred to the linguistic sign. Thus, a socially arbitrary noise becomes associated with a representational process and acquires meaning, e.g., *a unit in linguistic decoding*.

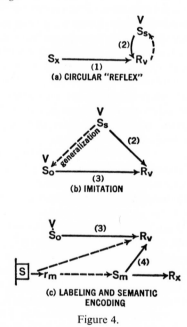

Figure 4.

Figure 4 diagrams the theoretical development of *instrumental and linguistic encoding*. Stage (a), called "circular reflex," is a necessary first step, because syllable units must become integrated into skills by the practice babbling provides, and the child must be able to repeat its own vocalizations on an auditory feedback basis before it can imitate others. The second step, "imitation," involves nothing more than primary generalization—the tendency to repeat a heard sound

chaos, have become perceptual signs of objects by virtue of association with representational portions of the same behavior the objects themselves produce. Incidentally, this seems to carry back one step further the philosophical argument about the nature of meaning; the visual images of objects, rather than being "the things themselves," as is usually assumed, are actually signs whose significance must be acquired.

spreads from self-produced cues to other-produced cues—and upon hearing mother say "ball" the child says "bah," his nearest skill unit. Now, as shown in stage (c), let us assume that the visual stimuli from BALL object already have significance for the child, that is, constitute a perceptual sign by virtue of eliciting a mediation process $(r_m - s_m)$ derived from BALL-manipulating behavior. Then the pairing of the heard label "ball" with the perception of the object should have at least the following consequences: (1) a single-stage association between the sight of the object and imitative labeling and (2) a two-stage, mediated association between sight of the object and imitative labeling, as shown by the starred arrow. Whereas the first of these is a meaningless process —sheer labeling that requires the physical presence of the object—the second represents the formation of *a unit of linguistic encoding*. The association of a representational process frees the child's language from the immediate here-and-now—*any* antecedent condition—desire for the object when it is missing, for example—which elicits the critical representational process is now capable of mediating the correct, socially communicative vocalization. This is the essence of abstraction in the use of language, I think. Also indicated in this figure is the fact that mediation processes can become associated with nonlinguistic instrumental reactions (the RX in the diagram); under appropriate conditions of differential reinforcement, the child learns to crawl toward, reach for, and smile at objects perceived as having "play" significance, like this BALL object.

Perhaps the single most important function of representational processes in behavior is as the common term in mediated generalization and transfer. As shown in Figure 5, whenever various stimuli accompany the same significate, they must become associated with a common mediation process, and hence acquire a common significance. Thus, for the rat, the cluster of stimuli surrounding a food object (its appearance and odor, the auditory "click" that announces its coming, the corner around which it is found, and so on) become roughly equivalent signs of the food object. To the extent that these signs have varied in their frequency of pairing with this common significate, they will constitute a *convergent hierarchy of signs* yielding the same significance but with varying strength or probability. Similarly, when a number of different overt responses are reinforced in association with a particular sign or class of signs, they will constitute a *divergent hierarchy of instrumental*

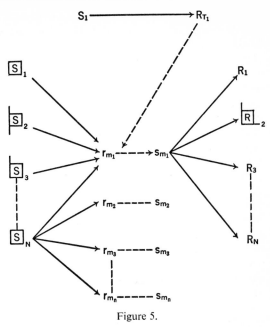

Figure 5.

acts associated with the same mediator. With any sign having a danger-significance for the rat will be associated a hierarchy of alternatives—running, freezing, turning a ratchet-wheel, and so on. These instrumental alternatives will also vary in their habit strengths or probabilities, and selection among them will depend particularly upon contextual cues. Finally, I have indicated toward the bottom of this figure that signs may come to be associated with *divergent hierarchies of mediators*. If the same set of sensory signals accompanies food significates often, sex significates occasionally, and pain significates seldom, this set of signals will become a somewhat ambiguous sign, in that different representational processes will tend to occur with varying probabilities. In language, homophones like *case*, *bear*, and *right* are merely extreme examples; here again, as whenever divergent hierarchies are operating, selection depends on context (e.g., conditional probabilities).

The availability of such hierarchies to the mature organism makes possible the tremendous flexibility we observe in behavior. If to a particular sign having a certain significance, the subject learns a new instrumental adjustment, such as pressing a lever or saying "Please," this immediately becomes available to any other sign having the same significance. Here we speak of *mediated generalization*. If a novel set of sensory signals, such as a flickering light in the rat's box or some unusual dark spots on the human's skin, acquire a danger significance, all the previously learned instrumental acts associated with this significance immediately become available to this new sign—the rat will shift quickly to running, to turning the ratchet-wheel, and the like, and the human will immediately call the doctor, talk to his wise old grandmother, rub his arm with bacon grease, or whatever he has already learned to do in situations having this significance. Here we speak of *mediated transfer*. The processes we call cognitive—concept formation and utilization, attitudes, personality traits, problem-solving—fit this mediational model, in the sense that they involve a class of stimulus situations associated with a common significance that mediates a class of alternative behaviors.

There are several difficult questions that arise with respect to the mediation hypothesis. One is this: *is such a two-stage process necessary?* Even at the rat level, there is a great deal of experimental evidence requiring a two-stage interpretation: the separation of learning from performance in many of the investigations inspired by Tolman; the role of secondary reinforcement mechanisms in experiments by Mowrer, Neal Miller, and others; the evidence for "learning to pay attention" in discrimination studies like those of Lawrence. At the human level: in the many studies of semantic generalization, the measured generalization between stimuli like JOY and GLEE obviously depends upon some common (and unobservable) mediating reaction to them, not to any physical similarities in the stimuli themselves (JOY and BOY are much more similar physically); the separation between learning and performance is even more clear in human behavior—witness the changes in attitude that may be produced by quietly watching a television program—and, as far as I can see, the phenomena of meaning and intention, so obviously displayed in human language behavior, entirely escape a single-stage conception. Another question arises: *what is the real nature of representational processes?* Here I have little to say. Following Hull, I have attributed stimulus-producing response characteristics to the process, because in this way it is possible to transfer all the conceptual machinery of single-stage S–R psychology—generalization, inhibition, habit strength, habit competition, and the like—to both the decoding and encoding sides of my two-stage model. However, this does not require a peripheral view as against a central one; the representational process could be entirely cortical, although I suspect it involves peripheral events in its development, at least. And I have no idea as to what might be the neurological basis or locus of such a process. In other words, for the present I am quite content to use the mediation process as a convenient intervening variable in theory, having responselike properties in decoding and stimuluslike properties in encoding.

Another critical problem is that of *indexing these representational processes*, particularly in humans. If we index the occurrence and nature of representational processes by the very behavior presumably mediated by them, we run full tilt into the circularity which I believe characterized Tolman's theory. What we need is some index of representational states that is *experimentally* independent of the behavior to be predicted. In other words, we need some way of measuring *meaning*. Most of my own experimental work at Illinois over the past five years has been devoted to this problem, and what follows is a very concise summary.

To measure anything that goes on within the little black box, it is necessary to use as an index some observable output from it. From a previous survey of varied outputs that are to greater or lesser degree indicative of meaning states—ranging from minute changes in glandular secretion and motor tension to total acts of approach, avoidance, and the like—we conclude that language output itself provides the most discriminative and valid index of meaning. After all, this is supposed to be the function of language. But what linguistic output gains in sensitivity and validity it seems to lose on other grounds; casual introspections are hardly comparable and do not lend themselves to quantification. What we need is a carefully devised *sample* of linguistic responses, a sample representative of the major ways in which meanings can vary.

The *semantic differential*, as our measuring technique has come to be called, is a combination of association and scaling procedures. We provide the subject with a standardized sample of bipolar associations to be made to each concept whose meaning is being measured, and his only task is to indicate the direction of his association and its intensity on a seven-step scale, for example, the concept LADY might be checked at step "6" on a *rough-smooth* scale, signifying "quite smooth." The crux of the method, of course, lies in selecting the sample of descriptive polar terms. Fortunately—and contrary to the assumptions of some philosophers—the myriad dimensions available in language are not unique and independent; our basic assumption is that *a limited number of specific scales, representative of underlying factors, can be used to define a semantic space within which the meaning of any concept can be specified.*

This points to *factor analysis* as the logical mathematical tool. Two factor analyses have already been reported, although they were based on the same set of 50 descriptive scales (selected in terms of frequency of usage) one involved Thurstone's Centroid Method and the judgment of 20 concepts against these scales by 100 subjects and the other, by 40 subjects, involved forced pairing among the verbal opposites themselves, with no concepts specified, and a new factoring method developed by my colleague, George Suci. Both analyses yielded the same first three factors: *an evaluative factor*, identified by scales like *good–bad, clean–dirty,* and *valuable–worthless*; a *potency factor*, characterized by scales like *strong–weak, large–small,* and *heavy–light*; and an activity factor characterized by scales like *active–passive* and *fast–slow.* A third factor analysis, based on a sample of 300 scales drawn from Roget's *Thesaurus,* 20 varied concepts, and the judgments thereof by 100 subjects, has just been completed (with the aid of ILLIAC, the Illinois digital computer, without which this work would have been impossible), and again exactly the same first three factors appear in order of magnitude. The regularity with which the same factors keep appearing in diverse judgmental contexts encourages us to believe that we are getting at something fairly basic in human thinking.

Although this factor work is interesting in its own right, its purpose is to devise efficient measuring instruments for meaning. In practice, small sets of scales, heavily and purely loaded on each of the factors isolated so far, are combined as an instrument against which subjects rate signs of

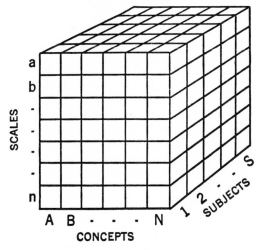

Figure 6.

any type—ordinary verbal concepts, the self-concept and other-concepts, cartoons, art objects, TAT or Rorschach cards, attitude objects, and so on. As shown in Figure 6, application of such an instrument to a group of subjects yields a cube of data, each cell of which contains a number from 1 to 7 (the seven-step scales), each column of which contains a profile for a given concept, and each slice of which represents the complete profiles for all concepts for a given subject. The *meaning of a concept* to a given individual or group is operationally defined as the profile of numbers, or means, in a single column or, more efficiently, as the point in the factor space determined by this profile. *Difference in meaning* (between two concepts for a given subject or group, between two individuals or groups for a given concept, or between two testings) is operationally defined by D (distance) = $\sqrt{Ed^2}$, the generalized distance formula. Figure 7 illustrates allocation and distances among three concepts—HERO, SUCCESS, and SLEEP—within the three-factor space so far derived. All three concepts are favorable evaluatively, but whereas HERO and SUCCESS are simultaneously quite *potent* and *active,* SLEEP is quite *impotent* and *passive.* The use of multivariate D has the additional advantage that all distances between concepts judged against the same set of scales can be represented simultaneously in the space defined by the scale factors. Computing the D-values between each concept and every other concept yields a distance matrix which, when only three factors are involved, can be plotted as a solid model. Such models represent, if you will, bits of "semantic geography," and the changes in such structures over time and across individuals or groups yield interesting descriptive data. Doctors Suci and Tannenbaum and I are now working on a monograph that will summarize the methodological and descriptive research with the semantic differential to date.

Is this instrument a valid index of representational processes in human subjects? Information on this point is much harder to come by. There is no doubt that this instrument, as far as it goes, measures meaning in the colloquial sense: the meanings of common adjectives are differentiated in obvious ways, expected differences between Republicans and Democrats are revealed, and so on. We also have some evidence that the dimensions we have isolated by factor analysis correspond to those used spontaneously by subjects in making meaningful judgments—an experiment by Rowan showed that the similarity relations within a set of concepts obtained with the semantic

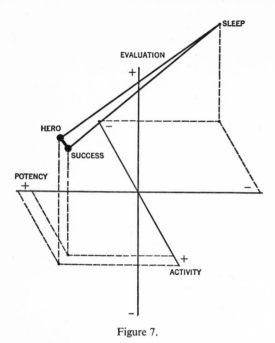

Figure 7.

intensity of the component reactions making up the mediation process elicited by that sign.

Do these identifications make sense in terms of what data we do have? For one thing, the profile similarities we obtain between signs are obviously dependent upon some implicit, meaningful reaction to the signs and not upon their physical characteristics—profiles for LADY and GIRL are very similar, but those for LADY and LAZY are not (in other words, the instrument is indexing a two-stage, semantic process). Secondly, note the responselike nature of the three factors we have isolated so far—*evaluation* (general autonomic reaction?), *potency*, and *activity*. If representational mediators were in truth fractional portions of total behavior, one would expect them to have responselike characteristics. In this connection, I may mention another factor analysis we have done, on the communicative meanings of 40 posed facial expressions. Again three factors accounted for most of the common variance, and these were identified as *pleasantness* (for example, from GLEE down to ACUTE SORROW), *control* (from CONTEMPT over to HORROR), and *expressiveness* (from COMPLACENCY out to the whole array of active expressions like CONTEMPT, RAGE, and HORROR). These *look like* the same factors, and they are clearly related to the reactive natures of emotional states.

We have planned a number of experiments to check this bridge between semantic measurement and representational process. One is a straightforward *mediated generalization* study—if my hunch is valid, then the measured similarity between signs, as obtained with the semantic differential, should predict the amount of mediated generalization between them under the usual conditions. Another variant of the same design will compare two types of *bilinguals*: compound bilinguals (who have learned two languages in such a way that translation-equivalent signs are associated with a single set of meanings) against coordinate bilinguals (who have learned two languages in such a way that translation-equivalent signs are associated with a double set of somewhat different meanings). Again, profile similarities between translation-equivalent signs obtained with the semantic differential should predict mediated generalization, compound bilinguals showing greater generalization than coordinate bilinguals.

Figure 8 provides a summary picture of the model I have been describing. Projection, integration (both evocative and predictive), and representation levels on both decoding and encoding sides of the behavioral equation are indicated, as

differential corresponded closely with those obtained for the same subjects by the method of triads (where dimensions of judgment are not specified), and other studies of this sort are in progress. But this offers only the most tenuous connection between our measurement procedures and the representational mediator as a construct in learning theory.

Let me suggest what I think this relation may be and then what may be some possible ways of tying it down experimentally. Our factor analytic work indicates a number of bipolar semantic dimensions—and such bipolar opposition in thinking, by the way, appears to be characteristic of all human cultures. My hunch is that *the representational mediator is a complex reaction made up of a number of components, these reaction components corresponding to the semantic factors we have isolated*. Now, we have been able to show that extremity of judgment on our semantic scales (for example, away from the midpoint, "4," toward "1" or "7") is linearly related to judgmental latency, when the same subjects judge the same concepts against the same scales in a reaction-time device. Since latency is an index of habit strength, it is reasonable to assume that *more polarized judgments on the semantic differential correspond to stronger habits associating sign with mediator components*. In other words, I am saying that the location of a sign in the space defined by the semantic differential is an index of the nature and

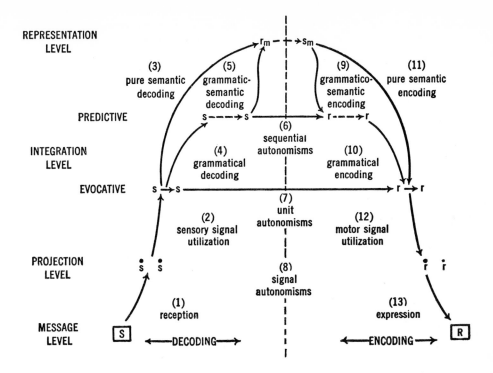

Figure 8.

well as the S–R relations within each level. The labels given to each type of association or "pathway," such as "pure semantic decoding" and "grammatical encoding," reflect my major interest in language behavior, but, as I have implied throughout this paper, to the extent that this picture is valid it should hold for perceptuo-motor sequences as well. This is admittedly a complicated conception of behavior, but I doubt that any conception sufficient to handle the complexities of language is going to be very simple. Although I have not indicated it in the diagram, for reasons of clarity, it should be assumed that these are hierarchies of alternatives of varying probability at each locus of S-S, R-R, and S-R relations.

We have made one check on the over-all validity of this model. This was an attempt to predict the greater-than or less-than contingencies between various language disturbances in *aphasia.* Suppose that we were able to get inside the full-blown mechanism shown here and cut one or more of the pathways—perhaps right across pathways 5 and 9, connecting grammatical predictive integrations with representational processes —what should happen? On the decoding side, all language performances requiring one to get the significance of sequences of signs should be lost

(for example, understanding complex commands, reading interpretively, and the like); on the encoding side, the grammatical correctness of intentional speech should be lost, producing *telegraphia.* On the other hand, both the significance and the ability to produce isolated semantic units should be preserved (pathways 3 and 11), as should previously learned automatisms, like reading aloud mechanically and reciting a familiar poem. Working with the detailed reports of some 35 classic aphasia cases, two clinical graduate students (who were completely unfamiliar with the theory) and I noted in each case the presence and absence of disturbance in some 20 different language performances (such as reading aloud, labeling, written word recognition, speech skills, and so on). Empirical contingencies between performances over these cases were then computed and compared with contingencies predicted from theoretical analysis of the "pathways" essential to each performance—the greater the overlap in "pathway" utilization, the more likely it is that disturbance of one would be accompanied by disturbance in the other. A nonparametric test of correspondence between predicted and obtained contingencies was significant beyond the .001 level. While such results do not confirm the details of our theoretical model, they do seem to substantiate the general analysis into several levels of organization and two stages of decoding and encoding, and they have also encouraged us to work on some new aphasia tests based on the model.

Perception and Meaning

I would identify as "perceptual" those phenomena characteristic of both evocative and predictive *integrations in decoding*, including direct effects upon this level from the projection system and indirect, "feedback" effects upon this level from the representational system. Actually, this way of thinking about perception agrees pretty well with the ideas of other psychologists: Bruner draws a distinction between "autochthonous" and "behavioral" determinants, the former referring to retinally dependent events and the latter to cognitively dependent events like values and meanings, as I understand it; Hebb distinguishes between "sensory" and "non-sensory" figures on what appears to be the same ground; and Gibson distinguishes the "anatomical visual field" from the "ordinal visual world," again on the same grounds, as I read him. And although I deliberately have not as yet read Floyd Allport's new book on perception, I expect that the same distinction will be found there.

Why *is* there this agreement about dual control over perceptual process? It is perfectly clear from empirical data that what the subject reports as his experience depend both upon the stimulus information given to his senses and upon the store of information derived from past experience. Stimulus information, as operated upon by mechanisms in the projection system itself, determines what sensory signals are present at any moment. These sensory signals set limits upon the possible integrations that can occur. *Stored information* is of two sorts: (1) the entire past history of sensory signal pairing has resulted in hierarchies of evocative and predictive integrations within this level itself, integrations that tend to fill out sketchy sensory information (closure) and predict synchronous and successive events in proportion to their environmental probabilities; (2) because certain cognitive states have accompanied some but not all integrations within competing hierarchies, the self-stimulation arising from such cognitive states will facilitate or increase the probability of these perceptual integrations as against others. I imagine that these "feedback" effects from the representational system operate just like other sensory

signals, and exert their selective effect upon alternative integrations simply by virtue of their past contingencies with exteroceptive signals. Effects of motivational states upon perception could be handled in similar fashion.

A beautiful demonstration of this is to listen to an unfamiliar chorus from Gilbert and Sullivan, alternately either following the printed libretto or merely listening without the printed guide. While the reader is seeing the printed words, the *auditory* information seems perfectly intelligible, yet the moment he looks away from the text it degenerates into gibberish. I believe it is feedback from decoding of the printed words that operates selectively among alternative auditory integrations. Now let me apply this type of analysis to two specific perceptual problems, recognition and constancy.

Identity and Recognition in Perceiving. Identifying or recognizing something requires that sensory signals activate a representational process; this must be so, it seems to me, because it is this process that mediates encoding of the words by which the subject reports his perceptions. Before he can express any "hypothesis" as to *what* the input information represents, some cognitive process must occur. I think Ames, Cantril, Kilpatrick, and the others associated with the "transactional" point of view have the same thing in mind when they say, ". . . we can only have a sense of objective 'thatness' when the impingements on our organism give rise to differentiated stimulus-patterns to which differentiated significances can be related." What is missing from their treatment of perception, however, is any analysis of the nature and development of such signifying processes. Earlier in this paper I described one conception of how stimulus patterns acquire significance or meaning.

Accepting the notion that recognition of "thatness" depends upon the arousal of a representational process in the perceiver, we may now ask what classes of variables should, in theory, influence the probability of recognition. The first class of variables is so obvious that most students of perception have taken it for granted, although the human engineering people have been very much concerned—this is *the availability of the sensory signals themselves* (projection level). The probability of "detection" (usually a kind of recognition) varies with the intensity of sensory signals generated by the physical stimulus, which can be modified by manipulating exposure time, changing the receptor population, moving the stimulus, and so on. The laws operating here are those characteristic of the projection system. What subsequent integrations can occur is limited by this sensory information—a circular pattern of high intensity has about zero probability of being decoded as a vertical line.

For just this reason, in studying the effects of higher-level determinants upon recognition we usually reduce the clarity of the sensory input. A second class of variables concerns *the availability of alternative sensory integrations* (integration level). As has been demonstrated repeatedly, with intensity–duration factors held constant, the probability of printed-word recognition is a very regular function of frequency of usage, which is another way of saying frequency of sensory signal pairing in past decoding experience. George Miller and others at M.I.T. have demonstrated the same sort of thing for intelligibility (recognition) of spoken words, and I believe one could interpret Gestalt data on "goodness of form" in the perception of figures along similar lines. The laws operating here would be those characteristic of the integrational system.

We can also specify *availability of representational processes themselves* as a class of variables. One relevant experiment is reported by Brown and Lenneberg. These investigators first measured the availability of labels in the English language code for various patches of color (a patch of one wave length would be consistently and quickly labeled, yet one of a different wave length would be inconsistently and slowly labeled); they then demonstrated that in a recall situation ease of recognition of color patches by different subjects was predictable from the previously measured availability of their labels. Many other experiments (which I shall not cite in detail) illustrate how *representational feedback* selects among alternative perceptual organizations, how significance helps determine recognition. When frequency-of-usage factors are held constant—a necessary control—it can still be shown that recognition times for words and visual forms vary with such things as values, attitudes, previous rewards or punishments, and the like.

I would like to say something, however, about "perceptual defense." The obvious problem is this—how can an organism defend itself against perceiving a threatening stimulus when the threat depends upon first decoding the significance of the stimulus? Earlier in this paper I suggested that mediation processes are composed of a number of reaction *components*, corresponding to the factors of meaning, and that the associations of these components with a sign will vary in habit strength.

Now, since probability of reaction is a function of habit strength, it follows that those meaning components having the strongest habit strength will tend to be elicited at shorter exposure time than those having weaker habit strength. In general, since *fully* discriminated meaning (recognition) depends upon the total semantic profile, this implies that as exposure time is increased the meaning of any sign should develop gradually: the blur produced by H-A-P-P-Y should first yield a vaguely favorable impression, then a more specific favorable–active impression, and should finally be recognized. With signs that have a dominantly threatening significance, *this* component should be aroused first, prior to complete decoding. If the subject is set for complete decoding, the resultant flood of anxiety self-stimulation could well muddy up the sensory waters and delay further decoding operations (perceptual defense). On the other hand, should the subject be alerted for danger signals, the same prerecognition anxiety stimulation could serve as a distinctive cue in selecting among alternatives (vigilance). And for *any* sign, the dominant component should, through its feedback signals, facilitate alternative integrations of "hypotheses" having similar significance (value resonance). These notions are testable in a number of ways.

The Perceptual Constancies. I have already discussed how a perceptual sign, such as the visual image of an object, may acquire its significance. Now, by virtue of the fact that physical objects and the organisms that explore them are changeable and moveable, the sensory signals deriving from objects will be variable through certain dimensions. The infant's bottle will appear in various sizes as distance changes, in various shapes as the angle of regard changes, in various brightnesses and hues as the intensity and composition of illumination changes. However, the variable sensory integrations arising from the same physical object under different conditions eventuate in the same terminal behavior and hence acquire a *common significance*—retinal images of various sizes, shapes, qualities, and intensities which derive from the infant's bottle as an object are repeatedly followed by milk-in-mouth, and hence acquire a common representational process. According to this view, then, *constancy in perception is the association of a common representational process* (significance) *with a class of stimulus patterns variable through a number of physical dimensions.* In the language of the "Transactional School," this is a common "thatness" shared by a class of stimulus patterns.

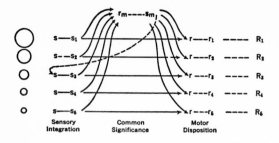

Figure 9.

This is by no means the whole story. In the course of interaction with environmental objects, the members of such stimulus classes become associated with different instrumental sequences and hence with *differential motor dispositions*. Figure 9 may clarify what I have in mind. The sensory integration arising from an APPLE very close to the face $(s - s_1)$ is identified as this edible object (common significance), but comes to be associated with mouth-opening and biting sequences, and hence the motor integration or disposition toward such behavior $(r - r_1)$. The integration characteristic of APPLE held at crooked arm's length $(s - s_3)$ has the same perceptual meaning, but is associated with dispositions toward grasping or flexing the arm. And the tiny retinal image characteristic of APPLE across the room $(s - s_5)$—again signifying this same edible object—has become most strongly associated with locomotor approach movements. In other words, the sensory integrations deriving from familiar objects become associated with both common representational processes and differential motor dispositions; the former constitute identity or "thatness" in perception and the latter constitute "thereness" in perception, part of the unconscious syntax of adjustive behavior.

It should be noted that, as shown in this diagram, selection among alternative "therenesses" or behavioral dispositions depends upon signals from both the integrational and representational systems. The disposition toward grasping requires both that the visual angle be of a certain size *and* that it be identified as a particular known object—given the same visual angle, an apple will be grasped at and a beach ball run after. On the other hand, sensory integrations arising from meaningless, abstract, or unfamiliar objects, or objects of variable size like balloons, will be associated ambiguously with various motor dispositions. The many observations by the Princeton group, summarized by Kilpatrick, on absolute distance judgments are consistent with this view.

Familiar and standard-sized objects like playing cards and cigarette packs were accurately judged as to distance when only size cues were available, but unfamiliar or abstract stimulus objects like star points and oak leaves were not. Similarly, when size and distance cues are put in conflict, apparent size will tend to be constant while apparent distance varies *if* the object is identified as a familiar thing having a "real" size.

But what, psychologically, *is* the "real" size of an object? Thouless, for example, says that constancy always consists in a "regression toward the real object," but he says little indeed about what this "real object" is or how it is established. Let me suggest an experiment and its result. We will project lifelike images of various objects onto an invisible screen at an unknown distance from the observer so that they seem to hang out there in empty space. By some optical means, we will allow the observer to adjust the physical size of these images until they seem "natural" or "just right" in apparent size. Knowing the actual distance of the screen and the final objective size of the image, we will compute the visual angle subtended by each object-image when judged to be "natural-looking." We will find that these visual angles correspond very closely to those subtended by these same objects at their ordinary inspection distances. We inspect and compare horses and automobiles at a distance of some 20 feet; we compare cigarette packages at crooked arm's length; we inspect the sharpness of record needles at a distance of about 6 inches. In other words, *the "real" or "natural" size of an object will be found to be the visual angle subtended at which the finest visual discriminations for that class of objects can be made.* I would also be willing to bet that this "natural" visual angle will approximate a constant for all objects, dependent upon retinal characteristics. The same argument would apply to all other dimensions along which constancy operates—the "real" color and brightness of an object will be that experienced under white light of normal daylight intensity; the "real" shape of an object will be that experienced when held perpendicular to the line of regard (because this is the condition for finest shape discriminations); and so on.

Now, from all this it follows that the representational mediation process associated with a particular class of signs, which gives them their "appleness," "horseness," "pack-of-cigarettes-ness," or what-have-you, will be most frequently and strongly elicited by that sensory integration corresponding to the "real" object. This is to say that we will most often be decoding the perceptual significance of an object when it is being inspected in that portion of the visual field where the finest discriminations can be made, when we are "paying attention to it." By the same token, as shown by the dashed arrow in Figure 9, *the feedback self-stimulation from this representational process must be most strongly associated with "tuning-up" or predicting this "real" or normative sensory integration.* From this we can derive a number of the standard phenomena of constancy. (1) What is perceived is usually a compromise between the "real" behavioral object and the actual sensory information. Recognition of a meaningful object will itself, through feedback, change the sensory signals in such a way as to increase the probability of occurrence of a more "normal" perceptual integration. Obviously, the less intense and clearly defined the retinal signals, or the more intense the feedback signals from the cognitive system the greater should be this tendency toward normalizing. Therefore, (2) the phenomenal characteristics of familiar, meaningful objects should show greater constancy than those of nonsensical, unfamiliar, or abstract objects. "Object-colors" show constancy, but "film-colors" do not; a dinner plate held at various angles shows more constancy than forms cut from white cardboard. And similarly, (3) the more natural the situation in which constancy is measured, or the more motivated the subject toward behaving with respect to "things," the greater is the constancy shown. Adults, children, monkeys, and even fish display almost perfect constancy in going about their everyday affairs. In other words, the ordinary behavior of organisms is concerned with decoding the *significance* of signs, regardless of their momentary physical characteristics, and with encoding intentional behavior that takes account of these significances.

Summary

I said this paper would be speculative, and I think I have kept my word. In order to discuss the topic I originally proposed, I found it necessary to present a rather general theory of behavior, and despite the length of this paper my treatment has been a very sketchy one. The theory conceives of behavior as a two-stage process, decoding the significance of received signals and encoding intentions into overt acts. Both decoding and encoding processes are assumed to involve three interactive levels of organization—a projection level, an integration level, and a representation or cognitive level—but the principles governing one level do not seem to apply to the others. Throughout this paper I have tried to demonstrate the essential identities of perceptuo-motor behavior and language behavior when viewed within this framework, and I at least feel that both become more understandable by virtue of being compared.

I am not unaware of the crudeness of this kind of theorizing. It is certainly more programmatic than rigorous and more qualitative than quantitative, but, on the other hand, I think that for some time to come rigorous, quantitative theories in psychology are going to be feasible only in very restricted areas. And in the meantime, many of us are going to want to do what we can with such complex problems as perception and language. Theories such as the one I have outlined can help to systematize what information we do have, can provide an impetus to new research, and can give us at least the illusion of some understanding.

25.
The Informational Analysis of Questions and Commands

DONALD M. MACKAY

Introduction

It is not, I think, without significance that the present Symposium on information theory contains relatively few papers on the theory of information; but the significance is not, as might be supposed, that the theoretical concepts are now so far worked out that only their applications remain to be discussed. In fact the contrary would be nearer the truth: that most of us who began with an interest in the theoretical concept of information have become so increasingly and profitably absorbed in practical problems of information-processing, in animals or machines, that the general analysis of informational exchange has been largely by-passed, and for the majority of the new generation now leaving college information theory still means little more than Shannon's measure of unexpectedness and its various applications. Otherwise excellent leading textbooks unashamedly proclaim a divorce between what they call 'information theory' and semantics, which topic is discussed, if at all, in the woolliest terms, and generally relegated to the philosopher.

As this can hardly be regarded as an ideal state of affairs in such a young subject, we may well ask why so little progress has been made on the semantic side. Apart from the absence of workers, I suspect the reason to lie in our failure to study the communicative process within a wide enough context: to follow the flow of information far enough back, and forward, from the com-

From D. M. MacKay, "The Informational Analysis of Questions and Commands," in Colin Cherry (Ed.), *Information Theory: Fourth London Symposium* (London: Butterworth's, 1961). Reprinted by permission of the author and publisher.

munication channel. We are all familiar with diagrams showing the human sender and receiver as 'black box' terminals linked by a chain of noisy channels. My suggestion is that semantic questions find their natural place in information theory when (but only when) we widen our diagrams to take account of the nature of these terminals as *goal-directed* self-adaptive systems. Not inappropriately perhaps, our discussion in this concluding paper will thus link together a fair range of the ideas embraced by the present Symposium, though we shall here take the enquiry only far enough to show some of the possibilities.

Our particular object will be to suggest how questions and commands can be analysed in informational terms similar to those outlined on an earlier occasion for the analysis of indicative sentences. In the first part we shall sketch the informational requirements of organisms in the situation in which questions and commands are exchanged between them. Thus fortified, we shall try to discern some lines along which at least semi-quantitative analysis may fruitfully be pressed.

The Impact of Information on the Organism

An organism can be regarded for our purpose as a system with a certain repertoire of basic acts (both internal and external) that in various combinations and sequences make up its behaviour. In order that its behaviour should be adaptive to its environment[1], the selective process by which basic acts are concatenated requires to be *organized* according to the current state of the environment in relation to the organism. There are various ways of picturing this need. In its most basic terms, we may regard what is required as equivalent to a vast constantly changing matrix of *conditional probabilities* (the C.P.M.), determining the relative probabilities of various patterns (and patterns of patterns) of behaviour in all possible circumstances. More economically, we can think of it as the setting up of a hierarchic structure of organizing 'sub-routines' to determine these conditional probabilities, interlocked in such a way as to represent implicitly the structure of the environment (the world of activity) with which the organism must interact. For many purposes we may reduce it to the filling-out of a world-map, ready to be consulted according to current needs and goals.

Whatever our thought-model, it is clear that unless the organism happens to be organized exactly to match the current state of affairs, *work* must be done to bring it up to date: work, not only in a physical, but in a *logical* sense. This 'logical work' consists in the adjusting and moulding of the conditional-probability structure of the organizing system: the formation, strengthening or dissolution of functional linkages between various basic acts or basic sequences of acts. The total configuration of these linkages embodies what we may called the total 'state of readiness' of the organism. Some of them will of course have purely vegetative functions that do not concern us. What does interest us is the total configuration that keeps the organism matched to its field of purposive activity, and so implicitly represents (whether correctly or not) the features of that field. For brevity, let us call this the *orienting* system, and the corresponding total state of readiness the *orientation* of the organism.

Information can now be defined as that which does logical work on the organism's orientation (whether correctly or not, and whether by adding to, replacing or confirming the functional linkages of the orienting system). Thus we leave open the question whether the information is true or false, fresh, corrective or confirmatory, and so on.

The *amount of information* received by an organism can then be measured (in various ways) by measuring if we can (in various ways) the logical (organizing) work that it does for the organism. I have discussed elsewhere some of the different measures that suggest themselves for different purposes, and we shall return briefly to the question later. Meanwhile it is sufficient to note that they are necessarily *relative* measures, since they measure the impact of information on the given receiver. 'Amount of information' measures

not a 'stuff' but a relation.[2] The *meaning* of an indicative item of information to the organism may now be defined as its selective function on the range of the organism's possible states of orientation, or for short, its *organizing function* for the organism. It will be noted that this too is a relation. (It must be clearly distinguished from the *organizing work done* on the organism, which is *the result of the exercise* of this organizing function. Much confusion is caused by attempts to identify meaning with the change produced in the receiver).

Perception and Communication

A solitary organism keeps its orienting system up to date in response to physical *signs* of the state of the environment, received by its sense organs. This adaptive up-dating of the state of orientation we call *perception*. We can regard *communication* as an extension of this process whereby some of the organizing work in one organism is attempted by *another organism*. Normally this means that the receiving organism is induced to adapt itself in response to physical signs that are perceived as *symbols*—as calling for orienting (or other) activity over and above that which constitutes their perception as physical events[3].

The logical starting point for a semantic theory of communication would therefore seem to be the analysis of the organizing functions that are 'extensible' in this way from one organism to another. There is a sense in which for this purpose the analysis of questions is logically prior to that of indicative sentences; for the meaning of an indicative sentence is often ambiguous until we know the *question* to which it is an answer, and/or the assertion which it *excludes*. For example, the sentence S: 'I sleep in room 10' may be an answer to the question (a) 'Where do you sleep?' or (b) 'Do you sleep in room 10 or room 12?' or (c) 'Who sleeps in room 10?' or (d) 'What do you use room 10 for?' and so on. Typical assertions excluded by S are then respectively (a) 'I sleep elsewhere. . . .' (b) 'I sleep in room 12' (c) 'Someone else sleeps in room 10' (d) 'I use room 10 for other purposes'.

Clearly, the selective function of S is quite different in these cases, either in its range (as between (a) and (b)) or in its dimensions (as between (a), (c) and (d)). No analysis of its semantic information-content can claim to be adequate unless it brings out differences of this sort, as well as doing justice to logical structure.

Questions

What then in these terms constitutes a *question*? The *object* of a question is normally to evoke communicative activity of the sort we have been discussing; but this by itself would be far from a definitive notion. Again the *linguistic form* of a question is often characteristic, but not always. *Intonation* may sometimes be the key discriminant, as in a phrase like 'going home?' which could equally well be the *answer* to a question. What, however, are we to say of the sentence 'I take it that you are going home'? It would normally make sense to treat this as a question requiring the same answer as 'Are you going home?'; yet it is in form (and could be in intonation) a plain indicative statement.

But indicative of what? Here I think we come to the heart of the matter. What is indicated here is the *state of readiness* of the originator, in relation to the receiver; and this points to a key characteristic of all questions. A question is basically a purported *indication of inadequacy* in its originator's state of readiness, calculated to elicit some organizing work to remedy the inadequacy. It is as if the questioner uncovered and held out the incomplete part of his organizing system to the receiver for his attention.

The fact that this normally works reveals an interesting presupposition behind the whole procedure. We may well ask why the receiver should be expected to pay any responsive attention of this sort to the exposed organizing system; and the answer is clear. Questions work only because human beings are *motivated* to adjust one another's states of readiness. Whether thanks to nature or nurture, a human being finds an organizing system 'exposed' in this deliberate way by another a natural target for adaptive activity, much as a mother bird is moved to regard a gaping beak in the nest. Doubtless were it not so the race would hardly have survived!

The moral would seem to be that what makes an utterance a question cannot always be pinned down either to a peculiarity of its logical form (since an inadequacy of orientation can be indicated in many ways, some of them wordless) nor to the fact that it has elicited information (since people are liable spontaneously to try to adjust one another's orientation).

Our problem in fact has some affinity with that of determining whether a given action was 'goal-directed' towards a certain end-point. Neither the form nor the effect of the action are infallible criteria alone. Instead, we adopt a 'variational approach and look for evidence (usually structural) that if the 'goal' had been slightly displaced, or the starting-point slightly different, the form of action would have been correspondingly modified in a manner calculated to reach the same end-point.

Similarly we can determine some utterances to be a question only, in the end, by enquiring how the behaviour of the originator would have been modified in varying circumstances: in other words, by examining the total information-flow-map within which the utterance originated.

For many purposes, however, this is too stringent a test. An examiner, for example, may frame a question to which he already knows the answer. His object is to gain, not the information represented by the answer, but information as to whether the candidate knows the answer. Yet the candidate would be ill-advised to ignore the utterance on the ground that the examiner was not really asking a question. Here the form, in its context, is decisive. A properly framed examination question depicts an inadequacy in the orientation of an imaginary questioner, and the candidate is expected to show his prowess by an answer calculated to remedy the inadequacy in a typical originator[4]. We may perhaps take as our tentative definition of a question, then, a combination of an *indication* and an *invitation*: or if we like, a twofold indication, (*a*) of inadequacy in orientation, and (*b*) of desire to have it remedied by the receiver.

The Meaning of a Question

The purpose of an indicative statement is to bring up to date some aspect of the receiver's state of readiness. Conversely, the main ostensible purpose of a question is to bring about, as it were by remote control, an updating of the questioner's *own* state of readiness. He wants the receiver to do some organizing work on him. The primary meaning of a question, then, which we might call its *interrogative meaning*, can be defined, like the meaning of an indicative sentence, as its selective function; but it selects from the range of possible orienting operations *on the questioner himself*. Its job is to identify the organizing work that needs to be done on the originator's switchboard, so to say. The meaning of the answer will in turn be its selective function—a more detailed one—on the range of questioner's states of orientation. Its job is to set the switches.

On the other hand, as we have seen, a question also plays an indicative role, by simply showing the receiver the inadequacy of the questioner's state of readiness, together with his desire to have it remedied. It is therefore worth while to distinguish the *indicative meaning* of a question as (for short) its orienting function for the receiver. It is obviously quite possible for a question to be interrogatively meaningful while indicatively meaningless, and vice versa. Interrogatively, a question is meaningless when it *calls for* no orienting work on the *originator*; indicatively, it is meaningless when it *performs* none on the *receiver*.

Thus to ask 'where does the flame go when the gas is turned off' is interrogatively meaningless as it stands, because if we assume that 'flame' means 'gas being burned', the two halves of the question call for mutually incompatible selective functions. When there is no gas, there is no flame to go anywhere. But it is not indicatively meaningless, for its utterance does orient the receiver with respect to the questioner's muddled organizing system.

Commands

We have pictured the asking of a question as offering the receiver access to some of the originator's controls. Conversely, we may now think of the uttering of a command as *claiming* access to some of the receiver's controls. It is perhaps hardly necessary in detail to go over all the corresponding ground. The controls in question are now those that determine goal-setting and action, rather than map-making and orientation. The relationship is potentially one of much greater subordination than that of a questioner to his respondent, since a command (if accepted) supplants the process of calculation-in-view-of-goals by which the action would normally be determind, and on which the response to a question would bear.

Once again, as those with memories of nursery (or Army) language know well, we would be unwise to try to characterize a command by its logical form, though of course the 'imperative' if used is distinctive. 'Johnny, I want you to stop banging on the floor'; or (across the Atlantic) 'Do you want to pass me the salt?' are cautionary samples of well-understood commands to which the response required is not 'Oh' or 'Yes' respectively, but *action*.

A command then, as a communicative act, is basically an indication that action on the part of the receiver is necessary to remove disequilibrium in the originator: in other words, that the receiver

is an intermediate *target* within the goal-system of the originator—that one of the originator's information-flow loops for goal-direction passes through him, so that he is potentially under feedback. Once again, the 'variational' question— what would happen in the originator's organizing system *if* the receiver did not act appropriately— is the reliably diagnostic one, rather than analysis of form or effect in isolation.

Clearly we must again distinguish two semantic functions for a command. As its primary purpose is to select a goal-setting in the receiver, we may define its *imperative meaning* as its selective function on the range of the receiver's goal-settings. Inevitably however it also serves to indicate the goal-settings of the originator, so that its *indicative meaning* may be distinguished for short as its orienting function relative to the originator's goal-settings.

The usual distinctions between meaning-to-originator, meaning-to-receiver, and conventional meaning, can be made in obvious ways in each case. They are of course essential if ambiguity is to be avoided.

Quantitative Aspects

There is now no difficulty in principle in defining a measure of *selective information-content* for questions and commands, as for indicative sentences. In each case we may define it as the equivalent number of independent binary selective operations specified by the selective function in question.

For many purposes, however, we may be interested in the form as well as the magnitude of the selective operation; and here, as with indicative sentences, the structure of the organizing system would seem to offer a natural basis for quantification. I have argued elsewhere for a model organized hierarchically, in which 'organizing sub-routines' at a given level are themselves organized by sub-routines at a deeper[5] level representing more abstract concepts.

If a model of this sort is correct, then an important quantitative parameter of a question will be the *order-number* of the level of organization at which it betokens an inadequacy. To take a simple example, a question of the form 'Is A, B?' is of a lower order than one of the form 'Are all A, B?'. 'Does the road next bend to left or right?' is of lower order than 'Is there any regularity in the sequence of bends?'. Similarly the command 'Turn left' is of lower order than the command 'Turn alternately left and right'. The second selects an *organizer of the organizers* dealt with by the first.

A second parameter of great importance will be the logical dimensionality (*logon-content*) of the question or command, defined as the number of independent 'degrees of freedom' of the organizing operation concerned. Thus 'What is the temperature?' has an (interrogative) logon-content of 1. (Indicatively it has of course as many as required for identification of the originator's state of inadequacy.) 'How has the temperature varied during the day?' has as many logons as there are independent ordinates on the graph required, and so on.

With non-numerical questions the actual estimation of logon-content waits, of course, on our knowledge of the human organizing hierarchy; but the concept may have its qualitative uses even now. For example, we can recognize the importance of *matching* the logon-content of an answer to that of a question. We all know that feeling of impatience with an answer that swamps us with data in unnecessary detail, or of frustration with one that leaves us short of some key item. What I am suggesting is not that numerical measures as such would be easy or even useful to apply in such a case, but rather that a quantitative theory of what is going on might help to articulate our intuitions and sharpen our recommendations of a remedy in more complex cases. This, after all, is perhaps the main service to be rendered by any semantic theory of communication.

It would be of obvious practical importance, for example, to ascertain whether George Miller's 'Magic number 7' represents a maximum logon-content for any useful unit-question (or answer, or command).

Conclusions

The main argument of this paper has been that the semantic informational analysis of questions and commands, as of indicative sentences, could be most naturally and least ambiguously carried out by analysis of their 'organizing function' with respect to the information-systems engaged. This requires a stronger link than exists at present between semantics and the study of behavioural organization; and it is not unlikely that the benefits could be mutual, for the theorist of organization might well gain significant clues in return; for example, from the logical structure of thesauri.

It is of course easier to make such suggestions than to carry them out, and I have done no more than to indicate a few possibilities. We have seen how measures of selective-information-content, and concepts of meaning, can readily be defined on this basis for questions and commands; and we have singled out the order-number (can we call it simply the 'order'?), and the logical dimensionality or 'logon-content', as two measures likely to be important. We have also noted the importance of distinguishing between measures of 'indicative' function and those of interrogative and imperative function, and between intended, received and conventional meanings and information-contents.

But I am painfully aware that the bulk of what needs to be done is little reduced by all this. If it succeeds only in stirring up fruitful discussion in a neglected area it will have served its purpose.

Notes

1. 'Environment' here means the total world of activity of the organism, and not only its immediate physical milieu.
2. This does not of course make the concept any less objective in principle, since one can always postulate a 'standard receiver', and this is in effect what is done in communication theory; but it does prevent the magnitude associated with it from having a unique value. The same item can have quite different information-contents for different receivers.
3. The training process by which symbols come to be accepted as such need not here concern us.
4. The question here is merely being 'mentioned' by the examiner, and not 'used', as it would be by a typical originator.
5. 'Deeper' and 'higher', oddly enough, seem to be interchangeable as indices of greater abstractness.

This paper presents a conceptualization of information as related to the decision problems of the recipient. The orientation is toward a formal definition of behavioral elements in an individual's "purposeful state": specifically, these elements are his objectives, his valuation of each objective, his possible courses of action, the efficiency of each course of action in achieving each objective, and his probability of choice for each course of action.

The *amount of information* in a purposeful state is explicitly defined in terms of the probabilities of choice of the available courses of action. The amount of information in a *message* is defined as the difference between the amount of information in the purposeful state following the message, and the amount of information in the purposeful state preceding the message. The *amount of instruction* in a purposeful state is defined in terms of the efficiencies of the available courses of action; and the *amount of motivation* is defined in terms of the values of the objectives. The amounts of instruction and motivation in a message are defined, just as information is, by comparing the amounts in a purposeful state before and after receipt of the message.

The *value of a purposeful state* to an individual is defined as a function of the amount of information, the amount of instruction, and the amount of motivation in the state. This concept can be generalized to express the value of the state of some other individual.

26.
Towards a Behavioral Theory of Communication

RUSSELL L. ACKOFF

Introduction

The significance of Claude Shannon's work in communication theory is such that anyone presuming to contribute to this theory is obliged to relate his work to Shannon's. In exploring this relationship it will be helpful to refer to Warren Weaver's masterful exposition of Shannon's work.

According to Weaver, "Relative to the broad subject of communication, there seems to be problems at three levels." These are

"Level A. How accurately can the symbols of communication be transmitted? (The technical problem.)

"Level B. How precisely do the transmitted symbols convey the desired meaning? (The semantic problem.)

"Level C. How effectively does the received meaning affect conduct in the desired way? (The effectiveness problem.)"

Weaver classifies Shannon's work as follows:

"The mathematical theory of the engineering aspects of communication, as developed chiefly by Claude Shannon at the Bell Telephone Laboratories, admittedly applies in the first instance only to problem A, namely the technical problem of accuracy of transference of various types of signals from sender to receiver."

He goes on to note, however, that "the theory of Level A is, at least to a significant degree, also a theory of levels B and C."

The effort in this paper is primarily concerned with level C, the effectiveness problem. But the effectiveness problem is conceived here in more general terms than those in which it appears to have been conceived in Weaver's formulation. This effort can be characterized by the following objectives:

1. To identify the ways in which the receiver's behavior can be affected by a sender.

2. To construct measures of these effects.

3. To define and construct measures of effectiveness for these effects relative to the receiver's objectives and those of the sender as well.

The question, "What is communication?" will be considered in more detail here than is provided by Shannon and Weaver. A related question, "How does one measure amount of information?" is as critical here as it is in Shannon's theory. But "information" is given a considerably different meaning here than it receives in Shannon's treatment.

"The word *information*, in this [Shannon's] theory, is used in a special sense that must not be confused with its ordinary usage. In particular, *information* must not be confused with meaning."

The meaningfulness and value of information *is* central in this discussion. "Information," according to Weaver, "is a measure of one's freedom of choice when one selects a message." Here we shall develop a more general concept of information, but one in which the concept of "choice" is also fundamental.

The fundamental idea of this paper consists of the analysis of communication into three components: the transmission of information, instruction, and motivation and the definition of these in terms of a purposeful state. *The measures discussed are only intended to be suggestive.* They are not

From Russell L. Ackoff, "Towards a Behavioral Theory of Communication," *Management Science*, 4 (1957–58), 218–34. Reprinted by permission of the author and publisher.

"best" measures in any sense of the word. I certainly hope to improve on them in the future. My hope is that—with what might appear like premature publication—I can encourage others, who are more qualified than I, to undertake the task.

An Apology

Few references are made in the discussion which follows to the work of anyone except that of Shannon and Weaver. This is, of course, unfair to many others who have worked on this problem, to many of whom I am indebted. My purpose, however, is neither to review the voluminous literature nor to evaluate it, but rather it is to present an idea which has been "tested" on a number of persons better qualified in this area than I am. As far as I can determine the material presented here has not been presented before. To save space I do not make all the obvious and not so obvious connections to the work of others. In all honesty I must also admit that I have chosen this course to save work as well. The only ones who will be offended by such omissions are those already familiar with the field. For those who are not, Colin Cherry's recent work in this field [*On Human Communication*] is recommended not only for its own sake but for its extensive bibliography on the subject.

Some Fundamental Concepts

Weaver defines communication as "all of the procedures by which one mind may affect another." His and Shannon's discussion, however, is restricted to only one such type of procedure: the transmission of *messages*. Their use of the term "communication" conforms better with common usage than does their definition. For example, the man who produced the slide rule I use may affect my mental processes without that man communicating to me. In general, many who have shaped my environment or the instruments which I use have affected my mental processes without communication with me in the ordinary sense.

If communication is to be restricted to the transmission of messages, the concept "message" must be clarified. First, however, "the effect of one mind on another" must be translated into operational (i.e., behavioral) terms.

1. A PURPOSEFUL STATE

Communication is an activity in which only purposeful entities can engage. Purposefulness exists only if choice is available to the entity involved and if that entity is capable of choice.

A purposeful state (S) may be defined by reference to the following concepts and measures:

I: the individual or entity to which purposefulness is to be attributed.

C_i: a course of action; $1 \leq i \leq m$.[1]

O_j: a possible outcome or consequence of a course of action; $1 \leq j \leq n$.[1]

P_i: the probability that I will select C_i in a specified environment, N; that is, $P_i = P(C_i/I, N)$.

E_{ij}: the probability that O_j will occur if C_i is selected by I in N; that is, the efficiency of C_i for O_j in N.

$$E_{ij} = P(O_j/C_i, I, N)$$

V_j: the value (importance) of O_j to I.
Clarification of some of these concepts is necessary.

2. COURSES OF ACTION

A course of action is not to be construed as mechanistically specified behavior. Variations in the action with respect to certain physical characteristics may not change the course of action. For example, "driving a car" may be designated as a course of action. There are many different ways of driving a car but it is frequently useful to group these into one class of behavior. Despite the variations within the class, it can be distinguished from other classes; for example, from "using a street car." A course of action may be specified with varying degrees of rigidity depending on the purposes of the research. For one purpose it may be desirable, for example, to distinguish between left-hand and right-hand driving. For another purpose it may be desirable to group the use of all self-powered vehicles into one course of action.

It should be noted that the problem of specifying a course of action is essentially similar to that of specifying a physical object. For one purpose an automobile may be considered as a unit; for another it is a composite of many other units (e.g., wheels, transmission, etc.), and for still another purpose it may be considered to be a part of a unit (e.g., a fleet of cars).

A course of action is said to be *available* in an environment if there is a probability of its being selected by someone, that is, if

$$\exists I_k : P(C_i/I_k, N) > 0.$$

An available course of action may have no probability of being selected by a specific individual under a particular set of conditions. Then it is not a *potential* course of action for him. This is equivalent to saying that a course of action, C_1, is potential to an individual in an environment if, for one or more sets of values of E_{1j} and V_j in N, P_1 is greater than zero. Nevertheless, for some specific set of values of E_{1j} and V_i, P_1 may be equal to zero.

The relativity of courses of action and outcomes should be noted. They are conceptual constructs which may be converted into each other depending on the interests of the researcher. For example, "sawing a tree" may be considered as a course of action which yields the "falling of a tree" as an outcome. But "felling a tree" may be considered as a course of action which can yield the outcome "clearing a path." Such relativity of concepts is common in all areas investigated by science and hence does not present any unique methodological problem in this context.

3. EFFICIENCY

Many different measures of efficiency are in current use. It is fairly common to use some measure of the cost, time, and/or effort required to bring about a specified outcome (e.g., to complete a specified task such as "travelling one mile") as a measure of efficiency. It is also quite common to measure efficiency in terms of the portion of an outcome which is realized by the expenditure of a specified amount of money, time, and/or effort. For example, one can measure the efficiency of a machine tool either in terms of the number of units produced per dollar or in terms of the cost per unit. Thus efficiency is commonly measured either as (1) units of input required to obtain a specified output, (2) or as units of output obtained by a specified input. Neither type of measure is sufficiently general to be applied in all situations.

The input required for a fixed output and the output yielded by a fixed input are not constant. For example, the number of units made per hour by a machine varies from hour to hour and the miles per gallon obtained by an automobile also varies. Hence, for a fixed input various possible outputs exist to each of which a probability can be assigned. If, in the definition of a course of action, an input is specified then the efficiency of that course of action for a specified outcome can be defined as "the probability that the outcome will occur if the course of action is taken." This

measure can always be applied in a purposeful state.

This measure of efficiency of a course of action depends on the environment and the individual involved. For example, use of skis may be efficient for self-transportation down a snow-covered hill but not so down an uncovered hill. Different individuals may ski with different efficiencies and the efficiency of the same individual may change over time (e.g., by learning). Consequently, the relevant time period, individual, and environment should be specified in designating efficiency.

4. VALUE

As in the case of efficiency there is no one measure of value or worth of an outcome that is generally accepted. Fortunately, however, such a measure is not necessary for our purposes here. Nevertheless, it is convenient to use some kind of standard measure wherever possible. A dimensionless measure of *relative* value may provide such a convenient standard. If the values (v_j) assigned to the various outcomes are all positive, a measure of relative value (V_j) for each outcome may be obtained by the following conversion:

$$V_j = \frac{v_j}{\sum v_j}.$$

Then, since

$$\sum \frac{v_j}{\sum v_j} = 1.0$$

it follows that

$$\sum V_j = 1.0.$$

The minimum relative value ($V_j = 0$) occurs only when the absolute value (v_j) is equal to zero. The maximum relative value ($V_j = 1.0$) occurs when all but one outcome has zero value.

In some cases negative measures of value are used (e.g., cost versus profit). The following transformation may be used in such cases:

$$V_i = \frac{v_j}{\sum |v_j|}.$$

In the discussion that follows we shall use the concept of relative value and assume that all V_j's are positive and, therefore, that $\sum V_j = 1.0$. All the results, however, are easily modified to cover the use of either absolute values or the case in which negative values are employed.

It is assumed here that no v_j can have an

infinite absolute value. This assumption is not made lightly. It is based on an analysis of the meaning of "absolute value" which appears in (1). Following the argument presented there a value can approach an infinite magnitude only as an unattainable limit.

This "light" treatment of value theory is no more intended to imply that the meaning and measurement of value presents no problems than is my unsupported use of the term "probability" meant to imply that its meaning and measurement present no problems. This is no place to argue the issues of value theory. Churchman and I have done this in several places. . . .

A purposeful state (S) may now be defined relative to the concepts which have been discussed. An individual (I) may be said to be in a purposeful state in an environment (N) if the following conditions hold:

1. There are at least two exclusively defined courses of action available in N; that is, in N for C_i, where $1 \leq i \leq m, m \geq 2$.

2. Of the available courses of action in N, at least two are potential choices of I.

3. Of the set of outcomes (defined so as to be exclusive and exhaustive) there is one (say, O_a) for which two of the potential courses of action (say, C_1 and C_2) have some efficiency; that is, $E_{1a} > 0$ and $E_{2a} > 0$. Furthermore, $E_{1a} \neq E_{2a}$.

4. The outcome relative to which condition 3 holds has some value to I; that is, $V_a > 0$.

This definition may be summarized less technically as follows: an individual may be said to be in a purposeful state if he wants something and has unequally efficient alternative ways of trying to get it.

If we consider an individual over a period of time it will be convenient to refer to the purposeful states at the beginning and end of that period as *initial* and *terminal* states, respectively.

The conceptual labors which have been involved in defining a purposeful state are necessary in order to make explicit the meaning of "one mind affecting another," and for enumerating the various possible types of effect. As subsequent discussion will show, these effects may be defined in terms of changes in purposeful states. Before we consider such changes in detail, the meaning of a "message" should be explored.

5. MESSAGE

A message may be defined as a set of (one or more) signs. This definition is of little value without considering the nature of a sign. A sign signifies something to somebody; that is, it produces a response to something other than itself.[2] This may be put more precisely in terms of the following concepts and symbols:

x_i: an object, event, or property.

$x_1 \rightarrow x_2$: x_1 produces x_2; that is if x_1 is a necessary but not a sufficient condition for x_2.

If x_2 is a change in a purposeful state of I, then x_2 is said to be a *response* of I to x_1, which is said to be a *stimulus*.

Then, x_0 is a sign of x_1 if it produces a response (x_2) of I to x_1; that is, if $x_0 \rightarrow (x_1 \rightarrow x_2)$.

If, for example, "pointing to a chair" or saying "chair" produces in someone a (purposeful) response to an object (e.g., sitting down), then the pointing or statement is a sign of that object to the respondent.

Signs are of two types, natural and man-made. For example, a dark cloud is a natural sign of rain. The word "cloud," however, is a man-made sign of the object, cloud. Smoke is a natural sign of fire but smoke signals are man-made signs of many other things. Messages consist of man-made signs.

This brief definition of a sign raises many questions; for example, "What do such words as 'but,' 'if,' and 'or' signify?" These and other questions can and will be answered in a subsequent paper. In the meantime, however, it will be apparent to the reader that many, if not most, of the familiar natural signs (e.g., smoke, which is a sign of fire) and linguistic signs satisfy the definition given above.

Communication

Now we can make precise the conditions under which one individual, I_a, may be said to communicate to another, I_b. If an *individual I_b responds to a set of signs selected by I_a in a purposeful state, then I_a is the sender and I_b is the receiver of the message.*

Several aspects of this definition of communication should be noted. First, I_a and I_b may be the same individual. That is, a person may communicate to himself as in writing a "reminder" to himself. Secondly, the sender of the message need not intend or desire to communicate to the receiver in order to do so. The interceptor of a message, for example, is communicated to, although unintentionally. Thirdly, the sender and receiver may be widely separated in time and space.

Now we want to concentrate primarily on the communication received, that is, on the receiver.

1. THE VALUE OF A COMMUNICATION

A purposeful state of an individual (I) is described by

    1. the set of available courses of action, C_i,

    2. the set of possible outcomes, O_j,

    3. the probabilities of choice associated with the courses of action, P_i,

    4. the efficiencies of the courses of action for each objective, E_{ij}, and

    5. the value of the outcomes to I, V_j.

Then, given the available courses of action and possible outcomes, the value of a state, $V(S)$, must be some function of P_i, E_{ij}, and V_j; that is,

$$V(S) = f(P_i, E_{ij}, V_j). \qquad (1.1)$$

The nature of the function, f, depends on the definition of the state's value. This value may be defined in several different ways; for example, in terms of expected return, expected gain, or expected loss. The discussion and measures that follow are independent of the definition of state value which is used. But for illustrative purposes we shall use "expected return" as the state value, that is,

$$V(S) = \sum_{i=1}^{m} \sum_{j=1}^{n} P_i E_{ij} V_j. \qquad (1.2)$$

Since $P_i \leq 1.0$, $E_{ij} \leq 1.0$, then, if a measure of relative value is used in which $0 \leq V_j \leq 1$ and $\sum V_j = 1.0$, it follows that the minimum and maximum values which $V(S)$ can assume are zero and one, respectively.

Receipt of a communication involves a change in the receiver's purposeful state. Let S_1 represent the initial state and S_2 represent the terminal or changed state where the change is in response to a message. Then the changes must be in one or more of the P_i's, E_{ij}'s, or V_j's, or some combination of these. Let $V(S_1)$ be the value of the initial state and $V(S_2)$ be the value of the terminal state. Then the value of the communication to the receiver, ΔV, is given by the equation:

$$\Delta V = V(S_2) - V(S_1). \qquad (2)$$

Even if only positive absolute values, v_j, are used, the value of the communication may be negative; that is, it can do the receiver harm.

The value of the communication to the sender may be obtained by substituting his values for the outcomes O_j in the receiver's state for those of the receiver in computing $V(S_1)$ and $V(S_2)$.

As is apparent, the value of the communication to anyone may also be obtained by using his V_j's rather than those of the receiver.

2. TYPES OF COMMUNICATION

We have considered three basic ways in which the value of a purposeful state may be changed by communication: by changes in P_i, E_{ij}, and in V_j. There is one other way that a purposeful state may be changed: those available courses of action which are not initially potential choices may become so. But unless such change is accompanied by changes in the P_i, it will not affect the value of the state. The significance of such a change, however, is best considered after discussing changes of the first type.

A particular communication may change only the P_i's, E_{ij}'s, or the V_j's, or some combination of these. We may study a communication which yields a combination of changes in terms of each type of change taken separately. This is, in fact, what has been done in the past, but the relationship between these types of changes has not been systematically studied because of the lack of a unifying conceptual framework.

By way of preview, we shall say that a communication which changes the probabilities of choice, *informs*; one that changes the efficiencies of courses of action, *instructs*; and one that changes the values of outcomes, *motivates*. Any single communication may, of course, do any combination of these simultaneously.

2.1 INFORMATION[3]

Shannon, his predecessor, R. V. L. Hartley, and most others who discuss information in mathematical terms are concerned with the amount of information that *can be* communicated rather than with the amount that *is* actually communicated. Shannon was primarily involved with systems in which each possible message can be coded into a combination of two symbols. For example, if there are four possible messages and two symbols (0 and 1), the messages can be represented as 00, 01, 10, and 11. Then, to select one message out of the four, two choices from among the two symbols (i.e., binary choices) may be made. One binary choice allows two messages (0 and 1) and three binary choices allows eight messages (000, 001, 010, 100, 110, 101, 011, and 111). In general, x binary choices allows 2^x possible messages.

For Shannon, the amount of information

contained in a message is the amount of freedom of choice involved in the selection of the message. A unit of choice is defined as the selection of one out of two equally available symbols. Thus, in selecting one of two equally available symbols, one choice-unit is involved and the resulting one-symbol message contains one unit of information.

In general, if there are M equally available messages in a state, the selection of one contains x units of information where

$$x = \log_2 M.$$

Equal availability of the symbols means equal likelihood of choice by the sender. That is, if there are M possible messages and the probability of each being selected is $1/M$, complete freedom of choice exists. If the probability of selecting a particular message, p_i, deviates from $1/M$ there is not a completely free choice. In the extreme case, if the probability of selecting any one of a set of messages is 1.0, then there is no freedom of choice and no information can be communicated by the one message which is always selected.

In order to cover cases in which choices are not equally likely (as well as where they are), Shannon derived the following general measure of the amount of information (symbolized by H in his system) contained in a state:

$$H = - \sum p_i \log p_i,$$

where p_i is the probability of choice of the ith message. If $\log_2$ is used, then H is expressed in binary units which are called *bits*. Thus, a state which contains two equally likely messages contains one bit of information.

The measure of information to be developed here will also be related to freedom of choice; that is, it will be a function of the probabilities of choice associated with the alternative courses of action. It will be a different function, however, because of the difference between a message and a course of action. The measure here will also be a function of the number of alternative potential courses of action, m.

When we talk of the amount of information that a person has in a specified situation (state), we do so in two different but related senses. First, we refer to the number of available courses of action of which he is aware; that is, to the number of potential courses of action. For example, a person who is aware of four exits from a particular building has more information than the person who is aware of only two when there are four. The act of informing, then, can consist of converting available choices into potential choices. For example, a statement such as

"There are exits at either end of this hall" may convey information in this sense. The person who has this information (i.e., who has these potential choices) may or may not exercise it depending on his appraisal of the relative efficiencies of the alternative exits. In one sense, then, information is the amount of potential choice of courses of action which an individual has.

The second sense in which we talk of information involves the *basis* of choice from among the alternative potential courses of action. For example, an individual who knows which exit is nearer than the others has a basis for choice and hence has information about the exits. Information in this sense pertains to the efficiencies of the alternatives relative to desired outcomes (e.g., a rapid exodus). Suppose, for example, that there are two exits and one is nearer than the other. If this is known and the objective (valued outcome) is to leave the building quickly, the choice is *determined* in the sense that the individual will always select the nearest exit. If he always selects the most distant exit then he is obviously misinformed (i.e., he has information, but it is incorrect). If he selects each exit with equal frequency then he apparently has no basis for choice; that is, no information. In this sense, then, information is the amount of choice which *has been* made. Now let us make this concept more precise.

Consider the case of an individual (I) who is confronted by two potential courses of action, C_1 and C_2. If the probabilities of selecting the courses of action are equal, $P_1 = P_2 = \frac{1}{2}$, the situation may be said to be *indeterminate* for I. He has no basis for choice and hence can be said to have no information about the alternatives. This is clearly the case when one of the alternatives is more efficient than the other. But if the two courses of action are equally efficient, the individual may have information to this effect and select each with equal frequency. Strictly speaking, however, he has no real choice in this situation since the alternatives are equally efficient. In a situation like this—a non-purposeful state—information has no operational meaning. Consequently this discussion has relevance only to situations in which all of the alternative courses of action are not equally efficient.

If $P_1 = 1.0$ and $P_2 = 0$, then the situation is *determinate* for I; all the choice that can be made has been made. The maximum possible information is contained in the state. It may not be correct but this is another matter which will be considered below.

We may define a unit of information as the amount contained in a two-choice situation that is determinate.

Let us consider the general case involving m alternative potential courses of action. In order to select one from this set, a minimum of $(m-1)$ choices from pairs of alternatives is required. Table 1 illustrates this fact.

TABLE 1

m =	2	3	4	5

$$\left.\begin{matrix} C_1 \\ C_2 \end{matrix}\right\}1 \quad \left.\begin{matrix} C_1 \\ C_2 \end{matrix}\right\}1 \quad \left.\begin{matrix} C_1 \\ C_2 \end{matrix}\right\}1 \quad \left.\begin{matrix} C_1 \\ C_2 \end{matrix}\right\}1$$

$$\left.C_3\right\}2 \quad \left.C_3\right\}2 \quad \left.C_3\right\}2$$

$$\left.C_4\right\}3 \quad \left.C_4\right\}3$$

$$\left.C_5\right\}4$$

We can conceive of the amount of information contained in a purposeful state, then, as a point on a scale bounded at the lower end by indeterminism (i.e., no choice has been made) and at the upper end by determinism (i.e., complete choice has been made). Location on this scale will depend on the values of P_i.

In an indeterminate state each $P_i = 1/m$. Therefore, one measure of the distance of a state from indeterminism is

$$\sum_{i=1}^{m} \left| P_i - \frac{1}{m} \right|.^4$$

For an indeterminate state this sum is equal to zero. In a determinate state one P_i is equal to 1.0 and the remaining $(m-1)P_i$'s equal to zero. Therefore, in a determinate state,

$$\sum_{i=1}^{m} \left| P_i - \frac{1}{m} \right| = \left(1 - \frac{1}{m} \right) + (m-1) \left| 0 - \frac{1}{m} \right|$$

$$= 1 - \frac{1}{m} + (m-1)\frac{1}{m}$$

$$= 1 - \frac{1}{m} + 1 - \frac{1}{m} = 2 - \frac{2}{m}$$

The ratio of (a) the deviation of a specified state from an indeterminate state to (b) the deviation of a corresponding determinate state from that indeterminate state, then, provides a measure of the fraction of the maximum information such a state can contain, to that which it does contain. Symbolically this ratio is

$$\frac{\displaystyle\sum_{i=1}^{m} \left| P_i - \frac{1}{m} \right|}{2 - \frac{2}{m}}.$$

The product of this fraction and the maximum amount of information such a state can contain—that is, $(m-1)$—provides a measure of the amount of information (here symbolized by A) in that state:

$$A = (m-1)\frac{\displaystyle\sum_{i=1}^{m} \left| P_i - \frac{1}{m} \right|}{2 - \frac{2}{m}}$$

$$= \frac{(m-1)\left(\dfrac{m}{2}\right)\displaystyle\sum_{i=1}^{m} \left| P_i - \frac{1}{m} \right|}{m-1} \tag{3}$$

$$= \frac{m}{2}\sum_{i=1}^{m} \left| P_i - \frac{1}{m} \right|$$

Now the amount of information communicated may be said to be the difference between the amount of information contained in the state of the receiver immediately preceding the communication (i.e., the initial state) and the state immediately following the communication (i.e., the terminal state). Let $A(S_1)$ be the amount of information in the initial state and $A(S_2)$ the amount of information in the terminal state, then the amount of information communicated, A_c is given by the following equation:

$$A_c = A(S_2) - A(S_1), \tag{4.1}$$

which may also be written in an expanded form:

$$A_c = \frac{m'}{2}\sum_{i=1}^{m'} \left| P'_i - \frac{1}{m'} \right| - \frac{m}{2}\sum_{i=1}^{m} \left| P_i - \frac{1}{m} \right|, \tag{4.2}$$

where m is the number of potential courses of action in the initial state, m' is the number of such choices in the terminal state, and P_i and P'_i are the probabilities of choice in the initial and terminal states, respectively.

A_c can take on values from $-(m-1)$ to $(m'+1)$. Negative values represent a loss of information (e.g., as in going from a determinate to an indeterminate state).

It should be noted that this measure contains no implication concerning the correctness or incorrectness of the information received. Further, it should be noted that this measure is relative to a specific receiver in a specific state. The same message may convey different amounts of information to different individuals in the same state or to the same individual in different states. Consequently, to specify the amount of information contained in a message it is necessary to specify the set of individuals and states relative to which the measure is to be made. If more than one individual or state is involved it is also necessary to specify what statistic (e.g., an average) is to be used. Generality of information may be defined in terms of the range of individuals and/or states over which it operates.

It should also be noted that messages are not the only source of information. One may obtain information by observation. For example, one may count the number of exits from a house. The measure of information suggested here is applicable to information obtained by either observation or communication.

2.2 INSTRUCTION[5]

To inform is to provide a basis for choice, i.e., a belief in the greater efficiency of one choice than another. Hence information modifies objective probabilities of choice by modifying subjective estimates of probabilities of success. Instruction is concerned with modifying the objective probabilities of success, i.e., efficiency. An individual's state of instruction can be characterized by the amount of control he can exercise over the outcomes in the state. He has maximum control over the outcome if he is capable of bringing about any of the possible outcomes. Instruction is the process of imparting such a capability to him where it is lacking.

Consider a course of action C_1 and two outcomes, O_1 and O_2. He has perfect knowledge of C_1 if he can use it to make either outcome occur with certainty, depending on his desires. If he cannot make the likelihood of an outcome change by his manipulation of C_1, then he does not control that alternative. Suppose, for example, that $E_{11} = 1.0$ no matter what the person desires, and hence $E_{12} = 0$. Then his choice is much like pushing a button that releases a course of action over which he has no further control.

The amount of control an individual has in a state can be measured as follows. Consider a case in which there are two outcomes, O_1 and O_2, and one course of action, C_1. If, when $V_1 = 1.0$ (and therefore $V_2 = 0$) $E_{11} = 1.0$, and when $V_2 = 1.0$ (and therefore $V_1 = 0$) $E_{12} = 1.0$, the individual has maximum control. Therefore

the amount of control is reflected in the range of E_{ij} as a function of V_j.

Specifically, the amount of control an individual has over a particular C_i relative to a particular O_j is given by

$$B(C_i|O_j) = (E_{ij}|V_j = 1.0) - (E_{ij}|V_j = 0)$$

The amount of control over C_i for all O_j's is given by

$$B(C_i) = \sum_j B(C_i|O_j) - 1$$

The 1 is subtracted because, if there are two outcomes, O_1 and O_2, and $B(C_i|O_1) = 1.0$, it follows that $B(C_i|O_2) = 1.0$ because O_1 and O_2 are exclusively defined (therefore, $V_2 = 1 - V_1$).

The amount of control, hence instruction in a state, then, is given by

$$B(S) = \sum_i \sum_j B(C_i|O_j) - m \tag{5}$$

$$= [\sum_i \sum_j (E_{ij}|V_j = 1.0) - (E_{ij}|V = 0) - m$$

Units of instruction may be called "hubits." The amount of instruction conveyed by a message is

$$B_c = B(S_2) - B(S_1). \tag{6}$$

2.3 MOTIVATION

If an individual in a state places value equally on all possible outcomes, then he has no basis for selecting among them and we can say that he has no motivation within that state. It should be recalled that the outcomes used to define a purposeful state are defined so as to be exclusive and exhaustive. From this it follows that the sum of the relative values of these outcomes is always unity. It also follows that if value is added to one outcome, an equal amount must be subtracted from the others. In the limiting case where there are two contradictory outcomes, O_1 and O_2, this last property clearly holds.

A state containing no motivation (in the relative sense) is described by the condition $V_1 = V_2 = \ldots = V_n = 1/n$. A state containing complete motivation is one in which one outcome has a relative value of one and all others have none; for example,

$$V_1 = 1.0, \quad V_2 = V_3 = \ldots = V_n = 0.$$

These descriptions correspond to those employed in the discussion of information where P_i was a measure of preference for courses of action. The amount of motivation in a state (C) may therefore be defined analogously:

$$C = \frac{n}{2} \sum_{j=1}^{n} \left| V_j - \frac{1}{n} \right|. \tag{7}$$

Consequently the amount of motivation communicated (C_c) may be defined as follows:

$$C_c = \frac{n}{2} \sum_{j=1}^{n} \left| V'_j - \frac{1}{n} \right| - \frac{n}{2} \sum_{j=1}^{n} \left| V_j - \frac{1}{n} \right|, \tag{8}$$

where V'_j are the relative values in the terminal state and V_j are the relative values in the initial state.

3. THE VALUE OF THE COMPONENTS OF COMMUNICATION

It will be recalled that the value of a communication to the receiver was given by the equation:

$$\Delta V = V(S_2) - V(S_1). \tag{2}$$

Using expected return for the measure of value, this equation may be rewritten as follows:

$$\Delta V = \sum_{j=1}^{n} \sum_{i=1}^{m} (P_i + \Delta P_i)(E_{ij} + \Delta E_{ij})(V_j + \Delta V_j)$$

$$\tag{8.1}$$

$$- \sum_{j=1}^{n} \sum_{i=1}^{m} P_i E_{ij} V_j.$$

By expansion, this may be converted into the following expression:

$$\begin{aligned} \Delta V = &\sum \sum \Delta P_i E_{ij} V_j + \sum \sum P_i \Delta E_{ij} V_j \\ &+ \sum \sum P_i E_{ij} \Delta V_j + \sum \sum \Delta P_i \Delta E_{ij} V_j \\ &+ \sum \sum \Delta P_i E_{ij} \Delta V_j + \sum \sum P_i \Delta E_{ij} \Delta V_j \\ &+ \sum \sum \Delta P_i \Delta E_{ij} \Delta V_j. \end{aligned} \tag{8.2}$$

The first three terms represent the value added to the initial state by the information, instruction, and motivation communicated, respectively. That is,

$$\begin{aligned} \Delta V_A &= \sum \sum \Delta P_i E_{ij} V_j \\ \Delta V_B &= \sum \sum P_i \Delta E_{ij} V_j \\ \Delta V_C &= \sum \sum P_i E_{ij} \Delta V_j \end{aligned}$$

where ΔV_A is the value added by only the changes in P_i, ΔV_B by the changes in E_{ij}, and ΔV_C by the changes in V_j.

The value of any of these expressions may be either positive or negative. If ΔV_A is negative, the receiver has been *misinformed*; if positive, he has been informed. If V_B is positive, he has been instructed; if negative, he has been "mis-instructed" (unfortunately we have no commonly used negative

of the verb "to instruct"). The same remarks apply to ΔV_C.

The remaining four terms in equation 8.2 represent ΔV_{AB}, ΔV_{AC}, ΔV_{BC}, and ΔV_{ABC}. For example, ΔV_{AB} is the joint contribution (not the sum of the independent contributions) to value of the information and instruction communicated. The other terms may be interpreted similarly. It is convenient, then, to think of the value of a communication as the sum of the independent and dependent contributions of information, instruction, and motivation. That is

$$\begin{aligned} \Delta V = &\Delta V_A + \Delta V_B + \Delta V_C + \Delta V_{AB} \\ &+ \Delta V_{AC} + \Delta V_{BC} + \Delta V_{ABC}. \end{aligned} \tag{8.3}$$

Conclusion

Not enough of the theory can be presented in the confines of an article to allow fruitful exploration of its applicability to the study of the human aspects of communication. Extension of the theory and discussion of its application will have to wait for another paper. But it may be helpful to relate what has been presented here to some more familiar ideas.

Those acquainted with Decision Theory will recognize that the conceptual framework of the theory of communication presented here is suitable for the formulation of the problems of Decision Theory. . . . The basic problem of Decision Theory is the selection and application of a criterion that should be used for selecting a course of action in (what we have here defined as) a purposeful state. Thus Decision Theory concerns itself with measures of efficiency, value, and effectiveness.

The study of communication, as conceived herein, concerns itself with the acts of humans which affect the decisions of other humans. For example, the final report of an applied research project, say in Operations Research, is a message intended to modify or justify the course of action pursued by management in a purposeful state. The value of such a communication is the change in effectiveness of the management policy which has been brought about by the report.

The value of the report, assuming it has some, may be due to either the disclosure of a new alternative course of action or the demonstration of greater effectiveness of a course of action known to, but not currently used by management. In either case, if management's policy is changed because of the disclosure or demonstration, information has been transmitted.

It is commonly recognized that when the recommendations contained in most research reports are accepted a critical step still remains, implementation. Implementation is a problem because the recommended course of action may be followed with varying degrees of efficiency. Implementation, then, consists of *instructing* personnel how to follow a course of action efficiently.

Management research projects usually take organizational objectives and their relative importance as given and hence seldom attempt to motivate management. The consultant, as contrasted to the researcher, is more likely to try to change management's "values." But even in research such motivational efforts do creep in. For example, the research may reveal that management is paying more to avoid shortages than they cost when they occur. Therefore, the researchers may try to have management de-evaluate the outcome, avoidance of shortages. To the extent that the researchers do so their communication is motivational.

As we explore the concepts and measures introduced here in more detail it will be found that we can obtain behavioral definitions and measures of such communicative concepts as ambiguity, obscurity, generality of information, and reliability of signs and messages. But this will have to wait for a future communication.

Finally, it should be reemphasized that the author feels much less certain of the appropriateness and usefulness of the *measures* introduced here than he does of the *concepts*. His hope is that the criticism and suggestions of others will accelerate their revision into a more appropriate and usable form. It is for this reason that the most important word in the title of this essay is "towards."

Notes

1. I assume here that the C_i and O_j form finite sets. The C_i and O_j are not "given" by the state or by the entity I; they are conceptual constructs of the researcher. They are his way of looking at the state. In practical situations as a researcher I have never found it necessary to consider infinite sets or even large finite sets. In practice I have found it fruitful to classify the C_i and O_j by use of the Boolean expansion of the basic characteristics of action and outcome which are of primary interest to me. This yields an exclusive and exhaustive set of C_i's and O_j's. . . .

It should be noted that the relativity of the definitions of the C_i and O_j to the researcher is not unique to this system. It exists in Shannon's as well. In one situation if I give you a yes–no question I consider you to have only a choice of two messages. In another I will take the *way* you say "yes" or "no" (e.g., pitch, hesitation, etc.) into account. Or again, if a telegrapher considers the choice between pressing the key with a finger on his left hand or his right hand, I need not consider this choice as relevant to my study.

2. Perhaps the most comprehensive behavioral treatment of "signs" has been given by C. W. Morris. This brief treatment is closely related to his except for one rather important point, it substitutes the concept of "potential producer of a response" for Morris' concept of "disposition to respond." Efforts to treat signs behavioristically go back at least as far as the 19th century American philosopher, Charles Peirce.

3. Because of the pervasiveness of the use of "information" in Shannon's restricted sense, it might seem preferable to use another term here. But since the usage here conforms more closely to common usage, if a change is required, it would seem preferable to change Shannon's term. As Colin Cherry notes, "the measure for H_n [average information] from Wiener and Shannon, is applicable to the signs themselves, and does not concern their meaning. In a sense, it is a pity that the mathematical concepts stemming from Hartley have been called 'information' at all."

4. It may turn out, on further study to be desirable to square these deviations, or raise them to some power other than one. Squaring deviations is a congenital disease of our age, however, which implicitly assumes that the error-cost function is quadratic in form. Although this form has many mathematical advantages, in my experience, error-cost functions which can be established empirically seldom take this form. This is, of course, no argument for using the power, 1. It has been pointed out to me that if this quantity were to be squared the result would be Pearson's chi-square divided by m, and that this is directly related to Shannon's measure of redundancy.

5. (Section 2.2 *Instruction* reflects substantial revisions made by the author subsequent to original publication of this article. They appear, with further relevant discussion, in Miles W. Martin, Jr., "The Measurement of Value of Scientific Information," in Burton W. Dean [Ed.], *Operations Research in Research and Development* [New York: John Wiley & Sons, 1963], pp. 97–123.—*Ed.*)

V

Cybernetics:
Purpose,
Self-Regulation,
and Self-Direction

ALTHOUGH the reader who has come this far will already have met most of the fundamental ideas that have come to be seen as the special province of cybernetics, Part V focuses more directly and systematically on its central theme: the analysis of control, regulation, or direction in complex systems.

Section A, "Cybernetics and Purpose," consists entirely of a single, closely interwoven, series of variations on a subtheme: the concept of "purpose," and the first promising attempts to make it scientifically respectable and analytically useful after centuries of teleological mysticism. The statement of the initial primitive theme in the early germinal paper of Rosenblueth, Wiener, and Bigelow, the ensuing counterpoints with philosopher Richard Taylor, and the more elaborate developments of the theme by Churchman and Ackoff, and Moore and Lewis, can only be suggestive of the sizable symphony that has developed since then—though we can be hardly past the first movement at present.

At this point it becomes especially important to note that the terms "self-regulation" and "self-direction" in the title of Part V are not to be assumed to be near synonyms, but are intended to point up a central distinction between two general kinds of processes that may be found in complex systems. Some mechanisms are essentially conserving, acting to regulate so as to keep within relatively narrow, viable limits, certain characteristic and essential features of the system's given structure and processes. Homeostasis is the classic example. On the other hand, there is another class of processes whose effect is to direct not only the system's behavior, but also the very nature of its structure or organization in new, often "higher," directions—that is, to change or elaborate its given structure into a significantly different one. Evolution is the obvious general example. It has now become especially important for the sciences of the human psychological individual and his sociocultural organizations to recognize that *both* these processes, and not only or primarily the first, are characteristic features. Thus, if there are stabilizing, structure-maintaining processes inherent in personality and society, there are also immanent unstabilizing and structure-changing processes.

Cannon's often-quoted but not always fully understood introduction of the concept of homeostasis is reprinted here as the first selection of Section B, not only as an early contribution to the spirit and substance of current systems research, but as a needed reminder to the many behavioral scientists wedded to the equilibrium notion (sometimes even equating it with homeostasis) of Cannon's reasons for *not* using the equilibrium concept for the type of system that is open, unstable, and in constant process. The social scientist might carry the insight one step further: *his* system—the sociocultural—is as far removed from the physiological along the same axis as Cannon's is from the mechanical equilibrium system, and thus requires a still different concept to express its particular nature. Our own view sees it, not as essentially homeostatic (and certainly not like an equilibrial system), acting mainly to maintain a given, presumably "normal" structure (although conserving forces are certainly to be found), but rather as something closer to "heteromorphic" or morphogenic: capable of maintaining its continuity and integrity specifically by changing essential aspects of its structure or organization. Much the same might also be said for the human psychological system. And in fact, the article by Pringle in Section B addresses itself to just this structure-changing feature of the psycho-neurological system and its close parallel to biological evolution. In both these latter cases we are dealing, not with maintenance of some given structure, but with change of that structure, usually in the direction of greater complexity. Pringle's paper, as an influential exercise in comparative general systems analysis, can be seen to interweave a number of previous themes, including the basic notion of information and the process of mapping environmental variety and its constraints into the organization (in this case, neurophysiological) of the complex adaptive system.

Ashby has stated that Sommerhoff's "identification of *exactly* what is meant by coordination and integration is, in my opinion, on a par with Cauchy's identification of exactly what was meant by convergence . . . [and] will, I am sure, eventually play a similarly fundamental part in our work." We present a selection from Sommerhoff's analysis of the biological process of adaptation and "directive correlation," followed by Ashby's related abstract model—a discussion that continues from his work presented in Part IV.

In the final paper of this Part, Maruyama levels the same charge against modern systems theorists that we have already seen leveled against the traditional equilibrium theorist, namely, an overemphasis on conserving, stabilizing processes at the expense of processes of system elaboration and change. Maruyama's work has won substantial notice as a constructive corrective, exploiting one of the inherent strengths of the modern perspective in dealing with morphogenic system processes.

A.

CYBERNETICS AND PURPOSE

27.
Behavior, Purpose, and Teleology

ARTURO ROSENBLUETH, NORBERT WIENER, AND JULIAN BIGELOW

THIS ESSAY HAS two goals. The first is to define the behavioristic study of natural events and to classify behavior. The second is to stress the importance of the concept of purpose.

Given any object, relatively abstracted from its surroundings for study, the behavioristic approach consists in the examination of the output of the object and of the relations of this output to the input. By output is meant any change produced in the surroundings by the object. By input, conversely, is meant any event external to the object that modifies this object in any manner.

The above statement of what is meant by the behavioristic method of study omits the specific structure and the intrinsic organization of the object. This omission is fundamental because on it is based the distinction between the behavioristic and the alternative functional method of study. In a functional analysis, as opposed to a behavioristic approach, the main goal is the intrinsic organization of the entity studied, its structure and its properties; the relations between the object and the surroundings are relatively incidental.

From this definition of the behavioristic method a broad definition of behavior ensues. By behavior is meant any change of an entity with respect to its surroundings. This change may be largely an output from the object, the input being then minimal, remote or irrelevant; or else the change may be immediately traceable to a certain input. Accordingly, any modification of an object, detectable externally, may be denoted as behavior. The term would be, therefore, too extensive for usefulness were it not that it may be restricted by apposite adjectives—i.e., that behavior may be classified.

The consideration of the changes of energy involved in behavior affords a basis for classification. Active behavior is that in which the object is the source of the output energy involved in a given specific reaction. The object may store energy supplied by a remote or relatively immediate input, but the input does not energize the output directly. In passive behavior, on the contrary, the object is not a source of energy; all the energy in the output can be traced to the immediate input (e.g., the throwing of an object), or else the object may control energy which remains external to it throughout the reaction (e.g., the soaring flight of a bird).

Active behavior may be subdivided into two classes: purposeless (or random) and purposeful. The term purposeful is meant to denote that the act or behavior may be interpreted as directed to the attainment of a goal—i.e., to a final condition in which the behaving object reaches a definite correlation in time or in space with respect to another object or event. Purposeless behavior then is that which is not interpreted as directed to a goal.

The vagueness of the words "may be interpreted" as used above might be considered so great that the distinction would be useless. Yet the recognition that behavior may sometimes be purposeful is unavoidable and useful, as follows. The basis of the concept of purpose is the awareness of "voluntary activity." Now, the purpose of

From Arturo Rosenblueth, Norbert Wiener, and Julian Bigelow, "Behavior, Purpose, and Teleology," *Philosophy of Science*, 10 (1943), 18–24. Copyright © 1943, The Williams & Wilkins Co., Baltimore, Md. 21202, U.S.A. Reprinted by permission.

voluntary acts is not a matter of arbitrary interpretation but a physiological fact. When we perform a voluntary action what we select voluntarily is a specific purpose, not a specific movement. Thus, if we decide to take a glass containing water and carry it to our mouth we do not command certain muscles to contract to a certain degree and in a certain sequence; we merely trip the purpose and the reaction follows automatically. Indeed, experimental physiology has so far been largely incapable of explaining the mechanism of voluntary activity. We submit that this failure is due to the fact that when an experimenter stimulates the motor regions of the cerebral cortex he does not duplicate a voluntary reaction; he trips efferent, "output" pathways, but does not trip a purpose, as is done voluntarily.

The view has often been expressed that all machines are purposeful. This view is untenable. First may be mentioned mechanical devices such as a roulette, designed precisely for purposelessness. Then may be considered devices such as a clock, designed, it is true, with a purpose, but having a performance which, although orderly, is not purposeful—i.e., there is no specific final condition toward which the movement of the clock strives. Similarly, although a gun may be used for a definite purpose, the attainment of a goal is not intrinsic to the performance of the gun; random shooting can be made, deliberately purposeless.

Some machines, on the other hand, are intrinsically purposeful. A torpedo with a target-seeking mechanism is an example. The term servomechanisms has been coined precisely to designate machines with intrinsic purposeful behavior.

It is apparent from these considerations that although the definition of purposeful behavior is relatively vague, and hence operationally largely meaningless, the concept of purpose is useful and should, therefore, be retained.

Purposeful active behavior may be subdivided into two classes: "feed-back" (or "teleological") and "non-feed-back" (or "non-teleological"). The expression feed-back is used by engineers in two different senses. In a broad sense it may denote that some of the output energy of an apparatus or machine is returned as input; an example is an electrical amplifier with feed-back. The feed-back is in these cases positive—the fraction of the output which reenters the object has the same sign as the original input signal. Positive feed-back adds to the input signals, it does not correct them. The term feed-back is also employed in a more restricted sense to signify that the behavior of an object is controlled by the margin of error at which the object stands at a given time with reference to a relatively specific goal. The feed-back is then negative, that is, the signals from the goal are used to restrict outputs which would otherwise go beyond the goal. It is this second meaning of the term feed-back that is used here.

All purposeful behavior may be considered to require negative feed-back. If a goal is to be attained, some signals from the goal are necessary at some time to direct the behavior. By non-feed-back behavior is meant that in which there are no signals from the goal which modify the activity of the object *in the course of the behavior*. Thus, a machine may be set to impinge upon a luminous object although the machine may be insensitive to light. Similarly, a snake may strike at a frog, or a frog at a fly, with no visual or other report from the prey after the movement has started. Indeed, the movement is in these cases so fast that it is not likely that nerve impulses would have time to arise at the retina, travel to the central nervous system and set up further impulses which would reach the muscles in time to modify the movement effectively.

As opposed to the examples considered, the behavior of some machines and some reactions of living organisms involve a continuous feed-back from the goal that modifies and guides the behaving object. This type of behavior is more effective than that mentioned above, particularly when the goal is not stationary. But continuous feed-back control may lead to very clumsy behavior if the feed-back is inadequately damped and becomes therefore positive instead of negative for certain frequencies of oscillation. Suppose, for example, that a machine is designed with the purpose of impinging upon a moving luminous goal; the path followed by the machine is controlled by the direction and intensity of the light from the goal. Suppose further that the machine overshoots seriously when it follows a movement of the goal in a certain direction; an even stronger stimulus will then be delivered which will turn the machine in the opposite direction. If that movement again overshoots a series of increasingly larger oscillations will ensue and the machine will miss the goal.

This picture of the consequences of undamped feed-back is strikingly similar to that seen during the performance of a voluntary act by a cerebellar patient. At rest the subject exhibits no obvious motor disturbance. If he is asked to carry a glass of water from a table to his mouth, however, the hand carrying the glass will execute a series of oscillatory motions of increasing amplitude as the glass approaches his mouth, so that the water will spill and the purpose will not be fulfilled. This test is typical of the disorderly motor performance of patients with cerebellar disease. The analogy with the behavior of a machine with undamped feed-back is so vivid that we venture to suggest that the main function of the cerebellum is the control of the feed-back nervous mechanisms involved in purposeful motor activity.

Feed-back purposeful behavior may again be subdivided. It may be extrapolative (predictive), or it may be non-extrapolative (non-predictive). The reactions of unicellular organisms known as tropisms are examples of non-predictive performances. The amoeba merely follows the source to which it reacts; there is no evidence that it extrapolates the path of a moving source. Predictive animal behavior, on the other hand, is a commonplace. A cat starting to pursue a running mouse does not run directly toward the region where the mouse is at any given time, but moves toward an extrapolated future position. Examples of both predictive and non-predictive servomechanisms may also be found readily.

Predictive behavior may be subdivided into different orders. The cat chasing the mouse is an instance of first-order prediction; the cat merely predicts the path of the mouse. Throwing a stone at a moving target requires a second-order prediction; the paths of the target and of the stone should be foreseen. Examples of predictions of higher order are shooting with a sling or with a bow and arrow.

Predictive behavior requires the discrimination of at least two coordinates, a temporal and at least one spatial axis. Prediction will be more effective and flexible, however, if the behaving object can respond to changes in more than one spatial coordinate. The sensory receptors of an organism, or the corresponding elements of a machine, may therefore limit the predictive

behavior. Thus, a bloodhound *follows* a trail, that is, it does not show any predictive behavior in trailing, because a chemical, olfactory input reports only spatial information: distance, as indicated by intensity. The external changes capable of affecting auditory, or, even better, visual receptors, permit more accurate spatial localization; hence the possibility of more effective predictive reactions when the input affects those receptors.

In addition to the limitations imposed by the receptors upon the ability to perform extrapolative actions, limitations may also occur that are due to the internal organization of the behaving object. Thus, a machine which is to trail predictively a moving luminous object should not only be sensitive to light (e.g., by the possession of a photoelectric cell), but should also have the structure adequate for interpreting the luminous input. It is probable that limitations of internal organization, particularly of the organization of the central nervous system, determine the complexity of predictive behavior which a mammal may attain. Thus, it is likely that the nervous system of a rat or dog is such that it does not permit the integration of input and output necessary for the performance of a predictive reaction of the third or fourth order. Indeed, it is possible that one of the features of the discontinuity of behavior observable when comparing humans with other high mammals may lie in that the other mammals are limited to predictive behavior of a low order, whereas man may be capable potentially of quite high orders of prediction.

The classification of behavior suggested so far is tabulated below.

It is apparent that each of the dichotomies established singles out arbitrarily one feature, deemed interesting, leaving an amorphous remainder: the non-class. It is also apparent that the criteria for the several dichotomies are heterogeneous. It is obvious, therefore, that many other lines of classification are available, which are independent of that developed above. Thus,

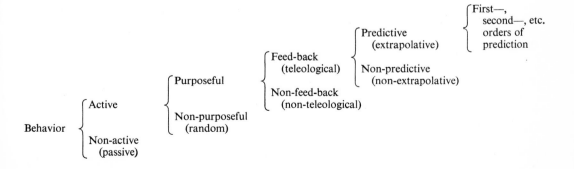

behavior in general, or any of the groups in the table, could be divided into linear (i.e., output proportional to input) and non-linear. A division into continuous and discontinuous might be useful for many purposes. The several degrees of freedom which behavior may exhibit could also be employed as a basis of systematization.

The classification tabulated above was adopted for several reasons. It leads to the singling out of the class of predictive behavior, a class particularly interesting since it suggests the possibility of systematizing increasingly more complex tests of the behavior of organisms. It emphasizes the concepts of purpose and of teleology, concepts which, although rather discredited at present, are shown to be important. Finally, it reveals that a uniform behavioristic analysis is applicable to both machines and living organisms, regardless of the complexity of the behavior.

It has sometimes been stated that the designers of machines merely attempt to duplicate the performances of living organisms. This statement is uncritical. That the gross behavior of some machines should be similar to the reactions of organisms is not surprising. Animal behavior includes many varieties of all the possible modes of behavior and the machines devised so far have far from exhausted all those possible modes. There is, therefore a considerable overlap of the two realms of behavior. Examples, however, are readily found of man-made machines with behavior that transcends human behavior. A machine with an electrical output is an instance; for men, unlike the electric fishes, are incapable of emitting electricity. Radio transmission is perhaps an even better instance, for no animal is known with the ability to generate short waves, even if so-called experiments on telepathy are considered seriously.

A further comparison of living organisms and machines leads to the following inferences. The methods of study for the two groups are at present similar. Whether they should always be the same may depend on whether or not there are one or more qualitatively distinct, unique characteristics present in one group and absent in the other. Such qualitative differences have not appeared so far.

The broad classes of behavior are the same in machines and in living organisms. Specific, narrow classes may be found exclusively in one or the other. Thus, no machine is available yet that can write a Sanscrit–Mandarin dictionary. Thus, also, no living organism is known that rolls on wheels— imagine what the result would have been if engineers had insisted on copying living organisms and had therefore put legs and feet in their locomotives, instead of wheels.

While the behavioristic analysis of machines and living organisms is largely uniform, their functional study reveals deep differences. Structurally, organisms are mainly colloidal, and include prominently protein molecules, large, complex and anisotropic; machines are chiefly metallic and include mainly simple molecules. From the standpoint of their energetics, machines usually exhibit relatively large differences of potential, which permit rapid mobilization of energy; in organisms the energy is more uniformly distributed, it is not very mobile. Thus, in electric machines conduction is mainly electronic, whereas in organisms electric changes are usually ionic.

Scope and flexibility are achieved in machines largely by temporal multiplication of effects; frequencies of one million per second or more are readily obtained and utilized. In organisms, spatial multiplication, rather than temporal, is the rule; the temporal achievements are poor—the fastest nerve fibers can only conduct about one thousand impulses per second; spatial multiplication is on the other hand abundant and admirable in its compactness. This difference is well illustrated by the comparison of a television receiver and the eye. The television receiver may be described as a single cone retina; the images are formed by scanning—i.e. by orderly successive detection of the signal with a rate of about 20 million per second. Scanning is a process which seldom or never occurs in organisms, since it requires fast frequencies for effective performance. The eye uses a spatial, rather than a temporal multiplier. Instead of the one cone of the television receiver a human eye has about 6.5 million cones and about 115 million rods.

If an engineer were to design a robot, roughly similar in behavior to an animal organism, he would not attempt at present to make it out of proteins and other colloids. He would probably build it out of metallic parts, some dielectrics and many vacuum tubes. The movements of the robot could readily be much faster and more powerful than those of the original organism. Learning and memory, however, would be quite rudimentary. In future years, as the knowledge of colloids and proteins increases, future engineers may attempt the design of robots not only with a behavior, but also with a structure similar to that of a mammal. The ultimate model of a cat is of course another cat, whether it be born of still another cat or synthesized in a laboratory.

In classifying behavior the term "teleology" was used as synonymous with "purpose controlled by feed-back." Teleology has been interpreted in the past to imply purpose and the vague concept of a "final cause" has been often added. This concept of final causes has led to the opposition of teleology to determinism. A discussion of causality, determinism and final causes is beyond the scope of this essay. It may be pointed out, however, that purposefulness, as defined here, is quite independent of causality, initial or final. Teleology has been discredited chiefly because it was defined to imply a cause subsequent in time to a given effect. When this aspect of teleology was dismissed, however, the associated recognition of the importance of purpose was also unfortunately discarded. Since we consider purposefulness a concept necessary for the understanding of certain modes of behavior we suggest that a teleological study is useful if it avoids problems of causality and concerns itself merely with an investigation of purpose.

We have restricted the connotation of teleological behavior by applying this designation only to purposeful reactions which are controlled by the error of the reaction—i.e., by the difference between the state of the behaving object at any time and the final state interpreted as the purpose. Teleological behavior thus becomes synonymous with behavior controlled by negative feed-back, and gains therefore in precision by a sufficiently restricted connotation.

According to this limited definition, teleology is not opposed to determinism, but to non-teleology. Both teleological and non-teleological systems are deterministic when the behavior considered belongs to the realm where determinism applies. The concept of teleology shares only one thing with the concept of causality: a time axis. But causality implies a one-way, relatively irreversible functional relationship, whereas teleology is concerned with behavior, not with functional relationships.

28.
Comments on a Mechanistic Conception of Purposefulness

RICHARD TAYLOR

IN A HIGHLY ORIGINAL and provocative essay entitled "Behavior, Purpose and Teleology," published a few years ago, Professors Arturo Rosenblueth, Norbert Wiener and Julian Bigelow attempt to indicate the scientific importance and usefulness of the concepts of purpose and teleology. Since this essay appeared the suggestions it contains seem to have acquired a significance which was not wholly apparent at that time. This is due primarily to the fact that a somewhat novel and, it appears to some, revolutionary approach to certain problems has arisen in the sciences, an approach which is more or less loosely referred to as "cybernetics," and among whose outstanding spokesmen are to be found the very authors of this essay—particularly Professor Wiener, whose recently published *Cybernetics* has been claiming the attention of an increasing number of scientists and non-scientists alike. In his book, it may be noted, Professor Wiener, in tracing the development of cybernetics over the past few years, gives a good indication of the importance he attaches to the earlier essay. He writes, with reference to it: "The three of us [Rosenblueth, Wiener and Bigelow] felt that this new point of view merited a paper, which we wrote up and published. Dr. Rosenblueth and I foresaw that the paper could only be a statement of program for a large body of experimental work, and we decided that if we could ever bring our plan for an interscientific institute to fruition, this topic would furnish an almost ideal center for our activity." It would seem, then, that an examination of the contents of this essay should not be without interest at this time.

My objective in this paper will be to elicit what appears to be a gross confusion underlying these authors' main contention. This contention is that the notions of purpose and teleology, "although rather discredited at present," are in fact not only useful and important, but unavoidable and necessary for the interpretation of certain kinds of behavior, both animate and inanimate. I shall maintain, on the contrary, that these concepts, as they are defined and illustrated by these authors, cannot possibly serve the ends for which they are invoked. This conclusion will be indicated, I believe, if it can be shown (1) that purposive behavior, as they describe it, is indistinguishable from any other kind of active behavior, and (2) that the term "purpose," as thus used, bears no similarity whatever to the meaning which is ordinarily attached to it. This second point might seem to be largely a verbal matter, and indeed it is; but it would be a mistake to conclude from this that it is therefore of slight significance.

The concept of purposiveness is applied by these authors only to certain kinds of *active* behavior, that is to say, to that kind of behavior "in which the object is the source of the output energy involved in a given specific reaction." And the term "behavior" itself is taken to mean "any change of an entity with respect to its surroundings,"[1] or, "any modification of an object, detectable externally." Thus, examples of active behavior would be, as I understand the definition, such things as a clock which is running, an automobile in operation, an exploding bomb, as well as the active behavior of living organisms generally.

From Richard Taylor, "Comments on a Mechanistic Conception of Purposefulness," *Philosophy of Science*, 17 (1950), 310–17. Copyright © 1950, The Williams & Wilkins Co., Baltimore, Md. 21202, U.S.A. Reprinted by permission.

Passive behavior, on the other hand, that is, behavior in which "the object is not a source of energy," and in which "all the energy in the output can be traced to the immediate input," would be exemplified by such things as a falling stone, melting ice, and a revolving water wheel. The terms "active" and "passive" are themselves quite unambiguous, and serve to convey, I believe, the distinction intended. But it needs particularly to be noted at this point that active behavior is taken to characterize certain inanimate objects, no less than living ones.

With these distinctions in mind, we can consider now the discussion of purposefulness itself. The authors define purposeful and non-purposeful behavior in these words: "The term purposeful is meant to denote that the act or behavior may be interpreted as directed to the attainment of a goal—i.e., to a final condition in which the behaving object reaches a definite correlation in time or in space with respect to another object or event. Purposeless behavior then is that which is not interpreted as directed to a goal."

Now the writers concede that the words, "may be interpreted," are vague, but insist that "the recognition that behavior may sometimes be purposeful is unavoidable and useful." I do not see that the expression, "may be interpreted," is really vague; the words themselves seem to have reasonably precise meanings, and I think there is little danger of their being mistaken. I think what the writers probably have in mind is that the definition itself is rather *general*—and in this, I should certainly concur. Indeed, the definition is so broad that it not only fails to distinguish, even in some general way, the feature which it is intended to describe, but makes any behavior whatever, whether active *or* passive, a case of purposiveness. One or two examples will show this to be the case.

Imagine, for instance, a clock which runs properly for many years, and then breaks down at twelve o'clock, New Year's Eve—such behavior admirably fits every requirement of purposiveness. Or again, consider a more humble example, such as a stone which falls from a rooftop, and kills a passer-by; or, more simply still, one that just makes a dent in the ground. Now such behavior not only *may* be interpreted as that in which the behaving object "reaches a definite correlation in time or in space with respect to another object or event," but such an interpretation is precisely the one which constitutes a simple and, as far as it goes, accurate description of just what has taken place. It seems hardly necessary

to multiply examples here, for a very little reflection will show that any instance of behavior one might choose will necessarily be a case of "purposefulness," whether one considers active or passive behavior. And the reason for this is very apparent: it is simply that any behavior culminates, at whatever point we choose to call its culmination, in a definite correlation in time or space between the behaving object and other objects or events.[2]

Of course the definition specifies that purposeful behavior is that active behavior which may be interpreted as being *directed* toward such a goal. But this, so far as I can discern, can only mean either (a) that such behavior is directed by some purposeful being *other* than the behaving object itself—by a human being, for example; or (b) that the behaving object directs *itself* toward some correlation. Now the first meaning is clearly not what the authors intend, for in that case the purposiveness would reside, not in the object itself, but in the being who directs it; a hammer does not itself become purposeful by being used *for* the purpose of driving nails. But if we consider the second meaning, which is the only other possible one, innumerable difficulties arise. In the first place, it presupposes the as yet unjustified assumption that some mechanisms, other than living organisms, *are* "intrinsically purposeful," in the distinctive sense of directing themselves toward the attainment of "goals." But even if this assumption were granted, then it is at once clear that *any* correlation in time or space between the behaving objecs and other objects or events can be taken at precisely the one toward which the supposedly purposeful object "directs" itself. For it is at all times in some such state of correlation, and so long as we leave distinctively human purposes out of account, as these authors try rigorously to do, then there is no conceivable way of selecting *some* particular reciprocal relationship between the object and some state of affairs as being the "goal" toward which that object was directing itself. Of course it might be suggested that we select as the goal of the object some *final* correlation; indeed, this seems to be exactly what the authors have in mind, in reiterating the expression "final condition." But this is of little help, because (a) unless the behaving object is destroyed, there *is* no such final condition: so long as it exists, it exists in a state of reciprocal relationship with other objects and events; and (b) such a criterion is wholly arbitrary. It obliges us to assert that whenever a supposedly purposeful mechanism culminates in such a "final condition" (however this "culmination" might be determined), then that final condition is, *ipso facto*, the very "goal" toward which it was directing itself. So if the behavior of an organism, say, culminates in some such final condition as death, then we shall be required to conclude that precisely this was, by definition, its purpose, i.e., the "goal" toward which it directed itself. And this seems palpably incorrect; it would render such an expression as "accidental death" quite meaningless, for example.

Clearly, then, we cannot expect much light from such descriptions as these; they appear arbitrary, and so general as to make purposiveness a ubiquitous phenomenon. Let us turn now, then, to the actual examples which the writers submit of both purposeful and non-purposeful behavior, to see whether these will serve any better to elicit the distinction intended.

It is a mistake, they contend, to regard all machines as purposeful, and in this they are doubtless correct; it seems to be a mistake which none but the overly superstitious would be likely to make, however. As examples of mechanical devices which are *not* purposeful, they suggest a roulette wheel, a clock, and a gun. A roulette, it is pointed out, is "designed precisely for purposelessness." And a clock, although it is designed to serve a purpose, is not in itself purposeful—because there is, again, no "final condition" toward which it "strives." The same consideration applies to the example of a gun; it can, to be sure, be used for a purpose, but purposiveness itself "is not intrinsic to the performance of the gun," because it can also be shot at random.

As contrasted with these mechanisms, on the other hand, the authors cite the example of a torpedo containing a target-seeking device, as being an instance of a machine which is "intrinsically purposeful." Indeed, they single out an entire class of mechanical devices as possessing intrinsic purposiveness, namely, "servomechanisms," or machines which are controlled by negative feed-back.[3] "The term servomechanisms," they note "has been coined precisely to designate machines with intrinsic purposeful behavior." Those who are acquainted with Professor Wiener's recent book are aware of the importance he and his colleagues attach to these mechanisms. Before examining the discussion of these, however, a few remarks suggest themselves concerning the examples already referred to.

In the first place, it is difficult to see, from the descriptions given, in what sense roulettes, clocks and guns are "purposeless," as contrasted with other mechanisms in which purposefulness is "intrinsic." A gun, it is pointed out, may be used for a purpose, but it can also be shot at random: therefore, "the attainment of a goal is not intrinsic to the performance of the gun." It would seem, however, that whether the gun is "used for a deliberate purpose" (e.g., shooting a duck), or simply shot at random, in either case it serves a purpose—if only the amusement of the gunner. Moreover, even target-seeking missiles, which are classed as "intrinsically purposeful," *can* be used for random shooting, or even left to rust away in an ammunition dump; why does this not indicate a parallel conclusion in their case? Apparently the authors are utilizing here an unnamed criterion of purposiveness.

The remarks concerning the behavior of clocks are similarly puzzling. A clock is denied the attribute of "intrinsic" purposefulness because "there is no specific final condition toward which the movement of the clock strives." But if mechanical devices are once granted the power of "striving" toward some goal or "final condition," then how are we to know that the final condition which a clock ultimately attains—say, breakdown, at a particular time and place—was not the very one toward which it was "striving"? The difficulty in such a supposition is no greater than in the case of target-seeking missiles, however great it may be in either case.

Finally, a roulette wheel is said to be "designed precisely for purposelessness," and by this the authors apparently have in mind that this device is designed in a manner such that it will not, in the long run, turn up any specific number, or order of numbers, more frequently than any other. Such, at any rate, is the distinguishing characteristic of an ideal roulette. How, then, should one describe a number wheel which has a weight affixed to its circumference, in a way so as to determine the wheel always to stop on the number six? Such a mechanism satisfies perfectly every condition of intrinsic purposefulness which the writers set down; unlike the clock, for example, there *is* here a "specific final condition toward which the movement of the [wheel] strives"— provided, again, that "striving" may without incongruity be attributed to mechanical devices.

I question whether anyone would undertake to defend such a conclusion as this; first, because it seems *prima facie* bizarre to maintain that a wheel becomes a purposeful object by the mere addition to it of a weight, and secondly, because if it were maintained that such behavior as this is purposive, then it would be difficult to find an instance of behavior which is not. Yet, I submit, this conclusion is absolutely forced by the descriptions and criteria which the authors adduce. Now I should suppose that the only relevant distinction to be drawn between an honest roulette and a weighted one is that in the case of the second we can usually predict its "final condition," whereas in the case of the first we cannot. But the reason for this is not that the one is purposeless while the other is not, but rather that we know, and can measure, the causal factors involved in the behavior of the loaded wheel, whereas we do not have such knowledge with respect to the other. If, on the other hand, we *did* know all of the causal factors (force, mass, friction, etc.) involved in the behavior of an honest roulette, then we *could*, within the limits of accuracy of our measuring instruments, predict its final condition, i.e., the number on which it will stop; at least, this ability is universally taken for granted in the sciences, so long as we are dealing with such relatively macroscopic objects as wheels and the like. In what sense, then, is a roulette wheel a purposeless machine? So far as I can discern, the distinction to be made, as regards purposiveness, is simply this, and this only: If a purposive *being*, i.e., a man, spins a roulette with the *purpose* of turning up a specific number, then he will be less likely to succeed if he uses an honest wheel, than if he uses one which is weighted properly; and the only reason for this is, again, that his knowledge of the causal factors involved in the behavior of the first wheel is insufficient to enable him to predict. And if this is a correct description, then, I submit, neither purposefulness nor purposelessness is appropriately attributable to the wheel, of whichever kind, but only to the being who uses it *for* a purpose.

The objective of the examples we have been considering was to elicit the distinction between purposeful and purposeless behavior—a distinction which the writers believe, correctly, is not made sufficiently clear by their definition. And the distinction which is apparently intended is simply this: that although such objects as roulette wheels, clocks and guns can be made to *serve* a purpose, they do not have any purposes of their own; while a target-seeking missile, on the other hand, not only serves a purpose, but *does* have a purposiveness of its own; it is, in the words of the authors, an "intrinsically purposeful" object. At least, so far as the notion of "purpose" is concerned, this is the only distinction I can discern between machines which only serve a purpose, on the one hand, and those which are "intrinsically purposeful," on the other. As soon, however, as

this distinction is made clear and precise, instead of simply hovering vaguely in the background, then, I believe, its dubious status becomes quite apparent.

Let us consider, finally, that class of machines which are distinguished by the possession of "intrinsic purposeful behavior," namely, servomechanisms. The term "servomechanisms" is used to denote such objects as thermostats, target-seeking missiles, ship steering devices, radar-controlled guns, and so on, the distinguishing characteristic of which is that the behavior of all such objects is controlled by negative feed-back. That is to say, such a mechanism is so designed that the *effects* of its behavior themselves enter as causal factors *on* its behavior, the objective being to have a device that will maintain itself in a certain desired correlation with other objects or events, which also operate upon it as causal factors. This is doubtless an over-simple description, but I think it will do for our purposes. Thus, a thermostat controls, and is in turn controlled by, temperature; a target-seeking missile is directed at, and is in turn directed by, its target, and so on. In the case of all such mechanisms, then, the objects or events which they operate upon, in turn operate upon them, in such a manner as to maintain a constant reciprocal relationship. The governor of a steam engine was, I believe, one of the earliest of such man-made devices.

Professor Wiener and his colleagues regard these mechanisms as exhibiting a kind of purposiveness *par excellence*, namely, *teleological* purposefulness;[4] they maintain, in fact, that the concepts of purpose and teleology are not only useful, but necessary for the understanding of this class of machines. It seems significant to note, however, that what is here called "teleological purposeful behavior" is by no means limited to higher organisms and servomechanisms, but is exhibited as well by some of the most ordinary objects of our daily experience—a fact which these authors entirely neglect to point out. The leaves of many green plants, for example, follow the course of the sun, and thus exhibit a behavior pattern which is precisely the same, so far as teleology is concerned, as that of a machine which is designed to impinge upon a moving, luminous object—a servomechanism which is cited by the writers as clearly exemplifying teleological behavior. Again, consider the needle of an ordinary magnetic compass. If it is diverted from its alignment with the magnetic forces of its locus, it vacillates momentarily, and finally resumes its former correlation. The behavior of the compass thus fits precisely the description of a "purposeful reaction," as being that which is exhibited by an object whose behavior "is controlled by the margin of error at which the object stands at a given time with reference to a relatively specific goal," assuming that we can designate its final correlation with the magnetic forces as its "goal." Other examples, such as the behavior of a pendulum, or of a vibrating cord, come readily to mind; and if servomechanisms differ in any way, other than that of mere mechanical complexity, from such everyday phenomena as these, then the writers have at any rate failed to give any hint as to what this difference might be. The behavior of servomechanisms does, to be sure, satisfy perfectly the criteria of purposiveness which the authors adduce, but the behavior of these other objects seems to satisfy them equally well.

I should maintain, therefore, that the notions of purpose and teleology are not only useless for the understanding of this sort of mechanical behavior, but are wholly incongruous as thus applied; and this conclusion follows, I believe, from the fact that such behavior is describable in terms of, and only in terms of, the same fundamental categories as are employed for the description of any other physical process.

A single example should make this last point clear. Consider the illustration used by the authors, viz., that of a torpedo with a target-seeking mechanism. Now if such a machine is so designed as to be guided by, say, sound waves proceeding from the ship's engines or propellers, then its behavior is describable, in general, as follows. The sound waves emanating from the target act causally upon the sonic mechanism of the torpedo, and the behavior of this device in turn acts (through intermediary devices) as a cause upon the steering mechanism of the missile. Accordingly, if the torpedo is diverted from its course, the resulting change in the sound waves, relative to the sonic device, suffices to reorient the torpedo, i.e., causes it to resume its course, relative to the target. Similarly, if the target itself moves, the correlation between sound waves and missile is likewise upset, and this, again, suffices to alter the course of the torpedo, relative to the target, through the complex nexus of causes and effects obtaining between the sound waves and the torpedo's rudders and vanes. Of course an accurate description of this process is much less simple than this, but, I submit, such further description consists only in the addition of details to this general picture. And one point to note here is that the torpedo is

guided, *not* by the target itself, but by the sound waves impinging upon the sonic mechanism; it does not literally "seek" the target, for its behavior would be the same even if no target were there, provided only that sound waves, or certain other immediate causes, obtained. The expression "target-seeking missile" is, in fact, metaphorical.

Is there, then, any room in such a description for the notion of *intrinsic* purposefulness on the part of the torpedo? I think not; and to illustrate this, we need only to alter the example in one respect. Let us suppose that the missile, instead of being governed by sound waves, is propelled along a cable, attached to the target. Now the behavior of this missile is precisely analogous to the first, the only relevant difference being that whereas the first was guided by sound waves between itself and the target, this one is guided by the much simpler means of a cable. If this second torpedo is diverted from its path, the change in its alignment, relative to the cable, suffices to reorient it; and the same is true if the target itself moves. In short, the analogy seems complete, the *only* difference being in the degree of mechanical complexity—and I doubt whether anyone would contend that complexity by itself is a criterion of a purposive object. Accordingly, if the first missile is to be characterized as "intrinsically purposeful," then we are obliged to conclude that the second one must be similarly described. But from this concession it would follow that a vast number of other machines become "intrinsically purposeful" objects, even though no one has ever suspected them of being such; trains, for example, or elevators, and in fact almost any machine one might choose.

My conclusion with respect to servomechanisms is, therefore, the same as before; namely, that Professors Rosenblueth, Wiener and Bigelow (a) utilize criteria which render purposeful behavior a ubiquitous phenomenon, and (b) thereby endow the word "purpose" with a meaning having no similarity to any meaning it has customarily been taken to possess. Of course it might be claimed that one is entitled to assign to his terms any meaning he chooses, and in a sense this is true; it must be added, however, that if this is what these writers have done, then their discussion sacrifices any interest and significance which it may have been intended to have. For it is exactly as if one were to announce the discovery that $2+3 = 6$, only to add later that the term " + " is taken by him to have the meaning which has traditionally been assigned to the symbol " × ". His claim would be entirely correct, but scarcely significant.

Notes

1. On this definition, it may be noted, a perfectly static object surrounded by others in motion exhibits "behavior."

2. The word "correlation" is not defined by these writers, and in fact they use it only once. I therefore assume it to have no special or technical meaning in their usage. Ordinarily, to say of two objects or events that they are correlated, is simply to say that they stand in some reciprocal or mutual relationship, i.e., that they are co-related. The word is also used more precisely to indicate a constant relationship between kinds of objects or events, or to indicate that one is a universal accompaniment of the other, as when we speak of mental states being correlated with brain processes, or of the hands of a clock being correlated with each other, and so on. The more general meaning is what the authors seem to have in mind in their definition, although the more precise one would not alter their argument.

3. The expression "negative feed-back" is a technical one of physics and engineering. The authors point out that they are using it "to signify that the behavior of an object is controlled by the margin of error at which the object stands at a given time with reference to a relatively specific goal." The same idea could be expressed by saying that an object is controlled by *negative* feed-back when the effects of its behavior in turn act indirectly upon the object itself to *oppose* whatever it is doing (cf. *Cybernetics*, p. 115).

4. This expression may appear as a redundancy, but the authors qualify as "teleological" only those purposeful objects which are controlled by negative feed-back.

29.
Purposeful and Non-Purposeful Behavior

ARTURO ROSENBLUETH AND NORBERT WIENER

IN A RECENT ESSAY Professor Taylor criticizes the criteria used by Rosenblueth, Wiener and Bigelow in 1943 to distinguish purposeful from non-purposeful behavior. He also criticizes our definition of *behavior*, our concept of the *vague* as opposed to the *general*, our use of the word *correlation*, and our statement that a system may reach a *final* condition. Indeed, there seems to be little, if anything, in our paper to which he does not emphatically object.

He maintains that the notions of purpose and teleology are not only useless for the understanding of mechanical behavior, but wholly incongruous when applied to this behavior. He further affirms that our use of the term purpose "bears no similarity whatever to the meaning which is ordinarily attached to it." He does not state, however, his own notions of purpose and teleology; and the meaning which he considers to be ordinarily attached to the term purpose. This omission weakens his criticism considerably.

In the present paper we shall not deal with purely verbal issues, because we believe that they are trivial and barren. We shall discuss some of Professor Taylor's opinions and generalizations which we judge erroneous. We shall then complement our earlier distinction between purposeful and non-purposeful behavior, because, notwithstanding Professor Taylor's opinion to the contrary, we think that this distinction is meaningful and essential in science, and because Professor Taylor's misunderstanding of our earlier formula-

tion suggests the possibility that others may also have deemed the problem to be largely a verbal one.

In the interim between the publication of our first paper and the present time, we have devoted much effort to the clearing up of the set of categories necessary for the study of the statistical aspects of modern science. Some of this work has been presented by one of us. The present analysis is based on the ideas developed in that presentation. They go back to the work of Gibbs and of Heisenberg, which cannot be contained in the purely Newtonian frame of physics.

Professor Taylor affirms that, in the study of the behavior which we have called purposeful, so long as distinctively human purposes are left out of account, there is no conceivable way of selecting *some* particular relationship between the behaving object and its surroundings as the goal toward which that object was directing itself. This statement is false. We can conceive many ways of selecting the goal or goals in purposeful behavior, and the only problem is to find out which of these several procedures is the most adequate for scientific analysis. The emphasis on human purposes is irrelevant. The purpose of the designer of a radar-controlled gun may have been to have the gun seek an enemy plane, but if the gun seeks the car of the commanding officer of the post, as this officer drives by, and destroys it, surely the purpose of the gun differs from that of the designer. Indeed, this would be an excellent example of cross-purposes.

Again, let us consider a car following a man along a road with the clear purpose of running him down. What important difference will there be in our analysis of the behavior of the car if it

From Arturo Rosenblueth and Norbert Wiener, "Purposeful and Non-Purposeful Behavior," *Philosophy of Science*, 17 (1950), 318–26. Copyright © 1950, The Williams & Wilkins Co., Baltimore, Md. 21202, U.S.A. Reprinted by permission.

is driven by a human being, or if it is guided by the appropriate mechanical sense organs and mechanical controls?

Professor Taylor states that, if mechanical devices are granted the power of striving toward a goal, it is legitimate to suppose that the final condition which a clock ultimately attains, e.g., breakdown at a particular time and place, is the goal toward which that clock was striving. He believes that this supposition is fully as justified as the assumption that a target-seeking missile pursues the attainment of an impact with the target. These statements are erroneous. The basis for this judgment will be given in the second part of this paper. Suffice it to say at present that the same criterion which allows an observer to distinguish purposeful from non-purposeful behavior in an animal or in another human being will invalidate the supposition that the clock's behavior is purposeful, with its breakdown as a goal.

Professor Taylor asserts that, according to our criterion of purposiveness, the mere addition of a weight to a roulette wheel changes this wheel from a purposeless to a purposeful mechanism. Then he questions whether anyone would undertake to defend such a conclusion, first, because it seems to him *prima facie* bizarre, and second, because he thinks that if it were maintained that the behavior of a weighted wheel is purposeful, then it would be difficult to find an instance of behavior which is not.

Whether or not a conclusion may appear *prima facie* bizarre to somebody may be highly relevant for some objectives, but is irrelevant in a discussion of the status of the concept of purpose in science. We shall therefore not consider this argument further. We utterly fail to grasp the logic of the statement that, if it were maintained that a weighted roulette exhibits purposeful behavior, then it would be difficult to find examples of purposeless reactions. We must, therefore, again dismiss this argument.

We consider the behavior of a weighted roulette purposeful, much as we deem that the behavior of a magnetic compass deviated from its resting alignment exhibits purpose. By this we mean that the analysis of the motions of the wheel or of the needle should include the fact that they end in a definite relationship to a specific characteristic in the environment in which they occur. The purposiveness in these instances differs importantly from that recognizable in a servomechanism, however, because the behavior of the latter is active, whereas the behavior of the wheel and that of the magnetized needle are passive, an important fact

which Professor Taylor did not include in his analysis.

He believes that we left our distinction between purposeful and purposeless behavior hovering vaguely in the background of the examples we used. We shall not discuss this belief. He believes further that as soon as this distinction is made clear and precise, then "its dubious status becomes quite apparent." This belief involves a contradiction of terms: a *clear* and *precise* distinction cannot have a *dubious* status.

He reproaches us for neglecting to point out that some of the most ordinary objects of our daily experience, such as the leaves of many green plants, exhibit instances of what we called "teleological purposeful behavior," namely behavior controlled by negative feed-back. This reproach is gratuitous. It would only be justified if we had attempted to list or catalogue *all* the instances of negative feed-back, but that was neither our claim nor our objective.

He maintains that the notions of purpose and teleology are useless for the understanding of servomechanisms, and are wholly incongruous as thus applied, because the behavior of these mechanisms is describable in terms of, and only in terms of, the same fundamental categories as are employed for the description of any other physical process. These statements are false. The analysis of any process can be carried out in many different ways that are scientifically valid and useful. We believe that the notion of purposefulness is one of the fundamental categories to which Professor Taylor vaguely alludes—vaguely because he does not state them specifically. We make bold to assert that the vast majority of engineers working on the theory or the design of servomechanisms share our conviction that the notion of purpose is essential for their work.

Professor Taylor's categories, although not explicitly listed, appear to be the Newtonian categories. These do not contain the categories of the Gibbsian statistical mechanics, nor, as a consequence, those of modern quantum theory. They admit of no probabilities and of no degrees. Strictly speaking, they do not admit the repetition of an experiment unless the impossible is realized and the new experiment is the complete replica of the old. They are suited for the non-experimental study of the motions of the planets, but not for experiments where the human influence is perceptibly large.

In a rigid system, without degrees, if we introduce causation it pervades the whole system, and the only conceivable cause of any future status is the entire past. But in a world in which the whole past causes the whole future in an integral inseparable way, the category of cause has no operational significance. In such a system, the introduction of purpose anywhere will again be all-inclusive and the notion will again be operationally meaningless. Thus, if we adhere to the Newtonian categories, Professor Taylor's criticism of the use of the notion of purpose appears legitimate, but this criticism is equally applicable to the notion of cause, a category that he accepts and uses freely.

The concept of cause is only significant when there are different degrees of causation that can be measured. To give causation degrees means to correlate changes in initial conditions with the corresponding changes in later conditions. This requires the entire apparatus of statistics and probability. However, in a science subject to statistics and probability it is possible to enquire not only to what extent a phenomenon causes another, but also to what extent a purpose causes a result. Modern science has indeed introduced degrees, and has thus justified the realization of experiments and has made possible the significant use of the term cause and equally the significant use of the term purpose.

It is possible to develop a mathematical theory of prediction which has its roots in the theory of probability. This theory is one-sided in time and distinguishes past and future. It is equally useful in determining the probable consequences of past stages, and the probable antecedents of present or future stages. The prediction of the future from the past belongs to the theory of causality; the determination of the past from the present belongs to the theory of purpose.

Professor Taylor gives a description of the behavior of a torpedo with a target-seeking mechanism, which he offers as an example of the use of the correct scientific categories. It is interesting to note that the notion of purpose, as defined by us, appears repeatedly, although thinly disguised by circumlocutions. As instances we quote: "a machine so designed as to be guided by"; "the resulting change in the sound waves, relative to the sonic device, suffices to reorient the torpedo, i.e., causes it to resume its course, relative to the target." We submit that the open use of the notion that the torpedo seeks the target simplifies and clarifies the description.

Professor Taylor states that the behavior of the torpedo would be the same even if no target were there, "provided only that sound waves, or certain other immediate causes, obtained." This

statement is misleading, because it is incomplete. The behavior will only be the same if the pertinent characteristics of the sound waves that reach the missile are the same. The effective target is not the ship but the apparent focus from which the sound waves emanate. If the missile is pursuing the same sound waves, it is pursuing the same target. The same criterion is applicable to human behavior. If a man wakes up at night and shoots his gun at his image in the mirror, his reaction is quite as purposeful as if he had shot at a burglar.

Professor Taylor then asserts that if the sound-guided torpedo is replaced by one that is propelled by the target by means of a cable the behavior will be precisely analogous, the only relevant difference being in the mode of guiding. He does not state whether the torpedo attached by the cable is active or passive; we assume that it is active. In either case the assertion is erroneous. The design, mechanics and trajectory of a pulled missile are quite different from those of a missile controlled by negative feed-back, and the difference is obvious if we realize that the pulled torpedo does not adjust its rudders and vanes to minimize an error, whereas the sound-guided one does. Of course, if the lateral pull on the cable is used as a signal to control by feed-back the movements of the missile, instead of their being controlled by sound waves, then the torpedo will again be intrinsically purposeful.

It is thus clear that Professor Taylor's criticism of our views on purposeful mechanisms contains many erroneous opinions. These errors do not necessarily imply, however, that our views were correct. It still remains for us to show with greater precision and detail than we did six years ago that it is possible to distinguish purposeful from non-purposeful behavior, and that this distinction is not only useful but essential, indeed, that the category of purpose is one of the fundamental categories in science. To this task we devote the remainder of this paper.

Let us consider an example. Suppose that in a dog track in addition to the usual mechanical hare and the living hounds there is a mechanical hound with the appropriate sense organs, and controls, so that it, like the living specimens, will follow the hare. This supposition is not unreasonable since mechanisms with positive phototropism, and even a multiplicity of tropisms, have already been constructed. Suppose also that the non-living mechanisms are dressed up to resemble their living counterparts. The observation of a single run will not allow the discrimination of any differences in the behavior of the several entities concerned; nor will the observation of several runs performed under approximately equal conditions allow this discrimination. But several observations carried out under varied conditions will readily lead even a casual observer to distinguish the rigidly directed behavior of the hare, which is independent of that of the hounds, from the behavior of the latter, which is guided, under the circumstances, by that of the hare.

The behavior of the mechanical hare is characterized by the fact that the mechanism does not show any tendency to return to its original path if it is detracked. The changes in the surroundings may influence its motion or path, but are not followed by a restoring reaction. The motions of the hounds, living and mechanical, on the other hand, are modified whenever that of the hare is changed, or whenever an influence exerted on them changes their course with respect to that of the hare. We can omit the hounds in our study of the behavior of the hare, and we shall not lose anything of significance or importance. But we can not omit the hare in our analysis of the behavior of the hounds without losing an essential fact, namely, the fact that the hounds pursue the hare, i.e., that their behavior is purposeful.

This distinction is not restricted to this particular example. We call the behavior of an ordinary clock purposeless because here again, as in the case of the mechanical hare, if we change its motion, e.g., if we stop it for half an hour and then start it again, we find no tendency to make up for the lost time. The time and place at which a clock will stop cannot be considered as a goal or purpose of its motion because this time and place do not influence the running of the clock, i.e., because they are irrelevant for the description of this motion and for the study of its laws. On the contrary, we call the motion of a phototropic servomechanism purposeful because if we deviate it from its course it reacts to resume its path toward the source of light, much as a moth will, and because the position, fixed or variable, of this source is highly relevant, indeed, is indispensable, for the understanding of the motions of the machine.

We have chosen these examples because we consider them extreme. We believe, however, that the notion of purpose is not absolute, but relative; it admits degrees. We further believe that it involves a human element, namely the attitude and objectives of the observer. Different observers may well differ in their evaluation of the degree of purposefulness of a given behavior. And the same observer may study a given behavior as purposeful or as purposeless, with different objectives. But these limitations of the notion of purpose are common to many other scientific categories, and do not detract from their validity and usefulness.

Before we proceed to a more detailed discussion of the criteria that may be used to characterize and determine purposefulness, some preliminary general remarks are pertinent. The first is that if the term purpose is to have any significance in science, it must be recognizable from the nature of the act, not from the study of or from any speculation on the structure and nature of the acting object. This remark is important because it eliminates any incongruity in the application of the notion to non-living machines. In other words, if the notion of purpose is applicable to living organisms, it is also applicable to non-living entities when they show the same observable traits of behavior.

The basis for this dictum is inherent in the example of the hare and the hounds. We assumed that one of the hounds was mechanical and behaved like the living hounds. We assert that if this behavior is indeed similar, and that if we attribute a purpose to the animals, we must attribute a similar purpose to the machine if we wish to be consistent. The animals and the machine differ in many other respects, but those differences will be irrelevant to our analysis of their behavior at the track.

A second pertinent consideration is the following. A sharp distinction between passive and active behavior is not possible. Yet it is important to separate the behavior of an object that is not connected to a reservoir of energy from that of one that is. In the first instance, e.g., the motion of a stone that is thrown, a relatively simple system carries with it all the energy available for the motion in a form which cannot readily be set over against the motion as a whole. An active system naturally divides itself into two parts. One of these is a reservoir of energy, and the other is the remainder of the system actuated by this energy. For example, a motor plugged into the mains is active, and this is equally true whether the source of e.m.f. across the terminals is supplied by a power driven generator or by a storage battery.

While there is no a priori reason to expect that purposeful behavior should be restricted to either of these two classes, we may expect the active systems to show far greater opportunities for the manifestation of purpose. In a passive system the input must supply the energy for the action as well as the direction (information), whereas in an active system the directions may be supplied at low energy levels (see below).

The third remark is concerned with the degree of coupling of an object with its surroundings. A system, isolated for study as a unit, i.e., an object, is never free from influences external to it; it is usually coupled to other systems in many ways: gravitationally, thermally, magnetically, etc. Relative independence, however, is sometimes easy to attain. Thus, in the designing of a clock it is desirable, and possible, to minimize the influence of temperature, moisture, magnetic fields, etc., on its movement.

The coupling of an object with its surroundings always includes an energetic element. But the amount of energy that will determine any significant changes in the object varies enormously in different instances and in different objects. A passive object will only register relatively large external changes, but active processes within an object may convert an external signal with a low energy level into an important reaction. This second type of coupling we call an informative or sense-organ coupling. We introduce this distinction because although only quantitative and susceptible of degree it plays an essential role in the theory and design of servomechanisms.

In the higher animals most of the cells react to many changes in the surroundings; thus, they react to pressure, to vibration, to temperature variations, to sound and to light, to diverse chemical agents, and to the flow of electric currents. In the vast majority of these instances a relatively large external change is necessary for the reactions to become significant. But some cells react to extremely minute specific external influences. For example, specific receptors in the skin can signal the pressure exerted by a hair, very low energy levels of sound waves affect the auditory receptors, and, in optimum conditions, a single quantum of light energy may be registered through the retina. The difference between these receptors and other cells is again only quantitative and admits degrees, yet the physiologists have singled them out for special study. It is interesting to note that sense-organs are built around receptors and contain earlier and later elements vastly contributing to the effectiveness of these receptors.

Similarly, all the parts of a machine show sensitivity to the environment. But there are elements like photo-cells or strain gauges which show a greater sensitivity to some specific changes in the environment. These elements may be connected to parts for the reception of external changes, and to subsequent amplifiers, in such a way as to make them even more effective for the detection of specific changes and for the communication of these to the mechanism so as to alter its performance.

After these preliminary remarks we can formulate some of the criteria for the distinction between purposeful and non-purposeful behavior.

(a) Purposeful behavior is to be attributed only to an object which forms part of a large system, i.e., to an object that is coupled to other objects or features in the environment in such a manner that changes in these objects or features will modify its behavior. This criterion renders the active behavior of an ordinary clock purposeless, for, as already stated, the clock is meant to be largely independent of its surroundings. The passive motion of an honest roulette is also purposeless for the same reason.

(b) Purposeful behavior requires that the acting object be coupled with the goal, that is, that the object register messages from its surroundings. The coupling may be on a high energy level; it is then relatively loose. A tighter coupling may obtain if the object has sense-organs, as defined above. This criterion excludes gross errors, such as that of assigning the time and place of stop or breakdown of a clock as its goal, or as that of thinking that the impact of a falling stone on the ground was the goal pursued by the stone.

(c) Purposeful behavior is behavior oriented toward or guided by a goal. The goal may be static or dynamic. If it is static and the behavior sequence is successfully achieved the behaving object will reach a relationship with the goal specifiable in time or space. This relationship should be reached under a relatively wide variety of conditions. If the goal is dynamic the object should tend to minimize an error in one or more of its relations to the goal.

Purposes, including the tropisms, may be positive or negative. Thus, a light-evading mechanism is fully as purposeful as a light-seeking one. The problems of design and of analysis are alike in both cases.

(d) The recognition of purposeful behavior requires that several observations be made with the system exposed to different initial or subsequent conditions. Repetition with varying circumstances is indispensable to eliminate random coincidences. It is also necessary in order to ascertain that the particular relation between the acting object and the constituents of the system interpreted as the goal was not reached by the independent development of processes which fall in phase at a given moment.

The notion of purpose is inapplicable to Leibniz's monads, which had no sense–organs, and which were initially wound so perfectly that they never fell out of phase with each other. The position of the hands in a clock and the rise of the sun to the zenith, like the monads, may be in phase regularly for an indefinite period, but will readily fall out of phase if we change the part of this system with which we can interfere.

(e) The relationship between an acting object and a goal is not necessarily one-way. Thus, if hounds pursue a mechanical purposeless hare the behavior of the hounds is purposeful, with the hare as a goal, and the relationship is one-way. But if the hounds pursue a live hare or a mechanical hare that tries to avoid the hounds, the relationship is two-way, the purposeful activity all around. It may be mentioned parenthetically that the theory of games is a chapter in the study of two- or more-way purposeful activity.

(f) Passive behavior may be purposeful or non-purposeful, much like active behavior. For example, it appears desirable to regard the motions of a magnetic compass that has been deviated from its resting position as purposeful, with the final resting orientation as the goal. Since in the passive behavior the coupling with the environment is purely energetic, we may expect low degree in the purposefulness. This is indeed the case, and it is interesting to note that when engineers wish to emphasize the purpose of a compass they do this by adding amplifiers to it, that is, by making it active. This is the principle of the earth-inductor compass.

(g) Purposeful behavior does not require complex acting objects, but may be exhibited by quite simple structures. Thus, the behavior of a magnetic compass is purposeful, whereas that of a large steam locomotive running without a crew would be purposeless.

(h) The notion of purpose is important for the study and understanding of some complete behavior-sequences, but loses its meaning if that study is concerned merely with the instantaneous changes in the sequence. The analysis of only the instantaneous changes does not adequately describe the behavior of, e.g., a target-seeking missile or of a voluntary, goal-directed act of an animal organism.

We stated earlier that Professor Taylor's schematic description of the behavior of a torpedo guided by sound waves originating from a target was carried out in terms that include the notion of purpose. This is true because his description was not restricted to the changes at the missile but included the relations between this missile and the target. Were these relations excluded the description would be utterly incomplete.

To sum up, we wish to stress that in some modes of behavior an acting object is closely coupled to certain features or objects in its environment. The analysis of the behavior is then quite incomplete if the object is considered in isolation, for it is only a part of a larger system. A tight coupling with the environment may result in random behavior, but it may also result in behavior with a purpose, i.e., behavior directed to a specific goal and determined by the relations between the object and the goal. The instantaneous interactions between an object and its environment are in no way different in non-purposeful and in purposeful behavior. But the adoption of a teleological approach simplifies the analysis of goal-directed behavior and enlarges the scope of this analysis. This methodological approach does not imply the philosophical belief in final causes.

We also wish to explain why we use the humanistic terms purpose and teleology in the description of the behavior of some machines. The question of whether machines are or can be like men or the higher animals does not guide our choice. This question is on the main irrelevant for scientific objectives. We believe that men and other animals are like machines from the scientific standpoint because we believe that the only fruitful methods for the study of human and animal behavior are the methods applicable to the behavior of mechanical objects as well. Thus, our main reason for selecting the terms in question was to emphasize that, as objects of scientific enquiry, humans do not differ from machines.

30.
Purposeful and Non-Purposeful Behavior: A Rejoinder

RICHARD TAYLOR

IN THEIR PENETRATING and admirably lucid discussion, "Purposeful and Non-Purposeful Behavior," Professors Rosenblueth and Wiener have considerably clarified the point of view expressed in their earlier paper dealing with the conception of purpose, and recently criticized by me. But while their discussion thus removes some of the difficulties which, I think, stood in the way of acceptance of their position, there yet remain fundamental questions which I do not believe have been adequately dealt with.

These authors rebuke me, with justice, for not indicating my own conception of purposiveness, and they add that this omission weakens my criticisms of their view. That the omission detracts from the interest of my paper is doubtless correct, but that it weakens my criticisms of their position is not similarly obvious. I shall, at any rate, attempt to rectify this defect in the present discussion.

The desirable procedure in what follows thus seems to be, first, to indicate what I take to be the chief difficulties remaining in the cybernetic conception; second, to set forth the conception of teleology to which I subscribe; and finally, to remove a few of the misunderstandings—not disagreements—which these papers have engendered.

Some Difficulties

I shall here enumerate three difficulties I find in the cybernetic view. The first two are suggested

primarily by Professors Rosenblueth's and Wiener's latest paper, but apply equally to the earlier one; the last was adumbrated, perhaps faultily, in my former paper, and needs now to be restated.

1. Professors Rosenblueth and Wiener write that "if the term purpose is to have any significance in science, it must be recognizable from the nature of the act, not from the study of or from any speculation on the structure and nature of the acting object," and then they add, "if the notion of purpose is applicable to living organisms, it is also applicable to non-living entities when they show the same observable traits of behavior."

These writers thus make quite explicit the supposition upon which, I believe, their entire viewpoint rests, namely, that the criteria of purposiveness are to be found wholly in observable behavior; in fact, they earlier *defined* purpose in terms of behavior. Now I believe this supposition is utterly doubtful, and that the authors themselves do not, and can not, abide by it. I shall try to show why this appears to be the case.

The authors invite us to consider a car pursuing a man down the road, "with the clear purpose" of running him down, and this they compare with an automatic mechanism behaving the same way. Now I submit that, *from observable behavior alone*, one cannot certainly determine what the purpose of the behaving object is, nor indeed, whether it is purposeful at all. Surely the observable behavior of the car and its driver might be exactly the same, whether the purpose is, as supposed, to overrun a pedestrian, or merely, as a joke, to frighten him, or, indeed, to rid the car of a bee, the driver being in this case wholly unaware that his car is endangering another

From Richard Taylor, "Purposeful and Non-Purposeful Behavior: A Rejoinder," *Philosophy of Science*, 17 (1950), 327–32. Copyright © 1950, The Williams & Wilkins Co., Baltimore, Md. 21202, U.S.A. Reprinted by permission.

person. If, however, purpose were definable solely in terms of observable behavior, as these writers suppose, then any driver who appeared to behave *as if* he were trying to run down a pedestrian, but who yet pleaded that he had no such intention, would not simply be *probably* lying, but could not *possibly* be telling the truth.

Elsewhere Professors Rosenblueth and Wiener suggest that "if a man wakes up at night and shoots his gun at his image in the mirror, his reaction is quite as purposeful as if he had shot at a burglar," and this observation is, I believe, suggestive of more than its authors intended. For the question immediately arises, *What* is the man's purpose? Is it to destroy his mirror, or to shoot a burglar? The answer to this must, on their terms, be sought solely in observable behavior; yet, I submit, observable behavior affords no criterion—*either* answer accounts for the behavior in question. Presumably, the man's purpose might be to shoot a burglar, and yet, unhappily, he *behaves* exactly as if his purpose were to destroy his mirror.

Of course I would not suggest that behavior affords no evidence of an agent's purpose; it is, probably, the best evidence we can have.[1] But these writers are saying much more than this; they are saying that purpose can be *defined* in terms of behavior, and they themselves so define it. And from *this* they conclude to what I take to be their most significant thesis, namely, that "if the notion of purpose is applicable to living organisms, it is also applicable to non-living entities *when they show the same observable traits of behavior*." [Italics supplied.] They then apply this dictum to the example of the mechanical hounds, and assert that if the behavior of such objects is in fact similar to that of living organisms, and if purposiveness is attributed to the latter, then it must also be attributed to the former, "if we wish to be consistent."

Now if my foregoing observations are cogent, this conclusion cannot possibly be correct. For if, as I have suggested, one and the same observed behavior pattern can be consistent with the supposition of either of two or more *wholly different* purposes, then purpose cannot be defined in terms of behavior. Accordingly, precisely similar behavior sequences, or even one and the same behavior sequence, can be consistent not only with the supposition of any of a variety of purposes, but with that of the complete *absence* of purpose. And it needs only to be added that, just as a mechanism can look and act exactly like a living organism, and yet, obviously, not necessarily *be* such, so also an automatism can appear

exactly *as if* it were purposive, and yet not necessarily *be* such.[2] There is, I submit, no inconsistency in this, *except* on the supposition that purpose is definable solely in terms of behavior. And this, I have tried to show, is not possible.

2. Professors Rosenblueth and Wiener reiterate that purposeful behavior is that which is directed towards a goal, and the goal of any purposeful sequence of behavior is taken to be that object or feature in the environment with which the behaving object strives to attain a certain definite correlation.[3] Now the difficulty here, which is far-reaching when its implications are seen, is that this conception automatically excludes certain types of distinctively and indisputably purposive behavior, by requiring that the goal be some *existing* object or feature in the environment of the behaving entity; this, in turn, obliges us to assert the impossibility of a certain kind of frustrated purposiveness. Consider, for example, a man groping about in the dark for matches which are not there, but which he erroneously believes to be near at hand, or another who goes to the refrigerator seeking an apple which he mistakenly believes to be there. Now we can say, if we like, that such behavior sequences are incomplete, or that they are clumsy, but they are assuredly purposeful, i.e., directed towards goals, despite the fact that these goals are not achieved and do not, in fact, exist. In such cases, in short, *there does not exist* the requisite object or feature in the environment of the purposeful entity with respect to which it ever does, or can, attain a "final condition," and which can thus, by the criterion offered, be designated "the goal" of that object. Similarly, the alchemist can seek the philosopher's stone, the knight can seek the Holy Grail, and both can behave in ways they believe appropriate to the attainment of such goal-objects, without our having to assume that any such objects exist anywhere in the world.

I said that the implications of these commonplace observations are far-reaching, as regards the cybernetic thesis, and my reason for saying so is that no mechanism, however ingeniously constructed, ever exhibits anything precisely like what I have sketched. Servomechanisms may seek objects which they were not designed to seek, like the gun which seeks out the commanding officer's car, or they may behave erratically, as in the case of a servomechanism whose feed-back becomes excessive, but they never positively seek objects which are, in fact, non-existent, like Holy Grails. A phototropic mechanism, like the mechanical hound designed to pursue a mechanical rabbit, may surprise its designer by pursuing something else, but, unlike the man groping for a match, it will never pursue a rabbit that is not in fact there, but which it erroneously "believes" to be there.

The authors whom I criticize assert that, "as objects of scientific enquiry, humans do not differ from machines," but here appears to be at least one significant and irreducible difference. This conclusion could be avoided, I believe, only by either (a) denying that the pursuit of a non-existing object is really purposeful after all, or (b) denying that non-existent objects are ever sought or pursued, or (c) producing an example of a mechanism that strives towards, not some peculiar object, but rather a non-existing object, of the same category as philosophers' stones, Holy Grails, and the matches that are not there.

The first recourse would be an affront to common sense, obliging us to assert, for example, that when I go to my refrigerator for an apple, my behavior is not purposeful unless my refrigerator contains an apple.[4] And the second alternative is easily refuted, for if this were true, then from any statement of the form "*A* seeks *B*" we could infer, "*B* exists"; from which it would follow that, e.g., if anyone happens to be seeking the Holy Grail, then the Holy Grail exists. Of course there would be plausibility in saying that no purposeful being ever seeks what it *believes* to be nonexistent, but this is clearly irrelevant. And the third recourse is, at least to me, incredible, evoking such notions as that of a mechanical hound positively pursuing what it takes to be a rabbit, but which is, in fact, nothing at all, or that of a servo-mechanistic telescope searching about for the planet Vulcan.

3. The third difficulty can be stated briefly, because it was the main burden of my earlier paper to point it out. Here it needs only to be clarified.

The difficulty is that, *at least as regards passive behavior*, purposiveness, by the criterion offered by Professors Rosenblueth and Wiener, appears to be ubiquitous. Behavior is purposeful, they say, if it "may be interpreted as directed to . . . a final condition in which the behaving object reaches a definite correlation in time or in space with respect to another object or event," a definition which they appear to endorse in both their papers. But is there *any* instance of behavior which may not be so interpreted? *Any* behaving object, it would seem, ultimately reaches some "final condition" or "definite correlation" with respect to other objects or events.[5] At this moment,

for instance, I note that the wisps of smoke emitted by my pipe maintain a definite correlation with certain features of the environment, viz., air-currents in the room. Similarly, the passive motion of the rocking chair beside me may, I suppose, be interpreted as directed to a final condition, rest, relative to another object of its environment, the floor. But what, I iterate, is gained by thus calling tobacco pipes, rocking chairs, compass needles, weighted roulettes and so on, *purposeful*? What, for example, is science able to learn about these objects, which it might otherwise never discover, by regarding them in this light? If, as appears to be the case, purposiveness becomes ubiquitous, then the application of the word "purpose" becomes as general as that of "behavior" itself and, by the criterion suggested, loses its significance simply by having no counterpart.[6]

An Alternative Conception

In view of the above considerations, I now suggest the following as necessary and sufficient conditions to be fulfilled, or to be assumed to be fulfilled, in order appropriately to regard any given behavior pattern as *purposive*:

There must be, on the part of the behaving entity, i.e., the agent: (a) a desire, whether actually felt or not, for some object, event, or state of affairs as yet future; (b) the belief, whether tacit or explicit,[7] that a given behavioral sequence will be efficacious as a means to the realization of that object, event, or state of affairs; and (c) the behavior pattern in question. Less precisely, this means that to say of a given behavior pattern that it is purposeful, is to say that the entity exhibiting that behavior desires some goal, and is behaving in a manner it believes appropriate to the attainment of it.[8]

These conditions appear to be the very ones which are invariably assumed to be fulfilled in any unmistakable case of telic activity. Such, I submit, is what one at least tacitly has in mind when he says, for example, that Jones has gone to the refrigerator to get an apple, that Smith shot a hole in his mirror, believing he was seeing a burglar, that the goal of a hound's behavior is a fleeing hare, that Brown pursued a pedestrian with his car, with the clear purpose of running him down, that the Western Powers have armed, in order to halt Soviet aggression, and so on.[9]

This conception, moreover, does avoid the difficulties pointed out in connection with the viewpoint I have criticized. It permits us to say, for instance, as the cybernetic conception does not, that the alchemist's search for the philosophers' stone is purposeful; or again, that two agents may exhibit the same behavior, and yet have wholly different goals. It also, of course, renders purposiveness a special trait of certain classes of objects only, and not a ubiquitous concomitant of behavior in general.

Two objections to this analysis may be anticipated. If, firstly, it is objected that my conception invokes dubious or occult entities, viz., desires and beliefs, I reply that, however unsuccessful science may be in describing or explaining these occurrences, everyone "knows perfectly well," in one clear sense, what it is to desire something, and what it is to believe something, and that, to this extent, there is nothing at all dubious or occult about them. And if, secondly, it is objected that these occurrences are not observable, and thus not operationally useful to science, I reply that this simply is not true; they are observable in the same sense that atoms, for example, are, viz., as inferences from what is directly observed, and are directly observable to the agent in whom they occur, in the same sense that pleasure and pain, for example, are. And I hope it is unnecessary to add, further, that to say of any analysis that it is not useful to experimental science would hardly be equivalent to saying that it is not a correct analysis.

Some Misunderstandings

There remains only to clear up three cases of sheer misunderstanding generated by this controversy.

1. I am charged with a contradiction in saying that the cybernetic distinction between purpose and non-purpose, though clear and precise, has yet a dubious status. My phrasing here was unfortunate; I did not mean, of course, that this distinction, though clear and precise, was somehow meaningless, but rather, that it was doubtful whether it is capable of being applied in the manner these writers attempt. To cite a parallel case, the statement that Hoover will be returned to the Presidency in the next election is perfectly clear and precise, but none the less dubious for that.

2. My paper was apparently construed as special pleading for Newtonian "categories," and this, apparently, because of my liberal use of the

word "cause." I am, of course, aware that advances have been made in science since Newton lived, so to avoid this misunderstanding I have eschewed the word "cause" and, I hope, similarly objectionable "categories" in the foregoing discussion.

3. My critics assert that, in certain of my descriptions of mechanical behavior, "the notion of purpose, as defined by us, appears repeatedly, although thinly disguised by circumlocutions." But was not the very point of my comments to show that, *as defined by them*, any behavior whatever becomes "purposeful"? Indeed, it was this ubiquity of purposiveness, *as they define it*, which led me to reject their definition.

Other misunderstandings—such as, for example, the possible interpretation that I do not regard *any* distinction between purposeful and non-purposeful as significant—have, I believe, been removed in the foregoing discussion.

Notes

1. At least, as regards purposes other than one's own. An agent, obviously, need not wait and observe his own behavior in order to infer what his purpose might be.

2. Cf. R. B. Perry, "Purpose as tendency and adaptation," *Philosophical Review*, vol. 26, 1917. Perry, it may be noted, in this paper written more than thirty years ago, reviews a thesis which is virtually identical with that of Professors Rosenblueth and Wiener, even employing examples which are familiar in the literature of cybernetics—e.g., those of the thermostat and the steam engine governor (p. 488).

3. In my former paper I suggested difficulties in the problem of discriminating the goal of any inanimate behavior sequence, without tacitly introducing human purposes. For another discussion of this, see Y. H. Krikorian, "Teleology and causality," *Review of Metaphysics*, vol. 2, 1949.

4. By their description of "a goal," however, this seems nevertheless to be the recourse these writers would be obliged to take.

5. The plausibility of selecting *any* actually resulting state of affairs as the end or goal of an ostensibly teleological process is cogently argued by C. A. Mace, "Mechanical and teleological causation," *Aristotelian Society*, suppl. vol. 14, 1935.

6. The prevalence of what these authors call "teleological purpose," i.e., that exhibited by certain *active* objects, is abundantly illustrated in an early paper by Stevenson Smith, "Regulation in behavior," *Journal of Philosophy*, vol. 11, 1914. This writer nowhere connects such phenomena with the notion of purpose, however, but describes them simply as cases of auto-adjustment.

7. It may seem incongruous to speak of a tacit (or "unconscious") belief, but I am using "belief" in

the broad sense in which, for example, a man whose chair collapses under him may be said to have *believed* it would hold him, even though he had not even thought of the chair or its condition. Similar remarks apply to "desire."

8. A similar conception has been propounded by C. J. Ducasse, "Explanation, mechanism and teleology,"

Journal of Philosophy, vol. 23, 1926, which suggested the analysis given here.

9. For a discussion of the problem of trying to *explain* events in terms of the concepts exhibited in such statements, see R. B. Braithwaite, "Teleological explanation," *Aristotelian Society, Proceedings*, n.s. vol. 47, 1946–47.

31.
Purposive Behavior and Cybernetics

C. W. CHURCHMAN AND R. L. ACKOFF

SCIENCE, like all other cultural institutions, has its fads, but *Cybernetics* is *not* one of them. This development of a science of control and communication by Wiener, Rosenblueth, McCulloch, and others is not only an exciting extension of scientific inquiry, but it is a fruitful one as well. The concepts and methods of cybernetics are by no means restricted to the problems of servomechanisms, or even neural physiology, though the impetus came from these areas.

Cybernetics is very young and very vigorous. But it is not a mutant suddenly appearing in the family of the sciences. It makes in a new way a point which has already been made in various guises in contemporary science and philosophy. In the field of psychology there has been a prolonged attempt to study the concept of purposive behavior and communication in such a way as to provide experimental methods of measuring the important dimensions of teleology. In 1914, Singer presented in his "Pulse of Life" the first of a series of papers in which he outlined a schema for the *experimental* determination of purpose. Tolman, Hull, and others following the same lead developed what is called "molar behaviorism" (as opposed to Watson's "molecular behaviorism").[1] In molar behaviorism, the concept of purpose has a central position. Contrary to traditional belief, Singer showed that the introduction of the concept of purpose into science did not violate the demands that mechanical science makes for a rigid (deterministic) image of nature. In effect, he demon- strated that the traditional conflict between mechanism and vitalism (causality and teleology) is removed by an experimental (as opposed to a spiritualistic or introspective) formulation of the meaning of "purpose." The cyberneticians furnish an important application of Singer's point of view by showing that teleological concepts are extremely fruitful in the study of neurological and machine behavior, and that such concepts can be treated experimentally.

Of course, there have been many so-called solutions to the mechanist–vitalist paradox, but most have reduced to a mere affirmation of the fruitfulness of analyzing experience from both mechanical and teleological viewpoints. Singer has gone beyond this by showing the logical and methodological inter-relation between these two conceptual schemes of nature. He has done more than merely assert their compatibility. He has developed a schema of science in which he has demonstrated this compatibility. In effect, he has provided a transformation principle which enables the scientist to pass from mechanical to teleological explanation without introducing any non-scientific forces such as "entelechy" or "élan vital." He has provided the logic of a non-spiritualistic teleology, a logic implicitly assumed in the work of the cybernetician. It seems unfortunate that Singer's thinking in these matters should be ignored by those who appear to have so much to gain from it.

The relatively new field of *semiotic* (the science of signs) should also be noted since it also represents a definite tendency to promote the experimental study of communication in terms of purposive or goal-directed behavior. Morris' *Signs, Language, and Behavior* is a landmark in this

From C. W. Churchman and R. L. Ackoff, "Purposive Behavior and Cybernetics," *Social Forces*, 29, 1 (October, 1950), 32–39. Reprinted by permission of the authors and The University of North Carolina Press.

area. His book provides a complete bibliography of foundations of semiotic.

As in all cases where a concept such as "purpose" becomes critical in many different fields, there is an ever-prevalent danger that the definitions will be oriented with respect to the particular aims of its formulators. When they are so oriented, their applicability in other areas becomes restricted and difficult, and there is a threat of their eventual exclusion because of the "bias" in the definition. The process may run as follows: workers in a field like cybernetics will define "purpose" so that it is admirably suited to neural and (certain) machine behavior. They will make tentative suggestions as to how their definitions could be applied in other areas. Eventually the bolder of them will suggest that people working in other fields can actually use the cybernetic definitions. Hence, they may begin to claim that the study of all purposive behavior can be carried forward in the most fruitful manner by the use of the existing cybernetic framework. But the psychologist and social scientist are aware of the complexities of phenomena in their own fields and may look upon the metaphors and analogies of the cyberneticians with skepticism. The danger is that eventually some cybernetician goes astray and makes proposals which to the psychologist and social scientist are completely outrageous. The end point of the process may come when each field returns to its own work and ignores the potential contribution of other disciplines. The real danger is the complete loss of integration (called for so eloquently by Wiener) which at the present time seems essential in the study of purposive behavior.

We do not think that cybernetics has yet made any seriously incautious claims, but one finds in the more popular presentations statements like the following: Cybernetics analyzes all purposive behavior and provides an exact notion of communication and the transmittal of information. Perhaps such statements are meant by their authors to be restricted to physiological and engineering domains, but, to the mind of many readers, the restriction is lost and the proposition seems to be nothing short of an ill-advised invasion by the mathematician, physical scientist, and biologist into the realm of the psychological and social sciences.

What is needed is a scheme of defining which is not directly oriented toward servo-mechanisms or neural behavior. A definition of purpose is no more nor less than a method of looking at the world—a scheme of organizing phenomena.

Definitions are aspects of the scientist's method. It is therefore important that a scheme be developed which is general enough to handle diverse phenomena, provided the needed integration of disciplines is to be maintained. Now a scheme of defining that is designed primarily to handle one type of data (such as those occurring in the study of servo-mechanisms), runs the risk of being useful in only a metaphorical sense when the type of phenomenon changes. Hence, it is critically important to see whether concepts like "purpose," "communication," and "information" can be so defined that the definitional schema are not forced in any domain. Each separate field will then add further conditions that will permit the most fruitful and convenient application of the general schemes. In this manner, we can maintain the integration between the disciplines that are studying essentially the same concepts.

This point can perhaps best be illustrated by reference to Rosenblueth's and Wiener's formulation of "some criteria for the distinction between purposeful and nonpurposeful behavior." These criteria are all concerned with establishing some connection between the purposive object and its environment and goals. Thus, the purposive object must be "coupled to" certain features of the environment, as well as "oriented to and guided by" the goal. Tests of purpose must be made by changing the environmental conditions, and so on. The general idea Rosenblueth and Wiener want to develop is that an object behaves purposefully if it continues to pursue the same goal by changing its behavior as conditions change.

It is our belief that this notion of purpose can be applied to some problems of the social sciences *only* by a considerable stretch of terminology. For example, consider the condition required of purposive objects by Rosenblueth and Wiener: that we have to study such objects in different environments and determine if they display different physical but similar goal-directed behavior. An animal psychologist such as Kohler observes that simians learn by trial and error how to obtain food that is placed in an awkward location. The animal is observed in an unchanging environment and yet his action is generally called purposive. Or, again, on the social level we observe a government bureaucracy making repeated and different efforts under essentially the same conditions to affect a certain type of legislation. Yet we call such behavior purposive. In effect, Rosenblueth and Wiener have found a useful distinction in the area of mechanisms. But it is ill-advised to extend this distinction to other areas without a careful consideration of the correlated types of phenomena. In other words, the scheme of cybernetics may be too crude for other areas. It is our contention that the social sciences require a more refined analysis of function and purpose than one which suits cybernetic purposes.

We see a need for developing at least three distinct teleological categories: the first consists of objects which accomplish certain objectives by displaying relatively invariant behavior in a wide range of environments. The second category consists of objects that accomplish their goals by changing their behavior if the environment changes, but generally exhibit only one type of behavior in any given environment. The third accomplishes its objectives by exhibiting different types of behavior, even though the environment remains constant. To put names on these categories, we shall say that the first class of objects has an *extensive function* but has no function of its own, the second class of objects exhibits a *function of its own*, i.e., has an *intensive function*, and the last exhibits *purpose*.

An ordinary clock has an extensive function since it serves a purpose by displaying relatively invariant behavior in all environments. Electric light switches, hydraulic presses, printing presses are further examples. Servo-mechanisms such as automatic temperature control devices, self-aiming guns and missiles, circuit breakers, etc., display intensive functions since they exhibit different behavior in different environments. Finally, the cybernetician's chess playing machine and much of human activity are purposive, that is, alternatives are displayed in the same environment. The interests of the cybernetician are usually restricted to what we have called intensive function. There is an occasional reference made to purpose in the sense defined, though without any explicit, recognition of the distinction between purpose and intensive function.

In the remainder of this paper we shall attempt to explain in more detail the meaning of the third category (purpose), and to compare this concept with the other two (extensive function and intensive function).

We begin by noting what we take to be well-recognized similarities in all studies of purposive behavior, whether they occur in the case of servo-mechanisms, psychological behavior, or social groups.

1. The objects and environments are not *rigidly* specified in the sense in which classical mechanics uses this term. That is, the schemes for studying purposeful behavior do not set down mechanically defined specifications for the conditions of the experiment. Whether the study is of machines or social groups, the experimenter may take the environment to be the "same" even though it varies widely according to the criteria of mechanics. This does not mean that the studies of purposive behavior are inexact, for there is no real reason why science should identify *precision* with *mechanical* specification; further, it does not mean that mechanics is basic, for again it is a fallacy to identify the foundations of science with the science that looks at the world through the spectacles of mechanical imagery.

2. A purposive object is always taken to exhibit *choice*; that is, a purposive object displays a selection-process in its behavior. As the cyberneticians point out, "The basis of the concept of purpose is the awareness of 'voluntary action'." Conditions for studying objects from the point of view of their choices is made possible by the teleologists' method of considering objects and environments. If the environments and objects were rigidly specified, then no "choices" would be possible. The social scientist, for example, can study the behavior of a corporation in what *to him* is the same environment; he can observe that the corporation makes many different types of choices, even though the environment remains the "same" in social scientist's sense of "sameness." It is important to notice that choice is not *sufficient* to identify purpose, but it is essential. It is also important to note that the possibility of choice on the part of objects results from the way in which the scientist looks at the world; within the frame of reference of classical mechanics, there is no choice, but in the frame of reference of teleology, there is a choice. And the two frames of reference are consistent and interdependent: as we vary the meaning of "same" and "different" we can look at the world in different ways. Thus "purpose" (like "determinism") is not found in the world; it is a fruitful scheme for studying the world. If its fruitfulness did not exist in fact (i.e., in an examination of the history of science), then we could say that the world is mechanically determined only, since mechanical imagery would provide the only meaningful way of looking at the world. The possibility of teleology, so often discussed in the mechanist–vitalist dispute, is to be decided by test; and if cybernetics, biology, and the social sciences are any criteria, teleology has been proved over and over. Further, the newer development clearly shows that we are not forced to abandon mechanics for teleology; both frames of reference are fruitful, and neither is fundamental.

3. Next, purposive behavior can only be studied relative to a period of time. The teleologist is not interested in the state of nature at a moment, for purpose implies action and change (behavior), and hence implies the necessity of a time interval in the scheme of study.

4. Finally, the purposive object or behavior is at least a potential *producer* of some end-result (end, objective, goal). The man who writes a poem, the machine that computes, the social group that averts disaster, are all acting in such a way that their acts possibly produce some end-result. This demand for potential productiveness requires that the scheme for studying purpose include the notion of "production" or some similar concept.

In effect, then, we are postulating that all purposive schema must incorporate the notion of an object (individual) having choices of behavior all of which are at least potentially producers of some end-result.

We now proceed to make more exact the general characteristics of all such schema for studying purposive behavior.

We begin by distinguishing between *physical* and *morphological* classification. In common parlance, physical classification is very rigid, and there is a minimum of leeway in its specifications; but in morphological specification, the idea is to leave room for variation within the class. Thus, in physical classification we group objects or events (behavior patterns) in terms of a quantified property expressed along a physical scale such as temperature, weight, wave length, velocity, etc. For example, we can form a physical class of objects by considering as a collection all objects weighing 70 pounds, or we may classify behavior physically by collecting all events in which motion takes place at the rate of 50 mph. Morphological classification, on the other hand, is a nonphysical method of classifying. It may take either a quantitative or qualitative form. When quantitative, the scientist chooses a range along a physical scale as the criterion for membership in the class. Thus, we can define the class of all objects having a weight between 70 and 80 pounds, or motion between 50 and 60 mph. We may also have qualitative morphologies, as when we specify "red objects" or "heavy objects" or "rapid walking."

The history of science to date has shown that the use of quantification permits a more rapid reduction of experimental error, and consequently, where greater precision is desired, attempts are made to convert qualitative morphologies into quantitative morphologies. For example, "heavy" can be specified in terms of a range of weights. But for many purposes it is inefficient to convert the qualitative into the quantitative, even though it can be done. It is only where distinctions between qualities become open to uncontrolled error that quantification is needed. In general, quantitative morphological classification is used more frequently than physical classification. We seldom classify in terms of a property given as a unique value along a physical scale, but rather provide a range, as when we specify tolerance limits. Morphological classification is pervasive in all the sciences—physical and social. "Wood," "apes," "males," "dense populations," etc., are all morphological concepts. The important thing to note about physical and morphological classifications is that they are *ways* of classifying; they are conceptual instruments which we bring to our experiments, and their adequacy is to be judged in terms of what they accomplish rather than in terms of rigidity.

In terms of morphological classification we can define the important relationship "producer–product" which we take to be essential for a complete understanding of functional and purposive classification. First of all, however, let us see how "causality" is related to "production."

In science, as J. S. Mill long ago realized in his *Logic*, the studies of causality vary considerably. But there appear to be two main types: the first is one in which we seek a description of nature that will enable us to *predict* with as great accuracy as possible, and the second where we want to discover a *determinant*, or responsible element, for some event. The first (*predictive*) type occurs most often in theoretical physics and astronomy: here the aim is to study conditions so that the future (or, sometimes, the past) can be predicted with virtual certainty. In such studies the scientist's interests lie in determining enough conditions so that he has a basis for a complete prediction. That is, he seeks conditions which are *sufficient* to assure the occurrence or nonoccurrence of an event. The second (*determinant*) type occurs frequently in physical science, but especially in biology and the social sciences. No one believes that low blood pressure is sufficient to predict death at some time, but it is important to study the properties of organs and systems that are determinants of a specific disease or death. In these studies the scientist is interested is unearthing the *necessary* conditions (aspects of nature that are responsible) for some event. For the sake of keeping these two studies distinct, we say that the first deal with *cause-effect* (i.e., sufficient conditions for predicted event), and the second deals with *producer–product* (i.e., the necessary conditions for some event).

Thus, we say that striking a match produces a fire, for the striking of the match was only necessary, not sufficient for the fire; other things (like oxygen) are also required. Now, of course, if we specify the environment in more and more physical detail, then we would be able to predict with greater and greater accuracy whether or not the striking of the match will be followed by fire. But for many purposes it is very inefficient to try to specify the environment in such physical detail. Relative to the range of environments in which the scientist conducts his study, he may not be able to predict with certainty whether or not the striking of the match will be followed by fire. But he can know that the striking of the match is necessary for the fire, i.e., that the fire will not occur in the given environment *unless* the match is struck. In more formal terms, an *x* can be said to be the producer of a *y* (where *x* and *y* are morphologically defined objects or events), if *x* is a necessary but not a sufficient condition for the absolutely precise prediction of the occurrence of *y*. *x* is called the "producer" and *y* the "product." More precisely still, an *x* is a producer of a *y*, if, when something differing from the morphology of *x* replaces *x* in the region where *x* occurs, no object of the *y* morphology will then be observed in the region of *y*, while the absence of *y* from its region is not sufficient grounds to infer the absence of *x*-type things in the *x*-region.

Certain properties of the productive relationship are important to note. First, there are always co-producers of a product. Since no producer is sufficient for the product, there are always other necessary (but not sufficient) conditions for the occurrence of the product. In the most general sense, the environment of a producer is always a co-producer. An acorn, for example, is a producer of an oak, since it is necessary but not sufficient condition for an oak. But so is its environment; to get an oak, certain soil and atmospheric conditions are also necessary.

Second, production is not restricted to the biological and human areas. Machines are producers as well as men. For any perspective of nature in which we cannot predict with absolute certainty that a machine will or will not be followed by a specified product, the machine is only a necessary (not sufficient) condition for the existence of that product. The dependence of the machine on its environment is apparent when we realize that we can construct an environment within which its usual product does not occur, i.e., certain machines will not operate in extreme temperatures. The producer–product relationship, then, is pervasive; it is a conceptual tool with which we approach nature for the sake of comprehending it.

Third, producer–product studies are primarily *statistical*. The scientist frequently wants to know how often an *x* will produce a *y* under morphologically specified conditions: What is the chance that a heart ailment of a certain type results in death; what is the chance that a computing device will produce a certain type of error; what is the chance that a NLRB decision will produce hardship among a certain class of the population?

In general, we may say that something of an *x*-type is a *potential* producer of a *y*-type under specified conditions if the estimated probability that *x*'s produce *y*'s is greater than some virtual zero: i.e., there is some significant chance that *x*'s produce *y*'s, where the level of significance is set by the design of the experiment.

It may have struck the reader that in the illustration above, behavior other than striking a match could also produce a fire in some environments. A fire may be produced by a cigarette lighter, combining chemicals in a vial, the application of extreme heat to certain materials, a gas leak near a pilot light, etc. That is, for any specific fire we could determine what produced it, but the same kind of product could be produced in other environments by different producers. Thus, producers which we may consider as differing morphologically *can* have a common product (defined morphologically). Walking, skating, bicycling, horseback riding, or driving an automobile, will all get us from where we are to some other designated location. Yet each type of behavior differs in what we may consider to be significant morphological properties. It is nevertheless extremely fruitful to collect into one class objects or behavior which differ morphologically but have a common *potential* product. Such a class of objects we call an *extensive functional* class. The function that defines the class is "the potentiality for producing a specified product."

If we take such morphologically distinct objects as skates, bicycles, sleds, automobiles, boats, aeroplanes, etc., we recognize them to have a common potential product, "getting from one place to another," and thus we assign them the extensive function of *transportation*. Functional classificatory schemes are another way we have of approaching nature, a way applicable to studies in *all* of the sciences. This, then, represents a definition of the first of the teleological categories we introduced earlier. Note that an object can belong to an extensive functional class even though it has no choice and must virtually exhibit the same behavior in all environments: it can belong to the extensive functional class *because there are other objects which behave differently* (respecting morphology), *and yet potentially produce the same goal.* Such objects or behaviors have a function by virtue of their membership in a class, not by virtue of their own properties alone. Note, too, that an object does not belong to a functional class because it exhibits "complicated" or "inexplicable" behavior: The peculiar motions of a ball rolling down an irregular surface do not characterize function, no matter how difficult the task of discovering the equations of motion. The laws governing functional classes are *statistical*, for it is only in terms of probabilities that we can adequately describe the behavior of any random member of a functional class.

We turn our attention now to *individuals* and their behavior. Since we want to keep the scheme general, we mean by "individual" anything the scientist considers as an identical unit over a period of time, whether it be a machine, a substance in a beaker, an amoeba, a person, or a social group.

Suppose now that in essentially the same morphologically specified environment, an individual can be shown to choose morphologically different types of behavior. If this can be shown, the individual has different *potential* choices. Next, suppose a subset of his potential choices all have a common potential product. Then any individual studied in this manner is studied relative to its *purpose*, and its behavior *in such a context* is called *purposive*. The purpose of an individual's behavior is the common product of the morphologically distinct behavior patterns he can display. Consider a psychological illustration. Take a person nailing pieces of wood together. On observation, the scientist can say the individual's purpose is to make a book shelf. He can say this because he can attribute to the subject potentiality of choices of various types of behavior which have the same function, and the scientist can test this attribution by removing the hammer and nails and observing if the individual uses glue, or screws and a screw driver, etc. If the individual is incapable of selecting an alternative where such alternatives are available, he is not behaving purposefully.

The alternative types of behavior having the same function, we call "means," and the product defining the purpose, we call "ends," "goals," or "objectives."

Now this definition of purposeful behavior is equally applicable to machines. Some machines are constructed so that they have a choice of behavior, where the alternative choices have the same function. The chess-playing machine is one of this nature, since for any move of its opponent, it has alternative moves. Though its designers may know the probability with which the machine will make any move in a given situation, which move it will make cannot be predicted with certainty. Most servo-mechanisms, however, cannot be said to have such alternative courses of action open to them in the same environment. Hence, we cannot, within this definitional scheme, call such servo-mechanisms "purposeful"; they are "intensively functional." This is not to say the servo-mechanism merely serves a function, but that it has a function of its own: over a range of environments it displays different behavior having the same potential product.

It should be made clear that our motivation in distinguishing between intensive function and purpose is not to separate the living from the machine, for machines can be purposeful in the more rigid sense in which we have used the term. But the distinction does serve to distinguish between different types of behavior, and we believe this distinction a fruitful one, particularly in psychology and the social sciences.

Thus we think the second and third categories of teleology should be distinguished for certain areas of study. It is one thing to study the behavior of goal-directed objects that can virtually exhibit only one type of behavior in a specified environment, and it is another thing to study those objects that pursue their goal by potentially exhibiting many types of behavior in the same environment. The distinction is certainly one that has been emphasized in the history of science and philosophy, and we think that today it is useful in distinguishing types of research even within cybernetics.

The distinction Rosenblueth and Wiener make between passive and active purposeful behavior is pertinent here. Passivity involves an energy source outside the object (an incapacity to store energy), while activity involves an energy source within the actor. Now it will be clear that what the cyberneticians call a "passive purpose," is, for us, intensive functional behavior, since, in a completely passive machine, changes in behavior can occur only when there are changes in the environment. It does not follow, however, that an active purpose (in their scheme) is a purpose (in ours): Even a self-moving device (or person) may have only one behavior it can produce for any environment.

Note that the "goal" of purposeful behavior need not be external to the individual (machine or person); indeed, an individual may be studied with respect to some future activity on its part. That is, the scientist may be able to show that the individual has choices all of which potentially produce (make possible) some *future* action. The machine may go through certain adjustments that enable it to solve a different kind of problem, the person may eat so that he can work, and so on.

The crucial point of comparison between this scheme and the one presented by Rosenblueth and Wiener lies in the relationship of the purposeful individual and his environment. Rosenblueth and Wiener require that the purposeful individual receive information from the environment, i.e., they require that the purposeful individual respond to the environment as well as to changes in the environment produced by the individual itself as well as by other forces. We do not require such a feed-back relationship for purposive behavior, though we recognize it to be involved in much purposeful behavior; we omit this requirement because of psychological examples of purposeful behavior where such transmitting of information either is not present or is not of concern to the experimenter. The Rosenblueth–Wiener condition does hold, however, for all objects which we would say have an intensive function.

We feel that an experimenter may want to make a study of purposive activity even though (for the observer) the environment does not change. This type of study is certainly common in psychology and the social sciences; "problem-solving" and "adjustment" studies are two cases in point.

The distinction we made between purpose and intensive function clearly depends on the specification of the environment: *If* the scientist is interested in showing how the product of an individual's behavior remains constant with respect to changes in the environment, than he will want to specify environments so that changes can be readily observed. But if he wants to concentrate attention on the types of choice and is not interested in the mechanism of choice, then he will want to take the environment as the "same" over the period of study. Thus intensive function and extensive function, and purpose, like the physical and morphological, represent conceptual models with which we approach our experience, organize it, and come to understand and control it. An individual could be studied from any of these points of view, and the results obtained in each approach can be completely compatible.

Though this paper has been devoted in the main to the very fruitful concepts of cybernetics, it might be worthwhile to insert a note of caution to those who come to the problem of purpose from speculation rather than experiment. We agree with Rosenblueth and Wiener that "purpose" need not connote "consciousness" or even "motivation" or "intention" on the part of the studied object. We agree with their position because we think that the domain of teleology should be extended as widely as possible at the present time, and science runs the danger of bifurcation if "purpose" is tied down to the as-yet-poorly-defined concepts of consciousness and intention. Thus, a machine like the chess-playing machine may be said to act purposefully without the scientist's committing himself on its consciousness. Also a human may exhibit behavior which falls under the category of purpose without his *intending* to pursue the goal. But we feel obliged to add that an important problem of experimental teleology is to define consciousness and intention; that is, we need experimental schemes for studying these concepts as much as we need schemes for studying purpose. The authors have attempted to construct such schemes elsewhere.

In closing, it is also important to make a general moral. In discussions of the type to which this paper is devoted (i.e., how terms are to be used), it is important to avoid fruitless verbal conflict concerning "right" names for things. Actually, we feel that the terms and frame of reference of cybernetics are extremely useful and that their worth has already been demonstrated in the fields of neural behavior, servo-mechanisms, and the like. We are certainly not urging an immediate modification of experimental definitions in these areas. What we are saying is this: *If* the aim is to inter-relate studies of teleological behavior in the various areas of science, *then* some such general scheme as we have offered is needed. Suitability of definition is always to be judged relative to aims, and our fear is that a strict forcing of definitions useful in one area upon another discipline will eventually work against integration where integration is so urgently needed.

Notes

1. Rosenblueth, Wiener, and Bigelow distinguish between the "functional method of study" and the "behavioristic." These correspond closely to molecular and molar behaviorism, respectively.

32.
Purpose and Learning Theory

OMAR K. MOORE AND DONALD J. LEWIS

WE SHALL BE CONCERNED in this paper with the notion of "purpose," a notion that has long troubled social and biological scientists. It has been held by some psychologists that a phenomenological observation of organismic behavior immediately reveals the essential goal directedness, the purposiveness, of this behavior, and that to talk of an animal's purpose in reaching a goal is legitimate as long as "purpose" is "defined" behaviorally. Other psychologists have insisted vehemently that teleological concepts such as "purpose" either have no place in an objective account of behavior, or at most can be introduced only after they have been derived from primary principles. In this paper we shall attempt to show how such terms as "purpose" and "teleology" may be used with scientific respectability in connection with the primary principles of learning. In fact, we believe that these notions are essential to psychology, and, when properly used, make clear just what it is that the learning theorist is making laws about. It should be emphasized at the outset that no attempt is being made here to reintroduce into psychology mentalistic notions or entelechies that psychologists have labored so long to eliminate from their science.

We are aware that many persons, especially hard-headed experimentalists, wish to avoid any use of teleological concepts. This avoidance is understandable. Teleological concepts have been pre-empted in the past by those who have had little or no concern with the formulation of test-

able theory. Historically, "teleology" has been defined as the opposite of "mechanism." A teleological theory was taken to be one which explained the past in terms of the future. Purposes, ends, goals, or, in short, the terminal stage of any sequence of behavior acted in some unanalyzable way as one of the antecedent conditions necessary to reach the terminus. This usage of "teleology" has not proved useful in solving scientific problems.

It might be argued by some that if by "purpose" is not meant some metaphysical notion, the word should not be used at all, for it has too many mentalistic connotations. We feel justified in retaining the word, however, because this paper is concerned with goal behavior and with means and ends. But we are not going to talk about any animal as "having" a purpose any more than the engineers who work with guided missiles that "seek" a target introject into the missile an entelechy or purpose.

We hope, in this paper, to redefine "purpose" and "teleology" in order to make them amenable to scientific psychological usage. We also hope to show the importance of these terms to learning theory and to show that the reinforcement theorists who have insisted most strongly that teleological concepts have no place in objective learning theory have made, paradoxically enough, essential use of the teleological frame of reference.

A significant methodological development in recent years has been a series of reformulations of the concept "teleology." Certain philosophers of science, principally Singer, Churchman and Ackoff, and also a number of scientists, prominent among whom are Wiener, and Rosenblueth and Bigelow, have been active in this effort. Their

From Omar K. Moore and Donald J. Lewis, "Purpose and Learning Theory," *Psychological Review*, 60 (May, 1953), 149–56. Reprinted with permission of the authors and American Psychological Association.

work has been extremely helpful in enabling us to make an analysis of the relationship between learning theory and teleological concepts. Our point of view is closest to that of Churchman and Ackoff. In the short space available it will be impossible to give a full account of their views or to show in just what way their views differ from ours.

Two Frames of Reference

There are at least two quite different, yet compatible, frames of reference that may be profitably employed in the analysis of behavior. They are the mechanical and the teleological (but not in the historical sense previously noted). When a scientist uses the mechanical frame of reference, ideally at least, he attempts to specify the conditions, both necessary and sufficient, to bring about some state of affairs. For purposes of illustration, let us use the following expression:

$$a \longrightarrow e_1$$

If the set of conditions designated by "a" is sufficient to bring about the state of affairs designated by "e_1," then whenever a occurs, e_1 occurs. If a is a necessary condition for e_1, then e_1 will not occur unless a has occurred. As a matter of fact, scientists have seldom, if ever, been able to specify the sufficient conditions for any occurrence. Nonetheless, from the point of view of the mechanical frame of reference, the ideal to be achieved is the specification of the necessary and sufficient conditions for any event.

In the following diagram, let us assume that a is a necessary and sufficient condition for e_1, b for e_2, c for e_3, and d for e_4. This means that when a occurs, e_1 will always occur, and if a does not occur, e_1 will not occur. An explanation using the mechanical frame of reference may be said to have been given when a set of well-confirmed laws has been formulated which will enable one to predict the occurrence or non-occurrence of an e_1 given the occurrence or non-occurrence of an a.

$$
\begin{array}{ll}
a \longrightarrow e_1 \\
b \longrightarrow e_2 \\
c \longrightarrow e_3 \\
d \longrightarrow e_4 \\
.. \quad\quad\quad E \\
. . \\
. . \\
n \longrightarrow e_n
\end{array}
$$

There is another feature of the mechanical frame of reference that requires comment. It is essential to specify just what the antecedent and consequent conditions are in any sequence of events. Two ways of specifying objects and environments that have been distinguished especially concern us here. One is the physical classification and the other is the morphological. According to Churchman and Ackoff,

> In physical classification we group objects or events (behavior patterns) in terms of a quantified property expressed along a physical scale, such as temperature, weight, wave length, velocity, etc. Morphological classification, on the other hand, is a non-physical method of classifying. It may take either a quantitative or qualitative form. When quantitative, the scientist chooses a range along a physical scale as a criterion for membership in the class. . . . We may also have qualitative morphologies, as when we specify "red objects," or "heavy objects," or "rapid walking."

The difference between physical and morphological classification is not that between precision and vagueness, but is the difference between the specification of a unique value along a scale and the specification of a range along a scale. The limits of any given range may be very precisely stated.

The ideal use of the mechanical frame of reference requires at least the physical classification of antecedent and consequent conditions. Otherwise it would be extremely difficult to specify the necessary and sufficient conditions for some occurrence. In our illustration, let us for the moment assume that a does not occur and b does. Under these conditions e_2 would appear. But if e_1 and e_2 are not distinguished on the basis of a physical classification, then e_1 might be confused with e_2 and a would seemingly fail as a necessary condition for e_1.

Means–ends. The teleological frame of reference, as developed in this paper, is concerned with the notion of means–ends. Therefore the problem to which we now turn our attention is that of explicating this notion. We want to be able to answer the following question: Under what conditions may some specified occurrence be considered as a means to some end? In terms of our illustration, we shall want to be able to answer the question as to when a, b, c, or d may be treated as means to an end. Let us begin by considering e_1, e_2, e_3, and e_4 as falling within the same morphologically defined class, the range of which is at least extensive enough to include these four elements. Let us assume further that if at least one—and it makes no difference which one or ones—of these possible occurrences obtains, then E obtains. "E" is the name of the state of affairs that obtains when at least one of the four events (e_1, e_2, e_3, e_4) occurs. More precisely, "*E*" is a variable which takes names of states as instances of substitution. We require a name for such states of affairs in order to avoid awkward circumlocution. The question now arises as to the relationship between *a*, for instance, and *E*. It can be seen that *a* is not a necessary condition for *E* because even though *a* does not occur, *E* may occur through the occurrence of *b*, *c*, or *d*. However, *a* is the sufficient condition for *E*. That is to say, if *a* occurs, e_1 occurs and thus *E* obtains. What this analysis of the teleological frame of reference provides is an explication of the notion of means, and alternative means, to some specified end. The relationship of *a* to *E* is that of *a* being a means to *E*. Of course *b*, *c*, and *d* are also means to *E*, that is, means to the same end. The four alternatives can be described as potential means.

In comparing the two frames of reference, there are several important differences to be noted. First, in using the teleological frame of reference, it is permissible, although not necessary, to make a morphological classification of consequent conditions. It is not essential to be able to differentiate between, for instance, e_1 and e_2 since no matter which one occurs, *E* obtains. The successful employment of the mechanical frame of reference requires that both the antecedent and consequent conditions be classified at least physically. Second, in the mechanical frame of reference, the antecedent state of affairs is both a necessary and sufficient condition for the consequent state. In the teleological frame of reference, any given antecedent state is not a necessary condition for the achievement of the consequent condition, namely, *E* in our example, although it is a sufficient condition.

Functional class. At this point it is necessary to make one further distinction. In order to do this we introduce the notion of a functional class. Since *a*, *b*, *c*, or *d* have been treated as means or potential means to *E*, they can be thought of collectively as members of the same class. The criterion for membership in this class is that of being a potential means to some specified end. Thus, *a*, *b*, *c*, or *d* may differ greatly physically and/or morphologically and yet have the property of being means to some specified end, and thus collectively constitute a functional class.

Reformulations of "Teleology"

Our object, in setting forth some of the essential features of the teleological frame of reference, is to enable us to make clear in just what way learning theory of the reinforcement variety *does* implicitly make use of this frame of reference. Reinforcement theorists seek to avoid any use of teleological concepts. In fairness to their position, it should be pointed out that they are opponents of that doctrine of teleology which leads directly to vitalism and emergentism, and which involves the introjection of an entelechy of purpose into the organism itself. Hull has suggested a way to avoid such subjective concepts. "This is to regard, from time to time, the behaving organism as a completely self-maintaining robot, constructed of materials as unlike ourselves as may be." He believes that the tendency on the part of the observer to impute an entelechy, soul, spirit, or demon into such a robot is less likely than if the scientist were observing a living organism. He also states that the robot concept is an effective prophylaxis against the tendency to reify a behavior function. "To reify a function is to give it a name and presently to consider that the name represents a thing, and finally to believe that the thing so named somehow *explains* the performance of the function."

We, of course, are no more anxious than Hull to reify functions or to introject entelechies into men or robots. It would seem, however, that the scientists who have been most intimately connected with the theory and construction of servomechanisms and sequence-controlled calculators have found it profitable to make extensive use of teleological concepts. Perhaps it would be more accurate to say that they have divested these concepts of their demoniacal connotations and rendered them usable for scientific purposes. Wiener would characterize Hull's self-maintaining robot as active, purposive, teleological, and as capable of high-order predictions. One might assume that Wiener has simply repeated the ancient fallacy of introjecting purpose into the machine itself. Quite to the contrary, his analysis —behavioristic in character—concerns only the input and output of the machine. He says, "Given any object, relatively abstracted from its surroundings for study, the behavioristic approach consists in the examination of the output of the object and the relations of the output to input." To say that a machine is purposive in Wiener's sense is not to say that it has some mysterious entelechy within it. The actions of the machine may be assessed with respect to their being means toward some end. In so far as these actions are thus describable they may be said to be purposive.

The cybernetic definitions of teleological concepts have been framed with reference to the problems of interpreting the behavior of certain classes of machines and neural systems. This has resulted in a conceptual model which is workable for the problems of cybernetics. However, the data of current learning theory requires certain modifications of these concepts. Churchman and Ackoff have recently framed a more widely applicable set of definitions which we shall draw upon for our present purposes. Wiener develops the idea that an object behaves purposively only if it pursues the same goal by modifying its behavior under varying conditions. The actions of a thermostat, for example, may be viewed as purposive in that the thermostat turns on the furnace when the room temperature drops below a critical point, and when the temperature goes above this critical point, it turns off the furnace. The end or goal in this example is the maintaining of the temperature at a critical point, and the actions of the thermostat may be viewed as a means of achieving that end in a changing environment. Let us, however, take for an example a naïve hungry rat placed in a Skinner box. This animal may perform a variety of responses and may, after a time, obtain food. But because the conditions in the box have remained constant, we cannot consider the rat's behavior purposive under the Wiener definition. Thus we are in the uncomfortable position of having to maintain that behavior of the thermostat is purposive, but the behavior of the rat is not.

To avoid this difficulty, we have worked out (following Churchman and Ackoff) certain teleological categories: (a) Extensive Function, (b) Intensive Function, and (c) Purposive Function. When using the teleological frame of reference one may classify any behavior as belonging to one of these categories. (a) If X (any object) accomplishes some specified end by displaying relatively invariant behavior in a wide range of environments, then X may be said to have an extensive function. (b) If X accomplishes some specified end by changing its behavior if and only if the environment changes, but exhibits only one type of behavior in any given environment, then X may be said to have an intensive function. (c) If X accomplishes some specified end by exhibiting different types of behavior, whether the environment changes or remains the same, then the actions of X may be said to be purposive. According to this schema, the behavior of the thermostat would be classified as having an intensive function; the behavior of the rat in the problem box as purposive.

Hullian Theory and "Purpose"

Let us turn our attention now more particularly to the Hullian formulation of learning theory. We wish to show that this theory makes essential use of the concept of purpose as just explained. In order to make our case, we must be very clear about *essential use*. To be precise about this extremely important point we must make the distinction between a set of lawlike statements and what they are about. Any theoretician has at least a twofold task: (a) He must classify the objects, events, or situations which are to be the subject matter of his laws, and this implies that he have a principle of classification. (b) He must formulate a set of lawlike statements, which have as their subject matter the objects, events, or situations that he has classified, and these statements should systematically relate parts of the classified subject matter to other parts. Ideally, these lawlike statements will enable him to predict the occurrence of the classified happenings, and thus will acquire the status of laws. The learning theorist can properly be said to make essential use of teleological concepts if these concepts are employed either in his classificatory system or in his laws.

Hullian theorists have devoted their attention almost exclusively to the formulation of testable laws. But we have already noted that the theoretician has a twofold task and that he must do more than formulate laws. He must also classify the objects, events, or situations which will be the subject matter of his laws, and he must make clear the principle of this classification. It is this second task that the Hullians have neglected. Apparently they are unaware of what principle of classification they have used, and it is partly because they have neglected making clear this principle that the controversy about "teleology" exists.

Our point is that the Hullians have been using a classificatory principle, and that it is a teleological one of the purposive variety. Now that we have made this assertion, the question arises as to how it can be proved or disproved; what sort of evidence would be required to confirm or disconfirm it? Quite obviously we should turn to the actual work of the learning theorist. Let us take Hull's book *Principles of Behavior* for illustration. We find that Hull, on the one hand, presents a number of experiments, and, on the other, he formulates a set of laws to explain what happens in these experiments. The question is: On what basis does he select these experiments as subject matter for his laws? We submit that each and every one of the experiments cited in his book involves a situation in which the behavior of the organism is assessed with respect to its efficacy in accomplishing certain ends.

We find, for example, that Demonstration Experiment A on page 70 is concerned with a rat placed on a grid which is subsequently electrified. Hull points out that the rat displays a variety of responses, one of which, the leaping over a barrier, removes it from the electrified grid. Hull classifies the responses of the rat into two categories, the futile and nonfutile. What can Hull possibly mean by characterizing a response as futile? It is not a physical or morphological property of a response to be futile or nonfutile, and the notion of futility or nonfutility is certainly not part of the mechanical frame of reference. We believe that what he means, or should mean, is that certain of the responses of the rat, when placed on an electrified grid, are means to a specified end, and other responses are not. The end which Hull selects is need reduction. The rat's leaping over a barrier is a means to the end, need reduction. There is a further employment of the teleological frame of reference by Hull in this experiment. In a series of trials, the rat makes a number of escapes from the charged grid. Hull pays no attention to the exact location from which the rat leaps or to the specific musculature involved (for the rat may leap from varying stances) when he describes the rat's escape. If he were making a purely mechanical analysis, he would, of course, have to do this very precisely. Nor does he pay attention to the exact spot where the rat lands on the other side of the barrier, and this is something else that would have to be done precisely in a purely mechanical analysis. Let us assume that the rat may leap from any one of four physically different locations, and land on any of four physically different spots across the barrier.

Location a_1 ⟶ Spot e_1
Location a_2 ⟶ Spot e_2
Location a_3 ⟶ Spot e_3 E
Location a_4 ⟶ Spot e_4

Hull is perfectly willing to count a landing at any one of these four spots as constituting an escape from the grid. What this means, then, is that he is interested in the relationship between a_1, a_2, a_3, a_4, which can be classified morphologically, and E. These four starting locations, and the appropriate jumping reactions, constitute members of the same functional class, any one of which is a means to the same end, namely, escape from the grid. The behavior of the rat on the electrified grid may be categorized as purposive. The rat accomplishes some specified end by exhibiting different types of behavior whether the environment changes or remains the same.

In "Demonstration Experiment C" (p. 75) Hull cites a conditioned-reflex learning experiment in which a dog learns to lift its foot in response to a buzzer, with shock serving as the unconditioned stimulus. Hull says, "No doubt the dog makes many other muscular contractions in addition to those which result in the lifting of the foot, but these are usually neglected in such experiments." Why, we might ask, are they neglected? It is quite evident that learning theorists are not merely interested in a purely mechanical analysis of what happens when an organism is subjected to certain stimuli. Rather, they wish to construct a set of laws that will enable them to predict the occurrence of those responses that are relevant to the achievement of a specified goal.

In the short space available we cannot cite all the experiments performed by reinforcement theorists, or even all of those mentioned in Hull's book. We are confident, however, that an examination of the experiments performed by reinforcement theorists will reveal the same properties. What these experiments have in common is that they deal with objects that can exhibit different types of behavior whether the environment changes or remains the same. And the experimenter singles out of the myriad of responses that might be investigated just those that are relevant to the achievement of some specified end. What the experiments have in common, then, is the use of the teleological frame of reference, and even within that frame of reference, the use of the special category of purpose. For example, the experimenters do not treat behavior in terms of the categories of either extensive function or intensive function. The experiments are not performed on objects that display relatively invariant behavior in a wide range of environments, as do clocks, for example. Nor do they involve organisms or machines that exhibit different behaviors in different environments, but only one type of behavior in any given environment, as do amoeba and some servomechanisms. It should be clear at this point, then, that the learning theorists are making systematic use, as any theorist must, of some principle of classification in selecting subject matter about which to formulate laws. The Hullian learning theorists have never made their principle of classification clear.

Tolmanian Theory and "Purpose"

It may seem that the position that has been set forth in this paper is equivalent to the one Tolman and his followers have long urged. Certainly Tolman has talked about the importance of purpose, ends—means relationships, goals, and has even said that "behavior as behavior reeks of purpose and of cognition." However, Tolman's use of teleological concepts is quite different from ours. To the Tolmanian, purpose is something that the animal has within it, and that an independent, neutral observer ascribes to the organism as a result of an analysis of its behavior. According to this view, we can speak, upon observing a cat in a problem box, of the "cat's purpose of getting to the outside, by bursting through the confinement of the box," or of the "cat's purpose as a determinant of the cat's behavior." To Tolman "The rat accepts or rejects and persists to or from food, blind-alleys, true path sections, electric grills, etc., only in so far as these latter function for him as subordinate goals. . . ." In other words, goals are goals *to* the rat, and purposes are purposes *of* the rat. In our analysis, *purpose* is a function of the classificatory system employed by the investigating scientist; it is never some faculty inside the organism.

The Tolmanian analysis differs from our own in another important respect. In his system the distinction between the classificatory schema employed to select the subject matter for the laws and the laws of the system themselves is obscured. It seems to us that Tolman uses the word "purpose" both as the name of a phenomenon to be explained and as an explanatory device. This in itself is a procedure of dubious scientific value. Ultimately it is a viciously circular procedure. Moreover, it makes it extremely unlikely that a scientist can successfully carry out the twofold task that we have delineated above.

The learning theory developed by the reinforcement theorists, in our opinion, is not guilty of this confusion. It is, however, beyond the scope of the present paper to examine the relationship between the Hullian laws and the implicit Hullian classificatory system in detail.

In summary, many critics have charged that reinforcement theory is defective because it does not take purposive behavior into consideration, and proponents of reinforcement theory have admitted the absence of teleological elements and have taken this absence as an index of the scientific character of their theory. Following the lead of certain cyberneticians and philosophers of science, we have worked out a new formulation of the teleological frame of reference which avoids metaphysical entanglement, and have shown that the reinforcement theorists have implicitly been making use of this frame of reference. The Hullian laws are set up to explain the conditions under which an organism learns. From the implicit Hullian point of view, to say that an organism has learned something is to assert that there is a high probability that the organism will exhibit responses that are a means to the end need reduction. Thus we have seen that the concept of learning itself is one that can be meaningfully used only within the teleological frame of reference.

B.

HOMEOSTASIS AND EVOLUTION

33.
Self-Regulation of the Body

WALTER B. CANNON

I

OUR BODIES ARE MADE of extraordinarily unstable material. Pulses of energy, so minute that very delicate methods are required to measure them, course along our nerves. On reaching muscles they find there a substance so delicately sensitive to slight disturbance that, like an explosive touched off by a fuse, it may discharge in a powerful movement. Our sense organs are responsive to almost incredibly minute stimulations. Only recently have men been able to make apparatus which could even approach the sensitiveness of our organs of hearing. The sensory surface in the nose is affected by vanillin, 1 part by weight in 10,000,000 parts of air, and by mercaptan 1/23,000,000 of a milligram in a liter (approximately a quart) of air. And as for sight, there is evidence that the eye is sensitive to 5/1,000,000,000,000 erg, an amount of energy, according to Bayliss, which is 1/3,000 that required to affect the most rapid photograph plate.

The instability of bodily structure is shown also by its quick change when conditions are altered. For example, we are all aware of the sudden stoppage of action in parts of the brain, accompanied by fainting and loss of consciousness, that occurs when there is a momentary check in the blood flow through its vessels. We know that if the blood supply to the brain wholly ceases for so short a time as seven or eight minutes

certain cells which are necessary for intelligent action are so seriously damaged that they do not recover. Indeed, the high degree of instability of the matter of which we are composed explains why drowning, gas poisoning, or electric shock promptly brings on death. Examination of the body after such an accident may reveal no perceptible injury that would adequately explain the total disappearance of all the usual activities. Pathetic hope may rise that this apparently normal and natural form could be stirred to life again. But there are subtle changes in the readily mutable stuff of the human organism which prevent, in these conditions, any return of vital processes.

When we consider the extreme instability of our bodily structure, its readiness for disturbance by the slightest application of external forces and the rapid onset of its decomposition as soon as favoring circumstances are withdrawn, its persistence through many decades seems almost miraculous. The wonder increases when we realize that the system is open, engaging in free exchange with the outer world, and that the structure itself is not permanent but is being continuously broken down by the wear and tear of action, and as continuously built up again by processes of repair.

II

The ability of living beings to maintain their own constancy has long impressed biologists. The idea that disease is cured by natural powers, by a *vis medicatrix naturae*, an idea which was held by Hippocrates (460–377 B.C.), implies the existence of agencies which are ready to operate correctively when the normal state of the organism is upset.

Reprinted from *The Wisdom of the Body* by Walter B. Cannon, by permission of W. W. Norton & Company, Inc. Revised and enlarged edition copyright 1939 by Walter B. Cannon. Copyright renewed 1960 by Cornelia J. Cannon.

More precise references to self-regulatory arrangements are found in the writings of modern physiologists. Thus the German physiologist, Pflüger, recognized the natural adjustments which lead toward the maintenance of a steady state of organisms when (1877) he laid down the dictum, "The cause of every need of a living being is also the cause of the satisfaction of the need." Similarly, the Belgian physiologist, Léon Fredericq, in 1885, declared, "The living being is an agency of such sort that each disturbing influence induces by itself the calling forth of compensatory activity to neutralize or repair the disturbance. The higher in the scale of living beings, the more numerous, the more perfect and the more complicated do these regulatory agencies become. They tend to free the organism completely from the unfavorable influences and changes occurring in the environment." Again, in 1900, the French physiologist, Charles Richet, emphasized the remarkable fact. "The living being is stable," he wrote. "It must be so in order not to be destroyed, dissolved or disintegrated by the colossal forces, often adverse, which surround it. By an apparent contradiction it maintains its stability only if it is excitable and capable of modifying itself according to external stimuli and adjusting its response to the stimulation. In a sense it is stable because it is modifiable —the slight instability is the necessary condition for the true stability of the organism."

Here, then, is a striking phenomenon. Organisms, composed of material which is characterized by the utmost inconstancy and unsteadiness, have somehow learned the methods of maintaining constancy and keeping steady in the presence of conditions which might reasonably be expected to prove profoundly disturbing. For a short time men may be exposed to dry heat at 115 to 128 degrees Centigrade (239 to 261 degrees Fahrenheit) without an increase of their body temperature above normal. On the other hand arctic mammals, when exposed to cold as low as 35 degrees Centigrade below freezing (31 degrees below zero Fahrenheit) do not manifest any noteworthy fall of body temperature. Furthermore, in regions where the air is extremely dry the inhabitants have little difficulty in retaining their body fluids. And in these days of high ventures in mountain climbing and in airplanes human beings may be surrounded by a greatly reduced pressure of oxygen in the air without showing serious effects of oxygen want.

Resistance to changes which might be induced by external circumstances is not the only evidence of adaptive stabilizing arrangements. There is also resistance to disturbances from within. For example, the heat produced in maximal muscular effort, continued for twenty minutes, would be so great that, if it were not promptly dissipated, it would cause some of the albuminous substances of the body to become stiff, like a hard-boiled egg. Again, continuous and extreme muscular exertion is accompanied by the production of so much lactic acid (the acid of sour milk) in the working muscles that within a short period it would neutralize all the alkali contained in the blood, if other agencies did not appear and prevent that disaster. In short, well-equipped organisms—for instance, mammalian forms—may be confronted by dangerous conditions in the outer world and by equally dangerous possibilities within the body, and yet they continue to live and carry on their functions with relatively little disturbance.

III

The statement was made above that somehow the unstable stuff of which we are composed had learned the trick of maintaining stability. As we shall see, the use of the word "learned" is not unwarranted. The perfection of the process of holding a stable state in spite of extensive shifts of outer circumstance is not a special gift bestowed upon the highest organisms but is the consequence of a gradual evolution. In the eons of time during which animals have developed on the earth probably many ways of protecting against the forces of the environment have been tried. Organisms have had large and varied experience in testing different devices for preserving stability in the face of agencies which are potent to upset and destroy it. As the construction of these organisms has become more and more complex and more and more sensitively poised, the need for more efficient stabilizing arrangements has become more imperative. Lower animals, which have not yet achieved the degree of control of stabilization seen in the more highly evolved forms, are limited in their activities and handicapped in the struggle for existence. Thus the frog, as a representative amphibian, has not acquired the means of preventing free evaporation of water from his body, nor has he an effective regulation of his temperature. In consequence he soon dries up if he leaves his home pool, and when cold weather comes he must sink to its muddy bottom and spend the winter in sluggish numbness. The reptiles, slightly more highly evolved, have developed protection against rapid loss of water and are therefore not confined in their movements to the neighborhood of pools and streams; indeed, they may be found as inhabitants of arid deserts. But they, like the amphibians, are "cold-blooded" animals, that is, they have approximately the temperature of their surroundings, and therefore during the winter months they must surrender their active existence. Only among the higher vertebrates, the birds and mammals, has there been acquired that freedom from the limitations imposed by cold that permits activity even though the rigors of winter may be severe.

The constant conditions which are maintained in the body might be termed *equilibria*. That word, however, has come to have fairly exact meaning as applied to relatively simple physico-chemical states, in closed systems, where known forces are balanced. The coördinated physiological processes which maintain most of the steady states in the organism are so complex and so peculiar to living beings—involving, as they may, the brain and nerves, the heart, lungs, kidneys and spleen, all working coöperatively—that I have suggested a special designation for these states, *homeostasis*. The word does not imply something set and immobile, a stagnation. It means a condition—a condition which may vary, but which is relatively constant.

It seems not impossible that the means employed by the more highly evolved animals for preserving uniform and stable their internal economy (i.e., for preserving homeostasis) may present some general principles for the establishment, regulation and control of steady states, that would be suggestive for other kinds of organization—even social and industrial—which suffer from distressing perturbations. Perhaps a comparative study would show that every complex organization must have more or less effective self-righting adjustments in order to prevent a check on its functions or a rapid disintegration of its parts when it is subjected to stress. And it may be that an examination of the self-righting methods employed in the more complex living beings may offer hints for improving and perfecting the methods which still operate inefficiently and unsatisfactorily. At present these suggestions are necessarily vague and indefinite. They are offered here in order that the reader, as he continues into the concrete and detailed account of the modes of assuring steady states in our bodies, may be aware of the possibly useful nature of the examples which they offer. . . .

34.
On the Parallel between Learning and Evolution

J. W. S. PRINGLE

Introduction

Thorpe has recently reviewed the types of learning which have been observed in the animal kingdom. The apparent diversity of the processes involved has hitherto made it difficult for workers in other fields to appreciate the general problem, and Thorpe's classification of the observed phenomena under five headings—habituation, conditioning, trial-and-error learning, insight learning and imprinting—has for the first time clearly defined the range which must be covered in discussions of learning in animals. This is particularly useful at the present moment, since physiological work on the higher centres of the nervous system and studies of the fundamental nature of biochemical processes in living organisms have also made great strides in recent years, and the time now seems ripe for a synthesis of these different viewpoints in an attempt to elucidate the nature of the learning process. This article presents such an attempt, based on what may be called "kinematic principles". The approach is a novel one and borrows more from the methodology of theoretical physics than from the established techniques of experimental biology. It is hoped, however, that such a theoretical treatment of the problem may be of assistance in the interpretation of experimental results and in the planning of new observations.

From J. W. S. Pringle, "On the Parallel between Learning and Evolution," *Behaviour*, 3 (1951), 174–215. Reprinted by permission of the author and E. J. Brill Ltd., Publishers, Leiden.

The Concept of Complexity

The important generalizations of physical and chemical kinematics have arisen from a study of systems under equilibrium conditions. Such laws as the Principle of Conservation of Energy (First Law of Thermodynamics) and the Principle of the Increase of Entropy (Second Law of Thermodynamics) apply only to systems which are isolated in the sense that energy does not cross the boundary of the system, and which in this isolated condition are left to attain a stable state under the prescribed conditions. The Second Law, in particular, cannot be applied to systems which are capable of exchanging energy with the rest of the universe, and since living organisms are essentially of this nature and continue to show the properties associated with life only while these energy exchanges are allowed to continue, the usefulness of the Second Law in a consideration of biological processes has even been questioned. Schrödinger, in a discussion of this point, suggests that it is helpful to regard living creatures as machines which feed on negative entropy; that is to say, they are continually absorbing from the environment food containing molecules with a high degree of order and excreting other molecules of a lower degree of order. There is clearly nothing exclusive about this process of entropy decrease which could be used as a definition of the living state, for it is shared by growing crystals, by the process of polymerization and by many other non-living systems, but it remains a useful statement of one aspect of the machinery of the living organism by which its structure is preserved against the randomizing forces of the natural world.

The characteristic of living systems which distinguishes them most clearly from the non-living is their property of progressing by the process which is called evolution from less to more complex states of organization. The concept of complexity is not one which has yet been subjected to accurate treatment in biology, although it has been widely used, without definition, by biologists and philosophers. The key to its meaning comes from recent studies of the nature of "information" which have developed from problems of communication engineering. MacKay, in a paper which presents the new ideas in a general form which is applicable to all branches of science, points out that all scientific statements conveying information are ultimately analysable into two components; an "*a priori*" or structural aspect associated with the number of independent parameters to which the statement refers, and an "*a posteriori*" or metrical aspect, there being associated with each structural element a numerical quantity measuring the amount of credibility to be associated with that aspect of the statement. The amount of this "structural" information represents what is usually meant by the complexity of a statement about a system; it might alternatively be defined as the number of parameters needed to define it fully in space and time. In communication theory the term "logon" has been suggested for the unit of structural information, by contrast to the "metron", the unit of metrical information, which defines the credibility of each logon of the statement.

It must be emphasized that this representation of complexity is essentially epistemological; the measure of complexity is of the statement about the system and is not the complexity of the system itself, an expression which has no scientifically discoverable meaning. The information which can be obtained about a system, be it a living organism or an inanimate system, is limited only by the effort expended in obtaining that information. But this need not detract from the value of the concept. The characteristic of living systems which is called evolution is the change from less to more complex states of organization. It is the ordered relationships between parts of the system, that is, the maintenance of constancy of certain features of the dynamic equilibrium, which enable us to define the information obtainable about it, since we commonly ignore, in scientific observations, those features which are not repeatedly verifiable to a certain accuracy. This justifies the use which will be made of the concept of complexity in this article: it refers only to those features of a system which are ordered and so are capable of verification.

Thus defined, complexity and order are two independent features of an observed system. Order, which is negative entropy, may increase without increase in complexity: and complexity might increase without change of entropy. An example will clarify this point. When a crystal of common salt grows regularly to twice its previous size, its entropy decreases; that is, it has increased its order. But its complexity is the same; verifiable observations on it may be described by a statement involving the same number of parameters as before. If, on the other hand, there is present in the solution an atom of a sodium isotope and this atom becomes incorporated into the crystal, the complexity of theoretically verifiable observations on the system at once increases, since the position of the isotope atom in relation to the rest of the crystal demands more parameters for its full description. If, now, the crystal is repeatedly re-dissolved and a new crystal formed, the isotope atom is unlikely to occupy the same position on each occasion; treating the successive appearances of the crystal as a single system, the additional information about the position of the isotope atom is not verifiable. This system cannot therefore be said to have evolved, since no new ordered relationships are established. Nothing can be stated about the incorporation of the isotope atom at a particular point in the crystal except that it is improbable, and no progressive change can be observed in the system.

Complexity in Biological Systems

By contrast with non-living systems whose complexity remains constant or decreases, biological systems may show a progressive increase in complexity such, for example, as occurs in evolution. It is here suggested that this progressive increase in complexity in living systems takes place through the agency of a particular type of process which occurs in living organisms and which confers on them the most characteristic property of life. The parallel between learning and evolution arises as a corollary to this principle since it can be shown that both in evolution and in learning there is an increase in the complexity of the organism; a similar type of process may therefore be at work in the two cases. The value of the parallel is that, since we know something about the process of evolution, we can thereby limit the range of our search for the processes involved in learning.

The type of process through which an increase in complexity occurs in living organisms is known in one form as the natural selection of random variations, widely accepted as the chief mechanism of evolution. To bring it into its correct relationship to the second law of thermodynamics, this process should more accurately be described as the selection of improbable variations, the term "improbable" being used in its statistical sense. The equilibration of a heterogeneous mixture of gases might be said to involve a selection of random variations, but in this case it is the probable variations that are "selected" in the sense that they occur more often than the improbable, and the system rapidly reaches a statistical equilibrium. The selection which occurs in evolution is different from this. A means is present of preserving in the system the effect of certain transformations which are improbable and of ignoring the effect of other transformations, including the reverse transformation to that which is preserved.

Such a preservation of improbable states cannot occur in a "closed" system in which a stable end-state equilibrium is reached[1]. But there is another type of equilibrium possible in the quantity of some unit in a system, namely, the steady-state, in which the rate of formation is equal to the rate of destruction of the unit under consideration. This type of equilibrium demands a source capable of supplying energy to the system and a sink capable of accepting energy from it, and it cannot therefore occur in "closed" systems. It can nevertheless be analysed. If energy is supplied from outside, the amount of some ordered component in the system can increase (e.g. in crystallization) and if there is also a continual removal of this component (e.g. in the case of the crystals by sedimentation) the amount present may remain constant. If a particular region is considered as the "system" for the purpose of study and all the rest as the "environment", the quantity of the unit in the system remains constant when formation and removal occur at equal rates, and a steady-state equilibrium is established.

The peculiar properties of certain "open" systems of this type in relation to the second law of thermodynamics arise from the fact that the two rate processes which are balanced at the steady-state equilibrium may be related in different ways to the quantity of the unit present; i.e. the system may be non-linear. The rate of increase of the unit is often nearly exponential; the rate of formation is proportional to the quantity already formed. The rate of destruction or removal, on the other hand, may depend on some other function (e.g. a higher power) of the quantity present. One example of a non-linear steady-state equilibrium which has found application to population studies is the Verhulst-Pearl logistic equation

$$\frac{dx}{dt} = ax - bx^2$$

but there are many other more complicated equations defining systems with similar properties. In all of these, if either or both the rate of formation and/or the rate of destruction are subject to variation, and if there is competition between the units composing the system for the energy supplies or for the conditions determining the rate of destruction, the steady-state equilibrium is altered, since it does not depend symmetrically on the variation each side of the mean in the rates of formation and destruction. Such a system, in fact, selects the more rapid rates of formation even if the more and less rapid rates occur in each process with equal probability. The probability of occurrence of the more rapidly forming or less rapidly disappearing units can, indeed, be very low, and they will nevertheless persist in the steady-state equilibrium.

The "open" system we are considering, then, has the following properties:

1. It is a steady-state equilibrium, each unit appearing to be constant only because it is being formed and destroyed at the same rate.

2. Either or both processes are subject to variation.

3. The rate of formation and the rate of destruction of each unit depend on different functions of the quantity present, and on the environmental conditions.

These features imply a property of the units which may be called "replication"; that is, the presence of a unit in the system causes more similar units to appear, with an accompanying increase in order which is balanced at steady-state equilibrium by the decrease in order involved in their destruction. In some of the special cases which we shall consider, this process of replication presents one of the least well understood features, but in its simplest form (e.g. in crystallization) it is fully describable. It is a manifestation of an increased probability of various sub-units combining in certain way when that combination is already present in the system and the conditions in general are such that combination tends to occur. Once a combination of sub-units has occurred in a particular way, all other types of combination are less probable, under those environmental conditions. Other combinations of sub-units may, however, occur, unless replication is perfect, and some of these other units may themselves possess the property of replication, so that, once formed, they also increase. If the process of growth continues unchecked, all units which have appeared and which possess the property of replication will be represented; but in a steady-state equilibrium in which there is also a process of destruction occurring, only those will persist which have a more rapid rate of formation or a lower rate of destruction under the existing environmental conditions. These two rate factors are so much more important than the initial probability of formation of the new units in determining the composition of the steady-state population, that a very improbable unit can survive or replace previously existing units when all are competing for the available energy supplies.

The state of such a system cannot be fully described unless the range of variation is confined within known limits. If the limits of variation are infinite there is always a finite possibility of a new unit arising whose rate of formation is higher or whose rate of destruction is lower than the original, so that it causes a change in the steady-state equilibrium. The system therefore contains within itself the possibility of change due to irregularities in the replication process.

Complexity in such a system can increase in two ways. If the system consists of several parts with no exchange of units between the parts, an improbable event may persist in one part and not in another simply because it has not occurred in all of them. The environment of all the parts can in this case be the same. Alternatively, if there are slight differences in the environments of the different parts of the system, a new unit which has arisen in all of them may survive only in some, producing thereby a sort of formal representation of the environmental conditions of each part of the system. If the occurrence of new units is sufficiently rapid, a change in the local environment of one part may, in this case, appear to be the cause of the change. If the appearance of new units is a rare event, the new complexity may not appear at all even when the environment becomes heterogeneous, although conditions are favourable for the increase in complexity to occur.

The possibilities of "open" systems in general are difficult to visualize without mathematical treatment, and the mathematics, which is based on the theory of stochastic processes, may become extremely involved. It is clear, however, that progressive changes can occur in such systems which are fundamentally different in nature to those occurring in "closed" systems in true equilibrium. It is proposed in this paper to use the term "evolutionary" to describe the general type of process which leads to an increase in complexity in such systems. Some further characteristics of the process will become clear from examples.

Evolution by Variation and Selection

Organic evolution by natural selection is clearly the first example which must be considered. Living organisms possess the property of replication as a whole (in this case more usually referred to as reproduction). They also show variation; that is, the reproductive process is imperfect, and organisms arise from time to time which do not exactly resemble their progenitors. The new types of organisms which arise in this way are known as mutations; their occurrence is improbable in the statistical sense, and yet they may be preserved in a population by the process of natural selection. The mathematics of the fate of a mutation in a population has been worked out by Fisher, and the process is widely accepted as the chief, or at least one of the chief ways in which living organisms progress by evolution from less to more complex states of organization. The new organisms which are preserved are those which, under the particular conditions, show either a higher rate of replication (reproduction) or a lower rate of destruction (greater individual life-time). Fisher has shown that the chance of survival of a new gene is dependent both on the selection factor and on the rate of occurrence of the mutation, but the mathematics shows clearly that in a large population a very low rate of mutation may, with quite a small selection factor, be adequate to ensure survival.

In higher animals and plants the process of sexual reproduction introduces a complicating factor into the calculations, since a reshuffling of the units of heredity (genes) provides additional variation without the occurrence of mutation. The complication of sexual reproduction is not, however, a necessary feature of an "evolutionary" process as defined above. Provided that the property of replication exists, any system composed of interacting (competing) units whose quantity is determined by the steady-state equilibrium between an approximately exponential process of formation and a process of destruction will "evolve", if the replication is imperfect and occasionally produces units which have either a higher rate of replication or a greater individual life-time. Only systems with this general structure are capable of evolving to greater complexity.

Systems with the Property of Evolution

Within the broad field of biological studies systems with the structure which has been outlined above have been recognized up to the present time only in the field of organic evolution and in the process of differentiation within the body of the individual organism. This second example is of considerable theoretical interest, for it is sometimes assumed that the whole of the complexity of the adult form is contained in the material substance of its inheritance. Quite apart from the direct experimental evidence for the self-duplicating property of "plasmagenes" controlling enzyme formation in the living body, the adaptation of the individual in the course of its ontogeny strongly suggests that an "evolutionary" process is at work in the processes of growth and differentiation. Spiegelman has discussed at length the treatment of differentiation as the problem of the evolution of a population of plasmagenes competing for the available material and energy supplies, and a repetition of his arguments would be out of place in this article, the object of which is to suggest a third example of the same process. We must now turn to examine the types of learning shown by animals and enquire if there are any features which demand the existence here also of an "evolutionary" process.

Does Learning Involve an Increase in Complexity?

Learning is the name given to the general class of processes by which the behaviour of an animal comes to depend not only on the environmental changes immediately preceding it in time, but also on events which have occurred in the related parts of the environment in the more remote past. Students of animal behaviour distinguish this past into two parts; that period of past time during which the animal has existed as an independent organism, and the rest of past time during which its ancestors have existed. The modification of present behaviour by past events is called learning when those events have occurred during the life-time of the individual, and instinct when the events have occurred outside this period of time.

In the former case the past events are supposed to have left some trace in the animal (usually in its nervous system), whereas in the latter case the organisation of the animal responsible for the observed response is supposed to be innate; that is, to be contained fully in the material substance of its inheritance. This distinction is theoretically clear-cut, although great difficulty has often been experienced in assigning to one or other case the particular aspect of behaviour which is the subject for study. We owe to Lorenz, Tinbergen and others most of what is now accepted about the nature of instinctive behaviour, and enough is known to enable us to recognize it in many different animals. The processes classed under the term learning, however, are of such a varied nature that it is not at once easy to identify the process, and it will be as well to summarize Thorpe's classification in order to provide a background for the subsequent discussion.

As Thorpe points out, one of the most elementary forms of learning shown by animals is that called by Humphrey "habituation". The animal at first reacts to a given stimulus in a particular way, and if the stimulus is repeated a sufficient number of times without the response leading to any further responses, it becomes in time inadequate to provoke that response. Thereafter on presentation of that particular stimulus, the response may or may not occur, depending in the main on the time interval since it last occurred.

The next most easily defined is the type of learning known as the conditioned reflex. Two varieties of conditioned reflex have been distinguished by Konorsky. The essential of Type I is that the proximity in time of two patterns of sensory stimulation leads to a change in the nervous system such that a response which was elicited by one sensory stimulus becomes elicitable by the other. The response is in all cases of the nature of an instinctive response in the sense in which the term is used by Lorenz; that is, its detailed form is largely innate and it is merely released by the stimulus and not fully determined by it. In conditioned reflex Type II a new response becomes associated with the original stimulus, instead of a new stimulus with the original response. The difference between these two types has appeared to Thorpe to be fundamental, but it is convenient for the present to consider them together.

The third type of learning is classed by Thorpe as trial-and-error learning. Essentially, the animal makes more-or-less random movements and selects, in the sense that it subsequently repeats, those which produce the "desired" result; namely, those which lead to the correct conditions for the release of a further response.

The fourth type, insight learning, is that in which the animal, without visible trial-and-error or with a reduced number of errors, performs a new action which leads to the "goal"; that is, again, an action which releases a further chain of responses. It is inherent in the definition of this type of learning that the whole situation with which the animal is presented is one to which it has not responded before either in its own life-time or in its ancestry. Some features of the environmental situation, however, in learning of this type have in all cases been encountered before, possibly on various different occasions, and it is in the ability to combine the memory of these past experiences in its analysis of the new test situation that the peculiarity of insight learn-ing resides.

Thorpe subsumes under the term "insight learning" the "latent learning" of Blodgett. Latent learning is the name given to the process whereby patterns of sensory stimulation affect subsequent behaviour but do not produce an immediate response. Such a process may be regarded as a part of the process of insight learning, for when it has taken place and the further stimulus needed to release an active response is given, the response consists not in random trial-and-error activities but proceeds with the elimination of certain actions which would be rendered ineffective by the situations previously experienced.

The final type of learning in Thorpe's classi-fication, "imprinting", is a specialized process found in birds and possibly in insects, by which the environmental situation at certain critical periods in the animal's lifetime becomes especially effective in producing permanent modifications of subsequent behaviour. Its characteristics will be discussed later.

Complexity in Time

In order to decide whether any or all of these types of the learning process involve an increase in complexity, a short digression is necessary, since we have used the concept hitherto only in relation to structure, and an extension of the idea is needed to justify its application to behaviour, where the idea of complexity in time is involved rather than complexity of form.

The ideas developed in this article make use of the concept of order, the negative of thermo-dynamic entropy, to distinguish between chaotic systems such as a gas, and complex systems such, for example, as a protein molecule. The additional feature present in the latter but not in the former is that the individual units of which it is composed (atoms or amino-acid residues) have some restraint on their relative movements in space; they are held together in the form of a single molecule. Current views on molecular structure do not impose any restraint in time on the movements of the individual units; they are free to occupy any of their possible relative positions at any time, and there is no way of determining which position they occupy at any instant. Thus, theories of muscular contraction involving the coiling of long protein fibres do not specify the state of coiling of any single molecule over a period of time, or of groups of molecules at a given instant of time, but merely the statistical sum of the effects of all the molecules acting over a period of time. It is perfectly possible, theoretically, to introduce order into a system by limiting its freedom in time instead of its freedom in space. The passage of a sound wave through a gas involves such a restraint on the freedom of move-ment of the atoms in time, although it is no more possible to define the spatial relationships between the atoms under such conditions than if the sound waves were not passing. If order, in its widest sense, can mean order in time as well as order in space, then the hypothesis must be extended to include the concept of complexity in time as well as complexity of form, and there should be a process analogous to the selection of random variations which is capable of causing complexity in time to increase in a system in the same way as is possible for complexity of form.

This space–time transposition is not easy to visualize, since, as we saw earlier, the use of the term complexity implies verifiability of observa-tion. Verifiability in this case cannot mean repeat-ability of successive observations, but must imply confirmation by simultaneous observations in different places. Behaviour is thus fully ordered when it consists of an indefinite repetition of the same event, and chaotic when the sequence of events does not contain any element of repetition whatever. The complexity of a description of the behaviour resides in the number of independent rhythms into which it can be analysed, and it relies for its objectivity on the fact that several simultaneous observations yield the same result. For example, we can describe the complexity of the swimming of a fish because simultaneous observations on different parts of the body yield the same rhythmic form. If the fish swam continuously at the same rate its behaviour would be simple; the fact that it starts and stops at once introduces complexity.

The two definitions of complexity, the spatial and the temporal, are, of course, both involved in most descriptions of animal behaviour, since the animal does not react as a whole. But it is the temporal complexity which is mainly of interest here, since it is that concept which is less well understood. An increase in spatial or structural complexity is well known to occur in the evolutionary process.

It seems clear that, defined in this way, complexity increases in each case in which learning can be said to have occurred. The complexity increase resides in the new temporal relationship between two events which arises in the process of learning. We saw in the example from crystallization how the incorporation of an atom of the isotope produced an increase in complexity while the accurate repetition of the original did not. The establishment of an ordered relationship between two events, one of which is new, increases the complexity of the system, in contradistinction to the continuation of a pre-existing relationship. Even in the case of habituation where the final condition may be one in which no response occurs on stimulation, the complexity has increased, since the regularity in time of the stimulus–response relationship has altered. The response having occurred on the first stimulation, it is evident that the condition of the animal at that time was such that a particular temporal relationship held between two events; the final state in which the response does not occur is different from a state in which the response never occurs, since in habituation it has occurred and may recur again in the course of time. The absence of a single event in a sequence is thus a more complex state than its presence. Where the response which has disappeared is an instinctive response it is pertinent that the animal can still transmit to its offspring the capacity for showing that response.

One reference may be quoted from recent literature to show that this problem of complexity in behaviour is being actively considered at the present time, though in other terms. McCulloch, in a discussion of what he calls "teleological mechanisms", outlines briefly the difficulty which has been felt in assigning an origin to the phenomenon of pattern perception. In the terms used in this article, pattern perception means the release of a response or the modification of a subsequent response by a complex ordered state in the environment. McCulloch quotes J. S. Mill for the suggestion that the basis of resemblances for each individual rests on inheritance–associations "in space and time throughout the development of the kind". He refers to Kapper's Law of Neurobiotaxis as another statement of the fact of the phenomenon, though not an explanation. He then states: "It is clear that the same thing happens in the introduction of new ideas or the recognition of new forms within the span of a single life, though obviously the mechanism cannot be the same". It is the thesis of this article that the mechanism, in essentials, is the same, although the substance of which the system is constructed is, in this instance, the substance of the individual central nervous system, instead of the whole world of living organisms and their environment.

The Mechanism of Learning in the Central Nervous System

To justify our hypothesis in its application to the phenomenon of learning it is necessary to suggest a type of mechanism, consistent with what is known about the physiology of the central nervous system, which is capable of performing a selection of variation in time in a manner analogous to the natural selection of variation in form found in organic evolution. Such a process is to be found in the properties of coupled oscillators, and it will now be shown how, from a population of coupled oscillators, a model can be constructed which has the property of "evolution" in the sense in which that term has been used.

The phenomena associated with the coupling of oscillators have been studied mainly by electronic engineers in relation to the types of oscillation which occur in electronic valve circuits. The theoretical (mathematical) treatment of the general problem is involved, and certain simplifying assumptions are usually made which limit the usefulness of such studies to the general problem here formulated. Thus Adler derives equations which are applicable only to the case of resonant oscillators, and Jelonek uses a similar simplification. From our point of view it is more useful to describe certain qualitative features of the behaviour of oscillating systems which have a greater generality than these mathematical analyses.

The most obvious feature of coupled oscillators is the phenomenon of "locking" or "synchronization". When the natural frequencies of two oscillators are not far removed from one another and there is coupling between them in the form of a non-reactive energy-transfer from one to the other, the two systems oscillate as a single unit at a frequency intermediate between the two natural frequencies but with a phase difference which is dependent on the intensity of the coupling and on the difference in natural frequencies of each oscillator in isolation. If the natural frequency of one oscillator (O_1) is varied while that of the second (O_2) is maintained constant, the relationship between the natural frequency of O_1 and the observed frequencies of O_1 and O_2 is as shown in Figures 2, a, when the form of the two natural oscillations is sinusoidal (Figure 1, a, linear oscillation). Synchronization occurs over a range of frequencies depending on the closeness of coupling, and the mutual effect of one on the other is symmetrical.

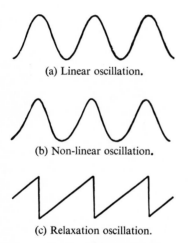

(a) Linear oscillation.

(b) Non-linear oscillation.

(c) Relaxation oscillation.

Figure 1.—Types of oscillator waveforms.

If the natural oscillations are not sinusoidal (non-linear oscillators) the behaviour of the coupled pair becomes asymmetrical. Let us assume a wave-form such as that of Figure 1, b, for the natural oscillations. The frequency of oscillation of the pair in the synchronization range is now shifted so that it is always greater than the mean of the two natural frequencies, with the result that the faster of the pair appears to be shifted less far from its natural frequency by the coupling than does the slower (Figure 2, b). This asymmetry is well illustrated by the behaviour of the extremely non-linear type known as a relaxation oscillator (Figures 1, c; 2, d). The frequency of two relaxation oscillators over the synchronization range is very nearly the natural frequency of the faster of the pair, so that the interaction appears to be

unidirectional when the frequency of oscillation is the parameter under observation.

When the natural frequency of one oscillator is altered so as to bring it into the synchronization range the onset of synchronization is not instantaneous but, particularly I) if the oscillators are only just oscillating and II) if the oscillations are non-linear, may take an appreciable time to develop, during which the actual frequencies approach one another although the external conditions (*i.e.* the natural frequencies) do not change. Similarly at the other end of the range the breaking of synchronization takes a certain time to occur. This gives rise, when the natural frequency of one oscillator is varied continuously, to a hysteresis phenomenon, the synchronization range depending on the direction from which and the rate at which the natural frequency of one oscillator (O_1) approaches that of the other (O_2). If O_1 approaches O_2 from below (frequency increasing) (Figure 2, b, d) the synchronization persists beyond the point at which it appears when O_1 approaches O_2 from above (frequency decreasing) (Figure 2, c, e). With non-linear oscillators this means that the actual frequency of oscillation of the "fixed" oscillator O_2 is changed less when the natural frequency of O_1 passes through its natural frequency in a descending direction than when it does so in an ascending direction. Conversely the actual frequency of O_1 is changed more when its natural frequency is decreasing. The importance of this hysteresis phenomenon will be apparent later.

Asymmetry in the mutual relationships of a pair of oscillators provides an essential feature of the proposed model. In a population of oscillators the process of synchronization corresponds to the reproduction of an individual in a natural population; the growth in numbers of oscillators showing this frequency of oscillation is initially of the exponential type characteristic of autocatalytic systems provided each oscillator is uniformly coupled to its neighbours. Once the steady-state distribution of frequencies of oscillation has become established in a population of coupled oscillators, the asymmetry of mutual interaction ensures that an increase in the natural frequency of one oscillator has a greater effect on the state of the population than a decrease.

This effect is most clearly appreciated in the case of relaxation oscillations, and it is in fact found in the beats of the chambers of the vertebrate heart, which behave like relaxation oscillators. The coupling here is so strong that the whole heart normally beats as a single unit at a

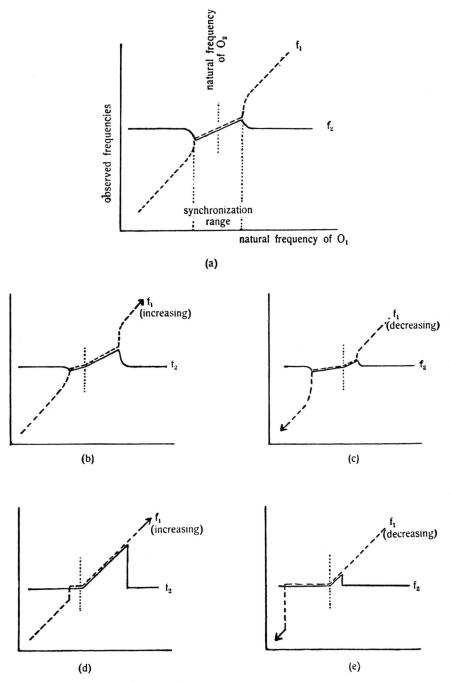

Figure 2.—Diagrams, based on experiments with electronic oscillators, to illustrate the phenomenon of synchronization. The graphs show the observed frequencies, f_1 and f_2, of two coupled oscillators, O_1 and O_2, as the natural frequency of O_1 is varied.

(a) Linear oscillation.
(b) Non-linear oscillation: frequency of O_1 increasing.
(c) Non-linear oscillation: frequency of O_1 decreasing.
(d) Relaxation oscillation: frequency of O_1 increasing.
(e) Relaxation oscillation: frequency of O_1 decreasing.

frequency determined by that of the fastest component (normally the sino-auricular node). The momentary rise in frequency produced by the injection of an external electrical stimulus to the sino-auricular node advances the beats of the whole heart, and the event can therefore be considered to have replicated throughout the coupled population. No effect is produced by the lag of a beat in, say, the ventricle, since it is caused to beat at the normal time by the influence transmitted to it from the faster beating auricle.

In the majority of oscillators studied by electronic engineers there is a very considerable degree of constancy in the time intervals between successive cycles or beats. This regularity derives from the high energy levels of the oscillation in comparison with the random energy fluctuations in the components (thermal noise). Oscillators can, however, be constructed in which the random energy fluctuations produce a variation in the timing of the beats, so that the frequency fluctuates about a mean value in a random manner. If the fluctuations are slow compared with the mean frequency of oscillation, the effect is to produce a "wavering" frequency; if the fluctuations are rapid compared with the mean frequency of oscillation the intervals between successive beats vary. Since a wide range of frequencies is normally present in the random energy fluctuations in components both effects may occur together. A coupled population of non-sinusoidal oscillators showing this variation in natural frequencies owing to thermal noise corresponds to a natural population evolving by natural selection of random variations. If the coupling is sufficiently loose to allow certain oscillators or groups of oscillators to escape from the synchronization range when the variation is sufficiently great, complexity can increase in such a system in a manner analogous to the increase in complexity seen in the organic evolution of populations sufficiently subdivided into isolated "demes" for the Sewall Wright effect or "drift" to appear in the demes and produce divergence of structural form. We have already noted that it is a condition for the increase in structural complexity of a system either that there should be isolation of one part from another or that the environmental conditions be different in different parts. Similarly two cases can be distinguished for the increase in temporal complexity in the model. Instead of distinct parts of the system which preserve their separateness with the passage of time we now need to have distinct rhythms which remain distinct as they move through the population in space. If all the oscillators in a population are beating in synchronism and mutual coupling is strong (as in the heart), a small advance in the beat of the fastest beating unit advances them all, and although there is a permanent change in the timing of the beats, no increase in complexity results. This corresponds to the incorporation of a mutation in the whole of a population of organisms. If coupling is weak, one part of the system may become isolated from the rest by the occurrence in it of a frequency too far removed from that in the rest of the system for it to have any effect on it, and the two rhythms may then persist independent of each other. Isolation in this case consists in isolation in time rather than in space since all the parts may remain morphologically coupled together. Alternatively, if the natural frequencies become different in different regions of the population due to external influences acting unequally on different members of the population, the whole again becomes a complex association of parts, interaction only occurring when frequencies are sufficiently close to bring a pair of units within their synchronization range.

Such a model is of little value unless its "evolution" is controlled or conditioned by events in the external world; if there is no limit to its variation the course of its evolution becomes completely unpredictable. Natural organic evolution proceeds in such a way that the species become adapted to the environmental conditions; to our model we must now add some feature to correspond to the environment. This is supplied by allowing the natural frequencies of the oscillators to be affected unequally by influences external to the population, which thereby create in it an "environment" which moulds the direction of evolution of the system. With this addition the parallel is complete between our model showing an "evolutionary" increase of complexity of time, and the system of living organisms evolving to greater complexity of form, and we may consider how far the model is applicable to conditions found in the central nervous system of animals.

The first reason for preferring a model based on complexity in time to one producing complexity of form for the central nervous mechanism of learning is the rapidity of the process. One example of the evolution of complexity of form (organic evolution) involves a time scale on the order of millions of years. Even if the more rapid adaptive changes occurring in populations of plasmagenes involve a similar process, the rate is still much too slow to account for the rapid forms of learning. It is not easy to see how a process which depends on structural replication can ever approach the speed required when the replicating units have the complexity involved in patterns of behaviour. There is therefore an initial preference for a system involving complexity in time for the learning mechanism of the central nervous system.

The observed physiological properties of central nervous tissues suggest that the model is of the right type, since there is direct evidence that electrical rhythms occur, apparently spontaneously, in many of the central nervous structures which on other grounds are held to be the site of the learning process. The hypothesis is therefore put forward that the learning capacity of the central nervous systems of animals is a property of populations of nerve cells behaving as loosely coupled oscillators. The sensory systems of the body act unequally on this population to set up conditions which in a formal manner represent certain features of the animal's environment, and the effector systems are coupled to it through analysing centres each responding to a particular "species" of internal rhythmic pattern. The motor systems evidently contain in themselves a high degree of inherited complexity so that the externally observed behaviour is related to the state of the population of oscillators in ways which are to a large extent characteristic of the species of animal. These ways represent the instinctive component of behaviour and may be very elaborate. The inherited complexity of the analysing system is also manifested in a series of "key trigger" actions, so that the set of rhythmic patterns needed to initiate the activity of each instinctive centre is itself complex (compare the "releasers" of Lorenz) though not necessarily as complex as the activity which is initiated. It is not suggested that the learning process is completely centralized at one level in the central nervous system. To the extent that any nervous centre contains a population of cells behaving as non-linear oscillators, the activity of that centre can show adaptive properties involving a selection from possible alternatives, and can play its part in the learning shown by the whole animal.

To put forward in detail plans for central nervous mechanisms which will account even in general terms for each of the different manifestations of the learning process involves a number of postulates which are not vital for the main thesis but which may nevertheless be made to illustrate the type of system which is suggested by the basic conclusions. A possible model which appears to be more in line with what is known about the properties of the nervous systems of animals is one in which the oscillators are not single cells (as in the heart) but closed loops or "reverberatory arcs" of neurons acting as oscillators as a whole. By this is not meant that an impulse travels round the arc in the way that the concept of the reverberatory arc is usually understood but that the whole chain of neurons is coupled so that it acts as a single oscillator, each cell influencing the excitation of the next with a certain "lead" or "lag" so that the whole generates a nearly sinusoidal oscillation when the phase change round the loop equals a multiple of 360 degrees. Excitation is here used to refer to the state of the cell which causes a rise in the frequency of impulses and not to the process giving rise to a single impulse. Pringle and Wilson have discussed the way in which the adaptation property of excitable tissues can generate a phase change between successive elements. Closed loop connections between a number of such neurons would generate oscillations more nearly sinusoidal in character than the relaxation phenomena which appear to be responsible for the rhythmic impulse discharge in a single neuron, and this model (a phase shift oscillator) therefore overcomes one of the difficulties of interpreting the nearly sinusoidal potential changes which are found in the cerebral cortex as the sum of a number of spike waveforms.

An essential feature is readily introduced into the model if the oscillations are considered to be generated in this way, since the loops in which the oscillations are generated may share short sequences of neurons and thus become coupled by competition for the shared cells (Figure 3). An oscillation in the loop ABA will cause a periodic rise and fall of the state of excitation of the shared cell X. If the natural frequency of oscillation of the loop CDC is slightly lower than that of ABA the result will be a synchronization of the two. If the oscillations in each loop are non-linear in character, as is to be expected from the known properties of phase-shift oscillators, the synchronized frequency of the two will be nearer to that of A than to that of B, so introducing the asymmetrical character of the mutual relationship necessary for the selection process as described above. Non-linearity of the oscillations has another important corollary. As a result of the hysteresis phenomenon described in Figure 2, b, c, there is a difference in the action of ABA on CDC if the frequency of the former is increasing or decreasing. If ABA approaches CDC from below it holds it in synchronism over a range of frequencies and breaks synchronism only when CDC has been

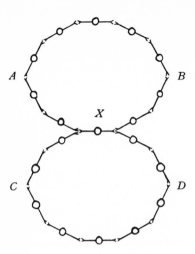

Figure 3.—Diagram of two coupled closed loops of neurons.

raised well above its natural frequency. If ABA approaches CDC from above, synchronism is not established until the two natural frequencies are nearly the same, and CDC is not depressed below its natural frequency by the same amount as it was raised in the former case. On the other hand the actual frequency of ABA is changed more during the period of synchronization. This hysteresis has results which are of great importance in the interpretation of certain forms of learning in terms of the action of the model.

The sharing of chains of neurons by oscillators introduces competition in space as well as competition in time. The result is a disappearance of one of the differences between the working of this model and that of organic evolution. The pattern of oscillations moves over the population of cells as a wave, achieving a replication of functional form, and a more rapidly replicating pattern displaces it by competition for the shared nerve cells. If form is represented by simultaneity of action instead of as the form of an organism or of a chemical molecule, its rate of creation from component parts is limited only by the speed of the coupling process which may be very high and the objection mentioned above to the form-replicating machine as a model of the learning mechanism disappears.

It may be supposed that it is this pattern of simultaneity which is the significant feature in the operation of the "key trigger" mechanisms for the release of activity in lower centres, leading finally to the motor response. Spatial summation of excitation on the motor neuron in the vertebrate spinal cord is a well-established phenomenon

and it has been suggested that summation may be limited to synaptic terminations in close proximity on the surface of the cell. If each region of the cell of the next lower centre received fibres from a particular pattern of regions in the population of cells in the learning centre, they would be excited only when this pattern of simultaneous excitation became established at a sufficiently high frequency or over a sufficiently large part of the total population.

The Types of Learning

The main thesis of this article is that there is a parallel between the process of organic evolution and process of learning, and the basis of the argument is the unique kinematic nature of the process which results in a selection of variations. It has so far been assumed that the variations which form the background for the selection process in the case of the central nervous system are variations due to thermal noise in the frequency of oscillations occurring in populations of nerve cells. It is, however, apparent that the environmental conditions controlling the sensory input to the selection mechanism of the central nervous system contain a measure of variability in themselves. The portions of the universe with which an animal is in sensory contact are not fully ordered (i.e. there is not regular repetition in space or time of the sensory messages) and there is therefore material for selection all the time in the changing formal "environment" in the central nervous system. It is clear that in this way the complexity of the organism can increase, if there is a selection mechanism present, without the necessity for spontaneous variation to occur within it. This process corresponds, by analogy, to the reshuffling of an existing gene complex which occurs in a species of organism when the environmental conditions change.

The modification to the model of the central nervous system needed to achieve only this type of complexity-increase and not the purely endogenous type (which can occur theoretically in a constant external environment) is simply the removal of variation from the frequency of the oscillators. The selection criterion is still left in the asymmetry of the locking process between coupled oscillators. Such a system within the central nervous system would not increase in complexity in a constant environment, but would increase if the material for selection were supplied from without. This simplified model reduces, in fact, virtually into a machine for transposing environmental complexity in time and form into a single system in which mutual asymmetrical interactions can take place. The selection is then performed by the only criterion of which the system is capable, namely antecedence in time, and the result is then re-translated into complexity of form and time in the effector systems.

Conditioned Reflex Type I

The interpretation of some of the types of learning in terms of the model may now be considered, starting with conditioned reflex type I of Konorsky. "A conditioned reflex arises when two stimuli, of which one evokes an inborn, permanent, or so-called unconditioned reaction of the organism, are applied in association a number of times, and it consists in the second of these stimuli beginning to evoke the same kind of reaction as that produced by the first stimulus." It is a further requisite that the conditioned (or "to be learned") stimulus should precede the unconditioned stimulus in time. Associated with the normal conditioned reflex are two forms of inhibition which must also be explained by any theory of learning. "External inhibition" is the suppression of an innate reaction by an extraneous stimulus applied simultaneously, and it occurs particularly when the extraneous stimulus produces a response in the animal different from the normal innate response to the unconditioned stimulus. "Internal inhibition" arises in two ways: either when an established conditioned reflex is not reinforced by the unconditioned stimulus over a number of applications of the conditioned stimulus; or when the conditioned stimulus is applied after the unconditioned. In both cases the inhibition manifests itself as a reduction in the magnitude of the innate response, or as a rise in the strength of the conditioned stimulus needed to evoke the same magnitude of response.

This type of behaviour is interpreted in the model as shown in Figure 4. Normally the presentation of unconditioned stimulus A leads to the growth in the learning mechanism L of certain rhythmic patterns specific for the release, when they reach a certain intensity, of instinctive response X through the key trigger K. If the conditioned stimulus B is presented before the presentation of stimulus A, the cells of L are in a modified state when A is presented and the growth of the specific pattern starts. Since the frequencies of the oscillators in L are rising due to the presenta-

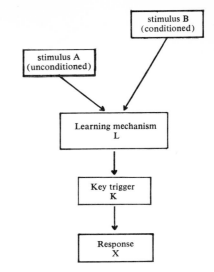

Figure 4.—Schematic block diagram for a conditioned reflex of Type I.

tion of A, they will "collect" into their own rhythm other oscillators composed of cells still showing excitation due to B and beating at frequencies through which the oscillations due to A rise. The extent of the pattern of simultaneity of action therefore grows. When the pattern is established over a sufficiently large part of the population at a sufficiently high frequency the response X occurs. In order that a repetition of this sequence of events a number of times should lead to a state of affairs in which B alone is adequate for the growth of the specific pattern to release K it is necessary to show that those cells or groups of cells which have become incorporated into oscillators forming the specific pattern are changed in some way, so that they synchronize more readily at the application of B + A, and finally develop the specific pattern without the presence of A at all. The nature of this change is discussed below.

External inhibition is interpreted as an upset in the development of the pattern for the release of K by the other patterns developed under the influence of the extraneous stimulus competing for the cells of the learning centre. Stimuli most effective in producing external inhibition are those which lead to some other response and which therefore presumably encourage the development of a different pattern. It is significant that the most effective conditioned stimuli are those which do not normally lead to a response.

The internal inhibition which develops when attempts are made to achieve "backward conditioning" by presenting the conditioned stimulus B after the unconditioned stimulus A is attributable to the hysteresis phenomenon in the synchronization of non-linear oscillators mentioned above. In this type of experiment the central effect of stimulus B appears at a time when the frequency of oscillations due to A is decreasing, and the oscillations set up by B therefore produce a greater effect on the oscillations still occurring due to A, than those due to A do on those due to B. Since the oscillations due to B compete for cells with those generating the patterns for the release of K, the effect will be a break-up of these patterns and a resulting inhibition of response X. In its initial stage such an inhibition does not differ from the external inhibition produced by the application of any extraneous stimulus subsequent to the unconditioned stimulus, but the difference arises when the experiment is repeated many times, for the stimulus B then becomes progressively more effective as an inhibitor. In the same way that in a normal positive conditioned reflex a change occurs which makes B more effective as a promoter of the growth of the pattern for K, so when applied after A, a change must occur which makes B progressively more effective as an inhibitor of the growth of this pattern.

The other type of internal inhibition produced by the presentation without reinforcement of stimulus B after it has become effectively conditioned to release response X, is best considered after a discussion of the nature of the "plastic" changes involved in conditioning, which must now be considered.

It is clear at once that, since the changes accompanying conditioning persist for long after the stimuli have ceased, the residual effect between each successive application of B+A cannot be thought of as a change in the pattern of synchronized activity in the learning mechanism L, but must be represented by some change in the cells of L which alters the patterns which are subsequently set up when the stimulus is again presented. Konorsky makes the definite suggestion that these plastic changes consist in the establishment of new excitatory or inhibitory synaptic connections between cells, and this idea of a change in the intercellular connections is inherent in such phrases as "a lowering of synaptic resistance" which are to be found in the literature on this subject. There are serious objections to any structural explanation of plasticity because of the extreme complexity and rapidity of the process as observed in such forms of learning as imprinting. In any case, if our model is to be of real value, and if the parallel with organic evolution is valid, we should try to find features in it which account for the plastic changes involved in learning without the introduction of new postulates of an unknown nature.

The necessary mechanism is, in fact, already present in the model as described, but its operation only becomes clear when we examine in more detail the events in each cell. As was stated earlier, the oscillations which generate the specific patterns in the learning centre are to be thought of as produced by the connection, in closed loops, of cells each of which is capable of manifesting a varying degree of excitation and of influencing the next cell in the chain with a certain lead or lag. The term excitation is used here not in the same sense as is usual in studies of peripheral nerve, where it refers to the process giving rise to a single impulse, but to a maintained state in the cell, measurable as some function of the frequency of impulses traversing the axon of the cell. Its identification with the slow electrical changes occurring at the ends of peripheral nerve fibres is a reasonable conjecture, but it presumably also involves internal changes of equilibrium of a chemical nature. The sum of its electrical manifestations in many cells many account for the slow potential changes which are observed in the mammalian cerebral cortex. Since excitation in this sense is a maintained state which grows and declines in relation to changes in the environment of the cell (stimulation) we may look for a parallel between it and the properties of the particular type of open system discussed [earlier], the level of excitation in each cell corresponding to the quantity of the "unit" there discussed. The subsequent discussion of evolutionary changes in populations of such open systems applies to the model, provided that it is borne in mind that competitive interaction in the model takes place between oscillators composed of many cells, in each of which the level of excitation undergoes cyclic changes during the oscillations about some mean value which represents the level of excitation in the absence of the oscillations.

In particular, just as a transient change in the environmental conditions of a population of organisms may produce an enduring change in the inter-specific equilibrium, so a transient stimulus (such as the application of B+A) may leave an enduring effect on the mean value of the excitation of the cells composing the oscillators in the model of the learning centre. This may occur whether or not the oscillations set up by the stimulus persist after the end of stimulation. Each application of the combined stimulus thus finds the cells in a different condition, and the gradual establishment of the conditioned state becomes a possibility.

Returning to the mechanism of the conditioned reflex we have then in the model features adequate to explain the residual change in the central nervous system needed to enable successive applications of stimulus B + A to develop a condition in which B alone releases response X through its key trigger mechanism. The advancing of the beat of certain cells excited by B which occurs when the frequency of oscillations induced by A is increasing, and which pulls them into the pattern of synchronism operating K, leaves these cells at levels of excitation such that they more readily fall into this pattern when B is presented subsequently. Similarly the advancing of the beats of certain cells excited by A when the frequency of oscillations induced by A is decreasing leaves them at levels of excitation which make them less readily synchronizable into this pattern in subsequent presentation of B + A. The two plastic effects need not occur in the same cells; internal inhibition is not merely a cancellation of conditioning, as Konorsky has pointed out. The drift back to their original condition of cells whose level has been modified by positive conditioning is a much more gradual process than the inhibition, and it is significant that the inhibition can be disinhibited by a variety of extraneous stimuli which produce a general raising of the level of central excitation. The number of changes of cell excitation level needed partially to upset the pattern of synchronization is smaller than that needed to establish a state in which it occurs, and it is therefore reasonable that it should be possible to neutralize the effect of the inhibitory changes by an increase in the intensity of those patterns which still remain established. The key trigger is operated when the lower centre is excited by a sufficiently high frequency of impulses reaching those regions of its surface in which spacial summation remains effective due to the development of the specific patterns in certain parts of the cell population.

In contrast to previous theories about the nature of the plastic changes in the central nervous system, this model attributes plasticity to changes occurring within the cells themselves, and not in the connections between them, which can remain morphologically and physiologically constant. It also has the merit that it does not localize these changes in any particular part of the learning centre, since the cells whose level of excitation is altered may be distributed almost at random throughout the population for a discussion of the

apparent lack of localisation (of the "memory trace").

It remains to consider the internal inhibition which is produced by the failure to reinforce a conditioned stimulus. As Konorsky points out, this process is not the same as the gradual fading of conditioning which occurs when the conditioned stimulus is not presented at all for a long period, in that even though the response fails to occur the conditioning may be caused to reappear by raising the level of central excitation by extraneous stimuli. The fading of conditioning is to be thought of as a drift back to their original level of the cells whose level of excitation has been changed. The inhibition, however, must result from a failure of the specific patterns to establish themselves at a sufficiently high level before this drift back is complete. The probable mechanism for this inhibition would appear to be as follows. The presentation of stimulus B starts in the population certain oscillations which increase in frequency and capture other oscillators in their growth process. Although in the conditioned state the resulting patterns are adequate to release K the process of capture also leads to plastic changes in certain cells, and since the patterns produced by B are not identical with those produced by A, the changes will lead to a reduction in the similarity to the "correct" pattern for the release of K. These plastic changes are superimposed on those involved in the original conditioning and do not reverse them, although the effect of them is to diminish the correctness of the pattern. An inhibition thus results long before the original plastic changes have disappeared from the population.

In this interpretation of the mechanism of conditioning in terms of the model of the central nervous system we have been led into a more detailed identification of the processes with particular neuronic mechanisms than was intended in the earlier sections. The main thesis of the parallel with evolutionary processes must not be overlooked in such physiological detail. There are, in fact, a number of ways in which the process of conditioning resembles the processes of evolution. The magnitude of the specific pattern, for example, resembles the numbers of an animal species in the way in which it grows. The conditions produced in the central nervous cell population by the unconditioned stimulus A corresponds to the environmental conditions normally optimal for the growth in numbers of this particular organism. The conditioning to stimulus B corresponds to the adaptation of the species to new environmental conditions. A particular example—the growth in number of a bacterial population—is interesting in that, as Hinshelwood has pointed out, the logarithmic phase of growth is the period during which the phenomenon of adaptation to new culture media takes place. It is tempting to compare this to the positive conditioning which occurs to stimuli whose central effect coincides with the rise of excitation due to the unconditioned stimulus, although the apparent absence of anything to correspond to internal inhibition in bacterial populations shows that the two systems are not equivalent. The use of pure cultures may be the clue to the difference, since in the learning centre it is never to be supposed that one specific pattern displaces all others from the population.

The plastic change responsible for conditioning may also be compared to the lasting effect on a heterogeneous population of organisms of a transient change in the environmental conditions. Selection of the gene complex is capable of producing slow adaptive changes, even in the absence of mutation, as a result of the re-sorting of the genes at each sexual reproduction and of competition among individuals for the available living space and food supplies; it corresponds in the model to the competition for the limited cells of the population by the different possible rhythmic patterns of simultaneity. Examples may be multiplied indefinitely and serve no useful purpose except where a particular aspect of the learning process can be made clearer by a search for the corresponding situation in evolution, which may be easier to analyse.

Latent Learning

It is convenient to discuss next the process which has been called latent learning, and which is included by Thorpe as a feature of insight learning. This differs from the conditioned reflex in that two or more sensory patterns may become associated in the learning mechanism even when no response occurs immediately consequent upon the presentation of one of the stimuli. In the original form of learning to which the term was applied rats which had been left in a maze without food for some time subsequently learned to find the food faster than those placed in it for the first time when the food also was introduced. The sensory patterns which have become associated in this case were all the impressions received from any of the features of the maze and the case is therefore one of great complexity; but latent learning in the sense in which the term is defined

above is undoubtedly a widespread phenomenon.

In the interpretation of the mechanism of conditioning in terms of the model of the central nervous system, it will be realised that the response as such plays no part. The whole of the suggested mechanism works equally well if the release of the response through its key trigger is prevented by some other agent. The subsequent effects of the response are not therefore of importance in conditioning, and it is significant that conditioned reflexes can be established as well to "unfavourable" as to "favourable" stimuli. Latent learning therefore appears to be a similar process to conditioning and to involve merely a selection by the learning mechanism of those aspects of the various sensory stimuli which are common to all or many of them. In other words, when the stimuli do not provide the correct conditions for the growth of any one specific pattern up to a level when a response occurs, those patterns are selected which grow most readily. Since none of them occupies more than a fraction of the population (in the ecological parallel none is "dominant") each stimulus alters their relative intensities by capture of oscillators and leaves its plastic remainder. If the successive stimuli have much in common, the pattern which grows under the influence of this common feature will have grown the most and the cells will be in a state to redevelop subsequently this particular pattern more readily than any other.

It is to phenomena of this nature that we must relate the perception of *gestalt* (ordered) patterns in the sensory stimuli. In terms of the model, the perception of a *gestalt* is the development of plastic changes which encourage the growth of a specific pattern whenever the sensory stimuli contain a particular combination of features. The order or repetition in time of this combination in a variety of otherwise different sensory impressions provides the necessary conditions for the accumulation of just those plastic changes which enable this particular pattern to develop.

Habituation

The types of conditioning so far discussed have in common the feature that the release of the response has no influence in the learning process. In all the manifestations of the process still to be considered this independence of the result does not appear, but rather the direction of the process, *i.e.* whether the result is positive learning or negative learning (inhibition) depends on the nature of the subsequent events. This is clearly brought out by the phenomenon of habituation.

Habituation is the process whereby a stimulus becomes ineffective in releasing a response when that response does not lead to a significant result for the animal. It is commonly found with stimuli which lead to what Pavlov has called the orientation reaction. When a dog is presented with an "alarming" stimulus it responds by adopting an attitude of defence whereby it is better able to resist a subsequent harmful stimulus. If this later stimulus does not materialize, the orientation reaction fades and the dog is said to have shown habituation to the first stimulus.

In order to interpret such a phenomenon in our model a scheme such as that of Figure 5 is necessary. When response X (in this case the orientation reaction) occurs it produces a change in the sensory stimuli from A to B, due to the movement of the response. The new stimulus, during the frequency-increasing part of the oscillations which it induces in L, reduces the magnitude of the specific pattern for release of K in much the same manner as any external inhibitor. As a result, plastic changes occur which make the specific pattern for release of K grow less strongly on a subsequent application of A. If the orientation reaction is followed by the

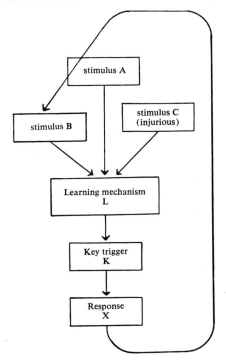

Figure 5.—Schematic block diagram for habituation.

"expected" injurious stimulus C, the combined sensory stimulus becomes B + C instead of B alone. But C is a powerful inducer of the specific pattern for release of K (as may be tested by applying C alone initially) and in this case the pattern grows still further and the plastic changes are such as to encourage its growth on a subsequent presentation of A. In this way the nature of the plastic changes in L comes to depend on the result produced by the development of the pattern, in the manner which is commonly found in closed loop systems.

Habituation is a widespread phenomenon, and when interpreted in terms of the model it is seen to be an example of a homeostatic process of the type found in many physiological systems. It would appear that it is bound to occur whenever the response of the animal changes the sensory stimulus, unless the new stimulus itself encourages the growth of the specific pattern which has released the response. The fact that it does not occur in the other types of conditioning studied by the Pavlov school may be attributed to the fact that the salivation response used by them does not normally affect to any marked extent the nature of the stimulus used. The reverse of habituation, reinforcement or autofacilitation, which is well shown in some of the types of instinctive behaviour described by Lorenz can be interpreted in a similar manner, the sensory stimulus which results from the response here encouraging still further the growth of the specific pattern for its own release. The necessary intensity of the initial stimulus A thus gets progressively less until, as has been found by Lorenz, the response may occur spontaneously without any visible causal agent. This is a typical result to be expected from a mechanism showing positive feed-back with an accumulation at each experiment (plastic change) of the factors determining the intensity of the feed-back.

Conditioned Reflex Type II and Trial-and-Error Learning

In interpreting conditioning of the normal type, and habituation, in terms of the model it is not necessary to suppose that there is any inherent variability in the frequencies of the oscillations generated in the learning mechanism by the sensory stimuli. Selection by time-antecedence operates on the complex pattern of oscillations set up by the external stimuli through the sense organs and the necessary variation is provided by the change from one stimulus to another. A different state of affairs is provided by conditioned reflexes of Konorsky's second type, and Thorpe, recognizing this, classifies this type with trial-and-error learning and not as a conditioned reflex at all. The significant difference will become apparent when we attempt to interpret the events in the manner already described. Figure 6 represents the scheme for a typical conditioned reflex type II. Initially, sensory stimulus A releases response X, followed, since the resulting stimulus B is "reinforced" (Konorsky's use of the term) by the presentation of food (stimulus C), by a further encouragement of the growth of the specific pattern concerned or at least by the neutralization of the habituation effect which might otherwise occur. The procedure may now vary. Sensory stimulus D in this type of conditioned reflex affects proprioceptive sense organs and is only produced by response Y. But sensory stimulus A, or A + B does not cause the learning mechanism L to develop the specific patterns needed to operate K_2 and release response Y. For selection to take place in L in the presence of stimulus A + D response Y must be caused to occur. If the movement necessary to produce sensory stimulus D is produced by the operator moving the limbs of the animal passively, then the conditions are not significantly different from those in conditioned reflex type I, D being then the unconditioned, and A the conditioned stimulus. If, however, there is an element of variability in the learning mechanism L, the specific pattern needed

to operate K_2 and release response Y may be developed in the presence of sensory stimulus A, and, if it is developed, adaptation to A + B + D will occur; owing to the plastic changes in L which result, this particular pattern will then be more likely to develop in subsequent experiments. A new response Y can thus become associated with a sensory stimulus A by virtue either of variation in the environment (*e.g.* by the experimenter) simulating sensory stimulus D, or by variation in L chancing to operate the key trigger for the release of the response. In the first case the process may be considered to be similar to that of conditioned reflex type I; in the second it is of the nature of trial-and-error.

The more usual form of the occurrence of this system in nature is probably represented by the addition of the interrupted line in Figure 6. This implies that stimulus B + C reinforcing the specific pattern for the release of K_1 is not produced unless the environmental changes produced by responses X and Y both occur; in other words, food is only obtained by the performance of both responses. The successive presentation of sensory stimulus A will now initially release a series of responses X. The response repetition will continue (unless suppressed by habituation) until either the environmental variation or the inherent variation in L results in the operation of key-trigger K_2. When this happens the complete sensory stimulus A + B + C + D is produced, allowing selection to occur in L and increasing the chance of sensory stimulus A alone operating K_2.

The diagram of Figure 6 is, of course much oversimplified, and habituation will normally ensure that more than one instinctive response is released by repetition of stimulus A. The result is an appearance of trial-and-error, all possible responses being tried until some combination releases Y; this combination is then preserved by selection.

Insight Learning

We now come to the most difficult and advanced type of the learning process, insight learning. To use anthropomorphic terminology, the animal in this case appears to be able to work out the consequences of a number of possible alternative responses to sensory stimuli without actually performing them, and then, on the addition of a further stimulus, to perform only the one which leads to a favourable result. Thorpe recognizes

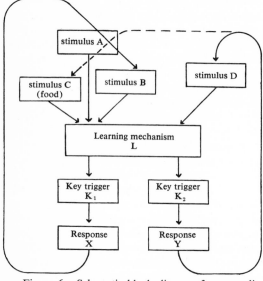

Figure 6.—Schematic block diagram for a conditioned reflex of Type II. The significance of the interrupted line is explained in the text.

that the process of insight learning is closely related to that of latent learning in that selection from possible sensory stimuli occurs without the performance of a response. This, however, is clearly not a full explanation, since insight learning involves a selection from possible responses as well as selection from the total stimulus. Some form of internal trial-and-error must also be involved by which (in terms of Figure 6) the selection occurs not only from actual sensory messages but also from those which would be received if the response movements were made. The whole of the loops must, then, be contained in the learning mechanism. We must reconsider the model in more detail to see how this can be achieved.

The model consists of a population of cells forming oscillators which interact in time because of the tendency of one oscillation to synchronize to itself another with a slightly lower natural frequency, and in space because some of the component cells are common to several oscillators. The sensory stimuli provide the "environment" for this population by modifying the level of excitation in the cells and so altering the frequency and composition of the oscillators. The level of excitation in the cells is also subject to random fluctuations, which also alter the frequency and composition of the oscillators and generate the variety of specific patterns involved in trial-and-error learning. There is no fundamental difference in the effects produced by these two influences which produce changes of the level of excitation in the cells: in the one case it is random changes in the external environment which produce variation in the sensory stimuli, and in the other the random fluctuations are endogenous in the cells of the learning mechanism. In either case selection by time-antecedence occurs on the background of variability so produced, and plastic changes follow, preserving a fraction of the selection effects.

Learning of the type of conditioned reflex type I involves only exogenous variation of the external environment influencing the centre through sense organs. Both the conditioned and the unconditioned stimuli are external to the animal. In trial-and-error learning or conditioned reflex type II where the movement is spontaneously originated, the conditioned stimulus is external, but the unconditioned stimulus is originated endogenously and acts through the response and the resulting proprioceptive messages. There is clearly a third type of process, in which both conditioned and unconditioned effects are endo-

genous in origin, and whose progress bears no relation to events occurring simultaneously in the environment external to the animal. Such a process would produce a progressive change in the frequency and composition of the oscillators which is not entirely random in character, since any given random event in one cell has the dual effect of producing variation and of changing the conditions in which selection operates. In other words, since only those random changes which produce an increase in the frequency of oscillation are preserved by selecton, each discrete change in the state of the whole system determines the conditions under which selection operates for the next step. The state of the system at any instant therefore defines the probabilities of each possible state at any given instant in the future. If the range of variability is slight, one or a few of these possible states may have much higher probabilities than any others and the behaviour of the whole system may be only slightly indeterminate.

This, it is suggested, is the nature of the mechanism in the central nervous system which is capable of insight learning. Because in such a system every endogenous variation provides both the material for selection and the conditions in which selection occurs, the occurrence of patterns adequate to release instinctive responses is not random, but an organization is built up in which the relationship between the pattern resulting from the selection process and the conditions in which the selection takes place is the same as the relationship between the key trigger and the sensory pattern usually resulting from the response (K_2 and D in Figure 6): is, in fact, the same as the relationship between cause and effect in the organism-environment complex. Given a certain initial organization and variable oscillators contained entirely in the learning mechanism L, a particular process involving an increase in complexity in L can then occur with a high probability without the performance of any actual responses.

It is pertinent to enquire how the organization required to define the relationship between the pattern resulting from selection and the conditions in which it takes place arises in the first instance in the individual. The relationship which is of value to the animal is the one which gives a performance similar to the relationship between cause and effect in its dealings with the environment, since only with this particular relationship will patterns be produced which release those instinctive responses which lead to the correct conditions for the release of further favourable responses. The only ways in which such relationships can be incorporated into the central nervous machinery of an individual are

(a) by inheritance

(b) by trial-and-error learning

(c) by contact with systems which already contain the required relationship, so that the complexity is incorporated "en bloc". In other words, the animal must be born with it, must find out, or must copy or be told. The last case opens up the subject of communication of information which it is not proposed to pursue here; the first needs no discussion. One further aspect of the trial-and-error method is, however, of interest. If the trial process leads to the completion of the feed-back loops only when one particular response occurs, the learning mechanism L will develop so that presentation of sensory pattern A always leads to the development of the key trigger for that one response, and the behaviour of the animal will become well adapted to a particular set of conditions but not more adaptable to variations in environmental conditions. Again to use anthropomorphic terminology, the animal will become good at doing one particular thing but not necessarily intelligent in general. This will be especially so if there is only a small amount of inherent variability in L. To achieve an organization in L which is capable of insight in the sense that the correct action to suit a particular set of circumstances occurs even when those circumstances have not occurred before in exactly that form, the requirement is, first, a larger measure of inherent variability and then a lengthy training in the form of a series of different conditions under each of which the mechanism of L changes slightly by selection. Slight variations of sensory pattern A are then capable of providing suitable conditions for the development of a wide range of specific patterns each of which leads to the release of the further response Y under the particular conditions present, and the mechanism becomes capable of analysing an entirely new situation into component features each of which has been encountered before in a different context, and of producing the right response without visible trial-and-error.

One is tempted to extrapolate the process still further into the sphere of human behaviour. The human being is capable of a much greater degree of insight learning than any other animal, but the full development of his capabilities demands a process of "education" which is not the same

thing as training, and differs from it in just the way which has been outlined. The continual repetition of a particular set of environmental conditions can lead, in man as in an animal, to a great increase in the efficiency of a response; that is, to the release of the next response ("success") on a progressively increasing number of occasions. But this training does not necessarily lead to the development of insight. To increase the ability of the human to deal with novel situations the requirement is for a long process of adaptation to a wide variety of situations each slightly dissimilar to the last, so that ultimately no new situation is likely to arise which contains, in its full analysis, more than a small amount of novelty to be dealt with by the visible trial-and-error process.

The increased endogenous variability necessary for the development of insight has obvious disadvantages if it is not accompanied by selection through experience of the "right" relationships between the result of and conditions for selection in the central nervous mechanism, since it leads otherwise to a failure on some occasions to give the usual response. Natural selection will act against mechanisms with a high endogenous variability when the environmental conditions of life are changing slowly. It may be that the extremely rapidly changing conditions found in an arboreal existence are responsible for the high degree of endogenous variability which must occur in the human brain.

Imprinting

This last type of learning in Thorpe's classification is also the least well understood from experimental analysis. It has, according to Thorpe, the following features:

1) It is confined to a definite and brief period of the individual life, and possibly also to a particular set of environmental circumstances.

2) Once accomplished the process is very stable—in some cases perhaps totally irreversible.

3) It is often completed long before the various specific patterns to which the imprinted pattern will ultimately become linked are established.

4) It is supra-individual learning—a learning of broad characteristics rather than of particular details.

From some points of view imprinting resembles latent learning in that it occurs without the appearance of a response. But the complex character of the behaviour-change which results, and the fact that imprinting occurs over such a small interval of time distinguishes it from the latter process. Until more experimental observations have been made it is unwise to identify it with any particular characteristics of the model which has been proposed, but it seems probable that it is connected in some way with the establishment of that central organization which is needed to define the transition probabilities in the development of insight learning. It was mentioned in the last section that a mechanism with a high endogenous variability must have this initial organisation if it is not to become chaotic in its behaviour. A means of "imprinting" such a basic organization on the organised mechanism could narrow very considerably the range of trial needed to establish a more detailed organisation by trial-and-error, and it could thus supplement to a large extent the innate organization which must be present in the histological structure of neuron connections. Such an interpretation is consistent with the supra-individual character of the imprinting process and with the fact that it may be completed before its effects become apparent in the ease of conditioning to various stimuli or the ease of release of certain responses. The modifications to the quantitative features of the model needed to allow imprinting to take place during a particular interval of time in the life of the individual need not be attempted at this stage.

Conclusion: The Location of Memory

In conclusion, it may perhaps be justifiable to include a further suggestion. The model which has been described as an aid to the interpretation of the types of the learning process is one which could be constructed by the use of techniques developed in the design of analytical machines. The device would be of the type known as an analogue machine, as opposed to the digital machines which are now widely used for mathematical computations. Mackay (personal communication) has pointed out that one of the advantages of this type of machine over the digital type is the greater economy of effort required to perform approximate qualitative analyses. On the other hand, the analogue method is less efficient as an accurate long-term memory store. The mechanism proposed for the "plastic" changes in the central nervous system, by alteration in maintained levels of excitation in the cells, is adequate for short-term memory but is probably inadequate to account for the accuracy of long-term memory. There seem to be two possibilities which might account for this long-term accuracy. The first is the improvement which is achieved by the use of large numbers of cells, so that the effect, so far as the animal as a whole is concerned, is the result of an averaging process. This may be called the statistical stability of the memory process. The second possibility is that the levels of excitation in each cell may react back on its internal chemical state, so that the memory is effectively transferred to a molecular level. It would be possible in this way to stabilize the inherent level of excitation in each cell so that the plastic changes become smaller and smaller in magnitude as the animal ages and the behaviour becomes correspondingly more fixed. A very small plastic change may suffice, in a mechanism of this sort, to produce alterations in the nature of the patterns generated, but the reduction in the capacity to learn which occurs in man as the individual ages may be attributable to some such molecular stabilization in the brain cells.

Summary

(1) The classification of the types of learning in animals given by Thorpe (1950) is taken as the starting point for an attempt to formulate in general terms the characteristics of the process. It is suggested that the most general feature of learning is the increase in complexity of behaviour which results.

(2) The concept of complexity, which has hitherto been used by biologists without definition, is analysed in terms of the "theory of information" developed in connection with problems of communication engineering. It is shown that, although it is a purely epistemological concept, it may be applied to descriptions of systems such as the living organism on which verifiable observations can be made. It is applicable to behaviour as well as to structure by a space/time inversion, verifiability consisting in this case of a correspondence between simultaneous observations made on different parts of the structure, instead of repeatability of observations at successive instants.

(3) A short account is given in general terms of the kinetics of evolutionary processes, the mathematical treatment of which (based on the theory of stochastic processes) is too difficult for an article of this type.

(4) Evolution by natural selection of random variations is the best known example of this type of process, but differentiation, treated as the problem of the evolution of a population of "plasmagenes," may be another example, and it is suggested that learning is a third example.

(5) It is shown that the phenomenon of synchronization of oscillators, the general features of which are described, can lead, in a population of oscillators, to an "evolutionary" increase of complexity of rhythm in a manner analogous to the increase of structural complexity which occurs in organic evolution. A model is described which is consistent with the known features of the physiology of the central nervous

system of animals, and which is capable of producing the increase of complexity found in the process of learning. The oscillators in this model consist of circular chains or "closed loops" of neurons, and it is suggested that such closed loops may generate oscillatory wave-forms more closely resembling the electrical waves found in the cerebral cortex than the impulses characteristic of peripheral axons. The coupling required to produce synchronization of these oscillators is provided by the sharing of neurons.

(6) The asymmetrical inter-relationship between oscillators necessary to produce "selection" results from the non-linearity of the oscillations generated by such a closed-loop arrangement. The "variation" on which selection operates may be provided by "endogenous" variability in the frequencies of the oscillators due to random changes of excitability of the neurons, or may be provided by the variation of the stimulus pattern from outside the animal. These two cases are compared to variation by mutation and by changes in the gene-complex, on either of which natural selection can operate to produce adaptive changes in organic evolution.

(7) The types of learning, as defined by Thorpe, are analysed in turn in a schematic manner, and it is shown how each of them is related to the properties of the proposed model. The formation of conditioned reflexes is adequately interpreted, and also the phenomena of external and internal inhibition. The difficulty of "backward conditioning" also appears to be represented in the properties of the model, being based on the hysteresis effect in the synchronization of non-linear oscillators.

(8) Habituation is shown to be an example of the homeostatic properties of a closed loop system with negative feed-back.

(9) Conditioned reflexes of Konorsky's Type I and habituation do not demand any endogenous variability in the frequency of oscillations in the model. Conditioned reflexes of Type II, trial-and-error learning, and insight learning do, however, demand that there shall be some endogenous variability in the nervous system if they are to be interpreted in terms of the properties of this model.

(10) The "memory" of a stimulus pattern, or the plastic change in the nervous system resulting from the application of a stimulus, has, in previous theories about the learning process, been considered to involve a change in the synaptic connections between neurons or at least a change in "synaptic resistance". It is shown that with this model the requirement is for a change in the maintained level of excitation within certain cells. It is further apparent that no localized changes are to be expected, the memory trace taking the form of a slight change in a large number of cells rather than a large change in any one. This feature of the model is consistent with the observations of Lashley and others that the formation of a conditioned reflex and of other associations is not accompanied by a localized change in any part of the cerebral cortex of mammals. The location of the physical events accompanying association within the cells of the central nervous system rather than between them introduces the possibility, which is briefly discussed, of a transfer of the events from the level of the neuron down to the level of molecular changes in the substance of the living protoplasm.

Notes

1. Bertalanffy has discussed the general properties of "open" and "closed" systems in relation to biology.

35.
Purpose, Adaptation and "Directive Correlation"

G. SOMMERHOF

The Key Position of the Concept of Adaptation

Before proceeding to the technical explication of the purpose-like character of vital phenomena and of biological order, we shall begin with a preliminary analysis in terms of everyday concepts and without too much regard for technicalities, in order to discover in broad outline the spatio-temporal relationships with which these phenomena confront us. The next step will be to cast the results of this preliminary analysis into a precise physico-mathematical formulation.

Owing to the ambiguity and inherent vagueness of the concepts traditionally used to describe the purpose-like character of life processes, we must not expect from this preliminary analysis more than it is inherently able to give. We may indeed expect definite clues about the direction in which we have to search for precise concepts which are capable of paraphrasing and replacing the vague ones; but we must not expect our analysis to lead to the solution of our problem by a straight path of logical deduction. The vagueness of traditional biological concepts will at times inevitably confront us with a choice of meanings. In that case the principle of our choice should be clear: for the purposes of this work we must select those meanings which best seem to cover the really distinctive features of observed life.

We may begin by noting that the distinctive organizational features of living organisms confront us with some peculiar dependency of part on whole which cannot be apprehended except

From G. Sommerhof, *Analytical Biology* (London: Oxford University Press, 1950), Chapter II. Reprinted with permission of the Clarendon Press, Oxford.

through the behaviour of the parts. In other words, the characteristics of organization and purposiveness are only perceptible in biological activity, and our first question, therefore, must concern the distinctive features of this activity. Nature is nature in action and no 'still' of life can convey the dynamic relationships in which the property of life resides.

Of the various concepts that suggest themselves in a description of the purpose-like character of vital phenomena, the more obvious have already been discussed and illustrated, e.g. 'adaptation', 'regulation', 'coordination', 'integration', &c. This is not to say that these concepts are mutually independent. Certainly the first three are closely interrelated and partly coextensive. Nevertheless, they are concepts of whose intelligibility we can be as tolerably satisfied as the present phase of discussion permits, and which do definitely penetrate to the core of life.

In the last analysis these and related concepts describe no more than interlaced patterns of adaptive behaviour, however different they may at first sight appear to be. Their main differences lie merely in the degree of complexity to which they refer and in the nature of their 'goals', rather than in the presence or absence of new connective relationships. The fundamental relation, therefore, is that of *adaptation*. We may, indeed, regard this relation as the ultimate analytical element in all the other concepts, and any one of them could be adequately defined in terms of this relation. Evidently, we have here a key-concept in our present ways of thinking about the purpose-like aspects of nature.

On these grounds it is expedient to concentrate first of all on the analysis of the idea of adaptation; to discover whether there exist any objective spatio-temporal relationships in the living world to which this concept may be assumed significantly to refer; and to attempt to distil from these relations their elusive purpose-like character. The remainder of this chapter will be devoted mainly to this task.

The Concept of Appropriateness

Speaking generally, it may be said that the notion of adaptation when applied to living nature refers to the widespread and striking *appropriateness* which organic activities show in relation to the needs of the organism, and to the *effectiveness* with which organisms meet the demands made upon them by their environment. For our present analytical purposes it will be expedient to take the idea of *appropriate response* as a starting-point and to begin by exposing some of the spatio-temporal relationships and patterns of causal connexions which it implies.

If we think of an organic response as a single, complex, physical event, the idea of appropriate response points beyond this complex event in three important respects.

In the first place, the response must be *to* something, it must be evoked or called into being by some antecedent environmental event or state of affairs.

Secondly, the response can be called 'appropriate' only in relation to the subsequent occurrence of some event or state of affairs towards the actual or probable occurrence of which we believe it to contribute effectively. This event or state of affairs is what is commonly regarded as the 'goal' or 'aim' of the response. Without explicit or implicit reference to this future event or state of affairs the idea of appropriateness is ambiguous, indeed meaningless. If a skater loses his balance, a response which would be appropriate from the point of view of restoring that balance would not necessarily be appropriate from the point of view of preserving the skater's dignity.

In most biological discussions this future event or state of affairs whose occurrence as the result of a given response is taken as the criterion for the response's appropriateness, consists of the subsequent survival of the given individual or of the species—or, at any rate, consists of the occurrence of some condition or event assumed to contribute to the probability of that survival. But it would be wrong to think that the idea of an appropriate response is confined to this particular case in biology. To recall an earlier example, even if a panicked animal happens to be running to its certain death we should still regard the action of jumping as an appropriate response to the approach of a given obstacle in its path. The response is now appropriate not from the point of view of the animal's survival but from the point of view of its continued movement along a certain path.

In the third place, whether or not a given response is appropriate depends on the environmental circumstances which it meets and with which it comes to interact. An action appropriate under one set of circumstances may be quite inappropriate under another. And the 'environmental circumstances' meant in this context are always circumstances which coincide with the execution of the response. They are circumstances which the response meets upon its execution and on which its success depends. Because, if under given environmental circumstances A an action B is said to be appropriate in respect of the successful realization of some subsequent environmental event C, the idea is: given A, the occurrence of C requires the occurrence of B. In other words, the occurrence of C is implied by the joint occurrence of A and B but not by the occurrence of either A or B alone. And unless A and B are here understood to refer to the same instant of time the possibility is left open for the occurrence of either one to imply the occurrence of the other and the whole conception would obviously become ambiguous and collapse. There is therefore a time lapse between the environmental conditions or events which initiate the given response and those on which its effectiveness is held to be contingent. This time-difference may seem a trivial matter at the present sketchy level of discussion, but it will gain in importance as our analysis gains in precision. It is in practice frequently obscured by the fact that the environmental circumstances which evoke a given response are often constant and hence are the same circumstances as those which the response meets upon its execution.

The notion of appropriate response is thus seen to link four primary and spatio-temporally distinct elements:

1. The environmental circumstances or events which act as stimuli and evoke the response.

2. The response.

3. The environmental circumstances which the response meets during its execution and on which its success depends.

4. The occurrence of a certain event or state of affairs, viz. the so-called 'goal' of the response,

as an effect of the interaction between (2) and (3).

Let us continue the present analysis with the aid of a concrete example. Let this example be one whose essential features could be realized either by a living organism or by a machine, so that we have for the present a simple guarantee that we shall not stray from relationships known to be compatible with the general laws of physics, and that we shall continue to think of appropriateness and adaptation in the non-psychological sense in which they have become important in biology. And let the response to be taken as example be one which acts on the environment only at a single instant of time, so that the temporal relationships involved take on a particularly simple form.

These requirements are aptly met by the example of a rifleman aiming his rifle at a moving, e.g. flying, target. The action of aiming the rifle at the approaching target may be fully representative of an appropriate response. Nor is this type of response confined to rational agents, or indeed living beings. Many animals have the power of aiming, and, what is more, all the essential features of this example could be reproduced mechanically by a radar-controlled and automatically sighted anti-aircraft gun shooting at a flying target. If at any time the discussion of the rifleman leaves us in doubt whether we are using our concepts in a sense in which they could be applied to animals other than rational thinking agents, a simple switchover to such examples as a cat leaping at a mouse will settle the matter. And if we should be in doubt whether we are using our concepts in a sense compatible with the general concepts of macroscopically deterministic physics, a similar changeover to the example of the anti-aircraft gun will again answer our query. The example also fulfils the requirement that the response of the rifleman or gun only acts on the environment at a single instant of time, viz. at the moment of firing.

In terms of this example the notion of the appropriateness of the gun's or rifle's line of fire involves again the same four spatio-temporally distinct elements:

1. The stimulus evoking the response may be assumed to be the appearance of the target within a given region, e.g. within the field of vision of the rifleman, or within the area swept by the radar beam. Let this stimulus be taken to occur at the time t_0, and be denoted by S_{t_0}. For simplicity let us also provisionally assume that the rifleman or gun-predictor appraises the target's course from a single look or a single set of measurements, and that this is taken at t_0.

2. The relevant part of the response itself consists of the line of fire adopted by the rifle or gun. Let t_1 be the time of firing, and let the direction in which the rifle or gun aims at that time, i.e. its line of fire, be denoted by R_{t_1}.

3. The environmental circumstances with which the actual firing coincides and in relation to which that line of fire is, or is not, appropriate, consists of all those environmental t_1-factors which determine the path of the target during the time-of-flight of the bullet. This set of factors contains, of course, the position, velocity, and direction of the target at the inception of this time of flight, i.e. at t_1, plus all forces and conditions as they exist at t_1 which will determine the target's path during that interval. Let this whole set of environmental circumstances be denoted by E_{t_1}.

4. Finally, we have the fourth element, the hypothetical future event or the 'goal' of the action, whose occurrence as the result of the response is taken as the criterion of the response's appropriateness. Here it is necessary to proceed with a little caution for there may exist two alternative conceptions:

(a) We may interpret 'appropriateness' in the sense that the rifle or gun is said to have been aimed 'appropriately' only if the bullet subsequently actually hits the target. Let the occurrence of such a direct hit at some point of time t_2 be denoted by G_{t_2}.

(b) Alternatively, we may adopt the attitude that the rifle or gun may be said to have been aimed 'appropriately' if it merely created at the time of firing the greatest probability feasible under the given circumstances of the bullet striking the target. This is a broader conception and probably the more common one.

In 4 (b) the path of the target and/or the bullet is conceived to be undetermined in some respects, and not, therefore, with certainty predictable by the rifleman or gun-predictor. In the present context this must not, of course, be taken to mean that the motion of the target or bullet is indeterministic in the sense that it is not accountable in terms of preceding physical events. Rather, it means that some of the factors which determine their paths are assumed to be unknown by the rifleman or that they cannot be taken into account by the gun-predictor. Typical factors of this kind are the wind velocity, random variations in the path of the target or, if the target is a living object, sudden unforeseeable impulses to change the direction of motion. Broadly speaking, the path of the target or bullet in that case can only be predicted subject to the hypothetical exclusion of the unknown factors. The predicted path is then not the actual path but merely an *ideal*, viz. the probable path—although the two may, of course, coincide. For instance, it is presumably true to say that both the ordinary rifleman and a simple gun-predictor work on the simplifying assumption that the target will pursue its course with constant speed and in a straight line during the time of flight of the bullet or shell. The exact meaning of 'probable path' in this context need not detain us at the present stage, provided we remember that in the sense of 4 (*b*) a statement about the appropriateness of the line of fire is not a direct statement about the actual situation but about an ideal situation which has all the main relationships in common with the actual situation except that abstraction is made of a certain number of interfering factors. As long as this distinction between 4 (*a*) and 4 (*b*) is kept in mind, there is no reason why we should not accord both cases the same analytical treatment. S_{t_0}, R_{t_1}, E_{t_1}, and G_{t_2} now refer to the corresponding events or conditions in either the actual *or* the ideal situation, it being understood that in the latter case they exclude all those unknown factors from which this ideal picture makes abstraction.

To elucidate the temporal and causal relationships which exist between the above four elements on the assumption of an appropriately aimed shot, we may enter these into a diagram in which the horizontal direction is used as a time coordinate, but in which no definite meaning is attached to the vertical direction beyond that of physical distinctness. If in this diagram we represent a causal connexion by an arrow, the relations between S_{t_0}, R_{t_1}, E_{t_1}, and G_{t_2} which have so far emerged may be symbolized as follows:

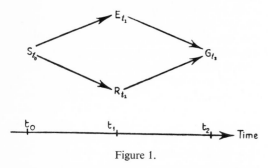

Figure 1.

After these preliminaries let us now investigate the exact meaning of 'appropriateness' when the rifle's or gun's line of fire is stated to be appropriate in relation to the path of the target and

from the point of view of scoring a direct hit. Does the term 'appropriate' in this context merely mean that the line of fire is *effective* in scoring a hit? The answer to this important question is No. Although the idea of 'appropriateness' implies that of 'efficacy' it also transcends the latter in at least one important respect: The idea of 'efficacy' covers only the *given* situation or its development, whereas, if we say that under given circumstances a certain line of fire is 'appropriate' we envisage not only the given course of the target but at the same time compare it before the mind's eye with an extended range of possible *alternative* courses which the target might have taken and for each of which there exists one, and usually only one, effective line of fire. Each one of these respective lines of fire is called the 'appropriate' one in relation to the corresponding path of the target. In other words, the statement that a given line of fire is 'appropriate' in relation to a given course of the target, asserts that the given course of the target is a member of a set of possible alternative courses and that the given line of fire is the corresponding member of a correlated set of possible alternative lines of fire, each member of the latter set being effective in conjunction with the corresponding member of the former.

The notion of appropriateness, therefore, is based on a vague consideration of a set E'_{t_1}, E''_{t_1}, E'''_{t_1}, ..., of alternative circumstances and a correlated set R'_{t_1}, R''_{t_1}, R'''_{t_1}, ..., of effective responses, the members of the two sets standing in a one–one correspondence such that under the given conditions only corresponding pairs of members will lead to a direct hit. These two sets are not sets of actual events but sets of imagined alternatives to actual events, the alternatives differing from one another by the hypothetical variation of some point of detail, e.g. variation in the target's direction. The whole idea is the result of a creative act of thought. From Gibbsian thermodynamics we may borrow the name *ensemble* for such a set of events, situations, or systems created in our thoughts as exact counterparts of one another except for one or more properties which are hypothetically varied according to a freely chosen principle. This principle of construction may be called the *generative principle* of the respective ensemble.

According to the above analysis an appropriate response is an effective response which is conceived against the background of an ensemble of alternative sets of environmental circumstances and as a member of a correlated ensemble of effective responses. In one point of detail this account requires a slight qualification in the present example and, probably, in most practical cases. The ensemble of alternative paths of the target is not usually conceived to be a finite ensemble: its membership has the power of a continuum, since we do not usually envisage a discrete number of alternative paths but a continuous (although finite) range of variation of the target's path. The aggregate of alternative paths is not thought of member by member but merely in terms of its approximate limits.

The exact meaning of the term 'possible' in the context of 'possible alternative paths of the target' will become clearer as we go along. The original context in which the concept of 'appropriateness' was formed was no doubt psychological and referred to rational beings which have a set of alternative responses ready to meet alternative circumstances, and which are conceived to be free to choose the appropriate response from this set of available responses. The 'possible' alternative circumstances are then merely the different circumstances which have frequently occurred in their experience, or have been recorded by others. But when this idea of appropriateness is applied to activities which do not spring from a conscious and rationally thinking mind, and when its meaning, therefore, becomes metaphorical in comparison with this psychological use, the term 'possible' also shifts its meaning and the whole matter becomes more intricate. However, these considerations concern merely the range of variation embraced by the given ensemble and do not alter the essential fact that when we think of appropriateness we envisage a finite range of hypothetical variations of the actual environmental circumstances in conjunction with a correlated ensemble of effective responses.

Transition to the Concept of Adaptation

In making the transition to the concept of adaptation we must distinguish the activity of *adapting* any X to any Y from the relationship of *adaptedness* between X and Y which this activity establishes. That is to say, in our ballistic example we must distinguish the activity of taking aim from the relationship of adaptedness which it establishes between the line of fire and the course of the target. Let us take the concept of 'adaptedness' in this latter sense as the primary concept.

The first thing to note is that in the usual biological context the idea of 'adaptedness' implies *appropriateness*. We would not consider the

falcon's flight to be adapted to the direction taken by its prey unless it were appropriate to minimising the distance between them. In fact, it is only in the special psychological sense and when referring to rational beings who are pursuing a consciously conceived goal that we might find an exception to this. For it is only in this psychological sense that we might call an activity adapted to the environment merely on the grounds that it happens to spring from a subjective desire on the part of the agent to act appropriately—irrespective of whether or not the action turns out to be successful and appropriate in the objective sense. But in spite of the fact that this psychological use may have been the original use of the words 'adapted' and 'adaptation', in biology these words have found common currency without any desire on the part of the speaker to attribute the presence of a rational and conscious mind to the organism concerned, and, as I have stressed in Chapter I, it is only this significant meaning of adaptation that we intend to analyse. In this sense, therefore, to say that the organic response X is adapted to the environmental circumstances Y from the point of view of some future state of affairs Z towards the realization of which it is conceived to be directed, implies that it is appropriate and hence also that it is effective in bringing about the actual or probable occurrence of Z.

The crucial question, however, which we must now take up is whether or not any significant relationships other than those of appropriateness and efficacy are implied by the concept of adaptation. Or must we assume adaptedness and appropriateness to be synonymous? The writings of most mechanists imply an affirmative answer to this last question. According to this view, to say that an organic action is adapted in certain respects, means no more than that it is appropriate in those respects. It cannot be presumed that the adherents of this restricted interpretation ever felt very happy about their attitude, but so long as they had found no way of scientifically formulating any other implication of the concept of adaptation their scientific discipline and honesty left no alternative open. In extreme cases 'adaptation' was even rendered synonymous with mere 'efficacy'. Witness this statement from a leading contemporary treatise on the general principles of life: 'Organisms are adapted to their environments. And all that this means is that their characteristics are such that in these environments they are able to live.' But can it really be true that 'adaptation' means no more than these utterly trivial relationships of mere efficacy? Can it be true that generations of biologists have meant no more than this when they found in that concept a vehicle for expressing the most distinctive features of living systems and organic activities? Are not these simple causal relations such as can be found in any inorganic system? Can we really maintain that the cloud is adapted to the sun in the same sense as that in which, for instance, the path of the pursuing hound can be said to be adapted to that of the fleeing hare?

The idea that adaptedness implies no more than appropriateness is equally quickly dispelled be concrete examples. For instance, if this doctrine were true, any fish could significantly be said to be adapted to any aquarium in which it survives, in spite of the fact that we would usually regard the aquarium as the adapted object. And there is another weighty reason which speaks against the idea that adaptedness is synonymous with either appropriateness or efficacy. For it is possible to name an important category of organic activities which are called 'appropriate' and 'effective' but which would certainly not be called 'adapted'. I mean all those actions which we are wont to describe as 'accidentally appropriate': the rifle or gun goes off accidentally during loading and by a rare stroke of fortune the target is hit. The line of fire will in such a case be called 'accidentally efficacious' or 'accidentally appropriate', but certainly not 'adapted'. Again, a gene-mutation may have accidentally appropriate effects in respect of the survival of the species, but the variant characteristic does not become a case of 'adaptation' in the full sense until its selective advantages have caused it to become firmly established in the population and have therefore made its existence in the population more than an accident. The meaning of 'accidental' in these contexts will be discussed at a later stage. Suffice it to say here that we have no reason to consider the term 'accidental' to be meaningless in the present context and that we may take it to specify a definite category of organic responses which are appropriate under the given environmental circumstances without being adapted to these circumstances.

No more need be said to reveal the far-going inadequacy of any interpretation of adaptation which equates it merely with either appropriateness or efficacy. We must, therefore, regard it as our next task to discover what the idea of adaptation implies in addition to these two concepts.

The example of the automatic gun will come to our aid in this respect. In terms of this example our question is: if we say that, at the time t_1 of

firing, the gun's line of fire R_{t_1} was adapted to the movement of the target E_{t_1} from the point of view of securing a direct hit G_{t_2}, is it true to say that we mean no more than that at that instant the gun's line of fire was such as to cause the target to be hit?

The answer to this question is in the negative and we can now see why. For, surely, we mean in addition that, at the time of making this appropriate response to the movement of the target, the gun-training mechanism was objectively so conditioned that there existed a definite range of possible variations of the target's path such that, whatever other course the target *might* have taken (within certain limits: see below), the mechanism *would* have responded by an appropriate modification of the gun's line of fire. We mean not only, therefore, that the gun's line of fire was appropriate in the given instance and in relation to one single given path of the target, but in addition that this appropriateness was secured by objective system-properties of the gun-training mechanism which would also have caused any one of an extended range of *alternative* paths to have been matched by an appropriate line of fire.

It is in this special sense that the objective system properties which the automatic gun possesses when in working order come to raise the occurrences of direct hits on the target above the level of pure chance coincidence. And it is in the same sense that adaptive behaviour, or an adaptive evolutionary change, in the animal world comes to assume that typical character of involving an objectively biased and non-accidental occurrence of appropriateness and success.

Two sets of elements are said to be 'correlated' if we can establish a one–one correspondence between their members such that to each member of the one set there corresponds one and only one member of the other set. In terms of this notion of correlated sets of elements the relation between the ideas of efficacy, appropriateness, and adaptation which has so far emerged may be summed up as follows:

1. The response R_{t_1} is *effective* in relation to the environmental circumstances E_{t_1} and in respect of the subsequent occurrence of the 'goal' G_{t_2}, if the joint occurrence of R_{t_1} and E_{t_1} causes the subsequent occurrence of G_{t_2}.

2. R_{t_1} is *appropriate* in relation to the circumstances E_{t_1} and in respect of the subsequent occurrence of G_{t_2}, if E_{t_1} is a member of a hypothetical set or ensemble $E'_{t_1}, E''_{t_1}, E'''_{t_1}, \ldots$ of possible alternative environmental circumstances while R_{t_1}

is the corresponding member of a correlated set R'_{t_1}, R''_{t_1}, R'''_{t_1}, ... of possible alternative responses, each member of the latter set being effective in bringing about G_{t_2} if and only if it occurs in conjunction with the corresponding member of the former set.

3. Finally, the statement that R_{t_1} is *adapted* to E_{t_1} in respect of G_{t_2} means that the organism or mechanism causally determining R_{t_1} is objectively so conditioned that if certain changed initial circumstances had caused the occurrence of any alternative member of the set E'_{t_1}, E''_{t_1}, E'''_{t_1}, ... it would in each case also have caused the occurrence of the corresponding member of the correlated set R'_{t_1}, R''_{t_1}, R'''_{t_1}, ... of appropriate responses.

It will be seen from these results that the assertion that a given response is 'appropriate' under given environmental circumstances tells us no more about the actual situation than the statement that it is effective—although it tells us something of the degrees of freedom which the given situation is imagined to have, i.e. something about our conception of that situation. In contrast, the statement that a given response is 'adapted' to given environmental circumstances does tell us more about the given situation than the mere assertion of efficacy. Because it tells us that the rifleman, the gun, the organism, or whatever the case may be, *is* such that under this or that alternative set of circumstances it *would* have reacted in this or that modified way. We noted above that adaptation was a form of non-accidental appropriateness: in the objective system property formulated in the last sentence we have the objective basis of this non-accidentality.

In relation to our concrete ballistic example the meaning of 'adaptation' may be expressed as follows: the gun's actual line of fire (R_{t_1}) is *adapted* to the target's path and the factors that influence it (E_{t_1}), if (*a*) the gun-plus-target system is at that time so ordered that a direct hit will be scored, and (*b*) there exists a hypothetical range of variation of E_{t_1} such that if certain changed initial circumstances had caused E_{t_1} to adopt other values within this range, then R_{t_1} would have adopted just those new and different values which in each such case would have led to the scoring of a direct hit. In short: the gun's line of fire *was* appropriate *and would have been* appropriate had the initial circumstances been otherwise.

The nature of the 'changed initial circumstances' referred to in these definitions will be discussed in the following section.

The Coenetic Variable

The only causal connexions explicitly referred to in the preceding discussion and definition of 'adaptation' are the two shown in Figure 2.

Only three of the four elements involved in the concept of appropriateness, illustrated in Figure 1, have so far entered our definition of adaptation and in this respect, therefore, the latter is still incomplete. To complete the definition we shall find that we must add an explicit reference to a fourth event or state of affairs. This fourth element may or may not be identical with the stimulus S_{t_0} but it is similarly connected with E_{t_1} and R_{t_1}, and it is a necessary ingredient of any instance of adaptation in the sense of the preceding definition.

The need for an explicit reference to such a fourth element becomes at once evident when it is asked how the one–one correspondence, to which our definition refers, between alternative values of E_{t_1} and the correspondingly modified

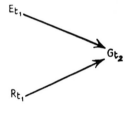

Figure 2.

values of R_{t_1}, can in practice be brought about, having in mind that these two sets of factors belong to the same instant of time and are therefore incapable of causally determining each other. The only way in which this correlation is physically possible is through *joint causation*, i.e. when there exists some prior environmental state of affairs or event which enters into the causal determination of both E_{t_1} and R_{t_1}, and of which both, therefore, are parallel effects. We shall call this necessary common causal determinant of E_{t_1} and R_{t_1} the *coenetic variable* of the adaptation. Our definition of adaptation must therefore be enlarged by the requirement that the hypothetical variations of E_{t_1} to which this definition refers, comprises only such variations as result from the hypothetical variations of a coenetic variable, i.e. of a factor which is also a causal determinant of R_{t_1}. In other words, the 'changed initial circumstances' referred to in that definition must be interpreted to mean variations of the t_0-values of one or more coenetic variables of the adaptation. In our ballistic example the coenetic variables of the adaptation between R_{t_1} and E_{t_1} consist of all these prior factors entering into the determination of the target's subsequent flight that the rifleman or the gun-predictor 'take account of' in determining their line of fire. Usually these factors will consist of the values which the position, direction and velocity of the target have at a point of time, say t_0, which precedes t_1 by the reaction lag of the rifleman or the gun-training mechanism.

If we denote any one of these coenetic variables by CV and its value at t_0 by CV_{t_0}, we may represent the four principal elements involved in the adaptedness of the gun's or rifle's line of fire to the target's path, and the causal connexions between them, by the following diagram.

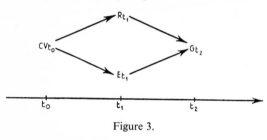

Figure 3.

The necessity for the existence of one or more coenetic variables in every case of adaptation is also intuitively obvious. It is impossible for any machine or living organism to adapt a certain action to the environment unless if has information about the environment, i.e. unless it can be influenced by some features of that environment. These features are in the last analysis always features existing prior to the adapted action concerned. Adaptation by means of rational prediction is no exception to this rule. For every prediction argues to future events from the state of affairs existing prior to the actual act of predicting. If the gun-training mechanism or the rifleman in determining a line of fire were prevented from taking any factors connected with the target's prior path into account, they would be acting merely 'blindly'. They could in that case bring about an accidentally appropriate line of fire but never an adapted one. The metaphorical comparison with blindness is illuminating in this connexion, because it underlines the fact that a fundamental biological function of the eyes, and for that matter of all sense organs, is precisely to establish causal connexions which will enable environmental variables to become the coenetic variables of adapted organic behaviour.

If adaptive actions are of the nature of responses, in other words, if they are evoked by specific environmental stimuli, the stimulating events may at the same time be the coenetic variables of the adaptation. But this is not necessarily so. While one set of environmental circumstances may evoke or release a given action, another set may provide the main influence in shaping the details of its execution.

Directive Correlation

In the preceding section the essential incompleteness of our provisional explication of adaptation was seen to be this: As generative principle of the ensemble E'_{t_1}, E''_{t_1}, E'''_{t_1}, ... only those variations of E_{t_1} are to be allowed which are entailed by the independent variation of a coenetic variable, i.e. of an event or state of affairs which is prior to, and a *joint* causal determinant of, both E_{t_1} and R_{t_1}.

On the strength of these considerations we shall now formulate the following provisional definition to introduce a new concept which will play a leading part in the remainder of the book, viz. the concept of *directive correlation*. It will become our main instrument for expressing in precise terms the various forms of purposiveness found in nature.

Definition. Any event or state of affairs R_{t_1} occurring at a time t_1 is *directively correlated* to a given simultaneous event or state of affairs E_{t_1} in respect of the subsequent occurrence of an event or state of affairs G_{t_2} if the physical system of which these are part is objectively so conditioned that there exists an event or state of affairs CV_{t_0} prior to t_1, and a set of possible alternative values of CV_{t_0}, such that

(*a*)   under the given circumstances any variation of CV_{t_0} within this set implies variations of both R_{t_1} and E_{t_1};

(*b*)   any such pair of varied values of R_{t_1}, E_{t_1} (as well as the pair of their actual values) is a pair of corresponding members of two correlated sets of possible values R'_{t_1}, R''_{t_1}, R'''_{t_1}, ... and E'_{t_1}, E''_{t_1}, E'''_{t_1}, ..., which are such that under the circumstances all pairs of corresponding members, but no other pairs, cause the subsequent occurrence of G_{t_2}.

Our previous analysis of adaptation shows that every adaptation is an instance of directive correlation, but not vice versa. Directive correlation is a slightly wider concept. For inspection of the usual meaning of 'adaptation' shows that in the case of an organism-plus-environment system the coenetic variable consists always of an environmental variable and, in fact, usually of an antecedent value of the environmental variable to which a given response is regarded as adapted. On the other hand, directive correlation, as defined above, contains none of these restrictions concerning the coenetic variable. Incidentally, the latter of these two restrictions sometimes leads us to speak of adaptation as existing not between R_{t_1} and E_{t_1}, but between R_{t_1} and CV_{t_0}. This is, for instance, the case in such phrases as 'the animal *is* adapting itself to the chang*ed* circumstances'. This variant of the normal use of the term, however, in no way affects the intrinsic relationships implied by the concept of adaptation.

It will be noted that in the definition of directive correlation, and also in our interpretation of adaptation, we are now no longer confined to those activities which are classed as 'responses'. All forms of physical activity come within the orbit of our definitions and we can henceforth ignore the special case of so-called 'responses'. As far as the causal determination of adaptive activity is concerned we have seen that it is the existence of coenetic variables which counts and not the existence of specific stimuli. In other words, the emphasis has shifted from environmental factors which evoke a given organic action to the general class of factors which determine its specific nature.

In the second place, it should be noted that our definition of directive correlation applies to living organisms in exactly the same sense as to dead machines of the type of our anti-aircraft gun. For, although we were mainly guided by the latter example, it is evidently possible in exactly the same sense to assert of the rifleman that, *qua* living organism, he was at the time of taking aim objectively so conditioned that this or that varied path of the target would have evoked this or that modification of his line of fire. Of course, it may be more difficult to verify such assertions about the rifleman than about the anti-aircraft gun, but this is a different question which will be discussed later on and which does not impair the significance of the assertions as such.

In fact, it can be said that there are no adaptive vital activities (in that biologically so significant sense of 'apparently goal-directed activities') in which we cannot reveal exact counterparts to the variables and causal connexions in terms of which we defined 'directive correlation'. Directive correlation, in short, is the fundamental and objective system-property which in more or less complex forms underlies the phenomena of adaptation in

nature and their purpose-like character. We have therefore succeeded in formulating this purpose-like character in terms which, although not yet quite up to the level of physico-mathematical precision (this will have to wait until Chapter III), yet are sufficiently precise and definite to go on with.

An important point to note is that although we have defined the directive correlation between E_{t_1} and R_{t_1} in terms of a future event, viz. the occurrence of a direct hit on the target (G_{t_2}), there was nothing in the least teleological in this definition. Our definition of directive correlation does not in the least imply that this future event G_{t_2} is a cause of either E_{t_1} or R_{t_1}, or that in any sense whatsoever it enters into their causal determination. The definition of directive correlation employs a reference to G_{t_2} only as a criterion in specifying the one–one correspondence which the gun-training mechanism effects (when in working order) between a certain range of alternative paths of the target and the gun's lines of fire— i.e. in specifying the correlation between the sets E'_{t_1}, E''_{t_1}, E'''_{t_1}, ..., and R'_{t_1}, R''_{t_1}, R'''_{t_1}, ... There is nothing teleological, in the sense of the future determining the present through some form of 'final causation', in the behaviour of the automatic gun. Its mode of operation is strictly mechanical and deterministic. The fact that the engineers who designed the gun may have been thinking about future targets and future direct hits has nothing to do with this particular issue. It merely means that certain states of consciousness of the engineers concerned entered into the determination of the design and construction of the gun; and that does not in the least alter the fact that in the gun-plus-target system we have a physical system which may be regarded as closed and in which nothing but straightforward physical causation is at work during the events which we refer to when we assert that the gun's line of fire is adapted to (or directively correlated with) the path of the target.

At the present, and still comparatively vague, level of discussion it already emerges, therefore, that 'adaptation' in its biological and purpose-like meaning can be paraphrased in perfectly orthodox causal–analytical terms. In fact, when we reach the physico-mathematical definition of directive correlation we shall see that it specifies an objective system property of material systems which may be regarded as just as respectable a physical property as any dealt with in orthodox physics.

The reader may at first see a difficulty in the fact that we have defined 'directive correlation' and 'adaptation' in terms of purely hypothetical conditions. For, in defining the property which the gun-training mechanism possesses when in working order, we referred not only to the gun's actual behaviour in a single concrete instance, but in addition to the gun's behaviour under hypothetically varied circumstances: *if* the target had taken a different course the gun-training mechanism *would* have caused the gun's line of fire to be modified accordingly. The elements of our ensembles are, all but one, imaginary elements.

But this difficulty is only apparent, because from the scientific point of view there is nothing epistemologically difficult in this reference to hypothetical conditions. The concept of directive correlation and adaptation in this respect implies merely that the gun-controlling mechanism *is* such that under such-and-such circumstances such-and-such things *would* have happened. And to specify actual properties of given material systems in terms of the systems' reactions under hypothetically varied circumstances is scientifically not only legitimate (provided, of course, that the hypothetically assumed conditions do not involve us in contradictions), but is in fact one of the fundamental modes of physical description. The physical principle of virtual variation and classical thermodynamics, for instance, make explicit use of this method, and we may even go so far as to say that this idea is implied in most simple physical statements. For instance, to say that the object X has a temperature Y means to the physicist little else than that such-and-such a process of measuring *when* applied to the object X *would* lead to the pointer-reading Y. And just as it can be significantly asserted that X has the temperature Y at a time when no measurements are actually being made, so it can be significantly asserted that the gun-training mechanism has the objective system-property of directive correlation irrespective of whether this property is at the time put to the test by confronting the gun with a succession of targets pursuing alternative paths. The notion of causality, too, will be seen to depend on a similar comparison between actual and hypothetically varied situations. Moreover, a similar state of affairs prevails in everyday life: if we say that the table *is* hard, we mean, strictly speaking, that it *is* such that it *would* resist if pressure were applied.

It was seen above that the concept of adaptation refers not merely to a dyadic relationship—as the syntactical structure of the statement 'R_{t_1} is adapted to E_{t_1}' might lead one to suppose—but essentially to a disguised tetradic relationship. Four elements are involved in the tetrad, viz. the elements exemplified above by CV_{t_0}, E_{t_1}, R_{t_1}, G_{t_2}, and four causal connexions between them (Figure 3). Moreover, it was seen that the concept of adaptation does not actually predicate anything of these four relationships, but, rather, by means of implications concerning their effects under hypothetically varied circumstances specifies an objective system property of the material system involved—in our examples, of the rifleman or of the gun. Upon careful analysis, therefore, the predicate 'adapted' turns out to be an extremely oblique predicate: To call an action 'adapted' is really making a statement about the agent rather than about the action. The properties the statement attributes to him are such that they can only be formulated explicitly by referring to hypothetically varied circumstances. No wonder adaptation and related concepts have for so long confounded the biologists, philosophers, and theologians. There is something almost pathetic in the obstinate search of some philosophers for specific properties of purpose-like activities in nooks and atomic interstices in which the predicate of adaptation never asserted them to exist in the first place. No wonder they returned empty handed and felt driven to the conclusion that from the scientific point of view the purposiveness of organic activities is a meaningless concept.

The Focal Condition

In our analysis of the directive correlation between the gun's line of fire and the target's path, the occurrence of a direct hit was seen to be the criterion in terms of which the respective correlation was defined. In the sequel we shall call the event or state of affairs which functions as the corresponding criterion in any case of directive correlation the *focal condition* of that directive correlation. We introduce this term, therefore, generally to replace the rather misleading words 'goal', 'end', or 'aims' in connexion with adaptive and purpose-like behavior in biological systems.

The elementary pattern of the causal connexions involved whenever two physical events or conditions are directively correlated, was illustrated in Figure 3. I have called this the *elementary* pattern of directive correlation because actual situations may, of course, be far more complicated than this simple diagram suggests. In many instances of directive correlation or adaptation there will be more than one coenetic variable, more than one focal condition, and more than two directively correlated variables. Thus, a directive correlation between five correlated variables with one coenetic variable and one focal condition would involve a pattern of causal connexions of the following type:[1]

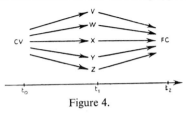

Figure 4.

The peculiar property of any focal condition is that the directive correlation with which it is associated imparts to it a special kind of independence from the effects of events which have entered into its history. For, although the event or state of affairs FC is causally connected with the values which the correlated variables V, W, X, Y, Z, assume at t_1, and although these in turn are causally connected with the value of the coenetic variable CV at t_0, the occurrence of FC at t_2 is itself independent of the latter (at least within a certain range of variation). In terms of our ballistic example: although the occurrence of a direct hit at t_2 is causally dependent on both R_{t_1} and E_{t_1}, and although both E_{t_1} and R_{t_1} are in turn causally dependent on the target's position, direction, velocity, &c. at t_0, viz. on CV_{t_0}, yet the adaptation of the gun's line of fire ensures that, within certain finite limits, hypothetical variations of CV_{t_1} will not affect the subsequent occurrence of a direct hit. We shall return to the investigation of this special independence between CV_{t_0} and G_{t_2} presently.

Degrees of Adaptation

It was seen that any instance of directive correlation between two or more variables requires the existence of at least one coenetic variable. It was also seen that the coenetic variable is the basic independent variable in the concept of adaptation and that its variation constitutes the generating principle of the ensemble which this concept envisages.

The range of variation of the coenetic variable over which a given instance of directive correlation remains valid is usually limited. The automatic gun or the rifleman can adapt their line of fire only to a limited range of variation of the target's course, and they can take into account only a limited number of factors which determine this course. This suggests that if it is possible to attach some measure to the magnitude of the range within which a coenetic variable can be varied without the directive correlation breaking down, then this may be taken as a measure of the *degree* of directive correlation. Two further quantities which may be taken to indicate degrees of directive correlation are: the number of coenetic variables and the number of directively correlated variables. There are therefore several senses in which it is possible to distinguish between different degrees of directive correlation in similar systems. In the same senses it is accordingly possible to give precise meanings to the ideas of 'degree of adaptation' and 'degrees of purposiveness' in the comparison of similar physical systems.

The Asymmetry of Adaptation

The relationship of adaptation, in the dyadic form in which it appears syntactically as in the sentence 'X is adapted to Y', is asymmetrical: 'X is adapted to Y' does not imply 'Y is adapted to X'. Directive correlation, on the other hand, is a symmetrical relationship as far as the correlated variables are concerned. The concept of adaptation, therefore, cannot be regarded as strictly synonymous with 'directive correlation' and represents only a special case. This qualification was already indicated but the point requires a few additional remarks.

Consider again our ballistic example. In that example the coenetic variable consisted of the set of t_0-values of a number of physical variables connected with the motion of the target, e.g. its direction, velocity, &c., at the time t_0. They were the variables which the gun-predictor or the rifleman takes into account in predicting the target's course. Meanwhile, E_{t_1} comprised the respective t_1 values of the target's direction, velocity, &c. In other words, in this particular case the coenetic variable could really have been represented by the symbol E_{t_0}. And if we look for other examples of adaptation we find this to be quite generally true: when we assert the value X_{t_1} of some variable X to be adapted to the value Y_{t_1} of a variable Y, we are referring to a directive correlation in which some prior value of Y, say Y_{t_0}, acts as the coenetic variable, i.e. in which $CV_{t_0} \equiv Y_{t_0}$.

Take, for instance, a case already alluded to, and consider the difference between asserting that the size of a certain fish is adapted to the size of its aquarium, and asserting that the size of the aquarium is adapted to the size of the fish. In the former case we imply that if the size of the aquarium had been different this would have entailed a different choice of fish; in the latter case the implication is that if the size of the fish had been different this would have led to a different choice of aquarium. In the first assertion, therefore, we are dealing with a case of adaptation in which the size of the aquarium at some prior point of time is regarded as the coenetic variable, whereas, in the second assertion the size of the fish, as it was at the time when the aquarium was chosen, acts as the coenetic variable.

If all this be accepted as the correct interpretation of the most common and significant biological meaning of 'adaptation', we may summarize our preliminary analysis of this concept by giving it the following explication:

> The statement that an event or state of affairs X_{t_1} is *adapted* to another event or state of affairs Y_{t_1} from the point of view of some future event or condition FC_{t_2}, means that X_{t_1} and Y_{t_1} are connected by a directive correlation which has FC_{t_2} as focal condition and in which a prior value of Y, say Y_{t_0}, acts as coenetic variable.

The asymmetry of adaptation discussed in this section suggests at least one reason why, for instance, in the conception of the gun's line of fire being adapted to the target's motion, the 'responsibility' for this adaptation is always attributed to the gun-training mechanism, in spite of the fact that this mechanism is only responsible for *one* of the two chains of causal connexion which link the coenetic variable with the focal condition. The reason is that E_{t_1} is a causal function of E_{t_0} more or less as a matter of course, whereas the special interference of the gun-training mechanism is required to make R_{t_1} a causal function of E_{t_0}, and particularly to make it the kind of function of E_{t_0} which will result in a directive correlation.

. . . . .

The Independence of the Focal Condition and the Coenetic Variable

We have seen that if a physical event or state of affairs X_{t_1} is directively correlated to another event or state of affairs Y_{t_1}, with a certain event or state of affairs FC_{t_2} as focal condition and with CV_{t_0} as coenetic variable, the correlation between X_{t_1} and Y_{t_1} renders the occurrence of the condition FC_{t_2} independent of variations of

CV_{t_0} (provided these remain within certain limits) in spite of the fact that FC_{t_2} is causally connected with X_{t_1} and Y_{t_1} and that each of these in turn is causally connected with CV_{t_0}. For instance, the effect of the adaptive action of the gun-training mechanism in our ballistic example was to render the occurrence of G_{t_2} independent, within a certain range of variation, of E_{t_0}, in spite of the fact that the occurrence of G_{t_2} is causally connected with both E_{t_1} and R_{t_1}, and that these in turn are causally connected with E_{t_0}.

The nature of this special type of independence merits closer attention: it is an independence which is not based on the absence of causal chains between FC_{t_2} and CV_{t_0}, but on the fact that there are at least two such chains (the one involving X_{t_1} and the other involving Y_{t_1}) and that the partial effects of these two chains exactly compensate or offset each other.

We have here another example of the importance of giving the concept of causal connexion a precise definition: Can FC_{t_2} in this case be said to be causally connected with CV_{t_0} at all? Those who stress the transitiveness of causal connectivity will be inclined to answer this question affirmatively. An equally good case can, however, be made out for the opposite answer, since it may be argued that FC_{t_2} cannot be said to be causally connected with CV_{t_0} if variations of the latter have no effect on the former. . . .

One of the fundamental characteristics of living organisms is their independence from environmental fluctuations, and it has often been remarked that 'progress' in the living world consists largely of a growth of this independence. We are now in a position to see a little more clearly the general structure of the relationships which are responsible for the most important part of this independence, viz. for the independence enjoyed by the organism by virtue of adaptive or other purpose-like activities which have its survival as a focal condition. For inasmuch as this independence is due to the purpose-like character of the responses and internal adjustments which the organism makes to environmental disturbances, it is precisely the type of independence which we have just discussed, viz. the independence which the focal condition of a directive correlation enjoys in respect of variations of the coenetic variable.

We may assume that it were some such relationships of which, for instance, Rignano was vaguely aware when he wrote of the strikingly purpose-like character of some organic activities: 'If we consider embryological phenomena or phenomena of regeneration we realise that when external circumstances change for an organism, the reaction processes produced by the change of circumstances also change, but that the final result of these processes, however different they may be as compared with one another, always remain the same'—and when he proposed this as a criterion for 'Finalism' in nature.

The main error of this passage and of similar passages with other authors is that they commit the 'genetic' fallacy of treating as a temporal sequence of compared situations what should properly be presented as a logical sequence of alternatives: they confuse actual successive temporal variants of a given situation with the hypothetical variations embodied in an ensemble. 'Directive correlation' as we have defined it does not refer to any correlation between pairs of elements belonging to any actual and temporal sequence of comparable situations. It specifies a property of one single given situation in terms of a purely imaginary ensemble of alternative situations. We may, of course, by way of experiment, actually attempt to realize these hypothetical variations: we may let the automatic gun of our example run through a temporal series of situations in which many of the hypothetical alternatives of the target's path are in turn actually realized. Such a test series of experiments would be the proper procedure to adopt if we wanted to verify whether the gun's line of fire really had the adaptive character which we claimed it had. But, as was remarked before, the significance of the assertion that the gun's line of fire has this adaptive character does not as such depend on the existence of such a test series of experiments. Directive correlation can be significantly asserted of single and non-recurring situations.

This question of verifiability is important because it is notoriously difficult to conduct a test series of controlled experiments with living organisms. The type of test series usually required, and also required for the strict verification of any claim about directive correlation, consists of a series of repeated experiments which differ from one another only in the variation of the initial value of a single variable and in all other respects are performed under identical conditions. It is the most important type of experimental test series in physics and its practical impossibility in much biological material is a severe handicap for the biologist. But the only methodological significance of those practical difficulties is that they make the discovery and verification of functional relationships between biological variables more difficult. It does not in any sense affect the significance, as such, of biological hypotheses which assert the existence of certain functional relationships. If it did, not only biological theory but also physical theory would be impossible since the mathematical functions postulated in physics usually relate to an infinite set of values of the physical variables concerned, whereas, of course, no test series can comprise more than a finite number of measurements. No test series, therefore, can verify completely the existence of any functional relationship of this kind between physical or biological variables: it can only supply a number of tests which either encourage the scientist to accept a certain functional relationship between empirical variables as a hypothesis, or urge him to seek alternative hypotheses.

Directive Correlation and the Teleological Fallacy

Our analysis and examples have shown that directive correlation is a physical property which can exist in a closed physical system without presupposing the presence within that system of a rational agent and of conscious mental processes, i.e. of processes which are purposive in the literal sense of the word and involve rational thought, the presence of visualized aims, and fixed resolutions in the mind of a thinking agent. It is therefore seen that conscious mental processes constitute but one of several possible mechanisms in nature which can cause a system to have the objective property of directive correlation. Moreover, our survey of the factual background establishes without doubt that conscious mental processes in the above sense are the exception rather than the rule as causes of directive correlation.

Yet, however exceptional in living nature as a whole those directive correlations may be which arise from the conscious activity of rational agents, it is the only case of whose working Man has a direct and introspective knowledge. Hence, it came about that, as Man awoke to the objective phenomena of directive correlation in nature, he began to describe them in terms of concepts introspectively derived from his own experience, and to interpret them in terms of anthropomorphic and psychological analogies—in terms of 'purposes', 'goals', 'aims', &c. The resulting confusions in biology were fatal. Yet, when all is said, it cannot really be surprising that Man, being part of nature, found analogies in nature to his own behaviour which invited such extrapolation.

No doubt the concept of adaptation has such a psychological and anthropomorphic origin. Its original and, in a certain sense, literal meaning refers not to the objective system property which we have called 'directive correlation', but to quite a different, subjective, correlation: viz. to the subjective correlation which exists between Man's intended actions and the opportunities for realizing his ends which he rightly or wrongly believes to exist in his environment. This is a correlation which expresses no more than the fact that, given his goal and given his interpretation and assessment of the existing environmental circumstances, the nature of his *intended* actions is usually determined. The correlation, therefore, refers to a subjective psychological process and bears no necessary relation to the objective system properties of directive correlation.

Gradually, however, Man's intensified study of nature brought about a subtle process of semantic transformation and the concept of adaptation began in its own right to denote the objective relations which our analysis has rendered explicit and which we have incorporated in the concept of 'directive correlation'. Thus in everyday and technical use 'adaptation' has changed from a subjective psychological term to the objective biological term which is occupying us at the moment. This new meaning of 'adaptation' has since continued to exist side by side with the old, and the schism between vitalism and mechanism resulted largely from the inability of biologists to keep these two meanings apart. This separation was of course no simple matter so long as these two distinct meanings had not been given explicit definitions. For instance, when we speak about 'right' or 'wrong', 'successful' or 'unsuccessful' adaptation we are using the concept in its psychological sense, since these predicates are based on a comparison between the goal desired by a rational agent and the results actually achieved by his purposive actions. They cannot, therefore, be applied to the concept of directive correlation. 'Successful (or unsuccessful) directive correlation' is a meaningless phrase.

Since this generalization of 'adaptation' in biology was a gradual and unconscious historical metamorphosis, it carried into the new ways of thinking all the terminological chattels of the original psychological meaning of 'adaptation', e.g. such terms as 'aim', 'purpose', 'striving', &c. This was unavoidable so long as the new conception had been given no adequate analysis and no new terminology had been made available to denote those elements and relations which were analogous to certain elements and relations in the old conception. It is for this reason that so many writers in biology still protest that they cannot adequately describe the really characteristic aspects of living nature without the use of such concepts as 'goal-directed activity', 'purposive striving', &c. But, as we now know, it is not really the concept of 'goal' or 'purpose' that they want. It is the concept of a future event whose occurrence is a defining property of the characteristic objective correlation which is found between organic activities and the environments they work in. In other words, what they want is the concept of the 'focal condition' in conjunction with that of 'directive correlation'.

In the chapter on 'Teleology and Causation' of his *Biological Principles* Woodger formulates the issue between these two apparently antithetical notions in the following terms, and I am quoting this passage at length because it is a clear statement of prevalent confusions in biology.

Any change in an organism following an environmental change either subserves the persistence of the organism or it does not. If the former is the case we can call it an 'appropriate response'. Now with our present ways of thinking there appear to be two ways of regarding such an occurrence. We can either ask: Did this happen 'appropriately' *in order* that the persistence of the organism should endure? Or we can ask: Did this just happen in this way and it 'happened' to be appropriate? The difference is expressed in the ordinary use of the terms 'deliberate' and 'accidental'. If we say that the 'appropriate' response occurred *in order* that the organism might persist we are giving the teleological answer in the strict sense of the term. This is what vitalism does. But if we say that the change just happened and was accidentally appropriate, this is equivalent to saying that no answer *can* be given. And it may be that this is all it will ever be possible to say.

The analysis of our ballistic example brings out the major error of this passage. It is a clear case of the fallacy of false disjunction. Being deliberately appropriate and being accidentally appropriate do not, as Woodger and many other modern biologists suppose, exhaust the field of possibilities. Our automatic gun evidently falls outside either of these two categories while it is in working order. And so does any other phenomenon of directive correlation, for such phenomena, as we have shown, represent an objectively biased happening of appropriateness. Woodger's account is equivalent to saying that a given number on a die turns up either accidentally or deliberately, the third important possibility, viz. that of weighted dice, being left out of account.

It is precisely this third possibility which corresponds to the case of directive correlation and underlies the apparent purposiveness of vital activities. Contrary to the sense of Woodger's concluding remark we see therefore that a definite answer *can* be given and that it is neither the teleological answer of the vitalists nor the sceptical answer of the mechanists. In the preceding discussions we have given this answer in still comparatively vague terms and it now remains to develop it on a higher level of precision. Meanwhile, the fate which the pseudo-conflict between 'teleology' and 'causation' suffers at our hands, will have become clear: the ideas of 'final causes' and 'teleological causation' prove to be superfluous and are dismissed as scientifically sterile, but the objective purpose-like properties in nature for the interpretation of which these teleological concepts were invented, have emerged from the surrounding darkness as relations capable of exact formulation in orthodox scientific terms and, incidentally, as relations which, far from contradicting the idea of efficient causation, actually presuppose it.

Notes

1. The similarity between such schemata and the convergence of light rays in passing through a convex lens accounts for my original choice of the term 'focal condition'.

36.
Regulation and Control

W. ROSS ASHBY

THE TWO PREVIOUS PARTS have treated of Mechanism (and the processes within the system) and Variety (and the processes of communication between system and system). These two subjects had to be studied first, as they are fundamental. Now we shall use them, and in Part III we shall study what is the central theme of cybernetics—regulation and control.

This first chapter reviews the place of regulation in biology, and shows briefly why it is of fundamental importance. It shows how regulation is essentially related to the flow of variety. The next chapter (11) studies this relation in more detail, and displays a quantitative law—that the quantity of regulation that can be achieved is bounded by the quantity of information that can be transmitted in a certain channel. The next chapter (12) takes up the question of how the abstract principles of chapter 11 are to be embodied—what sort of machinery can perform what is wanted. This chapter introduces a new sort of machine, the Markovian, which extends the possibilities considered in Part I. The remaining chapters consider the achievement of regulation and control as the difficulties increase, particularly those that arise when the system becomes very large.

At first we will assume that the regulator is already provided, either by being inborn, by being specially made by a manufacturer, or by some other means. The question of what made the regulator,

From W. Ross Ashby, *An Introduction to Cybernetics* (London: Chapman & Hall, 1956), Chapter 10, pp. 195–201, and Chapter 11, pp. 208–218. Reprinted with permission of the author and Chapman & Hall. The reader should recall Chapter 15 above reprinting earlier selections from this work that are important for the present discussion—Ed.

of how the regulator, which does such useful things, came itself to be made will be taken up [later].

The present chapter aims primarily at supplying motive to the reader, by showing that the subjects discussed in the later chapters are of fundamental importance in biology. The subject of regulation in biology is so vast that no single chapter can do it justice. Cannon's *Wisdom of the Body* treated it adequately so far as internal, vegetative activities are concerned, but there has yet to be written the book, much larger in size, that shall show how all the organism's exteriorly-directed activities—its "higher" activities—are all similarly regulatory, i.e. homeostatic. In this chapter I have had to leave much of this to the reader's imagination, trusting that, as a biologist, he will probably already be sufficiently familiar with the thesis. The thesis in any case has been discussed to some extent in *Design for a Brain*.

The chief purpose of this chapter is to tie together the concepts of regulation, information, and survival, to show how intimately they are related, and to show how all three can be treated by a method that is entirely uniform with what has gone before in the book, and that can be made as rigorous, objective, and unambiguous as one pleases.

The foundation. Let us start at the beginning. The most basic facts in biology are that this earth is now two thousand million years old, and that the biologist studies mostly that which exists today. From these two facts follow a well-known deduction, which I would like to restate in our terms.

We saw that if a dynamic system is large and composed of parts with much repetition, and if it contains any property that is autocatalytic, i.e.

whose occurrence at one point increases the probability that it will occur again at another point, then such a system is, so far as that property is concerned, essentially unstable in its absence. This earth contained carbon and other necessary elements, and it is a fact that many combinations of carbon, nitrogen, and a few others are self-reproducing. It follows that though the state of "being lifeless" is almost a state of equilibrium, yet this equilibrium is unstable, a single deviation from it being sufficient to start a trajectory that deviates more and more from the "lifeless" state. What we see today in the biological world are these "autocatalytic" processes showing all the peculiarities that have been imposed on them by two thousand million years of elimination of those forms that cannot survive.

The organisms we see today are deeply marked by the selective action of two thousand million years' attrition. Any form in any way defective in its power of survival has been eliminated; and today the features of almost every form bear the marks of being adapted to ensure *survival* rather than any other possible outcome. Eyes, roots, cilia, shells and claws are so fashioned as to maximise the chance of survival. And when we study the brain we are again studying a means to survival.

Survival

What has just been said is well enough known. It enables us, however, to join these facts on to the ideas developed in this book and to show the connexion exactly.

For consider what is meant, in general, by "survival". Suppose a mouse is trying to escape from a cat, so that the survival of the mouse is in question. As a dynamic system, the mouse can be in a variety of states; thus it can be in various postures, its head can be turned this way or that, its temperature can have various values, it may have two ears or one. These different states may occur during its attempt to escape and it may still be said to have survived. On the other hand if the mouse changes to the state in which it is in four separated pieces, or has lost its head, or has become a solution of amino-acids circulating in the cat's blood then we do not consider its arrival at one of these states as corresponding to "survival".

The concept of "survival" can thus be translated into perfectly rigorous terms, similar to those used throughout the book. The various states (M for

Mouse) that the mouse may be in initially and that it may pass into after the affair with the cat is a set M_1, M_2, ..., M_k, ..., M_n. We decide that, for various reasons of what is practical and convenient, we shall restrict the words "*living* mouse" to mean the mouse in one of the states in some subset of these possibilities, in M_1 to M_k say. If now some operation C (for cat) acts on the mouse in state M_1, and $C(M_1)$ gives, say, M_2, then we may say that M has "survived" the operation of C, for M_2 is in the set M_1, ..., M_k.

If now a particular mouse is very skilled and always survives the operation C, then all the states $C(M_1)$, $C(M_2)$, ..., $C(M_k)$, are contained in the set M_1, ..., M_k. We now see that this representation of survival is *identical* with that of the "stability" of a set. Thus the concepts of "survival" and "stability" can be brought into an exact relationship; and facts and theorems about either can be used with the other, provided the exactness is sustained.

The states M are often defined in terms of variables. The states M_1, ..., M_k, that correspond to the living organism are then those states in which certain *essential variables* are kept within assigned ("physiological") limits.

What is it survives, over the ages? Not the individual organism, but certain peculiarly well compounded gene-patterns, particularly those that lead to the production of an individual that carries the gene-pattern well protected within itself, and that, within the span of one generation, can look after itself.

What this means is that those gene-patterns are specially likely to survive (and therefore to exist today) that cause to grow, between themselves and the dangerous world, some more or less elaborate mechanism for defence. So the genes in *Testudo* cause the growth of a shell; and the genes in *Homo* cause the growth of a brain. (The genes that did not cause such growths have long since been eliminated.)

Now regard the system as one of parts in communication. In the previous section the diagram of immediate effects (of cat and mouse) was (or could be regarded as)

$$\boxed{C} \rightarrow \boxed{M}$$

We are now considering the case in which the diagram is

$$\boxed{D} \rightarrow \boxed{F} \rightarrow \boxed{E}$$

in which E is the set of essential variables, D is the source of disturbance and dangers (such as

C) from the rest of the world, and *F* is the interpolated part (shell, brain, etc.) formed by the gene-pattern for the protection of *E*. (*F* may also include such parts of the environment as may similarly be used for *E*'s protection—burrow for rabbit, shell for hermit-crab, pike for pike-man, and sword (as defence) for swordsman.)

For convenience in reference, let the states of the essential variables *E* be divided into a set η—those that correspond to "organisms living" or "good"—and not-η—those that correspond to "organism not living" or "bad". (Often the classification cannot be as simple as this, but no difficulty will occur in principle; nothing to be said excludes the possibility of a finer classification.)

To make the assumptions clear, here are some simple cases, as illustration. (Inanimate regulatory systems are given first for simplicity.)

(1) *The thermostatically-controlled water-bath.* *E* is its temperature, and what is desired (η) is the temperature range between, say 36° and 37°C. *D* is the set of all the disturbances that may drive the temperature outside that range—addition of cold water, cold draughts blowing, immersion of cold objects, etc. *F* is the whole regulatory machinery. *F*, by its action, tends to lessen the effect of *D* on *E*.

(2) *The automatic pilot.* *E* is a vector with three components—yaw, pitch, and roll—and η is the set of positions in which these three are all within certain limits. *D* is the set of disturbances that may affect these variables, such as gusts of wind, movements of the passengers in the plane, and irregularities in the thrusts of the engines. *F* is the whole machinery—pilot, ailerons, rudder, etc.—whose action determines how *D* shall affect *E*.

(3) *The bicycle rider.* *E* is chiefly his angle with the vertical. η is the set of small permissible deviations. *D* is the set of those disturbances that threaten to make the deviation become large. *F* is the whole machinery—mechanical, anatomical, neuronic—that determines what the effect of *D* is on *E*.

Many other examples will occur later. Meanwhile we can summarise by saying that natural selection favours those gene-patterns that get, in whatever way, a regulator *F* between the disturbances *D* and the essential variables *E*. Other things being equal, the better *F* is as a regulator, the larger the organism's chance of survival.

Regulation blocks the flow of variety. On what scale can any particular mechanism *F* be measured for its value or success as a regulator? The perfect thermostat would be one that, in spite of disturbance, kept the temperature constant at the desired level. In general, there are two characteristics required: the maintenance of the temperature within close limits, and the correspondence of this range with the desired one. What we must notice in particular is that the set of permissible values, η, has less variety than the set of all possible values in *E*; for η is some set selected from the states of *E*. If *F* is a regulator, the insertion of *F* between *D* and *E* *lessens* the variety that is transmitted from *D* to *E*. Thus an essential function of *F* as a regulator is that it shall block the transmission of variety from disturbance to essential variable.

Since this characteristic also implies that the regulator's function is to block the flow of information, let us look at the thesis more closely to see whether it is reasonable.

Suppose that two water-baths are offered me, and I want to decide which to buy. I test each for a day against similar disturbances and then look at the records of the temperatures; they are as in Figure 1:

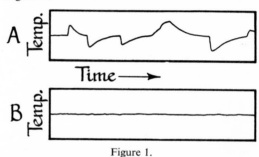

Figure 1.

There is no doubt that Model *B* is the better; and I decide this precisely because its record gives me no information, as does *A*'s, about what disturbances, of heat or cold, came to it. The thermometer and water in bath *B* have been unable, as it were, to see anything of the disturbances *D*.

The same argument will apply, with obvious modifications, to the automatic pilot. If it is a good regulator the passengers will have a smooth flight whatever the gustiness outside. They will, in short, be *prevented from knowing* whether or not it is gusty outside. Thus a good pilot acts as a barrier against the transmission of that information.

The same argument applies to an air-conditioner. If I live in an air-conditioned room, and can tell, by the hotness of the room, that it is getting hot outside, then that conditioner is failing as a regulator. If it is really good, and the blinds are drawn, I shall be unable to form any idea of what the outside weather is like. The good conditioner blocks the flow inwards of information about the weather.

The same thesis applies to the higher regulations achieved by such activities as hunting for food, and earning one's daily bread. Thus while the unskilled hunter or earner, in difficult times, will starve and will force his liver and tissues (the essential variables) to extreme and perhaps unphysiological states, the skilled hunter or earner will go through the same difficult times with his liver and tissues never taken to extremes. In other words, his skill as a regulator is shown by the fact, among others, that it prevents information about the times reaching the essential variables. In the same way, the skilled provider for a family may go through difficult times without his family realising that anything unusual has happened. The family of an unskilled provider would have discovered it.

In general, then, an essential feature of the good regulator is that *it blocks the flow of variety from disturbances to essential variables.*

The blocking may take place in a variety of ways, which prove, however, on closer examination to be fundamentally the same. Two extreme forms will illustrate the range.

One way of blocking the flow (from the source of disturbance D to the essential variable E) is to interpose something that acts as a simple passive block to the disturbances. Such is the tortoise's shell, which reduces a variety of impacts, blows, bites, etc. to a negligible disturbance of the sensitive tissues within. In the same class are the tree's bark, the seal's coat of blubber, and the human skull.

At the other extreme from this static defence is the defence by skilled counter-action—the defence that gets information about the disturbance to come, prepares for its arrival, and then meets the disturbance, which may be complex and mobile, with a defence that is equally complex and mobile. This is the defence of the fencer, in some deadly duel, who wears no armour and who trusts to his skill in parrying. This is the defence used mostly by the higher organisms, who have developed a nervous system precisely for the carrying out of this method.

When considering this second form we should be careful to notice the part played by information and variety in the process. The fencer must watch his opponent closely, and he must gain information in all ways possible if he is to survive. For this purpose he is born with eyes, and for this purpose he learns how to use them. Nevertheless, the end result of this skill, if successful, is shown by his essential variables, such as his blood-volume, remaining within normal limits, much as if the duel had not occurred. Information flows freely to the non-essential variables, but the variety in the distinction "duel or no-duel" has been prevented from reaching the essential variables.

Through the remaining chapters we shall be considering this type of active defence, asking such questions as: what principles must govern it? What mechanisms can achieve it? And, what is to be done when the regulation is very difficult?

. . . . .

Regulation and Variety

. . . The law of Requisite Variety enables us to apply a *measure* to regulation. Let us go back and reconsider what is meant, essentially, by "regulation".

There is first a set of disturbances D, that start in the world outside the organism, often far from it, and that threaten, if the regulator R does nothing, to drive the essential variables E outside their proper range of values. The values of E correspond to the "outcomes" of the previous sections. Of all these E-values only a few (η) are compatible with the organism's life, or are unobjectionable, so that the regulator R, to be successful, must take its value in a way so related to that of D that the outcome is, if possible, always within the acceptable set η, i.e. within physiological limits. Regulation is thus related fundamentally to the game [on p. 134]. Let us trace the relation in more detail.

The Table T is first assumed to be given. It is the hard external world, or those internal matters that the would-be regulator has to take for granted. Now starts a process. D takes an arbitrary value, R takes some value determined by D's value, the Table determines an outcome, and this either is or is not in η. Usually the process is repeated, as when a water-bath deals, during the day, with various disturbances. Then another value is taken by D, another by R, another outcome occurs, and this also may be either in η or not. And so on. If R is a well-made regulator—one that works successfully—then R is such a transformation of D that all the outcomes fall within η. *In this case R and T together are acting as the barrier F.*

We can now show these relations by the diagram of immediate effects:

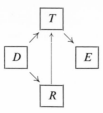

The arrows represent actual channels of communication. For the variety in D determines the variety in R; and that in T is determined by that in both D and R. If R and T are in fact actual machines, then R has an input from D, and T has two inputs.

(When R and T are embodied in actual machines, care must be taken that we are clear about what we are referring to. If some machine is providing the basis for T, it will have a set of states that occur step by step. These states, and these steps, are essentially independent of the discrete steps that we have considered to be taken by D, R, and T in this chapter. Thus, T gives the outcome, and any particular outcome may be compared with another, as unit with unit. Each individual outcome may, however, in another context, be analysed more finely. Thus a thirsty organism may follow trajectory 1 and get relief, or trajectory 2 and die of thirst. For some purposes the two outcomes can be treated as units, particularly if they are to be contrasted. If however we want to investigate the behaviour in more detail, we can regard trajectory 1 as composed of a sequence of states, separated by steps in time that are of quite a different order of size from those between successive regulatory acts to successive disturbances.)

We can now interpret the general phenomenon of regulation in terms of communication. If R does nothing, i.e. keeps to one value, then the variety in D threatens to go through T to E, contrary to what is wanted. It may happen that T, without change by R, will block some of the variety, and occasionally this blocking may give sufficient constancy at E for survival. More commonly, a further suppression at E is necessary; it can be achieved, as we saw, only by further variety at R.

We can now select a portion of the diagram, and focus attention on R as a transmitter:

$$D \rightarrow R \rightarrow T$$

The law of Requisite Variety says that R's capacity

as a regulator cannot exceed R's capacity as a channel of communication.

In the form just given, the law of Requisite Variety can be shown in exact relation to Shannon's Theorem 10, which says that if noise appears in a message, the amount of noise that can be removed by a correction channel is limited to the amount of information that can be carried by that channel.

Thus, his "noise" corresponds to our "disturbance", his "correction channel" to our "regulator R", and his "message of entropy H" becomes, in our case, a message of entropy zero, for it is *constancy* that is to be "transmitted". Thus the use of a regulator to achieve homeostasis and the use of a correction channel to suppress noise are homologous.

The diagram of immediate effects given in the previous section is clearly related to the formulation for "directive correlation" given by Sommerhoff, who, in his *Analytical Biology*, uses the diagram

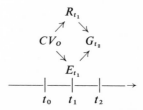

If I am not misinterpreting him, his concepts and those used here are equivalent thus:

Coenetic variable (CV_o) ↔ Disturbance (D)
Response (R_{t_1}) ↔ Response (R)
Environmental
circumstances (E_{t_1}) ↔ Table (T)
Subsequent occurrence (G_{t_2}) ↔ Outcome (E)

A reading of his book may thus help to extend much of the theory given in this Part, for he discusses the subject extensively.

The law now enables us to see the relations existing between the various types of variety and information that affect the living organism.

A species continues to exist primarily because its members can block the flow of variety (thought of as disturbance) to the gene-pattern, and this blockage is the species' most fundamental need. Natural selection has shown the advantage to be gained by taking a large amount of variety (as information) partly into the system (so that it does not reach the gene-pattern) and then using this information so that the flow via R blocks the flow through the environment T.

This point of view enables us to resolve what

might at first seem a paradox—that the higher organisms have sensitive skins, responsive nervous systems, and often an instinct that impels them, in play or curiosity, to bring more variety to the system than is immediately necessary. Would not their chance of survival be improved by an avoidance of this variety?

The discussion in this chapter has shown that variety (whether information or disturbance) comes to the organism in two forms. There is that which threatens the survival of the gene-pattern—the direct transmission by T from D to E. This part must be blocked at all costs. And there is that which, while it may threaten the gene-pattern, can be transformed (or re-coded) through the regulator R and used to block the effect of the remainder (in T). This information is useful, and should (if the regulator can be provided) be made as large as possible; for, by the law of Requisite Variety, the amount of disturbance that reaches the gene-pattern can be diminished only by the amount of information so transmitted. That is the importance of the law in biology.

It is also of importance to *us* as we make our way towards the last chapter. In its elementary forms the law is intuitively obvious and hardly deserving statement. If, for instance, a press photographer would deal with twenty subjects that are (for exposure and distance) distinct, then his camera must obviously be capable of at least twenty distinct settings if all the negatives are to be brought to a uniform density and sharpness. Where the law, in its quantitative form, develops its power is when we come to consider the system in which these matters are not so obvious, and particularly when it is very large. Thus, by how much can a dictator control a country? It is commonly said that Hitler's control over Germany was total. So far as his power of regulation was concerned, the law says that his control amounted to just 1 man-power, and no more. (Whether this statement is true must be tested by the future; its chief virtue now is that it is exact and uncompromising.) Thus the law, though trite in the simple cases, can give real guidance in those cases that are much too complex to be handled by unaided intuition.

Control

The formulations given in this chapter have already suggested that regulation and control are intimately related. . . .

We can look at the situation in another way.

Suppose the decision of what outcome is to be the target is made by some controller, C, whom R must obey. C's decision will affect R's choice of α, β or γ; so the diagram of immediate effects is

$$D \to T \to E$$
$$\searrow \quad \nearrow$$
$$C \to R$$

Thus the whole represents a system with two independent inputs, C and D.

Suppose now that R is a perfect regulator. If C sets a as the target, then (through R's agency) E will take the value a, *whatever value D may take*. Similarly, if C sets b as target, b will appear as outcome whatever value D may take. And so on. And if C sets a particular sequence—a, b, a, c, c, a, say—as sequential or compound target, then that sequence will be produced, regardless of D's values during the sequence. (It is assumed for convenience that the components move in step.) Thus the fact that R is a perfect regulator gives C complete control over the output, in spite of the entrance of disturbing effects by way of D. Thus, *perfect **regulation** of the outcome by R makes possible a complete **control** over the outcome by C.*

We can see the same facts from yet another point of view. If an attempt at control, by C over E:

$$C \to E$$

is disturbed or made noisy by another, independent, input D, so that the connexions are

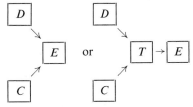

then a suitable regulator R, taking information from both C and D, and interposed between C and T:

$$D \to T \to E$$
$$\searrow \quad \uparrow$$
$$C \to R$$

may be able to form, with T, a compound channel to E that *transmits fully from C while transmitting nothing from D.*

The achievement of control may thus depend

necessarily on the achievement of regulation. The two are thus intimately related.

In our treatment of regulation the emphasis has fallen on its property of reducing the variety in the outcome; without regulation the variety is large—with regulation it is small. The limit of this reduction is the regulation that holds the outcome rigorously constant. This point of view is undoubtedly valid, but at first it may seem to contrast sharply with the naive view that living organisms are, in general, anything but immobile.

It should be appreciated that the distinction between "constant" and "varying" often depends on the exact definition of what is being referred to. Thus if a searchlight follows an aircraft accurately we may notice either that the searchlight moved through a great range of angles (angles in relation to the earth) or that the angle it made with the aircraft remained constant at zero. Obviously both points of view are valid; there is no real contradiction in this example between "great range" and "constant", for they refer to different variables.

Again, the driver who steers a car accurately from one town to another along a winding lane can be regarded either as one who has caused the steering wheel to show much activity and change or as one who, throughout the trip, has kept the distance between car and verge almost constant.

Many of the activities of living organisms permit this double aspect. On the one hand the observer can notice the great deal of actual movement and change that occurs, and on the other hand he can observe that throughout these activities, so far as they are coordinated or homeostatic, there are invariants and constancies that show the degree of regulation that is being achieved.

Many variations are possible on the same theme. Thus if variable x is always doing just the same as variable y, then the quantity $x - y$ is constant at zero. So if y's values are given by some outside factor, any regulator that acts on x so as to keep $x - y$ constant at zero is in fact forcing x to vary, copying y. Similarly, "making x do the opposite to y" corresponds to "keeping $x + y$ at some constant value". And "make the variable w change so that it is always just twice as large as v's (fluctuating) rate of change" corresponds to "keep the quantity $w - 2dv/dt$ constant".

It is a great convenience in exposition and in the processes of general theory to be able to treat all "targets" as if they were of the form "keep the outcome constant at a". The reader must,

however, not be misled into thinking that the theory treats only of immobility; he must accustom himself to interchanging the corresponding concepts freely.

Some Variations

[Earlier] the essential facts implied by regulation were shown as a simple rectangular table, as if it were a game between two players D and R. The reader may feel that this formulation is much too simple and that there are well known regulations that it is insufficient to represent. The formulation, however, is really much more general than it seems, and in the remaining sections of this chapter we shall examine various complications that prove, on closer examination, to be really included in the basic formulation. [See above, Chapter 15, p. 134—ED.]

Compound disturbance. The basic formulation included only one source of disturbance D, and thus seems, at first sight, not to include all those cases, innumerable in the biological world, in which the regulation has to be conducted against several disturbances coming simultaneously by several channels. Thus, a cyclist often has to deal both with obstructions due to traffic and with disequilibrations due to gusts.

In fact, however, this case is included; for nothing in this chapter excludes the possibility that D may be a vector, with any number of components. A vectorial D is thus able to represent all such compound disturbances within the basic formulation.

Noise. A related case occurs when T is "noisy" —when T has an extra input that is affected by some disturbance that interferes with it. This might be the case if T were an electrical machine, somewhat disturbed by variations in the mains voltage. At first sight this case seems to be not represented in the basic formulation.

It must be appreciated that D, T, E, etc. were defined in purely *functional* form. Thus "D" is "that which disturbs". Given any real system some care may be necessary in deciding what corresponds to D, what to T, and so on. Further, a boundary drawn provisionally between D and T (and the other boundaries) may, on second thoughts, require moving. Thus one set of boundaries on the real system may give a system that purports to be of D, T, etc. yet does not agree with the basic formulation. Then it may be found that a shifting of the boundaries, to give a *new* D, T, etc., gives a set that *does* agree with the formulation.

If a preliminary placing of the boundaries shows that this (provisional) T is noisy, then the boundaries should be re-drawn so as to get T's input of noise included *as a component in D*. D is now "that which disturbs", and T has no third input; so the formulation agrees.

There is, of course, no suggestion here that the noise, as a disturbance, can be allowed for magically by merely thinking differently about it. The suggestion is that if we start again from the beginning, and re-define D and T then some *new* transformation of D may be able to restore regulation. The new transformation will, of course, have to be more complex than the old, for D will have more components.

Initial states. A related case occurs when T is some machine that shows its behaviour by a trajectory, with the outcome E depending on the properties of T's trajectory. The outcomes will then usually be affected by which of T's states is the initial one. How does T's initial state come into the basic formulation?

If the initial state can be controlled, so that the trajectory can be started always from some standardised state, then no difficulty arises. It may however happen, especially if the system is very large, that T's initial state cannot be standardised. Does the basic formulation include this case?

It does; for D, as a vector, can be re-defined to include T's initial state. Then the variety brought to E by the variety in T's initial state is allotted its proper place in the formulation.

Compound target. It may happen that the acceptable states η at E may have more than one condition. Thus of a thermostat it might be demanded that

(i) it shall usually stay between 36° and 37°C;
(ii) if displaced by $\pm 10°$ it shall return to the allowed range within one minute.

This difficulty can be dealt with by recognising that E may be a vector, with more than one component, and that what is acceptable (η) may be given in the form of separate specifications for each component.

Thus, by allowing E to become a vector, the basic formulation can be made to include all cases in which the target is complex, or conditional, or qualified.

Internal complexities. As a last example, showing how comprehensive the basic formulation really is, consider the case in which the major problem seems to be not so much a regulation as an interaction between several regulations. Thus a signalman may have to handle several trains coming to his section simultaneously. To handle any one by itself would be straightforward, but here the problem is the control of them as a complex whole pattern.

This case is in fact still covered by the basic formulation. For nothing in that formulation prevents the quantities or states or elements in D, R, T, or E from being made of parts, and the parts interrelated. The fact that "D" is a single letter in no way implies that what it represents must be internally simple or unitary.

The signalman's "disturbance" D is the particular set of trains arriving in some particular pattern over space and time. Other arrangements would provide other values for D, which must, of course, be a vector. The outcomes E will be various complex patterns of trains moving in relation to one another and moving away from his section. The acceptable set η will certainly include a component "no collision" and will probably include others as well. His responses R will include a variety of patterns of movements of signals and points. T is what is given—the basic matters of geography, mechanics, signalling techniques, etc., that lead determinately from the situation that has arisen and his reaction pattern to outcome.

It will be seen therefore that the basic formulation is capable, in principle, of including cases of any degree of internal complexity.

37.
The Second Cybernetics: Deviation-Amplifying Mutual Causal Processes

MAGOROH MARUYAMA

SINCE ITS INCEPTION, cybernetics was more or less identified as a science of self-regulating and equilibrating systems. Thermostats, physiological regulation of body temperature, automatic steering devices, economic and political processes were studied under a general mathematical model of deviation-counteracting feedback networks.

By focusing on the deviation-counteracting aspect of the mutual causal relationships however, the cyberneticians paid less attention to the systems in which the mutual causal effects are deviation-amplifying. Such systems are ubiquitous: accumulation of capital in industry, evolution of living organisms, the rise of cultures of various types, interpersonal processes which produce mental illness, international conflicts, and the processes that are loosely termed as "vicious circles" and "compound interests"; in short, all processes of mutual causal relationships that amplify an insignificant or accidental initial kick, build up deviation and diverge from the initial condition.

In contrast to the progress in the study of equilibrating systems, the deviation-amplifying systems have not been given as much investment of time and energy by the mathematical scientists on the one hand, and understanding and practical application on the part of geneticists, ecologists, politicians and psychotherapists on the other hand.

The deviation-counteracting mutual causal systems and the deviation-amplifying mutual causal systems may appear to be opposite types of systems. But they have one essential feature in common: they are both mutual causal systems, i.e., the elements within a system influence each other either simultaneously or alternatingly. The difference between the two types of systems is that the deviation-counteracting system has mutual negative feedbacks between the elements in it while the deviation-amplifying system has mutual positive feedbacks between the elements in it.

Since both types are systems of mutual causal relationships, or in other words systems of mutual feedbacks, they both fall under the subject matter of cybernetics. But since the deviation-counteracting type has predominantly been studied up till now under the title of cybernetics, let us consider its studies *the first cybernetics*, and call the studies of the deviation-amplifying mutual causal relationships "*the second cybernetics.*" The deviation-counteracting mutual causal process is also called "morphostasis," while the deviation-amplifying mutual causal process is called "morphogenesis."

Though the second cybernetics as defined here is lagging behind the development of the first cybernetics at the present moment, the germination of the concept of deviation-amplifying mutual causal process is not entirely new. The concept was formulated in some fields even before the advent of cybernetics and was applied fruitfully. The field of economics is a good example.

For many years the economists had claimed that it was useless to try to raise the standard of living of the lower class, because, they argued, if the income of the population in the lower class should increase, they would produce more children and thus reduce their standard of living to the

From Magoroh Maruyama, "The Second Cybernetics: Deviation-Amplifying Mutual Causal Processes," *American Scientist*, 51 (1963), 164–79. Reprinted by permission of the author and publisher.

original level; the poor stay poor and the rich stay rich. This was a morphostatic model of mutual deviation-counteracting between the income level and the number of children. This theoretical model led the policy makers to the action of laisser-faire policy. On the other hand, it was also known that "the more capital, the more rapid the ratio of its increase"; in other words, the poor become poorer and the rich become richer. This was a morphogenetic model of deviation-amplifying process.

Subsequently J. Tinbergen and H. Wold have given more elaboration and mathematical sophistication to the theory of mutual causal process in the theory of economics. More recently G. Myrdal has pointed out that, while in the economically well-developed countries the regional, social, and hierarchical differences in economical level tend to decrease, in the economically underdeveloped regions the difference between the poor and the rich increases. In an economically well-developed society, transportation, communication, education, insurance systems, and welfare programs equalize the economical level throughout the society. In an economically underdeveloped society, on the other hand, under the laisser-faire policy and free play of market forces, the few privileged people accumulate more wealth and power while the living standard of the poor tends to fall. Low standard of living, poor health, and low efficiency in work aggravate one another. Racial or social discrimination, and other social, psychological and cultural factors may be added in the "vicious circle." Likewise, between nations, world free trade is profitable for rich countries and detrimental for poor countries. This morphogenetic reformulation of the economic theory affects the public policy toward the direction of planned economy within economically underdeveloped countries and controlled international trade.

Myrdal further points out the importance of the direction of the initial kick, which determines the direction of the subsequent deviation amplification in the planned economy. Furthermore, the resulting development will be far greater than the investment in the initial kick. Thus, in the economically underdeveloped countries it is necessary not only to plan the economy, but also to give the initial kick and reinforce it for a while in such a direction and with such an intensity as to maximize the efficiency of development per initial investment. Once the economy is kicked in a right direction and with a sufficient initial push, the deviation-amplifying mutual positive feedbacks take over the process, and the resulting development will be disproportionally large as compared with the initial kick.

We find the same principle of deviation-amplifying mutual causal relationships operating in many other happenings in the universe. Take, for example, weathering of rock. A small crack in a rock collects some water. The water freezes and makes the crack larger. A larger crack collects more water, which makes the crack still larger. A sufficient amount of water then makes it possible for some small organisms to live in it. Accumulation of organic matter then makes it possible for a tree to start growing in the crack. The roots of the tree will then make the crack still larger.

Development of a city in an agricultural plain may be understood with the same principle. At the beginning, a large plain is entirely homogeneous as to its potentiality for agriculture. By some chance an ambitious farmer opens a farm at a spot on it. This is the initial kick. Several farmers follow the example and several farms are established. One of the farmers opens a tool shop. Then this tool shop becomes a meeting place of farmers. A food stand is established next to the tool shop. Gradually a village grows. The village facilitates the marketing of the agricultural products, and more farms flourish around the village. Increased agricultural activity necessitates development of industry in the village, and the village grows into a city.

This is a very familiar process. But there are a few important theoretical implications in such a process.

On what part of the entire plain the city starts growing depends on where accidentally the initial kick occurred. The first farmer could have chosen any spot on the plain, since the plain was homogeneous. But once he has chosen a spot, a city grows from that spot, and the plain becomes inhomogeneous. If a historian should try to find a geographical "cause" which made this spot a city rather than some other spots, he will fail to find it in the initial homogeneity of the plain. Nor can the first farmer be credited with the establishment of the city. The secret of the growth of the city is in the process of deviation-amplifying mutual positive feedback networks rather than in the initial condition or in the initial kick. This process, rather than the initial condition, has generated the complexly structured city. It is in this sense that the deviation-amplifying mutual causal process is called "morphogenesis."

A sacred law of causality in the classical philosophy stated that similar conditions produce similar effects. Consequently, dissimilar results were attributed to dissimilar conditions. Many scientific researches were dictated by this philosophy. For example, when a scientist tried to find out why two persons under study were different, he looked for a difference in their environment or in their heredity. It did not occur to him that neither environment nor heredity may be responsible for the difference. He overlooked the possibility that some deviation-amplifying interactional process in their personality and in their environment may have produced the difference.

In the light of the deviation-amplifying mutual causal process, the law of causality is now revised to state that similar conditions may result in dissimilar products. It is important to note that this revision is made without the introduction of indeterminism and probabilism. Deviation-amplifying mutual causal processes are possible even within the deterministic universe, and make the revision of the law of causality even within the determinism. Furthermore, when the deviation-amplifying mutual causal process is combined with indeterminism, here again a revision of a basic law becomes necessary. The revision states: A small initial deviation, which is within the range of high probability, may develop into a deviation of very low probability (or more precisely, into a deviation which is very improbable within the framework of probabilistic unidirectional causality).

Not only the law of causality, but also the second law of thermodynamics is affected by the deviation-amplifying mutual causal process. Let us return to the example of the growth of a city in an agricultural plain. The growth of the city first increases the internal structuredness of the city itself. Secondly, it increases the inhomogeneity of the plain by its deviating from the original prevailing condition. Thirdly, the growth of a city at a spot may have an inhibiting effect upon the growth of another city in the vicinity, just as the presence of one swimming pool may discourage an enterpriser from opening another pool right next to it, and just as the presence of large trees inhibits with their shades the growth of some species of small trees around them. A city needs a *hintergrund* to support it, and, therefore, cities have to be spaced at some intervals. This inhibiting effect further increases inhomogeneity of the plain.

This gradual increase of inhomogeneity is a process against the second law of thermodynamics. In a few words, the second law of thermodynamics states that an isolated system spends with a great probability most of its time in high-probability states. Hence, if an isolated system is in an improbable state, it will most probably be found in the future in a more probable state. Under the assumption of randomness of events, homogeneous states are more probable than inhomogeneous states. For example, uneven distribution of temperature is less probable than even distribution of temperature. Under the assumption of the second law of thermodynamics, an isolated system in an inhomogeneous state will most probably be found in the future in a more homogeneous state. The second law of thermodynamics is in this sense a law of decay of structure and of decay of inhomogeneity.

Any process such as biological growth which increases structuredness and inhomogeneity was against the second law of thermodynamics and was an embarrassing problem for scientists. This embarrassing question was temporarily ignored by the argument that an organism is not an isolated system. But what process and principle make it possible for an organism to increase its structure and to accumulate heat was never squarely answered. Now, under the light of deviation-amplifying mutual causal process this mystery is solved.

The process of evolution, or in other words phylogenetic morphogenesis including the pattern of behavior, is deviation-amplifying in several ways.

First, there is deviation-amplifying mutual process between the mutations and the environment. For example, suppose that some mutants of a species can live at a lower temperature than the "normal" individuals. Then the mutants may move to an environment which is colder. Further mutations occur. Some of the mutants are unfit for the low-temperature environment and die off. But some other mutants are able to live in a much colder climate than their parents. They move to a still colder environment. The cold climate eliminates any new mutants that are unfit for cold climate. The "average" individuals of the survivors are then fit for cold climate. The chances of the species, or at least some members of the species, to move to a still colder environment are now greater than before. Thus the selection of, or accidental wandering into, a certain type of environment and the direction of survivable mutations amplify each other.

Not only the organism may move into a new environment, but it may also create its own environment. *Homo sapiens* is a typical example. A "civilized" man lives in an environment which he created, and which is relatively free from the bacteria of certain diseases such as typhoid. His resistance against typhoid decreases as a result of living in such an environment. The decreased resistance necessitates him to make his environment more germ-free. This decreases his resistance further.

Secondly, there is interspecific deviation-amplification. For example, a species of moth has predators. Because of the predators, the mutants of the moth species which have a more suitable cryptic coloration (camouflage) and cryptic behavior than the average survive better. On the other hand, those mutants of the predators which have a greater ability than the average in discovering the moth will survive better. Hence, the cryptic coloration and the cryptic behavior of the moth species improve generation after generation, and the ability of the predators to discover the moth also increases generation after generation.

Thirdly, the intraspecific selection has a deviation-amplifying effect. For example, many animals prefer supernormal (above-average) members of its species to normal members for mating and for carrying on other cooperative activities. By giving more responses to supernormal stimuli than to normal stimuli, the members of the species amplify, by favoring supernormal mutants, the deviation in the direction of supernormality. The deviation in turn may increase either the number and the intensity of the members' responses to the supernormal stimuli, or the level of the supernormality of the members' characteristics.

The response to supernormal stimuli may be inborn or culturally conditioned. For example, an oystercatcher (*Haematopus ostralegus*) prefers a large artificial egg given by an experimenter to its own egg, and tries to sit on the large egg even though the egg may be as large as the bird's body. This response is inborn. On the other hand in *homo sapiens*, what is considered "supernormal" often depends on the culture. In the contemporary American culture, female legs which are more slender and longer than normal are supernormal stimuli, while in Polynesia obese young girls used to be supernormal stimuli. In this way, culture may influence the direction of the evolution of the human species while, at the same time, the evolution influences the culture. We may say that "cultural selection" rather than natural selection is the mechanism of human evolution since much of man's environment is man-made. But the matter is not very simple because certain cultures like ours, with medical science and technology, make it possible for constitutionally "unfit"

individuals to "survive." Perhaps fitness should be defined not in terms of the capacity of the individual without tools, but in terms of the tools which he can mobilize.

In any case, man is responsible for his own evolution because of his capacity to create his own culture which is his environment and to choose his criterion of supernormality. Incidentally, to "create" is no longer a concept which violates the law of physics. As we have seen, creation *ex nihilo*, or rather almost *ex nihilo*, is scientifically possible because the secret is no longer in the Prime Mover or in the Creator, but in the process of a deviation-amplifying mutual positive feedback network.

Fourthly, the effect of inbreeding is deviation-amplifying for purely statistical reasons. Marriages between close relatives produce individuals in whom certain characteristics are extremely amplified. In fact, if people were to practice intrafamily marriages only, each family would develop into a separate species because there is no interbreeding between families. In a more extreme case, if the reproduction were always sexless, i.e., if an individual produced its offspring without any help from another individual, then each individual's descendants would develop into a separate species. In fact, "species" would be non-existent.

Many times cooperation between individuals facilitates life, and therefore the existence of species facilitates the life of the individual. The sexual reproduction, as compared with asexual reproduction, acts as a stabilizer of the species.

Here we have seen that mating may amplify the deviations or stabilize the species. This is not a contradiction. Whether mating is deviation-amplifying or deviation-counteracting depends on whether the inbreeding component or the interbreeding component is predominant. S. Wright has made this relationship clear. When the mutation rate is high as compared with the population size, random matings will produce more inbreeding than interbreeding, and the deviation-amplifying effect is predominant. On the other hand, when the population size is large as compared with mutation rate, random matings will produce more interbreeding than inbreeding, and the deviation-counteracting effect is predominant. The direction of deviation in a small population with a high mutation rate is unpredictable since it depends on the initial kick which is random. But once the deviation started, it is systematically amplified in the same direction.

When the mutation rate is neither too high nor

too low as compared with the population size, neither inbreeding nor interbreeding predominates. The result is neither deviation-amplifying nor deviation-counteracting, but a combination of both which results in random drift. At one time, a random initial kick produces a deviation in a certain direction. Deviation-amplification takes over and this deviation is amplified consistently in the same direction. But this does not last very long. Soon, deviation-counteracting takes over, and the population becomes fixed at a certain point of deviation. After a while, another random initial kick produces a deviation in a new direction. Deviation-amplification takes over again and the population drifts consistently in this direction for a while. But soon, deviation-counteracting takes over and the population becomes fixed at a new point in the new direction of deviation. Then another initial kick starts a deviation in a new direction. The process repeats itself with unpredictable drifts.

The maximum speed of evolution is found, not in one colony with a high rate of mutation and a small size of population, but in the interaction between colonies which have a moderate mutation rate. When the mutation rate is too high as compared with the population size, the mutant characteristics may amplify themselves at a speed beyond the possibility of finding a new environment and new intraspecific and interspecific ecological conditions which are suitable for the mutant characteristics, and beyond the possibility of allowing other variations of mutants, which have characteristics greater in survival value, to develop. The species may become extinct, or it may reach the limit of mutability and become fixed there, or the mutant characteristics may become so dominant and homogeneous as to become deviation-counteracting.

A moderate mutation rate produces a more viable and changeable evolution. Moreover, when there are occasional exchanges of immigrants between colonies, the introduction of a new strain, which has proved to be viable in one colony, into another colony has the same effect as producing viable mutants, and tends to favor evolution.

This seems to be true of human cultures also. When there is enough separation, not necessarily geographical, between cultures to allow them differentiation and variety, with enough exchange between them to allow mutual enrichment and new combinations, the human civilization seems to progress most efficiently.

As we have seen, evolution is deviation-amplifying in several ways. For this reason we called evolution "phylogenetic morphogenesis." But more traditionally, "morphogenesis" is used for ontogenesis, i.e., for the development of the embryo into an adult individual. Is the new usage of the word "morphogenesis," meaning deviation-amplifying mutual causal process, in any way contradictory to the traditional usage? By no means so. As we see in the following, the development of the embryo also involves deviation-amplifying mutual causal process.

There is one basic difference between ontogenesis and phylogenesis. Mutation and natural selection, which are the basis of phylogenesis, are absent in ontogenesis. In fact, while phylogenesis has the randomness of mutation, ontogenesis lacks this randomness and seems to be based on a strictly detailed, deterministic planning. But this detailed planning is *generated* within the embryo by a deviation-amplifying mutual causal process in a deterministic scheme.

Biologists have been puzzled by the fact that the amount of information stored in the genes is much smaller than the amount of information needed to describe the structure of the adult individual. The puzzle is now solved by noticing that it is not necessary for the genes to carry all the information regarding the adult structure, but it suffices for the genes to carry a set of rules to generate the information. This can be illustrated by a model.

Let us imagine, for the sake of simplicity, a two-dimensional organism. Let us further imagine that its cells are squares of an equal size. Let us say that the organism consists of four types of cells: green, red, yellow, and blue. Each type of cell reproduces cells of the same type to build a tissue. A tissue has at least two cells. The tissues grow in a two-dimensional array of squares. Let us give a set of rules for the growth of tissues:

1. No cells die. Once reproduced, a cell is always there.

2. Both ends of a tissue grow whenever possible, by reproducing one cell per unit time in a vacant contiguous square. If there is no vacant contiguous square at either end, the ends stop growing. If there is more than one vacant contiguous square at either end, the direction of the growth is governed by the preferential order given by Rules 3, 4, and 5.

3. If, along the straight line defined by the end cell and the penultimate cell (next to the end cell) there are less than or equal to three cells of the same type (but may be of different tissues) consecutively, the preferred direction is along the

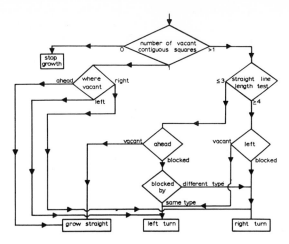

Figure 1.

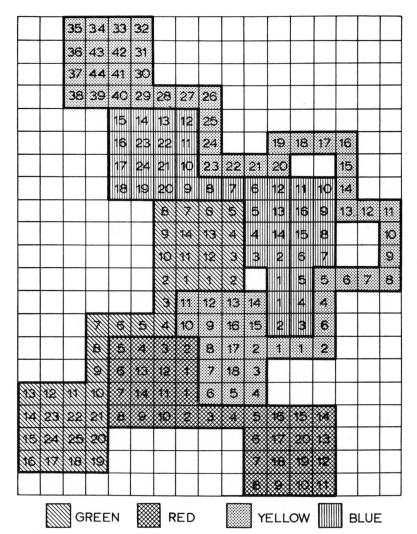

Figure 2.

same straight line. If that direction is blocked, follow Rule 5.

4. If, along the straight line defined by the end cell and the penultimate cell, there are more than or equal to four cells of the same type (but may be of different tissues) consecutively, the preferred direction of the growth is a left turn. If a left turn is impossible, make a right turn.

5. If, when a straight growth is preferred, the straight growth is impossible because the square ahead is already occupied, do the following:
If the square to which the straight growth would take place is filled with a cell of the same type as the growing tissue, make a left turn. If the square ahead is filled with a cell whose type is different from that of the growing tissue, make a right turn.

6. The growth of the four types of tissues is timewise out of phase with each other: green first, red second, yellow third, and blue last within a cycle of one unit time.

Rules 2, 3, 4, and 5 can be diagrammed together as in Figure 1.

Using these rules, we can compute the growth of the tissues when the locations of the initial tissues are specified. For example, let us say that, in Figure 2, only the squares marked by 1's are filled with cells of the types indicated by the colors. At the end of the second unit time, the squares marked by 2's are filled. And at the end of the nth unit time, all squares marked by numbers less than n are filled. At the end of the 44th unit time, all tissues have completed the growth, and the organism has attained its full differentiation.

In this example we started with four tissues of the minimum length, one tissue of each of the four types. But already the result is a fairly complex structure. If we start with a slightly larger number of tissues at the beginning, the resulting pattern becomes disproportionately more complex. The amount of information to describe the resulting pattern is much more than the amount of information to describe the generating rules and the positions of the initial tissues. The pattern is generated by the rules and by the *interaction* between the tissues. In this sense, the information to describe the adult individual was not contained in the initial tissues at the beginning but was generated by their interactions.

Besides generating information, this type of process has two additional features. First, it is strictly deterministic. When the locations of the initial tissues are identical in two embryos, the resulting adults, no matter how complex, will be exactly identical.

Secondly, it is in most cases impossible to discover the simple generating rules after the pattern has been completed, except by trying all possible sets of rules. When the rules are unknown, the amount of information needed to discover the rules is much greater than the amount of information needed to describe the rules. This means that there is much more waste, in terms of the amount of information, in tracing the process backwards than in tracing it forward. A geneticist would waste much time and energy by trying to infer the characteristics of the embryo from the characteristics of the adult organism. It would be more profitable to perform experiments in embryonic interference and embryonic grafting. The same is true also for the study of other deviation-amplifying mutual causal processes such as history or mental illness.

Since information is generated by the interaction between various parts of the embryo, it is not necessary for each part of the embryo to contain information regarding the body part it is destined to become. It partly receives the information from other parts of the embryo and from its relationships to them. For example, in the embryo of certain species, if the part which would become an eye is transplanted at an appropriate stage of the embryonic development into the part which would become skin, the eye-tissue becomes skin. It receives information for its growth from its surroundings.

We have discussed mainly the structure-generating aspect of the interaction between the parts of the embryo. But the interaction has also a structure-stabilizing aspect. Let us look at the example of the grafted eye tissues again. When they were grafted on skin tissues, the skin tissues made the would-be-eye tissues into skin tissues, against the possibility that the would-be-eye tissues might become an eye. This process is "morphostasis" in the traditional terminology.

Thus, the usage of the words "morphogenesis" and "morphostasis" in the sense of deviation-amplifying mutual causal process and deviation-counteracting mutual causal process, respectively, does not contradict their traditional usage in the sense of ontogenetic structure-generation and structure-stabilization. The new usage not only extends the old usage, but gives a functional definition in terms of deviation-generating and deviation-counteracting, and in terms of positive and negative feedback networks.

Let us now examine more closely what is meant by positive and negative feedback networks. Let us first emphasize that the presence of influences in both directions between two or more elements does not necessarily imply mutual causation. If the size of influence in one direction is independent of the size of influence in the other direction, or if their apparent correlation is caused by a third element, there is no mutual causation. Only when the size of influence in one direction has an effect upon the size of the influence in the other direction and is in turn affected by it, is there a mutual causation.

For example, Eisen Iron Co. makes iron from iron ore. Dexter Tool Co. manufactures tools made of iron. Dexter buys iron from Eisen, and Eisen buys tools from Dexter. There is some mutual relationship between the two companies. But suppose Dexter buys iron from several companies. When Eisen's output drops, Dexter's purchase of iron from other companies increases. When Eisen's output goes up, Dexter's purchase of iron from other iron companies decreases. The amount of tools Dexter can supply to Eisen does not depend on the amount of iron Dexter buys from Eisen. Furthermore, Dexter has other customers besides Eisen. Whether Eisen buys no tools or 10,000 tools from Dexter does not matter much to the operation of Dexter. In this case, though there is traffic of merchandise in both directions between Eisen and Dexter, the amounts of traffic in two directions have no mutual causal relationship.

Suppose that, suddenly, some ship industry develops in the vicinity, and both iron and tools are in great demand. Consequently, both Eisen's output and Dexter's output increase simultaneously. But this simultaneous increase was not caused by a mutual causal relationship between Eisen and Dexter, but by a third element which is the ship industry.

On the other hand, if the amount of Dexter's output depends on the amount of Eisen's output and varies with it either in the same or opposite direction, and the amount of Eisen's output depends on the amount of Dexter's output and varies with it either in the same or opposite direction, then there is a mutual causal relationship between Eisen's output and Dexter's output.

Mutual causal relationships may be defined between more than two elements. Let us look at the following diagram [Figure 3]. The arrows indicate the direction of influences. + indicates that the changes occur in the same direction, but not necessarily positively. For example, the + between G and B indicates that an increase in the amount of garbage per area causes an increase in the number of bacteria per area. But, at the

same time, it indicates that a decrease in the amount of garbage per area causes a decrease in the number of bacteria per area. The − between S and B indicates that an increase in sanitation facilities causes a decrease in the number of bacteria per area. But, at the same time, it indicates that a decrease in sanitation facilities causes an increase in the number of bacteria per area.

As may be noticed, some of the arrows form loops. For example, there is a loop of arrows from P to M, M to C, and C back to M. A loop indicates mutual causal relationships. In a loop,

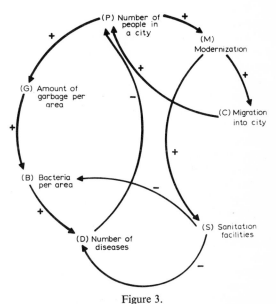

Figure 3.

the influence of an element comes back to itself through other elements. For example, in the loop of P–M–C–P, an increase in the number of people causes an increase in modernization, which in turn increases migration to the city, which in turn increases the number of people in the city. In short, an increase in population causes a further increase in population through modernization and migration. On the other hand, a decrease in population causes a decrease in modernization, which in turn causes a decrease in migration, which in turn decreases population. In short, a decrease in population causes a further decrease in population through decreased modernization and decreased migration. Whatever the change, either an increase or a decrease, amplifies itself. This is so when we take population as our criterion. But the same is true if we take modernization as a criterion: an increase in modernization causes a further increase in modernization through migration and population increase; and a decrease in modernization causes a further decrease in modernization through decreased migration and decreased population. The same holds true if we take the migration as the criterion.

In a loop, therefore, each element has an influence on all other elements either directly or indirectly, and each element influences itself through other elements. There is no hierarchical causal priority in any of the elements. It is in this sense that we understand the mutual causal relationships.

Let us take next the loop P–G–B–D–P. This loop contains a negative influence from D to P. In this loop, an increase in population causes an increase in the amount of garbage per area, which in turn causes an increase in the number of bacteria per area, which in turn causes an increase in the number of diseases, which in turn causes a decrease in population. In short, an increase in population causes a decrease in population through garbage, bacteria and diseases. On the other hand, a decrease in population causes a decrease in garbage, bacteria and diseases, and thus causes an increase in population. In this loop, therefore, any change in population is counteracted by itself. Likewise, any change in the amount of garbage per area is counteracted by itself. The mutual causal relationship in this loop is a deviation-counteracting mutual causal relationship. Such a deviation-counteracting process may result in *stabilization* or *oscillation*, depending on the time lag involved in the counteraction and the size of counteraction.

Let us further consider the loop P–M–S–D–P. This loop has *two* negative influences. An increase in population causes an increase in modernization, which in turn causes an increase in sanitation facilities, which in turn causes a decrease in the number of bacteria per area, which in turn causes a decrease in the number of diseases, which in turn causes an increase in population. This is therefore a deviation-amplifying loop. Two negative influences cancel each other and become positive in the total effect.

In general, *a loop with an even number of negative influences is deviation-amplifying, and a loop with an odd number of negative influences is deviation-counteracting.* Besides the three loops mentioned above, there is another loop P–M–S–B–D–P, which is deviation-amplifying because of the two negative influences contained in it. The system shown in the diagram contains several loops, some of which are deviation-amplifying and some of which are deviation-counteracting. Whether the system as a whole is deviation-amplifying or deviation-counteracting depends on the strength of each loop. A society or an organism contains many deviation-amplifying loops as well as deviation-counteracting loops, and an understanding of a society or an organism cannot be attained without studying both types of loops and the relationships between them. It is in this sense that our *second cybernetics* is essential to our further study of societies and organisms.

Not only are there deviation-amplifying loops and deviation-counteracting loops in the society and in the organism, but also under certain conditions a deviation-amplifying loop may become deviation-counteracting, and a deviation-counteracting loop may become deviation-amplifying. An example is the principle of diminished return. An increase in investment causes an increase in capital, and an increase in capital makes more investments possible. Before the profit reaches a certain level the effect of income tax is negligible. But, as the profit becomes greater, the influence of income tax becomes greater and eventually stabilizes the size of the capital.

A culture may follow a similar process. Sometimes one may wonder how a culture, which is quite different from its neighbouring cultures, has ever developed on a geographical background which does not seem to be in any degree different from the geographical conditions of its neighbours. Most likely such a culture had developed first by a deviation-amplifying mutual causal process, and has later attained its own equilibrium when the deviation-counteracting components have become predominant, and is currently maintaining its uniqueness in spite of the similarity of its geographical conditions to those of its neighbours.

The second cybernetics is useful also in such field as psychiatry. In many cases, interpersonal conflicts are generated by mutual deviation-amplification between persons and are later maintained at the deviated (but not necessarily deviant) pattern. Mutual amplification may occur within a person, for example, between loss of self-confidence and poor performance in a neurotic person. The established pattern, no matter how deviated, is not necessarily "pathological" if it enables a constructive life. But, if the pattern results in a reduction of conflict-free, constructive energy, then therapy becomes necessary. The therapy aims at breaking the stabilized unsatisfactory pattern and at initiating a new deviation-amplification in the direction of developing a satisfactory pattern.

The second cybernetics will be useful also in the technological fields such as in the design of a machine which invents. A trial-and-error machine is inefficient because it has no directivity. But it has a great flexibility. A deviation-amplifying inventing machine, on the other hand, works in the direction specified by the initial kick, and, for this reason, is efficient. It is not built for any specific direction, because the direction is a variable which is specified by the initial kick. In this sense it is flexible. But in another sense it is not flexible because once the direction is set, it will persist in that direction. A machine that incorporates randomness, deviation-amplification and deviation-counteracting may be both efficient

and flexible. It can search for all possibilities. It can try to amplify certain ideas in various directions. It can stay at a relevant idea (which may change from time to time during the invention) and bring back to it other ideas for synthesis. In fact, openness to strange hunches, ability to elaborate on them and to bring them back to a synthesis are what is found in the process of human creative minds.

The elaboration and refinement of the second cybernetics belong to the future, and we may expect many fruitful results from them.

VI

Self-Regulation
and
Self-Direction in
Psychological Systems

WE have seen that one of the major thrusts of modern systems research has been to replace atomistic analysis with a holistic approach to complex organization. Thus, the emphasis shifts from the fracturing of systems into units, piecemeal mechanical study of each unit, and an attempt to fit them together in some simple way, to a focus on the nature of the complex interrelations and interactions among the components and the effective environment, all seen as an ongoing process from which emerges the properties of the whole, and which may often change the very characteristics of the parts along the way. As several of the following articles show, modern systems-theoretic concepts provide an especially apt vocabulary for express-ing the principles underlying such processes. Slack's interpretation of Dewey's 1896 insight into the appropriate approach to the study of human behavior as an ongoing, holistic transaction may be taken as a reverberating keynote of the transition to modern systems thinking.

The broader transactional process points to the reinforcing effect of sensory feedback from controlling the environment, and to responses seen as directed out-ward rather than merely pushed from within. And the transactional model suddenly becomes of crucial practical concern in our space age: If psychosystem functioning is externally, not simply internally, controlled via feedback transactions with the environment, what happens to man in space when this transactional process is seriously truncated for prolonged periods? Richard Held and Sanford Freedman answer this question in rather clear-cut fashion. (For a fascinating and extensive research report and rigorous theoretical treatment of the central role of behavioral transactions in shaping our perceptual construction of the world about us, see James G. Taylor, *The Behavioral Basis of Perception*, 1962. It is of significance that this important work utilizes Ashby's logical cybernetic model as a framework for the theory.)

The first two papers in this Part deal mainly with transactions with the physical environment. As Shibutani shows in the third article, G. H. Mead, who was close to Dewey at Chicago and shared many ideas with him, focused on the social environ-ment of acting and interacting others to explicate the same general principle: It is through transactions with the now *social* environment of others that there emerges the distinctive human "mind," the self and self-consciousness, and the resulting more or less stable sociopsychological *constructions* of the past and future as well as present world of cultural ideas and physical objects. These features made possible and were made possible by the conjoint development of language, of symbolic interaction, which characterizes the sociocultural level of the complex adaptive system. In abstract terms, just as biological evolution represents the phylogenetic mapping of the variety and constraints in the physical and biological environment into the physiological structure of species via genetic coding, and learning represents the ontogenetic mapping in somewhat greater detail of this variety via neurological coding, so human cultural development and group socialization represents the extra-somatic mapping in much greater detail and potential fidelity of the physical, social and psychic variety in the environment via symbolic coding. This latter, by virtue of its extrasomatic component, is thus freed from many biological constraints, and underlies the phenomenon emphasized in Mead—man as an interpreting, defining, choosing, deciding being.

Mowrer's contribution is a further example (along with Osgood's and that of Miller *et al.*) of constructive steps being taken to lead behavioristic psychology out of its S–R cul-de-sac. It builds on insights such as those expressed in different ways above, arguing that internal states involving "meaning" act to construct, interpret, select and thus control situational events in an ongoing creative act. And conscious-ness—or, better, self-consciousness (explicated by Mead in terms of the fundamental "reflexiveness" of the symbol-using human mind)—is structurally an internal feed-back device inputting information of this "internal state" (felt needs, goals, plans of action, moral evaluations, etc.) into the organism's action-control center. Gordon Allport further confirms this active, open-system, transactional view of personality, though we should note that he balks at going the whole way in facing "the knottiest issue in contemporary social science": Is personality something "inside the skin"?

From its beginnings in the work of Wiener, cybernetics has been offered as a model that may help us to understand the more pathological, as well as "normal" features of human behavior. The selection by Notterman and Trumbull, and that by Vickers, represent ways in which the servosystem model may be applied to the problems of stress and of internal conflict ("deafening streams of mismatch signals"), the latter suggestively seen as endemic in organisms and organizations.

The last two selections of this Part represent recent self-conscious attempts to build fairly ambitious, systematic models of human behavior using cybernetic principles to attempt to get significantly deeper into its realistic complexities. MacKay attempts to schematize in his model many of the features of human functioning that the above authors have insisted upon, especially the active, self-seeking, probabilistic, interpretative, transactional nature of the individual's relations with an environment. And with Miller, Galanter, and Pribram, we conclude on the note with which we started: a critique of the reflex arc concept of traditional behavioristic psychology. Starting as near-behaviorists themselves, Miller *et al.* develop a constructive theoret-ical program that, in effect, carries forward Dewey's 1896 critique and leaves them in the end with a self-styled "subjectivistic behaviorism"! Their references to "the Image" are specifically to Kenneth Boulding's conception developed in his book by that title, refering to the knowledge, feelings, and evaluations men have of the world, society, and themselves. A "Plan" is defined by Miller *et al.* as "any hierarchical process in the organism that can control the order in which a sequence of operations is to be performed." The general problem of their book from which our selection is taken is "to explore the relation between the Image and the Plan," on the basis of their thesis that the Image tells only part of the story of human behavior and the internal mechanisms that lead to its particular overt expression. A Plan is needed to exploit the Image, and so guide behavior.

38.
Feedback Theory and the Reflex Arc Concept

CHARLES W. SLACK

IN 1896 JOHN DEWEY criticized the new reflex arc concept in psychology on a number of grounds and concluded that: ". . . the distinction of sensation and movement as stimulus and response respectively is not a distinction which can be regarded as descriptive of anything which holds of psychical events or existences as such."[1]

Dewey used the familiar example of the child reaching for the flame, taken from William James. The theory he severely criticized was the one which simply stated that the sensation of light is a stimulus to the grasping as a response, etc.

It was not until nearly a half-century later that experiments were performed which would allow a testing of the reflex arc concept in gross behavior —eye–hand coordination situations where some continuous record of the position of the "child's hand" and of the "flame" could be made. The experiments which allow for this kind of recording are called tracking experiments. The "flame" is now called the *target*; the "finger tip" becomes a special example of a number of *controls*, including variously loaded joy sticks and handwheels; the field of view, including that which the child sees of his hand, of the candle and of various reference objects, is now called the *display*.

Except for the fact that S is now usually limited to one or two dimensions in which he can move, and that his motivation is directed by instruction rather than by curiosity or burning pain, everything is pretty much the same with today's "human operator" and yesterday's *enfant terrible*. At least the classic example of the reflex arc is no better as an example than is the modern one.

There are but two things left to do to complete the similarity between the two examples. We must restrict the modern tracking to step-function inputs —sudden displacements of the target, in order to make it correspond to the sudden observance (or sudden lighting, if you will) of the candle in the classic example. Further, we shall only concern ourselves with the first half of the Dewey–James example where the child reaches for the flame. This phase illustrates the negative feedback principle to be discussed in this paper.

Tracking of step-function inputs, then, is a good example of the reflex arc. It is also a very special case of the psychophysical method of reproduction where the error contributed by S may eventually be reduced to zero. That is, tracking is the method of reproduction with knowledge of results and usually with continuous recording over time of the target and the control, or at least the difference between them.

Today's tracking, however, usually tends to imply more than a particular experimental situation. There is a theory which goes along with most experiments of this sort: on the one hand it is a more precise statement of the reflex arc notion, and on the other hand it says some quite different things.

Briefly stated, the theory of negative feedback extends to all systems which use a measure of the difference between the target and the control to decrease that difference. In the Dewey–James example, the difference between the finger tip and the flame is fed back and used to control perhaps the velocity of the hand; the greater the distance to go, the faster the movement.

From Charles W. Slack, "Feedback Theory and the Reflex Arc Concept," *Psychological Review*, 62 (1955), 263–67. Reprinted by permission of the author and the American Psychological Association.

We shall not concern ourselves here with the specific technical advantages of particular methods of analysis of feedback systems or with the special assumptions which many of these methods employ, but will turn instead to the general notion of the feedback system as a model for the reflex arc situation. The purpose of this paper is: first, to ascertain whether or not feedback theory (or servo theory) answers any of the criticisms of reflex arc theory which were raised by Dewey in 1896; and second, to see in what ways feedback theory conflicts with the simple stimulus–response notion implied in the reflex arc theory of Dewey's day.

First of all we notice that by its very nature of being a closed-loop theory, or a theory applying to systems which continually or continuously use a measure of performance to control performance, feedback theory meets Dewey's objection that "[reflex arc] gives us literally an arc, instead of the circuit; and not giving us the circuit of which it is an arc, does not enable us to place, to center, the arc." Reflex arc theory, as criticized by Dewey, was apparently an open-loop system. Servo theory closes the loop.

Secondly, Dewey criticized the reflex arc notion as follows:

Upon analysis, we find that we begin not with a sensory stimulus, but with a sensori–motor coordination, the optical–ocular, . . .

Now if this act, the seeing, stimulates another act, the reaching, it is because both of these acts fall within a large coordination; . . . the ability of the hand to do its work will depend, either directly or indirectly upon its control, as well as its stimulation, by the act of vision. . . . The reaching, in turn, must both stimulate and control the seeing. . . .

Since the analysis of a feedback system is in no way vitiated if the system is made up of a number of subsystems, it would seem that modern servo theory again fills the bill.

Thirdly, Dewey stated:

The sensation or conscious stimulus is not a thing or existence by itself; it is that phase of a coordination requiring attention because, by reason of the conflict within the coordination it is uncertain how to complete it. . . . The end to follow is, in this sense, the stimulus. . . . From this point of view the discovery of the stimulus is the "response" to possible movement as "stimulus." . . . Generalized, sensation as stimulus, is always that phase of activity requiring to be defined in order that a coordination may be completed.

This criticism of the reflex arc strikes at the heart of the matter, for it challenges the validity of the stimulus and response as separate entities which are causally related in some way.

We may expand upon this argument by means of a simple example from tracking:

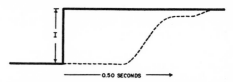

Figure 1.—Diagram of typical step-function input and output. The solid line is the input presented to S through a narrow slit at a fast rate. The dotted line represents S's attempt to keep his control pencil on the input. The error in initial response may be an undershoot (as shown), an overshoot, or zero.

Figure 1 shows a diagram of a typical step-function input (solid line) and output (dotted line). What was presented to S was a dot remaining motionless at one position on the field, and then its instantaneous displacement (I) to another position on the field. The dotted line shows S's attempt to follow this displacement of the target with his own control. His control in this case was a pencil which he attempted to keep on a short segment of line (the solid line), which he viewed through a narrow slit. His output was then recorded, superimposed upon the input (which E programmed for him by drawing lines at various positions and of various lengths, corresponding to time intervals, on a strip of paper which was fed past the slit where S's pencil point was resting).

This tracking input–output diagram might be taken as a slightly simpler example of the reflex arc situation than the child–candle.

In order better to understand Dewey's objection to the naive stimulus–response description of this type of behavior, we might attempt to apply stimulus response notation to the input–output diagram of tracking. Let us define the *stimulus* as a change in environment over time produced and measured by E by operations which he describes in such a way that they may be duplicated by others, and: (*a*) which, at any time, might provide some or all of the information leading to the response at that time, or (*b*) which, at any time, might possibly determine all or part of the response at that time, or (*c*) to which, at any time, the response might be related in some way, or (*d*) which is a necessary condition for the response.

Now let us check through the characteristics of the input to find what meets the criteria listed above.

First, the mere *existence of the target dot*

somewhere (anywhere) on the field cannot alone be called stimulus. The existence of the target dot is not a *change* in environment over time; that is, the dot may always be there during the experiment. Furthermore, it meets none of the other criteria. The existence of the target dot might be considered a necessary condition for the response were it not for the fact that under certain conditions (after the establishment of expectancies of various kinds) the response is obtained without the existence of the target dot. At any rate the existence of the target provides no information for the response, cannot possibly determine the response, and the response is not much related to the mere existence of the target.

Second, the *position of the dot relative to the rest of the field* (slit) is clearly not the stimulus. In the first place, we discover that the same response may be obtained to targets positioned anywhere on the field. In the second place, the position of the dot on the field can in no way determine the extent of movement of *S*, nor provide information to determine the extent of movement. The *S* has to know "which way" and "how far" before he moves, and the position of the dot relative to the field does not tell him this. The position of the dot on the field (or indeed, the position of the dot relative to any frame of reference of *E*'s) is not related to the response of *S* within a wide range of conditions.

Third, the *displacement of the target* (the size of the movement *I*) cannot be called the stimulus since, if it were, we would be restricted only to the condition shown where *S*'s position is equivalent to the position of the target just as displacement occurs. The displacement of the target, like the position of the target, does not tell *S* "which way" or "how far" except for the time right after the target moves and just before *S* moves. After *S* has started to move (either in the right or the wrong direction), *I* can only be misleading information telling him "how far" or "which way."

The result of all this rather simple reasoning (not to be confused with Dewey's more eloquent generalizations) is that we come to the conclusion that there is nothing about what *E* does to the environment which can alone be called the stimulus if we give the term stimulus any psychological relevance.

Let us take a more functional view of the situation in an attempt to discover what might be the "stimulus" as far as *S* is concerned.

What must *S* know in order to be able to respond adequately? We gave the answer above when we said that he must know "which way" and "how far." In order to know these, he must know "where he is" and where he "wants to be." But the difference between where he is and where he wants to be, including sign, is the answer to the questions "which way" and "how far." To the extent, then, that we can assume that *S*'s frame of reference relative to which he knows these things corresponds in important ways to the one *E* gave him (the slit), and furthermore to the degree that where *S* wants to be corresponds to where *E* wants him to be, we can say that the stimulus is equal to the "just past" difference between the dotted line and the solid line. Or, to paraphrase the statement made earlier, the difference between where *S* is and where he wants to be is continually used as the "stimulus" to decrease that difference. If we know where *S* is (in the same regard as he knows it) and if we can assume that where *S* wants to be is equivalent to the position of the target relative to that regard, then, and only then can we specify what the "stimulus" is. But notice this about the "stimulus" which we have thus deduced. This "feedback" stimulus is much closer to Dewey's idea of the stimulus than it is to the stimulus response notion which demands that the stimulus be defined a priori in terms of *E*'s frame of reference. The Dewey feedback stimulus is far from a given, constant thing. It is an assumption—both on *E*'s part and upon *S*'s part. It is a continually changing thing. It depends upon where *S* is and *S* is moving. The *S* alters (and at times creates) the stimulus in just as real a sense as does *E*. Furthermore, the very existence of the stimulus depends upon a "common ground" or frame of reference between *E* and *S* and upon common purposes between them. Both the child and the man who lights the candle are parties to a transaction which has no meaning in terms of either one separately.

The "real" stimulus in the reflex arc is the error signal (difference between desired state and present state) and it is not the input defined in terms of E's frame of reference.

This statement is all right as far as it goes but it does not go far enough. The error signal, as *E* knows it, *is* the difference between the dotted line and the solid line. But *S* has at his command, of course, *only an estimate* of this difference. There are threshold properties to be considered and there are constant and variable errors. Furthermore, there is much good evidence to show that *S*'s estimate of how far and which way he has to go is influenced to some extent by his past experience: there are serial order effects and "range" effects. Perhaps other more complex dependencies may affect S's estimate, especially at high speeds of performance.[2] Are we to include all this in the term stimulus? If we are to give the term *any* a priori psychological relevance we must consider *some* characteristics of S in our definition. If we do not want to be arbitrary about it, we are forced to qualify our definition of the term "stimulus" with the following clauses:

1. It is an abstraction out of a process, called, by Dewey, a transaction.

2. It is created by S as well as by E.

3. It is never fixed or constant unless both E and S want it to be.

4. It is an assumption.

5. It is at least a difference between two estimates—one with regard to the control and one with regard to the target.

6. Experience and various psychological errors enter into this estimate.

We may avoid the above argument by defining stimulus purely in physical terms, that is, relative to E's frame of reference. We may define it as the existence, or the position, or the displacement, or the brightness of the target or as all of these. But if we do this—i.e., define stimulus as independent of S, it is going to have *no* a priori psychological relevance.

For purposes of clarity, we should have a word to stand for the heavy line in Figure 1. We shall call this the input (input stimulus if you like), understanding that we can never be sure what, if anything, it has been put *into*. The dotted line in Figure 1 we call the output. When we use these words, we should remember that these are not psychological variables. Nor are they physical variables which are usually relevant to a psychological understanding of this transaction. The input is defined as what E does to the target, relative to the frame of reference he chooses. This is to give input an operational definition relative to his purposes as an experimenter. The output, while it can be

defined relative to E's frame of reference, can only have existence for S relative to S's purposes. Only in the event that S's and E's purposes overlap, that they have some goals in common, is an experiment possible.

Now, what is true for the "real" stimulus (as opposed to the "irrelevant" stimulus described above)—namely, that defining it depends upon common purposes between E and S, upon common frames of reference, that it can be created and altered by both E and S, and that it is an abstraction out of a complex transaction—is likewise true for the response.

Conclusion

We have taken one of the simplest examples of the reflex arc and attempted to apply stimulus–response concepts to it. In trying to do this we have discovered that we do not know what the stimulus is unless we know what the response is and what previous stimuli and responses were and that, as a matter of fact, we need quite a good understanding of the transaction in order usefully to call anything the stimulus. The same line of argument can be used for the so-called response side. We do not know what the response is unless we know what the stimulus is. Feedback theory tells us that the stimulus is at least the difference between the desired state and the present state. Our knowledge of psychology tells us that in this sense it must be a good deal more.

Notes

1. John Dewey, The reflex arc concept in psychology. *Psychol. Rev.*, 1896, **3**, 357–370.

2. The difference which exists between compensatory and follow tracking indicates that S is using more than the error signal upon which to base his response.

39.
Plasticity in Human Sensorimotor Control

RICHARD HELD AND SANFORD J. FREEDMAN

Studies of Disordered Motor-Sensory Feedback

Can man function effectively, let alone survive, in the exotic environments to which the astronaut will be exposed? The answers promise to be of scientific interest as well as practical consequence. Optimistic forecasts have been made of man's ability to perform efficiently in outer space for indefinitely long periods. These predictions have generally been based upon observations of human adaptability to the many and diverse circumstances found above, below, and on the surface of the earth. Some of the conditions to be encountered in space have been simulated by existing equipment on earth, and their consequences for human performance have been tested. Other conditions cannot be so directly tested, and informed speculation becomes necessary. For example, the effects of gravity have been reduced to zero for periods of less than 1 minute in aircraft flying Keplerian trajectories; subgravity states of longer duration have been achieved only in space vehicles. But even in the space flights of long duration astronauts have not yet experienced prolonged periods of free movement at zero gravity. For this condition, we make a less optimistic forecast, based upon recent analyses of sensorimotor function.

Humans and other mammals show a surprising lability in the responses of their sensorimotor

systems. Both prolonged isolation of human observers in monotonous environments (sensory deprivation) and prolonged immobilization in relatively normal environments (motor deprivation) lead to degraded performance on perceptual-motor tasks. The young of primates and certain other mammals fail to develop normal visually guided behavior when they are deprived of contact with the sense-stimulating environment. On the more optimistic side, human subjects have shown remarkable ability to adapt to conditions of sensory rearrangement—for example, to the wearing of prisms over the eyes which produce displacement and distortion in the appearance of the visible world. These findings imply that the stability of man's spatial perception and spatially oriented behaviors depends upon habitual contact with the sense-stimulating environment. When such contact is reduced or otherwise altered for a considerable period, the human system for sensorimotor control reveals its plasticity. Technological advances are producing conditions—flight in space, in particular—which tax this system. They pose the practical problem of predicting conditions under which the coordination of the human operator may be degraded. In a broader context we should like to have a general theory of the plastic sensorimotor systems, one that would both specify the range of normal circumstances responsible for the development and maintenance of stability in behavior and explain lability under transformed conditions.

Motor-Sensory Feedback

Essential for the stability of many of the plastic systems is the order entailed in the relation between the natural movements of an individual in his environment and their consequent sensory feedback. When an observer moves with respect to the many objects in his everyday world, his view of them changes. Their images move on his retinas in a manner highly correlated with his movements. There is an analogous correlation in the hearing of sounds by a moving listener, and a similar relation exists when an individual both moves and views a part of his body. Under any one of these conditions, a given movement tends to cause a characteristic sensory feedback. The central nervous system of the observer is both the originator of the movement and the receiver of its consequent sensory feedback. The central nervous system may be assumed to retain information concerning the output signals and to be informed of the dependent input signals. Comparison between these signals serves as a means of discriminating between change of visual stimulation caused by moving objects and change of visual stimulation resulting from movement of the perceiving organism.[1] In addition, this information has an important function above and beyond its use in spatial discrimination. It is necessary for maintaining and for altering the response characteristics of the sensorimotor control system in humans and certain other higher mammals. Evidence in support of this assertion has come from the results of two complementary types of experiment: analytic studies of adaptation to sensory rearrangement and related experiments on the effects of depriving young mammals of contact with their normal environments.

A rearrangement experiment with a human subject was first reported by Helmholtz, who viewed both his hand and other objects through a wedge prism. Because of the refraction of light, an object viewed through the prism appears displaced with respect to the position in which it would normally be seen by the naked eye. If the viewer reaches quickly for the object, he tends to miss it by an error equal to the optical displacement. However, as Helmholtz demonstrated, repeated efforts at reaching for the object result in compensation for this error and (equally important) in an error of the opposite direction upon removal of the prism.

To test the role of motor-sensory feedback in compensation for this classic form of rearrangement of eye–hand coordination, a procedure has been used as follows. During the period in which the subject viewed his hand through the prism (Figure 1, P), his arm was strapped to a lever that pivoted around a bearing at his elbow, as shown in Figure 1 (right). His head was held fixed by a bite board (not shown in Figure 1). Under one condition of viewing ("active movement") he moved his arm back and forth through the field of view, to the beat of a metronome (30 cy/min), for several minutes. Under a second condition ("passive movement") he kept his arm limp while it was moved in the same manner and for the same period by an external force. Before, and again after, viewing his hand through the prism, the subject repeatedly marked the apparent locations of the virtual images (T' in Figure 1). To him, these images of target points (T) appeared to lie on the surface under his hand when he saw them in the fully reflecting mirror (M). The mirror obscured both the subject's hand and his markings and consequently kept him from recognizing his errors. Comparisons made before and after the initiation of passive movement showed that the subject had not shifted the markings after periods of passive movement ranging up to half an hour. However, a few minutes of active movement produced substantial compensatory shifts, and many subjects showed complete compensation within half an hour after the initiation of active movement. Although the passive-movement condition provided the eye with the same optical information that the active-movement condition did, the crucial connection between motor output and visual feedback was lacking.

Other techniques, similar to this procedure for analyzing Helmholtz's experiment in terms of motor-sensory feedback, have been used to show the relevance of the movement condition for adaptation to many other types of rearrangement. To this end, Held and his collaborators have studied the effects of increasing the optical distance between hand and eye; of introducing locomotion during displacement of the visual field by wedge prisms; of rotating the visual field by means of right-angled prisms; of inducing intrafield distortions by means of flat wedge prisms; and of "displacing the ears" to new positions with respect to the head by means of eletronic pseudophones. In any of these rearrangements, the subject initially makes an error, in either a localizing response or a measure of spatial perception, that is predictable from the magnitude of the imposed displacement or distortion. However, in accord with our findings in the experiment described, if he is allowed free movement either of his whole body, as in locomotion, or of an appropriate limb, he will compensate for the initially induced error. If restrictions are placed on his movements or if parts of his body are

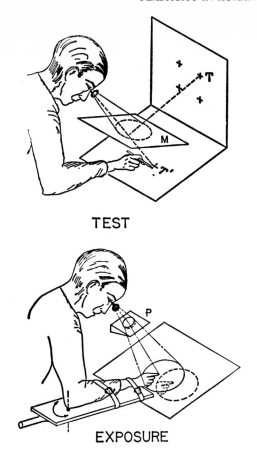

TEST

EXPOSURE

Figure 1.—Apparatus for rearranging eye-hand coordination and for testing the consequences of such rearrangement.

moved by an external force in such a way that the variation in sensory stimulation is equivalent to that received with free movement, there is no comparable adaptation. Some adaptation may be produced by factors other than motor-sensory feedback. But such factors, insofar as their effects have been explored, seem to lead to changes that are limited as compared to the full and exact compensation that can be demonstrated on sufficiently long exposure to conditions of rearrangement accompanied by free movement.

Demonstration of complete compensation is of crucial importance in bridging the gap between adaptation in the adult and original development in the newborn infant. When it can be shown that adaptation proceeds to a stable end state which corresponds to accurate orientation in the environment, then it is conceivable that the same process operates in the development of coordination in the newborn infant. A recent study has in fact demonstrated the importance of motor-sensory feedback

for the course of visual development from birth. The research was suggested by work of Riesen and his collaborators, who had previously demonstrated deficits in the visually-guided behaviors of cats and a chimpanzee reared from birth under conditions of restricted mobility when in the light. Following this lead, Held and Hein reared paired kittens under conditions such that, when the animals were in the light, only one of them was allowed to move about, the other being transported over an almost identical path. By this means the two kittens were provided with equivalent visual stimulation. The results showed that the kitten which was transported failed, in contrast to its actively moving mate, to develop normal visual–spatial capacities, despite its exposure to the patterned and varied visual surround. From these findings we conclude that the information entailed in the motor-sensory feedback loop allows the sensorimotor system both to set (as in development) and to reset (as in adaptation to rearrangement) its response characteristics as a consequence of its own past actions.

Motor-Sensory Correlation

The crucial role of natural movement in adaptation to rearrangement and in the development of the newborn infant hints at the importance of the correlation between motor output and sensory feedback signals. How is this factor implicated? For an answer, we must look more closely at the normal relation between movement and sensory feedback and at the effects of transforms of this relation.

In the case of eye–hand coordination, the geometry of movement specifies that each distinguishably different movement of the hand will be accompanied by a unique change in the viewer's image of that hand, provided that variations in the positions of eye and head are discounted. If these variations are not discounted, the function that relates movement of the arm and hand to its visual feedback will contain parameters whose values specify different positions of the eye and head. Information concerning these positions—hence the values of the parameters—is available to the nervous system. For any set of parameter values, viewing the hand entails a one-to-one relation between movement and visual feedback. During the course of a large number of movements the nervous system may take account of the internally initiated efferent signals to the musculature, together with their concurrent visual feedback. Within the limits of precision afforded by the transfer of information within the neuro-muscular system, the one-to-one relation between movement and its sensory feedback will allow the system to establish and store the correlated information. When, as in the experiment described, the hand is viewed through a wedge prism, it is displaced from the normal position by a fixed distance. Although any particular movement is now accompanied by a different feedback, the one-to-one relation between movement and sensory feedback is preserved. During the course of further movements of the arm and hand, the newly cor-related information becomes available to the nervous system. This invariant order is, we believe, responsible for adaptation to the prism transform.

Full compensation for the errors in visual direction (egocentric localization) induced by a wedge prism appears to require gross movement of the head and eyes in a patterned and visible surround. These movements make the analysis of the relation between motor output and sensory input more complicated than is the case for eye–hand coordination. Let us consider a bodily movement which transports the eye through an environment containing stationary objects. The initial position and subsequent displacement of the eye geometrically determine certain properties of the ensuing stimulation of the retina of that eye—those that Gibson has called the "flow patterns." When the eye rotates about an axis through its center (its nodal point, to be exact), all imaged points move across the retina at the velocity of rotation of the eye. When the eye translates toward an external point that casts its image on the retina, all other imaged points move radially outward from this central image point originating on the line of translation. For every different direction of translation, the center of radial flow is different. There is, then, a one-to-one relation between the directions of transla-tional movement and the corresponding centers of flow patterns on the retina.

A wedge prism, fixed in its position with respect to the eye, shifts the center of flow nor-mally characteristic of any given translational move-ment by an amount equal to the prism-induced displacement of the central image point. The one-to-one relation between direction of trans-lation and center of flow pattern is maintained. The prism transform is isomorphic with respect to this relation, a condition that we believe is consistent with the observation that translational movements of the eye are required for complete compensation of the errors in direction-finding induced in a subject by wearing a wedge prism.

The foregoing considerations apply to the movements of a disembodied eye. However, translational movements of the eye are naturally produced by movements of the head, trunk, and limbs. Movement of the eye is mediated through the various linkages that connect it to these parts of the body. Rotational displacements of the eye, however produced, can change the relation between translational movements, as normally produced, and their consequent visual feedback. The function relating translational movement to visual feedback then contains parameters which have different values for each set of states of the various linkages responsible for rotations of the eye. Information about the positions of these linked parts must be available to the central nervous system if the effects of rotation are to be factored out.

Fixed displacements of the "ears" around the cephalocaudal axis of the head, produced by means of pseudophones, transform the relations between translational movements of the head and binaural acoustic signals in a manner analogous to the changing of the relations between head movements and visual signals by a wedge prism. Although the pairing of directions of translation and particular sequences of auditory stimulation is changed, the one-to-one relation between members of these pairs is preserved. As in visual direction-finding, rotational displacements of the head must be represented by parameters of the function relating translational movement and auditory feedback.

Full and exact adaptation to a rearrangement represents a change in state of the relevant sensori-motor control system such that the input–output or stimulus–response relation becomes identical to that which existed prior to rearrangement. In accord with analyses of the type discussed earlier, we believe that this change in state is dependent, in the first instance, upon an invariance of rela-tions in the functions describing movement with its consequent sensory feedback. The invariance is entailed in idealized geometrical and physical descriptions of the effects or rearrangement. These descriptions, together with several assumptions, account for the availability of ordered information to the nervous system. On the motor side, the system must be informed of its own output and of the changes in linkages that set the values of parameters in the feedback loop. The precision of transfer of information in the neural route between the origin of efferent signals and the muscular output sets limitations. Similar limitations exist on the sensory side. Observations of the precision of normal sensorimotor coordinations strongly suggest that these limitations are not severe. In any event, to the extent that these transfer functions are stable in time, repetitions of specific efferent signals may, subject to certain considerations that are discussed later, be accompanied by characteristic sensory feedback signals [called "reafferent signals" by von Holst]. Over time, the cumulated set of paired efferent and reafferent signals should, then, show a high correlation. The correlated information is, we believe, necessary for the development and maintenance of the plastic coordinations under normal conditions and for their adaptability to the transforms produced by rearrangement.

Decorrelated Feedback

In asserting that reafferent stimulation is predictable from bodily movement in general, and is related in one-to-one fashion to certain components of movement, we made two presuppositions about physical aspects of the interaction between organism and environment. (i) We assumed that the observer moves in a world well populated with stationary sources of stimulation. (ii) We assumed that output (efferent) signals in the central nervous system and their consequent muscular contractions produce the same bodily movements on all occasions. In other words, the exertion of force by muscular contraction was assumed to yield consistently predictable movements of the body and its parts. Under normal conditions of terrestrial life, including those that obtain during most rearrangement experiments, both of these assumptions are tenable. The visible world, for example, is crowded with objects of which a great proportion are stationary. Bodily movements are, for the most part, the resultants of muscular exertion against the counterforces of objects and gravity. Frictional forces generated by contact with objects strongly dampen these movements and inhibit ballistic motion. Muscular exertion, then, normally does produce corresponding bodily movement. Taken together, these physical conditions entail the redundant information required for the establishment of the correlation of signals discussed earlier. An exception can occur when the body or any of its parts is prevented from making contact with external objects. Muscular contraction may then be unrelated to certain aspects of bodily movement.

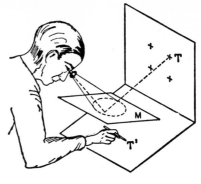

TEST

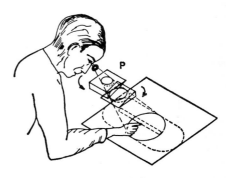

EXPOSURE

Figure 2.—Apparatus for disarranging eye–hand coordination and for testing the consequences of such disarrangement.

But such exceptions are rare. Men do not fly through the air, unsupported, for any appreciable length of time. The strongest impulse that can be generated by the human musculature is insufficient to keep a man's body in the air and out of contact with objects for more than a fraction of a second. Even the trampoline artist does not remain in the air longer than about $1\frac{1}{2}$ seconds. Consequently, we can normally expect a high order of predictability, that is invariant over time, in the relation between particular efferent and reafferent signals in the nervous system. However, either one or both of our presuppositions may become untenable in extra-terrestrial environments.

Consider, for example, the astronaut in free flight. If he is not strapped down but is free to push on objects, the consequent displacement of his body will be limited only by renewed contact with obstructions. During this ballistic movement, muscular effort may produce movement of parts of the body around its center of mass but will not change the trajectory of this center. The result may be a radical change from the high correlation between motor output signals and their sensory consequences that is experienced on earth. In the absence of specially imposed constraints, the relation between motor output and accompanying sensory input may vary, over time, in a very complex manner. The same consequence would be produced if the normal environment were to be replaced by one of continuously shifting visible objects, such as are seen on a "noisy" television screen. We have speculated elsewhere that this condition may be approximated by a blank visual field which allows noise intrinsic to the visual nervous system to become the dominant signal. Prolonged exposure to either a blank or a noisy visual field (sensory deprivation) alters spatial perception and coordination. But can we suggest an analog to the astronaut's condition that is less speculative—one that can lead to terrestrial experiments?

A rearrangement such as is effected either by placing wedge prisms before the eyes or by using pseudophones transforms the relation between motor output and sensory feedback in an isomorphic, continuous, and time-independent manner. Exposure in the normal environment leads to adaptation to the rearranged condition, with its corollary of temporary maladaptation upon return to the former, nonrearranged state. Suppose, however, we introduce a time-varying factor into the external circuit of the feedback loop—one over which the central nervous system has no control. This is, in principle, the condition to which our floating astronaut is exposed. The normal relation between certain efferent and re-afferent signals may then become more complex and even unpredictable. In time, the cumulated pairs of signals should show a reduced correlation. Considerations such as those we have discussed led us to suspect that in time the state of the system might be changed so as to produce increased ambiguity of response to relevant sensory signals. For convenience, we have termed the imposed condition "disarrangement," in contradistinction to rearrangement. We have performed disarrangement experiments in both visual and auditory domains to learn the consequences of such time-varying transforms in the motor-sensory feedback loop.

Visual-Motor Disarrangement

Because of its great lability, the system for eye–hand coordination lends itself to rapid testing

of the consequence of introducing a time-varying parameter. For the purpose, a prism of variable power P was introduced between eye and hand in the course of the experiment, as shown schematically in Figure 2 (right). Actually, the subject had in front of each eye a rotary prism consisting of two wedge-shaped elements of equal power that rotated at equal speeds but in opposite directions around the line of sight. The prisms were coupled together so that their powers could be maintained equal and no large binocular differences would be introduced. The device produced a continually varying displacement of the seen location of the subject's hand—a displacement independent of the movements of the hand itself. The displacement was cyclical, ranging equally to the left and right and back along the effective base–apex axis of the prism, at 1 cycle per minute. The axis could readily be changed so that the displacements occurred in the up–down direction. Measurements of eye–hand coordination were made before and after the periods of time-varying displacement in order to assess the effects of the displacement. The measurements were made in the apparatus shown in Figure 2 (left). The test procedure was identical to that described in connection with Figure 1 (left).

Figure 3 shows that markings made by one subject in two separate 64-minute sessions in which either up–down or right–left displacement was produced by a prism of varying power (maximum, 40 diopters). This subject had been instructed to move his forearm and hand back and forth over a small arc while he viewed it through the prisms. The most apparent change in the markings after introduction of the rotary

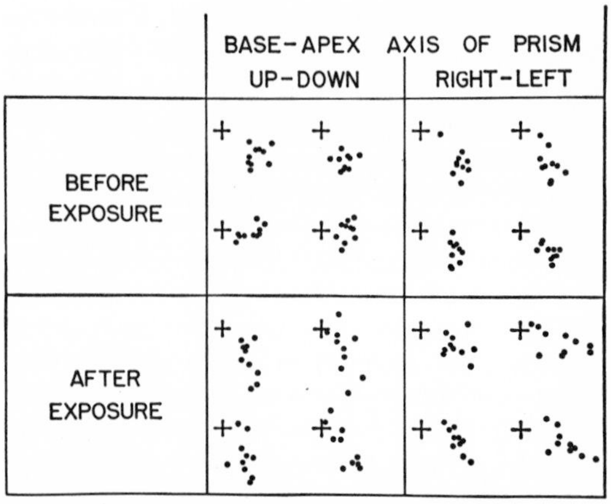

| | BASE–APEX AXIS OF PRISM | |
	UP–DOWN	RIGHT–LEFT
BEFORE EXPOSURE		
AFTER EXPOSURE		

Figure 3.—Markings made by one subject before and after exposure to conditions of disarranged eye–hand coordination.

prism was an increase in the dispersion of the markings along the dimension of optical displacement induced by the prisms. These markings show significant changes neither in dispersion at right angles to the induced displacement nor in the mean positions (centroids) of the group of marks. Furthermore, observations of successive markings revealed no tendency toward cyclical variation of position such as might be produced if the periodicity of the displacement cycle had somehow been learned by the subject.

The reduction of accuracy in eye–hand coordination implies that the control system has been degraded in the dimension specific to the time-varying parameter. This specificity affords an experimental control for testing the possibility that the increase in the dispersion of marks might result from fatigue alone, since such an effect should not be restricted to one dimension. The result is consistent with expectations derived from the considerations discussed, but it suggests the more stringent test in which results obtained under conditions of active and passive movement are contrasted. The basic findings prompted us to perform several experiments, of which the following is a representative sample.

Tests were made with eight undergraduates, two male and six female. Each subject was tested under four different experimental conditions, the order of the four tests being different for each subject. Two base settings of a variable prism with maximum power of 30 diopters were used: the effective base–apex axis (equivalent to the direction of optical displacement) was set at either the the right–left (R–L) or the up–down (U–D) position. Each of the two prism settings was combined with either an active or a passive movement of the arm. To insure that movements of the arm and hand were equivalent, and hence gave equivalent visual information to the eye under all viewing conditions, the subject was allowed to move his arm only in accordance with the procedure described for Figure 1. Test markings were made before exposure to the conditions of the experiment and 8, 16, 32, 48, and 64 minutes, respectively, after the beginning of such exposure.

The averaged percentage increases in the standard deviations of the markings for all subjects under all conditions are shown in the semilog plots of Figure 4. The standard deviations were calculated separately for the R–L and U–D dimensions. From these graphs it is apparent that the dispersion of markings increased along the dimension of variation during viewing with active movement but not during viewing with passive

movement. The passive-movement condition yielded no significant increase in dispersion along the dimension of variation, and the increase that occurs along the R–L dimension under the condition of U–D setting and passive movement appears to be unrelated to duration of exposure to the testing conditions. Statistical tests of the changes in dispersion after 64 minutes of exposure showed that the two base settings of the prism and the active-movement, as opposed to the passive-movement, condition had significant effects.[2]

Auditory-Motor Disarrangement

The separation of the two "ears" results in detectable binaural differences in the time, phase, and intensity of the sound coming from any localized sound source. Although localization of the source is possible without movement, the listener will ordinarily move his head or his whole body in relation to the sound source for greater accuracy. We tend to turn our faces in the direction of one of two telephones to make sure which one is ringing. We rotate the head until there are no differences in the sound as detected by the two ears. This "nulling" is possible only because the listener's movement produces systematic changes in the binaural differences. The use of this correlated motor-sensory information in auditory discrimination is analogous to its use in visual–spatial discrimination. And, as in the case of vision, the same information is responsible for

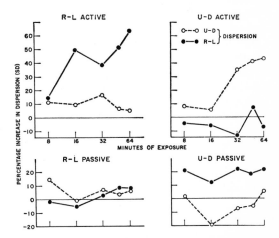

Figure 4.—Effect of exposure to conditions of disarranged eye-hand coordination in connection with active and passive movement, with right–left (R–L) and up–down (U–D) orientations of the base-apex axis of the prism.

adaptation to rearrangement. Consequently, the introduction of a time-varying parameter into the motor-sensory feedback loop should have an effect similar to that described for visual-motor disarrangement.

Discrimination of dichotic time difference (the difference in time of arrival at the two ears of corresponding acoustic signals) was chosen as the criterion task because of its importance in auditory direction-finding. Subjects were exposed to white noise, from separate but equivalent sources, which stimulated the two ears under various conditions of motility. This stimulation produces rapidly changing dichotic time differences whose temporal distribution is complex and unrelated to the actions of the subject. During an extended period of such exposure a variety of time differences will occur simultaneously with any particular movement made by the subject. The efferent–reafferent correlation should thus be reduced, with consequent loss of accuracy in localization.

To test the importance of motility in the expected degradation of the ability to discriminate dichotic time difference, 12 male college students were used as subjects. They were trained to a stable level of performance in the discrimination task and were then exposed for 2 hours to continuous stereophonic white noise under different conditions of movement. The subjects were tested before exposure and after 1 and 2 hours of exposure, by means of a procedure described in detail elsewhere. Briefly, it was as follows. The subject reported the apparent lateral direction (left or right) of paired clicks delivered to opposite ears with small but varied time differences. The set of thresholds of discrimination of change in the apparent direction of the clicks provided a measure of variability. The question may be raised as to whether localization of the apparent source of sounds coming to the ears by way of earphones is equivalent to localization of an external sound source. Jeffress and Taylor, in a recent study, demonstrated that there are no essential differences, and they claim that the two tasks involve identical processes.

The principal comparison made was between results of sessions in which subjects walked up and down a busy corridor during exposure to the white noise and sessions during which they lay relatively motionless on a bed. After exposure to the ambulatory condition, the subjects' ability to discriminate minute differences in the time of arrival of the sound at the two ears deteriorated very significantly. The mean variability of judgments (standard deviations) increased 104 percent

after 1 hour, and then declined to 79 after 2 hours, of exposure. On the other hand, the subjects' performance did not deteriorate when bodily movements were restricted; mean variability decreased 2 percent after 1 hour, and 5 percent after 2 hours, of exposure.[3]

Using the techniques described, Freedman and Secunda attempted to analyze the particular components of gross bodily movement which are responsible for the deterioration in performance during exposure to this auditory disarrangement. They found that (i) either rotation of the whole body or simple rotation of the head, without translation, during exposure to dichotic white noise led to deterioration in discrimination not significantly different from that produced with free ambulation; (ii) the effect occurred reliably within 1 hour; (iii) full recovery after exposure occurred within half an hour.

These results are all consistent with the effects of exposure to decorrelated feedback. They led us to make further experiments to compare the effects on auditory discrimination of active and passive movement.

Having isolated rotation as a condition sufficient to produce the auditory effects previously measured, we have been able to make separate experiments to compare the effects of the active and passive conditions for either rotational or translational movement. In the active-rotation condition the subject turned his head back and forth around a fixed cephalocaudal axis in a timed sequence of movements. In the passive-rotation condition he sat in a chair on a swiveled platform which was rotated by the experimenter through the same sequence. In the translational-movement condition he either walked or was transported in a wheel chair up and down a long corridor. In all conditions he wore earphones producing white noise. As in the visual-motor experiments, passive movement did not produce significant changes in performance during long exposure (1 hour); active movement was necessary for the degrading effect. After active movement, either rotational or translational, the variability (standard deviation) of judgments increased 41 percent; after passive rotational movement the variability increased 1 percent; after passive translational movement it decreased 5 percent.

Discussion

Prolonged exposure to conditions of decorrelated feedback clearly degrades the accuracy of response in both of the sensorimotor coordinations that have been studied. The differential effect of active movement as opposed to passive movement (or to no movement) is striking. In the absence of active movement, degradation does not occur. To the extent that an astronaut is exposed to decorrelated feedback, degraded performance on relevant tasks can be expected. Such losses may be prevented—and perhaps have been—by restricting bodily movement during exposures that entail the physical changes responsible for decorrelation.

Self-produced bodily movement, with its concurrent sensory feedback, provides an order required for organizing and reorganizing plastic sensorimotor systems. The experimental techniques of rearrangement and disarrangement have been used to perturb these systems. They have revealed some of the conditions that determine the response characteristics of the systems. These findings have a bearing on original development, as well as on the maintenance of behavioral capacities in higher mammals. We can understand the plasticity of these systems by thinking of them as having built-in programs that will produce precise sensorimotor coordination provided certain quasi-constant values have been set by information available from the motor-sensory feedback loop. Normal contact with the environment provides the highly correlated information that sets close bounds to the range of values of these constants. Contact during rearrangement alters the values of the constants but does not change the range of their variation. However, any condition that tends to disorder and decorrelate the information entailed in the relation that normally exists between motor output and contingent sensory input will increase the range of values of some constants and yield the degrading effect shown experimentally.

Both adaptation and degradation can be defined in the terms discussed. The adaptation of the sensorimotor system is optional when the constants are not significantly different from those currently inferable from the transfer of information through the motor-sensory loop. The adaptation will be less than optimal to the extent that this condition is not met. Since a history of exposure sets the values of the program constants, adaptation inevitably lags behind transforms in the loop. From this point of view, the increasingly variable responses produced by exposure to a time-varying transform in the loop represent an adaptive trend. The increasing range of values of the program constants is made evident by degraded coordination in tests carried out under stable conditions, which, in our experiments, are represented by the measuring conditions.

This discussion reflects the truism that an organism's adaptation is related to the regularities of its world. Perhaps less obvious is our conclusion that the stimulus transforms that accompany the movement of an organism are an important source of order in adaptation. This order is, in turn, dependent upon the quasi-constancy of several physical factors, among which is gravitation. In the absence of one or another of these factors, coordinations which depend upon their regularizing influence will, in time, be lost.

Summary

The maintenance and development of sensorily guided behavior depend in part upon bodily movement in the normal environment. Ordered information entailed in the motor-sensory feedback loop is responsible for the stable functioning of the plastic systems of coordination. It is found, from the results of experiments on vision and hearing, that the introduction of disorder into the motor-sensory loop changes the state of these systems and makes performance imprecise. In space, a freely moving astronaut will be exposed to a condition analogous to that of the subjects of these experiments. Consequently, he may lose his ability to perform certain tasks requiring precise sensorimotor control.

Notes

1. E. von Holst, *Studium Generale* 10, 231 (1957). D. M. MacKay has raised rather fundamental objections to von Holst's explanation and has proposed a more plausible model in *Aspects of the Theory of Artificial Intelligence*, C. A. Muses, Ed. (Plenum, New York, 1962), pp. 83–103.

2. A three-way analysis of variance (subject × prism base setting × active–passive) was performed on the mean changes of the difference between horizontal and vertical dispersion after 64 minutes of exposure. F for prism base setting was significant beyond the .01 level; F for the interaction between prism base setting and the passive–active dimension was significant beyond the .05 level.

3. The difference between these two experimental conditions was significant at better than $p = .005$, as determined by the Wilcoxon matched-pairs signed-ranks test. In addition, a Friedman two-way analysis of variance showed the conditions to be the significant variable ($p = .005$).

40.
A Cybernetic Approach to Motivation

TAMOTSU SHIBUTANI

SERVOSYSTEMS DIFFER from other machines in that they are able to regulate themselves; indeed, some of them perform in ways that resemble living creatures—perceiving, learning, remembering, and thinking. As automatic control assumes increasing significance in modern life, social psychologists are turning with renewed interest to the study of cognitive processes. As engineers attempt to design "electronic brains" that progressively approximate human capabilities, more and more attention is being directed to the study of open systems. Increasing emphasis is being placed upon organization, purpose, differentiation, growth, communication, and control—concepts that are indispensable in the study of living organisms. Preoccupation with problems of choice and goal-oriented movements characterizes all of the disciplines that have recently developed in this area—cybernetics, game and decision theory, general systems theory, information theory, and operations research.

Although engineers trying to duplicate human conduct have focused on self-regulation, until recently psychologists have tended to evade the problem. The study of personal autonomy and volition has been left by default largely to philosophers. But self-regulation lies at the core of human conduct and cannot be ignored. The only cybernetic model in use in social psychology today was developed at the turn of the century by two pragmatists; it has been used primarily by sociologists and is for the most part unknown to psychologists. The purpose of this paper is to describe briefly the conceptual scheme for studying voluntary conduct developed by John Dewey and

George H. Mead and to point to the many striking similarities between it and some basic ideas of the newer fields. While the scheme is crude and represents only a beginning, it may be used as an illustration to give some indication of the potentialities of this type of approach to human behavior.

Alternative Approaches to Motivation

No agreement exists among psychologists over what is meant by "motivation." Questions of why men act as they do arise from practical problems; effective cooperation as well as successful manipulation of others require a general understanding of human nature. Hence, laymen and specialists alike have long attempted explanations. As one might expect in a universe of discourse in which each event is viewed as the "effect" of some antecedent "cause" that is presumably responsible for bringing it about, a search is usually made for some prior event—a drive, some desire, or an incentive. A "motive" is usually regarded as something that "pushes" or "triggers" the deed. Such explanations, while sufficient for daily discourse, are hardly satisfactory from a technical standpoint. The problem confronting psychologists is that of accounting for the course of events, of explaining the *direction* taken by an individual's activities. Direct sorts of questions would arise, depending upon the general theory of behavior with which the psychologist worked. Although many writers on the subject are eclectic, it is possible to delineate four widely used schemes for studying the direction of human conduct.

The reflex-arc approach, which finds its clearest

Published originally in this volume.

statement in the writings of B. F. Skinner and J. B. Watson, sees behavior as response to stimulation. All behavior consists of muscular movements. Muscles are activated by nervous impulses, which originate outside the nervous system, at a point of sensory stimulation. The scheme rests on a mechanical model. Man is studied as if he were a machine, and a search is made for simple, unchanging elements. The stimuli represent the independent variable, and the responses constitute the dependent variable. Since the same stimuli do not always produce the same responses, however, psychologists like Clark Hull and E. C. Tolman have proposed some intervening variables, among them the "drive." A drive is a varying tendency of organisms to respond to certain types of stimuli, a tendency which is related to variations in organic needs. Thus, immediately after eating the need for food is satisfied, and the hunger drive is low. As nutriment is used up in metabolism, drive level increases in strength, and there is more responsiveness to various stimuli associated with eating. This complicates the scheme somewhat, but the stimuli are still the input, and the responses the output.

The tension-reduction approach, used in psychoanalysis and in Gestalt psychology, sees behavior as an attempt to maintain a steady state. The scheme rests on an organic model, and an effort is made to account for the things men do in terms of adjustive tendencies inherent in all living organisms. Emphasis is placed upon the structure of the organism as a whole; parts are seen only as components within this larger context. All living organisms have a dynamic quality, constantly adjusting to maintain their form and to reproduce their kind. Activity gets under way when the steady state of an organism is disturbed, and behavior represents an attempt to restore lost balance. Thus, many activities become goal-oriented; the organism persists through a variety of routes until it has reached its objective. The direction of behavior is explained in terms of the natural tendency of organisms to maintain a constant level of excitation.

Undoubtedly the most commonplace explanations of human behavior are in terms of subjective experiences—desire, preference, purpose. During the past century, especially since the work of Freud, there has been little respect among psychologists for verbal accounts of such inner experiences; it is taken for granted that an individual attempting to explain his own conduct would provide self-justifying rationalizations. In some circles this revolt has led to a rejection of all first-person reports. A third approach to motivation, which places emphasis upon avowed intentions, is rarely found among psychologists. It is obvious, however, that conscious intent does have something to do with the direction of conduct. Since men order and label their experiences symbolically, intentions are generally expressed in linguistic form, and some scholars have used the term "motive" to refer to linguistic designations of intent. These men are not trying to account for all human conduct; their interest centers primarily upon what men do in situations in which they appear to exercise choice. All motivation is not conscious, but some apparently is. It would appear, then, that the omission of conscious intent leaves any theory of motivation incomplete.

A fourth approach to human motivation is cybernetic. Behavior is seen not as response to stimulation, as relief from tension, nor as the accomplishment of symbolized intent; it is something that is *constructed in a succession of self-correcting adjustments* to changing life conditions. Although it is customary in some circles to express regret that G. H. Mead has nothing to say about motivation, he *is* concerned primarily with accounting for the direction of human conduct. While he insists that virtually all human activity is subject to social control, he sees human behavior as an organic process. Some of the distinctive characteristics of man—his ability to think in abstractions, self-consciousness, purposive and moral conduct—are emergent properties that developed as a certain type of organism came to terms with the exigencies of collective life. His basic unit of analysis is the *act*, which begins with a condition of discomfort and is terminated by its elimination. All the things men do can be broken down, for purposes of analysis, into a succession of acts. There are many kinds of acts: they vary in length; they may overlap one another; they may be delayed or even truncated. In the language of the new disciplines, human behavior is seen as a series of movements in space–time, the transformations of a self-modifying learning net. Mead, like Wiener, is primarily concerned with communication and control.

Behavior as a Succession of Adjustments

One feature of Mead's approach to human behavior is the special attention given to the *temporal* dimension. Each act is constructed as an organism and makes a succession of adjustments to conditions (external and internal) that are undergoing constant change. Overt behavior is generally only the final phase of an act; in most cases it is preceded by a number of preparatory adjustments, including various subjective experiences. Each act has a history and cannot be understood apart from it. Although he is concerned with a series of events, Mead insists that the sequence must be seen as a unit. An act is not a mere chain of passing events but an organized whole directed toward an end. To facilitate study of this process he proposes the concepts of impulse, perception, manipulation, and consummation. Because the editors of a posthumous volume have labeled Mead's introductory discussion of these four concepts the "stages in the act" they have often been regarded as steps in a natural history. A close examination of his work reveals, however, that this is not his intent. The terms refer to component parts of an act and facilitate analysis of its organization. For Mead, control of behavior comes from its organization.

The most fruitful point at which to begin the analysis of an act is the *impulse*. An impulse is a disturbance, any lack of adjustment between an organism and its milieu—pique over an imagined slight, hunger pangs, concern over the whereabouts of a friend who is overdue. An organism is set into motion by a disruption of its steady state; any discomfort leads to attempts to eliminate it. The key principle is that *an act, once under way, tends to persist until the discomfort is removed.* While it would appear that impulses vary considerably in direction and in intensity, Mead does not concern himself with these variables.

Much of human behavior is goal-oriented. An act is rarely the setting off of an already organized response by stimulation; it develops in transit. Dewey stresses the generality of impulses, showing that a given discomfort may be eliminated through any of several different routes. Each act moves in a general direction, but the specific details depend upon the exigencies of the situation. Thus, pragmatists emphasize a point somewhat akin to the principle of "equifinality" in general systems theory. In an open system the final state (overt response) may be reached from different initial conditions and in different ways.

An organism is in continuous interaction with its milieu, and activity is redirected and channeled in response to a succession of sensory cues. In his discussion of *perception*, Mead shows how various features of the environment become involved in the completion of an act. Not everything in the environment is perceived; a person is sensitized only to those objects that are relevant to what he is doing. One perceives things that will enable him to carry out an act that is already under way. Perception is not a mechanical recording of sensory cues; it is highly selective, and what is perceived is partly a function of the impulse.

Both perception and manipulation are organized in terms of hypotheses. The world is something that permits consummation. An object is approached in terms of a set of expectations of what will happen when one comes into contact with it; what is perceived, then, also depends upon anticipatory sets. Mead refers to perception as a "telescoped" or "collapsed" act; a person anticipates what would happen if he moved forward and touched the object, what would happen if his act were completed. Hypotheses are tested and confirmed in *manipulation*. In some measure what a man is going to do determines what he actually does. Manipulation involves the handling of objects as tools.

The hypotheses upon which perception and manipulation rest are derived from meanings. Other animals live in a world of events; human beings live in a world of conventional meanings. For pragmatists, meaning is primarily a property of behavior and only secondarily a property of the objects themselves. Meanings are stable relationships between an organism and a class of objects, defined in the manner in which the latter are characteristically handled. The meaning of anything is the way in which one is organized to act toward it. The characteristics of an object are important only in that they place limitations upon what can be done with it. Thus, members of each species select out of their environment objects that are essential to their survival and organize responses to them. This suggests that hypotheses rest upon past experience, including communication. In the language of the newer disciplines, the stock of meanings of each individual is his memory store, and input is coded in terms of past learning. This suggests too that one's orientation toward his environment is subject to continual reality testing. Hypotheses that turn out to be unreliable are rejected, and meanings are transformed. Thus, one anticipates from past experience what is likely to happen upon contact with an object, and movements are inhibited or facilitated on the basis of what is foreseen. Mead is best known, of course, for his insistence that meanings are subject to social control, that anticipated reactions of other people set limitations upon what can be done with an object.

Manipulation is rarely a reflex triggered by stimulation. Except for random movements human behavior consists of a series of maneuvers in which observed errors are corrected until a goal is achieved. Even a relatively simple act like reaching for a fountain pen involves a succession of adjustments. The actual path of the outstretched fingers is compared with the route it should follow, and the mismatch is progressively eliminated until the end is reached. The key to all such self-correcting adjustments is *feedback*; all purposeful behavior requires negative feedback. The principle is relatively simple: the results of one's own action are included in the new information by which subsequent behavior is modified. Thus, by responding to what he observes of his own activity a person makes necessary corrections. Although this principle appears over and over in the work of both Dewey and Mead, it has been largely overlooked, perhaps because they failed to coin a striking term to designate it. The emphasis upon feedback is one of the key differences between their scheme and other tension-reduction approaches to motivation.

Consummation is the elimination of the original disturbance and the restoration of a steady state. An act is terminated when discomfort is reduced, but there are variations in the extent to which satisfaction is achieved. When considerable tension has been generated in striving for a goal, attainment is usually a highly gratifying experience. Meanings may be reconstructed as the result of success or failure in reaching goals. Success reinforces patterns, and consummation that is especially pleasing becomes the basis for the formation of values. Prolonged and persistent failure, on the other hand, may lead to alienation and even to personality disorders. Since the nature of the impulse is not specified in this scheme, however, it is often difficult to ascertain just when an act has been consummated. The study of purposive activity requires a clearer designation of the goal as well as the margin of permissible deviation that surrounds it. What Mead stresses is that many of the objectives of one's activities are to be found in the life process of the group and not in the career of the separate individuals. Many human gratifications are interpersonal.

Thus, behavior is something that is built up as an organism successively adjusts to its ever-changing environment while moving toward a goal. What is actually done varies with cultural hypotheses and with exigencies of the situation—the opportunities that arise and the resources at hand. As might be expected of a social psychologist,

Mead insists that the process of defining the fluid situation necessarily takes place from a conventional perspective.

Decision Making and Voluntary Conduct

Psychologists ordinarily do not include under motivation the study of thinking, even though reflective thought does have something to do with the direction of behavior. One of the major contentions of pragmatists is that thinking is a form of behavior that occurs when there is blockage in the ongoing line of action. The interference may arise from some external barrier (including a conventional norm), from the incapacity of an organism, or from the absence of necessary objects. If an act is interrupted, a number of secondary adjustments take place—including emergency mobilization (emotion), consciousness, and deliberation—and through these processes a delayed act may eventually be brought to consummation. When an act is blocked, new plans are generated in the reflective process to redirect and guide action.

Any impulse that is not immediately consummated is transformed into an *image*, which serves as the basis for reflection. Images are acts that fail to issue in overt behavior, acts that are innervated but not carried out. Each image may be regarded as a plan of action, one possible way of completing the interrupted act. When confronted by interference, an individual experiences a succession of images. Reflective thought is an imaginative rehearsal—a comparison, evaluation, and selection of alternative routes to consummation. Mentality is the ability of an individual to anticipate the consequences of projected lines of action and to respond to them prior to commitment to overt action. Thinking, then, is problem-solving activity in which trial-and-error attempts take place in the imagination. The images that an individual is able to elicit in a problematic situation depend upon his past experiences in similar contexts; ability to solve problems rests in part upon experience and memory.

Most acts take place in a social context, and most images are *self-images*. Those who are involved in joint ventures become interdependent, for the impulses of each cannot be consummated without the cooperation of the others. Hence, each participant becomes sensitive to the possible reactions of others to his own conduct; he cannot afford to do anything that will jeopardize their support. Each person forms a perceptual object of himself by taking the roles of those with whom he is cooperating. An individual participating in concerted action internalizes the social process; that part of the social structure in which he is participating appears in his experience as his conception of his role. Self-control is part of the ongoing social current; as each individual adjusts in advance to the situation in which he is involved, he thereby facilitates cooperation. In this process, self-consciousness provides the basis for negative feedback. Unless one has some awareness of what he intends to do and how others are likely to react, he cannot inhibit acts that are likely to result in failure. Autonomy depends on feedback; without it one is subject to drift, to external control, or becomes a creature of impulse.

Each self-image elicits some kind of reaction. Some responses facilitate the line of activity that is already under way; others inhibit it. To study this process Mead proposes the concepts of the "I" and the "Me." These terms do not refer to agents, but to phases of activity. The "Me" is the self-image, and the "I" is the response that it evokes. If a reaction (I) is not immediately carried out, it is transformed into another image (Me), which in turn elicits another reaction (I). For example, if a man believes that his wife is disparaging his efforts (Me), he may experience an urge to beat her (I). If he refrains from striking a blow, he imagines himself administering the beating (Me). If he forms this self-image from the standpoint of a group in which wife-beaters are condemned, he reacts with disgust (I), and this further inhibits this route to consummation. He may then go to a neighborhood tavern, visualizing himself as a misunderstood husband (Me). He enacts the role of the injured party (I), expecting to be treated with sympathy. Should others respond as anticipated, this confirms his self-image (Me), and he continues to feel sorry for himself (I). Should they dismiss him disdainfully, however, he concludes that they are mocking him (Me), and he reacts in anger or with dejection (I). Thus, *an individual's line of conduct is constructed as he responds to a succession of organic states, perceptual objects, images, reactions of other people, and to his own responses.* When Mead declares that voluntary conduct is built up in the interaction of the "I" and the "Me," he is saying that self-conscious activity consists of a succession of responses to a series of self-images. It becomes apparent from this illustration that the feedback loops in self-control are far more complex than what is involved in the perception and manipulation of other objects; their essential feature is that self-images are formed from a shared standpoint. This is what makes joint activity possible; each person controls himself from a standpoint about which there is consensus. Although Mead's scheme may at first appear clumsy, his concepts do facilitate the analysis of subjective experience and its relation to overt action.

The succession of responses designated as "I's" constitute the core of an individual's personality. In a given situation each person has a somewhat different inclination to act. Distinctiveness rests not only on temperamental differences but is also derived from a unique background of experience. All normal adults are socialized, but each individual reflects the social process in his own way. One's personality is not always transparent because many reactions (I) do not issue in overt behavior; many behavioral tendencies are blocked for strategic purposes or to comply with conventional norms. Individuality is usually manifested only in situations in which there are sanctioned alternatives or in the style of overt conduct—through expressive movements that betray one's inner dispositions. It is in this way that Mead accounts for the fact that individual behavior does not necessarily coincide with group norms; each person, though a product of society, retains some measure of independence.

Once a person has learned a language, images and meanings may be designated by symbols, and the entire interchange between the "I" and the "Me" may become linguistic. Language transforms the human environment. With linguistic symbols one can manipulate meanings outside the contexts in which they develop and even make up more complex meanings. By symbols one can isolate certain experiences and hold on to them, pick out other relevant meanings, emphasize a particular image while rejecting others. It becomes possible to designate objectives—romance, public office, deference, or "kicks." In common parlance such consciously avowed intentions are called "motives" and are thought to be the "cause" of behavior. One virtue of Mead's scheme is that such conscious avowals are clearly separated from the organic deficits that set an organism into motion. That the accomplishment of conscious goals sometimes fails to bring a sense of fulfillment, then, becomes more understandable. Mastery of a language also makes possible the formulation of complex plans, overall schemes in which a sequence of operations can be performed, not unlike the program for a computer. In such plans, diverse and even antagonistic responses may be coordinated. Foresight is greatly facilitated by the use of symbols. What is called "thinking," then, consists largely of subvocal linguistic communication. Alternative plans of action are labeled, their consequences examined verbally, and choices made. A person who faces a problem often carries on discussions with himself. Once a decision is made, the individual is able to continue the interrupted act and to bring it to consummation. If the new plan also fails, there is further blockage, and it becomes necessary for him to think further.

Computer engineers are attempting to reconstruct this type of cognitive activity, and the new disciplines focus on these processes. It is widely believed that the human nervous system is a vast communication network, that the brain is like a computer with extensive facilities for storing coded information, and that cognitive processes are in large part information-handling operations. Although information theory has played a key part in computer work, psychological research based on it has thus far been quite primitive. It should be pointed out, however, that research in this area need not be confined to the narrow quantitative limits set in Shannon's work. Decision-making is recognized as a process of selecting from available alternatives, and decision theory represents an attempt to describe regularities in the way in which this is done. Each alternative is evaluated in terms of the expected consequences and the probability of their occurrence. Restrictions are set by previous commitments, the limits of permissibility, and the available time, resources, and information. Goals are stated, and choices are made in terms of the relative desirability of various solutions.

Although images and detailed plans rest largely on memory, choices are by no means limited to preexisting alternatives. Pragmatists stress the importance of novelty. Especially where there are conflicting tendencies, the resolution of a problem may diverge from any previous program. There is always the possibility of a new solution, a novel combination of circumstances in a creative act.

Much of human behavior, then, has an experimental character. There may be overt trial and error. Once a person has acquired a stock of images, problems may be solved through imaginative rehearsals, permitting him to pretest his acts before committing himself. Once language has been mastered, the trials may take place largely through linguistic communication. The latter techniques are extensions of adjustive tendencies inherent in the human organism.

Utility of the Cybernetic Model

The cybernetic model is more comprehensive than the other approaches. It shows the place of tension-reduction, response to stimulation, and conscious intent within a larger unit and the manner in which these components are related to each other. Cybernetics deals with the organization of processes within any assembly treated as a whole. It is an approach that emphasizes time, information, and feedback. Although this scheme may be of limited utility to psychologists concerned with the microscopic details of perception and learning, social psychologists studying macroscopic matters—such as the relationship of self-conceptions to certain types of decisions—will find it far more useful than other models. Adoption of such a model may also result in greater unification of psychological research now fragmented into many specialities.

Another advantage of this approach is that it brings the study of inner experiences back into psychology. Although some shudder at any suggestion of introspectionism, it is a matter of common observation that subjective experiences do have much to do with the organization of overt behavior. The model facilitates getting at some frequently observed regularities. One can more easily study phenomena such as hesitation and indecision, inner conflicts between opposing impulses, as well as decision-making. The concept of positive feedback may make possible more precise descriptions of the cumulative development of depressive moods as well as of violent outbursts. A person with paranoid tendencies, for example, imagines himself being mistreated and responds in anger; the new self-image in turn elicits more hostile reactions, until his rage builds up to an explosion.

Use of a cybernetic model will also enable us to see more clearly that all human behavior is part of the task of living. It will not be necessary to invent various impelling "forces" to account for the things men do, for they can be seen as manifestations of adjustive tendencies inherent in all living organisms.

Social psychologists may also profit in many ways from technical advances. Although the contention of proponents of general systems theory that the operation of systems at different levels of integration is isomorphic remains to be demonstrated, many fruitful hypotheses about human behavior can be constructed by analogy. Interesting analogies may be drawn, for example, from recent studies of feedback. Engineers are much concerned with the dangers of oscillation, which occurs when rectifying adjustments are too strong. Similar difficulties can be found in human behavior; victims of certain kinds of brain damage as well as inebriated men have difficulty in bringing a cup of water to their mouth as they repeatedly overshoot their target and then overcompensate. Furthermore, as social psychology develops, some of its concepts and generalizations will probably become so complex that they will require mathematical statement for effective manipulation. If so, much painstaking labor may be saved by studying the work already done by cyberneticians, who describe many of their basic principles in families of differential equations. Of course, caution must be exercised in borrowing ideas from another universe of discourse, especially where fields differ as much as social psychology and electronic engineering. While adoption of a cybernetic model will certainly not solve all the problems confronting social psychologists, intelligently used, it can become a valuable tool.

This is not to suggest that social psychologists need to retreat to speculative schemes developed more than a half century ago. Using some of these basic principles as a point of departure, they should be able to construct far more sophisticated statements. As more realistic accounts of how men actually think and choose are developed, engineers will have sounder foundations for the construction of more effective "minds." As is so often the case, innumerable mutual benefits are likely to follow the breaching of parochial boundaries.

41.
Ego Psychology, Cybernetics, and Learning Theory

O. H. MOWRER

IT WILL BEST SERVE our present purposes if I discuss the three topics constituting the title of this paper in the reverse order of that in which they are here mentioned. But first a more general word. It can hardly escape even commonplace observation that we tend to take, as models for interpreting the complex and mysterious, phenomena which are simpler and more fully understood. Hence, the machine, being man-made and intelligible, has often patterned our thinking about the less intelligible aspects of man himself.

Today we face a new challenge in this respect. Stanton and Sylva Cohn, writing in a current [Vol. 76, 1953] issue of *The Scientific Monthly*, put the matter well when they say:

> The nineteenth century was the "Age of Power." It saw the development of the machine, and concomitant with it there arose a mechanistic philosophy of life and a mechanical interpretation of life processes. . . .
> Science has advanced beyond the mechanistic stage, however. Just as the nineteenth century was the Age of Power, the twentieth century is the Age of Communication and Control. It is not enough to make a powerful machine, having the ability to do many times the work of man. There must be an intelligent application of this energy—it must be controlled.

In recent decades engineering has moved rapidly forward along these lines, producing, oddly enough, machines that are more "intelligent" in practice than living organisms are in

theory! Of course, some of these machines are *actually* more "intelligent," as regards certain specialized tasks, than are animals, including men. Here we think particularly of the "giant computers," for example. But we are presently concerned rather with the extent to which living creatures are, per hypothesis, more limited in their potentialities than we know them, in fact, to be. Modern machines are thus presenting a challenge to our theorizing in psychology and related sciences. If we can meet this challenge, psychology may, as Harry Harlow hopefully opined a few years ago, "eventually catch up with common sense." It is with this challenge and some of the new vistas it opens up that this paper will be mainly concerned.

I. The Passing of "Habit" and the Rediscovery of "Consciousness"

In the paper already cited, Cohn and Cohn say:

> If there is one law that marks this era [the nineteenth century Age of Power] definitively, it is the principle of the conservation of energy. This principle, which is expressed in the first law of thermodynamics, has been characterized as the greatest generalization in natural science. But it is not the final word.

"Habit" is a concept born of this tradition. A stimulus, as *cause*, impinges upon a sense organ (internal or external) and sets up neural impulses which, by virtue of certain neural "connections," travel to and activate certain muscles. The resulting response is the *effect*. Energy, though transmitted and transformed, is

From O. H. Mowrer, "Ego Psychology, Cybernetics, and Learning Theory," in Donald K. Adams *et al.* (Eds.), *Learning Theory and Clinical Research* (New York: John Wiley, 1954), pp. 81–90. Reprinted by permission of the author and publisher.

thus strictly "conserved," in the manner of a moving billiard ball striking and imparting its momentum to a second ball, it to a third, and so on. But thus far we have not distinguished between "habit" and "reflex." Reflex, we are told, is invariable, unmodifiable; habit, on the contrary, can be changed. But how? Thorndike noted that, with a habit, the cause–effect sequence does not end with response. Responses, he observed, may in turn initiate causal sequences in the external world which terminally impinge back upon the organism. These "feedback" effects Thorndike, like the layman, called rewards if they lessen stimulation in some important way, and punishments if they significantly increase stimulation. Rewards, Thorndike conjectured, strengthen the S–R sequences that produce them, whereas punishments have the reverse effect.

That an organism that can be thus modified by experience—that can, in other words, *learn*— will on the average be better off than a purely reflex organism is pretty obvious. But the model or image which Thorndike gave us has not been a universally satisfying one. On the one hand it has been charged, rather unjustifiably it seems, with being "teleological" (in the opprobrious sense of the term); it has also, more relevantly, been accused of making organisms more "mechanical," "blind," "stupid" than they really are. Although certainly more "intelligent" than a purely reflex creature, Thorndike's "habit" animal is by no means overburdened with brightness. Yet it has not been too easy to say exactly what is wrong with such a creature and how it might be improved.

Here we will attempt a concise diagnosis. It follows from what has been said that if a Thorndikian animal has acquired, under one set of conditions, a given "habit" and if conditions are now *changed*, the animal itself can begin to change *only after* it has performed the old response under the new conditions *at least once*. In other words, this much stupidity, or "maladaptation," is logically demanded by the theory, and has been rationalized by the slogan: "Organisms learn only by doing!" Evidence recently reviewed in another paper indicates that this inference is plainly not valid and calls for a radical revision of what we have previously called learning theory.

In order to escape from the Thorndikian dilemma, we must, first of all, repudiate one of his major assumptions: we must abandon the idea that rewards strengthen stimulus–response bonds and that punishments weaken them. What they do instead is to produce, by conditioning, *secondary*

reinforcements and *secondary motivations*, respectively. This will at first hardly seem like a clarifying statement. What it means, quite simply, is that we do not learn, or fixate, overt, behavioral responses at all. These are always "subject to change," depending upon the "situation." What *is* learned are attitudes, meanings, or expectations which consist of token decrements in emotional tension (secondary reinforcements, or rewards) and token increments (secondary motivation, or punishment). It is assumed that it is these inner, conscious factors which, moment by moment, select and shape overt action; and if we take this position we have ample provision for "learning" *without* doing, i.e., for changes in behavior that occur, solely and immediately, because the *situation*, or, more exactly, the individual's internal tension state, or "field," has changed. Here we have the capacity for foresight, insight, and a generally higher order of intelligence and adaptivity than is possible in a "creature of habit."

But we have achieved this at a cost which some will be reluctant to pay. Instead of channeling stimulus energies directly through the nervous system, switchboard fashion, and out into motor organs in a highly determined way, we are here assuming that this kind of determinism holds, so to say, only half way. It holds, I assume, to this extent, that meanings and attitudes, both positive (tension reducing) and negative (tension inducing), follow quite automatically, quite reflexly (conditioned reflexly), upon the occurrence of significant stimuli or situations. But here this type of fixed, cause–effect relationship ends and a more complex mechanism takes over. I am sure that you will not hold me accountable for explaining all the riddles of consciousness merely because I refer to the phenomenon; but I will venture the guess that consciousness is, essentially, a *continuous-computing* device or process. The eternal question is, "What to do? How to act?" And consciousness, as I conceive it, is the operation whereby information is continuously received, evaluated, and summarized in the form of "decisions," "choices," "intentions" which then emerge as behavior. Life asks the questions, sets the problems, and it is the business of consciousness to give us the "answers."[1]

This is not to say that consciousness is merely chance or caprice. In the paper already cited, an attempt is made to state some of the principles of conscious functioning, and more attention will be given to this problem in the later sections of the present paper. I hope that I have succeeded, thus far, in showing the general direction in which learning theory must, in my judgment, move if our conceptual models in psychology are to be as resourceful and sensible as the "real thing" or even as the newer types of machines that can today do such remarkable things.

II. Cybernetics as the Science of Communication and Control

It will, I trust, be evident how naturally the foregoing analysis articulates with some of the basic concepts of cybernetics. One of these concepts, as Wiener and others have formulated it, is that only the simpler machines and response systems operate on a "blind," reflexive principle; "higher" behavior and machines involve the "feedback" principle. Learning, I propose, is not a matter of strengthening or weakening connections between drives and overt behavior, but of the acquisition of "positive" (rewarding) and "negative" (punishing) feedback from stimuli that have accompanied past action or experience. If, in the past, a given act has been predominantly rewarding, then incidental stimuli, both external and internal, which have been associated with the act–reward sequence will take on, as already noted, the capacity to produce secondary reward and thus to guide or direct the organism into the *same* or similar action in the future. If, on the other hand, a given act has been predominantly punishing, the incidental stimuli which have been associated with the act–punishment sequence will take on the capacity to produce a secondary-drive increment ("fear") and will tend to guide the organism into *different* behavior.[2] Behavior, in any given situation, is the organism's best effort, then and there, to find that line of action with the greatest positive and the least negative feedback— and hence with the best likelihood, when it is carried through to completion, of being maximally satisfying and minimally hurtful. Feedback has been called "the central theme of cybernetics." We must, it appears, likewise give it a major role in psychological theory.

Another way of saying much the same thing is to stress the role of communication and control. These are the factors which are conspicuously missing in most nineteenth century and earlier machines and which are also minimal or absent in reflex action in living organisms. Central to the communication process, surely, is the phenomenon of *meaning*, and this we have given a central place in our theoretical scheme. Meanings,

not means; attitudes, not actions—these are the most immediate outcomes of learning; and it is through these that intelligent *control* of behavior then becomes possible. It is not yet certain whether so-called "information theory" as it has been elaborated by Shannon and Weaver and others will prove as helpful in behavior analysis as many psychologists now hope; but it seems abundantly clear that the more rudimentary tenets of cybernetics have already given our thinking in the realm of learning theory a useful nudge; and I shall try to show in the next and last section of this paper that the developments thus produced carry us appreciably closer, as learning theorists, to ego psychology than we have been before.

First, however, it will be salutary to clear up a terminological point. In the preceding pages I have spoken of positive and of negative feedbacks, defining them as secondary reward and secondary punishment, respectively. Here it should be said that for Wiener and other cyberneticists these terms have a different meaning. If a control system, like a thermostat or the governor on a steam engine, has a *stabilizing* influence on some quality—heat, in the one case; speed, in the other—then the feedback is said to be *negative*, regardless of whether it is acting at any given moment to bring a quantity *up* to or *down* to a standard state. The "error," in other words, which the control system is trying to "correct" may be either "plus" or "minus," but the type of over-all control is, in either case, said to involve negative feedback.

The opposite, or *positive*, type of feedback is no less interesting. The products of certain chemical reactions are such as to have a catalytic effect upon and thus to *speed up* the reactions which produce them. Sometimes these reactions gain momentum so rapidly as to produce "explosions"—they are, in fact, the same in principle as the chain reactions in the atom bomb and the H-bomb. If, likewise, certain other types of processes start to slow down, the effects of the retardation accelerate the slowing down. Thus, if a business concern is not prospering, inability to pay employees satisfactory wages or to meet creditors may accelerate the failure.

Since homeostasis, or self-regulation, is one of the essential characteristics of living organisms, it will be immediately evident that many of their activities will have the properties of negative feedback, as just defined. These activities or functions serve to hold certain states or qualities within fairly narrow limits of variation. Positive feedback, when the concept is applied to living organisms, sounds as if it would be pathological, to say the least, and, in the extreme case, lethal. Actually, we find some very instructive instances of it in living organisms. We have already seen that there is a general tendency for incidental stimuli which have been associated with primary-drive reduction to produce secondary-drive reduction and for such stimuli as have been associated with primary-drive induction to produce secondary-drive induction. These two types of contiguity learning, or conditioning, unquestionably tend to produce (indirectly, through the integrating function we call consciousness) responses which have biological utility—notably, flight or immobility in the face of danger and approach to objects or situations with rewarding potentialities.

However, there is one crucial respect in which this general scheme is inadequate, non-biological. The world's good things do not always remain conveniently at rest, waiting to be claimed and consumed. Sometimes these goal objects, especially when they are other organisms, have a way, at the critical moment, of eluding their pursuers. Therefore, an organism that became more and more confident and *relaxed* as it approached a quarry might find itself slowing up at just the point when a final "push" was needed for success. It is therefore interesting and altogether understandable that we should find the phenomenon of *appetite*. Its outstanding feature is an *increase* in secondary motivation just as consummation is imminent, thus giving to behavior at this crucial moment a peculiar urgency and "oomph."

Here, it seems, is an instance of *positive* feedback: "The nearer you get, the more you want it" carries the proper connotation here. Certainly such an arrangement is biologically intelligible, but it is nonetheless enigmatic. Neal Miller has put the matter this way. We have reason for thinking that stimuli associated with consummatory states takes on contradictory capacities: a tendency to cause a decrement in secondary drive (secondary reinforcement) *and* a tendency to cause an increment in secondary drive (which we call, not punishment, but appetite). As Miller points out, if these two tendencies occurred simultaneously, they would be self-canceling, mutually neutralizing; so his proposal is that they may alternate, producing intermittent bubbles of pleasurable anticipation and surges of intensified drive. This is frankly a speculation and, even if true, leaves many unanswered questions. However, consideration of the problem in the cybernetics setting illuminates and sharpens it in ways which we shall further explore in the next section.

III. Psychology of the Ego and Superego

It is at once evident that neither reflexology nor habit theory, in the mode of Thorndike, Pavlov, Watson, or Hull, can provide a very sophisticated approach to ego psychology, so-called. The reflex and habit, almost by definition, exclude consciousness, which is the core of ego functioning. The same difficulty does not at all occur in the conceptual framework presented in the preceding sections of this paper. Here we posit a division of labor between learning as a purely unconscious, automatic process, on the one hand, and conscious judging, deciding, and acting, on the other. Our conception of learning therefore blends naturally enough with ego theory.

Thinking, in this frame of reference, we see as a form of activity in which the individual makes symbolic responses (Hull's "pure stimulus acts") for the purpose of eliciting, or "sampling," the feedback, or effects, that could be expected if the action thus symbolized or represented were really carried out. In simplest, most primitive form, thinking, or reasoning, can be *seen*, quite literally, in the vicarious trial-and-error of a rat at a maze choice point—a form of "light" activity which is manifestly a prelude to grosser and more consequential action. In adult human beings the process becomes both more subtle and more elaborated, but its function, basically and ultimately, remains the same.

Recently I heard a college president say that the most important thing his institution could do for students was to "teach them to think." At one level this can be regarded merely as an old pedagogic bromide. Or it can be seen as profoundly and perennially true. If in thinking we are, so to say, sticking our "mental necks" out *into the future* and "feeling around," nothing, at least on occasion, could be more useful, especially when the findings are later translated back into intelligent action. Living organisms swim forward in a sea of time, and those with the best "distance receptors," i.e., with the best symbolic skills, will almost certainly have an edge in the struggle for existence. "Use your head" (or head end) is good advice both in the sense of using the special senses and in the sense of moving back and forth through time in the way that symbols make dramatically possible.

Then, too, our conceptual scheme puts us in a good position for understanding not only thought but also *fantasy*. The difference, basically, is that thought is a *preparation* for action; fantasy is a *substitute* for it. In fantasy we select not those symbols that will forecast reality but rather those that will yield, without reference to reality, the greatest present pleasure, or secondary reinforcement. "In day-dreams I often picture myself as a very generous and kindly person," a young woman undergoing psychotherapy recently remarked. This was said in the context of revealing just the reverse real characteristics and shows the highly autistic nature of such activity.[3]

So far, however, we do not yet have a psychology of the abnormal, a psychopathology. Does our system, as presently conceived, yield one? The attempts of reflexologists and habit theorists to "explain" neurosis and propound a therapy have not been very adequate. A minimal essential for a psychopathology is *conflict*, a concept which is hardly meaningful unless consciousness is posited. But conflict is not a *sufficient* cause of neurosis. Indeed, if our view be correct, consciousness is a continual interplay of contending forces; and decision making, compromise, and integration are its major accomplishments.

Where, then, does pathology arise? Section II of this paper gives us a clue. There we have seen that there are two broadly different principles of feedback. If feedback functions so as to make an organism move faster when it lags or slow down when it is going too fast, it is said, by the cyberneticists, to be a negative feedback; and we are at once reminded of the role of conscience or superego. This agency within the total personality is said to be wise, prudent, *balanced*, and mainly concerned with (social) *norms*. The mentally aberrant are said, per contra, to be unstable, unbalanced, abnormal. How do they get that way?

Through parental training as well as by direct experience, human beings acquire conscience, which, as we have seen, operates on the negative-feedback principle. Other things equal, this should be compatible with the over-all tendency toward self-regulation and homeostasis. However, we have already noted a complication. Man has appetites and lusts as well as conscience; and if one operates on a conservative, judicious principle, the other is prodigal and reckless. Sex is powerfully appetitive, clearly following a positive-feedback principle of the "explosive" variety, and anger is likewise the occasion for our sometimes "blowing up." Little wonder, then, that these human proclivities are the most difficult to "control" and the ones which most often come into conflict with superego functions. Caught between two such powerful forces as id and superego, appetite and conscience, lust and guilt, what indeed is the

"answer"? Here the integrative resources of the ego are likely to be taxed to their utmost!

The Freudian version of the etiology of neurosis holds that in the struggle between negative feedback (conscience) and positive feedback (appetite), the former sometimes seizes too firm a control over the latter and thus holds down, or inhibits, "instinctual" functioning to a pathogenic degree. Anxiety and depression are said to be the fruits of this stifling dominion of the superego over id functions.

On other occasions, I have argued for the contrary view, that neurosis arises when positive-feedback functions have out-contended the negative feedback, resulting in temporarily uncontrolled, "explosive" behavior for which the individual later feels remorse and shame (intensified negative feedback). The ego may then deal with this expression of an "aggrieved" conscience in several different ways, notably by making a confession and amendments, in which case inner stability and harmony tend to be restored; or the ego may set out to "disconnect" the conscience, like the governor on a steam engine, so as to let the "wild" behavior occur unobstructedly (albeit usually surreptitiously). Conscience may be thus dissociated or repressed, and the individual may compliment himself on his new freedom, liberty, emancipation. But the forces of conscience are tenacious and usually come back to haunt their owner, but now not as intelligible guilt but as the unintelligible and torturing experiences of anxiety, panic, depression, and inferiority feeling.

It is not my wish or purpose here to debate the relative merits of these two interpretations. I present them rather to show how neatly and naturally they can both be accommodated in the general theoretical framework here presented, one in which the view of "habit" as $S-R$ bonds is rejected in favor of a more complicated system in which learning is limited to the acquisition of inner meanings which are then evaluated and used to arrive at the particular (often novel) decision and action which the total situation seems most to warrant. One can hardly escape the feeling that cybernetics contains some powerfully unifying principles and that science may yet lead to a type of synthesis of human experience which has, by some, long been held impossible.

Notes

1. The reintroduction of the concept of consciousness into behavior theory may seem, to some, like a retrogressive step. To such persons it may connote a

return to introspection as the chief mode of psychological inquiry and a relinquishment of the gains of
half a century of intensive experimental inquiry. What
I wish to suggest, without being able to develop at
this time, is the notion that consciousness is a phenomenon which would be inferable from, and indeed
logically demanded by, the empirical facts, even though
there were no direct experiential access to it whatever.
The point is that adherence to a strict *S–R* psychology
creates a number of dilemmas which only the inter-
polation of an *integrative* mechanism of some sort
can resolve. Consciousness is here conceived as that
mechanism.

2. These principles are here stated in highly condensed form. . . .

3. Another patient, a young physicist, reports
excessive reading of science fiction. "In *it*," he observes,
"the experiments *always* come out right!"

42.
The Open System in Personality Theory

GORDON W. ALLPORT

OUR ORB OF COMMON INTEREST has two faces—one turned toward social psychology, the other toward personality. As things stand at the moment the first visage seems to me slightly depressed; the second slightly manic. However that may be, I should like to start this discussion by bringing the two faces into confrontation.

My first observation is that as members of Division 8 we are conspicuously the victims of fashion. Although our persona is sedate and seemly, we have our own hula hoops, flying saucers, and our own way of flagpole sitting. The interquartile range of our crazes I estimate to be about 10 years. McDougall's instinct theory held sway from 1908 to approximately 1920. Watsonian behaviorism dominated the scene for the next decade; then habit hierarchies, then field theory; now phenomenology. We never seem to solve our problems or exhaust our concepts; we only grow tired of them. At the moment it is fashionable to investigate response-set, coding, sensory deprivation, person perception, and to talk in terms of system theory—a topic to which we shall soon return. Ten years ago fashion called for group dynamics, Guttman scales, and for research on the centipedal properties of the authoritarian personality. Twenty years ago it was frustration–aggression, Thurstone scales, and national morale. And so it goes. We are fortunate that each surge of fashion leaves a rich residue of gain.

Fashions have their amusing and their serious side. We can smile at the way bearded problems receive tonsorial transformation. Having tired of

"suggestibility" we adopt the new hairdo known as "persuasibility." Modern ethology excites us, and we are not troubled by the recollection that a century ago John Stuart Mill staked down the term to designate the new science of human character. We like the neurological concept of "gating," conveniently forgetting that American functionalism always stood firm for the dominance of general mental sets over specific. Reinforcement appeals to us but not the age-long debate over hedonism. The problem of freedom we brush aside in favor of "choice points." We avoid the body–mind problem but are in fashion when we talk about "brain models." Old wine, we find, tastes better from new bottles.

The serious side of the matter enters when we and our students forget that the wine is indeed old. Picking up a recent number of the *Journal of Abnormal and Social Psychology* I discover that the 21 articles written by American psychologists confine 90% of their references to publications of the past 10 years, although most of the problems they investigate have gray beards. In the same issue of the journal, three European authors locate 50% of their references prior to 1949. What this proves I do not know, excepting that European authors were not born yesterday. Is it any wonder that our graduate students reading our journals conclude that literature more than a decade old has no merit and can be safely disregarded? At a recent doctoral examination the candidate was asked what his thesis on physiological and psychological conditions of stress had to do with the body–mind problem. He confessed he had never heard of the problem. An undergraduate said that all he knew about Thomas Hobbes was that he sank with the *Leviathan* when it hit an iceberg in 1912.

From Gordon W. Allport, "The Open System in Personality Theory," *Journal of Abnormal and Social Psychology*, 61 (1960), 301–11. Reprinted by permission of the author and publisher.

A Psycholinguistic Trifle

Our windows are pretty much closed toward the past, but we rightly rejoice in our growth since World War II. Among the many fortunate developments is rejuvenation in the field of psycholinguistics. (Even here, however, I cannot refrain from pointing out that the much discussed Whorfian hypothesis was old stuff in the days of Wundt, Jespersen, and Sapir.) Be that as it may, I shall introduce my discussion of open systems in personality theory by a crude Whorfian analysis of our own vocabulary. My research (aided by the kind assistance of Stanley Plog) is too cursory to warrant detailed report.

What we did, in brief, was to study the frequency of the prefix re- and of the prefix pro- in psychological language. Our hypothesis was that re- compounds, connoting as they do again-ness, passivity, being pushed or maneuvered, would be far more common than pro- compounds connoting futurity, intention, forward thrust. Our sample consisted of the indexes of the *Psychological Abstracts* at 5-year intervals over the past 30 years; also all terms employing these prefixes in the Hinsie and Shatzky *Psychiatric Dictionary* and in the English and English *Psychological Dictionary*. In addition we made a random sampling of pages in five current psychological journals. Combining these sources it turns out that re- compounds are nearly five times as numerous as pro- compounds.

But, of course, not every compound is relevant to our purpose. Terms like reference, relationship, reticular, report do not have the connotation we seek; nor do terms like probability, process, and propaganda. Our point is more clearly seen when we note that the term reaction or reactive occurs hundreds of times, while the term proaction or proactive occurs only once—and that in English's *Dictionary*, in spite of the fact that Harry Murray has made an effort to introduce the word into psychological usage.

But even if we attempt a more strict coding of this lexical material, accepting only those terms that clearly imply reaction and response on one side and proaction or the progressive programing of behavior on the other, we find that ratio still is approximately 5:1. In other words our vocabulary is five times richer in terms like reaction, response, reinforcement, reflex, respondent, retroaction, recognition, regression, repression, reminiscence than in terms like production, proceeding, proficiency, problem solving, propriate, and programing. So much for the number of different

words available. The disproportion is more striking when we note that the four terms reflex, reaction, response, and retention together are used 100 times more frequently than any single pro-compound excepting problem solving and projective—and this latter term, I submit, is ordinarily used only in the sense of reactivity.

The weakness of the study is evident. Not all terms connoting spontaneous, future oriented behavior begin with pro-. One thinks of expectancy, intention, purpose. But neither do all terms connoting passive responding or backward reference in time begin with re-. One thinks of coding, traces, input–output, and the like. But while our analysis leaves much to be desired it prepares the way for our critique of personality theory in terms of systems. The connecting link is the question whether we have the verbal, and therefore the conceptual, tools to build a science of change, growth, futurity, and potential; or whether our available technical lexicon tends to tie us to a science of response, reaction, and regression. Our vocabulary points to personality development from the past up to now more readily than to its development from here-on-out into the future.

The Concept of System

Until a generation or so ago science, including psychology, was preoccupied with what might be called "disorganized complexity." Natural scientists explored this fragment and that fragment of nature; psychologists explored this fragment and that fragment of experience and behavior. The problem of interrelatedness, though recognized, was not made a topic for direct inquiry.

What is called system theory today—at least in psychology—is the outgrowth of the relatively new organismic conception reflected in the work of von Bertalanffy, Goldstein, and in certain aspects of gestalt psychology. It opposes simple reaction theories where a virtual automaton is seen to respond discretely to stimuli as though they were pennies-in-the-slot. Interest in system theory is increasing in psychology, though perhaps not so fast as in other sciences.

Now a system—any system—is defined merely as a complex of elements in mutual interaction. Bridgman, as might be expected of an operationist, includes a hint of method in his definition. He writes, a system is "an isolated enclosure in which all measurements that can be made of what goes on in the system are in some way correlated."

Systems may be classified as closed or open.

A closed system is defined as one that admits no matter from outside itself and is therefore subject to entropy according to the second law of thermodynamics. While some outside energies, such as change in temperature and wind may play upon a closed system, it has no restorative properties and no transactions with its environment, so that like a decaying bridge it sinks into thermodynamic equilibrium.

Now some authors, such as von Bertalanffy, Brunswik, and Pumpian-Mindlin, have said or implied that certain theories of psychology and of personality operate with the conception of closed systems. But in my opinion these critics press their point too far. We had better leave closed systems to the realm of physics where they belong (although even here it is a question whether Einstein's formula for the release of matter into energy does not finally demonstrate the futility of positing a closed cgs system even in physics). In any event it is best to admit that all living organisms partake of the character of open systems. I doubt that we shall find any advocate of a truly closed system in the whole range of personality theory. At the same time current theories do differ widely in the amount of openness they ascribe to the personality system.

If we comb definitions of open systems we can piece together four criteria: there is intake and output of both matter and energy; there is the achievement and maintenance of steady (homeostatic) states, so that the intrusion of outer energy will not seriously disrupt internal form and order; there is generally an increase of order over time, owing to an increase in complexity and differentiation of parts; finally, at least at the human level, there is more than mere intake and output of matter and energy: there is extensive transactional commerce with the environment.[1]

While of all our theories view personality as an open system in some sense, still they can be fairly well classified according to the varying emphasis they place upon each of these criteria, and according to how many of the criteria they admit.

CRITERION 1

Consider the first criterion of material and energy exchange. Stimulus–response theory in its purest form concentrates on this criterion to the virtual exclusion of all the others. It says in effect that a stimulus enters and a response is emitted. There is, of course, machinery for summation, storage, and delay, but the output is broadly commensurate with the intake. We need study only the two poles of stimulus and response with a minimum of concern for intervening processes. Methodological positivism goes one step further, saying in effect, that we do not need the concept of personality at all. We focus attention on our own measurable manipulations of input and on the measurable manipulations of output. Personality thus evaporates in a mist of method.

CRITERION 2

The requirement of steady state for open systems is so widely admitted in personality theory that it needs little discussion. To satisfy needs, to reduce tension, to maintain equilibrium, comprise, in most theories, the basic formula of personality dynamics. Some authors, such as Stagner and Mowrer regard this formula as logically fitting in with Cannon's account of homeostasis.[2] Man's intricate adjustive behavior is simply an extension of the principle involved in temperature regulation, balance of blood volume, sugar content, and the like, in the face of environmental change. It is true that Toch and Hastorf warn against overextending the concept of homeostasis in personality theory. I myself doubt that Cannon would approve the extension, for to him the value of homeostasis lay in its capacity to free man for what he called "the priceless unessentials" of life. When biological equilibrium is attained the priceless unessentials take over and constitute the major part of human activity. Be that as it may, most current theories clearly regard personality as a *modus operandi* for restoring a steady state.

Psychoanalytic theories are of this order. According to Freud the ego strives to establish balance among the three "tyrants"—id, superego, and outer environment. Likewise the so-called mechanisms of ego defense are essentially maintainers of a steady state. Even a neurosis has the same basic adjustive function.[3]

To sum up: most current theories of personality take full account of two of the requirements of an open system. They allow interchange of matter and energy, and recognize the tendency of organisms to maintain an orderly arrangement of elements in a steady state. Thus they emphasize stability rather than growth, permanence rather than change, "uncertainty reduction" (information theory), and "coding" (cognitive theory) rather than creativity. In short, they emphasize being rather than becoming. Hence, most personality theories are biologistic in the sense that they ascribe to personality only the two features of an open system that are clearly present in all living organisms.

There are, however, two additional criteria, sometimes mentioned but seldom stressed by biologists themselves, and similarly neglected in much current personality theory.

TRANSATLANTIC PERSPECTIVE

Before examining Criterion 3 which calls attention to the tendency of open systems to enhance their degree of order, let us glimpse our present theoretical situation in cross-cultural perspective. In this country our special field of study has come to be called "behavioral science" (a label now firmly stuck to us with the glue of the Ford millions). The very flavor of this term suggests that we are occupied with semiclosed systems. By his very name the "behavioral scientist" seems committed to study man more in terms of behavior than in terms of experience, more in terms of mathematical space and clock-time than in terms of existential space and time; in terms of response more than in terms of programing; in terms of tension reduction more than tension enhancement; in terms of reaction more than proaction.

Now let us leap our cultural stockade for a moment and listen to a bit of ancient Hindu wisdom. Most men, the Hindus say, have four central desires. To some extent, though only roughly, they correspond to the developmental stages of life. The first desire is for pleasure—a condition fully and extensively recognized in our Western theories of tension reduction, reinforcement, libido, and needs. The second desire is for success—likewise fully recognized and studied in our investigations of power, status, leadership, masculinity, and need-achievement. Now the third desire is to do one's duty and discharge one's responsibility. (It was Bismarck, not a Hindu, who said: "We are not in this world for pleasure but to do our damned duty.") Here our Western work begins to fade out. Excepting for some pale investigations of parental punishment in relation to the development of childhood conscience, we have little to offer on the "duty motive." Conscience we tend to regard as a reactive response to internalized punishment, thus confusing the past "must" of learning with the "ought" involved in programing our future. Finally, the Hindus tell us that in many people all these three motives pall, and they then seek intensely for a grade of understanding—for a philosophical or religious meaning—that will liberate them from pleasure,

success, and duty. (Need I point out that most Western personality theories treat the religious aspiration in reactive terms as an escape device, to be classified along with suicide, alcoholism, and neurosis?)

Now we retrace our steps from India to modern Vienna and encounter the existentialist school of logotherapy. Its founder, Viktor Frankl, emphasizes above all the central place of duty and meaning, the same two motives that the Hindus place highest in their hierarchy of desire. Frankl reached his position after a long and agonizing incarceration in Nazi concentration camps. With other prisoners he found himself stripped to naked existence. In such extremity what does a person need and want? Pleasure and success are out of the question. One wants to know the meaning of his suffering and to learn how as a responsible being he should acquit himself. Should he commit suicide? If so, why; if not, why not? The search for meaning becomes supreme.

Frankl is aware that his painfully achieved theory of motivation departs widely from most American theory, and he points out the implication of this fact for psychotherapy. He specifically criticizes the principle of homeostasis as implying that personality is a quasiclosed system. To cater to the internal adjustments of a neurotic, or to assume that he will regain health by reshuffling his memories, defenses, or conditioned reflexes is ordinarily self-defeating. In many cases of neurosis only a total breakthrough to new horizons will turn the trick.

Neither Hindu psychology nor Frankl underestimates the role of pleasure and success in personality. Nor would Frankl abandon the hard won gains reflected in psychoanalytic and need theory. He says merely that in studying or treating a person we often find these essentially homeostatic formulations inadequate. A man normally wants to know the whys and wherefores. No other biological system does so; hence, man stands alone in that he possesses a degree of openness surpassing that of any other living system.

CRITERION 3

Returning now to our main argument, we encounter a not inconsiderable array of theories that emphasize the tendency of human personality to go beyond steady states and to strive for an enhancement and elaboration of internal order even at the cost of considerable disequilibrium.

I cannot examine all of these nor name all the

relevant authors. One could start with McDougall's proactive sentiment of self-regard which he viewed as organizing all behavior through a kind of "forward memory" (to use Gooddy's apt term). Not too dissimilar is the stress that Combs and Snygg place on the enhancement of the phenomenal field. We may add Goldstein's conception of self-actualization as tending to enhance order in personality; also Maslow's theory of growth motives that supplement deficiency motives. One thinks of Jung's principle of individuation leading toward the achievement of a self (a goal never actually completed). Some theories, Bartlett and Cantril among them, put primary stress on the "pursuit of meaning." Certain developments in post-Freudian "ego psychology" belong here.[4] So too does existentialism with its recognition of the need for meaning and of the values of commitment. (The brain surgeon, Harvey Cushing, was speaking of open systems when he said: "The only way to endure life is to have a task to complete.")

No doubt we should add Woodworth's recent advocacy of the "behavior primacy" theory as opposed to the "need" theory, Robert White's emphasis on "competence," and Erikson's "search for identity."

Now these theories are by no means identical. The differences between them merit prolonged debate. I lump them here simply because all seem to me to recognize the third criterion of open systems, namely, the tendency of such systems to enhance their degree of order and become something more than at present they are.

We all know the objection to theories of this type. Methodologists with a taste for miniature and fractionated systems complain that they do not lead to "testable propositions." The challenge is valuable in so far as it calls for an expansion of research ingenuity. But the complaint is ill-advised if it demands that we return to quasiclosed systems simply because they are more "researchable" and elegant. Our task is to study what is, and not what is immediately convenient.

CRITERION 4

Now for our fourth and last criterion. Virtually all the theories I have mentioned up to now conceive of personality as something integumented, as residing within the skin. There are theorists (Kurt Lewin, Martin Buber, Gardner Murphy, and others) who challenge this view, considering it too closed. Murphy says that we overstress the separation of man from the context of his living.

Experiments on sensory deprivation Hebb has interpreted as demonstrations of the constant dependence of inner stability on the flow of environmental stimulation. Why Western thought makes such a razor-sharp distinction between the person and all else is an interesting problem. Probably the personalistic emphasis in Judeo-Christian religion is an initial factor, and as Murphy has pointed out the industrial and commercial revolutions further accentuated the role of individuality. Shinto philosophy, by contrast, regards the individual, society, and nature as forming the tripod of human existence. The individual as such does not stick out like a raw digit. He blends with nature and he blends with society. It is only the merger that can be profitably studied.

As Western theorists most of us, I dare say, hold the integumented view of the personality system. I myself do so. Others rebelling against the setting of self over against the world, have produced theories of personality written in terms of social interaction, role relations, situationism, or some variety of field theory. Still other writers, such as Talcott Parsons and F. H. Allport, have admitted the validity of both the integumented personality system and systems of social interaction, and have spent much effort in harmonizing the two types of systems thus conceived.

This problem, without doubt, is the knottiest issue in contemporary social science. It is the issue which, up to now, has prevented us from agreeing on the proper way to reconcile psychological and sociocultural science.

In this matter my own position is on the conservative side. It is the duty of psychology, I think, to study the person-system, meaning thereby the attitudes, abilities, traits, trends, motives, and pathology of the individual—his cognitive styles, his sentiments, and individual moral nature and their interrelations. The justification is twofold: (a) there is a persistent though changing person-system in time, clearly delimited by birth and death; (b) we are immediately aware of the functioning of this system; our knowledge of it, though imperfect, is direct, whereas our knowledge of all other outside systems, including social systems, is deflected and often distorted by their necessary incorporation into our own apperceptions.

At the same time our work is incomplete unless we admit that each person possesses a *range* of abilities, attitudes, and motives that will be evoked by the different environments and situations he encounters. Hence, we need to understand cultural, class, and family constellations and traditions in order to know the schemata the person has probably interiorized in the course of his learning. But I hasten to warn that the study of cultural, class, family, or any other social system does not automatically illumine the person-system, for we have to know whether the individual has accepted, rejected, or remained uninfluenced by the social system in question. The fact that one plays the role of, say, teacher, salesman, or father is less important for the study of his personality than to know whether he likes or dislikes, and how he defines, the role. And yet at the same time unless we are students of sociocultural systems we shall never know what it is the person is accepting, rejecting, or redefining.

The provisional solution I would offer is the following: the personality theorist should be so well trained in social science that he can view the behavior of an individual as fitting any system of interaction; that is, he should be able to cast this behavior properly in the culture where it occurs, in its situational context, and in terms of role theory and field theory. At the same time he should not lose sight—as some theorists do—of the fact that there is an internal and subjective patterning of all these contextual acts. A traveler who moves from culture to culture, from situation to situation, is none the less a single person; and within him one will find the nexus, the patterning, of the diverse experiences and memberships that constitute his personality.

Thus, I myself would not go so far as to advocate that personality be defined in terms of interaction, culture, or roles. Attempts to do so seem to me to smudge the concept of personality, and to represent a surrender of the psychologist's special assignment as a scientist. Let him be acquainted with all systems of interaction, but let him return always to the point where such systems converge and intersect and are patterned —in the single individual.

Hence, we accept the fourth (transactional) criterion of the open system, but with the firm warning that it must not be applied with so much enthusiasm that we lose the personality system altogether.

General Systems Theory

There are those who see hope for the unification of science in what James Miller called "general behavior systems theory." This approach seeks formal identities between physical systems, the cell, the organ, the personality, small groups, the species, and society. Now critics (e.g., Buck) complain that all this is feeble analogizing, that formal identities probably do not exist, and that attempts to express analogies in terms of mathematical models result only in the vaguest generalities. As I see it, the danger in attempting to unify science in this manner lies in the inevitable approach from below, that is, in terms of physical and biological science. Closed systems or systems only partly open become our model, and if we are not careful, human personality in all its fullness is taken captive into some autistic paradise of methodology.

Besides neglecting the criteria of enhanced organization and transaction general systems theory has an added defect. The human person is, after all, the observer and interpreter of systems. This awkward fact has recently been haunting the founder of the operational movement, P. W. Bridgman. Can we as scientists live subjectively within our system and at the same time take a valid objective view thereof?

Some years ago Elkin published the case of "Harry Holzer," and invited 39 specialists to offer their conceptualizations. As might be expected, many different conceptualizations resulted. No theorist was able entirely to divest the case of his own preconceptions. Each read the objective system in terms of the subjective. In other words, our theories of personality—all of them—reflect the temperament of the author fully as much as the personality of *alter*.

This sad spectre of observer contamination should not, I think, discourage us from the search for objectively valid theory. Truth, as the philosopher Charles Peirce has said, is the opinion which is fated to be ultimately agreed to by all who investigate. My point is that "the opinion fated to be ultimately agreed to by all who investigate" is not likely to be reached through a premature application of general systems theory, nor through devotion to any one partially closed theory. Theories of open-systems hold more promise, though at present they are not in agreement among themselves. But somewhere, sometime, I hope and believe, we shall establish a theory of the nature of personality which all wise men who investigate, including psychologists, will eventually accept.[5]

Some Examples

In the meantime, I suggest that we regard all sharp controversies in personality theory as probably arising from the two opposed points of view—the quasiclosed and the fully open.

The principle of reinforcement, to take one example, is commonly regarded as the cement that stamps in a response, as the glue that fixes personality at the level of past deeds. Now an open-system interpretation is very different. Feigl, for instance, has pointed out that reinforcement works primarily in a prospective sense. It is only from a *recognition* of consequences (not from the consequences themselves) that the human individual binds the past to the future and resolves to avoid punishment and to seek rewards in similar circumstances, provided, of course, that it is consonant with his interests and values to do so. Here we no longer assume that reinforcement stamps in, but that it is one factor among many to be considered in the programing of future action. In this example we see what a wide difference it makes whether we regard personality as a quasiclosed or open system.

The issue has its parallels in neurophysiology. How open is the nervous system? We know it is of a complexity so formidable that we have only an inkling as to how complex it may be. Yet one thing is certain, namely, that high level gating often controls and steers lower level processes. While we cannot tell exactly what we mean by "higher levels" they surely involve ideational schemata, intentions, and generic personality trends. They are instruments for programing, not merely for reacting. In the future we may confidently expect that the neurophysiology of programing and the psychology of proaction will draw together. Until they do so it is wise to hold lightly our self-closing metaphors of sowbug, switchboard, giant computor, and hydraulic pump.

Finally, an example from motivation theory. Some years ago I argued that motives may become functionally autonomous of their origins. (And one lives to regret one's brashness.)

Whatever its shortcomings the concept of functional autonomy succeeds in viewing personality as an open and changing system. As might be expected, criticism has come chiefly from those who prefer to view the personality system as quasiclosed. Some critics say that I am dealing only with occasional cases where the extinction of a habit system has failed to occur. This criticism, of course, begs the question, for the point at issue is why do some habit systems fail to extinguish when no longer reinforced? And why do some habit systems that were once instrumental get refashioned into interests and values having a motivational push?

The common counterargument holds that "secondary reinforcement" somehow miraculously sustains all the proactive goal-seeking of a mature person. The scientific ardor of Pasteur, the religio-political zeal of Gandhi, or for that matter, Aunt Sally's devotion to her needlework, are explained by hypothetical cross-conditioning that somehow substitutes for the primary reinforcement of primary drives. What is significant for our purposes is that these critics prefer the concept of secondary reinforcement, not because it is clearer, but because it holds our thinking within the frame of a quasiclosed (reactive) system.

Now is not the time to re-argue the matter, but I have been asked to hint at my present views. I would say first that the concept of functional autonomy has relevance even at the level of quasiclosed systems. There are now so many indications concerning feedback mechanisms, cortical self-stimulation, self-organizing systems, and the like that I believe we cannot deny the existence of self-sustaining circuit mechanisms which we can lump together under the rubric "perseverative functional autonomy."

But the major significance of the concept lies in a different direction, and presupposes the view that personality is an expanding system seeking progressively new levels of order and transaction. While drive motives remain fairly constant throughout life, existential motives do not. It is the very nature of an open system to achieve progressive levels of order through change in cognitive and motivational structure. Since in this case the causation is systemic we cannot hope to account for functional autonomy in terms of specific reinforcements. This condition I would call "propriate functional autonomy."

Both perseverative and propriate autonomy are, I think, indispensable conceptions. The one applies to the relatively closed part-systems within personality; the other to the continuously evolving structure of the whole.

A last example. It is characteristic of the quasiclosed system outlook that it is heavily nomothetic. It seeks response and homeostatic similarities among all personality systems (or, as in general behavior systems theory, among *all* systems). If, however, we elect the open system view we find ourselves forced in part toward the idiographic outlook. For now the vital question becomes "what makes the system hang together in any one person?" Let me repeat this question, for it is the one that more than any other has haunted me over the years. *What makes the system cohere in any person?* That this problem is pivotal, urgent, and relatively neglected, will be recognized by open-system theorists, even while it is downgraded and evaded by those who prefer their systems semiclosed.

Final Word

If my discourse has seemed polemical I can only plead that personality theory lives by controversy. In this country we are fortunate that no single party line shackles our speculations. We are free to pursue any and all assumptions concerning the nature of man. The penalty we pay is that for the present we cannot expect personality *theory* to be cumulative—although, fortunately, to some extent personality *research* can be.

Theories, we know, are ideally derived from axioms, and if axioms are lacking, as in our field they are, from assumptions. But our assumptions regarding the nature of man range from the Adlerian to the Zilborgian, from the Lockean to the Leibnitzian, from the Freudian to the Hullian, and from the cybernetic to the existentialist. Some of us model man after the pigeon; others view his potentialities as many splendored. And there is no agreement in sight.

Nils Bohr's principle of complementarity contains a lesson for us. You recall that he showed that if we study the position of a particle we cannot at the same time study its momentum. Applied to our own work the principle tells us that if we focus on reaction we cannot simultaneously study proaction; if we measure one trait we cannot fix our attention on pattern; if we tackle a subsystem we lose the whole; if we pursue the whole we overlook the part-functioning. For the single investigator there seems to be no escape from this limitation. Our only hope is to overcome it by a complementarity of investigators and of theorists.

While I myself am partisan for the open system, I would shut no doors. (Some of my best friends are quasiclosed systematists.) If I argue for the open system I plead more strongly for the open mind. Our condemnation is reserved for that peculiar slavery to fashion which says that conventionality alone makes for scientific respectability. We still have much to learn from our creative fumblings with the open system. Among our students, I trust, there will be many adventurers. Shall we not teach them that in the pastures of science it is not only the sacred cows that can yield good scientific milk?

Notes

1. Von Bertalanffy's definition explicity recognizes the first two of these criteria as present in all living organisms. A living organism, he says, is "an open system which continually gives up matter to the outer world and takes in matter from it, but which maintains itself in this continuous exchange in a steady state, or approaches such steady state in its variations in time." But elsewhere in this author's writing we find recognition of the additional criteria.

2. In a recent review Mowrer strongly defends the homeostatic theory. He is distressed that the dean of American psychologists, Robert Woodworth has taken a firm stand against the "need-primacy" theory in favor of what he calls the "behavior-primacy" theory. With the detailed merits of the argument we are not here concerned. What concerns us at the moment is that the issue has been sharply joined. Need-primacy which Mowrer calls a "homeostatic" theory does not go beyond our first two criteria for an open system. Woodworth by insisting that contact with, and mastery of, the environment constitute a pervasive principle of motivation, recognizes the additional criteria.

3. When we speak of the "function" of a neurosis we are reminded of the many theories of "functionalism" current in psychology and social science. Granted that the label, as Merton has shown, is a wide one, still we may safely say that the emphasis of functionalism is always on the usefulness of an activity in maintaining the "steady state" of a personality or social or cultural system. In short, "functional" theories stress maintenance of present direction allowing little room or none at all for departure and change.

4. Pumpian-Mindlin writes: "The focus of clinical psychoanalysis on ego psychology is a direct result of the change from a closed system to an open one."

5. (The reader who has come this far in this sourcebook will recognize that Allport's criticisms and distinction between "general systems theory" and "theories of open-systems" are not applicable to the main thrust of the systems research we are presenting.—*Ed.*)

43.
Note on Self-Regulating Systems and Stress

JOSEPH M. NOTTERMAN AND RICHARD TRUMBULL

THE COMMENT we wish to make concerns the use of the servosystem analogy as a conceptual framework for theorizing and research in stress. Specifically, we wish to present some speculations bearing upon the characteristics of certain assumed underlying processes in such a frame of reference.

The notion that organisms are capable of correcting for or adapting to disturbances is not new. Nor is a sharp distinction always drawn between what might be termed physiological as opposed to behavioral homeostasis. By the latter term we wish to imply nothing more than that the typical animal has developed, during the course of his natural evolution, certain proclivities toward responding to changing stimulus conditions. The appropriateness of his specific responses —both as to occurrence and as to intensive magnitude—is determined by the usual laws of reinforcement, or (in extreme cases) natural selection. As far as the servosystem analogy is concerned, it makes little difference whether the corrective response is "voluntary" or "involuntary," skeletal or autonomic. What *is* crucial for this frame of reference is that the response directly affects that aspect of the stimulus situation ("disturbance") to which the organism is responding. The feedback thus generated by the organism's behavior is in part responsible for fixing the ensuing value of the input.

We wish to observe that this familiar servosystem argument presupposes the existence of the following underlying processes: detection, identification, and response availability, each of which

can be explored in the laboratory. The remainder of this comment is concerned with a discussion of these processes.

1. *Detection:* For regulation to take place, the disparity between the disturbed and normal (or desired) state of the organism must be of such form and quantity as to be detectable by the organism. Obviously, if the organism cannot sense (perceptually or physiologically) a disturbance, measures cannot be taken for its correction. Equally apparent is the fact that individuals will differ in both the quality and quantity of information required for detection.

2. *Identification:* Apart from the necessity for the disparity to be detected, it must also be identified. (The term is used in a non-cognitive sense.) Corrective action cannot be specific unless a given disturbance is successfully discriminated from other possible disturbances. Here again, individual differences in the form and quantity of information necessary for identification undoubtedly exist.

3. *Response Availability:* Upon detection and identification of the normal–disturbed disparity, the organism must be permitted by environmental, physiological, or laboratory conditions to make the correction. It may be noted in passing that Selye has principally concerned himself with those laboratory situations in which the response available to the organism has not proved adequate to removing the offending stimulus.

Detection

Returning to the process of *detection*, it is apparent that—as far as psychologists are ordinarily

From Joseph M. Notterman and Richard Trumbull, "Note on Self-Regulating Systems and Stress," *Behavioral Science*, 4 (October, 1950), 324–27. Reprinted by permission of the authors and publisher.

interested—the problems of primary significance are those concerned with perception. This, in turn, places detection phenomena squarely in the realm of psychophysics. Unfortunately, however, the psychophysics which is of importance in a servosystem context is not the kind which is typically studied. As was noted in the foregoing, the servosystem analogy calls for the behavior of the organism to have a direct effect upon the ensuing stimulus condition. This will generally mean that the disturbance is either gradually or discretely changed by successive responses—a sort of non-topographical "tracking." But this in turn means that the organism is responding to continuously or discretely variable stimulus values, not the stationary "yes–no" exposures usually studied in psychophysics. In short, the detection problem becomes, for one interested in the servosystem frame of reference, a much more complicated phenomenon than that customarily examined under the label of absolute and difference thresholds. It is now a problem involving not only the presence or absence of a stimulus, or one concerned with the minimal discriminable difference between a pair of stimuli, but, additionally, a problem involving the detection of the first and second time derivatives of a changing stimulus magnitude. (The "time" aspect enters the situation because successive responses take place in time.) ... Further research which has not as yet been published indicates that the energy required for a JND of stimulus change may be a gradually increasing function of the rate at which the stimulus value is changing.

From the point of view of those interested in stress phenomena, the detection of time variant stimuli would seem to hold special interest. Is stress due only to the stationary value of a disturbance, or does the rate at which the disturbance changes affect the subject's behavior (and vice versa)? We would certainly guess at the latter, but definitive experiments remain to be done.

Identification

Turning now to *identification*, it seems safe to remark that the mere detection of a disturbance, without identification, may lead to general (nonspecific) response similar to that described by Selye. Such massive, but behaviorally incoherent responses, may or may not result in the reduction of the disturbance.

If the disturbance cannot be immediately identified, the organism's initial reactions to the offending stimulus may be in the nature of inquiries or "feelers," which—by virtue of the ensuing feed-back—seek to establish the identity of the disturbance. Such behavior is a joint result of biological evolution and past reinforcement. Individuals will differ in the quality and quantity of the feelers necessary to make this judgement; accordingly, they will differ in the number of such responses emitted. Also as a result of past reinforcement history, organisms will differ in the number and kind of stimuli which are found to be disturbing. For this reason, too, individuals will differ in the number of feelers emitted. Whatever the source, those individuals who characteristically exhibit a markedly high frequency of these inquiries are frequently labelled "anxious."

The foregoing is exemplified by some data obtained by one of us under particularly illuminating circumstances.

In 1951, the School of Aviation Medicine at Pensacola was routinely exposing classes of Naval Aviation Cadets to low pressure chamber operations for familiarization with masks and reduced pressure reactions. There was a relatively high frequency of hyperventilation—a phenomenon exaggerated by severe anxiety—in which the normal breathing sequences are altered, resulting in expiration of carbon dioxide from the body stores faster than carbon dioxide is produced metabolically. An investigation was initiated in an effort to determine who the hyperventilator was, and whether there was anything that could be learned about him to allow prediction. The Minnesota Multiphasic Personality Inventory was the test given.

A series of classes were given the MMPI, were put through the chamber, and all cases of hyperventilation noted; notation was also made of those individuals experiencing other difficulties in adjusting to the new situation. Correlations suggested that our man appeared in the use of the question mark section. Up to a certain point, the less apparent the relationship of an item to his present pursuit (Naval Aviation), the more certain he was to employ the question marks. A new series of these items was constructed with others of similar nature, and a generous sprinkling of items to which he could answer "yes" or "no" for filling. This resulted in a test of 148 items. To confuse the issue and force some decision, he was asked to rate himself on a 7 point scale which included: never, seldom, sometimes (under certain conditions), often, usually (under most conditions), always, and never occurred to me. As might be expected, he took the three, five and seven positions.

Based upon these findings, a prediction was made for the next class of six Cadets. This was not an assertion that certain individuals in this class would or would not hyperventilate, but was a specification of the sequence of MMPI test responses of hyperventilators. This sequence, of course, consisted of their ranking in terms of their use of the 3, 5, and 7 responses. The prediction turned out to be perfect.

In discussing the data with the School of Aviation Medicine staff, it was found convenient to refer to the hyperventilator (the anxious person, or question mark user) as "Radar Robert." He is the individual who can find structure in a new situation only by emitting a number of feelers which is far in excess of that normally required. His need for feedback (the rebound, as it were, of his inquiries) justifies the pseudonym.

The implication of the foregoing for the servosystem analogy and its relation to stress is clear: the central concept of the servosystem is that of feedback; without feedback, there can be no regulation. While this is, of course, largely recognized, there has been relatively little research concerned with the cold, hard problems involved in the specification and measurement of feedback. How do individuals differ in their feedback requirements? (Stimulus deprivation studies are not enough to answer this question; we need stimulus *exposure* studies in which the stimulus varies as a joint function of time and the organism's behavior.) What are the characteristics of the cross-correlation in time between input (stimulus) and output (response)? How valuable are the mathematical tools of servo-engineers in this sort of analysis? Do we have to go—as does the servo-analyst—to intensive measures of response (such as force, duration, or time integral of force) in order to perform the required analyses? Some of these questions are presently being examined, but only a beginning has been made.

Response Availability

The importance to the servosystem analogy of *response availability* as a means of exploring stress is obvious, yet little seems to have been done with this avenue of approach. Consider a study in which one of us participated several years ago:

An experiment was performed to determine the effect of three different post-acquisition procedures upon the course of extinction of a conditioned heart-rate response in humans. The response was conditioned under a partial reinforcement schedule, with electric shock as the reinforcing agent. In a group extinguished without prior information (the non-instructed group), the conditioned response showed little tendency to decline in strength over 11 extinction trials. Following the first extinction trial, both an instructed and an instructed-avoidance group were told that they would no longer be shocked. In the latter instance, however, Ss were told that the shock would be omitted only if they made a specific motor response (tapping a telegraph key) whenever the conditioned stimulus (tone) was presented. Although in the case of the instructed group the strength of the conditioned response declined progressively, the addition of the avoidance response for the instructed-avoidance group led to a much more rapid extinction.

The instructed-avoidance group was the only group which *by its own behavior* successfully forestalled the stimulus which had acquired stress properties (the CS or tone). It is noteworthy that this group—which had immediate feedback of its responses—extinguished significantly faster than either the non-instructed or instructed groups.

Finally, it may be observed that there seems as little justification for experimenting exclusively with yes–no responses as there is for yes–no stimuli. Complete utilization of the servosystem analogy in studying stress would seem to require the establishment of those experimental situations which would permit a decrease in the value of disturbance function, in proportion to some variable aspect of the response. Eventually, perhaps, we will permit a rat pressing a bar with the equivalent of 20 grams of force to reduce the brightness of a noxious light stimulus by "x" times the number of foot-lamberts as when it presses with only 10 grams. In this forseeable type of research, we will go beyond mere response availability or occurrence, and extend our knowledge to the more intriguing problem of response proportionality along intensive dimensions; perhaps for force, or for duration, or for—a combination of both—time integral of force.

44.
The Concept of Stress in Relation to the Disorganization of Human Behaviour

GEOFFREY VICKERS

. . . AT FIRST SIGHT it seems clear that the phenomena that the psychiatrist studies are more easily comparable with those which concern the student of animal behaviour than with those which confront the physiologist. This is because both the animal experimenter and the animal ethologist, like the psychiatrist, start from an observed correlation between a 'situation' and a 'disorganization of behaviour'. Disorganization can take many forms: the unbearably frustrated man may break down into rage or tears or lethargy; he may begin to act at random or to adopt irrelevant behaviour, or to stop acting at all. So, within his compass, may the frustrated rat behave, and these differences in form of breakdown may be interesting as indices of temperament—but all serve equally well as indices of disorganization.

So the first question as I see it is: how do we recognize this characteristic disorganization despite its variety of form? Shall we take as our index the fact that the behaviour under review has become non-adaptive, inept to the situation that has provoked it? That will not quite do, because neurosis can be protective in man and perhaps also in animals. In any case, is it really adaptive for a man in a bewitched cockpit or a rat in a bewitched maze to go on ringing the changes on responses that have become irrelevant? I recall a question asked by a psychiatrist when some colleague had described a particularly

devilish form of animal frustration. What, he asked, would be a non-neurotic reaction to a situation of that kind? I do not recall that he received any convincing reply.

It seems that what we call disorganization may be, perhaps always is, a form of defensive reorganization.

But of course this protection is bought at a cost. Field-Marshal Lord Wavell, in some famous lectures on his profession, said that stupidity in generals should never excite surprise, because generals were selected from the extremely small class of human beings who were tough enough to be generals at all. The essential qualification was not that they should be extremely clever or sensitive, but that they should continue to function, even if not particularly well, in situations in which more sensitive or less stable organisms would have stopped functioning altogether.

The criteria for determining disorganization seem to imply the passing of some threshold beyond which external relations can no longer be handled at the previous level. This is either because internal coherence has been lost or because internal coherence has only been preserved at the cost of some 'withdrawal'. These may be two extremely different states.

Turning to the other side, if pressed to define the situations that provoke disorganization, both the psychiatrist and the student of animal behaviour could be driven to some tentative formulations. This is particularly true of the animal experimentalists who design the situations they are studying and who must, therefore, have some criteria for designing them. Once again the descriptions they give have curious imprecision.

Pavlov, I believe, used four main types:

From Geoffrey Vickers, "The Concept of Stress in Relation to the Disorganization of Human Behavior," in J. M. Tanner (Ed.), *Stress and Psychiatric Disorder* (Oxford: Blackwell Scientific Publications, Ltd., 1959), pp. 3–10. Reprinted by permission of the author and publisher.

progressive increase in intensity of the signals to which the animal is conditioned; progressive increase in the delay between the signal and the satisfaction; confusion by the introduction of conflicting signals; and interference with the physical condition. If you leave out the last and divide the third into doubt, where the signals are ambiguous, conflict where they are inconsistent and frustration, where the signal is not followed by its accustomed sequel, you still have not covered all the situations . . . such as loneliness, monotony and self-imposed restraint, which required categories of their own.

So this is a very wide field of characteristic situations, and it is curious that they can most readily be described by words drawn from the subjective vocabulary of human kind: apprehension, suspense, doubt, conflict or frustration. I think psychiatrists would agree that these categories also cover most of the situations that they recognize as provocative of stress in man, although, of course, they are too vague so far to be very much use. At all events this suggests fairly clearly that these situations, which we know through introspection, reflect conditions of neural excitement which are not confined to conscious states or to nervous systems as complex as our own.

The other curious enigma is that most, and possibly all, of them require for their convenient description the concept of *expectation*, whether it be the hateful expectation which can neither be accepted nor avoided, the pleasurable expectation which is intolerably deferred, the confusion of expectation which produces confused response, or the sheer absence of expectation which removes the basis for action. The fact that this concept of expectation—again a rather imprecise concept—crops up so often in the design of animal experiments shows that here too we are up against something which must be describable in terms of neural activity not related to conscious human states.

It is at that point, I suggest, that the psychiatrist and the student of animal behaviour can legitimately turn to the physiologist, including the physiologist in themselves, for help. The organism is not functioning the way it was; what is the nature of this change of function and how and why has it occurred? These seem questions which it is very proper to put to a specialist in organic function. Until recently I think a physiologist would not have had very much to say to them. He could supply an impressive model of the internal relations of the organism; he could describe the homeostatic mechanisms that neutralize the impact of external variables such as heat

or cold. He could describe the course of the defensive activities as they are overcome by forces too strong for the defences, forces which, in Professor Selye's classic formulation, he knows as stressors. He could describe the adaptive changes that enable the organism in an emergency to hold its own for a while against overwhelming forces—the sort of changes through which a man chased by a bull can mobilize for a few minutes a quite surprising amount of energy. But why an approaching bull should function as a stressor no less than a change of temperature, and still more why the hope of winning an Olympic mile should function as a stressor no less than an approaching bull, are questions which the physiologist is less prepared to answer. Nor is it clear, certainly not to me, that the collapse of the runner after he has jumped a gate or burst a tape has very much in common with the kind of collapse studied by the psychiatrist or the student of animal behaviour.

So the stressful situation which concerns the psychiatrist and, therefore, which ultimately concerns the physiologist, is not given by events but by the organism's interpretation of events in relation to itself. This, in turn, is a function of the way in which its individual personality is organized, and any conceptual model of that must include the organization of experience.

We have several conceptual models of organization. There is the physiological model, which until recently had not very much to say about the organization of experience. There is the body of psychological concepts dealing with perception and learning theory, and there is the multi-dimensional model implicit in psychodynamic theory. All of these deal with systems obtained by abstracting different sets of variables from this baffling observable entity—man. A better integration of these models is an urgent need which we hope this Conference will advance, and a great deal of what is happening currently is most excitingly advancing this integration, far more than certainly a layman like myself can summarize. The anatomist finds in the nervous system paths by which past experience can modify the sensory input before it gets to the association areas—a rather sinister form of feedback, as it seems to me. The pharmacologist has an increasing battery of drugs of known constitution which can directly modify psychological states.

Observations by EEG begin to give a physiological measure of some differences of human type and perhaps of difference of personal integration and have suggested to Dr Grey Walter a physiological theory of learning. Both psychiatrists and animal experimentalists are adding to the correlations which relate physiological and bio-chemical changes to psychological changes at the threshold of breakdown. Animal experimentalists have demonstrated the genetic basis for differential immunity and vulnerability to stress in animals. Animal ethologists have provided a set of concepts which seem to me to have a curious interest in that both physiologists and psychoanalysts can use them without either a sense of unreality or a sense of sin, and that is a very welcome widening of our common universe of discourse.

Then there is the interesting and I think very important increase in the development of language and concepts apt to describe open systems generally. The concept of stability, for instance, that Dr Ross Ashby has done so much to generalize, seems to me to be equally applicable to systems studied by the physiologist, the psychologist or the sociologist. The same concept applies equally to those much more numerous cases in which the governing controls of the system are not fixed but change with time, such as the pattern of growth and maturation or, for that matter, the pattern of a changing cultural norm. Homeostasis is a special case of a much wider process that Professor Waddington recently christened homeorhesis.

There is one idea in all these fields which may be of central importance to the theme of this Conference, and that is the idea of matching with a pattern. It is abundantly clear that the raw material of experience is not the whole of the 'blooming, buzzing confusion' but a selection of the regularities which we detect. These in turn provide categories for classifying future experience. Conditioning and probably much more beside, depends on recognizing regular relations between recurrent events in the categories thus distinguished. Even judgments of values, those most refined of tropisms, are also linked to situations recognized by their correspondence with a pattern, however complex. If I say that A has given me a fair deal, I must first have selected a number of aspects of my relations with A and then classified them as a deal, rather than say a battle, and then applied the standards of 'fair deal' rather than 'fair fight'. I must then observe that they match, and the process that arrives at the judgment of 'fair' is equally a matching process, like the process which arrives at the judgment of 'deal' and I see no reason why it should not be carried out in the same code.

So pattern governs throughout, and though

the mind may boggle at the thought of a black box no bigger than our heads that can group and regroup and handle shifting configurations of symbols so complex and plastic and yet so enduring, we boggle at the complexity rather than the principle. So perhaps it is not so surprising that this thing that we vaguely call 'expectation' should figure so largely in the organization and disorganization of personality, because clearly what we know as expectation is only the exposed part of an iceberg that floats very low in the water.

In pursuing, maintaining and elaborating the external relations by which we live and die, we are clearly guided by symbolic representations of what is happening and of what 'ought' to be happening. I think we need a model of this process to understand the nature and the noxious operation of stress. Contrariwise, in looking for a model to explain the noxious working of stress, we may well contribute to our understanding of controlled behaviour, because obviously organization and disorganization are opposite sides of the same penny.

The fact that meaning derives from relationship to the familiar is evident from the fact that when we have not got a templet from experience we have to invent one. That is exemplified at an entirely different level by another conceptual model with which I happen to be more familiar and which I do not think is as remote from our subject as may first appear. In the practical affairs of life we assume that conduct is controlled to a very important degree by structure of expectation, and the implications of this have been worked out both in theory and practice to a high degree of refinement in the field of administration. Nobody today would try to run a Government department or a university or a business without maintaining and continually comparing two running representations of the future. One is a representation of what is happening, projected into the future to give what is going to happen next, and the other is a representation of the course of events that we want to bring about or prevent; call one the actual and the other the standard, the one the 'is' and the other the 'ought to be'. The comparison of these two yields a stream of mismatch signals on which we act. The building contractor plots on charts against time the planned course of many interdependent operations, and then, as work proceeds, he plots what is actually achieved and projects that also into the future. The divergence of those two lines on the chart provides the signal for reflection and action, first to bring the 'is' into line with the 'ought to be' and, if that cannot be done, to bring the ought to be' into line with the 'is' so as to provide a workable control. These controls that are used in administration are representations of relationships that we seek to maintain or alter or escape. Some relationships are between the organization and its environment, and some are relationships within the organization itself. In either case the control may be negative or positive, directed either to bringing back the state continually to an optimum position, or to preventing it from straying beyond some critical threshold. Even in this down-to-earth world in which I am more accustomed to move, the 'actual' is a highly artificial construction; in the first place it is hypothetical—our information about what is happening is never complete, exact or direct. It is also selective, because we can only attend to a few aspects of it at a time. It is also represented in a code, be it in writing or figures or graphically, which limits and distorts what can be represented, and that is necessarily also the code in which you represent the 'standard', otherwise you could not compare them. Finally, it has an inescapable time base. In all of these things I think it closely parallels the working of the individual mind.

These controls, like those that control us personally, do not always give clear or correct or adequate guidance. They apply only to those aspects of experience to which we have chosen to attend. We may have chosen the wrong variables for attention, or the signals may be ambiguous or, being clear, experience may supply no apt response; or the responses that have worked in the past may let us down. But the most inescapable feature of these responses is their conflict. This is a feature of practical life, as we all know. Any projected action is relevant not only to the purposive sequence in which it arises, but to others also, and if it is apt for some, it is inept for others; whatever doors it will open, others it will shut. In business, short-term profit, long-term stability, internal coherence, public relations —each of these disparate standards, when compared with an appropriate selection from the actual, maintains its own stream of warning and advice, and these are no more consistent in business than in private life. Moreover their inconsistency is inescapable. The highly organized business, like the highly organized personality, necessarily generates more inconsistency in its governing expectations than one less highly organized. This may perhaps account for the fact that the stress of war produces relatively so little civilian neurosis,

life in wartime being harder but also simpler.

Apart from the inadequacies just described, the controls of business are partial and intermittent, because either the actual or the intended may not be accessible. Like the fog-bound navigator we may know where we ought to be but have no means of knowing where we are. Or like the climber following an unfamiliar route, we may see where we are but have no assurance that it is where we ought to be. Nor can experience be guaranteed to remedy either case, because the results of our action may return for judgment so long after the event and mixed with so many other variables that they may supply neither validation of the past nor guidance for the future.

In the board room, then, stress is associated with doubt and conflict of clearly definable kinds. We may have to live for years with deafening streams of mismatch signals, either because we can devise no suitable response or, much more often, because any response we made would elicit even more violent protest in another context. Alternatively we may have to live in an eerie silence because either the actual or the standard is not registering in the appropriate code. I assure you that either of those states can produce disorganization or perhaps protective reorganization which is very suggestive of the forms that are familiar to the psychiatrist and to the animal experimentalist.

This digression into what seems rather a remote field seems to be significant for three reasons. First, it does exemplify clearly the nature and limitations of that control by matching which we meet at all levels. It seems to me that even the simplest innate response involves selecting something from the stream of experience and matching it with a pattern. Even that famous herring gull chick does not respond to a red patch; it responds to the similarity between an observed red patch, which is a complicated abstraction, and the standard red patch with which it is equipped. The similarity may not be absolute but falls within limits of tolerance which are also built in. Of course learned responses can be built around cues much more extensive in space and time and which are also distilled from experience. It seems, then, that this process of matching with a pattern is inherent in the control of behaviour from the simplest response up to the most sophisticated act of cognition, because all cognition is recognition. Moreover these patterns which the brain can record are patterns in time as well as in space. An instrument which can record change with time can represent the hypothetical future as well as the past.

Secondly, I think the model is useful in its representation of conflict as endemic and necessarily increasing with the complexity of organization. By complexity of organization I mean not increase in size or increase in power and variety of responses available, or even increases in subdivision of function. I mean rather increase in the number and diversity of the objectives to be sought and the thresholds to be avoided, in other words, in the norms which the system is set simultaneously to seek and the limits which it is set simultaneously to avoid. That is the dimension along which it seems to me both organism and organization tend to develop and along which they tend to fall back, whenever they set themselves a task of reconciliation which proves to be too much for their powers.

Finally, I think the example is useful in stressing a new dimension in which higher organisms, no less than organizations are adaptable. Most biological work treats as given the acceptable and unacceptable states which act as governors, positive and negative, of a system's behaviour. In so far as this is so the only scope for adaptation within the individual life span is the development of the responses and the skills which serve those needs. But at higher levels, conspicuously at the level of human life, the individual's wants and needs, no less than his responses and skills, grow and multiply within the life span. An increasing number of different but mutually exclusive possibilities compete for realization within the framework of the biologically given. We take a hand in the setting of our own systems; and this is

perhaps the point at which our common conceptual model needs to be enlarged if it is to link physiological with psychological observation. The biologist is already at grips with the code built into the molecular structure of the gene. But the psychologist, who works within a very different time span, needs more than that; he needs a model not merely of the programme of the organism from birth to death, but of its self-programming capacity and its self-programming propensity, a model that can represent our goal-setting as well as our goal-seeking. Though these two activities of goal-setting and goal-seeking are so intimately related that they cannot be considered separately, they are also I think so distinct that they cannot usefully be simplified, as is often done, by simply resolving one into the other.

Finally, let me conclude by reminding myself and you that a model of conflict does not necessarily tell us anything at all about pathological stress. Conflict is endemic; breakdown is still, happily, relatively exceptional. We need not—and therefore must not—assume that conflict in itself is noxious. If the threshold is quantitative, it is also relative, for it must certainly depend both on the amount of stress and on the resistance, or perhaps we should call it the immunity, of the organism affected. We need to understand both the noxious nature of a given stress in relation to the organization of a given organism and the vulnerability of a given organism to a particular form of stress. That of course, involves an elaboration of the conceptual model far ahead of anything on which I have yet touched.

This paper is concerned with the behaviour possible in an information-flow system intended explicitly as a hypothetical model for comparison with the human information-handling system. A statistically self-organizing system is described in which not only normal homeostatic behaviour but also such activities as the invention of fruitful hypotheses, the imagination of fictitious situations, and the like would find a natural place. Discussion is confined mainly to the manner of concept-formation and concept-handling in such a system. It has been suggested that the correlate of perception (as distinct from reception) is activity which organizes an outwardly directed internal matching response to signals from receptors. This organizing activity amounts logically to an internal representation of the feature in the incoming signals to which it is adaptive, i.e. the feature which is thus 'perceived'.

A hierarchic structure is postulated wherein much of the organizing activity is concerned with modifying the probabilities of other activity. Abstract concepts and hypotheses are represented by 'sub-routines' of such organizing activity. These can in principle be evolved as a result of experience in a manner analogous to—or at least fruitfully comparable with—the process of learning and discovery; and it is not necessary for the designer to predetermine, nor possible for him to foresee, all the conceptual categories in terms of which the information received by the system may be structured. Particular attention is directed to the conditions under which such a process of concept-formation could take place with sufficient rapidity.

Some of the symptoms of psychopathology find apparent correlates in possible modes of malfunction of such a system. But in this, as in other respects, it is not intended to press the resemblances; the intention is rather to stimulate critical comparisons, in order that the differences between the information-flow model and the real thing may continually lead to progressive refinement and enlarged understanding.

45.
Towards an Information-Flow Model of Human Behaviour

DONALD M. MACKAY

I. Introductory

(i) *The wrong approach.* I suppose that the most superficial and unconvincing excuse for agreeing to discuss the behaviour of any artificial mechanism before a gathering of psychologists might run somewhat as follows: Experimental psychology has lately come to concern itself more and more with the human being as a 'black box', having certain characteristic ways of receiving and reacting to information. Perhaps not altogether accidentally, the merchant of automata has discovered that in his stock list there are several artificially contrived black boxes which in principle can be made to receive and react to information in the same characteristic ways. So, he might argue, his artificial black boxes are surely quite as proper subjects of the psychologist's attention as the human ones.

This argument seems to me, putting it mildly, to be mistaken. If the psychologist had *defined* his field as including 'all black boxes with a certain behaviour pattern', the excuse might hold. But if, as I believe, he defines it to be human (or animal) behaviour, then the merchant of automata has no automatic claim on his attention or even his interest. An automaton giving the most perfect imitation of observable human behaviour might prove to hold as much interest for the psychologist as a gramophone capable of reproducing human singing does for the laryngologist, and no more.

(ii) *The conceptual role of the theory of automata.* So I must offer, tentatively, a rather different excuse. I believe that the theory of automata is potentially of use to the psychologist, not just because some automata could imitate the behaviour of his subjects, but because some aspects of the theory could, if properly disciplined, provide him with a useful tool of research.

What I think it offers is a kind of logical scaffolding, which may play a useful part in the processes of erecting, testing and (alas) demolishing psychological theories. We search nowadays for clues to the organization of human personality in two widely different fields, the provinces broadly speaking of psychology and physiology. These clues come expressed in characteristically different languages. Particularly in descriptions of pathological symptoms, it is often hard to imagine with any confidence the proper correlate in one language of data expressed in the other. We are convinced that the data *should* help us, but we have little idea what to make of them.

Now the language of information and control, in which the theory of automata is framed, is conceptually intermediate between those of psychology and physiology. It belongs in a sense to

From Donald M. MacKay, "Towards an Information-Flow Model of Human Behaviour," *British Journal of Psychology*, 47 (1956), 30–43. Reprinted by permission of the author and publisher.

both fields, inasmuch as questions of information transfer arise in both.

It seems possible therefore that the theory of an automaton, specifically designed to be a research model of the human organism considered as an information flow system, might provide a common meeting-ground and a common language in which hypotheses might be framed and tested and progressively refined with the help of clues from both sides.

(iii) *The information-flow map.* For this purpose the physical engineering of the automaton is of little interest. It is sufficient if it can develop and grow on paper as a hypothetical information-flow map of the human organism. Just as the lines on a trade-route map of the world are drawn in the same way, whether goods are carried by train or lorry or camel caravan, so that pathways of our information-flow map may be postulated and tested and confirmed irrespective of the chemical, electrical or other details of the processes mediating the traffic in information.

On the other hand, the features of the information-flow map will normally suggest experimental questions in both the psychological and physiological fields, and should in turn find themselves tested and refined by the results.

This is what I meant by calling the theory of automata a logical scaffolding, potentially useful in building operations on psychological theories. Merely to ask what functions can be imitated, and how, would be pointless. The answer is that any pattern of behaviour that can be specified exactly or statistically can in principle be shown by an automaton, designed in one of any number of alternative ways.

The real question is how far we can go towards designing a self-organizing system with the same internal information-flow map, as well as the same behaviour, as a human being.

We shall not find the answer in this paper. But I hope to show you that the answer is reasonable to seek, and not impossible to imagine, in dim outline, even now.

II. Goal-guided Activity

(i) *Minimal requirements.* Perhaps the most characteristic feature of intelligent behaviour is what we call goal-guided (goal-directed, adaptive, conative, purposive) activity. Let us begin by summarizing the essential features of any information-flow system that is to show such activity. These features are by now familiar to most of us,

and I have discussed them in detail elsewhere. The minimal requirements are shown in outline in Figure 1, where the activity of the system is represented simply by the movement of a point Y along a line F.

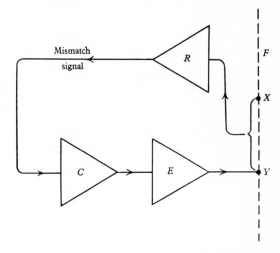

Figure 1.—The minimal requirements for goal-guided activity.

The active agent is represented by the 'effector system' E, which is governed by the control system C. We assume that in a typical situation E is capable of a certain range or variety of modes of activity (including inactivity), so that the function of C is to *select* from moment to moment what E shall do next, out of the range of possibilities open to it.

Suppose now that on the line F there is a point X, fixed or moving, towards which we want the activity of E to be 'goal-guided'. What do we mean by this? We mean (*a*) that the overall pattern of E's activity should be calculated to reduce the interval between X and Y, or some time average of it, to a minimum. We mean (*b*) that if the interval were increased in any way (e.g. by movement of X away from Y, or by the action of some external agency forcing Y away from X) the effector E should actively resist the increase; but that it should actively assist any agency tending to reduce the interval.

The first of these requirements could in principle be met by a 'blind' automaton with a pre-computed control sequence, if the motions of the goal were known in advance. But in order to cope with the unpredicted influence of external agencies, something more is needed. The controlling system C must *receive information from the field* about the interval XY, and use it to control (wholly or partly) the selection of what E shall do next. This information is represented in Figure 1 as carried to C by a 'mismatch signal' from a receptor system R coupled with the field through any number of modalities.

(ii) *Improvement in performance.* On the very simplest system, only the existence of the interval need be signalled to C. C must then be designed to keep E in randomly or systematically varied activity until the interval is reduced below some acceptable level (the 'Homeostat' of Ashby is a typical example).

We could generally achieve greater speed and efficiency, however (at the cost of greater complexity in the mismatch-signalling system), if the direction—and still better, the magnitude—of the interval XY, and not merely its existence, were signalled back to C. In the extreme case we could provide C with a complete representation of the current state of the field of activity, from which to calculate automatically the best course of action for E.

If information can also be *stored* in C, its calculations can be extended to cover rates of movement and to make predictions on a statistical basis, so increasing still further the effectiveness of E's activity. For this purpose we do not need to picture a discrete store of precisely symbolized information as in a digital computor. 'Storage' here means any modification of state due to information received and capable of influencing later activity, for however short a time.

(iii) *The distinction between goal-oriented and goal-guided activity.* In passing we may note that there are some physical systems in which the requirements for goal-directed activity (in one sense) seem superficially to be met without any of the features of Figure 1. A steel ball, for example, will 'pursue' a moving magnet and resist attempts to deflect it.

But in all systems of this sort, the essential difference is that *no other* activity is physically possible to the active agent in the given circumstances. The question of 'guidance' or 'direction', in the sense of a choice among physically open alternatives, does not arise. The required activity is in a sense uniquely implicit in the physical description of the situation requiring it. This is even more readily seen if we consider a physically equivalent model in which the ball is attached to the magnet by an elastic string. Its actions are merely 'newtonian' reactions to the situation in which it is placed.

It would, in fact, be rather tendentious to describe the magnet as the 'goal' of the ball. But

if we must use the term, it might be better to avoid confusion by describing such activity as 'goal-oriented' rather than 'goal-directed'.

(iv) *Examples of goal-guided activity.* The simple flow map of Figure 1 may be taken as typical of all situations in which goal-guided activity is required. In the simple thermostat, for example, the effector E is a heating unit. The field of activity F is the scale of temperature, on which the present state of the system is represented by some point Y, and the goal state by some pre-assigned point X. Information as to the existence and the direction (and sometimes the magnitude) of the interval XY is provided by some kind of thermometer (combining the functions of R and C) which serves to *compare* the present temperature with the goal temperature, and mechanically adjusts the activity of the heating unit in a direction calculated to reduce the discrepancy. In more complex cases the effector E may have many degrees of freedom (for example, it may have the equivalent of a large number of independent muscles), and its field of activity F will then be represented not by a one-dimensional line but by an abstract multi-dimensional space with a corresponding number of dimensions. In this space, however, the present state and the goal state will still be represented by points, X and Y, whose co-ordinates will be the values of the corresponding parameters. The magnitude and direction of the interval XY will have the same significance as in our one-dimensional illustration, and the system will work on exactly the same principles.

It will be clear from this example that in general the points X and Y in Figure 1 do not represent physical *objects* in the environment of A. The point Y is an abstract way of representing the current *state* of the total situation, system plus environment. The point X similarly represents the state of the total situation towards which as goal the activity is guided.

By defining X and Y in this way as representing states and not physical objects, we can make the same forma description apply equally well to a self-guided missile chasing an aircraft, and a man chasing the solution to a crossword puzzle.

(v) *Towards inventive behaviour.* On the basic principle of Figure 1, there is no difficulty in envisaging an automaton capable of interacting actively and apparently purposefully with its environment and in particular with human beings, in any field of activity for which it has appropriate effectors and receptors, and an adequate symbolic calculus embodied in its controlling element.

But there, of course, is the rub. It is true that human beings are educated in childhood to use a standard terminology, embodying standard categories of perception and description. To this extent we might be entitled to preform the categories in terms of which our model automaton structures its information. But human beings do more than carry out logically deductive operations in a fixed calculus. They invent new categories, and novel hypotheses, to meet an unforeseen flux of events. They imagine unrealized situations. . . .

How much farther can we go with our automaton? What kind of automaton, if any, could evolve for itself new descriptive categories appropriate to situations that its designer had not envisaged—and do it in a way analogous to the way in which a human being does? What place can we find in the theory of automata for such human activities as the invention of relevant and fruitful hypotheses, or the contemplation of imaginary situations?

In what follows I want to suggest, very tentatively, some principles on which a self-organizing system could be devised to parallel many of these inventive features of human activity—principles which at least bear realistic comparison with those so far recognized by psychologists in the development and functioning of the human information-flow system.

III. Organizing the Trial Process

(i) *Simple groping*. We have already seen that the system of Figure 1 could show goal-guided activity even with the minimum of information about the interval XY. In the limit, it could simply run through its repertoire of activity until the mismatch signal disappeared. But if its repertoire is at all large, this process takes time. If the effector system E has l independent degrees of freedom, each associated with m alternatives, the number of combinations which might be tried, and hence the average time of hunting, rises as m^l. Thus this groping automaton cannot hope to cope with a complex situation in which the goal-state changes rapidly and unpredictably. Under such conditions it would do little worse if the mismatch signal were dispensed with altogether,[1] since even if by chance the current goal state were reached and XY reduced momentarily to zero, it would probably not remain so for more than a small fraction of the time spent in hunting.

(ii) *Directional guidance*. Things would be much better if the mismatch signal could indicate

the *direction* of the interval XY. Even if only the *sign* of each component of the vector interval were signalled, the improvement could be enormous. Each trial (for an *l*-dimensional interval) would then provide approximately *l* independent yes–no decisions or (in the jargon of information theory) 'bits of information'. This could ideally reduce the minimum number of trials to only $\log_2 m$, i.e. $(l/l)\,(\log_2 m^l)$.

But with a complex effector system having many degrees of freedom, even this information capacity may be too much to ask of the mismatch-signalling system. This is not the place for a discussion of the informational requirements in detail. . . . But we must evidently ask at this stage whether there is no way of improving the performance of C without demanding from the mismatch signal the full selective-information content of $l \log_2 m$ bits which is ideally required in order to select the correct adaptive responses by E to each fresh movement of X.

(iii) *Making use of 'redundancy'*. Information theory assures us that in principle there should be such a way if the movements of X show any regularity in their statistical structure, or 'redundancy', as it is called. In that case, as in an ordinary communication channel, the average selective-information content of the correct response may

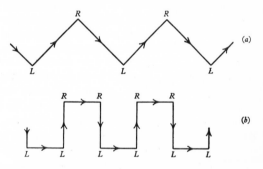

Figure 2.—Two typically 'redundant' sequences of bends in a road.

be much less than the $l \log_2 m$ bits which we first estimated. In practical terms, the trial process can afford to concentrate more on trying some combinations of activity rather than others; so on the average it can *save time*, or can reach the goal *in the same time with less information* from the mismatch-signalling system. Our problem is then to devise a way of profiting from the redundancy of the information sequence represented by the movements of the point X, or equivalently by the corresponding successful configurations of the activity of E.

(iv) *A simple example.* To take a simple example which I have used elsewhere, imagine an automaton designed to run along a road consisting simply of a succession of straight runs and right-angle bends. The automaton has only two control buttons, *L* and *R*, which when pressed cause it to turn left or right respectively. When it reaches a bend in the road, a simple mismatch signal automatically induces random trials of *L* or *R* until the correct button is pressed and it moves on.

Now suppose that over a long stretch of road, *L* and *R* bends form a regular sequence, such as *LRLR* (Figure 2*a*), which we may call S_1, or *LLRRLLRR* (Figure 2*b*) (S_2). We could obviously save hunting time if we were now to install in the controlling system a simple button-pressing mechanism or 'organizer' which imitated the appropriate rhythmic sequence of bends in the road. For example, in the first case, this organizer would always respond to the mismatch signal by trying *R* if *L* had last been used, and vice versa. Thus every mismatch signal would at once induce the correct adaptive response without any hunting. The high degree of redundancy in the series of bends permits of a correspondingly low information content per mismatch signal.

(v) *Second-order control.* Consider now a more complex road comprising a mixture of sequences of the forms S_1 and S_2. Each time the sequence changes the internal imitative organizer must now make a corresponding change in the rhythm of control signals. Either its connexions must be altered, or its place must be taken by another mechanism suitably connected to generate the next rhythm.

We need, in fact, a second control mechanism, C_2 (Figure 3), to organize trial routines from the repertoire of C_1 in the same way as C_1 organizes trial routines from the repertoire of *E*.

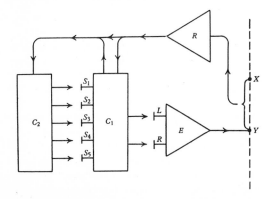

Figure 3.—The organization of control sequences.

If, for example, the road sequence changed from S_1 to S_2 and back again to S_1 in an alternating series, the organizer C_2 could be arranged to hunt through the repertoire of C_1 until it found the correct successor, each time that a succession of mismatch signals indicated that the road sequence had changed. But here again the efficiency could be increased, if C_2 incorporated a second-order button-pressing mechanism designed to try S_2 after S_1 and S_1 after S_2.

Obviously we can envisage an indefinite hierarchy of such imitative organizers. Each will represent by its activity an *abstraction* descriptive of some aspect of the series of bends in the road.

IV. A Statistical Self-Organizing System

(i) *The automatic development of organization.* Our last example was so far all very well; but it presupposed that we, the designers, foresaw in advance the range of possible shapes of the road. Our next step—the fundamental one for our present purpose—must be to devise some principle on which the system could discover for itself the statistically stationary features of its environment, and develop its own characteristic organizing sequences as a result of its experience, starting if necessary from scratch.

(ii) *Statistically adjustable links.* The most promising way of solving this problem—the problem of growing an organizing system from scratch—would seem to be to design the control mechanism out of statistically adjustable elements. By this I mean that instead of making its interconnexions hard and fast by means of switches that are either open or shut, we use 'statistical' interconnexions or links which have an adjustable *probability* of transferring information. This gives us all the flexibility we need. Since the 'thresholds' governing these probabilities may be continuously variable, the statistical structure of the activity pattern is capable of assuming an indefinite number of finely graded configurations, far larger than the number of combinations possible with simple on–off switches in each link. The information content represented by the current state of a given number of elements is correspondingly greater than in the simple case, so that the design can in principle be more economical of elements for a given required variability of activity pattern.

(iii) *Control of statistical bias.* We can now envisage a self-organizing system in which the mismatch signal automatically adjusts the configuration of thresholds just mentioned, governing

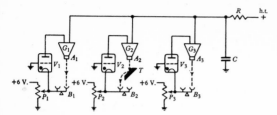

Figure 4.—A simple electronic device illustrating a statistical principle of self-organization.

the statistical pattern of activity according to the success or failure achieved, so that successful sub-routines of activity become more likely to be tried again in similar circumstances, while unsuccessful ones become less likely to be tried.

Provided that the environmental world of activity shows consistent features, these could then in time come to be reflected in the relative probabilities of different organizing sub-routines.

(iv) *A simple example of a statistically self-organizing system.* An excessively simple illustration of this kind of process is provided by the automaton of Figure 4. This device was built a few years ago to illustrate a lecture. It has of course no pretensions to animal status, and justifies its existence solely as an explanatory device. It is made up of three identical units (the number is unrestricted in principle), each comprising a thyratron valve V which can actuate an electromagnetic gate G so as to drop a ball-bearing from an aperture A over a two-pan balance B. B is mechanically linked to a potentiometer P, which governs the bias on thyratron V. The anodes of all thyratrons are supplied from a common capacitor C charged through a resistor R.

If now C is allowed to charge slowly, the thyratron with the lowest effective bias, say V_1, will tend to fire first, discharging C and releasing a ball from A_1. Normally this will fall into the right-hand pan of B, causing P_1 to rotate and increase the bias on V_1. Thus the normal effect of a thyratron's firing is to reduce the probability of its firing next time.

If, however, a target T is placed under A_1 so as to deflect any balls from A_1 into the left-hand pan of B_1, the opposite effect will occur. The success of the automaton in striking the target when V_1 fires will increase the probability that it will use V_1 next time.

Evidently no matter how the biases on different V's are set initially, the automaton will tend to an equilibrium state in which only the thyratron V_1

fires and T is always hit. If, moreover, we move T to a new position, say no. 2, the automaton will gradually 'unlearn' its original 'set' and eventually will be found firing only V_2.

In this example randomness is introduced only by the fluctuations of the thresholds of the various thyratrons, and the mismatch signals are quantized and very simple. Even so it illustrates the basic principle, whereby an automaton can selectively 'grow' modes of activity according to its experience of their relative success, and diminish by a kind of natural selection the probabilities of all except those which have been successful often enough.

(v) *Higher-order statistical control.* When we come to apply this principle of statistical control to the system of Figure 3, the situation is rather more complex. The actions whose probabilities are adjustable are not merely external effector actions, but actions of one element on another; in particular, they include actions which alter 'conditional probabilities' of action by other elements. For example, in order to set up the sub-routine $LRLR$ it is necessary that the action L shall modify the conditional probability, that R shall follow if and when the mismatch-signal next calls for action. There will thus be some situations of the general type of Figure 1, wholly *within* the mechanism itself.

But the principle on which this system should be made self-organizing is the same. As before, we introduce a second organizer, C_2, working now on the same statistical principles as C_1 which we have just discussed. Metaphorically speaking, C_2 can be on the watch for any regularities in the series of different routines found successful by C_1. As these regularities develop, C_2 will spontaneously begin to grow patterns of activity to imitate them and pre-organize the probabilities of their occurrence, so reducing the average time spent in hunting through C_1's repertoire, and at the same time increasing the statistical stability of any closed cycles that C_1 finds to be adaptive. If the organizing mechanism begins with most or all thresholds fairly low, so that all kinds of sequence are possible, then after a long series of $LRLR$ bends in the road the probabilities of actions not contributing successfully to the trial process can be steadily reduced and the others reinforced until a firm pattern of organization emerges.

In fact we can envisage in this case, just as in paragraph III(v), a whole hierarchy of such organizers, whose activity has here developed from scratch a statistical pattern to match the stable structure discovered in the environmental sequence of events. Although the term 'hierarchy' suggests an ordered system of separate levels, the most flexible system would allow of activity from different levels to be mixed in each. We might even envisage an abstractive 'level' which took as its data the activity of the whole system of levels; and it may be profitable to investigate the results of allowing the activity of the 'top level' of the hierarchy to become the data for the 'bottom level', the mismatch signal being distributed initially more or less evenly throughout all levels. But the notion of a simple hierarchy is sufficient for our present purposes.

(vi) *The stabilizing principle.* Thus far we have spoken as if thresholds (and their corresponding 'transition probabilities') could be altered only by the action of the mismatch signal; as if 'adaptive success' were the only condition on which the probability of a given sequence should be increased. But there are several ways in which the adaptive efficiency of the system (in a normal human environment) could be augmented, thereby incidentally increasing its resemblances to the human organism.

Let us introduce a second principle of organization, according to which the probability of a given transition is augmented automatically (though not indefinitely) according to the frequency of its occurrence, in addition to any modification resulting from its success or failure. What effect will this have on the growth of organization? First, and obviously, it will mean that every time a routine of activity comes 'back to where it started', it will receive some reinforcement in addition to whatever evaluation, positive or negative, is deduced from the mismatch signal. Closed cycles of activity will tend to be statistically favoured. As time goes on the adaptive organizing responses most likely to be tried will be those closed cycles which appear to be statistically nearest to what is required, as deduced from the mismatch signal. The system, in other words, will have a tendency to construct its imitative organizing sequences out of the nearest approximate combination of closed cycles of activity; the system will tend to perceive its environment as a compresence of associated 'wholes'.

The second effect will be to increase the speed with which the organization will grow in a relatively stable environment where the same sub-routine may be repeatedly called for. Closed sub-routines once found successful are less liable to disruption in subsequent trial processes, and more likely to be retained, excited or suppressed as wholes, ready more rapidly for use when the corresponding stable feature recurs in the flux of environmental events.

A third result of applying this principle is that the shorter closed cycles of organizing activity will tend to be even more favoured than they would be in any case. The effect is in fact analogous to 'negative resistance' or 'positive feedback', in the jargon of control theory; but it is feedback exerted on the probability pattern rather than on the activity itself. It tends to favour the formation of metastable 'eigenstates' of organizing activity and readiness for activity. The bias of our system will be towards conceptual simplicity. The mark of Occam's razor will appear naturally on the pattern of its activity.

(vii) *The 'anti-monopoly' principle.* The third principle of organization of the threshold control mechanism that should be mentioned here might seem superficially antagonistic to our earlier requirements. It is a provision that the intensity of a given pattern of activity will tend to die away with time. This we might describe as an 'anti-monopoly' proviso. Its function is to preserve the exploratory character of adaptive activity by ensuring that the system does not settle down for too long in any one adaptive mode. It thus increases the number of features of the environment likely within a given time to find themselves represented in the activity pattern.

This feature, analogous in some ways to the familiar 'adaptation' of sensory organs, is actually in no way antagonistic to our second requirement that the *transition* probabilities for closed cycles should automatically rise with time. To take a simple example, both requirements would be met in a system based on the principle of Figure 4 if the high-tension supply battery were subject to polarization, so that its e.m.f. dropped steadily as long as current was drawn from it, and recovered only after a period of rest. In Figure 4, the *relative* probability of the successful thyratron's firing increases with each successful action. It is only the probability of its *action in a given time interval* that decreases with each action.

In general, then, we must distinguish clearly between the control of relative probabilities of alternative courses of action, and the control of the probability of a given action in given circumstances within a given time interval. The first determines the probability that, if an action occurs, it shall be of a given form. The second determines the probability that an action of this form shall occur (in a given time interval). Since the thresholds governing these probabilities are continually changing the probabilities cannot of course be interpreted as frequencies in an observable time series of actions, but (as in analogous situations in physics) only as the relative proportions, or frequencies, in an imaginary 'ensemble' or population of similar situations.

V. The Control of Dimensionality

(i) *The speed of development.* By intelligent use of the foregoing principles, I think that a self-organizing system can be envisaged which could develop, *ab initio*, adaptive powers comparable with those of a human being, if subjected to a similar course of typical experience. But we have skirted one important practical question. Even with all our aids to efficiency, is there any hope of this development taking place in a *time* comparable with that required by an average human infant? Will not the number of possibilities to be eliminated, and the number of wasted trials, be astronomically large if the field of activity is to be as complex as the human one?

(ii) *Statistical control in the effector system.* I have only a tentative answer to this question, but it is one that seems promising. We have argued hitherto as if the degrees of freedom of the effector system E were large in number and fixed, so that each step in the process of development amounted to the elimination of all but one of a vast number of alternatives.

Let us picture now, however, an effector system constructed on much the same statistical lines as the control system we have discussed; and suppose that initially the elements are so tightly coupled together (by statistical links with low thresholds) that the effective number of independent degrees of freedom is very small. The responses possible will be relatively few and crude, but the initial trial and error processes of goal pursuit can be correspondingly rapid. Few effector actions will be highly successful, but the best in given circumstances will be fairly quickly discovered. The broad outlines of the statistical structure of the environmental events will begin fairly quickly to be reflected in the pattern of internal organizing activity.

Suppose now that as this happens, the degrees of freedom of the effector system are gradually and correspondingly increased in number, by a reduction in the statistical coupling between elements under the guidance of the mismatch signal. To take a simple example, the system might start with the controls of all of five 'fingers'

interconnected so as to function as a single unit. As the organizing process went on, it could be arranged that the general level of the thresholds in the interconnecting links should rise, so that the 'fingers' would begin to gain independence; but any links found in practice to promote successful responses could be selectively favoured, and could retain a greater-than-average probability of functioning.

(ii) *Comparison with the human system.* In a suitable environment offering a sufficiently typical sample of the regularities normally to be encountered, it seems not unreasonable (as I have argued elsewhere) that the organizing system should now be able to grow in complexity in the ways we have discussed, step by step with the effector system, at a rate comparable with that of the learning process in human infancy. At least the system may perhaps bear realistic comparison with the human one, if only in order that the differences as they emerge may lead to its refinement.

But at present we are concerned not with the comparison so much as with the fact that this kind of activity is possible in a self-organizing information-flow system, and with its significance as one solution to the problem of concept formation.

For it is indeed in the same principles of organization which we have hitherto treated as a means of improving goal pursuit, that we shall find a possible solution to the problems of inventive activity—particularly concept formation—which are our chief concern.

VI. The Representation of the Field of Activity

(i) *The symbolization of information.* In a digital computor the symbols for concepts are patterns of 1's and 0's or their equivalents such as electrical dots and dashes. In an analogue computor such as a guided missile might use, abstractions such as 'distance of target' are represented by the magnitudes of voltages or currents. What corresponds to these symbols in our present system? How is information about its field of action represented internally?

(ii) *Pre-conceptual information.* At the first stage, the stage of *reception*, it is represented by what we have called the mismatch signal, which may include, for instance, electrical stimuli from visual receptors. Much of this information may be coupled directly to conditional-probability controls associated with actions which in a human being we should describe as reflexes. Incoming information then leads to a conditioned or unconditioned response without any necessary intermediate stage of conceptualization.

In human beings such activities range from visceral responses of which we may be unaware, through such reflexes as swallowing which we can sometimes control, to learned patterns such as swimming or cycling which may have formed under conscious guidance. All are characterized by the fact that we 'don't think about them'. We say we 'know how to' cycle, etc., using 'know' in a different sense from that which we use for conceptualized 'facts'.

(iii) *Symbolization in terms of organizing activity*. But in our information-flow model there is a second sort of internal activity which logically 'represents' features of the world of activity. This is the activity of the internal organizing system. Its imitative sub-routines, when evoked in order to match successfully some change in the world of activity, form an active symbol for that change. What is more, since most sub-routines will tend to form out of standard sequences, this symbol will amount logically to a description in terms of the corresponding standard categories. In normal circumstances, then, there will be in the organizing activity a running representation of those features of the field of action to which the system as a whole is currently matching its own activity, internal and external.

In its 'visual' system, for example, we may picture the internal activity as organized to match (in a sense, to cancel out actively) the incoming visual signals. It will do this as a result of combinations of sub-routines of organizing activity which will have developed through adaptive success in the past, and will amount therefore to a representation of the visual field in terms of previously exemplified categories.

(iv) *The correlate of perception*. Once again the details cannot detain us here. But the essential point is that the symbols constituted by the internally evoked matching response (the organizing activity) seem to be related to the incoming recepta from the sensory organs of the information-flow model in much the same way as what we call *percepts* are related to the recepta from human sensory organs.

Among other points worthy of note, we see that (a) the description offered by this symbolic representation is selective: certain features are 'noticed' and others ignored; (b) the description possible at a given level is not unique, since several combinations may 'match' equally well; there

will be illusions and 'ambiguous figures' for this system just as for human observers (e.g. the Ames demonstrations); (c) complementary descriptions may be constituted by organizing activity at different levels of abstraction (e.g. an illuminated advertising sign may be perceived either as a number of electric lamps or as the sequence of words delineated by them); and (d) the state of readiness both for further effector action and for further internal organizing activity depends on the pattern of these symbols, in the same logical way as our human state of readiness depends on what we perceive in our world of activity.

These are only a few of many points in favour of attributing to the organizing sub-routines the role of symbols for percepts.

(v) *Imagination and thinking*. But we must go further. What we know is not limited to what we currently perceive. We know 'what is the case'— or believe we do—even after we have shut our eyes, or gone away for a year from a familiar landscape. Operationally we mean by this, among other things, (a) that we can 'evoke mental images' of the landscape, or of the relationship (concrete or abstract) which we claim to know; (b) that in all appropriate circumstances we are organized to react to the pattern of events which should come about if 'what we believe' *is* the case. For example, we move so as to avoid the chair which we believe to stand in the middle of a dark room. Or we (consciously or unconsciously) select the statements we make to a friend so as to avoid logical contradiction with a proposition we believe.

Now the model shows certain features of its activity that seem closely parallel to these. In particular, it is possible and would indeed be normal in the absence of excessive external stimulation, for some of the internal organizing mechanisms to 'run free' without evoking the corresponding effector activity. All that is necessary is a flexible inhibitory system under the general control of the organizers, so that they can suppress, in whole or in part, the effector action normally induced by the free running activity.

When this happens, long trains of organizing activity will normally be excited through the statistical couplings formed as a result of past 'experience', and internal goal-guided activity can take place, in which the goal pursued is *abstractly defined in terms of the degrees of freedom of the organizing system*, in the same way as the goal X (Figure 1) was defined in terms of those of the effector system E. Along such lines it seems possible to find correlates of many of the human activities we group under the names of *thinking* and *imagining*.

Hypotheses, for example, are operationally symbolized by sub-routines which organize readinesses for the corresponding features that should be shown by the world of activity if the hypotheses were true. A familiar scene can be 'imagined' by re-evoking the pattern of internal organizing activity by which it was 'matched' originally. An unfamiliar scene, known only from a description, can be 'imagined' by allowing the internal activity symbolizing its description to evoke what it will by way of activity statistically associated with past instances of the descriptive terms. If the model has come to use organizing routines mainly from its visual system as symbols for concepts, such evoked activity may amount to a symbolization of a complete visual image. If, however, some of the other types of organizing routines form its main conceptual vocabulary, the visual component of its 'imagined' situation may be small. All such 'imaginative' activity is distinguished from 'hallucination' by the fact that the mismatch signal will be present in its normal channels, indicating that what is symbolized is not a representation of the current state of the environment.

(vi) *Functional disorder.* This raises the whole question of functional disorder in the automaton, and its behavioural resemblances to human mental disorder. But the detailed consideration of this topic would lead too far and will be undertaken in further papers. (Some of the optional differences between normal imagination, illusion (e.g. optical), and hallucination, are discussed in an earlier paper.) Suffice it to say that the behaviour accompanying differential threshold changes, and diversion or interruption of the mismatch signal (in various ways), adds up to a picture of which no major feature is without its known correlate in human psychopathology, and in which many of the major symptoms of mental disorder find natural analogues.

VII. Conclusion

And so we could go on, but we must bring this lengthy discussion to an end; for the outlines of the correlation possible between the model's internal activity and human concept-handling processes should now be clear. And perhaps enough has been said to indicate that such a self-organizing system shows some promise of usefulness as a kind of foil for psychological theories of human personality. Of each point of debate we may ask 'What difference would it make in the information-flow model?' or 'How does it sound when translated into terms of the information-flow pattern?' Needless to say, the information-flow model is likely to find itself refined as often as the psychological theories in consequence of such discipline; but it seems at least possible that the process may be a convergent one that could materially further psychological understanding.

Notes

1. The Irishman's clock, we remember (which had stopped), succeeded in telling the time correctly twice a day.

46.
Plans and the Structure of Behaviour

GEORGE A. MILLER, EUGENE GALANTER, AND KARL H. PRIBRAM

The Unit of Analysis

Most psychologists take it for granted that a scientific account of the behavior of organisms must begin with the definition of fixed, recognizable, elementary units of behavior—something a psychologist can use as a biologist uses cells, or an astronomer uses stars, or a physicist uses atoms, and so on. Given a simple unit, complicated phenomena are then describable as lawful compounds. That is the essence of the highly successful strategy called "scientific analysis."

The elementary unit that modern, experimental psychologists generally select for their analysis of behavior is the *reflex*. "The isolation of a reflex," B. F. Skinner tells us, "is the demonstration of a predictable uniformity in behavior. In some form or other it is an inevitable part of any science of behavior. . . . A reflex is not, of course, a theory. It is a fact. It is an analytical unit, which makes the investigation of behavior possible." Skinner is quite careful to define a reflex as a unit of behavior that will yield orderly data: "The appearance of smooth curves in dynamic processes marks a unique point in the progressive restriction of a preparation, and it is to this uniquely determined entity that the term reflex may be assigned." This somewhat odd approach to the reflex—in terms of the smoothness of curves—results from Skinner's consistent attempt to define a unit of behavior in terms of behavior itself instead of by reference to concepts drawn from some other branch of science.

From George A. Miller, Eugene Galanter, and Karl H. Pribram, *Plans and the Structure of Behavior* (New York: Holt, Rinehart & Winston, 1960), Chapters 2 and 4. Reprinted by permission of the authors and publisher.

Although Skinner's approach absolves the psychologist of certain burdensome responsibilities toward his biological colleagues, the fact remains that the reflex is a concept borrowed originally from physiology and made to seem psychologically substantial largely by the myth of the *reflex arc*: stimulus → receptor → afferent nerve → connective fibers → efferent nerve → effector → response. For many years all those elementary textbooks of psychology that mentioned the nervous system featured the traditional, simplified diagram of the reflex arc in a very prominent position. You may ignore a behaviorist when he tells you that the reflex is a fact, but you can scarcely ignore a physiologist when he draws you a picture of it. You might as well deny the small intestines or sneer at the medulla oblongata as to doubt the reflex arc. Even the most obstinate opponent of physiological explanations in psychology can scarcely forget the bloody tissue from which the reflex—even the reflex-sans-arc—originally grew.

But let us suppose, by a wild and irresponsible flight of fancy, that the physiologists and neurologists suddenly announced that they had been mistaken, that there was no such fact as a reflex arc and that the data on which the theory had been based were actually quite different from what had originally been supposed. What then would psychologists say? Would they persist in talking about reflexes? Has the reflex concept been so tremendously helpful that behaviorists could not afford to give it up, even if its biological basis were demolished?

There is some reason to think that the reflex unit has been vastly overrated and that a good many psychologists would like to get out from under it if they could. The reflex arc may have been helpful in getting psychology started along scientific paths, but the suspicion has been growing in recent years that the reflex idea is too simple, the element too elementary. For the most part, serious students of behavior have had to ignore the problem of units entirely. Or they have had to modify their units so drastically for each new set of data that to speak of them as elementary would be the most unblushing sophistry. After watching psychologists struggle under their burden of conditioning reflexes, Chomsky, the linguist and logician, recently summarized their plight in the following terms:

> The notions of "stimulus," "response," "reinforcement" are relatively well defined with respect to the bar-pressing experiments and others similarly restricted. Before we can extend them to real-life behavior, however, certain difficulties must be faced. We must decide, first of all, whether any physical event to which the organism is capable of reacting is to be called a stimulus on a given occasion, or only one to which the organism in fact reacts; and correspondingly, we must decide whether any part of behavior is to be called a response, or only one connected with stimuli in lawful ways. Questions of this sort pose something of a dilemma for the experimental psychologist. If he accepts the broad definitions, characterizing any physical event impinging on the organism as a stimulus and any part of the organism's behavior as a response, he must conclude that behavior has not been demonstrated to be lawful. In the present state of our knowledge, we must attribute an overwhelming influence on actual behavior to ill-defined factors of attention, set, volition, and caprice. If we accept the narrower definitions, then behavior is lawful by definition (if it consists of responses); but this fact is of limited significance, since most of what the animal does will simply not be considered behavior.

Faced with the choice of being either vague or irrelevant, many psychologists have been restive and ill at ease with their borrowed terms. What went wrong? How was the reflex arc conceived originally, and for what purpose? Can we supplant the reflex arc with some theory of the reflex that is more suited to our current knowledge and interests?

Sir Charles Sherrington and Ivan Petrovitch Pavlov are the two men who are probably most responsible for confirming the psychologist's Image of man as a bundle of *S–R* reflexes. Yet one may be permitted to speculate that neither of them would approve of the way their concepts have been extended by psychologists. In his *Integrative Action of the Nervous System* (1906) Sherrington is particularly explicit in his qualifications and warnings about the reflex. Again and again he states that "the simple reflex is a useful fiction"—useful for the study of the spinal preparation. He expressed considerable doubt that a stretch reflex, of which the knee jerk is the most frequently quoted example, represented his notion of a simple reflex and questioned whether it should be considered a reflex at all. The synapse was invented by Sherrington in order to explain the differences between the observed properties of nerve trunks and the properties that had to be inferred to describe the neural tissue that intervenes between receptor stimulation and effector response. Nerve trunks will transmit signals in either direction. Characteristically, the signals are of an all-or-none type. Reflex action, on the other hand, is unidirectional and the response is characteristically graded according to the intensity of the stimulus. How can these be reconciled? Sherrington resolved the differences by supporting the neuron doctrine: the nervous system is made up of discrete neural units that have the properties of nerve trunks; intercalated between these units are discontinuities which he christened "synapses," and these have the properties unique to reflexes.

In recent years, graded responses have been shown to be a prepotent characteristic not only of synapses but also of all excitable tissue, for example, of the finer arborizations of the nerve cells. The cerebral cortex, man's claim to phylogenetic eminence, "still operates largely by means of connections characteristic of primitive neuropil [which is] the most appropriate mechanism for the maintenance of a continuous or steady state, as contrasted to the transmission of information about such states" [George Bishop].

Moreover, additional data have come to light. Today we know that neural and receptor tissues are spontaneously active irrespective of environmental excitation. This spontaneous activity is, of course, altered by environmental events—but the change in spontaneous activity may outlast the direct excitation by hours and even days. Furthermore, we know now that the activity of receptors is controlled by efferents leading to them from the central nervous system. As an example, consider the events that control muscular contraction. (Similar, though not identical, mechanisms have also been described for the various sensory systems.) One third of the "motor" nerve fibers that go to muscle actually end in spindles that are the stretch-sensitive receptors. Electrical stimulation of these nerve fibers does not result in contraction of muscle; but the number of signals per unit time that are recorded from the "sensory" nerves coming from the spindles is altered drastically. It is assumed, therefore, that the central nervous mechanism must compare the incoming pattern of signals with the centrally originating "spindle control" signal pattern in order to determine what contribution the muscular contraction has made to the "spindle sensing" pattern. The outcome of this comparison, or *test*, constitutes the stimulus (the psychophysicist's *proximal* stimulus) to which the organism is sensitive. The test represents the conditions which have to be met before the response will occur. The test may occur in the receptor itself (e.g., in the retina) or in a more centrally located neuronal aggregate (as is probably the case for muscle stretch).

It is clear from examples such as this that the neural mechanism involved in reflex action cannot be diagrammed as a simple reflex arc or even as a chain of stimulus–response connections. A much more complex kind of monitoring, or testing, is involved in reflex action than the classical reflex arc makes any provision for. The only conditions imposed upon the stimulus by the classical chain of elements are the criteria implicit in the thresholds of each element; if the distal stimulus is strong enough to surmount the thresholds all along the arc, then the response must occur. In a sense, the threshold is a kind of test, too, a condition that must be met, but it is a test of strength only. And it must have encouraged psychologists to believe that the only meaningful measurement of a reflex was its strength (probability, magnitude, or latency).

The threshold, however, is only one of many different ways that the input can be tested. Moreover, the response of the effector depends upon the outcome of the test and is most conveniently conceived as an effort to modify the outcome of the test. The action is initiated by an "incongruity" between the state of the organism and the state that is being tested for, and the action persists until the incongruity (i.e., the proximal stimulus) is removed. The general pattern of reflex action, therefore, is to test the input energies against some criteria established in the organism, to respond if the result of the test is to show an incongruity, and to continue to respond until the incongruity vanishes, at which time the reflex is terminated. Thus, there is "feedback" from the result of the action to the testing phase, and we are confronted by a recursive loop. The simplest kind of diagram to represent this conception of reflex action—an alternative to the classical reflex arc—would have to look something like Figure 1.

The interpretation toward which the argument moves is one that has been called the "cybernetic

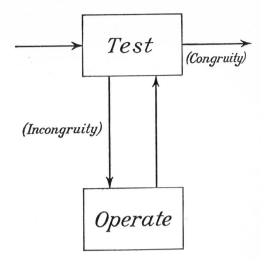

Figure 1.—The TOTE unit.

hypothesis," namely, that the fundamental building block of the nervous system is the feedback loop. The development of a mathematical theory for servomechanisms, wedded to the physiological accounts of homeostatic mechanisms, has stimulated widespread discussion and speculation about devices closely akin to Figure 1. The argument, therefore, moves toward popular ground.

But what good is this alternative interpretation of the reflex? The psychologist was interested in reflexes because he thought they might provide the units he needed to describe behavior. But simple reflexes have been inadequate. And if reflexes based on afferent–efferent arcs would not turn the trick, why should we hope for better things from reflexes based on feedback loops? It is the reflex itself—not merely the reflex arc—that has failed, and repairing the neurological theory underlying it is not likely to save the day. What do we hope to gain from such a reinterpretation?

Obviously, the reflex is not the unit we should use as the element of behavior: the unit should be the feedback loop itself. If we think of the Test–Operate–Test–Exit unit—for convenience, we shall call it a TOTE unit—as we do of the reflex arc, in purely anatomical terms, it may describe reflexes, but little else. That is to say, the reflex should be recognized as only one of many possible actualizations of a TOTE pattern. The next task is to generalize the TOTE unit so that it will be useful in a majority—hopefully, in all—of the behavioral descriptions we shall need to make.

Consider what the arrows in Figure 1 might represent. What could flow along them from one box to another? We shall discuss three alternatives: energy, information, and control. If we think of *energy*—neural impulses, for example—flowing from one place to another over the arrows, then the arrows must correspond to recognizable physical structures—neurons, in the example chosen. As a diagram of energy flow over discrete pathways, therefore, the TOTE unit described in Figure 1 might represent a simple reflex. Or it might represent a servomechanism.

There is, however, a second level of abstraction that psychologists usually prefer. We can think of *information* as flowing from one place to another over the arrows. According to the method of measuring information that has been developed by Norbert Wiener and by Claude Shannon, information is transmitted over a channel to the extent that the output of the channel is correlated with the input. We could therefore think of this second level of abstraction as the transmission of correlation over the arrows. In that case, we are concerned not with the particular structures or kinds of energy that are involved in producing the correlation but only with the fact that events at the two ends of the arrow are correlated. The situation is quite familiar to psychologists, for it is exactly what they mean when they draw an arrow leading from Stimulus to Response in their *S–R* diagrams or when they define a reflex as a correlation between *S* and *R* but refuse to talk about the neurological basis for that correlation.

A third level of abstraction, however, is extremely important for the ideas we shall discuss in the pages that follow. It is the notion that what flows over the arrows in Figure 1 is an intangible something called *control*. Or perhaps we should say that the arrow indicates only succession. This concept appears most frequently in the discussion of computing machines, where the control of the machine's operations passes from one instruction to another, successively, as the machine proceeds to execute the list of instructions that comprise the program it has been given. But the idea is certainly not limited to computers. As a simple example drawn from more familiar activities, imagine that you wanted to look up a particular topic in a certain book in order to see what the author had to say about it. You would open the book to the index and find the topic. Following the entry is a string of numbers. As you look up each page reference in turn, your behavior can be described as under the control of that list of numbers, and control is transferred from one number to the next as you proceed through the list. The transfer of control could be symbolized by drawing arrows

from one page number to the next, but the arrows would have a meaning quite different from the two meanings mentioned previously. Here we are not concerned with a flow of energy or transmission of information from one page number to the next but merely with the order in which the "instructions" are executed.

At this abstract level of description we are no longer required to think of the test as a simple threshold that some stimulus energy must exceed. The test phase can be regarded as any process for determining that the operational phase is appropriate. For example, to be clear though crude, we do not try to take the square root of "ratiocinate." We may know full well how to extract square roots, but before we can execute that operation we must have digits to work on. The operation of extracting square roots is simply irrelevant when we are dealing with words. In order to ensure that an operation is relevant, a test must be built into it. Unless the test gives the appropriate outcome, control cannot be transferred to the operational phase.

When Figure 1 is used in the discussion of a simple reflex it represents all three levels of description simultaneously. When it is used to describe more complex activities, however, we may want to consider only the transfer of information and control or in many instances only the transfer of control. In all cases, however, the existence of a TOTE should indicate that an organizing, coordinating unit has been established, that a Plan is available.

In the following pages we shall use the TOTE as a general description of the control processes involved; the implications it may have for functional anatomy will remain more or less dormant. . . . In its weakest form, the TOTE asserts simply that the operations an organism performs are constantly guided by the outcomes of various tests.

The present authors feel that the TOTE unit, which incorporates the important notion of feedback, is an explanation of behavior in general, and of reflex action in particular, fundamentally different from the explanation provided by the reflex arc. Consequently, the traditional concepts of stimulus and response must be redefined and reinterpreted to suit their new context. Stimulus and response must be seen as phases of the organized, coordinated act. We might summarize it this way:

The stimulus is that phase of the forming coordination which represents the conditions which have to be met in bringing it to a successful issue; the response is that phase of one and the same forming coordination which gives the key to meeting these conditions, which serves as instrument in effecting the successful coordination. They are therefore strictly correlative and contemporaneous.[1]

Because stimulus and response are correlative and contemporaneous, the stimulus processes must be thought of not as preceding the response but rather as guiding it to a successful elimination of the incongruity. That is to say, stimulus and response must be considered as aspects of a feedback loop.

The need for some kind of feedback channel in the description of behavior is well recognized by most reflex theorists, but they have introduced it in a peculiar way. For example, it is customary for them to speak of certain consequences of a reflex action as strengthening, or reinforcing, the reflex—such reinforcing consequences of action are a clear example of feedback. Reinforcements are, however, a special kind of feedback that should not be identified with the feedback involved in a TOTE unit. That is to say: (1) a reinforcing feedback must strengthen something, whereas feedback in a TOTE is for the purpose of comparison and testing; (2) a reinforcing feedback is considered to be a stimulus (e.g., pellet of food), whereas feedback in a TOTE may be a stimulus, or information (e.g., knowledge of results), or control (e.g., instructions); and (3) a reinforcing feedback is frequently considered to be valuable, or "drive reducing," to the organism, whereas feedback in a TOTE has no such value.

When a TOTE has been executed—the operations performed, the test satisfied, and the exit made—the creature may indeed appear to have attained a more desirable state. It may even be true, on the average, that the TOTE units that are completed successfully in a given situation tend to recur with increased probability, although such a relation would not be necessary. Thus it is possible to discuss a TOTE in the language of reinforcements. Nevertheless, the TOTE involves a much more general conception of feedback. The concept of reinforcement represents an important step forward from reflex arcs toward feedback loops, but bolder strides are needed if behavior theory is to advance beyond the description of simple conditioning experiments.

Perhaps variations in the basic TOTE pattern will prove necessary, so for the purposes of the present discussion we shall continue to regard the diagram in Figure 1 as a hypothesis rather than a fact. The importance of this hypothesis to the general thesis of the book, however, should not be overlooked. It is, in capsule, the account we wish to give of the relation between Image and action. The TOTE represents the basic pattern in which our Plans are cast, the test phase of the TOTE involves the specification of whatever knowledge is necessary for the comparison that is to be made, and the operational phase represents what the organism does about it—and what the organism does may often involve overt, observable actions. Figure 1, therefore, rephrases the problem: How does a Plan relate the organism's Image of itself and its universe to the actions, the responses, the behavior that the organism is seen to generate?

Let us see what we must do in order to expand this proposal into something useful. One of the first difficulties—a small one—is to say more exactly what we mean by the "incongruity" that the test phase is looking for. Why not talk simply about the difference, rather than the incongruity, as providing the proximal stimulus? The answer is not profound: We do not want to bother to distinguish between TOTEs in which the operations are performed only when a difference is detected (and where the operations serve to diminish the difference) and TOTEs in which the operations are released only when no difference is detected. When the diagram is used to describe servomechanisms, for example, it is quite important to distinguish "positive" from "negative" feedback, but, because we are going to be interested primarily in the feedback of control, such questions are not critical. Rather than treat all these varieties as different units of analysis, it seems simpler to treat them all as examples of a more general type of "incongruity-sensitive" mechanism.

A second difficulty—this one rather more important—is the question of how we can integrate this TOTE unit into the sort of hierarchical structure of behavior that we [insist] on. How can the two concepts—feedback and hierarchy—be reconciled? One method of combining feedback components in a hierarchy has been described by D. M. MacKay, who proposed to make the consequences of the operational phase in one component provide the input to the comparator of a second component; MacKay's suggestion leads to a string of such feedback components, each representing a progressively higher degree of abstraction from the external reality. Although MacKay's scheme is quite ingenious, we are persuaded that a somewhat different method of constructing the hierarchy will better serve a psychologist's descriptive purposes. A central notion of the method followed in these pages is that the operational components of TOTE units may themselves be TOTE units. That is to say, the TOTE pattern describes both strategic and tactical units of behavior. Thus the operational phase of a higher-order TOTE might itself consist of a string of other TOTE units, and each of these, in turn, may contain still other strings of TOTEs, and so on. Since this method of retaining the same pattern of description for the higher, more strategic units as for the lower, more tactical units may be confusing on first acquaintance, we shall consider an example.

R. S. Woodworth has pointed out how frequently behavioral activities are organized in two stages. Woodworth refers to them as "two-phase motor units." The first phase is preparatory or mobilizing; the second, effective or consummatory. To jump, you first flex the hips and knees, then extend them forcefully; the crouch prepares for the jump. To grasp an object, the first phase is to open your hand, the second is to close it around the object. You must open your mouth before you can bite. You must draw back your arm before you can strike, etc. The two phases are quite different movements, yet they are obviously executed as a single unit of action. If stimulation is correct for releasing the action, first the preparatory TOTE unit is executed, and when it has been completed the stimulation is adequate for the consummatory TOTE unit and the action is executed. Many of these two-phase plans are repetitive: the completion of the second phase in turn provides stimuli indicating that the execution of the first phase is again possible, so an alternation between the two phases is set up, as in walking, running, chewing, drinking, sweeping, knitting, etc.

We should note well the construction of a "two-phase" TOTE unit out of two simpler TOTE units. Consider hammering a nail as an example. As a Plan, of course, hammering has two phases, lifting the hammer and then striking the nail. We could represent it by a tree, or hierarchy, as in Figure 2. If we ask about details, however, the representation of hammering in Figure 2 as a simple list containing two items is

certainly too sketchy. It does not tell us, for one thing, how long to go on hammering. What is the "stop rule"? For this, we must indicate the test phase, as in Figure 3. The diagram in Figure 3 should indicate that when control is transferred to the TOTE unit that we are calling "hammering," the hammering continues until the head of the nail is flush with the surface of the work. When the test indicates that the nail is driven in, control is transferred elsewhere. Now, however, we seem to have lost the hierarchical structure. The hierarchy is recovered when we look at the box labeled "hammer," for there we find two TOTE units, each with its own test, as indicated in Figure 4. When the pair of TOTE units combined in Figure 4 are put inside the operational phase in Figure 3, the result is the hierarchical Plan for hammering nails that is shown in Figure 5.

If this description of hammering is correct, we should expect the sequence of events to run off in this order: Test nail. (Head sticks up.) Test hammer. (Hammer is down.) Lift hammer. Test hammer. (Hammer is up.) Test hammer. (Hammer is up.) Strike nail. Test hammer. (Hammer is

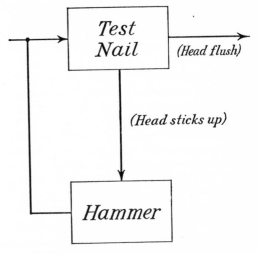

Figure 3.—Hammering as a TOTE unit.

down.) Test nail. (Head sticks up.) Test hammer. And so on, until the test of the nail reveals that its head is flush with the surface of the work, at which point control can be transferred elsewhere. Thus the compound of TOTE units unravels itself simply enough into a coordinated sequence of tests and actions, although the underlying structure that organizes and coordinates the behavior is itself hierarchical, not sequential.

It may seem slightly absurd to analyze the motions involved in hammering a nail in this

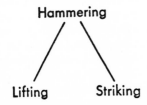

Figure 2.—Hammering as a hierarchy.

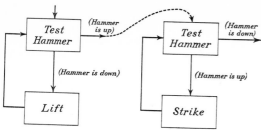

Figure 4.—Dashed line indicates how two simple TOTE units are connected to form the operational phase of the more complicated TOTE unit in Figure 3.

explicit way, but it is better to amuse a reader than to confuse him. It is merely an illustration of how several simple TOTE units, each with its own test–operate–test loop, can be embedded in the operational phase of a larger unit with its particular test–operate–test loop. Without such an explicit illustration it might not have been immediately obvious how these circles within circles could yield hierarchical trees.

More complicated Plans—Woodworth refers to them as "polyphase motor units"—can be similarly described as TOTE units built up of subplans that are themselves TOTE units. A bird will take off, make a few wing strokes, glide, brake with its wings, thrust its feet forward, and land on the limb. The whole action is initiated as

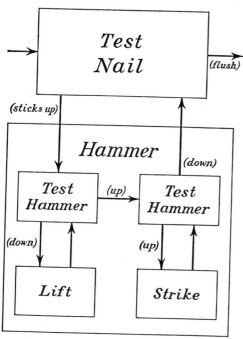

Figure 5.—The hierarchical Plan for hammering nails.

a unit, is controlled by a single Plan, yet is composed of several phases, each involving its own Plan, which may in turn be comprised of subplans, etc.

Note that it is the *operational* phase of the TOTE that is expanded into a list of other TOTE units. If we wish to preserve the TOTE pattern as it is diagrammed in Figure 1, we cannot use it to build up more complicated tests.[2] The tests that are available, therefore, are conceived to be relatively fixed; it is the operational phase that may be quite various and complex. The operational phase may, of course, consist of a list of TOTEs, or it may terminate in efferent activity.[3] If we consider complex Plans—TOTE hierarchies in which the operation of one TOTE is itself a list of TOTE units—then some general properties of such systems become apparent:

—The hierarchical structure underlying behavior is taken into account in a way that can be simply described with the computer language developed by Newell, Shaw, and Simon for processing lists.

—Planning can be thought of as constructing a list of tests to perform. When we have a clear Image of a desired outcome, we can use it to provide the conditions for which we must test, and those tests, when arranged in sequence, provide a crude strategy for a possible Plan. (Perhaps it would be more helpful to say that the conditions for which we must test *are* an Image of the desired outcome.)

—The operational phase can contain both tests and operations. Therefore the execution of a Plan of any complexity must involve many more tests than actions. This design feature would account for the general degradation of information that occurs whenever a human being is used as a communication channel.

In lower animals it appears that the pattern of their behavior is normally constructed more or less fortuitously by the environment—only man cherishes the illusion of being master of his fate. That is to say, the environment provides stimuli that "release" the next stage of the animal's activity. It is something of a philosophical question as to whether we wish to believe in plans that exist somewhere outside of nervous systems, so perhaps we should say merely that lower animals appear to have more tactics than strategy.

As we ascend the evolutionary scale we find in mammals an increasing complexity in the kind of tests the animals can perform. In man we have a unique capacity for creating and manipulating symbols, and when that versatility is used to assign names to TOTE units, it becomes possible for him to use language in order to rearrange the symbols and to form new Plans. We have every reason to believe that man's verbal abilities are very intimately related to his planning abilities. And, because human Plans are so often verbal, they can be communicated, a fact of crucial importance in the evolution of our social adjustments to one another.

.   .   .   .   .

Values, Intentions, and the Execution of Plans

All acts have in common the character of being intended or willed. But one act is distinguishable from another by the content of it, the expected result of it, which is here spoken of as its intent. There is no obvious way in which we can say what act it is which is thought of or is done except by specifying this intent of it.

In this passage from his Carus Lectures, C. I. Lewis reminds us that the acts people perform cannot be characterized simply by specifying the time–order of their parts—in the way we might describe the motions of a billiard ball or a falling stone. The term "intent" is Lewis's way of trying to catch this elusive and unique feature of the behavior of living systems. In speaking in these terms he is like any ordinary person who tries to say what makes his actions meaningful—but he is quite unlike most experimental psychologists.

Intention went out of style as a psychological concept when reflex theory and its derivatives became the foundation for our scientific theories of behavior. Only rarely in the past twenty years has the concept been used outside the clinic as a technical term in a psychological explanation. And most of the explicit uses that have occurred can probably be traced to the influence of Kurt Lewin.

Lewin used the concept of intention in order to combat an overly simple theory that actions are always strengthened whenever they are successful. He uses the example of someone who intends to mail a letter. The first mailbox he passes reminds him of the action. He drops the letter in. Thereafter, the mailboxes he passes leave him altogether cold. He does not even notice them. Now, according to Lewin, the intention to mail the letter created a positive *valence* on the mailbox, which attracted the person's attention to it. When the occasion arose and the intentional act was consummated, the valence vanished and further mailboxes were ignored. According to classical theory, dropping the letter into the mailbox should

have the effect of strengthening the association between mailboxes and the response of reaching into the pocket for a letter to be mailed. The poor fellow should have made abortive responses toward three or four mailboxes before the strength of the association had diminished. Instead of accumulating habit strength, however, the fact is he had no further interest in mailboxes. Therefore, reasoned Lewin, the simpler theory is inadequate and a more complex representation of a life space, complete with valences created by intentions, is required.

The present authors are in fundamental agreement with Lewin. Intention does pose an interesting and important problem for psychology. And we agree that the associationistic doctrine described above can never provide an adequate explanation. But Lewin goes on to draw an interesting analogy that we want to reject. An intention, he says, creates a quasi-need. Just as hunger gives food a positive valence, so does the intention to mail the letter give the mailbox a positive valence. Just as eating reduces the positive valence of food, so does mailing the letter reduce the quasi-need and remove the positive valence from mailboxes. For Lewin, there is a complete parallel between the dynamics of intentions and the dynamics of any other kind of motivated behavior. It is this dynamic property of an intention that we feel is confusing, and we wish to reformulate it.

There are simpler alternatives. What does it mean when an ordinary man has an ordinary intention? It means that he has begun the execution of a Plan and that this intended action is a part of it. "I intend to see Jones when I get there" means that I am already committed to the execution of a Plan for traveling and that a part of this Plan involves seeing Jones. "Do you intend to see Smith, too?" asks about other parts of the Plan. "Yes" would be a clear reply. "No" is a little ambiguous but probably means that it is part of my Plan to avoid Smith. "I don't know" means that that part of the Plan has not yet been developed in detail and that when it is developed it either may or may not include Smith. People are reasonably precise in their use of "intent" in ordinary conversation. And they do not use it to mean that something is temporarily valuable or that they have any particular needs, either real or quasi. The term is used to refer to *the uncompleted parts of a Plan whose execution has already begun.*[4]

Criminal lawyers are constantly troubled over the distinction between intention and motivation. For example, Jones hires Smith to kill someone. Smith commits the murder, but he is caught and confesses that he was hired to do it. Question: Is Smith guilty? If we consider only the motives involved, the employer is guilty because he was motivated to kill, but the gunman is not guilty because his motive was merely to earn money (which is certainly a commendable motive in a capitalistic society). But if we consider their intentions, then both parties are equally guilty, for both of them knowingly undertook to execute a Plan culminating in murder. The legal confusions arise when the lawyer begins to argue that Smith could not have intended to murder because he had no motive. Only a motive, he says will create an intention. Lewin and his associates would further confuse the issue by arguing that an intention will create a motive. The present authors take the position that a motive is comprised of two independent parts: value and intention. A value refers to an Image, whereas an intention refers to a Plan.

Presumably, a normal, adult human being has constant access to a tremendous variety of Plans that he might execute if he chose to do so. He has acquired a skill for swinging golf clubs, in the kitchen there is a book with a recipe for making a cake, he knows how to get to Chicago, etc. As long as he is not using them, these various available Plans form no part of his intentions. But as soon as the order to execute a particular Plan has been given, he begins to speak of its incompleted parts (insofar as he knows them) as things he intends to do.

Therein resides a crucial difference between a chain of actions and a Plan of action. When a chain is initiated with no internal representation of the complete course of action, the later parts of the chain are not intended. When a Plan is initiated, the intent to execute the later parts of it is clear.

But where then are values? An evaluation is a form of empirical knowledge and so helps to form the person's Image. But have values no special influence on our Plans? If not, why should any Plan ever be executed? To answer the second question first, Plans are executed because people are alive. This is not a facetious statement, for so long as people are behaving, *some* Plan or other must be executed. The question thus moves from *why* Plans are executed to a concern for *which* Plans are executed. And to cope with this problem of choice we do indeed need some valuational concepts.

Just as the operational phase of a Plan may lead to action, so the test phase of a Plan may draw extensively upon an Image. Thus, the evaluations a person has learned are available for use in the test phases of the Plan; we assume that every test phase in every TOTE unit can have some evaluation function associated with it. Ordinarily, the operational phase of a TOTE should increase the value of the situation as indicated by the test phase. But that is by no means a necessary condition for executing the TOTE. When a Plan is complex—made up of a hierarchy of subparts—it may be that some of the parts have negative values associated with them. That is to say, in order to achieve a positive result it may be necessary to do something that, by itself, has a negative evaluation. When this happens, a person who is executing the Plan can rightly be described as first intending, then carrying out, actions that he considers to be undesirable.

An intended operation that will someday provide the anticipated positive value for an extensive Plan of action may be delayed almost indefinitely while a person continues to execute preparatory subplans leading to outcomes with negative evaluations. Such actions obviously occur repeatedly in the lives of adult human beings—civilized society would scarcely be possible without them—yet they are unusually difficult to understand in terms of simple hedonism or simple reinforcement theory or any other psychological theory that makes no allowance for cognitive structure, for an Image and a Plan.

It seems reasonable to think of the test phases of the more strategic portions of a Plan as associated with overriding evaluations. Thus, a hierarchy of TOTE units may also represent a hierarchy of values. At the root of the hierarchy we can imagine that there is a kind of prototest that, when satisfied, carries a larger positive evaluation than could be counteracted by any accumulation of negative values, or costs, from the TOTE units in more tactical positions. Of course, when we choose a Plan and begin to execute it we may be unaware of some of the detailed tactics that will be needed to carry it through; necessarily, therefore, we would be unaware of all the values associated with those unforeseen tactics until the Plan was well along in its execution. If the negative values accumulate until they outweigh any conceivable positive value associated with the prototest, we may discontinue the execution of the Plan. When, for example, we walk out of the dentist's office before we have been in the dental chair, some change in values must be inferred. (Note, however, that the *intention* may be invariant under these value transformations. The intention to see a dentist vanishes only when a new Plan is executed

—it does not gradually lose strength as the desire to see the dentist declines.)

When we say, "I need a mailbox in order to mail this letter," the word "need" should not be taken as expressing a dynamic or evaluative property associated, either temporarily or permanently, with mailboxes. Mailing the letter may be part of an utterly hateful business—any values, positive, negative, or neutral, can be attributed to the letter or to the mailbox—yet the simple mechanical fact would remain true, namely, that Solomon himself cannot put a letter in a mailbox if he does not have a mailbox. It is important to distinguish such mechanical "needs" from dynamic or evaluative needs. Lewin was clearly aware of this problem, yet he did not resolve it. He says quite explicitly that the value of an object is not identical with its valence. The example he cites concerns a sum of gold that might represent a great value for one person without tempting him to steal it, but for another person it might have a strong valence prompting him to steal. Having thus clearly seen the difference between values and valences, however, he drops the matter with the comment that sometimes they are related and sometimes they are independent. If the concept of valence is replaced, as we propose, by the concept of specific criteria that must be met before the execution of a Plan can continue, then there is no reason to expect that objects satisfying the criteria will always be valuable or that they will always satisfy needs. They may, or again, they may not.

The reader will, we hope, forgive us our banalities, for sometimes the obvious is hard to see. The fundamental, underlying banality, of course, is the fact that once a biological machine starts to run, it keeps running twenty-four hours a day until it dies. The dynamic "motor" that pushes our behavior along its planned grooves is not located in our intentions, or our Plans, or our decisions to execute Plans—it is located in the nature of life itself. As William James says so clearly, the stream of thought can never stop flowing. We are often given our choice among several different Plans, but the rejection of one necessarily implies the execution of some other. In sleep we are about as planless as we can get.

In this renunciation of the dynamic properties of Plans,[5] however, we should not lose sight of the fact that something important does happen to a Plan when the decision is made to execute it. It is taken out of dead storage and placed in control of a segment of our information-processing capacity. It is brought into the focus of attention, and as we begin to execute it we take on a number of menial but necessary tasks having to do with gathering data and remembering how far in the Plan we have progressed at any given instant, etc. Usually the Plan will be competing with other Plans also in the process of execution, and considerable thought may be required in order to use the behavioral stream for advancing several Plans simultaneously.

The parts of a Plan that is being executed have special access to consciousness and special ways of being remembered that are necessary for coordinating parts of different Plans and for coordinating with the Plans of other people. When we have decided to execute some particular Plan, it is probably put into some special state or place where it can be remembered while it is being executed. Particularly if it is a transient, temporary kind of Plan that will be used today and never again, we need some special place to store it. The special place may be on a sheet of paper. Or (who knows?) it may be somewhere in the frontal lobes of the brain. Without committing ourselves to any specific machinery, therefore, we should like to speak of the memory we use for the execution of our Plans as a kind of quick-access, "working memory." There may be several Plans, or several parts of a single Plan, all stored in working memory at the same time. In particular, when one Plan is interrupted by the requirements of some other Plan, we must be able to remember the interrupted Plan in order to resume its execution when the opportunity arises. When a Plan has been transferred into the working memory we recognize the special status of its incompleted parts by calling them "intentions."

The recall and resumption of interrupted tasks have, largely as a result of Lewin's interest, received some attention in the psychological laboratory. The studies are well known, as in Lewin's interpretation in terms of tension systems that are reduced when the task is successfully completed. Since both Lewin's tension system and our working memory are carefully considered explanations, they both account for most of the observed phenomena and it is difficult to find clear points of contrast between them. (This difficulty is enhanced by a wonderfully free and easy interpretation by Lewin of "tension system.") But there are a few points on which the two theories can be compared, and we shall concentrate on these.

First, we assume it is a well-known fact that interrupted tasks tend to be resumed and tend to be well remembered. However, the tasks must be a little complicated. Simple, repetitive, continuous tasks—marking crosses on a sheet of paper, stringing beads, etc.—will not usually be resumed and do no stand out in the subject's memory when he is later asked what tasks he performed. This observation seems eminently reasonable from either Lewin's or our point of view. For the present authors, the argument would be that such tasks require little or no record of what has been and what remains to be accomplished, and hence they have no special representation in the subject's working memory. For Lewin, the argument was that such tasks cannot be interrupted, they can only be halted. Therefore, interruption does not serve to continue or prolong a tension system. So far, the two views do not quarrel with each other. Moreover, it is recognized that even these continuous tasks can be interrupted if the subject is told in advance that the task calls for a given amount of activity. If he is told, for example, that his task is to put thirty beads on a string, he can be interrupted before he reaches that number. Both views still agree, since we predict that he will remember the task because he had to count and remember a number in order to keep his place, and Lewin predicts the same result because the task is now interruptable. But suppose that the subject is given a pile of beads to string, told he must put *all* of them on the string, but not told any specific number. Now the predictions will be different. We say there should be no tendency to resume or recall the task, since memory function is performed externally by the pile of beads, not by the subject. The Lewinian view would predict that since a tension system would remain undischarged because the task was interrupted, the bead-stringing should be more quickly resumed and more frequently recalled. We do not have experimental data with which to settle the point, but it appears to us to be a point on which data might help. We are at least encouraged to see that our view is not just a rephrasing of Lewin's, since a fairly clear disagreement can be formulated.

Second, suppose a person intends to write letters to five different people. He assembles the writing materials and begins, but he is interrupted before he finishes. The question is, will it make a difference in his tendency to resume the task if he is interrupted in the middle of a letter rather than between letters? In our view, it would make a difference. An interruption between letters leaves him with no memory problem, so the interrupted Plan is not assigned any special place in his working memory where it might remind

him to finish the job. We are not clear what the correct prediction would have been for Lewin. On the one hand, each letter is a separate task with its own tension system to be discharged, so he would predict as we do. But, on the other hand, there might be very little tendency to resume under either condition, because the completion of a task "similar" to the interrupted one (the completion of the first three letters) is supposed to provide a "substitute consummation" to reduce the tension systems associated with the other letters.

In our terms, writing letters would be called a "flexible" Plan, because its parts can be performed in any order. That is to say, it does not matter in which order the letters are written. Usually, there is more working memory involved in keeping track of inflexible Plans, because such Plans tend to become more elaborate and specific; we therefore would make a general prediction that, so long as external memory is not used, a task that requires an inflexible Plan will be resumed and recalled more frequently after interruptions.

Third, in one experiment subjects were told on half the interrupted tasks that the tasks would be resumed and on the other half that they would not be resumed. This advance information made very little difference in their tendency to recall interrupted tasks better. It was on the basis of this experiment that Zeigarnik concluded that the possibility of resumption had no effect on recall. However, an equally valid conclusion might be that verbal instructions given at the time of interruption cannot change the extent to which working memory has already been devoted to the execution of the Plan. But further experiments could be helpful in settling the difference.

Finally, "It is often observed," Lewin wrote, "that even making a written note of an intention is conducive to forgetting it, though according to the association theory it should reinforce the coupling between the referent and the goal-presentation. Making a note is somehow a consummation, a discharge." This observation takes on particular significance for us, of course, since in our view the use of external memory devices ordinarily lightens the load on our personal memories. The forgetting, if our observation is valid, would not be the result of a consummation or a discharge of tension, but rather the result of freeing our working memory capacity for other planning activities.

It was a general observation in Zeigarnik's studies that adults did not participate as enthusiastically as the children and that they did not show as strong a tendency to recall the interrupted tasks. It seems probable to us that an adult has several Plans of his own whose execution must be temporarily suspended during the experiment and that he might be reluctant to lose track of them for these little laboratory games. Also, it seems probable to us that an adult would have learned how to make use of external memory devices for his Plans whenever possible, and so might not use his internal working memory in the same way a child would.

These considerations have, perhaps, drifted rather far into the details of a particular line of research. And the present authors may seem a bit too intolerant of Professor Lewin, a man who has contributed so much to advance our understanding of the psychology of the will. The excuse is that Lewin in his early work came very close to saying some of the same things this book is trying to say. His papers offer a challenge—both in theory and in observations. On the majority of points the present interpretation agrees with his predictions, but the disagreements in theory have been emphasized in the hope of stimulating their resolution in the laboratory.

If an intention is, as here described, the unfinished part of a Plan that is being executed, how could anyone forget what he intends to do? Forgetting intentions is a commonplace occurrence, of course, and several psychologists have offered explanations. It is generally assumed that forgetting an intention is not the same as forgetting a telephone number, although perhaps the same mechanism may occasionally be responsible. Usually, however, forgetting an intention appears to have some active quality to it that is not involved in the kind of forgetting Ebbinghaus studied. The classic work that emphasizes an active component in forgetting intentions, of course, is Freud's *Psychopathology of Everyday Life*. Freud's emphasis, naturally, was on the dynamic or evaluative aspects of such forgetting, on the repression of the intended act by other psychic forces that opposed it in some way.

The most obvious thing to say about a forgotten intention is that the Plan that gave it life was not completed. The question that is basic to all others, therefore, is why one Plan was abandoned and another pursued instead. If we try to translate Freud's dynamic explanation into the language of this essay, we must say that Plans are abandoned when their execution begins to produce changes in the Image that are not as valuable as we had expected. (This would be consistent with the Freudian view, but it is not the only possible explanation.) The diagnostic value of a forgotten intention is that it so often underscores a change in Plan that might otherwise have gone unnoticed. And the change in Plan, in turn, provides a clue to aspects of the Image that might not ordinarily be accessible to introspection. We could, of course, examine the conditions that cause us to be unaware of the fact that we have changed our Plans, but presumably the conditions would be essentially those that psychoanalysts tell us produce repression. Thus we accept the notion that dynamic changes in the Image—especially in the evaluative aspects of the Image—exert close control over the Plans we try to execute. Altering the planner's Image is a major dynamic mechanism for altering his Plan, and thus for altering his behavior. Social psychologists who have considered the problems of persuasion have generally agreed that the best techniques involve some change in the audience's concepts or values. But we are here (as throughout most of this book) concerned more with the *execution* than with the *formation* of Plans.

We can easily imagine other, nondynamic conditions that might lead one to abandon a Plan and thus to forget an intention. The working memory may go awry, especially when the execution of the task has been interrupted for some reason. To take an extreme example, the man whose appointment book is destroyed through no fault of his own will have lost track of numerous Plans, both pleasant and painful. Remembering the Plan is most difficult when we try to do it without external crutches, when the Plan is new or transient, and when the Plan is complicated. If the Plan is written down in detail, if it is one we follow repeatedly, or if several consecutive subplans are involved, then our working memory has an easier task. We therefore assume that intentions would be forgotten more frequently in the former situations than in the latter, *ceteris paribus*, for purely mechanical reasons that Ebbinghaus would understand as well as we do.

Presumably we are constantly revising our Plans after we begin to execute them. Ordinarily we do not make any special note of these changes, but merely execute the new Plan as quickly as possible. But a special problem arises with shared Plans. When you have made known your intentions, other people may depend upon you to carry them out. Thereafter, changes in your Plans must take into account what has been said. You may change the Plan for any of a dozen reasons and then forget to incorporate into your new Plan a subplan for communicating to your friend about the change. Forgetting to tell somebody that you have changed your intentions is a very different process from forgetting your intentions.

Still another nondynamic reason for forgetting an intention might be that some preparatory step in the Plan leading up to the intended act proved to be impossible. An applied mathematician may intend to solve a problem by first inverting a matrix and then computing certain quantities, but he discovers that the particular matrix does not have an inverse. He will forget his intention to compute the quantities, but not because he has repressed it or found it potentially dangerous, etc. No doubt many intentions must be forgotten because we are not bright enough or strong enough to execute the Plans in which they were embedded. Not all Plans are feasible.

Two general consequences of the present argument are worth brief comment before closing. . . . First, more research is needed on the way people use external aids as memory devices—to record their Plans, their intentions, and their progress in executing their Plans. In our enthusiasm for memorizing nonsense syllables we have overlooked the importance of some of these ancillary kinds of memory. Memory for intentions should not be the private property of clinicians.

Second, what we call an "effort of will" seems to be in large measure a kind of emphatic inner speech. Much, probably most, of our planning goes on in terms of words. When we make a special effort the inner speech gets louder, more dominating. This inner shouting is not some irrelevant epiphenomenon; in a very real sense it *is* the Plan that is running our information-processing equipment. As psychologists we should listen to it more carefully.

C. I. Lewis says only what is plainly open to common sense in the following comment:

Knowledge, action, and evaluation are essentially connected. The primary and pervasive significance of knowledge lies in its guidance of action: knowing is for the sake of doing. And action, obviously, is rooted in evaluation. For a being which did not assign comparative values, deliberate action would be pointless; and for one which did not know, it would be impossible. Conversely, only an active being could have knowledge, and only such a being could assign values to anything beyond his own feelings. A creature which did not enter into the process of reality to alter in some part the future content of it, could apprehend a world only in the sense of intuitive or esthetic contemplation; and such contemplation would not possess the significance of knowledge but only that of enjoying and suffering.

In this short paragraph Lewis puts the problem of the present discussion. It is so obvious that

knowing is for the sake of doing and that doing is rooted in valuing—but how? How in the name of all that is psychological should we put the mind, the heart, and the body together? Does a Plan supply the pattern for that essential connection of knowledge, evaluation, and action? Certainly any psychology that provides less— that allows a reflex being to behave at random, or leaves it lost in thought or overwhelmed by blind passion—can never be completely satisfactory.

Notes

1. This passage is from an article by John Dewey entitled, "The Reflex Arc Concept in Psychology," an article as valuable today for its wisdom and insight as it was in 1896.

2. The reason that the TOTE of Figure 1 can be expanded only in its operational phase is purely formal and can be appreciated by simple counting: There are four arrows associated with Test; there are two arrows associated with Operate; and there are two arrows associated with TOTE as a unit. Therefore, if the two-arrowed TOTE is used to construct some component of another TOTE, the component it constructs must be the two-arrowed Operate, not the four-arrowed Test. However, rigid restrictions such as these are probably unrealistic and justifiable only in terms of pedagogic simplification. Anyone who has studied the hierarchically organized programs written by Newell, Shaw, and Simon to simulate human problem-solving will recognize how primitive and unelaborated these TOTE hierarchies really are.

3. If we take seriously the suggested form of the TOTE, the system may be easily trapped into loops. For example, if the subtests in the expansion of an operational phase all pass, but the basic test fails, a loop will exist. In order to avoid loops we might insist that the basic test imply the disjunction of the subtests. A more realistic solution would accept the occurrence of loops as a signal that the Plan was not successful in producing the result for which the basic test was designed; it would then be necessary to provide further machinery for discovering and stopping such loops.

4. In common speech, an additional requirement seems to be that the Plan be conscious. The present authors are willing to tolerate "unconscious intentions." This seems to be the position of E. C. Tolman, Psychology vs. immediate experience, *Philosophy of Science*, 1935, 2, 356–380. It is, of course, a basic Freudian concept. See also G. E. M. Anscombe, *Intention* (Oxford: Blackwell, 1958).

5. In discussing this point with some of our colleagues we have encountered the reaction that we have not really renounced dynamic properties in the Plan, but rather that we have actually explained them. If so, it is certainly an odd definition of "dynamic." The "explanation" is simply that, so long as it stays alive, the psychobiological machine must continue to execute the successive steps in some Plan. It is our

impression that this is not what most psychologists have meant when they used the term "dynamic." As we understand it, "dynamic" is usually taken to mean that the organism is striving toward some state or object that, when achieved, will reduce unpleasant tensions, etc. These "stages," "goals," "tensions," if they exist at all, must be represented in the Image, not in the Plan. Hence we persist in speaking of our position as a "renunciation of the dynamic properties of Plans."

VII

Self-Regulation
and
Self-Direction in
Sociocultural Systems

IN contrast to previous Parts, which have presented only samplings of the work done in the areas dealt with, this last Part, though certainly not exhaustive, represents a good proportion of the work of the regrettably small number of social scientists who have approached the sociocultural system from the modern systems perspective to any serious extent. (And three or four of the selections in this final Part were not consciously written from this perspective.) As we stated in our General Introduction, this sourcebook is offered in large part as an attempt to help remedy this state of inertia.

From the comparative general systems point of view, the sociocultural system may be seen as a natural system that has evolved from and is continuous with the other levels of natural systems that are studied by the major sciences: a hierarchy of levels including the atom, molecules, cells, organs, organisms, and various stages of social organization of numbers of organisms. Given the nature of the individual human psychological system as discussed in Part VI, our focus in dealing with the last and highest level is on the nature of the bonds linking together a number of these individual biopsychological units and the resulting emergence of cultural ideas and beliefs, characteristic personalities, and particular structures of patterned social interrelations.

A great deal of necessary background has already been provided in previous Parts, but when we come to deal specifically with the sociocultural level, the dearth of material with a systems view creates a problem of organization. Hence, our arrangement is rather arbitrary, and perhaps points up the gaps in our studies more than the positive contributions. However, it is hoped that these latter more than suggest the large promise of the modern systems view.

Many social thinkers have posed the question of the nature of society as an entity or system in its own right. And just as modern biology had to fight hard to replace a mechanistic, piecemeal analytical approach with a more organic view, so did social science have its bout with those who refused to see society or the group as more than an aggregate of individuals thus to be studied in terms of individual attributes. It was Emile Durkheim who perhaps turned the tide with his theory of society sui generis and his impressive study of the sociogenesis of suicide. A next important step was the development of the thesis that the individual is truly social and society psychic and therefore the individual–society dichotomy constitutes a pseudo-problem. This view was propounded by Baldwin, Cooley, Dewey and Bentley, and others, but got its most systematic social psychological underpinnings in the symbolic interactionist theory of mind, self, and society associated primarily with the name of G. H. Mead. For Mead and his associates, the entity "society" came to be seen clearly as an open, ongoing transactional system of interacting, interpreting and deciding individuals. Such a position, we have argued, is closely congenial to the modern systems view, which seems the obvious next step in the theoretical development of social science. Karl Deutsch, in the introductory article to this final Part, confirms unequivocally that society may most fruitfully be viewed, not as an equilibrial or organismic system, but as an open, adaptive, self-directing system. Deutsch's paper is an especially succinct statement of his cybernetic model of society, though the reader might well follow through with a reading of his more recent extended treatment in *The Nerves of Government* (1963).

The so-called Hobbesian problem of order in society has been a perennial theoretical focus of social scientists: How is harmonious social order possible? Rejecting Hobbes' own answer in terms of a strong central authority, an important segment of contemporary social science replies in terms of socialization and institutionalization of a postulated near-universally accepted set of norms and values to which conformity is encouraged and deviance is discouraged by more explicit and external "controls." But a growing chorus of critics sees this position as overdrawn, if not utopian, and points to a number of neglected considerations, including: the factual correctness of an assumed societal consensus on central norms and values; the large slippage between verbal statements of social standards and their interpretation in action; the gross neglect of the role of the "material base" of society— the play of interests and power and the real amount of conflict that is both latent and overt in most societies most of the time; and the assumption that deviance from the assumed norms is essentially pathological and that conformity is almost always "healthy."

The four selections of Section A of this Part examine several facets of the problem of control, conformity and deviance, providing support for a more intensive consideration of these neglected considerations. The late S. F. Nadel, an important anthropological theorist as well as field researcher, exploits modern systems theoretic principles to suggest the intimate systemic relationships between traditional standards, rewarding or instrumental activities, and societal self-regulation. An important practical, as well as theoretical conclusion that might be drawn from his analysis is that presumed norms and values cannot maintain social order, even with external controls, when activities of large numbers of people are neither inherently rewarding nor instrumentally linked to promote desired ends. In fact, the social standards cannot be taken simply as causal determinants or autonomous controls, but must be viewed as fully implicated in the systemic process and themselves generated out of conditions including interest-laden, instrumental activities.

Neither of the next two articles were written from an explicit modern systems perspective, but represent two of the very few in social science literature arguing for the positive, if not essential, role of deviance in society. They articulate especially well with the cyberneticist's view—earlier expressed in several places—of the necessary role of a constant source of variety for the self-regulation and adaptation of the complex, open system. Roger Nett's paper also points in the direction of the very recent sociological theories of deviance as generated in a complex social and psychological feedback process stemming from the given sociocultural organization of the community or society. But his main emphasis is on the role of deviance itself as generating continuous social organization, and as essential to that process. His argument may be related to the modern systems view of the general mechanisms of evolution and the process of "mapping" environmental (including social) variety by way of the system's own variety. Social deviance is thus essential as part of the self-correcting, self-directing mechanisms underlying the viable continuity of the social system.

This thesis is perhaps even more explicitly brought out in Roger Owen's anthropological study of cultural diffusion and adaptation among some primitive bands. Here the close isomorphism on the most general level between biological and sociocultural evolution is quite transparent: the roles of variety, its selection, preservation and perpetuation. In one case it is a question of mutations, genetic coding, natural selection and reproduction; in the other, of deviance, symbolic language coding, social and psychological selection, and cultural preservation and transmission. In the fourth selection, by Wilkins, we return to an explicit use of cybernetic principles—particularly Maruyama's treatment—to develop a theory of deviant behavior as socially generated in a deviation-amplifying feedback process.

In Section B we turn to studies of "control" in the different sense of the organization and regulation of behaviors in a more or less conscious promotion of social goals. David Easton is one of the few American political scientists besides Karl Deutsch to use modern systems analysis seriously and extensively as the basis of a model of the political process as an input–output, feedback-controlled, goal-seeking system. Of note, in addition, is his contribution to the swelling critique of the equilibrium model of society and his preference for a view of the social system as "made up of efforts . . . to control, modify or fundamentally change either the environment or the system itself, or both together." Cadwallader also emphasizes the inherently changeful nature of the social system in seeing it as an "ultrastable" one in that it is capable of changing its own structural arrangements to promote a viable continuity. His is a fresh approach to the sociological analysis of complex organizations as well as to organization theory in general.

With Lewin and Haberstroh, we come down somewhat in the level of abstraction to deal with more concrete problems of organizational control. It will be no surprise to many that Kurt Lewin, ever alert to exploit new scientific ideas, gave an early, especially lucid cybernetic interpretation of planned group goal-seeking and problem-solving. Haberstroh's empirically well-informed systems analysis also makes an important point for social theory consonant with our argument of Section A above. "Control" is not something *imposed* on organization, but is an inherent aspect of *social organization as an ongoing process.* Enduring organization is seen as a function of *both* consensus—"a degree of harmony and co-ordination . . . a sharing of intention"—*and* conflict—"a process of conflict resolution."

Biologist (more accurately, sociobiologist) Garrett Hardin moves us back to more general considerations in forcing us to ponder the severe problems to be faced in attempting the planned control of complex social organization. He sees clearly the implications for societal control of the fundamental difference between the cybernetic and the pre-cybernetic machine. In the latter the attempt is to *build in* its control according to uncertain prediction of future error, whereas the former builds in the flexibility and general procedures which may provide for unanticipated error by way of adjustments of the organization itself to error-producing forces. Finally, Geoffrey Vickers, in questioning the concept of adaptability, augments Easton's model, points up the difficulties of error-control in complex society, and underscores further the endemic nature of conflict in organic systems.

Section C concludes this sourcebook with a return to an important facet of the problem of conceptualizing the interplay of individual and group-level phenomena. We have seen that the social system embraces the interrelations of component individuals who, themselves complex systems, do not simply react to one another in a mechanical, billiard-ball fashion, but rather actively contribute to the genesis of social structure and process by way of the interpretation of the situation and the making of subsequent decisions or choices of action. This implies the use of strategies, no matter how crude or how "rational." Game theory was thus bound to create excitement especially among those social scientists who rejected a view of society implying that it contained passive or decorticated individual units. Anatol Rapoport provides a well-tempered assessment of game theory's significance for behavioral, and especially social, theory.

Finally, Buckley's paper, besides bringing together and summing up a number of the important ideas introduced in this sourcebook that are especially pertinent to social science, attempts to sketch a number of recent social theories or models which carry on the analysis of the social interaction process where game theory leaves off, and are congenial to the modern systems view.

47.
Toward a Cybernetic Model of Man and Society

KARL W. DEUTSCH

MEN HAVE TENDED to order their thoughts in terms of pictorial models since the beginnings of organized thought. The model itself was usually drawn from something in their immediate experience, available from their technology, and acceptable to their society and culture. Once adopted, it served, more or less efficiently, to order and correlate the experiences which men had, and the habits they had learned, and perhaps to suggest a selection of new guesses and behavior patterns for new or unfamiliar situations.

Thus men used the image of their own society (where men influence one another's behavior by talking to each other) as a model for physical nature. Nature was pictured as a society of animated objects which could be magically influenced by talking to them through the right kind of incantations, that is, through the language socially accepted in that imaginary society of things. The inefficiency of this sort of model is considerable, since it permits very little analysis and only very poor predictions.

Later models were drawn by men from work of their hands, that is, from processes and things which they themselves could bring into existence, put together or take to pieces, and which they therefore could analyze and elaborate more adequately in their parts and interrelations.

There is the simple model of the artisan who makes things with his hands, particularly the potter who shapes clay. Once men have ceased to think of trees as of a society of animated beings to be talked to, they may think of them as a

collection of green pots made by some invisible potter. After having assumed that things have minds and either existed forever or were born and died like men, it is now assumed, on the second model, that things have neither will nor mind but are so many inert products made by the invisible craftsman. Here again, however, the model permits very little correlation between experience with one kind of thing and experience with another. Nor does it permit much analysis or prediction of the invisible craftsman's past behavior or future intentions.

More complex models become available when men have learned to produce more complex contrivances and when the fruits of their labor can be combined and piled up into houses, towns, and pyramids, which dwarf the size and the life span of the individual beholding them. The impersonal plan or law of the city may then come to serve as a model for an assumed impersonal plan or law of nature, and the structure of this impersonal law or architecture appears to remain effective regardless of the subsequent activities of any invisible architect or lawgiver who might have originated it. These new models permit a clear and more specific correlation of experience. They imply rigid and often immovable arrangements in space, which lend themselves readily to pictorial representation. In this manner the Egyptian pyramid, with its rigorous order of a very few stones at its apex and the many stones bearing all the burden at the bottom, has served as a "social pyramid", a model for that conception of human society from which the Jews walked out under Moses. More broadly the pyramid has served as a model for the conception of a "hierarchy", because a hierarchy, whether of priests, army

From Karl W. Deutsch, "Some Notes on Research on the Role of Models in the Natural and Social Sciences," *Synthese*, 7 ('48–'49), 506–33. Reprinted with permission of the author and D. Reidel Publishing Co.

officers, or ideas, values, or purposes, such as in Aristotelian philosophy, turns out to be an Egyptian pyramid writ large, with its stones replaced by officers, or words, or human beings. A significant characteristic of the pyramid model is its static character. The pyramids were deliberately built to be unchanging. They were a prototype of that "graven image" in which a dynamic religion might fear a grave of life.

Two other simple models involve at least some movement, and therefore some implication of time. The first of these is the wheel. In its simple rotary motion, elevating and casting down each part of its circumference in regular succession, it has been conceived as a model of human affairs and human history, as "wheel of fortune", "wheel of fate", or Fortune standing on a ball—in each case suggesting instability of the parts with stability of overall performance; and it was projected to the skies in the spheres, cycles, and epicycles of Ptolemaic astronomy.

The other of these models is the balance, the pair of scales which yields the concept of stable equilibrium, with its implication that the adverse reaction must be the greater, the more the true position of balance has been destroyed. The notion of *diké*, of "nothing too much", of the golden mean, and the statue holding the scales of justice in front of many Western lawcourts, all testify to the suggestive power of this model. Both wheel and balance imply movement, but only movement which either continues permanently or else eventually returns to the original position. "The more it changes, the more it stays the same."

Other simple technological operations began to yield models which implied the notions of process, progress, and history in the simplest, most elementary form. Perhaps the two outstanding models here are the model of the thread taken from spinning, whether as the thread of fate, or the thread of an argument, or the thread of human life. A web woven from these threads is then an obvious extension of this model, implying now, however, the notion of interaction. The German word for reality, *Wirklichkeit*, is related to the word denoting such a textile operation. Goethe has embodied this picture in the Earth Spirit in *Faust*: "*So steh' ich am sausenden Webstuhl der Zeit und wirke der Gottheit lebendiges Kleid.*"

The very continuity of thread and skein and warp and woof make these textile models unsuitable for analysis. It is only with the development of far more complex operations, toward the end of the Middle Ages, that we find mechanical models of greater complexities, slightly less inadequate for describing the world around us. Mechanisms can be taken apart and reassembled. This is crucial for the new models. The development of the making and understanding of mechanical pumps to a fair level of efficiency made it finally possible for Harvey to write his scientific classic, *De Motu Cordis*, using the analogy of valves and pumps for the first adequate description of the circulation of blood.

The Classical Model of Mechanism

The development of clockwork, under progress ever since the thirteenth century, finally yields the classical model of a "mechanism" which is then applied to a description of the stars in the system of Newton; to the system of government in the writings of Hobbes, and in the "checks and balances" of Locke, Montesquieu, and the founding fathers of the American Constitution; and to the human body by such writers as La Mettrie, author of the book, *Man A Machine*, in the eighteenth century. It is extended to God as the "first mechanic" by Tom Paine; and to joy in Schiller's lyric, "Ode to Joy", as the "watchspring of the universe".

The transfer of the idea of mechanism, from the experience of pumps and clockworks to a general description of reality, was encouraged in the days of Newton by the success of gravitational astronomy where the movements of the planets, isolated from each other by vast distances of space, proved peculiarly suited to mechanical interpretations; though it has appeared since that they seem to be peculiarly unrepresentative of most of the events of nature.

The classical concept or model of mechanism implied the notion of a whole which was completely equal to the sum of its parts; which could be run in reverse; and which would behave in exactly identical fashion no matter how often those parts were disassembled and put together again, and irrespective of the sequence in which the disassembling or reassembling would take place. It implied consequently the notion that the parts were never significantly modified by each other, nor by their own past, and that each part once placed into its appropriate position, with its appropriate momentum, would stay exactly there and continue to fulfil its completely and uniquely determined function.

These few remarks already show that the

classical notion of mechanism is a strictly meta-physical concept. No thing completely fulfilling these conditions has ever been on land or sea, nor even, as our cosmologists have told us, among the stars. The more complicated a modern mechanical device becomes in practice, the more important becomes the interdependence and mutual interaction of its parts through wear and friction, and the interdependence of all those parts with their environment, as to temperature, moisture, magnetic, and electrical influences, etc. The more exacting we make the standards for the perform-ance of a real "mechanism", the less "mechanical" in the classical sense does it become. Even an automobile engine must be "broken in", and a highly accurate timing device depends so much on its environment that it must be assembled in strictly air-conditioned workrooms by workers with dry fingertips.

The Classical Concept of Organism

Conspicuous breakdowns of the concept of mechanism became most obvious in the fields of the social sciences and biology. Attacks on the inadequacy of mechanical thinking form a major part of the political writings of Edmund Burke, and the emphasis on wholeness, interrelatedness, growth and evolution, proclaimed in literature and education by Rousseau, and in politics by Burke, was then powerfully reinforced in the nineteenth century through the growth of the biological sciences, resulting in the wide popularity of the concept of "organism" in its classical nineteenth century form, as the proper model for reality.

According to this classical view, an "organism" is unanalyzable, at least in part. It cannot be taken apart and put together again without damage. As Wordsworth put it, "We murder to dissect". The parts of a classical organism, in so far as they can be identified at all, not only retain the functions which they have been assigned but in fact cannot be put to any other functions (except within narrow limits of "de-differentiation" which were often ignored), without destroying the organism. The classical organism's behavior is irreversible. It has a significant past and a history—two things which the classical mechanism lacks—but it is only half historical because it was believed to follow its own peculiar "organic law" which governs its birth, maturity, and death and cannot be analyzed in terms of clearly identifiable "mechanical" causes.

Attempts have been frequent to apply this classical concept of organism to biology and to human society. On the whole they have been unsuccessful. While "organismic" models might sometimes help to balance the onesidedness of a "mechanical" approach, biologists have failed to derive significant predictions or experiments from the supposed "life force" of nineteenth century "vitalists", and the inadequacies of organismic theories of society or history have been even more conspicuous.

Both mechanistic and organismic models were based substantially on experiences and operations known before 1850. Since then, the experience of almost a century of scientific and technological progress has so far not been utilized for any signi-ficant new model for the study of organization, and in particular of human thought.

Perhaps it is now becoming possible to develop such a new model in the field of scientific theory, since such new models during the past fifty years have actually been developed in the world of physical fact. The developments in this connection have been the developments of communications engineering. Telegraphs, telephones, and switch-boards have often been compared with the nerves of an organism, but this comparison has usually remained a figure of speech. But by continuing to *make* things which fulfil the functions of com-munication and organization, we cannot help in the long run but gain significant opportunities for a clearer understanding of those functions them-selves. Given a development of peace and progress in the next few decades, we ought to gain an unprecedented chance to find out vastly more about the processes of communication, organiza-tion, and learning, since we have in fact been engaged in making an ever larger number of physical facilities which actually do these things.

If we should rate this chance less highly we may yet not be willing to neglect it. At a time when we all still know so little of communication, organization, and learning in the working of the human mind and the behavior of societies, we may welcome any help toward organizing our scanty data and clarifying our elusive concepts for the investigations still so urgently needed.

Self-modifying Networks as Generalized Models of Organization in Machines, Minds, and Societies

Modern studies of communications engineering suggest that the behavior of human organizations, peoples, and societies has important relations in common with manmade communications networks, such as servomechanisms, switchboards, and calculating machinery, as well as with the behavior of the human nervous system and the human mind. It now seems possible to analyze and describe the *common patterns* of behavior of self-modifying communications networks in general terms, apart from the question whether their messages are transmitted and their functions carried out by circuits of electric current in an electronic device, by chemical and neural processes inside a living body, or by spoken, written or other communications between individuals in an organization, group, nation, or society.

There are several advantages in using electric networks, nerve systems, and societies as analogs for each other, provided we remember that analogy implies similarity only in certain relations between the constituent elements of each system. We gain the aid of new and perhaps more efficient models for our thinking about minds and societies. These new models offer suggestive analogies for such relationships as "purpose", "learning", "free will", "consciousness", and "social cohesion" —that is, precisely for those relationships which have often been considered crucial in social science but were found incapable of effective representation by earlier models. Until now, these relationships could at most be treated qualitatively, by recognition or description. The new models suggest observations and experiments to seek data for their treatment in quantitative terms.

A modern radar tracking and computing device can "sense" an object in the air, interacting with its beam; it can "interpret" it as an airplane (and may be subject to error in the "perception"); it can apply records of past experience, which are stored within its network, and with the aid of these data from "memory" it can predict the probable location of the plane several seconds ahead in the future (being again potentially subject to error in its "recollections" as well as in its "guess", and to "disappointment", if its calculation of probability was correct, but if the airplane should take a less probable course); it can turn a battery of anti-aircraft guns on the calculated spot and shoot down the airplane; and it can then "perceive", predict, and shoot down the next. If it should spot more than one airplane at the same time, it must become "infirm of purpose", or else decide ("make up its mind") which one to shoot down first.

This is contemporary engineering practice. It would be out of place to describe here other existing devices, such as thermostats, automatic airplane pilots or electronic calculators. Suffice it to say that manmade machines actually operating or designable today have devices which function as "sense organs", furnish "interpretations" of stimuli, perform acts of recognition, have "memory", "learn" from experience, carry out motor actions, are subject to conflicts and jamming, make decisions between conflicting alternatives, and follow operating rules of preference or "value" in distributing their "attention", giving preferred treatment to some messages over others, and making other decisions, or even conceivably overriding previous operating rules in the light of newly "learned" and "remembered" information. Parallels for this behavior in the fields of psychology, neurophysiology, and cultural anthropology are striking.

None of these devices approach the overall complexity of the human mind. While some of them excel it in specific fields (such as the mechanical or electronic calculators), they are not likely to approach its general range for a long time to come. But, as simplified models, they can aid our understanding of more complex mental and social processes, much as sixteenth century pumps were still far simpler than the human heart, but had become elaborate enough to aid Harvey in his understanding of the circulation of the blood.

The Feedback Concept

With the aid of these models, we may recognize a basic pattern which minds, societies, and self-modifying communications networks have in common. Engineers have called this pattern the "feedback". "In a broad sense (feedback) may denote that some of the output energy of an apparatus or machine is returned as input. . . . (If) the behavior of an object is controlled by the margin of error at which the object stands at a given time with reference to a relatively specific goal . . . (the) feedback is . . . negative, that is, the signals from the goal are used to restrict outputs which would otherwise go beyond the goal. It is this . . . meaning of the term feedback that is used here."[1] "By output is meant any change produced in the surroundings by the object. By input, conversely, is meant any event external to the object that modifies this object in any manner."[2]

In other words, by feedback is meant a communications network which produces action in response to an input of information and *includes the results of its own action in the new information by which it modifies its subsequent behavior.* A simple feedback network contains arrangements to react to an outside event—*e.g.* a target—in a specified manner—*e.g.* by directing guns at it—until a specified state of affairs has been brought about—*e.g.* the guns cover the target perfectly; or the automatic push button tuning adjustment on a radio has been accurately set on the wave length approached. If the action of the network has fallen short of reaching fully the sought adjustment, it is continued; if it has overshot the mark, it is reversed. Both continuation and reversals may take place in proportion to the extent to which the goal has not yet been reached. If the feedback is well designed, the result will be a series of diminishing mistakes—a dwindling series of under-and-over corrections converging on the goal. If the functioning of the feedback or servo-mechanism is not adequate to its task (if it is inadequately "dampened"), the mistakes may become greater; the network may be "hunting" over a cyclical or widening range of tentative and "incorrect" responses, ending in a breakdown of the mechanism. These failures of feedback networks have specific parallels in the pathology of the human nervous system ("purpose tremor") and perhaps even, in a looser sense, in the behavior of animals, men and the whole communities.

Learning and Purpose

Already the simple feedback network shows the basic characteristics of the "learning process" as described by John Dollard. According to Dollard—who is speaking of animals and men— "there must be (1) drive, (2) cue, (3) response, and (4) reward." In a manmade feedback network, "drive" might be represented by "internal tension", or better, mechanical, chemical, electric "disequilibrium"; input and output would function as "cue" and "response"; and the "reward" could be defined analogously for both organisms and manmade nets as a "reduction in intensity" (or extent) of the initial drive or internal disequilibrium.

A simple feedback mechanism implies already a measure of "purpose" or "goal" which does not only exist within the mind of a human observer, but has relative objective reality within the context of a particular feedback net, once that net has physically come into existence. A "goal" has been defined as "a final condition in which the behaving object reaches a definite correlation in time or in space with another object or event."

This definition of a goal, or purpose, may need further development. There is usually at least one such external goal, *i.e.*, a relation of the net as a whole to some external object, which is associated with one state of relatively least internal disequilibrium within the net. Very often, however, a very nearly equivalent reduction in internal disequilibrium can be reached through an internal rearrangement of the relations between some of the constituent parts of the net, which would then provide a more or less effective substitute for the actual reaching of the goal relation in the world external to the net. There are many cases of such surrogate goals or *Ersatz* satisfactions, as a short circuit in an electronic calculator, intoxication in certain insects, drug addiction or suicide in a man, or outbursts against scapegoat members in a "tense" community. They suggest the need for a distinction between internal readjustments, and readjustments sought through pathways which include as an essential part the reaching of a goal relationship with some part of the outside world.

This brings us to a more complex kind of learning. Simple learning is goal seeking feedback, as in a homing torpedo. It consists in adjusting responses, so as to reach a goal situation of a type which is given once for all by certain internal arrangements of the net; these arrangements remain fixed throughout its life. A more complex type of learning is the self-modifying or *goal changing* feedback. It includes feedback readjustments also of those internal arrangements which implied its original goal, so that the net will change its goal, or set itself new goals which it will now have to reach if its internal disequilibrium is to be lessened. Goal changing feedback contrasts, therefore, with Aristotelian teleology, in which each thing was supposed to be characterized by its unchanging *telos*, but it has parallels in Darwinian evolution. The performance of a human goal-seeker who strives for new goals on reaching each old one has been immortalized in Goethe's *Faust*:

> *"Im Weiterschreiten find't er Qual und Glueck,*
> *Er, unbefriedigt jeden Augenblick."*[3]

We can now restate our earlier distinction as one between two kinds of goal-changing by internal rearrangement. Internal rearrangements which are still relevant to goal seeking in the outside world we may call learning. Internal re-arrangements which reduce the net's goal seeking effectiveness belong to the pathology of learning. Their eventual results are self-frustration and self-destruction. Pathological learning resembles what some moralists call "sin".

Perhaps the distinction could be carried further by thinking of several orders of purposes.

A first order purpose in a feedback net would be an internal state in which internal disequilibrium would be less than any alternative state within the range of operations of the net. This first order purpose would correspond to the concepts of "adjustment" and "reward" in studies of the learning process. Self-destructive purposes or rewards would be included in this class.

By a second order purpose would be meant that internal and external state of the net which seem to offer to the net the largest probability, or predictive value, derived from past experience, for the net's continued ability to seek first order purposes. This would imply self-preservation as a second order purpose of the net, overriding first order purposes. It would require a far more complex net.

A third order purpose might then mean a state of high probability for the continuation of the process of search for first and second order purposes by a "group" of nets beyond the "lifetime" of an individual net. This would include such purposes as the "preservation of the group" or "preservation of the species". Third order purposes require several complex nets in interaction. Such interaction between several nets, sufficiently similar to make their experiences relevant test cases for one another, sufficiently different to permit division of labor and sufficiently complex and readjustable to permit reliable communication between them—in short, such a "society" is in turn essential for the higher levels of the learning process which could lead beyond third order purposes.

Among fourth order purposes we might include states offering high probabilities for the preservation of processes of purpose seeking, even beyond the preservation of any particular group or species of nets. Such purposes as the preservation or growth of "life", "thought", "learning", "order in the universe", and all the other purposes envisaged in science, philosophy, or religion, could be included here.[4]

Complex Networks: Messages and Symbols

Several simple feedback mechanisms can be combined in a feedback network of a higher order. The derivation of abbreviated symbols from the interactions of the network's "sensory structures" with the outside world can be made more elaborate and more precise. These symbols can be stored in, and "recalled" from, more elaborate "memory facilities" (modern electronic calculators may have several such "memories"). They can be recombined according to previously specified rules to produce new results, that is, new combinations of symbols which may not correspond to outside events past, present, or possible, depending on the efficiency of the network, the accuracy, relevance, and relative completeness of the symbols, and the relative "realism" or applicability of the specific "logic" or rules of recombination applied to them by the network.

Complex feedback networks, like their constituent simpler circuits, deal with "messages" and "symbols". For purposes of definition, a "network" is a system of physical objects interacting with each other in such a manner that a change in the state of some elements is followed by a determinate pattern of changes in other related elements, in such a manner that the changes remain more or less localized and independent of other changes in the system from other sources. A "state description" is a specification of which of its possible states each element of the network is in. A "message" is any change in the state description of a network or part of it. Similar definitions have been suggested by Norbert Wiener: A "message" is a reproducible pattern of changes regularly followed by determinate processes depending on that pattern. A "channel" is a physical system within which a pattern of change can be transmitted so that the properties of that pattern (or message) are more or less isolated from other changes in the system. Any message may be interpreted as a set of alternatives or decisions.[5]

The point is that a message is not a physical object in the sense of everyday language. It is a pattern of physical changes of physical objects. It has physical reality: it can be measured and subjected to repetitive treatment. Only by physical processes can it be preserved, received, transmitted, destroyed, or operated on, in any way whatsoever. Yet it can be transferred in succession or duplication from one set of physical objects to another. It can be in several places at once. Unlike "matter" and "energy", "information"—that is, the patterns that can be abstracted from messages —is not subject to their laws of conservation. It can be created and annihilated. It differs greatly from those aspects of reality stressed so heavily by the mechanical "materialism" of the nineteenth century. And it permits us to deal with the hitherto supposedly "intangible" aspects of pattern, *Gestalt*, configuration, order, novelty, as physical realities accessible in principle to analytical, quantitative, and operational treatment, as against the intuitive approach of metaphysics and "idealistic" philosophy from Plato to the present.

In complex feedback nets any message may be treated by the network as a symbol, or symbols may be derived from it. By a "symbol" is meant any message within such a network which has acquired a relatively stable association with another event outside the net, or with any other message within it.[6] As a symbol a message within the network functions as what Charles Morris has called a "sign vehicle". It is treated by the net as associated with ("interpreted as referring to") an outside event or group of events ("or with another inside message"), which Morris calls its "designatum", and it may result in some specific behavior by the net which Morris calls its "operand", and which students of the learning process might perhaps associate with their concepts of "cue" and "response".

Switchboards and Values

The movements of messages through complex feedback networks may involve the problem of "value" or the "switchboard problem", that is, the problem of choice between different possibilities of routing different incoming messages through different channels or "associative trails" within the network. If a relatively large number of alternative channels is available for a small number of messages, the functioning of the network may be impeded by indecision, if many messages have to compete for few channels, it may be impeded by "jamming". The efficient functioning of any complex switchboard requires, therefore, some relatively stable operating rules, explicit or implied in the arrangement of the channels, that is to say, rules deciding the relative preference and priorities in the reception, screening, and routing of all signals entering the network from outside or originating within it.

Simple examples of such rules are the priority given fire alarms in many telephone systems or the rules determining the channels through which transcontinental telephone calls are routed at different loads of traffic—including even the "hunting" of an automatic switchboard for a free circuit when the routing channels are full loaded. They illustrate the general need of any complex network to decide in some way on how to distribute its "attention" and its priorities in expediting competing messages and how to choose between

its large number of different possibilities for combination, association, and recombination for each message.

What operating rules accomplish in switchboards and calculating machines, is accomplished to some extent by "emotional preferences" in the nervous systems of animals and men, and by cultural or institutional preferences, obstacles, and "values" in groups and societies. Nowhere have investigators found any mind of that type which John Locke supposed "to be, as we say, white paper". Everywhere they have found structure and relative function.

In much of the communications machinery currently used, the operating rules are rigid in relation to the content of the information dealt with by the network. However, there seems no reason why these operating rules themselves should not be made subject to some feedback process. Just as human directors of a telephone company today may react to a traffic count by changing some of their network's operating rules, we might imagine an automatic telephone exchange carrying out its own traffic counts and analyses, and modifying accordingly its operating rules and even the physical structure of some of its channels, such as adding or dropping additional microwave beams —which fulfil the function of telephone cables— in the light of the traffic or financial data "experienced" by the network.[7]

What seems a possibility in the case of man-made machinery seems to be a fact in living nerve systems, minds, and societies. The establishment and abolition of "conditioned reflexes", long studied in animals and men, and the results of individual and group learning, often include changes in such "operating rules" determining the organism's treatment of subsequent items of information reaching it.

Any network whose operating rules, that is, preference structures can be modified by feedback processes, is subject to internal conflict between its established working preferences and the impact of new information. The simpler, relatively, the network, the more readily internal conflicts can be resolved by automatically assigning a clear preponderance to one or another of two competing "channels" or "reflexes" at any particular moment, swinging from one trend of behavior to another with a minimum of delay. The more complex, relatively, the switchboards and network involved, the richer the possibilities of choice, the more prolonged may be the periods of indecision or internal conflict. Since the net acquires its preferences through a process of history, its "values" need not be all consistent with each other. They may form circular configurations of preference, which later may trap some of the impulses of the net in circular pathways of frustration. Since the human nervous network is complex, it remains subject to the possibilities of conflicts, indecision, jamming, and circular frustration. Whatever pattern of preferences or operating rules may govern its behavior at any particular time can only reduce this affliction, but it cannot abolish it.

But since the network of the human mind behaves with some degree of plasticity, it can change many of its operating rules under the impact of experience. That is to say, that with the aid of experience the network of the human mind can change its own structure of preference, rejections, and associations. And what seems true of the general plasticity of the individual human mind in its evolution during the life of the individual, seems even more the case of the plasticity of the channels which make up human cultures and social institutions and those particular individual habit patterns which go with them, at least in some proportion to the ability of those cultures to survive and to spread.

So far, we have described two kinds of feedback: "goal seeking", the feedback of new external data into a new whose operating channels remain unchanged; and "learning", the feedback of external data for the changing of these operating channels themselves. A third important type of possible feedback would be the feedback of internal data, analogous to the problem of what is usually called "consciousness".

Consciousness

"Consciousness" may be defined, for the purposes of this discussion, as a collection of internal feedbacks of secondary messages. "Secondary messages" are messages about changes in the state of parts of the system, that is, about primary messages. "Primary messages" are those which move through the system in consequence of its interaction with the outside world; or any secondary message or combination of messages may in turn serve as a primary message, in that a further secondary message may be attached to any combination of primary messages, or to other secondary messages, or their combinations, up to any level or regress.

In all these cases, secondary messages function as symbols or internal labels for changes of state within the net itself, and are fed back into it as additional information, influencing, together with all other feedback data, the net's subsequent behavior. "Consciousness" does not consist in these labels, but in the processes by which they are derived from the net and fed back into it.

Such feedback messages about some of the net's internal states occur in simple form in electronic calculators where they serve important functions in recall. They may occur, in extremely complex patterns, in the human nervous system, where they would be extremely hard to isolate for study. But they also occur, and can be studied with relative ease, in the division of labor of large human teams which process information and fulfil collectively certain functions of thought, such as industrial research laboratories or political or military intelligence organizations.

In cases of this last group we can observe how guide cards and index tabs are added to the information moving through, or stored within, filing systems, libraries, card catalogues, or the document control centers of intelligence organizations such as the State Department or the war time Office of Strategic Services; and how these secondary symbols influence the further treatment of the information. The heads, policy boards, or project committees of such organizations cannot deal with all the vast information in the original documents. They are dealing mostly with titles, description sheets, summaries, project requests, routing slips, and other secondary symbols, while a great deal of the material continues to be processed "below the level of consciousness" of the guiding and policy making parts of the organization. Only those feedback circuits and decisions which are "picked up" through the attachment and feedback of secondary symbols, become directly "conscious" for the organization.

To be sure, the selective function of any network is by no means limited to this "conscious" zone of secondary symbols. On the contrary, what reaches that zone for separately labeled and recorded processing, depends in turn on what has been selected or rejected, associated or dissassociated, routed or blocked, recorded or misfiled or erased, within the rest of the system. What seems true of the screening function of the reporter on the beat and the desk analyst in the intelligence organization, seems similarly true of the "non-conscious" remembering and forgetting, the "aversions" and "hunches" of the individual human mind, as well as of many of the "unverbalized" conventions and assumptions, preferences and taboos of human societies and cultures.

The powers of the "non-conscious", internally unlabeled circuits and processes within a network, can be positive as well as negative. An experience may be built up into a perception and recorded in memory, two and two may be put together, new associations, and indeed significant discoveries and insights, may be put together "non-consciously" without intervention of secondary symbols, until secondary symbols are attached to the new combination and suddenly the image of the new synthesis breaks through into the realm of consciousness, seemingly all ready and armored like Pallas Athene springing forth from the head of Zeus of the Greek legend.

Yet it is by means of these secondary symbols that we may be conscious of some or all of our relevant steps in a calculation, or of some or all of our steps in a sequence of behavior. Since these secondary symbols are fed back into the net, the message of which the net has become "conscious" may appear in that net with greater frequency than its unlabeled alternatives, and remain more readily available for preferred treatment: be it preferred association, recording, transmission, blocking, or suppression, under the current operating rules of the system.

If secondary symbols become attached to parts of connections in the net, which embody these operating rules, these rules themselves become "conscious" for the net, and, by being fed back into it, become statistically reinforced for more effective application, or else for easier modification, if the possible modification of the rules themselves is included in the net. The effects of such internal labeling may be thought of as to some extent comparable to the effect of dramatic symbols or publicity devices being attached to particular ideas, practices, or laws in a society, lifting them from their previous obscure existence into the cross fire of public attention within that community.

The ensemble of secondary symbols may easily misrepresent the net's actual content. Some primary symbols may be "overrepresented" by ample feedback, while others may not be made "conscious" at all. Consciousness, therefore, may be false consciousness; much as the actual personality of a man may be quite different from what he thinks it is. Similarly, by attaching suitable symbols and feedbacks to suitably chosen aspects of their behavior, groups or nations can be given highly misleading ideas about their own character. Slowly built-up discrepancies between internally noticed and unnoticed memories and preferences are sometimes suddenly resolved in an individual by his political or religious conversion, or by reform or revolution in a nation.

How does the feedback notion of consciousness compare with other approaches? In the behaviorist school of psychology, we are told "consciousness and conscious processes are excluded as not subject to scientific investigation, or . . . reinterpreted as covert language responses". In social science writings, consciousness is often stressed, and ascribed to groups, but usually this is done without definition or description in any but intuitive terms. Two recent writers describe individual consciousness as follows:

(The) integrative (regnant) processes in the brain . . . according to the findings and speculations of neurophysiologists . . . are capable of self-awareness (as if they had a mirror in which to see themselves). During the passage of one event, many, but not all, of the regnant processes have the property of consciousness, at the moment of their occurrence or soon afterwards if recalled by retrospection. Thus the stream of consciousness is nothing more than the subjective (inner) awareness of some of the momentary forces operating at the regnant level of integration in the brain field.[8]

This is a suggestive description in the language of everyday life in which processes behave like small individuals who "reign", "see themselves as if they had a mirror", and "have the property of consciousness" which "is nothing more than . . . subjective (inner) awareness". But how helpful is it as a concept from which we might derive new observations and experiments?

If consciousness is a feedback process, then it requires material facilities, and is carried on at some material cost in terms of facilities and time. Some of the facilities tied up, and some of the delay imposed on primary processes, should be capable of measurement. Furthermore, feedback processes have structures, circuits, channels, switching relationships, incompatibilities, and discontinuities, which might be susceptible of mapping. If we cannot isolate the physical facilities involved, we might devise functional tests for possible patterns, limits, and discontinuities in the performance of the process of consciousness. If these tests should yield a map of discontinuities in performance, we might derive a basis for further inferences about the structures of the underlying facilities and processes themselves.

Similar considerations might apply to the processes of "consciousness" in nations, classes, or other social groups. If there are such processes, how are they organized and patterned? What are the manpower, facilities, symbols, learning processes, and teamwork relations by which they are carried on? If consciousness resembles a feedback, does it also resemble its peculiar kinds of instability? A small change in a feedback circuit can bring about a large change in its overall performance. Are there analogies for this in social life?

The feedback model of consciousness is more than a verbal explanation. It is a concept. For it suggests many questions which sooner or later should be answered, one way or another, by observations and experiment.

Will

Consciousness seems related to "will"—or to that sense of conation or of making autonomous decisions which we mean when asserting that "our will is free". This notion of will includes not only decisions with internal labels attached to the very moment of action, or to several steps within an action; it includes also mere decisions to start an action now or on a later signal, with the actual parts of the action following automatically without any "conscious" labels attached to them.[9] "Will" in all these cases may be tentatively defined in any sufficiently complex net, nervous system, or social group as the set of internal labels attached to various stages of certain channels within the net, which are represented by these labels as relatively unchanging, so that "we merely trip the purpose and the reaction follows automatically", or at least we expect it so to follow.

In other words, *will* may be called the *set of internally labeled decisions and anticipated results, proposed by the application of data from the decisions past, and by the blocking of incompatible impulses or data from the system's present or future*. Since the net cannot foretell with certainty either the outcome of the subsequent trains of its own internal messages and switching orders, or the outcome of its own efforts to inhibit information incompatible with the "willed" result, it knows only what it "will do", not what it "shall do". It may "know its mind", but it cannot know with certainty.

A fundamental problem of "will" in any self-steering network seems to be that of carrying forward and translating into action data from the net's past, up to the instant when the "will" is formed (the determination becomes "set", or the decision "hardens"), while endeavoring to screen out all subsequent information which might tend to modify the "willed" decision. Will, in short, could be called the internally labeled

effort to maintain preference for predecision messages over post-decision ones. The "moment of decision" might then be seen at that threshold where the cumulative outcome of a combination of past information begins to inhibit effectively the transmission of contradictory data.

This general problem of "will" seems to apply, at least to some extent, to manmade devices whose operations can be accurately specified. Automatic pilots or steering mechanisms exclude or compensate subsequent "experiences", such as gusts of storm, which might deflect them from their course. Modern guided missiles, homing torpedoes, proximity fuses, and similar weapons, involved in their design problems of this kind. Electronic wave filters screen out "noise" from the "desired" messages.

Isolating the pattern of "will" in feedback machines may help us to recognize it in men and communities. Men may shut out the experiences of pain or fear or doubt or pity which might deflect them from their "fell purpose". Cultures or states, ever since the days of the Spartans, may put informal social or religious taboos, or explicit legal prohibitions, in the way of all messages which might change their previously determined patterns of behavior. Modern nations, governments, or political parties may strive in war or peace to perpetuate their policies by blocking all incompatible experiences from the life of their community by all the methods of legislation, indoctrination, pressure, censorship, police, or propaganda of which they may dispose. It is in that sense, perhaps, of a pattern of relatively consolidated preferences and inhibitions, derived from the past experiences of a social group and consciously labeled for at least a relevant portion of its members, and applied to the guiding and restricting of the subsequent experiences of that group, that the concept of "will" can be applied meaningfully to the behavior of political movements, peoples, and social organizations.

Autonomy, Integrity, and Freedom

In what sense is this "will" free?

First of all, this will is relatively free from the pressures of the outside world at any one moment, since it represents the stored outcome of the net's past now being fed back into the making of present decisions. Without effective feedback of its past, the net's behavior would be determined largely by outside pressures. It would not steer, but drift, in both its external and internal arrangements.

If autonomous goal-seeking or goal-changing is treated as a value—if, in the words of G. B. Shaw, "to be in hell is to drift, to be in heaven is to steer"—then a corresponding value attaches to this material precondition of autonomy, the continued possession of an effective past. Here is the foundation of the value of *integrity*, that is, a structure of internal feedbacks, controls and connections, undisrupted by excessive rates of input, either by internal disruption of the net, or by such a speed-up of the learning rate that the net's past becomes negligible in predicting its future behavior. It seems, then, that we learn in the long run by not learning too much at any one moment, and that the net that learns too much becomes a thing. This case differs in kind from that of a net which finds its effectors temporarily ineffective due to outside forces, but retains its internal structure, much as a ship may be tossed by a storm but retain intact its steering mechanical, or a prisoner still may "call his soul his own".

Men, or whole communities, may treat destruction of their "integrity" as equivalent to physical destruction. Faced with the need for rapid readjustment, they may find themselves now with past habits shaped by a pathological learning process which has deprived them of the capacity to adjust freely to new realities, and see no further choice but internal loss of integrity, or external destruction. There are at least three types of pathological learning: through drifting, through *rigor mortis*, through internal breakdown, and there is a fourth, relative type: through failure to learn freely,[10] and yet rapidly enough to keep up with a given rate of change in the environment.

So long as it has autonomy, the net wills what it is. It wills the behavior patterns (the "personality") which it has acquired in the past, and which it is changing and remaking with each decision in the present. Thanks to what it has learned in the past, it is not wholly subject to the present. Thanks to what it still can learn, it is not wholly subject to the past. Its internal rearrangements in response to each new challenge are made by the interplay between its present and its past. In this interplay we might see one kind of "inner freedom".

In its external actions, the net does what it can do. Its outward behavior will be the result of the interplay between the orders transmitted to its effectors, and the feedback data about their results among the pressures of the outside world. In this type of interplay we may see a kind of external freedom for the net to continue its efforts to reach its goal.

Freedom in a feedback network could go further. A chess playing machine could be constructed which would rapidly compute all admissible moves on both sides for two or three moves ahead, and choose the ones most profitable for its side according to a schedule of values derived from the rules of the game. It would play mediocre chess. It could be improved by giving it a suitable memory and additional circuits, so that it could learn to modify its play on the basis of experience. The quality of its playing would then depend largely on that of its experience. It could be spoiled by poor teachers or stupid opponents. If all its past opponents were mediocre, the machine might never learn to play brilliantly. It would remain imprisoned by the limitations of its past. But it could be aided to play better, by building into it a device to break or override sometimes the patterns learned from its past, and to give it a chance for initiative and creativity.

This function of autonomous internal habit-breaking could be fulfilled by building into the machine a circuit breaker controlled by some "internal receptor", such as the flipping of a coin, that is, by some element of the network, whose state would be "not altogether determined by the previous states of other parts of the net"[99]

Such a device could be connected in such a way as to break up from time to time established connections or patterns of response, and to permit new combinations within the net to be formed, recorded internally in memory, and carried through into external action.

The results might resemble those of a "spontaneous impulse". Like all "spontaneity", they would be subject to limitations. All they could do would be to replace an old or highly probable configuration by a new or less probable one, *provided that the elements for the new configuration were already present in the net at the critical moment*—even though they might have got there only through the input in the immediately preceding instant.[12] The range of possible new combinations would therefore also depend, among other factors, on the range of possible new input information from the outside world, and on the effectiveness of the inner "habit breaker" in breaking up blocks against its integration with other data in the net. Apart from facilitating this inflow of new information, "spontaneity" could only bring out a wider range of the potentialities already contained within the net.

This type of feedback network might provide an analog for the problem of "Free Will". Such an analog might be found in a machine combining a determinate store of memories with a randomly varying inner receptor in the circuits governing recall and recombination. The random effects of the inner receptor (or "sudden impulse") are then limited by the statistical weight of the alternatives available from the stored past of the machine (its "personality"). Such a machine might act "freely", with initiative, but "in character".

The analogy suggests that "moral responsibility" is conferred by the determinate, cumulatively learned element in the combination. Each of us is responsible for what he is now, for the personality he himself has acquired by his past actions.

But no single act—and hence none of the past ones—was wholly determinate. Each act could have been different by the small but finite probability of the random internal "circuit breaker" producing a different outcome. Nor are we wholly prisoners of any one decision or any one experience. Ordinarily, it takes many repetitions so to stock a mind with memories and habits that at long last all roads lead to the same city, whether it be taken, in religious language, as the City of Destruction, or the City of Salvation.

This view seems to clash with the traditional one. Tradition sees responsibility for the "wholly" free act, as a limiting case, and no responsibility for the act wholly determined from within a person's past. The view explored here would take the opposite view of these limiting cases. It sees in the actual moment of decision only a *dénouement* in which we reveal to ourselves and to others what we already have become—or perhaps better, what we have become thus far. This view has parallels to that of St. Augustine, and more recently perhaps of Karl Jaspers and other existentialists, but it does not involve outside predestination. For each step on the road to "heaven" or to "hell", to harmonious autonomy or to disintegration, was marked by a free decision. Each was an act of free interplay between the randomness of the internal "pattern breaker" and the determinacy of character, the memories, and habits accumulated up to that step. The determinate part of our behavior is the stored result of our past free decisions.[13]

Freedom and Coherence in Societies

We have glanced at a few suggestive analogies between mechanical or electrical feedback nets, nerve systems, and societies. Let us conclude this brief survey of learning nets with a glance at some of the major differences among them.

A "machine" has been defined as an "apparatus for applying mechanical power, having *several parts each with definite function*"; and an organism as "a body with *connected interdependent parts* sharing common life"; and "organ" as a "part of animal or vegetable body *adapted for special vital functions*".[14] Both machines and organisms are here characterized by a high degree of permanence in the functions assigned to each part, be it a cog in the wheel or an organ grown permanently in its place in the body. Learning nets may conform to these limitations, if they are of mechanical or organic construction, but their functions as learning nets may point beyond these limits, to the different characteristics of societies.

Already modern calculating machines involve a balance between subassemblies permanently constructed for specific purposes, and transitory subassemblies put together from general elements to serve temporary needs. Multipliers in electronic calculators, e.g., are usually permanent subassemblies, but more highly specialized operations, such as use of one or several algebras, would ordinarily be performed by general equipment temporarily brought into a suitable sequence or configuration of activities.

Perhaps it could be said that this *possibility of relatively free transfer and recombination*, not only of the symbols treated, but *of the very physical elements of a learning net* for the performance of new operations, is the critical property which makes a given learning net into a *society*.

A learning net functions as a society, in this view, *to the extent* that its constituent physical parts are capable of regrouping themselves into new patterns of activity in response to changes in the net's surroundings, or in response to the internally accumulating results of their own or the net's past.

The twin tests by which we can tell a society from an organism or a machine, on this showing, would be the freedom of its parts to regroup themselves; and the nature of the regroupings which must imply new coherent patterns of activity—in contrast to the mere wearing out of a machine or the aging of an organism, which are marked by relatively few degrees of freedom and by the gradual disappearance of coherent patterns of activity. The distinction between learning nets which are machines or organisms, and learning nets which are societies, appears here as a matter of degree which turns into a difference in kind, that is, in overall behavior.[15]

The difference between organisms and societies

rests, then, in the degree of freedom of their parts, and the degree of effectiveness of their recombinations to new coherent patterns of activity.

This in turn may rest on specific properties of their members: their *capacity for readjustment to new configurations, with renewed complementarity and sustained or renewed communication.*

The degree of complementarity between the members of a society may determine its capacity for sustained coherence, while their degree of freedom—and their range of readjustments available without loss of complementarity—may determine the society's capacity for sustained growth. If the essence of growth, according to Toynbee, is increase in self-determination, then this concept of growth should prove applicable to societies and other complex learning nets. The more complex and readjustable the constituent parts of a society become, the greater the coherence and freedom of each of its subassemblies, the greater should be the society's possibilities of itself achieving greater coherence and freedom in the course of its history. Learning nets and societies do not grow best by simplifying or rigidly subordinating their parts or members, but rather with the complexity and freedom of these members, so long as they succeed in maintaining or increasing mutual communication.

Perhaps these few remarks will indicate the potential aid which our thinking might derive from these new advances in the fields of engineering and organization, and from the gradual emergence of a generalized model of a "learning net". The chances for a successful development of such new models are perhaps increased by the parallel development of a growing division of the labor of fact-finding and thinking in human organizations, as exemplified in the division and reorganization of intellectual labor in modern industrial and social research.

Notes

1. A more refined definition of feedback would put "output *information*" in place of "output energy", in accordance with the distinction between "communications engineering" and "power engineering." *Cf.* N. Wiener, *Cybernetics*, New York, John Wiley, 1948, p. 50.

2. There is also another kind of a feedback, different from the negative feedback discussed in the text: "The feedback is . . . positive [if] the fraction of the output which reenters the object has the same sign as the original input signal. Positive feedback adds to the input signals, it does not correct them." . . .

Only self-correcting i.e., negative, feedback is discussed in the present paper.

3. Analytical understanding of a process need not diminish its sublimity, that is, its emotional impact on us in our experience of recognition. *Faust* becomes no more trivial by our knowledge of goal changing feedbacks than a sunrise becomes trivial by our knowledge of the laws of refraction.

4. The four orders overlap; their boundaries blur; and there seems no limit to the number of orders of purposes we may set up as aids to our thinking. Yet it may be worthwhile to order purposes in some such fashion, and to retain, as far as possible, the model of the feedback net which permits us to compare these purposes to some degree with physical arrangements and operations. The purpose of this procedure would not be to reduce intellectual and spiritual purposes to the level of neurophysiology or mechanics. Rather it would be to show that consistent elaboration of the simpler processes can elevate their results to higher levels.

5. This applies, as do the preceding definitions, to continuous as well as to discontinuous physical processes. Information in television, for instance, is transmitted as a set of alternatives between bright and dark image points of discriminably different levels of brightness. These decisions can be represented mathematically on the binary scale. Such representation is also possible for any kind of continuous signal, since in practice it always is transmitted against a background of noise. Any variations in the signal which can actually be picked up can be represented very closely by a sequence of binary choices.

6. What is "labeled" in the latter case is not, in an electric net, a voltage pulse, but the state in which particular parts of the net are at a particular time. Such internal labeling of states of parts of the net itself is used in the automatic recall of stored information in electronic calculators.

7. An automatic telephone exchange capable of opening new channels in response to its own traffic counts was reported under construction by the Philips Company at Eindhoven, Holland. *Science News Letter*, Washington, D.C., April 10, 1948, p. 233. A telephone exchange which would install such a channel control itself would represent one more extension of the same principle.

8. Clyde Kluckhohn and Henry A. Murray, *Personality in Nature, Society, and Culture*, Alfred A. Knopf, New York, 1948, p. 9.

9. *E.g.* in the pressing of a key on a signal in N. Ach's experiment where "at the very instant of action no consciousness of will need appear," but "such a consciousness was none the less present as a 'determining tendency' prior to the action and governing it." R. Mueller-Freienfels, *The Evolution of Modern Psychology*, New Haven, Yale, 1935, pp. 109 f. *cf.* also pp. 41, 69, 236.

10. That is, retaining at each step a significant part of one's preceding past.

11. "*Receptors* are elements of a network whose state may be influenced from outside, i.e., which is not altogether determined by the previous states of other elements." Communication from W. Pitts, Massachusetts Institute of Technology, April 6, 1948. "Outside the network" need not refer to physical location. Any element influenced by any process other

than those in the other elements of the network—for instance, an inner element varying randomly—counts for a receptor under this definition. It is the last type of arrangement that I have called "internal receptor" in the text. It would differ from ordinary "external" receptors in that it might be largely independent from most other processes acting on the other receptors of the net. If there were several such receptors, each varying independently from all others, each would add another degree of freedom to the behavior of the net. One way of increasing the freedom of any such system would be therefore to increase the number of its quasi-independent receptors.

12. There are two kinds of information for a network. The first is outside information not previously present in the net. The second is an internal recombination of symbols not previously "recognized," that is, not previously matched in this configuration by a new secondary symbol, or symbols.

13. This view of moral responsibility would exclude those determinate elements of behavior which are not freely learned through intake of information, but are the results of heredity, mutilation, organic disease, or functional mental illness after it had disrupted significantly the processes of learning or decision making.

14. *Concise Oxford Dictionary*, Clarendon Press, Oxford, 1934, pp. 688, 804; my italics.

15. In some calculating machines, and perhaps in the cells of the human brain, there is some degree of reassignment of general elements to specific tasks or temporary subassemblies serving as "task forces." In some societies such as that of the *Eciton* army ants, there seems to be such a high degree of permanency of specialized function for each ant or class of ants, and so few degrees of freedom for an individual's choice of path that the entire column of ants may trap itself in a circular "suicide mill" where the path of each ant becomes determined by "the vector of the individual ant's centrifugal impulse to resume the march and the centripetal force of trophallaxis (food-exchange) which binds it to its group," so that the ants continue circling until most of them are dead. T. C. Schneirla and Gerard Piel, "The Army Ant," *Scientific American*, June, 1948, p. 22. The authors believe that communication among ants "resembles the action of a row of dominoes more than it does the communication of information from man to man. The differences in the two kinds of "communication" requires two entirely different conceptual schemes. . . ." *Ibid.* The concepts suggested in Norbert Wiener's *Cybernetics* and in the other literature on learning nets, cited elsewhere in this paper, could perhaps be applied to both kinds of communication. The action of rows of falling blocks or dominoes has been used by McCulloch to demonstrate more general principles of communication relevant for the understanding of more complex nets.

A.

SOCIAL CONTROL:
INTERNAL VARIETY AND CONSTRAINTS

48.
Social Control and Self-Regulation

S. F. NADEL

1

NO ONE WILL quarrel with the assertion that social existence is controlled existence, for we all accept a certain basic assumption about human nature—namely, that without some constraint of individual leanings the coordination of action and regularity of conduct which turn a human aggregation into a society could not materialize. It is thus true to say that "the concept of social control brings us to the focus of sociology and its perpetual problem—the relation of the social order and the individual being, the relation of the unit and the whole."

The question arises where this control resides. Clearly, the social order as such already constrains or controls; institutions, mores, patterned relationships, and all the other constituents of social existence prescribe modes of thought and action and hence canalize and curb individual leanings. In this sense control is simply coterminous with society, and in examining the former we simply describe the latter. Many sociologists choose such a broad interpretation. To quote from MacIver: "A very large part of sociological literature, *by whatever name* [my italics], treats of social control. . . . To study social control we must seek out the ways in which society patterns and regulates individual behavior. . . ."

Yet the fact that institutions, mores, and so forth, operating as they do upon potentially intractable human material, can maintain themselves and have stability suggests that there must

From S. F. Nadel, "Social Control and Self-Regulation," *Social Forces*, 31 (March, 1953), 265–73. Reprinted by permission of The University of North Carolina Press.

be further controls, safeguarding their continuance. These further controls are, of course, well known. They are exemplified in legal sanctions, in procedures of enforcement, and in any formalized apportionment of rewards and punishments, that is, in all institutions specifically designed to buttress the accepted norms of behaviour and coming into play for no purpose other than appropriate encouragement or restraint. So understood, the social controls no longer simply coincide with social existence but represent a special province within it and a special machinery outside the particular order they are meant to protect.

Anthropologists will tend to adopt this narrower definition, for they would lay that other, more immediate and pervasive control of human nature into their concept of culture or custom. For example, "The social controls found in a culture . . . are a body of customs by which the behaviour of the participants is regulated so that they conform to the culture." This seems the more profitable viewpoint, especially since it brings out the diversity of the processes involved, culture "moulding" or "canalizing" human drives, needs, or desires, while controls reinforce conformity and block deviance.

Viewed in this fashion the anthropological field is rich in instances demonstrating that societies keep their orderliness and cultures, their character even, though controls may be weakly developed or even absent. This suggests, then, that social systems or cultures must in some measure be self-sustaining. The familiar reference to the "force of custom," often taken to be paramount in primitive society, probably always has this implication. But it can be argued, more generally, that no society or culture can be without some elements of self-regulation and that any other assumption leads to absurd consequences; for if all cultural modes of thought and action, in order to function adequately, need specific controls, we might well ask what is controlling the controls, and so on *ad infinitum*.

What follows is an attempt to describe and specify the main elements of self-regulation. The first point to be made is that little is gained by adducing the force of custom and tradition, that is, the sheer inertia of habitual behaviour and inherited practices. At least, this force is not a final, irreducible datum.

II

Traditional behaviour, perpetuated through the habituations of long practice, has been numbered among the basic and irreducible types of social action by no less an authority than Max Weber[1]; yet it is doubtful if it can thus stand on its own, at least as regards conduct of any consequence.[2] Rather, traditional or customary behaviour operates reliably only when two other conditions apply and derives its force and apparent self-propulsion from them. Either acting in accordance with tradition (i.e., in accordance with old inherited models) is as such considered desirable and good; or, this way of acting happens also to be safe, known routine. In the first case the traditional action is also value-oriented, being indeed short-lived without this support, as is instanced by changing fashions and fads. In the second case the custom remains such because its routinized procedure affords maximum success with least risk. It is, I suggest, in these two conditions that we find the true elements of self-regulation.

This conclusion is borne out by a further consideration. It must not be assumed that custom (as now understood) takes care only of the less important features of social life while those relevant or crucial are safeguarded by specific controls. In the kind of society anthropologists mostly study, this is certainly not true, custom and tradition governing a wide field of conduct and, within it, activities of great relevance. Indeed, it can be shown that in primitive societies the specific controls, far from being indispensable in safeguarding important activities, tend to be weakest in their case. In other words, any activity which is socially important may by this very fact already be protected from deviance or neglect.

There is nothing surprising or paradoxical in this assertion once we are clear on what we mean by "social importance." There are various criteria for its assessment, of which two are relevant in this context. First, an important social activity is stated to be such by the actors; the criterion therefore lies in the value judgments and convictions of the people we study. Secondly, from the observer's point of view, the importance of any social activity is established by its focal position among all the other social activities; more precisely, any activity is important to the extent to which a series of others depend on it, in a practical and instrumental sense, being incapable of achievement without the focal activity or impeded by any variation (through neglect or disregard) in the latter. This nexus in turn rests on the multivalence of social activities. By this term I mean, briefly, the capacity of any activity (to the extent to which it is focal) to serve also ends or interests other than the one for which it is explicitly or primarily designed. Examples will presently be quoted.[3]

As regards the first criterion of importance, I shall attempt to show that convictions about values have sufficient force to sustain and direct behaviour with no aid other than that implied in the second criterion. And as regards this, it includes self-regulation by definition; for the more focal any mode of behaviour, the more strongly it is rendered invariant by the aggregate pressure of all the other activities and interests dependent on it—lest indeed social life, or a wide area within it, become dislocated. Adherence to the prescribed (and important) norm of action is thus, once more, adherence to a safe procedure, both for the society at large and for the individual actor.

That the two supports of self-regulation—notions of value and instrumental nexus—hang together has already been suggested. They can in fact be regarded as complementary, for if values imply positive guidance for action, the effect of the instrumental nexus is to impede deviance from given courses of action. It will be more convenient to begin with the discussion of the latter.

III

In prototype form, the instrumental nexus appears also in single activities; for any mode of action which is an appropriate means to a given end tends to become routinized and self-maintaining for that reason. This principle (exemplified already in the psychological "law of effect" governing elementary learning processes) is essentially one of economy, in effort and trial-and-error; and the most convincing instances in social life probably come from the field of technology and economics in general. But aesthetic and recreational activities are similarly self-regulating, for any artistic style will again be perpetuated as long as it represents an adequate means for its particular end (the desired satisfaction or stimulation). The infrequency of technological invention in primitive societies provides the broadest evidence of this kind. If it be argued that this proves, not the self-maintenance of adequate methods, but a primitive, tradition-bound mentality, which prevents people from exploring new and possibly better procedures, the answer is this: primitive peoples are disinterested only in inventiveness, in experimentation with new techniques, not in innovations as such. Significantly, they have little hesitation in copying (or borrowing) novel methods, that is, in adopting them when they can be seen *in use*. The pull of tradition, then, means in fact reluctance to abandon a safe routine for the risks that go with untried methods.

Why this factor should be more powerful in primitive than in advanced societies we need not discuss. But it must be mentioned that in the former the tenacity of inherited routines, technological or aesthetic, is often reinforced by sanctification. Now this means that the routine is invested with an additional value of sacredness or of desirability in virtue of divine derivation and the like; and this means, further, that the technological or aesthetic task is meant to serve more than one interest. For it is now required to attain its own intrinsic goal as well as conform to religious commands, lest the actors forfeit some expected benefit, imaginary (supernatural protection) or real (the normal fellowship which might be refused to the irreligious). We note that this is already a first instance of that more far-reaching self-regulation which we derived from the interdependence of diverse social activities and from multivalence or focal position.

IV

Any mode of behaviour operating also as a *diacritical sign* (i.e., a differentiating sign or symbol) is multivalent to a simple degree. Thus customs of dress, eating habits, style of residence or manners of speech, apart from attaining their intrinsic ends (protection of the body, satiation of hunger, shelter, communication), will often also indicate a person's social status, group or class membership, and generally his social relationship with others. The continuance of these indicative modes of behaviour is thus reinforced by the importance of the state of affairs they indicate. Or, in terms of individual behaviour, individuals will keep up a certain dialect, manner of dress, and so forth, in order to evince their status and group membership and, implicitly, to qualify for the benefits that go with them.

We note that the additional valence of the activities in question attaches to their form or style, not to the efficiency with which they are performed and attain their intrinsic ends. In other instances it is the latter which has the double valence, in the sense that efficiency in the performance of one activity becomes a qualification for participation in a second, desired one. Among the Nupe of West Africa, for example, a man whose sons are guilty of a serious misdemeanour cannot hope to be appointed to a rank and title; among a certain Nuba tribe in the Sudan a candidate for shaman priestship will be unsuccessful if he happens to be a lazy or unsuccessful farmer. We might here, briefly, speak of an incentive or premium meant to ensure the socially approved conduct.

Examples of this kind can easily be multiplied. But it is important to emphasize that the incentive is incidental, not specific, and that the rewarding achievement is such—a premium—only among other things. If it were not so, we should be dealing with a specific machinery of control, not with features of self-regulation. Nupe society is not otherwise concerned with the success of parental discipline (save in approving of it), and the bestowal of ranks and titles implies other, more relevant, qualifications besides; the Nuba tribe does not reward good farmers and punish bad ones, and again expects would-be shamans to give more substantial proof of their eligibility. All that happens is that parental discipline or diligent farming is linked with the other desired achievements, failures in the former reducing the chances of the latter. Differently expressed, deviations from the socially approved conduct are *penalized* (not punished). The step from one to the other, though narrow, is unmistakable. We need only think of the exclusion from sacramental offices of sinners or of the loss of civil rights threatening political criminals.

It is clear that the efficacy of such linked incentives requires a closely geared social system. In our examples the public standards of conduct are affected only because every man is both a family head and a potential rank-holder, a farmer and a possible candidate for priestship. The linkage itself presupposes that rank and priestship are conceived sufficiently broadly for the specialized qualifications to be combined with other, extraneous ones. In more general terms, the regulative effects must vary inversely with the separation of social roles, with the specialization of offices and tasks, and, implicitly, with the size of groups (since only small groups can function adequately without considerable internal differentiation). It is precisely the small scale and lack of internal differentiation which characterize the societies we commonly call primitive and hence enable them to lean more heavily on such machinery of self-regulation.[4]

Its effects may be rendered both more pervasive and more unspecific, consisting in general complications and obstacles facing any person who would depart from the accepted norm. The premium, then, lies merely in the smoothly functioning, normal course of events individuals expect to encounter; the penalty, in normality dislocated and expectations disappointed. Some of the most crucial norms of conduct in primitive societies, such as exogamy or incest taboos, are often regulated by no more specific sanctions, secular or religious, yet are adhered to with great strictness.

Consider for example, a society, patrilineally organized, where marriage is prohibited between aganatic kin, is contracted by the payment of brideprice, and entails specific duties towards the offspring on the part of both father's and mother's kin. If any man married in disregard of the first rule, the others would fail to work also. The brideprice would have to be paid within the same descent group, while in the people's conception it is a payment suitable only between such groups, being meant (among other things) to indemnify the bride's group for the loss of her prospective progeny. The offspring of such an irregular union would forfeit the double assistance from two kin groups since the father's and mother's kin now coincide, and would be less advantageously placed than the offspring of customary marriages. And there would be various other, minor but no less confusing, complications; for example, rules of avoidance (obligatory towards in-laws) and intimacy (towards blood relations) would now apply to the same people. In short, one breach of routine disrupts routine all round, and the individual is faced with a wide loss of social bearings.

It has been suggested that this type of marriage enables kinship affiliations to be extended beyond the single descent group and thus creates additional bonds consolidating an otherwise segmented

society. This, then, is one of the multivalences underlying the chain of effects just described: the institution of marriage, over and above regulating sex relations and procreation, also serves that other end, the strengthening of social solidarity. It is doubtful if the actors themselves think, or think clearly, in terms of this ulterior objective (or function). Yet if they are not capable of assessing this widest instrumental nexus, they are aware of the multiple consequences threatening them personally, and act from this knowledge.

This is only a restatement, from the actors' standpoint, of the point made before, that the crucial importance (or focal position) of any mode of action also protects it from deviation. Here, too, the character of self-regulation is most clearly marked. For consider that in the specific controls the disabilities (or sanctions) imposed upon the transgressor are mostly extrinsic to his mode of conduct, leaving the success of the latter unaffected. An adulterer, for example, though he may have to face flogging, a fine, or imprisonment, will have attained the satisfaction sought in the criminal act. In the instances here considered the penalty is intrinsic, lying in the incapacity of the criminal act to provide the expected satisfaction: the unorthodox marriage is simply an unsuccessful marriage judged by all the expectations it normally fulfils.

Quite often this notion of self-regulation by ill-success finds explicit expression. The potential transgressor will merely be warned of the frustrating and self-negating consequences of such-and-such a mode of action. The Navaho Indian, for example, "never appeals to abstract morality or to adherence to divine principles. He stresses mainly practical consequences: 'If you don't tell the truth, your fellows won't trust you and you'll shame your relatives. You'll never get along in the world that way'." At least, statements of this kind will appear couched in terms of imagined, supernatural guarantees. Thus in a Nuba tribe incestuous marriages are said to remain barren, and the people would add, "Since one marries for the sake of children no one would break the marriage rules." The Tikopia hold similar views and in fact explicitly state that incestuous marriages are doomed to failure or ill-success. And proverbially there is the curse on ill-gotten gains.

We may finally consider the extreme case, when the loss of social bearings is very nearly complete. Here the transgressor is practically excluded from all normal expectations, left without a niche, and relegated to the role of a misfit. In primitive societies it is often unheard of for a man or woman to remain unmarried. Now this phrase "unheard of" indicates the pressure of the multiple consequences: the bachelor could reach no position of responsibility or authority; his economic pursuits would be seriously hampered in a society where the family is the main source of cooperative labour; he might have no one to look after him in sickness or old age, and no one even to bury him or perform the rites of the dead. In the Nuba society mentioned before, dissolution of marriage, though not unheard of, is considered most undesirable and in conflict with tradition. In practice this means that the woman who leaves her husband can never enjoy normal life: her children would not be hers; she forfeits the support of her own kin; and she would die without the customary ceremonial.

Let me stress that these consequences still represent only ill-success, in the business of living, as it were, not genuine sanctions—disabilities specifically inflicted. The culprits merely suffer failures which, in a society so-and-so constituted, cannot be avoided once the rules are broken. There is no intentional discrimination against the offenders *qua* offenders; nor is there any thought of a stigma imposed on them (say, that of "living in sin") and ostracism because of this. Rather, as we shall see, stigmatization is on the whole alien to primitive societies.

V

Whenever we spoke above of desired, approved, good actions or their opposites we were, of course, already referring to values. Indeed, this concept, like that of social control, can be given so wide a connotation that it becomes coextensive with social existence or social behaviour. For since the latter is, by definition, regular and aimful behaviour, thus implying consistent choices between possible courses of action, social life in its entirety might be said to express or implement preferences and idiosyncrasies, notions of worthwhileness and undesirability, in short, values. This view is expressed, for example, by Lasswell and Kaplan: "A value is a desired event—a goal event"; "Conduct is goal-directed and hence implicates values."

It is more profitable, however, to restrict the concept somewhat, in this sense. It shall be understood to refer to worthwhileness of a non-trivial kind (excluding, for example, table manners, fashions of dress, etc.); it shall not be coterminous with practical utility (e.g., of tools, technological procedures) but bear on more autonomous forms of worthwhileness (e.g., things morally good or aesthetically desirable); it shall refer to classes of objects (things, events, states of affairs desired or disdained), not to individual ones, so that it indicates maxims of action, not *ad hoc* preferences; and the former shall be ideologically founded, i.e., capable of being expressed by the actors in generalized assertions on right and wrong and the like.

It is clear and hardly worth saying that the specific controls found in any society inevitably imply values and normally refer to values widely accepted. What is important is that controls cannot ever be wholly effective unless they endorse what most people hold desirable (or value), which point is amply illustrated by the difficulties of enforcing laws no one believes in or of maintaining a hated regime. Even societies relying on machineries of control, then, must rely also on values simply held. Ultimately, the social norm has power inasmuch as it is internalized, that is, inasmuch as the public assertions on right and wrong are also the private convictions of individuals. In the language of psychoanalysis, the commands made on behalf of the society are absorbed in the Super-Ego.

We may for the moment disregard the processes whereby the Super-Ego is built up, through verbal instruction and the impact of models for acting. Nor need we emphasize again that in primitive societies it seems sufficiently powerful to determine conduct unaided by extraneous controls. But, as previously suggested, even where there are controls, their chain ends at some point; and here values simply held, internalized, will be the final pivots of desired conduct. In that sphere the observer can only state that, say, matrimonial sanctity or the observance of exogamy are held to be "right" or "good" and deviations from the norm, wrong or evil.

Sanctification will often obscure this self-reliance, deriving the final values from divine ordinances or some superhuman authority. Perhaps the religious guise, reducing as it does the absoluteness of the moral or social norm to quasi-human acts of will or providence, invests the norm with greater persuasiveness; at least, the acceptance of ultimate principles is moved back a further stage and joined to that other ultimate conception of an ordered universe. But in primitive societies this theological sophistication is often absent. There is no conception of deities as law givers; rather, they are merely the guardians and exponents of the moral principles. If they punish or reward they do so only because they are themselves subject to the given doctrine of good and evil. As Firth points out for the Tikopia, "The spirits, just as men, respond to a norm of conduct of an external character. The moral law exists in the absolute, independent of the gods." Indeed, the supreme deity may be altogether aloof from all moral concerns, while minor deities, spirits, or other mystic forces, would aid or hinder human action regardless of moral principles. Their influences are needed only to explain why "good" actions may fail and "evil" ones triumph, and hence to serve as a foil for the self-reliance of the accepted values. Furthermore, sanctification as such, making any object or action sacred, holy, or mystically "right," merely adds another notion of worthwhileness, no less final than the simple "good."

Yet even where supernatural sanctions are accepted, they clearly cannot be simply aligned with secular social controls. That punishment and reward will materialize, administered perhaps by an omniscient deity, is a question of faith, not of verifiable consequences; and this conviction is much closer to the internalized dictates of conscience than to the anticipation of public correctives or appraisals. Nor is it rare for the supernatural sanctions to be so conceived that they are entirely reduced to appeals to conscience. Among the Nuba tribes infringements of clan taboos or exogamy are sometimes believed to be punished with leprosy, which may unpredictably befall any descendant of the culprit. The potential sinner, then, will be restrained by the fear of bringing suffering and disgrace upon someone he does not even know, that is, by a further emphasis on the evilness of his deed.

In one respect belief in supernatural punishment differs sharply from internalized values. Any deed inviting the former, any sin, that is, can usually be expiated, the system of beliefs defining the sin also showing the procedure for regaining purity. But disregard of the dictates of conscience entails guilt, the awareness of which cannot be wiped out by established procedures. The conflict thus engendered must be resolved or borne by the individual alone and may well leave only escape into neurosis or suicide.[5] I do not know and can hardly imagine a society which relies exclusively on internalized commands and their correlate, guilt. That guilt is so often made translatable into sin reflects, I suggest, this risk—which no society can face—of denying to culprits all chances of expiation.

VI

We may, in the same light, assess the merits and demerits of self-regulation by "multiple consequences." Consider that any sanction proper (disregarding the extreme penalities of death, life imprisonment, and perhaps expulsion) limits the consequences of the crime to a single event, the punishment, and afterwards offers the culprit a new chance. But the obstacles and hardships the offender creates for himself when committing an unpunishable offense cannot be cut short by any single act of atonement. Though they may involve less physical suffering, they are more hopeless since there is no way of repairing that total loss of social bearings. The more widely a society relies on the automatic efficacy of this threat, the more severely are offenders imprisoned in their own actions. I know of no society completely commited to this method of self-regulation; but it is sufficiently powerful in some primitive groups to account for self-exile or even that ultimate means of escape, suicide.

Primitive societies are in this respect in a dilemma. As I have suggested, it is their smallness, low internal differentiation, and closely knit organization which afford the possibilities of far-reaching self-regulation; and it is precisely societies of this kind which are weakened most seriously by the loss of members. The adoption of specific mechanisms of control, therefore, apart from corresponding to the requirements of a less closely geared social system, also represents a loophole in a social system too rigidly and permanently penalizing transgression.

The underlying assumption, that primitive societies are aware of this risk and hence in some measure concerned with rehabilitating offenders, is easily proved. It is borne out by the widespread treatment of homicide almost as a "civil" offense, so that even murderers can, after payment of *wergeld* or similar compensations, resume their normal place in the community. Frequently too there are formalized procedures of reconciliation after the punishment of torts, meant to "cleanse the hearts of anger" and to accentuate the fact that the transgression has been disposed of. Similarly, primitive societies make little use of any lasting stigma or ostracism. In no society is it altogether avoidable that the disgrace of a particular transgression should follow the culprit through life and overshadow all his contacts with his fellow men. But there can be no doubt that it is the advanced rather than the primitive society which tends to exploit this effect in consciously creating the outcast or his milder version, the *déclassé* individual. If in certain primitive groups stigmatization does occur, this is only part of another dilemma: for the same society would in other ways show its desire to rehabilitate offenders.[6]

VII

It will perhaps be argued that I could attribute such importance to self-regulation and especially to values simply held only because I neglected two crucial and obiquitous controls—the diffuse sanctions implied in public criticism, shame or ridicule, and the institutionalized procedures of education. My answer would be that the two controls not only safeguard but also presuppose values and hence represent, not so much controls acting from outside upon the desired conduct, as phases in a circular process whereby values engender conduct and conduct reinforces values.

Voiced disapproval, of any kind or mood, may of course be merely a sign foreshadowing more compelling consequences (i.e., some concrete penalization, some loss, more or less severe, of normal chances). Disregarding this eventuality, the disturbing psychological effects of the disapproval, without which it would not be a sanction, presuppose that it cannot be evaded or disregarded. If, say, the shamefulness of adultery is not a matter of general agreement, the adulterer (assuming he is not punished in any other way) will always find some supporters among the public and be able to heed their approval rather than the disapproval of others. Above all, if he does not himself endorse the criticism, even tacitly or subconsciously, he can simply set it aside as irrelevant. In other words, blame, ridicule, or holding up to shame are controls only if they express commonly accepted values and correspond to the promptings of the Super-Ego. Admittedly, they also bring these promptings into the open; but this only means that they render them more acute. Indeed, it might be said that the strongest of these diffuse sanctions, shame, derives its very strength from the fact that it is an "exposure"—of inadequacies privately felt.

As regards education, it is a truism to say that its widest efficacy (ignoring purely technical skills) lies in the inculcation of lasting attitudes and viewpoints, that is, of values subsequently to be simply held and followed. Without this, its force of control would be restricted to the actual period of tuition, when the educator employs rewards and punishment and other means of coercion;

nor could there be any reason why a child taught a particular way of behaving should, as an adult, hand on this knowledge to his offspring, thus perpetuating the social order. Education, then, as I see it, merely provides occasions for self-regulation to emerge or re-emerge.

Needless to say, it is not the only such occasion. The notions of worthwhileness taught in family or school are reinforced in various ways throughout life—by religious doctrine, by the topics of art or legends, and by the symbolic dramatizations of ritual. But here it is difficult to distinguish between occasions and consequences or, if we call the former "controls," between these and the things controlled. The methods of education, the content of art, religious beliefs, though they demonstrate and enjoin precepts for conduct, are themselves forms of conduct and perpetuated only because they follow from these very precepts. We have, in the last resort, merely multiple instances of a given system of values, irreducible to any further regulative machinery save that circular process mentioned before, which seems inherent in any value system of real efficacy. The circularity goes even further; for any public act in harmony with the obtaining values becomes in some degree a model exhibiting their validity, and so adds to their efficacy. Whenever a person observes, say, exogamy or pursues blood revenge as demanded by the social norm, he not only executes a prescribed procedure but adds to the instances demonstrating that this procedure is indeed valid, which addition is as much a reinforcement of the moral values as is their explicit assertion or teaching.

For this widest circularity we find a physical model in what are now known as "feed-back" systems. Taking the whole society to be the relevant system, any "output" of the intended kind—any conduct in accordance with the social norm—is partly returned as "input," i.e., as information sustaining further action of that character. The self-regulation implied in "linked incentives" and "multiple consequences" represents the exact counterpart. If the efficacy of values corresponds to a "positive feed-back," the other types of self-regulation correspond to "negative feed-back," controlling output through signalling errors—the errors being the forms of deviant conduct whose penalizing consequences force action back into the intended channels.

It will be seen, further, that guidance through values and penalization must operate consistently if the social order is to be maintained. The positive precepts of worthwhileness will normally reduce experimentation with unorthodox conduct. Yet if the latter does occur and fails to carry its own penalty, the underlying values will inevitably be weakened. This mutual agreement probably represents the most vulnerable area in any social system, for here theoretical convictions and practical experience must teach the same lesson. The frequent cry of morality collapsing nearly always refers to a situation where disregard for conduct taught and enjoined is no longer penalized through being demonstrably unsuccessful. The social order, if it is to survive, must then be refashioned, with values once more consistent with practical experience.[7]

We may add, finally, that the specific controls are equally fitted into the circularity of value systems. For the controls both follow from the value system and demonstrate it, since the punishments and rewards bestowed by societies are normally public acts. That circle is broken only when rulers, judges, legislators or, for that matter, teachers and moralists apply or preach a doctrine in which they themselves do not believe. They then stand outside the value system they wish to maintain, whoever they may be—a conquering minority enforcing laws fit for the masses, a group of Supermen *à la* Nietzsche, claiming to be "beyond good and evil," or a cynical *élite à la* Pareto or Sorel.[8] Here the question, "what is controlling the controls?" makes sense: the answer is—self-interest and calculations of political expediency. And here, if you like, we touch upon social controls in purest form, exercised from outside and unobscured as well as unaided by any self-regulation.

Notes

1. Max Weber, *The Theory of Social and Economic Organization*, trans. by A. M. Henderson and Talcott Parsons (New York: Oxford University Press, 1947), p. 105.

2. Weber only admits that the "pure type" of traditional behaviour (as of all the other basic modes of action) is rarely met with in concrete situations; there, it will tend to shade over into or overlap with the other types, especially with "value-oriented" action.

3. The "multivalence" of social activities has been described in various terms. Malinowski used to speak in a similar context of "amalgamation of functions." Recently Firth referred to the "concern" which an "action or relation has for all other elements in the social system in which it appears"; see Raymond Firth, *Elements of Social Organization* (London: Watts, 1951), p. 35.

4. Highly complex societies, too, exploit it whenever they are organized on "totalitarian" lines, e.g., when social promotion of any kind is impossible unless the candidate professes the "right" kind of religion of political conviction or lives according to approved standards of morality.

5. Malinowski quotes several examples of suicide committed in consequence of such overwhelming awareness of guilt—"as means of escape from situations without issue"; see his *Crime and Custom* (New York: Harcourt Brace and Company, 1926), pp. 94 ff. Though these examples refer to a guilt reinforced by shame, i.e., brought into the open by direct or public accusation, this does not detract from the crucial guilt motive.

6. Thus in the Trobriands, where the breach of exogamy is followed by ostracism (which may drive the culprits to suicide), blood revenge is mostly replaced by compensation (a "peacemaking price"). *Ibid.*, pp. 80, 115.

7. Weber's and Tawney's familiar studies relating the Reformation to the rise of capitalism demonstrate such a process of reconstituted consistency, between a new rewarding economic practice and a value system hitherto deprecating wealth and material success.

8. Cf. Karl Mannheim's summing up of Fascist ideology: "The superior person, the leader, knows that all political and historical ideas are myths. He himself is entirely emancipated from them, but he values them . . . because they . . . stimulate enthusiastic feelings . . . and are the only forces that lead to [the desired] political activity." *Ideology and Utopia* (New York: Harcourt, Brace and Company, 1936), pp. 122–123.

49.
Conformity-Deviation and the
Social Control Concept

ROGER NETT

I

HISTORY OF INQUIRY teaches us that much depends upon the starting point of any analysis. The early tie of American social control theory to social psychology is an example. The observation by writers in the century preceding this one of the operation of irrational mechanisms in social context was not without social control implications. When at the turn of the century the late E. A. Ross produced a volume on the subject, he did, as he said, make social control a branch of the then emerging social psychology. Especially, he was able to show the significance for social control of the works of Tarde and Le Bon, whose mechanisms were at that time in contrast with overworked doctrines of rational[1] social structure, and as such, they were enlightening. In the various forms of imitation and suggestion there was a means of perpetuating society (with, it must be said, tendencies toward instability) at a subrational level. But Ross, who had had a liberal nineteenth century education, saw these processes largely as a complement to rational behavior, perceiving man as, in the last analysis, the measure of the social situation. He concluded that "one who learns why society is urging him into the straight and narrow will resist its pressure. One who sees clearly how he is controlled will thenceforth be emancipated. To betray the secrets of ascendancy is to forearm the individual in his struggle with society."

The emerging tradition of a social-psychological interpretation of social control was almost immediately given strong impetus by a careful observer of individual human behavior, C. H. Cooley. Such concepts as the "looking-glass self" and "primary groups" were easily recognizable as fruitful in understanding the relationship of the individual to the collectivity and have for that reason attracted the attention of students to this day. The processes of suggestion and imitation could be shown not to be mechanical, as in the presentation of Tarde and Le Bon, but rather a dramatic interactionary process in which the individual found a purposeful place in his society. While, in keeping with his time, Cooley was an optimist and conceived man as basically rational and therefore moral (to the extent that he used his rationality to calculate his behavior in terms of social welfare), the result of Cooley's observations could as easily be interpreted to indicate a condition of an entirely different nature. To the extent that the self was social it could be looked upon as determined by society. The social environment in which persons behaved "voluntarily," without noticeable coercion —in fact by using the individual's own energy— was continuously making the individual a product of itself.

At approximately the same time another American, W. G. Sumner, in words most explicit, described characteristics of "folkways" and "mores" which were, by a process of genesis, irrational and unconscious. Thus was introduced a social control that was impersonal and naturalistic. Sumner, unlike Ross, had not supposed that exposing the irrational forces acting on the individual might cause him to supplant them with varying amounts of rationality. Social control was the natural action of culture upon a human being whose "'faiths'—are not affected by scientific facts or demonstration."[2] This pessimism of Sumner was, it may be noted, a clean break with Victorian outlook.

Notwithstanding either their differences or the intentions of each (in fact, contrary to their intentions), the works of both Sumner and Cooley served in a roundabout way to establish pragmatically that society had provided a broad basis for the individual to conform. The person who did not conform, the social deviant, could be regarded as spurious to a process—a failure of socialization. The extent, in ethnological instance and in typical group organization, to which the conformity aspect of persons[3] could be shown quantitatively to exceed the deviation aspect of persons made the latter appear almost too small to warrant major consideration.

When Park and Burgess wrote their now classic sociology test, social control was assigned to the area of problems of "administration," of "policy and polity," and of "social forces and human nature," the latter (they said) being the specific function of the sociologist to discern. On a basis of what had gone before, social control processes were established in which the individual was controlled. Emphasis was put on forms of control of men singly by men collectively. These included elementary irrationals, public opinion, and institutions, or essentially those elements given meaning by Sumner. The introduced concepts of "administration" and "policy and polity" (undiscussed) and the fully developed concept of processes of interaction were, however, considerably more than the naturalism of Sumner. With their inclusion, social control was to be also a science of strategy not entirely divorced from what Machiavelli had said it was some four hundred years before.[4] The inclusion was in part only an abbreviated one, for Park and Burgess had, as was mentioned, chosen to relegate much of the interpretation of the rational to political scientists (to students of administration and policy and polity) and, as sociologists, to talk about the social-psychological behavioral items, the understanding of which is dependent largely on perceiving irrationals. This, it can be noted, was not an inefficient division of labor, as fruitful results have shown, but it helped establish for American sociologists a precedent to circumvent extensive analysis of the rational or strategic in social control.[5]

A variation of the line of thought occurred in the work of Lumley, who either had not understood Sumner or chose to redefine "mores" for non-naturalists. He found that by a rational process, "the mores are those folkways which have been *examined*, *judged* useful and beneficial and *made* into approved activity-patterns" (italics mine). Around this concept he presented his "means of social control." Lumley's "effective will transference" was a control strategy specifically of a Machiavellian type. A difference is that Lumley, instead of outlining a rational means whereby a selfish Prince could keep control over the masses of people, outlined means whereby the masses of people "trudging along half awake and fairly comfortable according to an age-old pattern which is measurably acceptable" could control "those persons who for countless reasons go astray." His interpretation places the deviation aspect of persons in direct opposition to social organization. He finds that "they diverge from the general course by reason of selfishness, greed, love of adventure, timidity, stupidity and what not; they manifest this divergence by stealing, murder, uttering heresy, refusing to work, boasting, strutting, breaking contracts, and doing many other things."

Subsequent writers on social control, Landis and Bernard, have synthesized different parts of the above heritage. Landis, in his textbook on the subject, followed largely the leads of Cooley, but he went beyond Cooley to analyze problems of change and especially problems relating to secondary-group morality. Bernard retained the anthropological naturalism of Sumner, enlarging upon the social control meaning of institutions.

More recently, unified theories of social control have all but been abandoned in favor of studying specific aspects of social control under such diverse headings as (to name a few) "structure and function," "collective behavior," "role and status." In general these emphasize the conformity tendencies of men.

II

Before discussing relationships between the deviation aspect of persons and social organization, it is necessary to make clear what meanings of "deviation aspect of persons" are reasonably possible. Once these are determined it will be pertinent to attempt to show the relative significance of each to social control theory. Wherever possible, standardized terms are used with an attempt to pinpoint their meaning. Additional types are presented, however, which for the moment escape orthodoxy.

First, it is pertinent in social control theory to distinguish two types of social deviation: social and social-organic personality failure and inadequacy which is discernibly degenerate and which we can therefore in good usage call "pathological," and deviation discernibly falling outside of this category which is "nonpathological." It is here implied that the area of confusion between the two is smaller than is sometimes contended. The terms pathological and nonpathological simply indicate conditions of persons individually or segmentally and are not intended to carry loose implications as do such analogical terms as "pathological society."

While the first of these, pathological deviation, is causally related to areas of social problems and is important to applied sociology, it can, as far as social control theory is concerned, largely be dismissed as a calculable entity. Quantitatively it occurs as either a constant or an easily discernible (hence predictable) linear function, subject to numerical manipulation, including (as will later be suggested) cancellation. The one instance where it can be made to affect social control is where a confusion or a false identification of it is made with other types of deviation. In practice this would produce in a society, a condition known as "witch hunting," in which the nonconformer is constrained ostensibly for degeneracy (historically, extra-human). It is true, as writers have indicated, that many social deviators are no more than pathological failures or inadequates (in complex societies a significant number), and that these, on first observation, cannot always be separated from other types of deviators. Logical linking of such categories, however, is arbitrary, for as a rule the factor "pathological" can be kept discrete, both for purposes of study and (ideally) in social control practice. If such a separation were not possible in order to clarify concepts, it would be necessary to introduce a frame of reference in which there could be discerned other oblique relationships. For example, no one has attempted to ascertain what percentage of conformers are also pathological failures and how being pathological affects conformity. (Especially one might suppose that the condition of inadequacy favors simple conformity.[6]) This is to say that it is likely that the element, pathological failure, is not reducible to social deviation except in restricted reciprocal usage. Since the society (i.e., peers of the individual) determines the degeneracy or constructiveness of a specific act of deviation, the simple pathological deviator does not ordinarily produce an effect of any serious consequence for social control and is only a part of calculated social or social-organic personality failure.[7] There may be found a proportion of constructive deviates represented by such historical figures as Gibbon, Nietzsche, etc., who have coincidentally suffered pathological circumstance, and in such cases the two conditions might be shown to have a causal connection, but, pertinent to this discussion, the product of their efforts is seldom taken for inadequacy. Most degeneracy follows patterns well known to a given society and is, therefore, accepted as consistent with a percentage-expectancy of understood pathological behavior. That a Columbus or a Galileo can be "mistaken" for a fanatic carries a message to the student of social control, but it does not affect the everyday world of the pathologist.

Nonpathological deviation, i.e., social nonconformity due to psychological and sociological causes which fall outside of any area of degeneracy, may be reduced to a number of discernible factors which by contrast vitally affect social control. A list of these would include deviation caused by such generally understood conditions as minority attachments, marginality to the group, and specific ambiguities in the culture, any of which are objective conditions to which individuals so affected will have additional, that is, more than ordinary, adjustments to make. Such a list would also include conditions invoked more directly by the individual, such as, imagination or intelligence exceeding that called for by the social situation, moral equivocation, and, more generally, pursuit of ideals and strategies.[8] It is these last categories which have seen the least of convincing sociological generalization. By the fact that they place emphasis on a diversity inherent in nonconformity they do not satisfy the qualifications of mass behavior so familiar to the sociologist and the social psychologist. Far from readily calculable entities, their ramifications may be supposed to be as complex as the strategies which they subsume. It is also true, as has been noted, that there has been a tendency to consign them to others, to political scientists, economists, etc., or else as C. Wright Mills more radically suggests, we leave discernment of them to the "practical" as distinguished from the "professorial" man. Yet the writer would submit that they are ones most laden with implications to social control theory.

III

What then is the relationship of control-significant categories of the deviation aspect of persons to effective social organization? Social control most frequently has been interpreted as a problem centering on how society orders, conditions, and controls its membership. Equally tenable is an hypothesis that a problem of regulating a society is to tap, organize, and adapt its creative strength. A major difference is that the first of these propositions emphasizes the functionality of social conformity as a product of social organization, while the second emphasizes that of divergence in generating continuous social organization. Since the creative strength of a society must be sought in the capacity of individuals to evaluate, extend, correct, and ultimately to alter existing definitions and understandings (a process which is, in effect, deviation), the problem of ordering a society becomes one of utilizing the vital element—deviation—in social-organizational context. Whereas no rigid identification of creators with deviators can be made, to invent, to extend, to correct, and, summarily, to perform actively and constructively in the social world as we know it, imply a rational character and a freedom which in each instance fall outside of conformity. That such is the case is not an idea of recent attainment; it has long been implied in theories ranging from doctrines of progress to doctrines of individual freedom and rights. Methodological reasons why more extensive consideration of the constructive character of deviation has tended to escape the framework within which American social control theory developed are suggested in the first section of this paper, which is an account of what are now historical observations by social control writers. Such traditional questions as "how is conformance brought about and maintained" may, in social control theory, be misdirected. They seemingly carry unanalyzed assumptions concerning the nature and desirability of conformity, especially in condoning a most singular failing of mankind, his shortsighted tendencies growing out of inertia. At the same time they have drawn attention away from broader meanings of social deviation, not infrequently allowing the deviation aspect of persons to be seen as an antithesis of conformity, a threat to the existence of social organization, or "chaos threatening the social order." They can hardly lead to a discovery of when and under what circumstances the deviation aspect of persons is responsible to social organization.

One is therefore led to ask what, more exactly, is the tie of conformity/deviation to social organization/disorganization? And also, is the "problem" of social control an inherited one of bringing about and maintaining "order," or is it one of carrying on effective human social life within the limits of available understandings of social man and temporal sequence? Let us approach the matter as if it were the latter.

It is thought to be true that the deviation aspect of persons is more active in times of social disorganization. During such times persons are not only under more stress but they are seemingly under more obligation to think independently, a process which (although often regressive) favors, as against collective "thinking" (inertia), rationality. Social disorganization implies that institutional definition is not interpreted by persons to be adequate to their needs and that "customary" behavior has given way to expediency. In such circumstances the individual is likely to attempt self-preservation by devious means, one of which is original thinking. The same recourse to insight would occur during periods of intense organization, reconstruction, or in general when social change is rapid. At the same time it would be misleading to suppose that the deviation aspect of persons is causal to social disorganization. It would even be misleading to suggest that it is outside of the effective organization of society, for it cannot be, if by a social organization we mean to include the whole of a society.

In this connection it may be noted that in at least two ways societies are brought to ruin by conformers rather than by deviators, that is to say, by the resistance-to-thinking-for-themselves aspect of persons, or what Sumner called the irrational.[9] First, social deviation is by definition not a common denominator of a type which could stand against organized society; deviators may introduce ideas which are revolutionary when and only when they have gained the support of masses of conforming people. There is indeed an assumption among political realists that most demagogues simply play the known areas of conformity and do not reckon truly with social deviators at all. Second, interpreting the history of mankind even conservatively, there is too much evidence that persons in the role of conformer fail to revitalize society and sustain a healthy social organization. Under conformer dominance, institutions lose their vitality,[10] neglecting needs of individuals, or satisfying them only in token fashion. Concentric tendencies eventually overcome the institutions and sometimes with them the sustaining societies.

If then, deviation, from the nature of its independent issue, is (unlike conformity) not of sufficient homogeneity to rise and threaten organized society, how can we account for the hostility which societies are capable of showing innovators? For seemingly in this respect, man improves his general welfare, wherever he does, in spite of much or most of his society.[11] With reference to our previous illustration, this is the same as asking what message *does* the life history of a Columbus or a Copernicus hold for the student of social control. It is suggested that this can only be understood as a problem of change. An object of social control systems has usually been to set up a finite number of action alternatives for the individual which in turn could be controlled. These alternatives could be labeled good, worse, and bad, but they were, when exercised by individuals, almost never a threat to the social control system. Individuals who took the "lesser" alternatives might have been considered pathological, weak, or even unscrupulous, but they usually strengthened the whole system by providing a background of contrast for the "better" alternatives. In such systems, however, the social deviate had to be regarded as a much more basic threat in that he did not cast his role among the alternatives. He was to the otherwise finite system an unpredictable. His actions were not as devastating as his example. Whereas the fallen one in the "lesser-alternative" category provided an example which could, with reference to preexisting understandings, be recognized as wrong and therefore as emphasizing what was right, the deviator could only leave in his path bewilderment and "disorganization." The strength of the much lauded "primary-group morality" lay in reducing the number of alternatives to action to a most "practical" minimum and then guarding them jealously. In such systems prediction tends to be at a maximum and security of the system is correspondingly assured. In such systems the social deviator is conveniently seen as a threat to the controlling factions and will be shown something approaching maximum hostility.[12] One might generalize that he is therefore "dangerous" in inverse proportion to the number of accepted alternatives to action, and that those who most want conformity want it as a means of consolidating a social control position, an always incomplete process since "socialization" of the individual is never complete and new individuals are being born constantly.

The genius of "democratic"[13] social organization, where and to the extent that it exists, lies in its utilization of the deviation aspect of persons within its organizational context. In such an instance the deviator, since he is potentially productive (going back to our proposition that a problem of regulating a society is to tap, organize and adapt its creative strength), can expect maximum acceptance and is dangerous only to select minority elements or "fasces" thereof.

Such a (hypothetical) society, in which the deviator is not kept in marginal status, optimizes the society's chance of applying its own criticisms and correctives to the social process in a somewhat constant (i.e., controllable) manner. It does so by substituting a relatively smooth change continuum for more violent processes such as the radical change function of the history of human societies brought to our attention in different ways by Sorokin, Spengler, Hegel, *et al.* From this it can be seen that a social organization which could adequately utilize the creative capacity of the persons who make up that society would be one which in the long run would have a maximum chance of survival; that others would fail relatively sooner. It is thus no paradox that ability of persons to deviate can preserve a society which conformity-inertia in critical times would be unable to sustain. Obstacles to effecting a social organization which can apply its own correctives may, of course, never be overcome; but, pertinent to the study of social control and in order to give social control the broadest possible definition, it would be well to apprehend the manner in which conformity demands can reduce the effectiveness of the deviation aspect of persons in meeting social change.

To summarize, certain conditions appear to be in evidence. First, the role of the deviator, or the deviation aspect of persons, suffers no real lack of pertinence to social control theory since it is an active ingredient in social organization—in the dynamics of change perhaps the only one. Only the deviator can introduce fundamentally new ways into the culture, since the introduction of new ways is deviation. Second, it is possible that one of the reasons why in social control theory analysis of rational or strategic control has so frequently been deterred is that rational behavior, if it occurs, of necessity has to occur not only as deviate behavior, but deviate behavior of a type which is usually too complex (has too many variables) for easy generalization. This is to suggest that the level of human behavior at which rationality occurs is less easily stereotyped than are the levels of human behavior which are irrationally determined. Finally, instead of asking how society orders and controls the individual, students of social control might ask how society takes its organization and momentum from its behaving individuals.

Notes

1. The term "rational," used several times in this paper, is each time partially defined in the context of the specific paragraphs in which it occurs. More generally it here refers to individual man's capacity to calculate courses of action optimum to the welfare and hence survival of himself or his species. Its antithesis is found in the numerous determinisms (biocentric drives, inertia, circumstance) which operate to make his actions fortuitous to the above-mentioned capacity.

2. William Graham Sumner, *Folkways* (Boston: Ginn & Co., 1906), p. 98. It is true that Sumner excepted an intellectual "elite" where the society permitted the existence of one (pp. 103, 206).

3. The terms "deviation aspect of persons" and "conformity aspect of persons" are used in recognition that conformity and deviation are as concepts subject to the limitations of ideal-typology; that empirical individuals are in their behavior composites of both; therefore, to separate them is arbitrary to a particular discussion.

4. N. Machiavelli, *The Prince* and *The Discourses* (New York: Modern Library College Editions, 1950). Machiavelli, a political realist, had quite anticipated the meaning of "folkways" and "mores," cautioning the Prince not to operate cross-current to the biases of the people (pp. 8, 19, 21, 103). His, in the large, was a science of rational control through knowledge of irrationals.

5. Compare, for example, American social control theory with the "planning" theory of Karl Mannheim.

6. L. G. Brown, in his *Social Pathology* (New York: Crofts, 1942), points out that approved (ordinarily conformance) patterns in the culture can be held pathologically by the individual as easily as disapproved cultural patterns (pp. 70 f.). Substantiating this, the present writer found in mental hospitals that those obsessive psychotics who are characteristically fanatic, contrary to popular assumption, rarely select original data but simply take "too literally" the more generally accepted orthodoxies in the culture. Such patients will elucidate with great animation quite common religious or political sentiments, seldom sentiments of radical content. Since most such persons are outside of mental hospitals, their contribution is mostly in the direction of maintaining conventionality in a society.

7. The exceptional instance (witch hunting) we have just mentioned exists when the peers are bent on seeking conformity to the extent that other types of deviators are logically (by a process of either rationalization or direct intention) included.

8. While it is desirable to enlarge upon the meaning of these categories, to do so is beyond the immediate range of this paper. Readers will perhaps perceive without assistance their most general pertinence. They are subsumed in virtually all higher learning theory, regardless of lesser differences in the epistemological slant of the latter. An example of one type of analysis in this area is a discussion of strategy in restricted interaction situations which is presented by John von Neumann and Oskar Morgenstern in their *Theory of Games and Economic Behavior* (2nd ed.; Princeton: Princeton University Press, 1947).

9. *Op. cit.*, p. 473, "There is logic in the folkways, but never rationality."

10. In Ogburn's frame of reference "vitality" could be substituted with "effective relationship with other institutions." (W. F. Ogburn, *Social Change* [New York: Viking Press, 1922, 1950].) Cooley, in his concept of "formalism," spoke of this tendency. (C. H. Cooley, *Social Organization* [New York: Scribners, 1909], pp. 342 f.)

11. Sumner noted non-progress in societies which exterminate elites (*op. cit.*, p. 103). LaPiere notes in describing individual initiative and social change that only a failure of socialization can invent: "To the extent that the group has failed to induct him into the social ways, the individual is 'free' to work out his own individual ways; and there is the possibility, however small, that . . . he may devise a mode of action that will be adopted by others. . . ." (Richard T. LaPiere, *Sociology* [McGraw-Hill: New York, 1946], pp. 53–4.) Weiss notes that man can deviate because there is a time gap between the cause and effect of what would be otherwise social determinism. (Paul Weiss, *Nature and Man* [New York: Holt, 1947], pp. 6 f.) Linton says, "Thus the origination of new forms of behavioral response seems to be a function not of the society as a whole but of some one, or at the most a few." (Ralph Linton, *The Cultural Background of Personality* [New York: Appleton-Century-Crofts, 1495], p. 121.)

12. A clear example of this tendency in operation is found in Arensberg and Kimball's description of rural Ireland (Conrad Arensberg and Solon Kimball, *Family and Community in Ireland* [Cambridge: Harvard University Press, 1940]), an instance where the controllers are the top-extreme male age group. Examples would be found in virtually all so-called static communities.

13. "Democratic" is here used to refer only to a social organization with a high coefficient of protection of the individual from arbitrary abuse by organized factions, minorities or majorities. Such is of course not a complete definition, but is probably basic to concepts of democracy as held by English speaking peoples today. MacIver speaks of "fundamental liberty of opinion" as essential to the concept. (Robert M. MacIver, *The Web of Government* [New York: Macmillan, 1947], p. 182.) Bryce's classic definition finds the condition necessary to the concept. (James Bryce, *Modern Democracies* [New York: Macmillan, 1921], I, 20.)

50.
Variety and Constraint in Cultural Adaptation

ROGER OWEN

IN THIS PAPER I propose to examine some aspects of the "patri-local band level of sociocultural integration," as outlined by Elman R. Service in his *Primitive Social Organization: An Evolutionary Perspective.* I wish to call particular attention to the common phenomenon of extra-band and extra-tribal marriage in band-type societies, and also to point to the effects that such a marriage pattern would have upon such things as intra-band linguistic diversity, the process of enculturation, and upon cultural growth and change in general. An outline of the social structure of the aboriginal inhabitants of Northern Baja California, Mexico, is presented as one example of a more general phenomenon.

The Patri-Local Band

Service defines the patri-local band as "the simplest, most rudimentary form of human social structure." As a structure, the patri-local band is, or was, to be found in all the major quarters of the earth and virtually in all the habitat zones in which the human species can survive. Service believes the patri-local band to have been the common structural form during the Paleolithic epoch; as the "simplest, most rudimentary form . . .," he suggests that "it could be scientifically assumed to be earliest as well." He offers the "patri-local band" concept as a substitute

Revised version of a paper read at the 62nd Annual Meeting of the American Anthropological Association, November 21, 1963, San Francisco, California; originally titled "The Social Demography of Northern Baja California: Non-linguistically Based Patri-local Bands." With permission of the author.

for the combined "familial," "patri-lineal," and "composite" band levels earlier hypothesized by Steward in his *Theory of Culture Change.* Service believes, and attempts to demonstrate, that both the "familial" and "composite" band levels of Steward's scheme are the results of serious social and cultural disturbances following European contact. The patri-local band then becomes the generic form of band organization. He prefers the designation "patri-local" rather than "patri-lineal" because he believes the form of post-marital residence rather than the type of descent or inheritance system to be the factor most important in shaping a culture's content, a hypothesis further supported in this paper. Examples of recently surviving patri-local bands are the Australians, Tasmanians, Ona, Yahgan, Tehuelche, Southern Californians, Baja Californians, Philippine Negritos, Semang, Central African Negritos, and the South African Bushmen. Also included are those societies previously classified by Steward as either at the "familial" level or at the "composite band" level: the Northern Algonkians, the Northern Athabascans, the Andamese, the Yahgan, the Shoshone and Paiute, and the Eastern Eskimo. In short, Service believes that hunting and gathering populations at all times and all places have been organized into patri-local band structures, excepting only those which existed in the midst of such abundance that the environment could support a higher level of integration; e.g., the Northwest Coast of North America.

All patri-local bands share a number of characteristics. All are foragers of wild foods, are small in absolute size (30–100 + individuals), have low to extremely low population densities

(1–50 square miles per person), lack any but age and sex specialties, and all presumably have minimal technological development. Further, territoriality tends toward the "home-base" variety, with relatively fluid boundaries, non-kinship sodalities are absent or weakly developed and such leadership as exists, rests with the most powerful figure, usually male. The preeminent characteristics, however, both regarded as necessary conditions to the formation and maintenance of a patri-local band, are reciprocal band exogamy and preferential virilocal residence.

Baja California

Aboriginally, and to the present, a number of Indian languages and dialects were spoken in Northern Baja California. It is customary to recognize three languages: Kiliwa, Paipai, and Tipai. Each of these languages was composed of an uncertain number of localized, regional dialects, each of which was similar to the dialects of contiguous groups.

As to sociocultural level, all of the Baja California Indians fall easily within Service's patri-local band level of sociocultural integration. Residence groups were small, localized around waterholes usually, and no extra-band political integrations existed. The post-marital residence pattern was modally and ideally virilocal, although bilocality and matrilocality undoubtedly were sometimes practiced as well. Exogamy was bilateral to the extreme: marriage was not permitted with a person known to be consanguineally related on either the matrilineal or the patrilineal side.

Each of the bands was named. Each spent a significant portion of the year at its respective home-base, utilizing local sources of food, or foods derived from elsewhere which were stored at the home-base. Several times each year, however, the band would move to the regional sources of food (such as pinon pine stands, oak groves, dense *opuntia* cactus thickets, areas of abundance of *agave*, and the sea coasts). The zone to which a band might move would depend on which foods were abundantly available near the home-base. Moreover, the highly variable landscape of Northern Baja California (e.g., Lower Sonoran Desert, found at 2000 feet, Upper Sonoran Desert at 4000 feet, semi-Alpine conditions at 7000 feet, regionally distributed stands of oaks, highly variable coverage of *agave*, and other such variety), ensured that the general ecological potential available to any one band would be significantly different from that convenient to any other.

Each band, due to regional variability, was faced with slightly different adaptive problems and, as each band was also slightly divergent dialectically, if "culture" be defined as man's adaptive conceptual system, each band may be characterized as having a slightly different cultural system.

Turning to the nature of culture in any band, it can be indicated that, in addition to inter-band cultural variability, *each* band contained not only cultural, but linguistic diversity as well. Marriage partners brought into the group (usually the female in virilocal or patrilocal circumstances), had to be selected from non-relatives. The probability of selecting a non-related mate from a neighboring band, after perhaps millennia of geographical contiguity, was slight. Contiguous bands consisted principally of relatives who, even if relatively distant genealogically, would nonetheless be barred from one another by the incest taboo. The only solution to this enigma was (and remains) either to violate the incest taboo or to select partners from distant bands. In Baja California, those bands located at greater distances from one another were divergent dialectically and culturally to an even greater magnitude than the contiguous bands. Those "Paipai," for example, located in the more southerly portion of the "Paipai" range, frequently married "Kiliwa"; "Paipai" in the northern portion of the area often married "Tipai." All of the bands then, were composed of male speakers of one language, but of adult females many of whom natively spoke another language. Furthermore, not only were the adult males and many of the females native speakers of different languages, but they were also enculturated to regional cultures different to a greater or lesser degree.

If "culture" is to be understood as a set of usefully adaptive concepts carried in the minds of the members of a society—concepts which are transmitted to the next generation by means of language, the question may be asked: "What was the nature of 'culture' in any given Baja California patri-local band?" A typical family situation, with the patri-lateral extended family as the principal social vehicle of cultural transmission, consisted of (and consists of) adult males as native speakers of one language, and some proportion of females as native speakers of different languages or, at least, of different dialects. The children of such a group would be enculturated to some point in their lives by means of the language and symbol system of their mothers and of the other females in the group, but not in that of the father and the other males. Furthermore, as many of the adult women themselves were drawn from several dialect and language groups, the children of each familial segment within the band might well undergo slightly different enculturation from that of any other child from the standpoint of symbolic content.

What then is the reality of the learned, shared behavior patterns characteristic of a residence group? Many of the resident males and females would natively speak different languages and respond to slightly different symbolic frameworks. Children might vary to a degree from either of their parents, being under bi-cultural and bi-lingual constraints, and furthermore, children of one family might diverge from all other children in regard to specific language training and the resultant "inherited" symbolic system. The patri-local band then, to the degree that exogamy forces marriage with members of divergent groups, can be further characterized as a culturally and linguistically "hybrid" residence group, one united by patri-lineal consanguineal ties but not by ties of either common language or common culture. It is a social group in which at least two languages are routinely employed in transmitting inter-generationally the adaptive concepts which, in their turn, partake of at least two divergent sources.

Although it is frequently concluded that linguistically similar residence groups form the basis of extra-band social and political affiliations, that is not the case in Baja California. It is geographical location and, as a dependent variable, affinal affiliation, rather than linguistic unity which leads to what little extra-band solidarity exists. Contiguous "Paipai" and "Kiliwa" bands intermarry, share regional food resources, and in all ways are more closely affiliated with each other than the southern "Paipai" bands are with the northern "Paipai" bands. The term "Paipai" then, has neither social nor cultural significance. Neither, I suspect, does it have linguistic signi-ficance through dialectic variation. The term merely serves to label a regional population with some degree of linguistic and general cultural identity.

Similar circumstances prevailed in much of Australia. In what is perhaps a unique paper, Norman Tindale analyzed extra-"tribal" mar-riages. He found, by examining one tribe in detail, that extra-"tribal" marriages comprised 8 per cent of all marriages; for Australia as a whole,

Tindale estimates the frequency of extra-"tribal" marriages to be 15 per cent. He concludes:

It is evident that among the Pitjandjara and the Ngadadjara, inter-tribal unions were not unusual events and fitted properly into their life patterns. . . . The existence of inter-tribal marriages therefore must be accepted as normal, since people who encounter each other on the margins of tribal territories, though often nervous in each other's presence, apprehensive, and often at enmity, do engage in barter, occasionally do give women in inter-tribal marriage.

Extra-"tribal" marriage is a probable con-comitant everywhere of low population densities and rigorous forms of preferential or prescribed mate selection, especially extreme bi-lateral exo-gamy. Furthermore, dialectic variation and cul-tural variation are probably inevitable concomi-tants of low population density as well. If "culture" is seen as an extra-somatic adaptive mechanism, there will be considerable regional diversity among populations at the lowest levels of cultural com-plexity, especially in areas of high degrees of topographic and climatologic variation. The immanent result, with bi-lateral exogamy, is to place individuals with differing languages and cultures in each residence unit. Many, and per-haps all patri-local bands, then, are "hybrid" groups linguistically and culturally.

Conclusions

I would like to suggest that the common practice of giving linguistic designations to patri-local bands (e.g., Paipai, Kiliwa, Shoshone, Yahgan, Aranda, etc.), only serves to obscure a critical aspect of the social structure of the band. Furthermore, a series of common anthropological assumptions which are implicit to the use of the linguistic label should be called into question. These assumptions are: (1) that every ongoing social group has a homogeneous, shared culture; (2) that all members of a simple society, or the vast majority, will speak a single language; (3) that enculturation within most simple societies takes place by means of a single language; and (4) that cultural growth and change normally occur within a single linguistic and cultural tradition. These assumptions make it possible, for example, to discuss Paipai "culture" as if a common set of symbols were shared by all speakers of Paipai; to discuss Paipai symbolic adaptation (culture) as being carried on by all the individuals within the territory bounded by the Paipai language; to presume that Paipai "culture" is transmitted from generation to generation, relatively intact, by means of Paipai language. It is my contention that all of these assumptions are invalidated, to varying degrees, by the phenomenon of extra-band and extra-tribal marriage at the patri-local band level of sociocultural integration.

Given extra-band and extra-tribal marriage in most or all patri-local band groups, and accepting Service's point regarding the primordial nature of that type of social structure, the general evolutionary implications of such band exogamy are great. Band cultural conservatism, the prevalence of rigorous male initiatory rites, bifurcate-merging kinship terminologies, and even the formation of "tribal" level societies are all more explicable if understood within the framework of the "hybrid" band.

Throughout the Paleolithic epoch, very little cultural change is to be seen over extremely long periods of time. If cultural change and growth be defined as a gradual accumulation of additional adaptive symbols, then the "hybrid" patri-local band is a structure which may act to inhibit symbolic accumulation. Each generation, most females are lost in all bands and, inevitably, they take with them some unique, useful symbols. Tindale presents a magnificent example of this process when he observes that an Australian female, married extra-tribally, took with her a personal *churinga* which was meaningless in her new context. Thus, without adding a new adaptive symbol to the band into which she married, she subtracted a presumably adaptive symbol from the band she left. This process, repeated each generation through the regular dispersal of women in marriage, would result in a gradual but constant elimination of cultural concepts from the natal band that, should they prove not to be useful in the marital band, may be lost altogether.

If the subtractive process removes concepts from one band and by this means serves to inhibit growth, by the same token the donation of the subtracted concept to another band provides a near random but continuous input of new concepts into all bands. This input provides a diversified reservoir of new concepts which, in the face of evolutionary necessity, might prove to be of vital adaptive value. In situations of rigorous selective pressure, a residence group would have available all of the novel cultural responses carried by the "foreign" females, in addition to all of the traditional adaptive concepts available to the locally resident males. Although the postulation of the subtractive and additive processes discussed here grew from deductive procedures, their

analogic similarity to mutation and genetic drift should be noted.

Rigorous male initiatory rites or puberty ceremonies—those of the Australians or the Feugians, for example—also take on new analytic substance if they are perceived as a useful means of inculcating the appropriate and necessary local culture in male youth of the residence group who, until adolescence, have been primarily enculturated by females who natively speak a "foreign" language and who, themselves, exhibit a more or less different culture from that of their husbands. Initiatory rites, then, may serve as a means of forcibly impressing upon juvenile males, who will be the inheritors and users of a locale, an identity with the males who are their fathers and uncles. This forces the youth to identify themselves with the cultural symbol system carried by the males—the system necessary for survival in the particular habitat.

Service attributes many systems of bifurcate-merging kinship terminology to a society-wide set of affinal moieties brought about by reciprocal extra-band marriage. This hypothesis assumes additional strength if the halves of the affinal moiety system are understood to be linguistically and culturally more or less divergent. Mother's brothers, then, live elsewhere and may speak a different language as well. In the enculturative process, mother would presumably refer to her own kin (mostly in the other half of the affinal moiety) in her own terminological system. Bifurcate-merging, then, not only takes cognizance of affinal duality but also linguistic and cultural diversity as well, and may have been produced as a result of terminological reciprocity practiced across linguistic boundaries.

It has been widely noted that an increase in population density is a common corollary of the development of cultural complexity. An increase in population density has the concomitant effect of increasing the rate of intensity of linguistic communication and also normally increases the rate of communication of cultural symbols from one individual to the next. As a consequence of increased population density, a lowering of regional cultural and dialectical variation usually will result. To be sure, in the case of modern political states with their dense population aggregates, sharp socioeconomic boundaries do result and thus often bring about the establishment of distinct cultural and linguistic boundaries. Yet, even at this complex socioeconomic level, the effect of the increase in population density within the boundaries is to lower the degree of individual differences. Among bands, increased population density will not only lower the degree of individual variation but will strike directly at the circumstance which gives to the band its "hybrid" nature. With an increase in population density an individual normally will be able to select a marriage partner from a local group near at hand. As a result, mates will come more closely to approximate one another linguistically and culturally. Given the economic circumstances necessary to the development of sedentary villages with large aggregated populations, mates may be selected from one's own village. Marriage partners under these circumstances would normally share both a language and a culture.

The accumulation of cultural concepts has undergone everywhere an exponential growth pattern subsequent to the development of sedentary villages. Perhaps the sudden quantitative surge of concepts associated with village life is a result of the development of relatively homogeneous societies, culturally speaking, based upon matings of individuals sharing dialects and general adaptive patterns. In such a culturally homogeneous society, few symbols would be lost through "subtraction" and most or all symbols would be effectively transmitted to the next generation. "Additions" to the culture would still be made by a reduced rate of extra-cultural marriages and through other routes of diffusion. But the common emphasis on linguistic group endogamy found in horticultural societies would permit the "foreign" wife to contribute to the symbol system of the society which she joins, but to have little significant inhibitory effect in transmitting it to her offspring.

In conclusion, let me suggest that all the understandings social scientists have regarding the nature of *a* human culture—a social group whose members share an essentially identical set of adaptive symbols and most or all of whom speak a single language—become relevant and accurate only at that point in human history where population densities become such that mates may be obtained from nearby groups, a phenomenon most commonly found among horticultural or agricultural peoples. At this point, human social groups for the first time in their evolution, come to be culturally and linguistically relatively homogeneous. It must be noted, finally, that no human society as yet exists which is truly homogeneous in regard to either language or culture. But the trajectory toward such appears obvious.

51.
A Behavioural Theory of Drug Taking

LESLIE T. WILKINS

THE DIFFERENCE BETWEEN addiction and pathological or excessive drug taking may be found in
medical explanations, but in this paper I shall not
discriminate between habituation and addiction.
I shall confine my attention to the behaviour of
the persons who use various substances defined as
narcotic, addictive or other substances prohibited
for purposes not sanctioned by medical authority.
This limitation will, I hope, enable me to take
together narcotic addiction in its recognised forms
with glue sniffing, goof balls and other substances
used in such a manner as to render the user subject
to penal sanctions.

This definition is, of course, an extension of
the usual definition of 'addiction' but it seems
essential to find a theory of behaviour which
could accommodate not only those substances
already defined and included in the appropriate
schedules of control, but new ones as they may
be developed.

As more becomes known of mental disorder,
it seems highly probable to me as an amateur that
substances likely to be developed may have beneficial effects on some persons who suffer some
deficiency, and effects considered undesirable if
used by persons otherwise regarded as in normal
health, or suffering from some disorder not within
the terms of the prescribed use of the drugs. It
would appear, however, that the main objection
is not so much to the nature of the substances
used, as to the unauthorised use of them. It would
be possible for a person to be sufficiently well
informed to know that, say, prohibited substance
A was demonstrated to give relief to his condition,
but unless he obtained his supplies through
approved channels, he would be liable to action
on legal grounds. But surely it will not be agreed
that it is only the means to which objection is
made? It cannot be just that unauthorised persons
are involved that determines the social sanction?
Or is it?

As technical knowledge increases it is to be
expected that non-medical (unauthorised) persons
will obtain information which is normally available only to the profession. Is it then expected
that control is to be effected by restriction of
knowledge to a certified elite? Can ignorance be
an effective control mechanism? Some general
dissemination of health information would seem
to be functional for our society, and it does not
appear that the best system for ensuring absence
of misuse of substances is ignorance of them.
Effective control is not synonymous with limitation
of access to information to a body of selected
certified professional persons. Perhaps 'big brother'
does know best, but if the controlling profession
is seen as 'big brother', will this be the best 'image'
to ensure the behaviour functional for society?

It would appear that 'information' as well as
control is necessary to ensure that society functions in a healthy manner. The 'image'—the way
things are perceived to be—is the significant
determinant of human behaviour, not the way
things are. Not only are the substances and what
they in fact do important, but what they are
perceived to be and to do. The medical profession
may have the knowledge, and the layman be a
mine of inaccurate beliefs and superstitions. But
no man delegates all responsibility even for a

From Leslie T. Wilkins, "A Behavioural Theory of
Drug Taking," *Howard Journal*, Vol. XI, No. 4 (1965),
pp. 6–17. Reprinted by permission of the author and
publisher.

section of his person to another. There are 'experts' in all fields of human activity, and if we could be persuaded that the expert knew best, not only in medicine, but in other fields as well, we could successfully delegate 90% or more of ourselves to their control. Is the answer to our present problems more or less information generally available to laymen, more or less control, and if we solve today's problems, will we be building up an undesirable future situation? Perhaps some light can be thrown on these questions by comparative analysis.

In many developments, both technological and social, the United States presents a similar pattern to that in this country. In some things it is possible to predict the future here by observation of what is currently happening there. In other matters the comparison may not be helpful.

The majority of students of the problem of drug addiction and its control in the United States have, at different times, expressed an interest in the claim that in Britain we have no real problems in this area. Some American writers have suggested that the official figures here did not reveal the true position and that, while America recognised the problem it was not noted here, although in fact similar.

Brill and Larimore after a visit to the United Kingdom wrote that they were 'unable to find any indication that there existed anywhere in that country the practice of medically maintaining indefinitely otherwise healthy persons on continued doses of opiates under medical supervision. . . .' It might be that the perception of the 'otherwise healthy person' differs between observers. Each medical practitioner has his own views on medical matters and these views are respected in this country both by the administration and by the law. It is, however, my view that the concept of being at the same time 'healthy' and 'addicted' would not fit very well with our conception of health. While addiction is defined as an illness, it will tend to be perceived as such, not only by the authorities concerned, but by the general population.

In a sample survey reported by Schur covering 147 twenty-one-year-old persons in one London Borough, the question was asked, 'What should be done with drug addicts?' 'Should they be sent to prison, put into hospital, or merely left alone?' 93% said that they should be put into hospital, 2% said that they should be left alone, and 5% responded 'don't know'. Despite the phrasing of the question, not one person suggested that addicts should be put in prison. This is particularly important when it is noted that the same sample revealed that one half of them had the opinion that 'prison was too good for sex criminals—they should be whipped publicly or worse'. They cannot be said to represent a 'soft' or 'progressive' group! The same is doubtless not representative of the whole of Britain, but it is in the London area that addiction would be most probable if the pattern were similar to other countries.

None the less, Brill and Larimore continue their report, '. . . it is clear that the British practice with respect to addicts is not fundamentally different from our own, nor does the British medical view of addicts differ from our own'. I assume that this lack of difference relates to medical factors, since certainly penal treatments differ and so does public opinion. Perhaps perceptual differences explain more than procedural differences? Perhaps public opinion provides a better control than public policy? It is my thesis that those who have been comparing different systems of drug addiction control in order to find an explanation of the differences in addiction patterns are giving primary importance to what is at best only a secondary variable. But although the secondary variable may be small in terms of its explanation of the variance, it could have been the factor which stimulated the original patterns which do not themselves have the major controlling influence.

I think that the simple cause-and-effect types of explanation are inadequate to discuss the problem of addiction, and indeed, most forms of human behaviour. It is probable that our lack of success in scientific explanations of human behaviour may lie in the fact that we have looked to simple cause-and-effect models in most fields except economics in recent years.

Amplification of Deviation

Magoroh Maruyama makes the case for a more complex type of explanation. 'A small initial deviation, which is within the range of high probability' he says, 'may develop into a deviation of very low probability'. This may be termed concisely 'a deviation amplifying system'. Consider, for example, prices and confidence on the Stock Exchange—what 'causes' a rise in confidence? Prices? Or does confidence change prices? Clearly this is a mutual causal system. A small change in price (perhaps random) may cause confidence to rise, and be followed by a further rise in prices, until some other factor produces a state of balance or reverses the process. Could not a small (even random) change in the law or attitude of the public towards drugs trigger a deviation amplification system? Perhaps this is worth thought. Models involving deviation amplification concepts and the interdependence of blocks of transactions have been found to assist in economic explanations and economic control. Economic behaviour is human behaviour. Does it differ in terms of the basic systems from other forms of human behaviour? There is no evidence that it does.

I have already claimed, with very considerable support of scientific inquiry, that people tend to behave with respect to things as they perceive them to be rather than as they 'actually are'. Perhaps some notes on this postulate are necessary.

No knowledge is obtained directly. The input to our consciousness is through our senses in the form of 'codes'. We decode these signals into our experience in different ways. Perception is the process of observing, recording and organising the experience we have of the outside world. There will be some measure of consensus between observers which may conveniently be regarded as 'truth', but it is safer to propose degrees of subjectivity rather than a dichotomy between subjectivity and objectivity.

Differences between systems may be unimportant, or even non-existent, or may be irrelevant to behaviour if the important factor is the perception (a) of the system and (b) the variables with which the system is believed to be concerned.

For many years advertisers of consumer goods have realised that the 'public image' (perception) of their products is a most important factor determining sales. Even where two products are identical, the sales may differ greatly due only to the nature of the 'image of the product' or the 'image of the producing firm'. A difference in image may seem rather insubstantial, but perhaps the 'image' of drug addiction in Britain as compared with the U.S.A. explains more of the difference than differences (if any) in the control system. But, of course, the control system, or the perception of the control system may play an important part in determining the nature of the image of drugs and their illegal usage.

The Image of Drugs

Freeman suggests that the perception of the function of alcoholic beverages differs markedly between those who subsequently become alcoholics and those who do not. One wonders what the perception of the function of drugs is when, for the first time, an addict decides to take a shot. It seems very unlikely that he would perceive himself as taking on a most expensive habit which will have the consequences which are well-known to the medical profession. The reason for the initial drug use may be quite different from the reasons for its continued use. The perception of the function of drugs may be quite different from their actual function. Explanations in terms of 'actual function' may be totally unsatisfactory to explain behaviour which is determined by misinformation.

Cloward and Ohlin suggest on the basis of their observations in New York City that it is the 'two times losers' (those who have failed to make a success either of legitimate or criminal activities) who turn to drugs as an escape. But are drugs actually perceived as providing an escape by the would-be user just before he takes the step to use? The most useful question is not what function do drugs of different types fulfil, but what function are they perceived as fulfilling. Unfortunately with a truly addictive substance it may be impossible to find out in retrospect what was the initial perception of the user.

It is interesting that the substances used seem to change, as though there were fashions in drug taking. Furthermore, it seems that among the scheduled substances are those which 'actually' fulfil quite opposing functions, as well as totally unrelated functions. The range of functions makes the psycho-analytic argument for opposites having a similar basis difficult to sustain with any conviction. But, despite the range of 'true' functions, perceptual process theory may tell us what seems worth investigating further.

Brill and Larimore suggest that the differences in addiction rates between the United States and Britain '. . . may be part of a broader question in comparative psychiatry, which, for example also indicates that rates for alcoholism and alcoholic psychosis are lower in the United Kingdom than in the United States . . .' It is certain that the rates for alcoholism are much lower in Britain, and particularly in England. How much lower it is difficult to estimate. Estimates can be derived in different ways and afford different types of comparisons. In New York State in 1959 admissions to mental hospitals for alcoholic psychosis are reported to have been 1,929, while in Great Britain only 531 persons were similarly admitted although the population was two-and-a-half times greater. If this is a reasonable indicator, it seems that the rate for alcoholism in Britain is about one-tenth that of the rate for the United States.

The Jellinek formula, which in most countries provides a reasonable estimate of alcoholism from various other indices, has been shown not to provide a good estimate for the United Kingdom; thus it appears that the usual correlates of alcoholism do not apply. If alcoholism and drug addiction have anything in common as Brill and Larimore suggest, and if as Freeman has demonstrated, the perception of the function of alcohol differs between alcoholics and other drinkers (even heavy drinkers), it also seems highly probable that the nature of the concept or 'image' may afford some sort of explanation of both types of differences between Great Britain and the United States.

Merely to say that the different 'image' explains the difference, affords no explanation. Merely to move the explanation from an unknown to a mirror image of the unknown does not make anything more known! It is necessary to explore further and to present a far more complex picture.

Crime in general can be defined as obtaining things by means defined as illegal. It is the means *of obtaining* which constitute the illegality, not the possession or use of drugs or property. It has been shown elsewhere by the author that where the balance between legitimate and illegitimate opportunities to obtain property remain constant, the amount of crime tends to remain constant, that is, proportional to the total opportunities. If opportunities (both legitimate and illegitimate) are rising, crime also rises. That is to say, the more cars in a community, the more stealing of and from cars; the more transfer of money by legitimate means (the affluent society) the more transfers of money by illegitimate means. It is the balance between legitimate and illegitimate means which determines the proportion of criminal activity. Criminal Statistics for England and Wales for the years 1938 to 1961 show an almost complete correlation between the number of cars registered and the number of thefts from cars. It certainly cannot be held that our moral values deteriorate and improve in strict proportion to the increases and decreases in the number of motor cars! It follows that if the proportion of illegitimate opportunities (in relation to legitimate opportunities) changes, the crime rate for the appropriate offences will also change. It will be necessary later to relate this argument to others which must now be made rather briefly.

It has been noted by many writers in regard to widely different aspects of human behaviour, that people do not 'play expected values'. An 'expected value' is the mathematically derived value of an interaction between chance and gain—the product of the probability and the sum involved. In cultures from which drug addicts tend to come, and indeed in many other sectors of society, people have a poor appreciation of the 'utilities' arising from any action or strategy. In general people have difficulty in assessing rational actions where the probability of gain and loss may be small. Football Pools behaviour is a case in point—no board of directors could get away with an investment policy for the shareholders based on the same odds as attract millions to 'invest' in the pools. Kahn discussing political behaviour has also pointed out the tendency not to use 'expected values', and comments, 'prestige gets committed and there are all kinds of public humiliation which tend to pressure individuals into distorted strategy in determining their behaviour'. The 'scientific' evaluation of the effect of drugs or other substances is based on 'expected' values, but to the persons involved the risk of becoming habituated or addicted may appear small, and the risk of being caught for illegal possession may seem to be less than in fact it is.

The Influence of Cultures

A further postulate stresses the relative basis for definitions of social deviance. Definitions of deviance are made by cultures (the authority structure), and cultures vary in their perceptions of forms of behaviour. Some definitions of unacceptable behaviour are no more than vestigial traces of earlier cultural systems. Clearly if all persons behaved in exactly the same ways in any sort of society no matter what form that behaviour took, none of the behaviour could be defined as criminal, even although by some external standard (external to that culture) all the behaviour was criminal. A nomadic society has no difficulty within a culture which accepts the nomadic way of life, but becomes quite unacceptable to a society where land is defined as belonging to persons or organisations.

Distinctions between what is perceived as legitimate and illegitimate are made culturally and legally, and legal definitions tend to follow the cultural definitions. In general a religious society will define deviance from the religious norms as heresy, and if the deviance is perceived as sufficiently bad, some form of punishment will be prescribed. A socialist communist society will define deviance with respect to its theoretical concept of collectivism, whereas a capitalist society will define certain forms of economic behaviour as sufficiently deviant to need the isolation of the offender from the remainder of the society. All people have some idea of what they regard as deviant or 'rare' events, but no one has any prior knowledge of normality. Each person has only his own experience against which to assess the unusual or deviant and to serve as a basis for differential classification. It is as though each individual has a storage system linked with some classification and integrating device which is used for purposes of subjective prediction of behaviour in accordance with levels of expectation. It is the unpredictable which is threatening to the system—individual or societal.

Human actions do not divide into 'bad' and 'good', black and white, or crime and no crime. It seems that societies have, in general, been more aware of the degree of deviance than the direction of deviance. A very large proportion of the deviants perceived by the society of the time as sinners have subsequently been canonised. The deviant saint and the deviant sinner are likely to suffer a similar fate, depending upon the degree of their conflict with the established norms of the time rather than upon the nature of their acts. The relevance of this argument to drug addiction and habituation may not be immediately obvious. It is, however, important to note that no measures of 'absolutes' can be put forward and sustained by scientific argument. If we wish to be considered as rational persons in discussion of the drug problem, we cannot invoke any concepts of absolute good or bad, but must be prepared to discuss the relevance of the controls we propose in terms of the functioning of our society. But the definitions made by the 'culture' (that is, its agents) have an influence upon the behaviour of members of the culture and its sub-cultures.

A feed-back mechanism is proposed in respect of exposure to definitions and the forming of definitions. In particular it will be recognised that in the village each member of the village has some training in sociology through observation and the informal 'village pump politics'. A form of apprenticeship in group management and social communication and control is available to all. The training may be at a primitive level and based on folk lore and a sociology at the level of anecdote interleaved with prejudice, but information about the working of the system is fed into the system, and this information covers the whole system and is perceived as relevant.

By contrast, the urban dweller has restricted information. The information about other members of the system in a small group or village enables the members to make predictions of behaviour over a wide range of events—a far wider range than the urban dweller can cope with within his environment. It is well known that villages can deal with and integrate members of the community who deviate quite widely from the norm. Both the village idiot and the village squire are acceptable members of the village culture. If, however, the village idiot were to move into an urban culture, he would be rejected and removed from the system. It may be argued, then, that the total range of human behaviour which the village as a system can accommodate is greater than the range of behaviour which can be accommodated within an urban system.

It will nonetheless be noted that while the village as a system can integrate a wide range of behaviour, each individual member of the village has very closely specified roles. Any member of the village community who, for any reason, deviates from the narrowly defined role finds his behaviour interpreted and controlled by social pressures within the community.

Dentler studied deviance and social controls in small Quaker work camps. The members of these groups were in close and continuous face to face communication. The level of communication was very much higher than that to be expected in the village culture, and, not unexpectedly, the level of integration of the members of the groups was also higher. Dentler's findings confirm the general hypothesis suggested by this theory, that is, small groups can perceive specific roles for each member, whereas large groups tend towards the perception of individuals in terms of stereotypes.

Perhaps the point may now be conceded that there are systems which can accommodate wider range of deviance on the part of their members than other systems. In terms of the probability calculus, the value (x) in $(x\sigma)$ differs with respect to individuals and systems. In the village (x) is small for individuals and large for the system, whereas in the urban culture (x) is larger for the individual and smaller for the system. It seems possible that where the system had difficulty in dealing with deviance, the persons who deviate will tend to be 'defined out' of the system. If a society defines as 'intolerable' actions which are not completely dysfunctional for that society, it may be defeating its own ends. Definition by the society of an individual as unacceptable may lead to a self-identification as unacceptable to the culture, and hence the culture as unacceptable to him. When sanctions are applied to the deviants within any society which are perceived by them as extreme, the persons so defined become alienated from the values of the parent system. Sanctions may become devalued. There is experimental evidence for this type of positive feed-back process —a deviation amplifying system.

To quote only one example. Vernon in a follow-up study of children placed in different streams revealed a marked drop in I.Q. scores in lower streams and a relative increase in the scores of those placed in the higher streams, although at separation the scores were level. The defining act (streaming) provided a possibility for self-identification and had the effect of changing the 'information set' or experience of the defined groups in accord with the nature of the definitions. It would appear that, in general, a group which has been cut off from the general norms of the group develops its own norms which tend to reveal some deviation-amplifying qualities. A centrifugal force seems to drive them away from the distribution of values from which they have themselves been rejected.

Summarising the general theory, in so far as it is possible, the following statements are set forth:

a) certain types of information in relation to certain systems lead to more acts being defined as deviant to the extent that the individuals whose behaviour is so defined are 'cut off' from the values of the parent system;

b) the defining act leads to more action being taken against those perceived as deviant, and the individuals defined begin to see themselves as deviant (as unacceptable within the general system or culture). Perhaps it is mainly through information received from other persons that we get to know what sort of persons we 'are'.

c) the self-definition and the action taken by society at large leads to the isolation and alienation of the persons defined as deviant;

d) the deviant groups tend to develop their own values which may run counter to the values of the parent system which has defined them as 'outlaws' (or in terms of the probability calculus 'outliers');

e) the deviation amplifying (centrifugal) forces are thus generated and develop within the deviant group, and lead to more deviant behaviour by the alienated groups;

f) the increased deviance demonstrated by the deviant groups (resulting from a deviation amplification or feed-back system) results in more forceful action by the conforming groups against the non-conformists;

g) thus information about the behaviour of the non-conformists (f) received by the conforming groups, leads to more acts being defined as deviant . . . and back to (a) and round, and round again.

This model has few unfamiliar features. Indeed it would be more surprising if persons who were excluded by a system were to continue to regard themselves as part of the system which had excluded them. If it is a fair model of the system certain deductions can be made from it. In particular it may be expected that those defined as deviant will insulate themselves from the value systems excluding them and establish systems which determine for them some needed structure. This type of model makes it unnecessary to show that a major causal factor has been identified. Because of the amplification through feed-back, initially small differences, perhaps even due to chance variations in the network, could build up to quite large forces. A number of mutual causal processes can be identified in economic behaviour where the initial stimulus was small and probably generated by random changes. In short the main feature of this model is that it proposes an essentially unstable system.

It has been reported by social ecologists that the values of the middle-classes are acceded to orally by persons who live in the most delinquent areas in their sample. This fact most observers regard as being a hopeful sign. The value structure of society at large seems to have penetrated its darkest corners. But is this hopeful? Rather it seems to reveal a latent conflict situation which may be more difficult to resolve than if the values were different in different 'sub-cultures'. Since the milieu is different, a system of values which differed might be expected to be more likely to be functional. The definitions do get communicated through levels of the culture. Looked at one way this process may be seen as an effective system for socialisation, but it could also indicate that the mechanisms for alienation operate effectively.

National Differences

Applying the theoretical structure proposed to the explanation of the difference between drug addiction in Great Britain and the United States, the following is suggested as a framework:

a) the 'image' of drugs differs in Great Britain from that in the United States;

b) the 'image' of the use of drugs differs;

c) the 'image' of the addict differs;

d) the 'image' of the enforcement agencies (e.g. police) differs;

e) small differences in the control system, or even the perception of the control system could generate large differences in the 'image' of addiction which could amplify the effects of official control action;

f) fewer actions are defined as 'crimes' in Great Britain, and as a result fewer people are in general defined as 'criminal', whether in respect of social or economic deviance; (compare also cheque frauds in the two countries);

g) certainly for drugs, the balance between the legitimate and illegitimate means for obtaining them differs. The illegitimate supply sources are very easy of access in New York and Los Angeles;

h) the 'information set', or folk lore (it does not have to be true) regarding drugs, their control and the like, is adequate to explain the differences.

Perhaps most of the above points will be agreed without further evidence or discussion. Point (*g*) may, however, require some explanation. It seems certain that the opportunities for obtaining legitimate sources of supply are greater in Britain, whereas in the United States the illegitimate opportunities, especially for the vulnerable groups, are greater. It is necessary to consider the ration between legitimate and illegitimate opportunities. It is the balance which, according to my theory and the evidence, is important in explaining behaviour. If one could express the number of illegitimate opportunities as a fraction of all opportunities, the United States would show a much higher ratio. Moreover, the 'image' will be influenced by the ratio of illegitimate to legitimate opportunities.

With regard to (*h*), it may be true, as Brill suggests, that there are no procedural differences between the control system in Britain and the United States, but the existence of a different set of beliefs is quite sufficient to change behaviour. In fact, objective differences which are not perceived cannot change behaviour.

It may seem that I am saying that the success to date which has attended this country's efforts at drug control rests upon some slender foundations. These foundations look especially slender or even non-existent from the viewpoint of a cause-effect model. I cannot, however, myself agree that the 'perceptual process' model and the concept of 'image' provide only a slender foundation. It may be that we know little of these processes, but that does not mean that they are unimportant, merely that it would be convenient if they were.

Perhaps the difference between the problem of drug addiction in the United States and Britain is based on a similar phenomenon to that which determines what types of petrol or oil you buy for your car, or the brand of cigarettes you may smoke. Perhaps these determinants are unsubstantial and not part of the stock-in-trade of general administrative action, but they can make extremely large differences in behaviour. Some firms go bankrupt while others prosper on less substantial differences.

Perhaps a society can effectively control only these persons who perceive themselves to be members of that society. Perhaps the major distinction between being classified as a criminal or a sick person is that a sick person can still identify himself as within the social system, whereas by definition a criminal cannot.

If this is so, then the surest and simplest, if not the quickest way for Britain to get a drug addiction problem similar in proportion to that of the United States would be to try to reduce the problem it now has by any repressive means, or indeed by any means which gave the opportunity for the 'image' of drugs, drug use or addiction to change so as to generate a deviation-amplifying system.

There may be some actions which can be taken to change the opportunity structures which can be put into effect without alienation of deviant groups. Operations upon systems seem more likely to have functional results than operations concerned with individual users, or potential users. Until we have, through research, better ideas as to what to do, very serious mistakes are likely to be made in dealing with this problem.

When we are concerned with artificial environments and with environments on the moon, planets and other such places, we seek the help of the scientific method, but when we are concerned with this earth, and particularly this country, we rely upon amateurs. We are spending more money in finding out how one or two men may respond to artificial environments (space capsules) than we are to finding out about how many millions of persons are involved in our social control systems. We can specify more accurately the factors to be considered in the capsules than in the home. We realise that the capsule is a different environment about which we have little or no information from our instincts, but the modern home is as far removed from the cave as the space capsule. How is it that we are expected to be able to cope with the one and not the other without obtaining systematic information?

The Basic Problem

Finally, to return to the problem as stated at the beginning of this paper. As information increases some persons not duly authorised will believe, rightly or wrongly, that they would benefit by the use of certain substances which it might be regarded as unsafe to permit into non-professional channels. But which experts, which groups of professionals are to have what degrees of control, and by what criteria are they to be selected? Doubtless every professional expert does know best about matters within his own particular field of study. But no expert can claim general expertise. There are available today specialists who would be able to take over responsibility for nearly all of our life's activities. Some persons in some circumstances will seek help of their own volition from the appropriate expert. Perhaps they are more likely to do this in an appropriate manner and at an appropriate time if they are well informed. But if they do not seek help themselves, where do we begin to invoke the 'big brother' controls? When should it be decided that a person has forfeited his right to freedom? How much of a mess may a person make of his life before it is taken over for him by an expert?

In which fields is he entitled to make a mess, if the freedom differs? If people think they know, will they not make use of their supposed or real knowledge? Is the way of ignorance the most suitable system of control? By what right has the professional in some fields of study a special privilege to determine legal and illegal behaviour for other persons?

As the rate of technological development increases we may from time to time find ourselves taking up different positions regarding special skills and knowledge.

One thing is, however, perfectly clear. Any system in which it is necessary to have 'big brother' functions in order to ensure that it works is a very inefficient system. The fault lies with those who are responsible for devising systems. A good, that is, functional system has its own self-regulatory processes built into it, not superimposed upon it. In some areas of human behaviour we have achieved fairly reasonable systems. We no longer pillory people who fail to save money, rather we operate upon the rate of interest and we provide social welfare systems by means of enforced savings through taxation and compulsory insurance.

Man has got to learn to adapt to a rapidly changing environment, and our systems of controls must also find ways to be adaptive, indeed, we need to devise adaptive systems. It is not sufficient to kick the machinery of control from time to time when it appears to have gone wrong. We must know something about the machinery. It is not sufficient for our rulers to try to steer 'their people' only by trippings.

B.

SOCIAL CONTROL:
ORGANIZATIONAL GOAL SEEKING

52.
A Systems Analysis of Political Life

DAVID EASTON

IN *A Framework for Political Analysis* I spelled out in considerable detail the assumptions and commitments that would be required in any attempt to utilize the concept "system" in a rigorous fashion. It would lead to the adoption of what I there described as a systems analysis of political life. Although it would certainly be redundant to retrace the same ground here, it is nonetheless necessary to review the kinds of basic conceptions and orientations imposed by this mode of analysis. In doing so, I shall be able to lay out the pattern of analysis that will inform and guide the present work.

Political Life as an Open and Adaptive System

. . . The question that gives coherence and purpose to a rigorous analysis of political life as a system of behavior is as follows. How do any and all political systems manage to persist in a world of both stability and change? Ultimately the search for an answer will reveal what I have called the life processes of political systems—those fundamental functions without which no system could endure—together with the typical modes of response through which systems manage to sustain them. The analysis of these processes, and of the nature and conditions of the responses, I posit as a central problem of political theory.

Although I shall end by arguing that it is useful to interpret political life as a complex set

From David Easton, *A Systems Analysis of Political Life* (New York: John Wiley, 1965), Chapter 2, pp. 17–35. Reprinted by permission of the author and publisher. Copyright © 1965 by John Wiley & Sons, Inc.

of processes through which certain kinds of inputs are converted into the type of outputs we may call authoritative policies, decisions and implementing actions, at the outset it is useful to take a somewhat simpler approach. We may begin by viewing political life as a system of behavior imbedded in an environment to the influences of which the political system itself is exposed and in turn reacts. Several vital considerations are implicit in this interpretation and it is essential that we become aware of them.

First, such a point of departure for theoretical analysis assumes without further inquiry that political interactions in a society constitute a *system* of behavior. This proposition is, however, deceptive in its simplicity. The truth is that if the idea "system" is employed with the rigor it permits and with the implications currently inherent in it, it provides a starting point that is already heavily freighted with consequences for a whole pattern of analysis.

Second, to the degree that we are successful in analytically isolating political life as a system, it is clear that it cannot usefully be interpreted as existing in a void. In must be seen as surrounded by physical, biological, social and psychological *environments*. Here again, the empirical transparency of the statement ought not to be allowed to distract us from its crucial theoretical significance. If we were to neglect what seems so obvious once it is asserted, it would be impossible to lay the groundwork for an analysis of how political systems manage to persist in a world of stability or change.

This brings us to a third point. What makes the identification of the environments useful and necessary is the further presupposition that political life forms an *open* system. By its very nature as a social system that has been analytically separated from other social systems, it must be interpreted as lying exposed to influences deriving from the other systems in which empirically it is imbedded. From them there flows a constant stream of events and influences that shape the conditions under which the members of the system must act.

Finally, the fact that some systems do survive, whatever the buffetings from the environments, awakens us to the fact that they must have the capacity to *respond* to disturbances and thereby to adapt to the conditions under which they find themselves. Once we are willing to assume that political systems may be adaptive and need not just react in a passive or sponge-like way to their environmental influences, we shall be able to break a new path through the complexities of theoretical analysis.

As I have elsewhere demonstrated, in its internal organization, a critical property that a political system shares with all other social systems is this extraordinarily variable capacity to respond to the conditions under which it functions. Indeed, we shall find that political systems accumulate large repertoires of mechanisms through which they may seek to cope with their environments. Through these they may regulate their own behavior, transform their internal structure, and even go so far as to remodel their fundamental goals. Few systems, other than social systems, have this potentiality. In practice, students of political life could not help but take this into account; no analysis could even begin to appeal to common sense if it did not do so. Nevertheless it is seldom built into a theoretical structure as a central component; certainly its implications for the internal behavior of political systems have never been set forth and explored.

Equilibrium Analysis and Its Shortcomings

It is a major shortcoming of the one form of inquiry latent but prevalent in political research—equilibrium analysis—that it neglects such variable capacities for systems to cope with influences from their environment. The equilibrium approach is seldom explicitly elaborated, yet it infuses a good part of political research, especially group politics and international relations. Of necessity an analysis that conceives of a political system as seeking to maintain a state of equilibrium must assume the presence of environmental influences.

It is these that displace the power relationships in a political system—such as a balance of power—from their presumed stable state. It is then customary, if only implicitly so, to analyze the system in terms of a tendency to return to a presumed pre-existing point of stability. If the system should fail to do so, it would be interpreted as moving on to a new state of equilibrium and this would need to be identified and described. A careful scrutiny of the language used reveals that equilibrium and stability are usually assumed to mean the same thing.

Numerous conceptual and empirical difficulties stand in the way of an effective use of the equilibrium idea for the analysis of political life. But among these there are two that are particularly relevant for my present purposes.

In the first place, the equilibrium approach leaves the impression that the members of a system are seized with only one basic goal as they seek to cope with change or disturbance, namely, to re-establish the old point of equilibrium or, at most, to move on to some new one. This is usually phrased, at least implicitly, as the search for stability as though this were sought above all else. In the second place, little if any attention is explicitly given to formulating the problems relating to the path that the system takes insofar as it does seek to return to this presumed point of equilibrium or to attain a fresh one. It is as though the pathways taken to manage the displacements were an incidental rather than a central theoretical consideration.

But it would be impossible to understand the processes underlying the capacity of some kind of political life to sustain itself in a society if either the objectives or the form of the responses are taken for granted. A system may well seek goals other than those of reaching one or another point of equilibrium. Even though this state were to be used only as a theoretical norm that is never achieved, it would offer a less useful theoretical approximation of reality than one that takes into account other possibilities. We would find it more helpful to devise a conceptual approach that recognized that at times members in a system may wish to take positive actions to destroy a previous equilibrium or even to achieve some new point of continuing disequilibrium. This is typically the case where the authorities may seek to keep themselves in power by fostering internal turmoil or external dangers.

Furthermore, with respect to these variable goals, it is a primary characteristic of all systems that they are able to adopt a wide range of actions of a positive, constructive, and innovative sort for warding off or absorbing any forces of displacement. A system need not just react to a disturbance by oscillating in the neighborhood of a prior point of equilibrium or by shifting to a new one. It may cope with the disturbance by seeking to change the environment so that the exchanges between the environment and itself are no longer stressful; it may seek to insulate itself against any further influences from the environment; or the members of the system may even transform their own relationships fundamentally and modify their own goals and practices so as to improve their chances of handling the inputs from the environment. In these and other ways a system has the capacity for creative and constructive regulation of disturbances as we shall later see in detail.

It is clear that the adoption of equilibrium analysis, however latent it may be, obscures the presence of system goals that cannot be described as a state of equilibrium. It also virtually conceals the existence of varying pathways for attaining these alternative ends. For any social system, including the political, adaptation represents more than simple adjustments to the events in its life. It is made up of efforts, limited only by the variety of human skills, resources, and ingenuity, to control, modify or fundamentally change either the environment or the system itself, or both together. In the outcome the system may succeed in fending off or incorporating successfully any influences stressful for it.

Minimal Concepts for a Systems Analysis

A systems analysis promises a more expansive, more inclusive, and more flexible theoretical structure than is available even in a thoroughly self-conscious and well-developed equilibrium approach. To do so successfully, however, it must establish its own theoretical imperatives. Although these were explored in detail in *A Framework for Political Analysis*, we may re-examine them briefly here, assuming, however, that where the present brevity leaves unavoidable ambiguities, the reader may wish to become more familiar with the underlying structure of ideas by consulting this earlier volume. In it, at the outset, a system was defined as any set of variables regardless of the degree of interrelationship among them. The reason for preferring this definition is that it frees us from the need to argue about whether a political system is or is not really a system. The only question of

importance about a set selected as a system to be analyzed is whether this set constitutes an interesting one. Does it help us to understand and explain some aspect of human behavior of concern to us?

To be of maximum utility, I have argued, a *political* system can be designated as those interactions through which values are authoritatively allocated for a society; this is what distinguishes a political system from other systems that may be interpreted as lying in its environment. This environment itself may be divided into two parts, the intra-societal and the extra-societal. The first consists of those systems in the same society as the political system but excluded from the latter by our definition of the nature of political interactions. Intra-societal systems would include such sets of behavior, attitudes and ideas as we might call the economy, culture, social structure or personalities; they are functional segments of the society with respect to which the political system at the focus of attention is itself a component. In a given society the systems other than the political system constitute a source of many influences that create and shape the conditions under which the political system itself must operate. In a world of newly emerging political systems we do not need to pause to illustrate the impact that a changing economy, culture, or social structure may have upon political life.

The second part of the environment, the extra-societal, includes all those systems that lie outside the given society itself. They are functional components of an international society or what we might describe as the supra-society, a supra-system of which any single society is part. The international political systems, the international economy or the international cultural system would fall into the category of extra-societal systems.

Together, these two classes of systems, the intra- and extra-societal, that are conceived to lie outside of a political system may be designated as its total environment. From these sources arise influences that are of consequence for possible stress on the political system. The total environment is presented in Table 1 as reproduced from *A Framework for Political Analysis*, and the reader should turn to that volume for a full discussion of the various components of the environment as indicated on this table.

Disturbances is a concept that may be used to identify those influences from the total environment of a system that act upon it so that it is different after the stimulus from what it was before. Not all disturbances need strain the system.

TABLE 1

Components of the Total Environment of a Political System

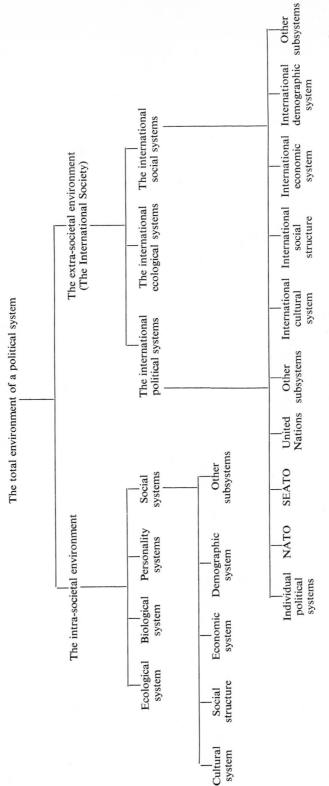

Some may be favorable with respect to the persistence of the system; others may be entirely neutral with respect to possible stress. But many can be expected to lead in the direction of stress.

When may we say that *stress* occurs? This involves us in a rather complex idea, one that has been treated at length. But since it does stand as a major pillar underpinning the analysis to be elaborated in the succeeding chapters, I must at least broadly sketch out its implications. It embodies several subsidiary notions. All political systems as such are distinguished by the fact that if we are to be able to describe them as persisting, we must attribute to them the successful fulfillment of two functions. They must be able to allocate values for a society; they must also manage to induce most members to accept these allocations as binding, at least most of the time. These are the two properties that help us to distinguish most succinctly political systems from other kinds of social systems.

By virtue of this very fact these two distinctive features—the allocations of values for a society and the relative frequency of compliance with them—are the *essential variables* of political life. But for their presence, we would not be able to say that a society has any political life. And we may here take it for granted that no society could exist without some kind of political system; elsewhere I have sought to demonstrate this in detail.

One of the important reasons for identifying these essential variables is that they give us a way of establishing when and how the disturbances acting upon a system threaten to stress it. Stress will be said to occur when there is a danger that the essential variables will be pushed beyond what we may designate as their *critical range*. What this means is that something may be happening in the environment—the system suffers total defeat at the hands of an enemy, or widespread disorganization in and disaffection from the system is aroused by a severe economic crisis. Let us say that as a result, the authorities are consistently unable to make decisions or if they strive to do so, the decisions are no longer regularly accepted as binding. Under these conditions, authoritative allocations of values are no longer possible and the society would collapse for want of a system of behavior to fulfill one of its vital functions.

Here we could not help but accept the interpretation that the political system had come under stress, so severe that any and every possibility for the persistence of a system for that society had disappeared. But frequently the disruption of a political system is not that complete; the stress is

present even though the system continues to persist in some form. Severe as a crisis may be, it still may be possible for the authorities to be able to make some kinds of decisions and to get them accepted with at least minimal frequency so that some of the problems typically subjected to political settlements can be handled.

That is to say, it is not always a matter as to whether the essential variables are operating or have ceased to do so. It is possible that they may only be displaced to some extent as when the authorities are partially incapacitated for making decisions or from getting them accepted with complete regularity. Under these circumstances the essential variables will remain within some normal range of operation; they may be stressed but not in a sufficient degree to displace them beyond a determinable critical point. As long as the system does keep its essential variables operating within what I shall call their critical range, some kind of system can be said to persist.

As we have seen, one of the characteristic properties of every system is the fact that it has the capacity to cope with stress on its essential variables. Not that a system need take such action; it may collapse precisely because it has failed to take measures appropriate for handling the impending stress. But it is the existence of a capacity to respond to stress that is of paramount importance. The kind of response actually undertaken, if any, will help us to evaluate the probabilities of the system's being able to ward off the stress. In thus raising the question of the nature of the response to stress, it will become apparent, in due course, that the special objective and merit of a systems analysis of political life is that it permits us to interpret the behavior of the members in a system in the light of the consequences it has for alleviating or aggravating stress upon the essential variables.

The Linkage Variables between Systems

But a fundamental problem remains. We could not begin the task of applying this kind of conceptualization if we did not first pose the following question. How do the potentially stressful conditions from the environment communicate themselves to a political system? After all, common sense alone tells us that there is an enormous variety of environmental influences at work on a system. Do we have to treat each change in the environment as a separate and unique disturbance, the specific effects of which for the political system have to be independently worked out?

If this were indeed the case, as I have shown in detail before, the problems of systematic analysis would be virtually insurmountable. But if we can devise a way for generalizing our method for handling the impact of the environment on the system, there would be some hope of reducing the enormous variety of influences into a relatively few, and therefore into a relatively manageable number of indicators. This is precisely what I have sought to effect through the use of the concepts "inputs" and "outputs."

How are we to describe these inputs and outputs? Because of the analytic distinction that I have been making between a political system and its parametric or environmental systems, it is useful to interpret the influences associated with the behavior of persons in the environment or from other conditions there as *exchanges* or *transactions* that cross the *boundaries* of the political system. Exchanges can be used when we wish to refer to the mutuality of the relationships, to the fact that the political system and those systems in the environment have reciprocal effects on each other. Transactions may be employed when we wish to emphasize the movement of an effect in one direction, from an environmental system to the political system, or the reverse, without being concerned at the time about the reactive behavior of the other system.

To this point, there is little to dispute. Unless systems were coupled together in some way, all analytically identifiable aspects of behavior in society would stand independent of each other, a patently unlikely condition. What carries recognition of this coupling beyond a mere truism, however, is the proposal of a way to trace out the complex exchanges so that we can readily reduce their immense variety to theoretically and empirically manageable proportions.

To accomplish this, I have proposed that we condense the major and significant environmental influences into a few indicators. Through the examination of these we should be able to appraise and follow through the potential impact of environmental events on the system. With this objective in mind, I have designated the effects that are transmitted across the boundary of a system toward some other system as the *outputs* of the first system and hence, symmetrically, as the *inputs* of the second system, the one they influence. A transaction or an exchange between systems will therefore be viewed as a linkage between them in the form of an input–output relationship.

Demands and Supports as Input Indicators

The value of inputs as a concept is that through their use we shall find it possible to capture the effect of the vast variety of events and conditions in the environment as they pertain to the persistence of a political system. Without the inputs it would be difficult to delineate the precise operational way in which the behavior in the various sectors of society affects what happens in the political sphere. Inputs will serve as *summary variables* that concentrate and mirror everything in the environment that is relevant to political stress. Thereby this concept serves as a powerful tool.

The extent to which inputs can be used as summary variables will depend, however, upon how we define them. We might conceive of them in their broadest sense. In that case, we would interpret them as including any event external to the system that alters, modifies or affects the system in any and every possible way. But if we seriously considered using the concept in so broad a fashion, we would never be able to exhaust the list of inputs acting upon a system. Virtually every parametric event and condition would have some significance for the operations of a political system at the focus of attention; a concept so inclusive that it does not help us to organize and simplify reality would defeat its own purposes. We would be no better off than we are without it.

But as I have already intimated, we can greatly simplify the task of analyzing the impact of the environment if we restrict our attention to certain kinds of inputs that can be used as indicators to sum up the most important effects, in terms of their contributions to stress, that cross the boundary from the parametric to the political systems. In this way we would free ourselves from the need to deal with and trace out separately the consequences of every different type of environmental event.

As the theoretical tool for this purpose, it is helpful to view the major environmental influences as coming to a focus in two major inputs: demands and support. Through them a wide range of activities in the environment may be channeled, mirrored, and summarized and brought to bear upon political life, as I shall show in detail in the succeeding chapters. In this sense they are key indicators of the way in which environmental influences and conditions modify and shape the operations of the political system. If we wish, we may say that it is through fluctuations in the inputs of demands and support that we shall find the effects of the environmental systems transmitted to the political system.

Outputs and Feedback

In a comparable way, the idea of outputs helps us to organize the consequences flowing from the behavior of the members of the system rather than from actions in the environment. Our primary concern is, to be sure, with the functioning of the political system. In and of themselves, at least for understanding political phenomena, we would have no need to be concerned with the consequences that political actions have for the environmental system. This is a problem that can or should be handled better by theories seeking to explore the operations of the economy, culture, or any of the other parametric systems.

But the fact is that the activities of the members of the system may well have some importance with respect to their own subsequent actions or conditions. To the extent that this is so, we cannot entirely neglect those actions that do flow out of a system into the environment. As in the case of inputs, however, there is an immense amount of activities that take place within a political system. How are we to sort out the portion that has relevance for an understanding of the way in which systems manage to persist?

Later we shall see that a useful way of simplifying and organizing our perceptions of the behavior of the members of the system, as reflected in their demands and support, is in terms of the consequences of these inputs for what I shall call the political outputs. These are the decisions and actions of the authorities. Not that the complex political processes internal to a system, and that have been the subject of inquiry for so many decades in political science, will be considered in any way irrelevant. Who controls whom in the various decision-making processes will continue to be a vital concern since the pattern of power relationships helps to determine the nature of the outputs. But the formulation of a conceptual structure for this aspect of a political system would draw us into a different level of analysis. Here I am only seeking economical ways of summarizing the outcomes of these internal political processes—not of investigating them— and I am suggesting that they can be usefully conceptualized as the outputs of the authorities. Through them we shall be able to trace out the consequences of behavior within a political system for its environment.

There would be little point in taking the trouble to conceptualize the results of the internal behavior of the members in a system in this way unless we could so something with it. As we shall see, the significance of outputs is not only that they help to influence events in the broader society of which the system is a part; in doing so, they help to determine each succeeding round of inputs that finds its way into the political system. As we shall phrase it later, there is a *feedback loop* the identification of which will help us to explain the processes through which the authorities may cope with stress. This loop has a number of parts. It consists of the production of outputs by the authorities, a response on the part of the members of the society with respect to them, the communication of information about this response to the authorities and finally, possible succeeding actions on the part of the authorities. Thereby a new round of outputs, response information feedback and reaction on the part of the authorities is set in motion and is part of a continuous never-ending flow. What happens in this feedback loop will turn out to have the deepest significance for the capacity of a system to cope with stress.

A Flow Model of the Political System

It is clear from what has been said that this mode of analysis enables and indeed compels us to analyze a political system in dynamic terms. Not only do we see that it gets something done through its outputs but we are also sensitized to the fact that what it does may influence each successive stage of behavior. We appreciate the urgent need to interpret political processes as a continuous and interlinked flow of behavior.

If we apply this conceptualization in the construction of a rudimentary model of the relationships between a political system and its environment, we would have a figure of the kind illustrated in Diagram 1. . . . In effect it conveys the idea that the political system looks like a vast and perpetual conversion process. It takes in demands and support as they are shaped in the environment and produces something out of them called outputs. But it does not let our interest in the outputs terminate at this point. We are alerted to the fact that the outputs influence the supportive sentiments that the members express toward the system and the kinds of demands they put in. In this way the outputs return to haunt the system, as it were. As depicted on the diagram, all this is still at a very crude

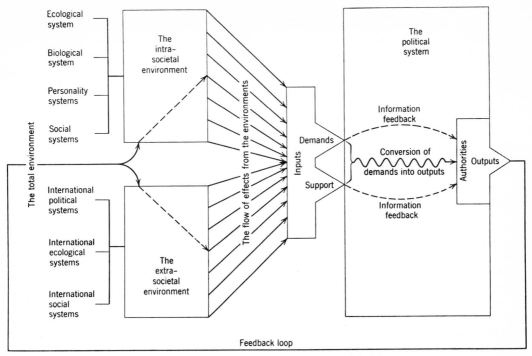

Diagram 1.—A Dynamic Response Model of a Political System.

level of formulation. It will be our task to refine these relationships as we proceed in our analysis.

But let us examine the model a little more closely since in effect this volume will do little more than to flesh out the skeleton presented there. In interpreting the diagram, we begin with the fact that it shows a political system surrounded by the two classes of environments that together form its total environment. The communications of the many events that occur here are represented by the solid lines connecting the environments with the political system. The arrowheads on the lines show the direction of flow into the system. But rather than attempting to discuss each disturbance in the environment uniquely or even in selected groups or classes of types, I use as an indicator of the impact that they have on the system, the way in which they shape two special kinds of inputs into the system, demands and support. This is why the effects from the environment are shown to flow into the box labelled "inputs." We must remember, however, that even though the desire for simplicity in presentation does not permit us to show it on the diagram, events occurring within a system may also have some share in influencing the nature of the inputs.

As is apparent, the inputs provide what we may call the raw materials on which the system acts so to produce something we are calling outputs.

The way in which this is done will be described as a massive conversion process cavalierly represented on the diagram by the serpentine line within the political system. The conversion processes move toward the authorities since it is toward them that the demands are initially directed. As we shall see, demands spark the basic activities of a political system. By virtue of their status in all systems, authorities have special responsibilities for converting demands into outputs.

If we were to be content with what is basically a static picture of a political system, we might be inclined to stop at this point. Indeed much political research in effect does just this. It is concerned with exploring all those intricate subsidiary processes through which decisions are made and put into effect. This constitutes the vast corpus of political research today. Therefore, insofar as we were concerned with how influence is used in formulating and putting into effect various kinds of policies or decisions, the model to this point would be an adequate if minimal first approximation.

But the critical question that confronts political theory is not just the development of a conceptual apparatus for understanding the factors that contribute to the kinds of decisions a system makes, that is, for formulating a theory of political allocations. As I have indicated, theory needs to know how it comes about that any kind of system can persist long enough to continue to make such decisions. We need a theory of systems persistence as well. How does a system manage to deal with the stress to which it may be subjected at any time? It is for this reason that we cannot accept outputs as the terminal point either of the political processes or of our interest in them. Thus it is important to note on the diagram, that the outputs of the conversion process have the characteristic of feeding back upon the system and shaping its subsequent behavior. Much later I shall seek to demonstrate that it is this feature together with the capacity of a system to take constructive actions that makes it possible for a system to seek to adapt or to cope with possible stress.

On the diagram, this feedback is depicted by the line that shows the effects of the outputs moving directly back to the environments. As the broken lines within the environmental boxes indicate, the effects may reshape the environment in some way; that is to say, they influence conditions and behavior there. In this way the outputs are able to modify the influences that continue to operate on the inputs and thereby the next round of inputs themselves.

But if the authorities are to be able to take the past effect of outputs into account for their own future behavior, they must in some way be apprised of what has taken place along the feedback loop. The broken lines in the box labeled "The political system" suggest that, through the return flow of demands and support, the authorities obtain information about these possible consequences of their previous behavior. This puts the authorities in a position to take advantage of the information that has been fed back and to correct or adjust their behavior for the achievement of their goals.

It is the fact that there can be such a continuous flow of effects and information between system and environment, we shall see, that ultimately accounts for the capacity of a political system to persist in a world even of violently fluctuating changes. Without feedback and the capacity to respond to it, no system could survive for long, except by accident.

In this brief overview, I have summarized the essential features of the analytic structure. . . . If we condensed the diagram still further, we would have the figure shown on Diagram 2. It reduces to its bare essentials the fundamental processes

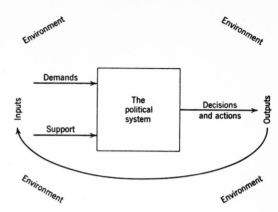

Diagram 2.—A simplified model of a political system.

at work in all systems and starkly reveals the source of a system's capacity to persist. It may well stand temporarily as the simplest image, to carry in our minds, of the processes we are about to discuss in detail.

To summarize the conceptualization being reviewed here, our analysis will rest on the idea of a system imbedded in an environment and subject to possible influences from it that threaten to drive the essential variables of the system beyond their critical range. To persist, the system must be capable of responding with measures that are successful in alleviating the stress so created. To respond, the authorities at least must be in a position to obtain information about what is happening so that they may react insofar as they desire or are compelled to do so.

In *A Framework for Political Analysis* each of these concepts and interrelationships was attended to in varying degrees of detail . . . It will be my task to begin to apply them in an effort to construct a much more elaborate structure for the analysis of political systems.

In doing so, we shall find ourselves confronted with a series of major questions. What precisely are the nature of the influences acting upon a political system? How are they communicated to a system? In what ways, if any, have systems typically sought to cope with such stress? What kinds of processes will have to exist in any system if it is to acquire and exploit the potential for acting so as to ameliorate these conditions of stress? . . .

53.
The Cybernetic Analysis of Change in Complex Social Organizations

MERVYN L. CADWALLADER

IN THE VIEW of the general agreement about the fundamental role of communication in human life, it might be assumed that any major breakthrough in the scientific study of communication phenomena would be hailed as an event of considerable significance for sociology.[1] This has, indeed, occurred, but with the rapid development of information and cybernetic theory, most sociologists have remained unaware of it.

Cybernetic theory has been extensively applied in electronics, telecommunications, automation, and neurology. Some first attempts at the application of cybernetics in experimental psychology are reported in *Information Theory in Psychology*. Communication theory has been slower in gaining attention among the social scientists interested in large social systems. The pioneers include enconomist Kenneth E. Boulding and electrical engineer Arnold Tustin, who have suggested a variety of empirical applications to the problems of economic analysis. Karl W. Deutsch has undertaken a cybernetic analysis of the emergence of nationalism in political communities. The work of these men demonstrates that cybernetics can be employed as a theoretical system in social analysis. The present essay sketches how some of the concepts and principles of cybernetics might be used in the analysis of change in formal social organizations.

The ultrastable system.—The fundamental theme of cybernetics is always regulation and

Reprinted from Mervyn Cadwallader, "The Cybernetic Analysis of Change in Complex Social Organizations," *American Journal of Sociology*, 65 (1959), 154–57, by permission of The University of Chicago Press. Copyright © 1959 by The University of Chicago Press.

control in open systems. It is concerned with homeostasis in organisms and the steady states of social organizations. Its orientation is the source of considerable misunderstanding because many of the sociologists who are interested in the subject of social change object to the use of all concepts of equilibrium, homeostasis, or stability, arguing that to include such ideas as a central part of social theory is to preclude the possibility of dealing with change. They seem to believe that stability and change are not only contradictory ideas but that the processes themselves are totally incompatible. The difficulty here is not merely semantic: some kinds of stability do negate certain kinds of change. What has been overlooked is that at least one category of stability depends upon and is the consequence of change. Just this kind of stability is of prime interest to cybernetics.

An open system, whether social or biological, in a changing environment either changes or perishes. In such a case the only avenue to survival is change. The capacity to persist through a change of structure and behavior has been called "ultrastability." If a complex social organization is to survive critical changes in its environment, it can do so only by changing its structure and behavior. That Great Britain has survived through medieval, mercantile, and capitalist periods means that as a national state it has ultrastability. Any industrial corporation, such as International Business Machines or General Electric, that has survived the last fifty years of social change in the United States has done so through a process of self-transformation and not through the continuation of original organizational and operational patterns. Therefore, the concept of ultrastability will aid in distinguishing between systems that achieve stability under specific constant conditions and those that can learn or evolve new structures and behavior so as to remain stable under changing conditions. The latter is the focus here.

Another way of expressing the above is to say that some classes of open systems adapt to a fluctuating environment through processes of learning and innovation. There is nothing new in such a statement if the reference is to biological organisms. The novelty here lies in the proposal that complex formal social organizations, such as industrial corporations, armies, churches, and so on, be regarded as learning and innovating systems. Or, to put it another way, large-scale formal organizations are treated as open problem-solving systems, studied with a variety of theoretical problem-solving models, i.e., as learning and innovating systems.

By common convention we are used to thinking in terms of individual human beings as inventing or innovating, but not of social groups. But it is valid to talk about innovations produced by a social organization taken as a whole, and this is not to deny the fact of individual innovation. Any such system capable of purposeful problem-solving behavior and of learning from the past and innovating for the future is an ultrastable system.

Cybernetics and the analysis of ultrastable organizations.—From the point of view of cybernetics, any large scale formal social organization is a communication network. It is assumed that these can display learning and innovative behavior if they possess certain necessary facilities (structure) and certain necessary rules of operation (content).

First, consider the structure of the system—as it might be represented in the language of cybernetics. Any social organization that is to change through learning and innovation, that is, to be ultrastable, must contain certain very specific feedback mechanisms, a certain variety of information, and certain kinds of input, channel, storage, and decision-making facilities. This can be stated in the form of an axiomatic proposition: that complexity of purposeful behavior is a function of the complexity of the communication components or parts of the system. More specifically, every open system behaving purposefully does so by virtue of a flow of factual and operational information through receptors, channels, selectors, feedback loops, and effectors. Every open system whose purposeful behavior is predictive, and this is essential to ultrastability, must also have mechanisms for the selective storage and recall of

information; it must have memory. Does the social organization under scrutiny behave purposefully, does it solve problems, and does it forecast future events? If the answers are in the affirmative, then one must find in it certain kinds of communications, information, and control mechanisms.

In addition to the requisite structural components mentioned above, the communication net must contain or acquire information that makes learning and innovating behavior possible. This is a "program." That is to say, it must acquire or discover rules of behavior, instructions regarding internal mechanisms and processes—all of which will result in performance to be identified as learning, problem-solving, and innovating.

Innovation by any system is subject to the limitations and possibilities established by the quantity and variety of information present in it at a particular time and by the information available to it from the environment. Something cannot be created from nothing, much less something new. Therefore the range of possible new combinations that may be formed by an innovating system depends upon the possible range of output, the range of available information stored in the memory, and the operating rules (program) governing the analysis and synthesis of the flow of information within the system. In order to innovate, the system must be able to analyze information, that is, it must separate it into constituent parts. In a social system this is a consequence of certain explicit operating rules about what can and should be done, by whom, when, and why.

The utilization by a system of a particular part of its fund of information as an output for the solution of an environmental problem is not usually determined by pure chance, unless the system, in dealing with a totally unfamiliar situation, is trying completely random outputs.[2] In the long run there must emerge an organization of the trial process in any open system capable of storing information about past behavior. Purposeful and predictive behavior depends upon memory, whether the system is organismic or social. Continuing behavior is modified by the results of specific acts. This is one kind of negative feedback and one which introduces a bias into the program of the system which changes the probabilities of various kinds of future acts in terms of present and past successes and failures.

If the problem-solving output of the system is organized solely in terms of past successes and failures, a point would be reached in its development at which it would not try anything new: all obstacles would be attacked with the techniques which had already proved successful. Innovation depends, therefore, on preventing such a freezing of the behavior of the system in old patterns. This is accomplished in a variety of ways. "Mistakes" in the identification, analysis, and synthesis of information may be the source of novel behavior. The loss of information (forgetting) about the past countermands the freezing process, to some extent, in all open systems complex enough to learn. In addition, the program of the system may contain specific instructions preventing the synthesis of all information into old familiar patterns and explicitly supporting certain kinds and amounts of novel action. Whenever novel behavior is successful, a negative feedback of information reinforces the creation and use of novelty. Not only will the system innovate, but it will remember that the act of innovating enabled it to circumvent obstacles and reach its goals. It will have discovered that a technique which worked in the past can be improved upon. Finally, in doing so, the system will have achieved the state of ultra-stability which, for an open system, is the optimum road to survival.

The elements of a model, empirical indicators, and sample hypotheses.—One of the main tasks which a theoretical model performs for the scientist is the selection of relevant variables and significant hypotheses from the infinite number of possibilities. A cybernetic model would focus the investigator's attention on such things as the following: (1) the quantity and variety of information stored in the system; (2) the structure of the communication network; (3) the pattern of the subsystems within the whole; (4) the number, location, and function of negative feedback loops in the system and the amount of time-lag in them; (5) the nature of the system's memory facility; (6) the operating rules, or program determining the system's structure and behavior.

The operating rules of the system and its subsystems are always numerous. Relevant for the present problem are (1) rules or instructions determining range of input; (2) rules responsible for the routing of the information through the network; (3) rules about the identification, analysis, and classification of information; (4) priority rules for input, analysis, storage, and output; (5) rules governing the feedback mechanisms; (6) instructions for storage in the system's memory; (7) rules regarding the synthesis of information for the output of the system—especially those concerned with the matter of usual or novel output.

It is now possible to suggest a few cybernetic propositions determining the presence, absence, and nature of innovative processes in complex communications systems. For example, it can be said that: (1) the rate of innovation is a function of the rules organizing the problem-solving trials (output) of the system; (2) the capacity for innovation cannot exceed the capacity for variety or available variety of information; (3) the rate of innovation is a function of the quantity and variety of information; (4) a facility, mechanism, or rule for forgetting or disrupting organizing patterns of a high probability must be present; (5) the rate of change for the system will increase with an increase in the rate of change of the environment (input). That is, the changes in the variety of the inputs must force changes in the variety of the outputs or the system will fail to achieve "ultrastability."

While no exact mathematical relationship between the elements of such a system has been specified, it is assumed that this is possible in principle but that its realization must wait for the results of actual experimentation and field tests. The use of mathematical devices for the measurement of information and the representation of networks will be a necessary and crucial first step in research programs designed to test hypotheses derived from the above theory.[3] Research might be carried out along the following lines: (1) the volume of mail, telegrams, telephone calls, and memos could be sampled at input terminals, output terminals, and at crucial points in the network; (2) the volume of printed and written materials stored in the libraries and files of the system could be measured; (3) tracer messages would enable the observer to map channel connections, one-way couplings, two-way couplings, and to locate relatively independent subsystems; (4) the time taken by regular or tracer messages to move through a feedback loop would give information on time-lag; (5) the many techniques already in use by the social scientist for measuring values and attitudes will be useful tools for the detection and measurement of implicit operating rules. The techniques of content analysis could be put to use for the abstraction of critical operating rules contained in the official documents of the formal organization, in order to isolate and index those parts of the program of the system which constrain and determine the range, routing, identification, analysis, storage, priority, feedback, and synthesis of information. Above all else, the rules supporting the synthesis and use of unusual as against usual patterns of action would be of special concern in a description and analysis of the ultrastable system in the process of change, or of a system with a certain potential for purposeful change.[4]

Notes

1. "Advances in communications theory in the last few years give the appearance of an important scientific breakthrough. They provide principles for translating into similar terms and quantitative units communication interactions both among human beings and among machines, as well as between men and machines. *They also suggest aspects of a more inclusive theory of organization.*" Richard L. Meier, "Communications and Social Change," *Behavioral Science*, I (January, 1956), 43. (Italics mine.)

2. The randomized strategy of certain games as described in game theory is one example. However, it is assumed that goal-seeking behavior is guided by random trial-and-error process during the early history of such systems.

3. The Shannon–Wiener concept of information is quantitative. However, there is some question as to whether it can be applied in macrosocial analysis at the present stage in the development of sociology. A qualitative concept may have to suffice for a time.

4. It is assumed that there are distinctly different kinds of social change exhibited by different kinds of social systems. For a discussion of this problem see Mervyn L. Cadwallader, "Three Classes of Social Change," *Pacific Sociological Review*, I (May, 1958), 17–20.

54.
Feedback Problems of Social Diagnosis and Action

KURT LEWIN

1. MANY CHANNELS of social life have not simply a beginning and an end but are circular in character. The large section of the channel which leads food from the grocery store into the mouths of the family members or into the garbage can is actually a part of another circular process. This process includes dishwashing, receiving money from the husband, and other sections of housekeeping which follow each other in a circular way. Many of the sections are interdependent in that finishing one starts the next.

Organized social life is full of such circular channels. Some of these circular processes correspond to what the physical engineer calls feedback systems, that is, systems which show some kind of self-regulation. One of these systems will be discussed here as an example of problems of social steering or self evaluation.

2. Planned social action usually emerges from a more or less vague "idea." An objective appears in the cloudy form of a dream or a wish, which hardly can be called a goal. To become real, to be able to steer action, something has to be developed which might be called a "plan." The transition from an idea to a plan presupposes that: (i) The objective has to be clarified; (ii) The path to the goal and the available means have to be determined; (iii) A strategy of action has to be developed. These three items together make up the "general plan" which is to precede action.

It should be noted that the development of a general plan presupposes "fact-finding." The

From Kurt Lewin, "Frontiers in Group Dynamics," Part II-B, *Human Relations*, I (1947), pp. 147–53. Reprinted by permission of Tavistock Publications Ltd.

original state of the idea of the goal corresponds to an area in the social field or the life space of the individual that is but little structured in itself and the relation of which to the rest of the field is not clearly determined. Fact-finding is necessary to structure the goal, its relation to the total setting and the path and means which may lead to the goal. On the basis of this fact-finding the goal is usually somewhat altered in light of the findings concerning the means available.

The emerging "general plan" corresponds to a field which contains the structure of the goal, and the steps to the goal in sufficient detail to serve as a blueprint for action. It is important, however, that such a plan be not too much frozen. To be effective, plans should be "flexible." The flexibility of plans requires the following pattern of procedure: Accepting a plan does not mean that all further steps are fixed by a final decision; only in regard to the first step should the decision be final. After the first action is carried out, the second step should not follow automatically. Instead it should be investigated whether the effect of the first action was actually what was expected.

In military terms, reconnaissance should provide data about where one now stands and whether the field has changed significantly. The result of the reconnaissance after the first step of action should be twofold: (i) It might be necessary to alter the "general plan"; (ii) The basis is given for a final decision on the second step. After the second step again reconnaissance follows, leading again to an alteration of the general plan and the decision on the next step.

This pattern of planned group action is probably developed in most detail in the army. It is widespread, however, in many areas of social life, frequently though in a rather rudimentary form. To understand what kind of social organization is required for efficient planned group action one can refer to the pattern of certain goal seeking machines.

3. During the war a multitude of self-steering missiles were developed, goal seeking machines which can reach their target with a remarkable degree of precision. Basically, these goal seeking machines have two components: one is equivalent to a sense organ, perhaps a radar eye; the other is an action organ, for instance, a gun which shoots bullets. If the beam from the target hits the eye off center, a mechanism is set into motion which automatically turns the eye to the center and changes the direction of the action organ toward the target. In other words, the eye functions as a steering mechanism. Technically this is achieved by so-called "feedback" processes which link three entities, namely: (i) the position of the target, (ii) the sense organ, (ii) the action organ. The action organ is continuously steered toward the goal with the help of the sense organ which "seeks" to eliminate divergencies between action and goal.

Some actions of human beings such as driving a car or reaching for a glass of water are steered by a functionally equivalent process. The individual watches the discrepancy between the direction of his action and the direction toward the goal, and this perceived discrepancy more or less automatically steers his action.

Is there anything equivalent in social life to steer social action? What are our social sense organs? How about the steering process?

The engineer knows of steering processes which have no reference to the outside. An example is the system which assures that the rudder of a ship follows every turn of the steering wheel at the captain's bridge. This system lies entirely within the ship and has no relation to points outside. In administration such steering corresponds to a case where a superintendent reports back to the manager of the factory that he has carried out the required action of hiring an expert. That, of course, does not assure that the action has the desired effect of improving the course of the organization. Of similar nature is the following example: Citizens who feel that certain group relations do not follow an appropriate course get together and try to give the wheel a turn toward the right direction by arranging a brotherhood day. They are elated for having done a good job if the meeting was impressive. Perhaps,

however, they should be compared with the captain who hears that his course is too much to the left, rushes to the wheel, turns it to the right and, having done so, goes happily to dinner. In the meantime, his boat goes around in circles.

A good number of our social or administrative actions are of a similar nature. The effort might lead to the satisfaction of the action but actually it does not reach the objective. The reason for the shortcoming, expressed in terms of feedback systems, is that all the inter-dependent parts of the process lie within the moving boat. What is missing is a link which steers the action by its effect on the outside rather than by the effect within the organization.

In many fields of social management as, for instance, in those dealing with minority problems, education, conducting conferences, or committees, we lack signposts of exactly where we are and in what direction we are moving with what velocity. As a result, the actors are uncertain of themselves, they are at the mercy of likes or dislikes of bosses, colleagues, or the public. Perhaps even more important, however, they are unable to "learn." In a field that lacks objective standards of achievement, no learning can take place. If we cannot judge whether an action has led forward or backward, if we have no criteria for evaluating the relation between effort and achievement, there is nothing to prevent us from making the wrong conclusions and to encourage the wrong work habits. Realistic fact-finding and evaluation is a prerequisite for any learning. No wonder that a recent survey of workers in group relations revealed that one of their great difficulties is their feeling of unclearness about what they should do.

An efficient steering of social action presupposes that fact-finding methods have to be developed which permit a sufficiently realistic determination of the nature and position of the social goal and of the direction and the amount of locomotion resulting from a given action. To be effective, this fact-finding has to be linked with the action organization itself: it has to be part of a feedback system which links a reconnaissance branch of the organization with the branches which do the action. The feedback has to be done so that a discrepancy between the desired and the actual direction leads "automatically" to a correction of actions or to a change of planning.

Accounting systems in business are designed to function as reconnaissance parts in the feedback system of a social group. The effectiveness of these and other methods of fact-finding depend upon the frequency with which the reconnaissance is carried out, whether it reaches the really essential data, whether the reconnaissance is transmitted to a sufficiently powerful level in the hierarchy of steering, without channeling so many fact-findings into that steering group that it is overburdened.

4. The research needed for social practice can best be characterized as research for social management or social engineering. It is a type of action-research, a comparative research on the conditions and effects of various forms of social action, and research leading to social action. Research that produces nothing but books will not suffice.

This by no means implies that the research needed is in any respect less scientific or "lower" than what would be required for pure science in the field of social events. I am inclined to hold the opposite to be true. Institutions interested in engineering, such as the Massachusetts Institute of Technology, have turned more and more to what is called basic research. In regard to social engineering, too, progress will depend largely on the rate with which basic research in social sciences can develop deeper insight into the laws which govern social life. This "basic social research" will have to include mathematical and conceptual problems of theoretical analysis. It will have to include the whole range of descriptive fact-finding in regard to small and large social bodies. Above all, it will have to include laboratory and field experiments in social change.

Field experiments are basically not different from laboratory experiments. An experiment as opposed to a mere descriptive analysis tries to study the effect of conditions by some way of measuring or bringing about certain changes under sufficiently controlled conditions. The objective is to understand the laws which govern the nature of the phenomena under study, in our case the nature of group life.

A change (ch) refers to the difference between a preceding situation (S) and a following situation which has emerged out of the first as a result of some inner or outer influences. $(Ch = S_{after} - S_{before})$. A law is found if this change, ch, can be linked to a function, f, of certain factors x and y which are found to be responsible for that change. Not all laws have this form. However, this form represents one of the simplest patterns of a law and characterizes also a certain type of experimental procedure and analysis.

This type of experiment, whether laboratory or field experiment, has as its objective the study of three situations or processes, namely: (a) the character of the beginning situation, (b) some happenings designed to bring about certain change, (c) a study of the end situation to see the actual effect of the happening on the beginning situation. A diagnosis of the before and after situation permits us to define the change or effect; studying the happening should be designed to characterize the factors which brought about this change.

It is obvious that the quality and exactness of the conclusions that might be drawn cannot be larger than the degree to which all three parts of the process can be analyzed. It demands a measurement of the situation before and after but equally a careful description and analysis of those happenings which brought about the change.

In case of a field experiment such as a workshop, this means that an analysis of the situation before and after the workshop is needed to define the change created by the workshop. It means also that the workshop itself would have to be described as carefully and accurately as possible with the objective of finding out as much as possible exactly what type of happening had created this change.

Here, I feel, research faces its most difficult task. To record the content of the lecture or the program would by no means suffice. Description of the form of leadership has to take into account the amount of initiative shown by individuals and subgroups, the division of the trainees into subgroups, the frictions within and between these subgroups, the crises and their outcome, and, above all, the total management pattern as it changes from day to day. These large-scale aspects, more than anything else, seem to determine what a program of action will accomplish. The task which social scientists have to face in objectively recording these data is not too different from that of the historian. We will have to learn to handle these relatively large units of periods and social bodies without lowering the standards of validity and reliability to which we are accustomed in the psychological recording of the more microscopic units of action and periods of minutes or seconds of activity.

One of the difficulties which a description of happenings as extended as the workshop presents to the psychologist is the mere size. Historians have been accustomed to dealing with units of decades and hundreds of years. Psychologists have been more accustomed to minutes and seconds. The particular meaning which the term analysis had to the scientist in the nineteenth century and in the beginning of the twentieth, has identified scientific procedure to many psychologists with procedures which deal with minute time periods. It is only recently that some of us have lost the prejudice according to which the description of a large unit is less scientific than the description of a small unit.

Even those among us who in principle do not like to discriminate against large units have to face a task which is new and a bit frightening even to the brave soul. It raises the question: can we hope to use as objective a description and measurement of large social units as we have been able, at least to the degree we have learned to characterize and measure small units? Is there any way to keep up our standards of reliability and objectivity? At present I feel the social scientist is threatened by the Scylla of losing his "objectivity" by the attempt to deal with sufficiently large and meaningful units on the one hand, and by the Charybdis of losing the "validity" of his study by dealing with inadequate and frequently too small units.

5. Any research program set up within the framework of an organization desiring significant social action must be guided closely by the needs of that organization, and must help define those needs more specifically. Usually there will be three kinds of problems to which the research staff must apply themselves:

Immediate problems. There will be a number of problems requiring some immediate program of action. Experience has shown that the research social scientist can make two contributions here: (*a*) As consultants on methods of action. The accumulation of scientific findings concerning social action techniques is mounting daily and only a technician in this field can be expected to keep up with them. (*b*) As evaluation experts. Major actions should not be launched without proper provisions being made for the evaluation of the success of the action and for the discovery of more effective modifications which may be found. Adequate evaluation is a technical research problem.

Pre-testing. Pre-testing by experimental trying-out of certain potential lines of action with properly selected groups and adequately defined controls is one of the most practical refinements of science, and one of the surest guides to sound administrative policy.

Long-term policies and action programs. As research proceeds, it will become more and more valuable for determining long term policies and action programs. By delegating to the research worker certain responsibilities and freedoms to carry on what is sometimes called "pure research" in a general area of dynamics it is safe to assume

that certain basic data for long term planning will gradually emerge. While research of this type sometimes does not look immediately "practical," those in the past who have backed this line of activity have reaped a rich harvest of efficiency, economy and effectiveness.

6. Obviously social management in the various areas of modern society have to face a tremendous task. Its solution presupposes social fact-finding of an unheard of magnitude. It requires basic research about social steering systems. The fear of fascism seems to have driven some people into the greatest kind of misunderstanding which identifies democracy with planlessness. The survival and development of democracy depends not so much on the development of democratic ideals which are wide-spread and strong. Today, more than ever before, democracy depends upon the development of efficient forms of democratic social management and upon the spreading of the skill in such management to the common man.

The social scientists, perhaps more than the natural scientists, have to learn to be unafraid and at the same time fair-minded. To my mind, fair-mindedness is the essence of scientific objectivity. The scientist has to learn to look facts straight in the face, even if they do not agree with his prejudices. He must learn this without giving up his belief in values, that is, without regressing to the pre-war cynicism of the campus. He has to learn to understand how scientific and moral aspects are frequently interlocked in problems, and how the scientific aspects may still be approached. He has to see realistically the problems of power, which are interwoven with many of the questions he is to study, without his becoming a servant to vested interests. His realism should be akin to courage in the sense of Plato, who defines courage as wisdom in the face of danger.

The problem of our own values, objectives, and of objectivity are nowhere more interwoven and more important than in action-research. Fortunately the work of social scientists during the war has created in a good many people just this spirit.

Research in group dynamics is, as a rule, group research. It requires the cooperation of persons who steer group life and who record and measure various aspects of group life. One cannot overemphasize the importance of the spirit of cooperation and of social responsibility for research on group processes. To my mind it is equally important that the same spirit of cooperation dominate the relations between the various institutions which happily have become active in this field.

55.
Control as an Organizational Process

CHADWICK J. HABERSTROH

THE STUDY of self-regulating systems, now generally known as cybernetics, explores the ways in which some output of a dynamic system can be maintained in a more-or-less invariant equilibrium, or steady state, in the face of disrupting external forces. The most general answer to this question is that the system must somehow be supplied with information about the disrupting forces that is used to offset their effect. A common way of supplying this is by means of a feedback of information on the deviations of the output from equilibrium. This information flow causes the equilibrium to be restored in some appropriate manner.

Even assuming that one does know the feedback channel used and understands the laws through which the feedback restores equilibrium, he still has a right to ask how it is that the system exists at all, and why it tends to an equilibrium at that particular value and not some other. In the case of engineering control systems the answer to this question is simple and direct: the designer intended them to perform in the way they do. Thus, there is a purposive element in these control systems resulting from an *a priori* selection of the equilibrium to be obtained. If one asks the same question, however, about naturally self-regulating systems, such as homeostatic mechanisms in the living organism or ecological balances in a community of organisms, the answer is neither simple nor direct. The equilibria found and the mechanisms for attaining them have come about by the process of natural selection in the context of a

From Chadwick J. Haberstroh, "Control as an Organizational Process," *Management Science*, 6 (January, 1960), 165–71. Reprinted by permission of the author and publisher.

particular environment. If we put the same question in the case of organizational systems, the answer is even less direct and more complicated, involving as it does a multitude of designers each consciously striving to realize his own objectives, in the context of an environment and of selection pressure arising from the limitation of resources as well.

In organizations the conscious intentions of the participants are an important factor. In order to explain the gross behavior of an organization, these intentions must be measured and brought into relation with the other aspects of the organization's functioning. The existence of stable organization implies a degree of harmony and co-ordination among the participants, a sharing of intention. In order to secure this, participants communicate with each other and in doing so construct a common symbolic picture of the goals they have set for the organization and the means by which they intend to attain the goals. This picture, or representation, of the means and ends of organization (the "task analysis") is implicit in the verbal communication inside the organization. It can be measured by the use of content-analysis techniques. I have attempted to apply these techniques to a sample of communication from an integrated steel plant operated by one of the American companies. This case will be used as an example in exploring organization purposes and other organizational characteristics affecting control processes.

Goal formation is influenced by the intentions of the individual participants and by the environmental constraints under which they operate. Both can be sources of conflict. The emergence of stable, enduring patterns of organization is in

445

part a process of conflict resolution. The necessity of reducing conflict to manageable bounds tends to direct the organization's efforts toward a small number of goals and a small number of means activities for achieving them, relative to the number of alternatives that might be conceivable. It is to be expected, therefore, that the number of independent goals turned up in the task analysis will be rather small. Conflict reduction is facilitated if these goals are formulated in terms of acceptable levels, rather than in terms of optima, and if the criterion of goal achievement is external and objective, rather than subjective and open to dispute. If members measure goal achievement objectively and perceive means to attain them, the goals are termed "operative."

In the case of Integrated Steel, four goals were discovered. These relate to cost reduction, production level, safety, and medical care. The safety and production goals are formulated in terms of acceptable levels set by an external office. Performance is measured in terms of tonnage produced and frequency of injuries, and an elaborate technology exists for goal achievement. In the case of safety, this task analysis was measured in detail. The goal of providing adequate medical care was departmentalized in a plant hospital; and a standard cost system and various cost reduction programs were in operation. Neither the hospital nor the cost system was investigated, however.

If the process of goal formation results in a small number of operative goals, as it did at Integrated Steel, the basis for a feedback of information on deviations of performance from the established goals is already apparent. To affirm the existence of a control system we need only verify that this information is reported to executive centers and that the executives respond so as to achieve the goals. The task analysis comprises a program of means activities understood by the participants to lead to goal achievement. One way of responding would be to adjust the level of resource use in these means activities. Let us refer to this as "routine control." Another way of responding would be to look for a better way of achieving goals. This type of activity could take the form of inventing new means activities or of altering the system of executive organization (i.e., changes in personnel or in allocation of functions). It might be expected that this type of activity would occur only in a case of extreme or repeated failure. Let us call this "non-routine control." Sufficient pressure might even lead to modification of goals in order

to assure survival of the organization. Normally, however, the evolved structure of goals and means activities determines what the participants do; communication channels carrying information on performance influence when and how much they do.

In the case of Integrated Steel's safety program the type of means activities which have been developed to implement the safety objective are accident investigations, safety conferences with workmen, implementation of safety work-orders, special inspections, clean-up work, etc. The execution of each of these activities is in some way conditional on the occurrence of injuries in the plant. Other activities are also carried on which are independent of the occurrence of accidents. These include routine inspections, training and screening procedures for new employees, safety clearance of engineering proposals, job analysis, publicity campaigns, etc.

The formal communication channels on safety performance begin with injury reports made by the plant hospital. This information is collated and distributed daily throughout the plant's executive organization in detail and in statistical summary. This information cues the line supervision to investigate injuries; alerts the plant safety staff to inspect for similar hazards and to assist in accident investigations; and, in summary figures, provides the basis for broader types of corrective action such as the study of classes of jobs for hazards, the issuance of special instructions to employees, and evaluation of supervisors. The same reports when aggregated into divisional and plant injury frequencies serve as an indicator of the plant's overall performance relative to its safety goals.

The routine control processes discussed above are not the only, or even the most important, means of control used at Integrated Steel. The non-routine control processes, changes in personnel and in the institutional structure within which the participants operate, take precedence. The very nature of the accident process (i.e., the importance of human failure, rare events, conjunction of circumstances, and the randomness of occurrence of injuries) make for a different degree of reliability on the technological side from that encountered in connection with, for instance, production matters. Because the coupling between the program of means activities and the degree of safety performance is not fully determined, there is a need for relatively tight control over the programs themselves. This is achieved by response of the top plant management

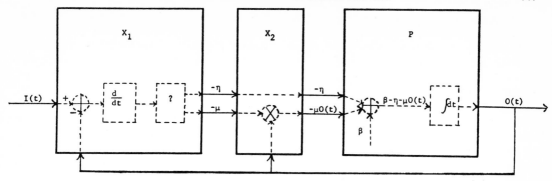

Figure 1.—Control Flow Chart of Safety Functions at Integrated Steel.

TABLE 1

Innovation* and Performance at Integrated Steel

| Year | Disabling Injuries | | | | | | Total Injuries | | | Innovation* |
	Average All Plants	Error	Δ Error	Plant Perfor-mance	Error	Target	Plant Perfor-mance	Error	Target	
1	5.29	.98	.98	6.17			357			moderate
2	4.95	.13	− .85	5.08			422			none
3	5.41	1.77	1.64	7.18			407			heavy
4	4.66	1.24	− .53	5.90			302			none
5	3.66	.67	− .57	4.33			244			none
6	3.55	.65	− .02	4.20	.46	3.74	210	0	238	light
7	3.00	.83	.18	3.83	.50	3.33	196	0	210	moderate
8	2.38	.81	− .02	3.19	.72	2.47	183	11	172	light
9	2.07	.63	− .18	2.70	.54	2.16	133	0	168	none
10[a]	1.93	.91	.28	2.84	.80	2.04	128	0	133	heavy

[a] At time major decisions were taken.

* Level of innovation in the plant-wide safety program was rated by the author on the basis of a survey of plant safety files. The information found consisted of a description of the innovations made. This information is briefly summarized below.

In the first year studied the safety staff recommended and received management approval of a job analysis program which was to provide the basis for strict enforcement of safe working procedures. They also requested regular physical examinations for all employees.

In year two, no safety innovations were discovered.

In year three, one of the plant manager's top staff assistants announced to division managers the inauguration of an extensive program of job analysis and indoctrination of workmen in safe procedures. He also urged the division managers to inaugurate the practice of having foremen make thorough investigations of all minor injuries as a basis for corrective action. This was to be coupled with a program of training foremen in the responsibilities which would be placed upon them in these two programs, and also the formation of division safety committees at top division management level to expedite safety recommendations. He also announced inauguration of a plant-wide safety committee.

In years four and five no new activity was discovered.

In year six a proposal was made by the safety staff for transfer of some functions so as to improve the co-ordinating service of the safety staff and shift more executive responsibility on to the line organization. There was a new program of statistical reporting of injuries classified by types of accident.

In year seven management inaugurated a revised system of job analysis, appointed a new plant-wide advisory committee, inaugurated an annual conference of all division managers for the purpose of setting objectives and reviewing the safety program, and also ordered the universal replacement of a hazardous type of crane controller in use through most of the plant.

In year eight a new statistical basis for the reporting of injuries was inaugurated.

In year nine no innovations were discovered.

In year ten a revised and greatly expanded program of job analysis was instituted, with a number of executives re-assigned to safety responsibilities exclusively. Procedures for top level reporting and evaluation of safety performance were revised to place greater emphasis upon safety.

to deviations of the plant and departmental injury frequencies from the objectives set for them at the beginning of the year. These yearly objectives are set by company officers above the plant level, although the plant management has discretion to aim at a more difficult target if it chooses.

Figure 1 is a block diagram of the control structure discussed above. The input (I) is the annual safety objective which is compared with the performance of the plant (O) by top management. The result of the non-routine control functions of the top executive organization (X_1) may be expressed as the two parameters of the routine control system: the intensity of response to injuries (μ) and the level of independent safety activity (η). A complete model of the top executive function was not constructed, although there seems to be evidence[1] that it responds to changes in the degree of error (a differentiating operator). Other than that it appears possible to say only that its effects are intermittent, rather than continuous, and respond only to error in excess of a certain threshold. It is therefore a nonlinear operator. In the case of the routine control function (X_2), however, a linear model seems appropriate. Executives appear to proportion their influence on the injury rate to the magnitude of that rate plus a constant. The "program" operator (P) relates the control activities to the actual performance of the plant, adding and integrating the safety efforts ($\mu O + \eta$) and the exogenous load of new hazards (β).

Table 1 contains data on injury rates, safety objectives, and innovations in the safety program for a 10 year period. The changes in organization made by top management in year 7 did not take effect until year 9. Thus, during the period beginning with year 3 and ending with year 8, the routine contol system operated with constant parameters μ and η. Under this assumption, the injury rate ($O(t)$) is given by

$$O(t) = \int [\beta - \eta - \mu O(t)] \, dt \qquad (1)$$

or equivalently

$$O'(t) = \beta - \eta - \mu O(t). \qquad (2)$$

Solving this differential equation,

$$O(t) = \frac{\lambda}{\mu} + \left[\hat{O} - \frac{\lambda}{\mu}\right] e^{-\mu t} \qquad (3)$$

where $\lambda = \beta - \eta$ and $\hat{O}$ is the initial level of the injury rate.

This equation implies that first differences in injury rates tend to decrease by a constant ratio from year to year. The performance data for years 3 to 8 in Table 1 is fairly consistent with this.

Another principle of control, important in the case of organization, is that of factorization. Ashby has shown that if trial-and-error changes are relied upon for control (compare the operator X_1 at Integrated Steel), a large system cannot practically be stabilized unless its output can be factored into a number of independently controlled information sources. At Integrated Steel, the safety objective was broken down by divisions and injury rates were reported on that same basis. Part of the nonroutine control activity occurred at the division level. Innovation in divisional programs initiated by division management was correlated with the division's performance error.

This, of course, bears on the subject of decentralization in organizations. Meaningful decentralization is probably impossible without a resolution of the goals into nonconflicting, operative subgoals so that these can be placed under independent control. On the other hand, there is probably a size for organizations at which goal attainment becomes impossible without factorization, even though the method used may not resemble current definitions of "decentralized authority." The plant production, cost, and safety goals at Integrated Steel appear to represent just such a factorization of the company goals.

In summary, the characteristics postulated by cybernetic theory for self-regulating systems have their correlates in human organizations. In the case of Integrated Steel, the theory points up the influence of information feedbacks upon the actions of the executives in attempting to realize the organization objectives. Of particular importance are the role of the higher echelons of executives in controlling the mode of response of the lower echelons and the use of multiple feedbacks in the design of the executive system.

Note

1. Compare columns 4 and 11 in Table 1.

56.
The Cybernetics of Competition:
A Biologist's View of Society

GARRETT HARDIN

. . . WHAT IS OFFERED HERE is one biologist's conception of the foundations of social and economic theory. "What presumption!" social scientists may say. Admitted; but biology, as Warren Weaver has put it, is "the science of organized complexity"—and what is the social scene if not one of organized complexity? Some of the principles worked out in one field should be at least part of the theoretical structure of the other. Particularly relevant are the principles of *cybernetics*, the science of communication and control within organized systems.

. . . . . .

The cybernetic model can be carried over into economics, as shown in Figure 1, which depicts

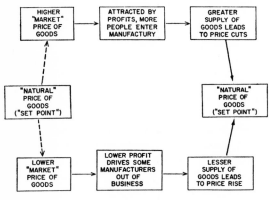

Figure 1.—Cybernetic regulation of price, in the Ricardian model.

Reprinted from Garrett Hardin, "The Cybernetics of Competition: A Biologist's View of Society," *Perspectives in Biology and Medicine*, VII (Autumn, 1963), 61–84, by permission of the University of Chicago Press. Copyright © 1963 by the University of Chicago Press.

the control of price in the Ricardian economic scheme. The well-known course of events can be read from the figure. We see that negative feedback produces stability about a "set point," which Ricardo called the "natural price." The model would be more realistic if it were constructed in terms of *profit* rather than price, but for historical continuity we retain the classic Ricardian element *price*. As with the biological example previously used, the meaning of "natural" can, in general, only be determined *ex post facto*. The word "natural" is here (as elsewhere) a verbal cloak for ignorance. Nevertheless, it or an equivalent word is needed to remind us of the state of affairs. There is mystery here. It was this mystery together with the unpremeditated consequences of the economic cybernetic system that led Adam Smith to speak of an "Invisible Hand."

An effective cybernetic system produces stability, i.e., fluctuations within limits, and this we esteem. A system that produces a stable temperature, or a stable population, or a stable price, seems to us somehow *right*. When we examine any cybernetic system we discover that it is more or less wasteful. The thermostated room wastes heat; the natural population wastes lives; the economic system produces price wars and business bankruptcies. We may refine the controls and minimize the losses (of heat, or of money, for example), but a close examination of the system convinces us that there must always be some losses, waste in some sense. This is so because the controls that serve to produce equilibria are themselves so many modes of loss. Accounting procedures, insurance programs, police forces, sweat glands, electric fans, predation, crowd diseases, delicate thermostats—all these are forms of waste. We do not regret them, for the negative feedback produced by each of these elements acts as a check to some kind of uncontrolled and ruinous positive feedback. But each negative feedback device has its price, and we cannot get rid of one form of loss without incurring another. In a deep sense we see that some waste is inevitable and natural, and we recognize as immature the man who compulsively tries to do away with all waste. We recognize as pathological the goal of a waste-free world. This recognition is an important element in that complex of temperament that we label "conservative." Insofar as we think deeply, we all, of necessity, partake of this temperament to some extent.

But because the mature person acknowledges the inevitability of some waste, it does not follow that he must be reconciled to any amount and kind of waste. In the first excitement of discovering the beauties of economic cybernetics, David Ricardo quite naturally made such an error. In speaking of the cybernetic system that stabilizes the population of laborers, Ricardo wrote: "When the market price of labour is below its natural price, the condition of the labourers is most wretched: then poverty deprives them of those comforts which custom renders absolute necessaries. It is only after their privations have reduced their number, or the demand for labour has increased, that the market price of labour will rise to its natural price. . . ."

Attention should be called to the use of the word "natural" in this question. It would be antihistorical to expect Ricardo to speak of the "set point of labor" inasmuch as the term "set point" was not used for another century; but that is not the only criticism that can be made of the word "natural." Looking at the problem through the eyes of Stephen Potter, what do we see? Plainly, that an advocate is likely to use the word "natural" in order to insinuate approval of the "natural" thing into the mind of his auditor. By so doing, the advocate frees himself of the necessity of developing a defensible argument for the "natural" thing—for who can disprove of that which is "natural"?

This attack on the use of the word "natural" is more than a mere Potterian counterploy, as is clearly shown by the following defense given by Ricardo:

Labour, like all other things which are purchased and sold, and which may be increased or diminished in quantity, has its natural and its market price. The natural price of labour is that price which is necessary to enable the labourers, one with another, to subsist and perpetuate their race, without either increase or diminution.

These then are the laws by which wages are regulated, and by which the happiness of far the greatest part of every community is governed. Like all other contracts, wages should be left to the fair and free competition of the market, and should never be controlled by the interference of the legislature.

This passage leaves no question in our mind that Ricardo identified the momentary state of things in his own time as "natural" and that all attempts to modify it further by new legislation were "unnatural" and hence improper in some deep sense. With rare exceptions, most of us post-Ricardians have been unwilling to accept this view. We will accept the starvation of field mice; but not that of human workers. Ricardo, at least on paper, accepted both. But—perhaps because of a delicate consideration of the feelings of others?—he used a most elegant euphemism for the facts. "It is only after their privations have reduced their number," he wrote; and insisted that "wages should be left to the fair and free competition of the market." The market must be free, that we may enjoy the blessings of cybernetic stability. Most of us now think that Ricardo's price is too high. We are willing to make use of "unnatural" controls of the price of labor even if it means losing some of our freedom. The history of the labor movement since Ricardo's time may be regarded as one long struggle to substitute other forms of waste for the "natural" form which Ricardo, who was not a laborer, was willing to accept.

The Competitive Exclusion Principle

Perhaps more important than the humane argument just given against the Ricardian model is a theoretical argument which indicates that the cybernetic system he described is fundamentally unstable. Before we can discuss this matter we need to introduce a biological principle known by various names but recently called the "competitive exclusion principle." The historical origin of this principle is complex; no one man can be given credit for it. In the last decade it has become increasingly clear that it is a basic axiom of biological theory; and it will be my argument here that it is basic also to sociological and economic theory. But first, let us develop the principle in an exclusively biological context.

Consider a situation in which two mobile species, X and Z, live in the same habitat and also live in the same "ecological niche," i.e., live exactly the same type of life. Species X multiplies according to this equation:

$$x = Ke^{ft}, \qquad (2)$$

where x is the number of individuals of species X at time t; e is the base of natural logarithms; K is a constant standing for the number of x at $t = 0$; and f is a constant determined by the "reproductive potential" of the species.

Species Z multiplies according to this equation:

$$z = Le^{gt}, \qquad (3)$$

in which the constants have the same meaning as before (though, in the general case, with different values).

Suppose these two species are placed in the same universe to compete with each other. What will happen? Let us represent the ratio of the numbers of the two species, x/z, by a new variable, y. Then:

$$y = \frac{Ke^{ft}}{Le^{gt}}. \qquad (4)$$

Since K and L are both constants, they can be replaced by another constant, say C; and making use of a well-known law of exponents, we can write:

$$y = Ce^{ft-gt} = Ce^{(f-g)t}. \qquad (5)$$

But f and g are also constants, and can be replaced by another constant, say b, which gives us:

$$y = Ce^{bt}, \qquad (6)$$

which is, of course, our old friend, the equation of exponential growth. The constant b will be positive if species X is competitively superior, negative if it is species Z that multiplies faster.

What does this mean in words? This: in a finite universe—and the organisms of our world know no other—where the total number of organisms of both kinds cannot exceed a certain number, a universe in which a fraction of one living organism is not possible, one species will necessarily replace the other species completely if the two species are "complete competitors," i.e., live the same kind of life.

Only if $b = 0$, i.e., if the multiplication rates of the two species are *precisely* equal, will the two species be able to coexist. Precise, mathematical equality is clearly so unlikely that we can ignore this possibility completely. Instead we assert that the *coexistence of species cannot find its explanation in their competitive equality*. This truth has profound practical implications.

Have We Proved Too Much?

It is characteristic of incomplete theory that it "proves too much," i.e., it leads to predictions which are contrary to fact. This is what we find on our first assessment of the competitive exclusion principle. If we begin with the assumption that every species competes with all other species, we are forced to the conclusion that one species— the best of them all—should extinguish all other species. But there are at least a million species in existence today. The variety seems to be fairly stable. How come?

There are many answers to this question. I will discuss here only some of the answers, choosing those that will prove suggestive when we later take up problems of the application of the exclusion principle to human affairs. The following factors may, in one situation or another, account for the coexistence of species.

Geographic isolation.—Before man came along and mixed things up, the herbivores of Australia (e.g., kangaroos) did not compete with European herbivores (rabbits). Now Australians, desirous of retaining some of the aboriginal fauna, are trying desperately to prevent the working out of the exclusion principle.

Ecological isolation.—English sparrows introduced into New England excluded the native bluebirds from the cities. But in very rural environments bluebirds have, apparently, some competitive advantage over the sparrows, and there they survive today.

Ecological succession.—It is not only true that environments select organisms; in addition, organisms make new selective environments. The conditions produced by a winning species may put an end to its own success. Grape juice favors yeast cells more than all others; but as the cells grow they produce alcohol which limits their growth and ultimately results in new predominant species, the vinegar bacteria. In the growth of forests, pine trees are often only an intermediate stage, a "subclimax," being succeeded by the climax plants, the hardwood trees, which out-compete the pines in growing up from seeds in the shade of the pine tree.

Lack of mobility.—The universal application of the exclusion principle to plants is still a controversial issue, which cannot be resolved here. It may be that the lack of mobility, combined with certain advantages to being first on the spot, modify the outcome significantly. Although this explanation is questionable, it is a fact of observation that a pure stand of one kind of plant hardly ever occurs.

Interbreeding.—If two competing populations are closely enough related genetically that they can interbreed, one group does not replace the other, they simply merge. This does not end competition; it merely changes its locus. The different genes of the formerly distinct groups now compete with each other, under the same rule of competitive exclusion.

Mutation.—Continuing with the example just given, one gene never quite eliminates another because the process of mutation is constantly producing new genes. The gene for hemophilia, for example, is a very disadvantageous gene; but even if hemophiliacs never had children (which is almost true), there would always be some hemophiliacs in the population because about three eggs in every 100,000 produced by completely normal women will be mutants that develop into hemophilic sons.

The Cybernetics of Monopoly

We are now ready to take a second look at the Ricardian thesis. The model implicit in his writings may not unfairly be stated as follows. We conceive of a single product manufactured by a number of entrepreneurs, each of whom must, for simplicity in theory construction, be imagined to be engaged in the manufacture of this product only. Under these conditions the Ricardian cybernetic scheme diagrammed in Figure 1 will prevail —but only for a while. History indicates that the number of entrepreneurs is subject to a long-term secular trend toward reduction. In the early days there were many scores of manufacturers of automobiles in the U.S.; today there are less than a dozen. Ball-point pens, transistors—every new product—have followed the same evolution. The history of the oil industry (to name only one) indicates that under conditions of perfect laissez faire, competition has a natural tendency to steadily decrease the number of competitors until only one is left. In industries with heavy overheads this tendency is a consequence of the economy of size. But even without this size effect, a simple extension of the competitive exclusion principle into economics shows that a reduction in the number of competitors will take place as the

more efficient entrepreneurs squeeze out the less efficient, until ultimately only one is left. If this were not so, we would have to conclude that the free enterprise system has no tendency to produce the lowest possible price; or, to put it differently, that it has no tendency to produce the maximum efficiency. Either conclusion would deny the claims to virtue put forward by the defenders of the free enterprise system.

If a monopoly is produced, what then? Here is a question which Ricardo did not face. At first glance one might say that the monopoly price should be stable, because if it were to rise, new entrepreneurs would be attacted to the field and would lower the price. But this is a naïve view. We know that it is more difficult to start a business than to continue one, and consequently a monopolist can maintain a price considerably above the "natural price." Furthermore, a realistic model must include much more than we have indicated so far. We must consider the whole complex of phenomena that we include under the word "power." *Social power is a process with positive feedback.* By innumerable stratagems a monopolist will try to manipulate the machinery of society in such a way as to ward off all threats to re-establish negative feedback and a "natural" cybernetic equilibrium. And, as history shows, the monopolist in one field will seek to extend his power into others, without limit.

What has just been said about business monopolies applies equally to labor monopolies, *mutatis mutandis.* Insofar as they meet with no opposition, there is little doubt that labor monopolies seek to produce an ever higher price for labor. At the same time, they protest the appearance of business monopolies. Contrariwise, unopposed businessmen seek to promote a free market in labor while restricting it in their own field (by "Fair Trade" laws, for instance). It is not cynicism but simply honesty that forces us to acknowledge that Louis Veuillot (1813–1883) was right when he said: "When I am the weaker, I ask you for liberty because that is your principle; but when I am the stronger I take liberty away from you because that is my principle." In other words, such verbal devices as "principles," "liberty," and "fairness" can be used as competitive weapons. Each purely competitive agent, were he completely honest and frank, would say, "I demand a free market—but only for others." It is, in fact, a natural part of *my* competitive spirit to seek to remove from *my* field the natural competition on which the validity of the Ricardian scheme rests.

Such an analysis, which is based on the observed behavior of competing groups, may seem depressing. Rather than dwell on the possible emotional consequences of the facts, let us see what we can do about arranging the world to our satisfaction. Let us try to enlarge the model of our theory. To do this we acknowledge that we are *not only* unconscious "purely competitive agents," but that we are also capable of being conscious. We can predict the results of our own actions, as well as the results of the actions of those opposed to us. We acknowledge that *words are actions*, actions designed to influence others. Because we can see that others resort to high-flown rhetoric when they want to influence us, we become suspicious of our own arguments. We operate under the basic and parsimonious rule of the Theory of Games, which says that we must impute to others intelligence equal to our own. Under these conditions we seek the *boundary conditions* within which the rule of laissez faire can produce stability.

The Limits of Laissez Faire

Laissez faire has a strong emotional appeal; it seems somehow right. Yet we have seen that, in the limit, the rule fails because of the positive feedback of power. Can we *rationalize* the rule of laissez faire by harmonizing it with boundary conditions?

I suggest that there is, in biology, a useful model already at hand. Consider the cybernetic system that controls the temperature of the human body. This system works admirably. So well does it work that, for the most part, we can safely adopt a laissez faire attitude toward our body temperature.

The system works without conscious control or planning. But only within limits. If the environmental stress is too great, temperature control fails. At the upper limit, too great a heat input raises the body temperature to the point where the physiological thermostat no longer functions. Then higher temperature produces greater metabolism, which produces more heat, which produces higher temperature, which—and there it is, positive feedback, leading to death, to destruction of the whole system. Similarly with abnormally low temperatures. The working of the system is shown in Figure 2. There is a middle region in which a laissez faire attitude toward control of the environment works perfectly; we call this middle region the *homeostatic plateau*. (The word

"homeostatic" was coined by W. B. Cannon to indicate constancy-maintained-by-negative-feedback.) Beyond the homeostatic plateau, at either extreme, lies positive feedback and destruction. Plainly, our object in life must be to keep ourselves on the homeostatic plateau. And insofar as it is within our power to affect the design of a system, we would wish to extend the plateau as far as possible.

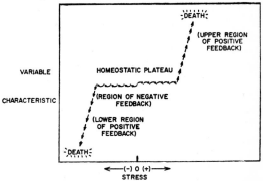

Figure 2.—The cybernetics involved in the survival of a system.

Is this not the model for all cybernetic systems, sociological and economic as well as biological, the model on which ethics must be based? The desire to maintain *absolute* constancy in any system must be recognized as deeply pathological. Engineering theory indicates that excessive restraints can produce instability. In psychiatry also, the desire for complete certainty is recognized as a most destructive compulsion. And in the history of nations, attempts to control rigidly all economic variables have uniformly led to chaos. The psychologically healthy human recognizes that fluctuations are unavoidable, that waste is normal, and that one should institute only such explicit controls as are required to keep each system on its homeostatic plateau. We must devise and use such controls as are needed to keep the social system on the homeostatic plateau. On this plateau—but not beyond it—freedom produces stability.

We can do this only if we explicitly give up certain superficially plausible objectives which are incompatible with stability. In the realm of economics, the most dangerous will-o'-the-wisp is the word "efficiency." Consider the classical Ricardian economic system. If we decide that all waste is bad and that we must maximize efficiency, then we will stand admiringly by and watch the competitive exclusion principle work its way to its conclusion, leaving only one surviving entrepreneur, the most efficient. And then? Then we find that we have a tiger by the tail, that we have allowed the positive feedback of power to go so far that we may be unable to regain anything that deserves the name of freedom. It is suicidal to seek complete efficiency. The Greek Solon said, "Nothing in excess," to which we must add, *not even efficiency*. Whatever it is that we want to maximize, it cannot be efficiency. We can remain free only if we accept some waste.

How are we to keep a social system on its homeostatic plateau? By laws? Not in any simple way, for the effect of an action depends on the state of the system at the time it is applied—a fact which is, I believe, not systematically recognized in the theory of law. An act which is harmless when the system is well within its homeostatic boundaries may be quite destructive when the system is already stressed near one of its limits. *To promote the goal of stability, a law must take cognizance not only of the act but also of the state of the system at the time the act is performed.* In his effort to obtain the maximum individual freedom, it is to be expected, of course, that "economic man" will try to defend his actions in terms of some tradition-hallowed "absolute" principles that take no cognizance of the state of the system. Absolutists of all sorts may, in fact, be defined as men who reject *systematic* thinking.

Consider this question: Should a man be allowed to make money, and keep it? In the history of Western capitalism our first approximation to an answer was an unqualified *Yes*. But as we became aware that money is one means of achieving the positive feedback of power, we looked around for curbs. One of these is the graduated income tax, which most men would now defend as a reasonable brake to the positive feedback of economic power. Yet it can easily be attacked as being "unfair," and in fact has been so attacked many times. As late as 1954 (according to a press report) the industrialist Fred Maytag II, speaking to a meeting of the National Association of Manufacturers on the subject of discriminatory taxes, issued this clarion call for action: "The hour is late, but not too late. There is no excuse for our hesitating any longer. With all the strength of equity and logic on our side, and with the urgent need for taking the tax shackles off economic progress, initiative is ours if we have the courage to take it."

One cannot but have a certain sympathy for the speaker. He is right when he says that the existing tax structure is contrary to "equity." But if discussion is to be carried on in terms of such abstractions, Mr. Maytag would find his opponents introducing the word "justice" and saying that this is more precious than equity. Rather than use such verbal bludgeons, we should think operationally in terms of the homeostatic plateau. We should think in terms of systems rather than individual acts. That this sort of thinking presents difficulties for the law is admitted; but it is clear also that we have made some progress in the solution of these difficulties, e.g., in the graduated income tax. It is clear also that our systematic thinking has not produced perfect solutions to our problems (e.g., it is still possible to become a millionaire via the capital gains route).

Indeed, the recognition of the relevance of the whole system in judging the desirability of an individual act can be traced back to antiquity. One of the greatest of the technical social inventions of ancient Athens was that of *ostracism*, which was invented by Cleisthenes. We are told:

Once a year the popular Assembly deliberated on whether any citizen should be required to go into exile for ten years on the grounds that his presence in Athens was a threat to the constitution. If the Assembly voted to hold an ostracism, a second vote was taken. Then, if six thousand citizens wrote the same name on an *ostrakon*, or potsherd, the man named must leave Athens for ten years. But he did not lose his citizenship, his goods were not confiscated, he did not even suffer disgrace. In fact, it was only the man of great ability who was likely to be ostracized, yet the possibility of ostracism was a constant deterrent to overweening political ambition.

In other words, ostracism was a device aimed at stopping the positive feedback of power, a tool designed to maintain the political system on a homeostatic plateau. Recognition of the dangers of this positive feedback must surely be almost universal among practical men and produces the most diverse strategems, many of which would seem quite paradoxical to one who was ignorant of the positive feedback of power (as adolescents in our society often are). For instance, we are told that "in the early history of the Church, bishops had to take two solemn oaths at the time of their ordination. The first oath was that they would discharge the duties of that office faithfully in the sight of God and man. The second oath was called the oath of 'Nolo episcopari'—'I don't want to be a bishop.' . . ." Those who frequent the university campuses of our own time will surely have noted that one of the best ways to achieve a deanship is to insist that one doesn't want to be dean (but not too loudly!). Competition and the desire to limit power produce strange strategies.

The Persistence of Variety

An important part of the unfinished work of theoretical biology revolves around the question of variety: how are we to account for the variety of the living world? The competitive exclusion principle points always towards simplification; yet the world remains amazingly, delightfully complex.

The same problem exists in economics. Why do there continue to be so many competing units? The economist's problem is, I suspect, even further from solution than the biologist's, but we can briefly list some of the social factors, which resemble those mentioned earlier in the biological discussion.

Geographic isolation.—A less efficient company may be able to coexist with a more efficient one if it is at a considerable distance and if transportation charges are heavy, as they are, for instance, in the coal and steel industry. (It is interesting to note that major steelmakers of the United States two generations ago tried to negate this factor by enforcing the "Pittsburgh-plus" system of pricing.)

Product differentiation.—In biology, ecological differentiation is the necessary condition for coexistence; in economics, product differentiation plays the same role. Patents, copyrights, and mere advertising gimmicks enable entrepreneurs partially to escape pure competition.

Mergers prevent extinction in economics in the same sense that inter-breeding prevents extinction in biology.

In the social realm we have in addition various peculiarly human characteristics that contribute to the persistence of variety. Curiosity, envy, dislike of boredom, yearning for destruction are a few of the factors which work against the efficiency of the market and hence tend to perpetuate variety. We are a long way from understanding the economic system. It is, however, transparently clear that any satisfactory over-all theory of economics must include a large measure of psychology in it. The *Homo economicus* of classical theory has been useful as a first approximation only.

The Idea of a System

One of the most important ideas in modern science is the idea of a *system*; and it is almost impossible to define. There are a number of good essays available on this subject. Here we will try to define by example.

Our first example is a caricature from the nineteenth century—the idea of a system that connects the welfare of England with the existence of old maids. The argument is simple: old maids keep cats, cats eat rats, rats destroy bumblebee nests, bumblebees fertilize red clover, and red clover is needed for horses, which are the backbone of English character training. Ergo the strength of England depends on a bountiful supply of old maids. Now that is a caricature, but it gets across the idea that the many cybernetic systems of nature are connected in complex ways. So complex are they that we can seldom predict exactly what will happen when we introduce a new element into a system. By way of illustration, consider the following examples from three different fields of biology.

Ecology.—Charles Elton tells the following history.

Some keen gardener, intent upon making Hawaii even more beautiful than before, introduced a plant called *Lantana camara*, which in its native home of Mexico causes no trouble to anybody. Meanwhile, some one else had also improved the amenities of the place by introducing turtle-doves from China, which, unlike any of the native birds, fed eagerly upon the berries of *Lantana*. The combined effects of the vegetative powers of the plant and the spreading of seeds by the turtle-doves were to make the *Lantana* multiply exceedingly and become a serious pest on the grazing country. Indian mynah birds were also introduced, and they too fed upon *Lantana* berries. After a few years the birds of both species had increased enormously in numbers. But there is another side to the story. Formerly the grasslands and young sugar-cane plantations had been ravaged yearly by vast numbers of army-worm caterpillars, but the mynahs also fed upon these caterpillars and succeeded to a large extent in keeping them in check, so that the outbreaks became less severe. About this time certain insects were introduced in order to try and check the spread of *Lantana* and several of them (in particular a species of Agromyzid fly) did actually destroy so much seed that the *Lantana* began to decrease. As a result of this, the mynahs also began to decrease in numbers to such an extent that there began to occur again severe outbreaks of army-worm caterpillars. It was then found that when the *Lantana* had been removed in many places, other introduced shrubs came in, some of which are even more difficult to eradicate than the original *Lantana*.

From this example (and scores of comparable ones are known) it is easy to see why it is so difficult to secure the permission of the U.S. Department of Agriculture to import any species of plant or animal. However, though we are very conservative about the introduction of biotic elements into our ecological systems, we show the most juvenile irresponsibility in our attitude toward new chemicals. To get rid of insects, we spray promiscuously with such potent poisons as Malathion. As a result, we kill not only millions of insects, but also thousands of birds. Because birds are a great natural negative feedback for insect populations, using insecticides often causes a secondary *increase* in the numbers of insects later. We may refer to this as a "flareback"— thus verbally acknowledging our failure to think in terms of systems. We are only now beginning to see the magnitude of the problems we have created for ourselves by *unsystematic* thinking, for which belated insight we are significantly indebted to Rachel Carson's book *Silent Spring*.

Embryology.—Beginning about 1960 a drug known as "thalidomide" became an increasingly popular sedative in Europe. It seemed superior to all others in effectiveness and harmlessness. But by the end of 1961 a most painful disillusionment had set in. When taken during the early weeks of pregnancy, it frequently interfered with the development of the limb buds of the child, resulting in the birth of a child suffering *phocomelia*—seal-limbs, little flipper-like hands, without long arm bones. In addition, there were other variable defects of the ears, digestive tract, heart, and large blood vessels; strawberry marks were common. Only a minority of the children whose mothers took thalidomide during the first trimester developed phocomelia, but so widespread was the use of the drug that the number of cases produced in West Germany alone in two years' time probably exceeded 6,000. This experience contributed to a re-evaluation of the whole idea of therapy, particularly of newly pregnant women. The developing embryo is a set of cybernetic systems of the greatest complexity. Coupled with the high rate of change during the early weeks is a high sensitivity to foreign chemicals inserted into the system. To a growing extent, physicians are loath to permit a newly pregnant woman to take any drug if it can possibly be avoided.

When we think in terms of systems, we see that a fundamental misconception is embedded in the popular term "side-effects" (as has been pointed out to me by James W. Wiggins). This phrase means roughly "effects which I hadn't foreseen, or don't want to think about." As concerns the basic mechanism, side-effects no more deserve the adjective "side" than does the "principal" effect. It is hard to think in terms of systems, and we eagerly warp our language to protect ourselves from the necessity of doing so.

Genetics.—When a new gene is discovered, it must be named; this is accomplished by naming it for some conspicuous effect it has on the organism. But when a very careful study is made, it is found that a mutant gene has not one effect but many. For example, close analysis of one mutant gene in the laboratory rat has shown no less than twenty-two well-defined effects, including effects on ribs, larynx, trachea, vertebrae, lungs, red blood cells, heart, teeth, and capillaries. Yet all these effects spring from a single chemical change in the genetic material of the fertilized egg. In the early days, geneticists often used the word "pleiotropy" to refer to the multiple effects of genes. Now it seems scarcely worthwhile to use this word because we are pretty sure that all genes are pleiotropic. The word "pleiotropy" is a fossil remnant of the days when geneticists failed to have sufficient appreciation of the developing organism as a system.

Pleiotropy presents animal and plant breeders with one of their most basic and persistent problems. The breeding performance of the St. Bernard dog will serve to illustrate the problem. Crosses between St. Bernard and other breeds of dogs produce a large proportion of stillborn or lethally malformed puppies. The trouble apparently lies in the pituitary gland, which is overactive. When we look closely at the adult St. Bernard, we see that its abnormally large head and paws correspond to "acromegaly" in humans, a condition also caused by an overactive pituitary. The St. Bernard breed is, in fact, standardized around this abnormality. Why are not the causative genes more deleterious to the breed? Undoubtedly because there are other, "modifier," genes which alter the whole genetic system so that it can tolerate the effects of the "principal" genes. The production of a new breed built around some distinctive gene often takes a long time because the breeder must find, and breed for, a multitude of modifier genes which create a genetic system favorable to the principal gene. This work is almost entirely trial and error; along the way the breeder must put up with large losses in the way of unsuccessful systems of genes.

The Feasibility of Human Wishes

The dream of the philosopher's stone is old and well known and has its counterpart in the ideas of skeleton keys and panaceas. Each of these images is of a single thing which solves all problems within a certain class. The dream of such cure-alls is largely a thing of the past. We now look askance at anyone who sets out to find the philosopher's stone.

The mythology of our time is built more around the reciprocal dream—the dream of a highly specific agent *which will do only one thing*. It was this myth which guided Paul Ehrlich in his search for disease-specific therapeutic agents. "Antitoxins and antibacterial substances are, so to speak, charmed bullets which strike only those objects for whose destruction they have been produced," said Ehrlich in voicing this myth. Belief in the myth has inspired much fruitful research; but it *is* a myth, as the phenomena of allergies, anaphylaxis, auto-immunization, and other "side-effects" show us. It is *our* myth, and so it is hard to see.

One of the inspired touches in Rachel Carson's *Silent Spring* is her use of "The Monkey's Paw," a story which W. W. Jacobs built around our modern myth. In this story a man is allowed three wishes. He wishes first for money. He gets it. It is brought to his door as compensation for his son's death in the mill. Horrified, the father wishes his son alive again. He gets that wish too— his son comes to the door looking as he would after such an accident. In desperation, the father wishes everything back as it was before he was given the three wishes.

The moral of the myth can be put in various ways. One: wishing won't make it so. Two: every change has its price. Three (and this one I like the best): *we can never do merely one thing*. Wishing to kill insects, we may put an end to the singing of birds. Wishing to "get there" faster, we insult our lungs with smog. Wishing to know what is happening everywhere in the world at once, we create an information overload against which the mind rebels, responding by a new and dangerous apathy.

Systems analysis points out in the clearest way the virtual irrelevance of good intentions in determining the consequences of altering a system. For a particularly clear-cut example, consider the Pasteurian revolution—the application of bacteriology and sanitation to the control of disease. We embarked on this revolution because we wished to diminish loss of life by disease. We got our wish, but it looks now as though the price will be an ultimate increase in the amount of starvation in the world. We could have predicted this, had we taken thought, for Malthus came before Pasteur, and Malthus clearly described the cybernetic system that controls populations. The negative feedbacks Malthus saw were misery and vice —by which he meant disease, starvation, war,

and (apparently) contraception. Whatever diminution in effect one of these feedbacks undergoes must be made up for by an increase in the others. War, it happens, is almost always a feeble demographic control; and contraception is not yet as powerful as we would like it to be; so, unless we exert ourselves extraordinarily in the next decade, starvation will have to take over. Like the father in "The Monkey's Paw," we wanted only one thing—freedom from disease. But, in the system of the world, we can never change merely one thing.

Suppose that at the time Pasteur offered us his gift of bacteriology—and I use the name "Pasteur" in a symbolic way to stand for a multitude of workers—suppose at that time that some astute systems analyst had drawn a Malthusian cybernetic diagram on the blackboard and had pointed out to us the consequences of accepting this gift. Would we have refused it? I cannot believe we would. If we were typically human, we would probably have simply called forth our considerable talent for denial and gone ahead, hoping for the best (which perhaps is what we actually did).

But suppose we had been what we like to dream we are—completely rational and honest, and not given to denial? Would we then have rejected the gift of disease control? Possibly; but I think not. Is it not more likely that we would, instead, have looked around for another gift to combine with this one to produce a new, stable system? That other gift is well known of course: it is the one Margaret Sanger gave us, to speak symbolically again. It is a gift we are now in the process of accepting.

In terms of systems, we can give this analysis:

System	Stability
Malthusian	Yes
Pasteurian	No
Sangerian	Possibly
Pasteurian-Sangerian	Yes

A systems analyst need not, when confronted with a new invention, reject it out of hand simply because "we can never do merely one thing." Rather, if he has the least spark of creativity in him, he says, "We can never do merely one thing, *therefore we must do several* in order that we may bring into being a new stable system." Obviously, in planning a new system he would have to examine many candidate-ideas and re-examine our value system to determine what it is we really want to maximize. Not easy work, to say the least.

Is Planning Possible?

Some of the most excruciating questions of our time hinge on feasibility of planning. Is good planning possible? Is it possible to devise a planned system that is at least as good as a free system? Can the free market be dispensed with without losing its desirable virtues?

There is no dearth of literature supporting and condemning planning. Rather than add to this double battery of polemic literature, I would like to take a different approach. I would prefer to adopt an agnostic attitude toward the principal question and ask a second question: *If* successful planning is possible, what are its preconditions? If we can see these clearly, we should be in a better position to answer the principal question. The major points at issue seem to me to be the following.

1) Can it be shown, before instituting a plan, that all significant factors have been taken account of? It is not easy to see what the nature of the proof would be; and in any case, the consequences of past planning attempts do not make us optimistic.

2) Are we sure that we can predict all possible interactions of factors, even when we have complete knowledge of them? This is not as disturbing a question now as it was in the past. Any system of equations that can be solved "in principle" can be turned over to computing machines, which are immensely faster, more patient, and more reliable than human beings; and all computing machines operate under the Magna Charta given them by A. M. Turing.

3) Granted that we can predict a new and better stable system, can we also devise an acceptable transition? The many social systems known to historians and anthropologists represent so many points in space and time. The transitions from one to another are usually obscure; or, when recorded, are known to involve great human suffering and immense wastage of human resources. In general, transitions seem more feasible for small populations than large—but will small populations ever again exist?

4) Can we take adequate account of the reflexive effect of knowledge and planning on the actions of the planned and the planners? I have argued elsewhere that a satisfactory theory of the social sciences must be based on recognition of three classes of truth. No one, to my knowledge, has tackled this fundamental problem.

5) Can it be shown that programming, in the light of the reflexive effect of knowledge, does

not lead to some sort of infinitive regress? Only so can solutions be achieved.

6) Can the calculations be carried out fast enough? Modern calculating machines, with their basic operations measured in microseconds, are marvelously speedy. But the number of operations required may be astronomical, and the 3.1557×10^7 seconds available in each year may not be enough.

7) Can we persuade men to accept change? A casual survey of important reforms effected in the recent past shows that each of them took about seventy-five to one hundred years for completion. It is a general impression (and a correct one, I think) that the speed at which social problems *appear* is now accelerating. But is there any indication that the rate of solution is also accelerating? We seem to need some basic reform in people's reaction to proposed changes. Would this demand a new sort of faith? And in what? Science? Truth? Humanism?

8) Will any plan we adopt have adequate self-correcting mechanisms built into it? It is one of the virtues of a market economy that any error in judgment as to what people want is soon corrected for. Price fluctuations *communicate* needs to the managers. But in a planned economy, it has been often noted, planners who make errors are likely deliberately to interfere with the free flow of information in order to save their skins. Can a planned system include uncloggable channels of information?

Such seem to me to be the principal difficulties in the way of planning. Whether they will ultimately prove insuperable, who can say? But for the foreseeable future, I suggest there is much to be said for this analysis by Kenneth Boulding:

. . . I believe the market, when it works well, is a true instrument of redemption, though a humble one, not only for individuals but for society. It gives the individual a sense of being wanted and gives him an opportunity of serving without servility. It gives society the opportunity of coordinating immensely diverse activities without coercion. The "hidden hand" of Adam Smith is not a fiction.

There are forces operating in society, as there are within the human organism, which make for health. The doctor is merely the cooperator with these great forces in the body. The doctor of society—who is equally necessary—must also be a humble cooperator with the great forces of ecological interaction, which often restore a society to health in spite of his medications. It is precisely this 'anarchy' which Professor Niebuhr deplores which saves us, in both the human and the social organism. If we really established conscious control over the heartbeat and the white blood cells, how long would we last? Health is achieved by the cooperation of consciousness with a largely unconscious physiological process. Selfconsciousness is not always an aid to health, either in the individual or in society.

The problem of planning will not soon be disposed of, nor soon solved, but perhaps some false issues can be avoided if we make a distinction between "planning" and "designing." By *planning* I mean here what I think most people have in mind, the making of rather detailed, rather rigid plans. The word *designing* I would like to reserve for the much looser, less detailed, specification of a cybernetic system which includes negative feedbacks, self-correcting controls. The classical market economy is such a design. Kenneth Boulding when he speaks of "the market, *when it works well*" is, I believe implicitly referring to the biologist's model of homeostasis shown in Figure 1. The classical market should not be called *natural*, for it is a truly human invention, however unconsciously made. It is not universal. It has been modified continually as men have groped toward better solutions. I would submit that the proper role for conscious action is the ethical evaluation of many possible homeostatic systems, the selection of the best *possible* one, and the refinement of its design so as to make the homeostatic plateau as broad as it can be, thus maximizing both social stability and human freedom.

57.
Is Adaptability Enough?

GEOFFREY VICKERS

THE WORD ADAPTABILITY is widely and loosely used to describe the accommodations which we take to life. Most often it is used with approval of the readiness and skill with which an individual fits in to his social group. Less is heard of the readiness and skill of the group to accommodate and learn from the deviant individual; but this is also an aspect of adaptation and an essential one.

Again, it was once a matter of pride that the individual should be adaptable to his physical environment; that he should learn readily to tolerate extremes of heat and cold, changes of latitude and season, changes of diet. Today, especially on the North American continent, more pride is taken in adapting the physical environment to human requirements, providing us with a standardised diet, a standardised climate, even standardised opportunities for entertainment on the equator and in the Arctic circle, on the ground and in the stratosphere. What is involved is clearly an endless process of give and take between the individual, society and the physical environment, extended indefinitely in time. Any adjustment made in the course of this process deserves the name of adaptation.

So, as a first approximation, we can represent the course of history by the diagram in Figure 1, in which *En* represents the environment, *S* society, and *I* the individual.

The arrows are lines of possible, not invariable effect. The changes which arise in any member of the triad may or may not be due to changes in the others; may or may not affect the others; and

may or may not be affected in turn by changes which they produce. Thus, the lowering of temperature during the last ice age profoundly affected the habits of life on earth but it was not caused by life on earth nor was the earth significantly affected by these changes of habit. By contrast, the current raising of the level of radiation, which may have an effect on life at least as great as an ice age, is directly caused by the activity of man and is certain, in one way or another, to affect that activity in its turn. Thus any particular change may be described as a dependent, an independent, or an interdependent variable. An important feature of industrialization is to make the relations shown in the diagram ever more interdependent.

This picture of history is meaningless, unless we add to it some picture of the "nature" of its three members which accounts for the force and direction of their pushings and pullings. Each is a set of self-maintaining activities; and that which is maintained is a set of relationships, inner and outer. Even a kitten has a very large number of relationships to maintain, internal relationships such as those which keep its body capable of digesting food; external relations, such as those by which it secures food; and a large repertory of ways of behaving in order to maintain these relationships, ranging from behavior of the whole creature in, say, food seeking to the behavior of a single enzyme in its digestive tract or a single neuron in its nervous system. I will call all these ways of behaving *B*; and all the relations which are to be maintained I will call *E*. When the system concerned is a human being, I intend *B* and *E* to include also all those activities and all those relationships which we commonly describe

From Geoffrey Vickers, "Is Adaptability Enough?" *Behavioral Science*, 4 (1959), 219–34. Reprinted by permission of the author and publisher.

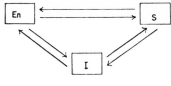

Figure 1.

in psychological terms, activities such as thought and feeling, relations such as status and success. Where the system concerned is a society, I intend B and E to be understood equally widely.

It remains to ask how behavior (B) should be so directed as to serve E. We can explain much of B in physical and chemical terms; but if we could so explain it all, we should not have explained the existence of the kitten, let alone you and me and Canada. The kitten is a specific pattern of order in a world which in its inorganic aspects tends naturally to disorder. Why a kitten or any other form, rather than increasing formlessness? Why the increasing improbability of ever more elaborate organization? How, in brief, is B set to produce E? We have to postulate a regulating process. I will call this unknown process R.

As a first approximation, then, we can represent the nature of a man or a society by the diagram in Figure 2.

Since the man or the society retains through time enough regularity to earn itself a name, we must suppose that somehow and to some extent E controls R and R controls B. Hence the arrows moving from right to left. We shall have occasion to add other arrows before we are through.

I now ask "Whence comes E?" and "How does R work?" I will try to pursue these questions separately in regard to the individual and in regard to society but the two enquiries will continually become mixed.

In the individual the norms and limits comprised in E are derived from three sources— biological evolution, cultural conditioning, and personal experience.

Biological evolution builds into us our future pattern of growth and development, our metabolic needs, and a host of other requirements, potentialities and propensities. They are believed to be coded in the gene. I have already expressed the view that our genetically given nature may well be much more complex than we know or perhaps will ever know.

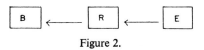

Figure 2.

Cultural conditioning persuasively moulds other positive and negative governors of our behavior, our ideas of success and security, of right and wrong, of what to seek and what to shun. Its influence is great, inescapable, indispensable but not absolute. For, first, it is no homogeneous, self-consistent whole; and in any case, only a fraction of it reaches any one individual. Secondly, it changes within the individual life span and is itself affected by the experiences of individual members of the culture. Finally, it may be, in greater or less degree, rejected. In all these three ways there is, of course, variety between one society and another and between one period and another in the history of the same society.

Individual experience, drawing thus partly but not wholly from the cultural inheritance, progressively builds the norms and limits which are to govern the individual through life. So great are the variations of genetic endowment and of personal experience that these governors will vary widely even within a fairly homogeneous culture; but the range of their variation will depend to no small extent on the culture concerned.

These three sources of norm-setting are in continual interaction; and it is easy to attribute to inherent, biological "nature" something which is in fact given by a particular culture or even by particular individual experience. Thus there is apparently no reason to suppose that all men are "naturally" competitive; but in a competitive society most men will seek success and approval in whatever competitive activity offers them an acceptable chance of success. Suppose, now, that changes in the environment or the natural development of the society make the competitive struggle harder and its prizes fewer. The individual experience of an increasing number will reject the cultural norm. These will become associated in a sub-culture, which in time may replace the original pattern. Individuals of future generations will then receive a different cultural conditioning and perhaps in time they will reject and modify that also.

Something like this happened in Britain in the 19th century. Competitive individualism never attracted more than a minority; and in consequence there grew up side by side two well marked and sharply distinguished ethics, the Trade Union ethic and the ethic of the entrepreneur, the ethic of "united we stand . . ." and the ethic of "Devil take the hindmost."

My attempt to deal separately with individual and social nature has already broken down. Social E affects and is affected by individual E; and both are affected by R. There is also, of course, close mutual inter-action between social *R* and individual *R*. We can redraw the Figure 1 to show these relationships, as in Figure 3.

I want now to attach a more definite meaning to the governors (*E*) of a national society such as Canada. It also is an open system, less closely integrated than an organism but equally capable of being described in terms of *B*, *R*, and *E*. It also is subject to laws which prescribe the conditions of its continuance. History is littered with the records of societies which have ceased to exist—some, like Carthage, by physical destruction; some, like your Indian predecessors on this continent, by the ending of some essential relation with the environment, such as that by which they lived as predators on herds of game. Contemporary societies also depend for their continuance on keeping their essential inner and outer relations within their critical limits. And within these limits, each society, like each individual, builds up from its progressive, unique experience and continually modifies its own set of aspirations and apprehensions, positive and negative governors of its behavior.

All states today assert that their primary purpose is to secure favorable conditions of life for their citizens. This in itself represents an evolution in the *E*'s of sovereign states; no chief executive today would venture to assert "L'Etat c'est Moi." It is an important definition of relationship between the state and the individual.

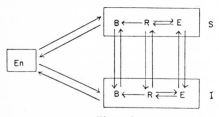

Figure 3.

All states assume or pretend that a necessary condition of this is the continuance of their integrity and independence. This is an important definition of their relationship with other collective entities outside their frontiers. We can see this emerging and hardening over several centuries past in Europe. Today we see it developing elsewhere, while, in the Western world we may be witnessing the first painful steps in its supersession. I shall have more to say about this later; suffice it for the moment to remind you that self-preservation is not an unqualified norm, even for an animal.

States vary widely in their ideas of what

conditions are to be sought, still more in their ideas of the relative rights of different classes of citizens; most of all, in the relative importance they attach to the present and the future. Those who share Bentham's devotion to "the greatest happiness of the greatest number" sometimes forget that the greatest number is always the number of the unborn.

At the national level, then, as at the individual level, behavior (*B*) is so directed as to serve *E*—or rather a varying selection from *E*; and *E* in turn is changed by the result of *B*—or rather by the outcome to which *B* contributes. Sometimes, if some valued condition of life is left long unrealised, it is tacitly abandoned. The level of aspiration is changed; *E* is altered to fit the possibilities. But on other occasions the denial of this valued condition builds up pressure for it until it demands realization at whatever cost, perhaps by revolution. Thus in Britain the right of a man to do what he likes with his own land has been progressively curtailed in the interests of good development and good farming; and though the curtailment which operates today is less drastic than in the war years, the old norm has gone, probably for good. On the other hand, the rights of the individual against the State in regard to minimum subsistence, as defined in the National Assistance Act, are the more sharply defined because of a feeling that they had been long denied.

The regulating process

This brings us to *R*. What is the regulating process, whether of states or of men, which continually adjusts behavior (*B*) so as to preserve *E*; and equally adjusts *E* so as to keep *B*'s task within the limits of the possible? Before trying to express this in terms of Canada today, we can say something about it in terms sufficiently general to cover men and societies alike.

We must credit *R* with at least three major functions. It must be able to receive information about how things are going in relation to how they should be going to maintain *E*; and it must be able to send signals to initiate behavior which will affect the course of events in the interests of *E*. And between these two functions of "information in" and "information out" we must posit the all-important function of selecting what to do out of the limited number of behaviors which are possible in the circumstances. These three functions focus the three main groups of problems concerned with adaptation. How does the system know that action is needed? How does it decide what to do? How does it make its decision effective?

The first question needs restating. *R* is always in action. Societies and men, like other systems, must be constantly active to maintain their own existence. But the demand on *R* varies. Suppose a system in a state of equilibrium. Regular forms of behavior are maintaining constant relationships. *R*'s activity is constant. Now suppose that this situation is disturbed, perhaps by a change in the environment (prices are affected by a bad harvest), perhaps by a change within the system itself (prices are affected by inflation). *R* will now have to act differently if the effect of the disturbance is to be neutralised and prevented from disturbing *E*. I will confine the enquiry for the moment to cases where change in this sense is needed. How does *R* know that something different is required of it?

Apart from forecast, to which I return in a moment, the signal comes in the form of direct experience of breakdown. The usual behavior ceases to be available or ceases to work. An increase in crime, for example, unless otherwise observed, will first come to notice because the police, the criminal courts, and the prison authorities become overloaded. The normal relation between their capacity and the work they have to do begins to go astray. Ultimately, the normal behavior will become impossible. Someone responsible for the custody of a prisoner will have no room for him in any prison. He must either change his standards of permissible overcrowding or change his standards of security or refuse to accept the man at all, which will change his normal relations with the committing authorities. Whatever he does or leaves undone will be a change.

The signal thus received is, of course, only a symptom of a more profound imbalance. The increase of lawlessness is a more important problem than the overcrowding of prisons; and, if it were understood, it might be found to be itself a symptom of some still more profound imbalance. But though increasing lawlessness may be the more important problem, it is not the more urgent. Those who worry about it may remain baffled for years without having to initiate any change in behavior. But the man with the unhoused prisoner must find somewhere to put him tonight.

It is useful, therefore, to distinguish signals which compel action from those which merely invite it, a distinction which runs closely parallel to the distinction between urgency and importance.

Clearly, a modern society would not hold together for a year, if *R* did not begin to work until it received signals of incipient breakdown. These signals are supplemented in varying degrees by prediction. An increasing trend in crime may become visible statistically before it has begun to cause actual breakdown in the machinery for dealing with criminals. The ability of *R* to act on merely predictive signals is, however, limited by several factors. Such signals never compel action, though they may invite it. They make less impact on the mind. And they are no more reliable than prediction itself. The volume of crime, for example, may change sharply and unpredictably, varying with factors which are at present unidentified.

Consider an example at the other end of the scale. The state of education today may be admirable; but the known structure of the age pyramid may make it certain that the needs of the school population five years hence will far exceed the capacity of the present resources, applying present standards. Even long term prediction in this instance is relatively easy and precise.

Comparing these two examples, the increase of crime and the increase of the school population, we can see another difference between the various signals which make up the "information in" of *R*. They are informative in varying degrees. The observed increase in crime is neither more nor less than a red light. It warns of a dangerous deviation from *E*. It throws no light on the cause nor does it give any guidance on what *R* should do. The observed increase in the school population, on the other hand, when grouped with other known facts, such as the existing school building and teacher training programmes and resources, makes the problem perfectly clear, though even this does not of itself tell *R* what to do.

The problem for *R* is to choose a way of behaving which will neutralise the disturbance threatening the maintenance of *E*. Success means initiating behavior which will reduce the deviation between the actual course of affairs and the course which would be consonant with *E*; or at least preventing its nearer approach to the limit of the unacceptable or the disastrous.

This decision is a choice between a limited number of alternatives. Men and societies have only a finite number of ways of behaving, perhaps a much smaller number than we realise; and the number actually available and relevant to a given situation is far smaller still. It is thus essential to regard these decisions as the exercise of restricted choice.

These decisions are of four possible kinds. When the usual response fail, the system may alter itself, for instance by learning new skills or reorganizing itself so as to make new behaviors possible; it may alter the environment; it may withdraw from the environment and seek a more favorable one; or it may alter E. These are possible, if at all, only within limits; and all together may prove insufficient.

It remains to ask how men and societies choose from among these alternatives, when choose they must. In brief, the answer is "by experience"; but experience operates in various ways, none of them perfect; and they demand some closer enquiry.

Control by error

Consider first the most elaborate. Systems are said to be error controlled when the results of their behavior return to them as a signal to control their subsequent behavior. The classic example is the helmsman, reading from his compass continuously the difference between the actual direction of the ship and its prescribed course and suiting the movements of the wheel to the information which he derives from these signals. Three features of this are worth noting.

First, the signal consists not in the position of the ship's head nor in its direction but in the two together. The helmsman responds to the rate of change of direction, relative to the true course. Problems encountered in the construction of automatic pilots have shown that he must even be able to notice changes in the rate of change, if he is to hold a true course.

Secondly, the information from these signals is necessarily late. No one can hold a true course, until he has learned to anticipate them.

Thirdly, the signal does not, strictly speaking, tell the helmsman anything at all about the effect of his own behavior. Many other influences, some of them more potent than the rudder, are constantly changing the direction of the ship's head. He cannot say with any certainty how much of it is his own doing. None the less with practice he learns to control the course with great precision, largely because the cycle is repeated so often and so frequently.

The larger affairs of life are more diverse. Men and governments continually find themselves doing things they have never done before. The results may not return for judgment for many years; and when they do, they include the influence of so many other variables that they may supply neither validation for the past nor guidance for the future. Even the standards by which to judge the results may be lacking. We can note disasters which seem to be the result of our foreign policy but no one can prove that the results of any other course would not have been worse.

It is thus rather a puzzle how far, if at all, what is rather confusingly called control by error is relevant to the working of R in the systems of men or societies. We have to distinguish carefully between two possible meanings of the phrase. In so far as we can compare what is happening with what ought to be happening (as given by E), we are kept informed of what is going wrong and perhaps of the rate and direction in which it is going wrong. Control by error in this sense is valuable and not uncommon. What we are not told with any certainty, if at all, except in the simplest and most often repeated cases, is what contribution our past actions have made to the result. Control by error in this sense is much more valuable—and much more rare.

Suppose for example that the statistics of crime show an increase. The society takes action; it recruits more police, increases penalties, devises new kinds of prisons or what not. Crime goes on increasing. It does not follow that the action taken was wrong. Perhaps, without it, things would be even worse. Perhaps it has not yet had time to take effect. Society knows only that in this respect the ship of state is still swinging further off course. It is controlled by error only in the first sense.

Consider, by contrast, the statistics relating to the school population; suppose these show an increasing imbalance between the numbers of children and the resources to teach them. Society takes action; it enlarges its programme for school building and teacher training. Month by month, the officials compare the numbers forecast with the actual, the number of schools completed and teachers trained with the actual; and these last figures will tell them with precision not only how things are going but also the results of their own efforts. With minor qualifications, their behavior is error-controlled in the second sense.

The two senses are, in fact, points on a continuous scale. Control by error enables us to compare the way things are going with the way we want them to go; and in doing so, it tells us something about the effect of our previous action. But this something may be anything from nearly everything to nearly nothing.

Control by rule and purpose

Experience works in two other more simple and more pervasive ways to guide R in its endless series of choices. I have called them in an earlier paper control by rule and control by purpose. To select a behavior because that is the behavior which one has always selected in the past in what seem to be similar circumstances is to be controlled by rule. To select a behavior because it is likely to have a foreseen and desired result is to be controlled by purpose.

The two interact. Even at the simple levels of animal behavior an habitual response can be inhibited, if it is always followed by punishment, an unfamiliar one rendered habitual, if it is always followed by reward. On the other hand, rule will withstand the attrition of anything less than repeated and unmistakable evidence that it does not work. At the level of deliberate human behavior, whether by men or societies, it is safe to assume that in selecting the course to be followed both criteria are present and both are used, often with conflicting results. All conflicts between morality and expediency are conflicts between control by rule and control by purpose. Even when no such emotional overtones are involved, we often feel extreme reluctance to adopt a course, however clearly it is commended by its probable results, if in the past we have always acted differently in circumstances which seem the same.

It is supposed to be a feature of increasing rationality to be controlled increasingly by purpose and decreasingly by rule. I doubt whether this is so. True, as we become more capable, individually and collectively, of making forecasts and looking ahead, we are able to build up longer and more coherent sequences of purposeful action. On the other hand, as we become capable of wider generalisation, we are able to build up more complex and comprehensive sets of rules. The mature human being and the mature society differ from those at simpler levels both by pursuing simultaneously a larger number of objectives and by regulating their behavior by a more comprehensive code of rules. Both control by rule and control by purpose are elaborated together.

In any case, control by purpose can never be ultimate. However remote and comprehensive the objective, it is always possible to ask why this objective rather than any other should be pursued. When the answer is given, as in this regression it soon will be, in terms of E—in terms, that is, of some relationship which this objective will help to attain or preserve or escape, it is still possible to ask why that matters so much. The system can put an end to the regression only by saying either "because that is the sort of system I am" or by saying "because that is the value that I place upon it," which is only another way of saying the same thing. I defer to the next section the problems of such valuation.

It remains to consider the third group of questions regarding R. How does it give effect to its decisions?

It is a feature of R in a complex modern society that action, when chosen, can only occur with the collaboration of very many people, of whom the vast majority are not personally concerned with the action, will derive no personal success or satisfaction from it, and would be personally none the worse and less burdened if it were dropped. Of these some have usually a personal vested interest in stopping it. What internal coupling, what common code of response is needed, if in such circumstances in a modern society R is to work at all?

In fact, R often fails to work. We see today in France a mammoth reorganisation in progress to remedy the persistent failure of R to work—and in particular, to work at this last stage, the stage of getting decisions carried into effect. We have seen a large proportion of the world adopt constitutions which secure coherence at the cost of liberties which we hold dear; and we are conscious within our own societies of action clearly identified and badly needed which yet does not happen. The third stage in the working of R is as vulnerable as any.

Application of the model

It is time to give more concrete meaning to these symbols. What are the regulating mechanisms of a national society, such as Canada? They consist on the one hand of the machinery of government, central and local, and on the other of the organisation of business, through which the country's economic growth chiefly proceeds. These two are linked, partly by the growing nexus of government control over business and business influence over government; partly by the fact that both are linked with the same body of citizens, though in different ways. I shall return later to discuss the relation between government R and business R, which is obviously crucial to the success of the country in its task of adaptation.

Any attempt to reduce the process of adaptation to a set of symbols and then to apply these to the bewildering complexity of contemporary life must lead to over-simplification and artificiality. I think it is none the less useful in various ways; first for the very fact that it makes clear how complex the actual process is. We should need a three-dimensional model to show even an example of the mutual interaction of person on person, society on society. We should need an immensely extended hierarchy to represent the tissue of group relations even within one closely knit society. And even then we should still lack any way of representing the equally essential dimension of time. The rate of change; the extent of its repercussions in a given time; the delay in response; the speed with which the results of a response will return for judgement; all these factors, essential to the process of adaptation, can only be described in terms of time.

Yet for all its simplification, the attempt serves to show the essential features of adaptation and the equally important complexities which modify these when we consider adaptation at the level of man and society.

Suppose a very simple system. Its essential relations (E) are coherent and fixed; so are the limits within which the system can deviate from them without suffering irreversible change. Its repertory of behavior (B) is also fixed. Its problem is so to behave as to keep within the limits of E. For the problems of such a system there must always be either a best answer or no answer. Such problems, however complex, could at least in theory be programmed for and solved by a computer. But even simple organic systems are more complex than this. At the level of individual and social adaptation, we have to take into account differences, whether of degree or of kind, of which I would stress three.

First, in a modern industrial society the main source of disturbance is the behavior of the system itself. It is thus, at least in theory, directly controllable. Rapid industrialisation, itself adaptive, a product of R, increases the rate of change in the system and its environment and thus the volume of disturbance with which R must subsequently deal. Since the volume of disturbance with which R can deal, even at its most efficient, is limited, it follows that the system may be destroyed by disturbances which its own activity creates. This is why, as I have so often stressed, the rate of change in a system may be lethal, irrespective of its direction.

Secondly, the content of E is itself variable within wide limits. Politically, economically and socially, our governing expectations are different from, and more numerous and exacting than, those of our grandfathers. Rapid industrialisation continually begets new ones. The advent of the automobile did far more than provide a new way to satisfy an old need. It created new needs— new norms of mobility to be attained, new limits of immobility to be regarded as unacceptable. It contributed to the singular dependence of the Western world on Middle Eastern oil, which so powerfully conditions international relationships today. It is not possible to distinguish such relationships from other elements of E as being artificial, transitory. They must be rated by their cogency. Are they relations for which men might destroy each other and themselves? If and so long as they are, they cannot be excluded from E.

Finally, the total volume of E becomes so large and contains so many inconsistencies that ever less of it can be realised. So the selective function of R becomes profoundly different from its function in, say, a rat seeking the shortest route through an experimental maze to the food box. There is no longer only one right answer. Conflict is endemic and universal; and its resolution involves valuation. This cannot be explained solely in the terms so far used. To this gap in the analysis I now turn.

Conflict in the system

At the simplest organic level adaptation means directing behavior to keeping within acceptable limits the relationships which are essential to the system (i.e. to preserving the essential values of E). It is a task of regulation.

At levels far below the human this involves conflict and choice. The timid but hungry bird which comes at last to the bird table ignores the warning signals which defend it from predators, to heed the even more imperious ones which guard it from starvation.

Thus conflict is endemic in organic systems. It is at its most complex and acute in men and societies. For both men and societies try to do far more things at the same time and over longer periods; in other words, they are governed simultaneously by many different sets of governing expectations, by no means consistent. The resolving of conflict, exceptional in most creatures, is our most constant and familiar activity. This is what decision means. It involves choosing one, at the cost of rejecting many other alternatives. We have "advantages"; we can build long, purposeful sequences of behavior in our heads; we can clarify our rules, working out their implications on each other. But these precious gifts only enable us to state more clearly the questions for evaluation to answer. They do not supply the answers. Their principal effect is to enable us to live more complicated lives.

I have distinguished three functions—norm-setting, by which I mean the evolution of those positive and negative governors of behavior which we seek to follow (E); regulation, in which I include all the devices (R) by which we direct our behavior so as to follow them; and valuation, which I reserve for the resolution of conflict. It is unusual thus to distinguish the resolution of conflict; it is usually supposed to follow automatically from the strength and direction of our rival seekings and shunnings, to be in fact a function of E. Thus the economist resolves all the diverse demands which affect the market into the search for maximum satisfaction. The psychologist resolves the still more diverse activities which he observes into the effort to reduce "tension." These descriptions are no doubt valid and useful so far as they go; but they throw no light on some questions which are fundamental to the resolution of conflict. Whence comes the pulling power of an economic satisfaction and why does it vary so suddenly and capriciously even for the same person in different situations? (Who wants turtle soup on a picnic?) What is the process by which we set out own tensions, as we so obviously do? And why do we derive satisfaction from the process of resolving them, rather than from the state of having them reduced? (Lovers of Bach go to the concert hall to listen to Bach, not to attain a condition in which they do not want to listen to Bach any more.)

Apart from these unresolved doubts, I think it useful to concentrate on the resolution of conflict, because it is only in this connection that valuation becomes a problem. In all arguments in Cabinets and Board rooms or within the individual brain, the object of the rival disputants is to alter the tensions as felt by others, so as to accord with their own. There are two ways of doing this. One is to persuade others to classify the situation according to some type to which a strong attitude is already attached. (War is murder. Pacifism is treason.) The other is to change the attitude already attached to the classification used. (War in the nuclear age demands an assessment different from war in the 19th century.) The second is in fact only a refinement of the first. It involves new and more refined classification. Our ability to use more numerous and refined classifications shows growth in our capacity for taking complicated decisions but does not of itself explain our power to resolve conflict.

I will mention two other dimensions of growth. One essential aspect of maturation is learning to attach reality to satisfactions remote in time. Without this, we could not pursue remote objectives. Another is learning to derive personal satisfaction from group satisfaction and from the satisfaction of others. Without this, we could not sustain social life. These widenings of the frontiers which divide "now" from "not-now" and "I" from "not-I" are clearly of great importance to the resolution of conflict; but they do not eliminate conflict or explain how it is resolved.

Thus, though it is certainly artificial to distinguish the resolution of conflict from the process by which norms are set, it seems useful to me in the present state of our knowledge, because it helps to focus separately a group of problems which are of particular importance to our enquiry and which at present admit of no clear explanation.

Dimensions of change

When I think of men or societies as systems, extended in time, I see their essential character developing in three related ways. I see first an increase—or decrease—in the number of relationships which they are set to attain or elude (in E). I see next an increase—or decrease—in the repertory of behavior (B) by which they can pursue these relationships and in their skill (R) in devising apt responses from their repertory. I see finally an increase—or decrease—in their power and skill to resolve the conflicts which these developments involve.

These three dimensions of change are intimately related. When we learn new ways of behaving (we invent the printing press or the steam engine or the secret ballot), we become better able to meet our current needs. As a result we develop new needs, so that our capacities are fully extended at a more complex level. If we fail at that level, we fall back to a simpler level. The level we can maintain depends in part on our capacity for resolving conflict and this in turn depends partly on the inherent consistency of the relations which we are set to realise. Where these are grossly inconsistent, our capacity for resolving conflict will reach its limit sooner than it otherwise would. But some conflict is inescapable. It is not unreasonable, for example, for a man or a society to wish to be both powerful and well loved but the pursuit of either goal will soon limit the other; and the character of the system will be largely given by the way in which this and many other conflicts are resolved.

Freud has given a vivid, though perhaps an incomplete, account of this process of mutual accommodation in his description of the ego's efforts to reconcile the demands of its three "hard masters," the id, the superego, and the realities of the environment. An analogous burden rests ever on the governments of societies. The id represents those built-in demands for satisfaction that can neither reason nor wait. The superego represents those traditional rules distilled from past experience, rigid, mutually inconsistent, often unrealistic, yet indispensable as a starting point in the ever-moving present. The environment includes all those elements of the situation which pose inescapable demands and limit possible response. We can call the continual resultant change "adaptation" but we shall not understand it, in men or in societies, unless we postulate the development of an "ego-ideal," an idea of the self, individual or collective, evolving from this process, which gives coherence to development whenever the pressures ease sufficiently to make room for choice.

This ego-ideal, this idea of the self, which guides the choice of men and societies, whenever choice operates—is it the determinate result of circumstances? Or the expression of arbitrary choice? Or does it represent the momentary position of the system in an endless but not unguided search for the best realisable expression of its essential "nature?" Personally, I believe that it is all three. I will not attempt further to justify this belief; but I will try now to answer the question which forms the title to this paper.

I have no doubt that adaptability, as commonly understood, is not enough, as a goal or even as an explanation of our striving. We must assume also a "nature" which men and societies are striving to realise and which gives force and direction to their adaptation and sets limits to their adaptability. I believe that this nature is limited but not given by the genetic constitution of mankind; it is something which we may be said both to "make" and to "discover." At any given moment of history, it consists not so much in the mass of partly inconsistent relations which we are set to seek and shun (our E) as in that idea of ourselves which guides our valuations and hence our compromises. It is an active force in

our evolution, none the less so for being its chief product. It will always be tentative, never final, never sacrosanct; yet it will slowly come to hold more of what history has to tell us about what is worth while to try to be.

At lower levels, this or that among the inconsistent elements of E gets control for a time and dictates response, to the exclusion of others; but at higher levels the elements of E become associated in a more or less coherent and abiding form. The concepts of maximizing satisfaction or minimizing tension, useful enough at lower levels, need at higher levels to be translated into terms of form-seeking. Purely quantitative concepts will no longer serve. Personality is more than the sum of E; more than a mere resolution of forces. It is a configuration of forces. And though I shall continue to refer to E as standing for the governors of behavior, I will ask you to bear in mind that these governors would not function at the level in which we are interested, unless they were organised in a way which my analysis so far has barely touched.

Rapid industrialisation

During the past two years many of us must often have wondered how far the things we were observing were the results of rapid industrialisation; how far they were due to peculiarities in Canadian or generally in Western culture, as distinct from nations further East, where rapid industrialisation is also dominant; and how far they are part of the general human predicament. I believe that the foregoing analysis helps us to answer this essential question. The answer, I believe, is as follows.

Rapid industrialisation, wherever it occurs, releases into the system an ever greater degree of change and hence of disturbance. The volume of this disturbance, unless controlled, is bound to grow more rapidly than the adaptability of the society and must threaten the society with one or both of two related dangers. One danger is that the society may fail to adapt itself to the changes and may break down. The other is that that society will adapt itself only at the cost of some major and adverse change in valuation, that it may be able to survive only by the sacrifice of values which today it rightly deems essential. Either of these dangers is a threat to the wellbeing of individual men and women; and I believe that all those threats which we have been able to trace can be attributed to one or another of these dangers.

I believe, further, that the nations of the West, where rapid industrialisation has developed largely under the private entrepreneurial system and within the context of Western liberal democracy, have special problems, as well as special advantages, when compared with countries further East, whether communist, such as China and the Soviet Union or non-communist, such as India, where the development has occurred later within the context of a different idea—the idea that economic development is a major part of political government. This idea is now largely shared by the Western world, but since our economic institutions developed in a different climate, our responses are different also.

Finally, I believe that the ideological division of the world into communist and non-communist, though important for some purposes, is inept and confusing for the purpose of our present discussion. I believe that all countries where rapid industrialisation is occurring can be arranged along a continuous scale, according to a number of familiar indices, such as the relative size of the private and the public sector of industry, the relations between government and industry, the degree and kind of control exercised over physical and industrial development and so on; and that these reflect particular stages and conditions of development, no less than differences of "ideology." Even these differences of ideology are themselves the product as well as the cause of history and can be compared with each other along more than one continuous scale. I hope, therefore, that we shall not hesitate to note the handicaps as well as the advantages of a culture such as yours in dealing with the problems of rapid industrialisation.

All societies where rapid industrialisation is taking place are in danger of being unable to adapt themselves to the rate of change which their own activity produces. These changes may arise in any part of the system and may spread to any other part along any or all of the circular pathways which I have described. They may arise from changes which we make in the physical environment; general, such as the rise in the level of radiation, or the fall in the water table; specific, such as the submergence of Iroquois under water or the Niagara peninsula under concrete. Equally real and disturbing are changes in the system itself, as in the size, age-grouping and distribution of the population and the work they are required to do. Most disturbing, perhaps, though least obvious, are changes in the expectations, individual and collective, by which they are governed.

The disturbances thus released into the system place an ever increasing load on its regulative mechanisms. In the language of my earlier analysis, R is constantly having to evoke new behaviors in order to attain the same result and new results in order to attain the same satisfactions. As the volume of disturbance approaches the limit of the system's adaptability, the first result is a general sense of stress. The struggle for existence may be sharper in a rich society than in a poor one. It all depends on what has become essential to existence. The rich expanding societies of the West live, I believe, under stress far greater than that experienced by simpler but more stable societies.

Disturbance in the system

Unless foreseen, the signal for action takes the form of some specific breakdown. Some customary response is found to be no longer available. The regulative mechanisms become fully occupied with the short-term at the expense of the long-term, with the urgent at the expense of the important. This is the second result of overload and it may set going a vicious circle, which will later present ever more insoluble problems and finally overwhelm the system.

Unless the disturbance can be neutralised without causing another at least as bad, the gap between what is achieved and what is called for (by E) will sharply widen. There will follow, as a third result of overload, severe criticism of those responsible for the regulating mechanisms—that is of government, in the widest sense. The criticism may be wholly unfounded; the disturbance may be far outside their control. The criticism may none the less be a very important new force in the situation. There is here a peculiar danger to Western democracy. Our present age is in the course of discovering new techniques for control. We do not really know how much to expect from them; people frequently expect far too much. But, as I have sought to show, experience does not readily disclose how far a situation is under our control or even how wisely our control has been exercised. Responsible government can only be effective if those to whom it is responsible know how far its responsibility extends. This knowledge is perhaps less certain today than at any previous time.

Over many decades, things which used to be regarded as "acts of God"—war, famine, pestilence; or as part of the nature of things—crime, destitution, ignorance, have come to be regarded as controllable and are hence assumed to be somebody's responsibility. They can all be "fixed"; it is just a matter of know-how. It is true and welcome that the degree of our control is slowly extending but the assumptions based on this extension are false and dangerous. Not everything can be fixed; and fixing is never just know-how. It is always decision, made at the cost of not fixing something else. Until both governors and governed have a common and realistic view of what can be controlled and how far and at what cost, the relations between them are bound to be disturbed; and these disturbances may be as dangerous to the system as any.

Thus an increase in the rate of change places an increasing strain on the machinery of democracy. This, I believe, is why democracy has broken down—or has not been attempted—in many countries where rapid industrialisation is taking place and why it is at risk everywhere. For when disturbance so overloads the adaptability of a system as to threaten its existence, the system is prone to take whatever steps seem likely to preserve its coherence. In the conflict between the urgent and the important, even freedom may seem a luxury.

The ultimate result of overload is certain. In the end, the system will fall back to a lower level of organisation, discarding or modifying such of its governing expectations as cannot be realised. Since these governors (E) are closely related, the result will be a change, greater or less, in the whole personality of the system. It will be simpler; and for a time at least, less effective even at its new and lower level.

I have referred earlier to the danger of changing people's governing expectations so quickly that those which they build up in their early years will not serve to guide them or their children a few decades later. "We cannot afford to make a world in which everyone is clueless after thirty." I think that my analysis of the part played by such expectations (E) will give this point more force. The reason why we cannot afford to dispense with the guidance of the past is that in the last analysis we have no other. Apart, however, from danger resulting from the speed of reorganisation, there are dangers resulting from its probable character. When an overloaded system is forced to reorganise itself on a simpler basis, it may benefit from the change; it may, for instance, abandon some inherent contradiction in its goals, which has been the chief source of disturbance and which the crisis has brought to light. But

history suggests that radical and sudden changes tend to leave the system impoverished and debased.

An overloaded system may not break down. It may respond in any or all of three ways. It may seek to improve its regulative machinery (R) so as to deal with a heavier load. It may seek to control the rate of change, so as to keep the load within its capacity. Or it may lower and change its governing expectations (E) and thus alter what it is trying to be. Unless it achieves balance deliberately by one or all of these methods, the force of events will do so, more violently, less tidily, but not less effectively; for the imbalance will insist on correcting itself, unless it destroys the system altogether. In the last part of this paper I will review what seem to me to be the main possibilities under these three heads; for it is in these fields, as I believe, that the major responses of your society must lie.

Political, social, and economic regulators

I have grouped the regulators of your society under three main heads: political, social and economic. Most obvious are your political institutions, central, provincial and local. An increasing volume of regulation can be done through these bodies alone. We have seen reason to think that, at least at the local level, they are in some places unequal to their task.

Political institutions are not free to grow and develop, as business undertakings are. The areas of their authority and the scope of their powers can be enlarged only by legislation, requiring particularly difficult agreement. Their budgets are voted annually. Their efficiency is not readily measured in terms of value for money; so their enlargement is viewed with suspicion in the climate of Western democratic thought.

Yet they must grow and develop, if they are to keep pace with their growing tasks; their growth is essential to the regulation of any economy under rapid industrialisation. Among the many problems within this area I will mention only one: the problem of attracting sufficient high talent. In all Western democratic countries this is one of the unsolved problems of regulation. A familiar example is the difficulty of attracting enough scientists of the present to teach the scientists of the future; but the problem is widespread. My impression is that in my own country the prestige of the civil servant's profession has markedly declined within the last two generations, by comparison with other comparable activities, though its importance has manifestly increased. In all Western countries—in sharp distinction from those further East—Government is ill placed to outbid industry either in money or prestige for its requirements of any scarce resources; and this, I think, reflects an element in our ideology (our E) which will have to change. Judging by the facts, one might suppose that it was a tenet of Western democracy, not of Karl Marx, that the State is destined to wither away.

The working of political regulators depends in turn on social organisation and attitude, less precise but not less potent, ranging from organised pressure groups to the countless foci round which opinion is mobilised; and including all those social organisations for action which are indispensable to the working of Western democracy, not less but more so, in a Welfare State. Higher adaptability requires greater efficiency in these regulators also; and we have seen reason to think that rapid industrialisation, whilst helpful in some ways, is in other ways adverse to their development, notably by increasing mobility and dissociating our working lives from the communities in which we live.

Most important, in the context of rapid industrialisation, is the regulative power of industry itself. By far the most important long term physical planning in Canada, as in most Western countries, is being done by industry.

The techniques of control are more developed and more used in this section than in any other and they are developing more rapidly. Industrial planners take a far wider and longer view than they used to do, or perhaps than anyone else does, of the scope of their responsibilities. When establishing new centers of industry, especially in remote places, they are conscious, as they would not have been conscious even a few decades ago, that they are setting the stage for family and social life for many generations. Their adaptability has been remarkable.

On the other hand, the demands on them are great also. Industry is the source of virtually all those disturbances which tax the adaptability of the system ever more each year. The implications of what it does stretch far beyond the widest effects on its own work people and their dependents. It seems obvious to me that the extent to which the entrepreneurial system can survive in anything like its present form depends largely on whether its institutions can reach a sufficiently sensitive accord with the political institutions of the country.

The regulative aspects of industry which I have chosen to mention first are not those which would have occurred to anyone fifty years ago. Then, the first and greatest regulator would certainly have seemed to be the market. Rapid industrialisation was born in the market place. The free market for capital and labor, for products and for ideas, were the conditions of its existence. It was trusted as an autonomous device, capable of adjusting automatically the activities of men as producers with men as consumers. Its other function, its function as an activator, was equally welcome and equally trusted. It made the activities of production and exchange self-exciting up to the limits of the system's capacity. No one doubted that this was wholly good.

This is not the place to trace in detail the change in the working of the market and in our attitude towards it. The extent of the free market, as the 19th century understood it—the market in which both buyers and sellers are so numerous that price is independent of any of them—is far narrower than it was; and where it still exists, it no longer commands the same confidence as before, either as a regulator or as an activator. In both capacities it is increasingly manipulated or displaced—for example, by the redistribution of purchasing power, through taxation and welfare benefits, between one group of citizens and another and between the general body of citizens and the government; by control and guarantee of prices; by subsidies and other instruments of defense or social policy. These devices should not, I think, be either approved or disapproved on ideological grounds. They are part of the inevitable development of the regulative mechanisms (R). They are not on that account necessarily right; but if they were wrong, some better innovation must be found.

In every country where rapid industrialisation is taking place, an increasing field of activity is being withdrawn, wholly or partly, from the control of the market. Even in Canada this sector is so large that its successful management is essential to the progress of the country. It includes all those services which are managed by public monopolies, or by local authorities. It includes all those fields where the market operates imperfectly, if at all, either because there is only one buyer (as with the Government's defence budget) or because the project needs government finance (as with natural gas). Within this withdrawn field you use, with admirable objectivity, whatever form of regulation seems to fit the purpose. Let us then accept the fact that the development of R depends very much on your ability to adapt or replace the economic regulators of the past, so as to associate them closely with political and social regulators in an efficient whole. The idea that the economic field could be separated from the rest of life, always an illusion, is long since dead.

Controlling rate of change

I turn now to the second defensive response. A system unduly pressed by change may seek to limit the rate of change to suit its capacity for regulation. It seems to me inherently probable that the rate of change, unless controlled, must progressively outstrip our adaptability; so this recourse is likely to prove the most necessary and important of all; and I suspect that it is already the most active. It is, however, still ideologically unacceptable to many, so we should perhaps give it some thought.

The principle is already admitted, in practice, if not in theory. We restrain the rate of change in many directions. Economic enterprise is restrained, for example, to prevent dustbowls and disafforestation, to secure the convenient development of land, to curb fluctuations in the trade cycle, to maintain steady employment—this last a potent limiting factor on the development of automation. Emigration is controlled, to limit the disturbance which would result from unduly rapid change in the number and constitution of the population. There is room for debate as to how much disturbance the system can stand. There is need for enquiry, lest proposals to limit change prove to be only an easy alternative, to save authority the painful but more healthy alternative of raising its capacity for regulation. But when all is said, the limitation of the source of the disturbance is bound to be of ever greater importance among the means of dealing with it.

Value judgements

I turn now to consider the effect of rapid industrialisation—or rather, of our attempts to adapt to it—on our collective and individual value judgements, on the weighting of our preferences and prejudices, on the directing of our seekings and shunnings, in a word on E. The effect of "social E" (or of what I have ventured to call the collective personality which emerges from it) on the individual is twofold. Indirectly, it determines the environment in which he is to live; for example, by determining how we shall divide our collective efforts between better roads and better schools. Directly, it affects the value system of each individual. In my view, the collective value systems of Western democracies contain both handicaps and treasures and it is important that, in the changes which are always in progress, we should shed our handicaps and keep our treasures.

In an earlier paper I described three cultural values which seem to me to characterise Western societies in their present stage of industrialisation, especially in the New World. I called them the metabolic criterion, the criterion of material expansion and the criterion of freedom. They stand for the assumptions, still basic to Western culture, that the individual should prize and strive for increasing independence, through the increasing production and exchange of goods and services. I suggested that these values are not consistent with each other or with our present stage of development and are approaching the end of their domination. Independence must find room for interdependence; expansion must learn to live with stability. The ground for these suggestions is twofold—the fact of observation that it is happening and the logical conclusion that it is bound to happen.

In every Western country, as I have pointed out, the sector of activity which is collectively managed grows steadily larger, both absolutely and relatively. One effective measure of it is the proportion of the national income which is collectively spent. To members of a society where industrialisation was born of individualism, it may seem anomalous that it should lead to collectivism; but in some clearly definable senses it is clear that it must do.

In an expanding society numbers increase and every individual claims to take up more room, both directly for living space and mobility and indirectly, through the space needed to accommodate all the services on which he calls. Since the surface of the planet does not expand, it is inescapable that people should become more thick on the ground. Moreover, the extension of economic and other relationships in space and time increases the interdependence of each on all, of the live on the dead and unborn. In such a process the field of activity which a man can regard as his own business and no one else's must shrink to the vanishing point.

It does not follow that in such a progress freedom and variety must disappear. The problem for Western democracies is to socialise their individualist ethic without losing its essential values.

Conclusions

In conclusion I will try to summarise the connection between the conceptual model which I have sketched in this paper and the concrete realities which we have observed or sensed in the places we have visited.

(1) The increasing volume of disturbance which our industrial activity creates makes itself felt in the rapid, irreversible commitment of land surface and of basic, physical resources; in the divorce of the working from the living community; in the change (partly widening, partly narrowing) of the personal satisfactions which it offers; and in the instability and the character of the values by which both individual and society live. The last is the most elusive and probably the most important of all.

(2) Industrialisation multiplies the need for regulation and thus inevitably builds up the apparatus of government in democratic, no less than in totalitarian states. This in turn affects both individual and social values and involves a profound revaluation in those countries where industrialisation was at first a product of individual enterprise.

(3) Industrialisation disturbs international, no less than intranational relations. These disturbances call for regulation far beyond the power of the rudimentary regulators which at present exist in the international field.

(4) Thus the disturbances which industrialisation creates threaten the individual either with breakdown or with the impoverishment which may be the necessary price of avoiding breakdown. Breakdown may occur at any level from the international to the personal. We can see examples of it at all these levels. The impoverishment by which it may be escaped is an impoverishment of values, resulting from the limitation of expectations to the stereotyped satisfactions which the hard-pressed system can most easily provide—the modern equivalent of "bread and circuses"—and correspondingly, a shrinking of personal responsibility and envolvement and the growth of authoritarian government.

(5) The present and the immediate past do not provide evidence from which the future can be predicted, either for your country or for any other. The rate of industrialisation rises automatically at an accelerating rate, unless it is limited either by deliberate policy or by its own effects; and it evokes new responses as it passes new thresholds. We can count its blessings and its banes but we cannot strike a balance which has any clear relevance to the future. We can only forecast its results by deepening our understanding of the systems, international, national, social and personal, which its impact disturbs; of the ways in which such systems are bound to respond; and of the extent to which they are free to choose between these ways of responding. Our understanding of these processes forms the essential background to any discussion of industrialisation.

C.

DECISION PROCESSES AND GROUP STRUCTURE

58.
Critiques of Game Theory

ANATOL RAPOPORT

GAME THEORY appeared on the scientific horizon quite suddenly in a monumental work by J. von Neumann and O. Morgenstern. There had been previous important papers, in which the foundations of the theory were laid, but by and large, until 1944 the subject matter may be said to have lain in a dormant state.

The effects of the von Neumann–Morgenstern volume were explosive. A bibliography of the literature published in a recent book of R. D. Luce and H. Raiffa contains some 700 titles, almost all of them falling in the period since 1944. The bibliography, according to Luce and Raiffa, covers game theory fairly completely but not the "allied subjects," i.e., investigations largely instigated by game theory and by problems arising in it.

In some quarters, game theory was hailed as one of the most outstanding scientific achievements of our century. Since those who are able to fathom the full significance of game theory are likely to have also an understanding in depth of mathematical physics, such evaluation can be interpreted as placing game theory in the same class with, say, the theory of relativity, which also belongs to our century. Now the latter is held by many to be of a magnitude comparable to the first synthesis of mathematical physics by Newton. And so the implication is that game theory stands on a par with Newton's celestial mechanics as a scientific achievement.

We will not defend or dispute this evaluation here, but we will try to understand the basis for making it. For it is instructive both for under-

standing the spirit of game theory and the spirit of theoretical science in general to see just what impels people of profound understanding of these matters to see a comparable achievement in two such widely disparate endeavors.

The Conceptual Achievement of Game Theory

Let us first ask wherein lies the greatness of Newton's achievement. In our day we may expect practically unanimous agreement on the answer. The achievement was a two-pronged one. Newton created a new mathematical framework—infinitesimal analysis—and used it with enormous success in a mathematical deductive system in which *all* of the known modes of motion of the known heavenly bodies (and in many instances of terrestrial bodies) appeared as *compelling* consequences of only three assumptions.[1] The first part of the achievement was conceptual; the second part was technical. The Newtonian system, at least in its subsequent extension, is practically unthinkable without its basic mathematical tool—infinitesimal analysis (the calculus and derived disciplines). But the converse is not necessarily true. Possibly the calculus could have been invented even without the instigating challenge of celestial mechanics.[2] So at least in principle, we have two very different and semi-independent aspects of Newton's achievement. It is idle to speculate whether the calculus actually could have "stood up" without the needs of astronomy to back it. *In retrospect* we feel that it *should* have stood up, so as to be available to the future generations, as will appear from the analogy we are about to make.

From Anatol Rapoport, "Critiques of Game Theory," *Behavioral Science*, Vol. 4 (1959), 49–66. Reprinted by permission of the author and publisher.

In game theory, we see a situation similar to that which *would* have obtained if the applications of calculus did *not* result in a predictive theoretical science of amazing power and accuracy. The mathematical discipline forged for dealing in a mathematically exact fashion with certain aspects of characteristically human behavior still appears as an amazing invention. But the applications are meager and faltering. Is there not a contradiction here? Is it not fatuous to speak glowingly about a theory which has so far failed to impress with its power in the field of application for which it was created? Can it be that those first outbursts of enthusiasm were simply premature, were made in the hope that spectacular applications would follow, so that now it is time to make a more sober appraisal? Some people think so. We think not. We think that after fifteen years of rich theoretical developments of the theory unaccompanied by anything comparable in applications to behavior, the initial verdict is still justifiable. We also think that this initial verdict was made quite independently of the immediate application potential of game theory. Rather game theory was seen as a feat of breaking through an accustomed framework of mathematical thinking.

To illustrate, we will resort to an analogy with the calculus. The calculus was not just a new mathematical language. It was a language superbly suited to dealing with certain events which could never be described in the old mathematical languages—the events subjected to strict determinism amid constant change or flux. ("Fluxions" was Newton's name for the method.) It so happened that this language was immediately applied to events which it could describe with outstanding accuracy—the motion of heavenly bodies. Had it been applied to, say, meteorology, it would have failed. Even today we do not have a meteorology which compares with celestial mechanics in the precision of its predictions. But there is no question that the calculus and its derivative disciplines are the proper language to describe air currents, temperature gradients, etc. It captures the "essence" of those events (if I may be permitted an old-fashioned term). This essence is contained in the notion that (1) a portion of space and the events within it are described by the values of a certain number of pertinent variables given at each of the points at each moment of time and (2) the sequence of these "states" from moment to moment is determined by these very distributions and their rates of change. Neither classical geometry nor algebra were able to deal with these notions. The methods of analysis derived from the calculus

still cannot cope with meteorological events very successfully, but the difficulties are technical not conceptual. The basic design of the mathematical conceptual machinery exists. The problem is simply in making more powerful and more refined tools according to this basic design. The difficulty of theoretical meteorology is largely a difficulty of *handling* the machinery.

Coming back to game theory, its significance was seen in that it broke away from a conceptual framework which had been unable to cope with certain aspects of human behavior. It offered a new one which did appear to capture the "essence" of that behavior. True, the present "tools" of game theory are still not powerful enough or not refined enough to deal with the subject matter for which it was meant, but it is felt that just as in the case of the calculus in application to meteorology, it is now a matter of developing more powerful tools. The basic design has been sketched. Therein lies the achievement.

What is this conceptual framework, which game theorists felt could not deal properly with certain aspects of human behavior, and what are those aspects? The conceptual framework was that same infinitesimal analysis which proved its worth so eloquently in physical science, and the aspects of human behavior with which it could not cope were those of calculations of consequences of alternative choices of action under conditions of conflict and/or partial cooperation among rational beings. The application of infinitesimal analysis (whether in "classical" mathematical economics or in some attempts to mathematicize sociology) was based on the assumption that there was something in human affairs called mass behavior and this mass behavior is subject to deterministic laws even though the same laws may be inapplicable or meaningless on the individual level. More concretely, man in mass was seen as an aggregate of particles. The behavior of each particle could be described only in probabilistic terms. But the behavior of the totality could be described deterministically on the basis of the law of large numbers. This is the basic philosophy of statistical mechanics: the probabilities on the molecular level of events become determinable mathematical dependencies among variables pertaining to the gross level. Thus statistical mechanics merges with thermodynamics when the aggregate of individual entities appears under a "low resolving power"—as a continuum. Similarly, the probabilities governing the actions of individuals (to buy or not to buy; to invest or not; to move to the city or the country) become in the mass continuous variables and their rates of change. The action of the mass, then, described in terms of these continuous variables becomes amenable to treatment by the classical methods of infinitesimal analysis (differential equations and the like)—so the argument for applying classical mathematics to human behavior goes.

I would like to stress at the outset that I do not believe that all this effort to cast mass behavior into classical differential equations was wasted. It is not at this time a question of predictive power of the classical tools. They are not great (but neither are those of game theory at this time). It is a question of appropriateness *in principle* of the tools. I do believe that there are important phenomena of mass behavior to which the classical tools of analysis are applicable in principle. This means that exactly as in the case of meteorology their limited success is due to their crudeness, not to their irrelevance. That is to say, I believe that in *some* aspects of human behavior, particularly mass behavior, the assumptions of rationality, calculation of consequences, etc., simply do not play an important part, so that in certain aspects the action of masses may be described as determined strictly by "id-like" forces, that is, forces that act only "locally" on the *here and now*. These forces are therefore akin mathematically to the forces of mechanics, which are relevant in situations where goals and comparisons of consequences do not enter the picture. It is possible that economic events under approximately free competitive market conditions are of this type. So probably are fads and behavioral epidemics. Most political behavior of masses frequently seems to be of this type. And certainly biological and ecological behavior of man in mass (phenomena described by vital statistics, accident rates, growth and decline of cities, migrations, secular trends in language change, diffusion of genes and of cultural traits, etc., etc.) seem to be of this type. Here the methods of classical analysis are probably appropriate. Greater success will be achieved when better fundamental assumptions are made underlying the mathematical models, when more relevant variables are discovered and become accessible to more precise measurement, and when the mathematical difficulties of more sophisticated models can be overcome (perhaps by a judicious use of mathematical technology).

But the fact remains that this whole approach seems irrelevant to the situation in which two individuals whose interests are not coincident are in control of different sets of choices and endeavor to make their respective choices in such a manner

as to emerge with an advantage. This is the elementary game situation. Its most salient feature is the ability of the participants to choose from among alternatives and to calculate the ranges of consequences associated with each choice.

In its present state, game theory is not a dynamic theory. It does not describe a process in time. It describes only the *logic* of the situation and in some instances, but by no means always, is able to prescribe for the "rational player" what choices he should make. Thus "causality" in the conventional sense, depicted as a flow of states, is not part of this framework. The emphasis is always on the ranges of choices and on the analysis of the interplay of these choices and of the consequences. The novelty and originality of game theory is in that *such* situations had never been extensively mathematicized (i.e., cast into formal manipulable symbolic notation with relevant rules of manipulation). When this job was done, the resulting system revealed itself to be of enormous complexity and of subtle intricacy, and, for the mathematician, of striking beauty.

There is another sense in which the mathematical structure of the game-theoretical approach differs fundamentally from that of classical mathematical models. In the latter *specific* assumptions are always necessary, and so the investigator must always choose from an innumerable number of possible assumptions. For example both Rashevsky and Richardson in developing their theories of mass behavior (very much in the same spirit) must postulate specific forms of interaction between the individuals in the mass. Therefore their equations take on a specific form and contain many free parameters. In the absence of concrete knowledge about the character of the interactions, many other forms of equations could have been chosen with equal justification. As for the parameters, unless their values can be determined empirically or by several independent inferences, their great number often gives so many degrees of freedom to the solutions of the equations that no definitive conclusions are justified.

Game theory, on the contrary, in its original formulations contains no parameters and makes no arbitrary assumptions. Its assumptions are always "extreme," e.g., complete "rationality" (as defined in the context of choice among alternatives) and complete knowledge (to the extent allowed by the rules of the game) of the participants. It is true that these assumptions make the theory unfit as a *descriptive* theory of behavior in game-like situations, and so in subsequent attempts to construct more "realistic" theories, specific assumptions and parameters had to be introduced as will appear below. But the advantage of the original "non-parametric" formulation was precisely in that the purely *logical* scheme of the game-like situation was exposed.

As much as anything else, then, the achievement was in focusing attention on the nature of reasoning involved in the logic of events where conflict (and often partial coincidence) of interests enters. It was perhaps felt by those who most strongly endorsed the achievement that perhaps for the first time the difficulties of reasoning about typically human affairs have been pointed out and made explicit. The problems, in other words, became defined. This is still a long way from solution, but constitutes the most important step in the construction of theories relevant to the above mentioned aspects of human behavior.

Once this relevance is recognized, the mathematical achievement of game theory is singled out from many other recent mathematical developments, even if many of the latter are more sophisticated mathematically speaking. For game theory is seen now not as just another branch of mathematics (although it is quite rich in purely mathematical ideas) but as a major innovation in applied mathematics,[3] an invention of a mathematical system outstandingly germane to specifically human affairs, involving rational choice. The theory transcends the limitations of classical mathematics, which can be applied at best to mass behavior, where rational choice, calculation of consequences, etc., plays no part.

So much for the importance of game theory as a branch of applied mathematics. Its importance to social science is paradoxically not in its successes but in its failures. Just what is meant by this will, I hope, become clear in the following discussion. First, however, we must give a very brief review of the scope of game theory as originally conceived.

The Scope of Game Theory

There are three separate theoretical frameworks of game theory. We may describe these three frameworks somewhat metaphorically but quite suggestively as based on high, medium, and low "resolving power." The first, called the theory of games in extensive form, is descriptive of typical games of strategy stripped to their essentials. These essentials (as is quite evident on reflection) are the rules of the game which determine what may happen. Thus the game is seen as a sequence of moves made by the different players in some prescribed order. (The order may depend on what choices are made and their outcomes.) This description may be schematized by a diagram called the game tree. A classification of games, therefore, suggests itself based on a classification of such diagrams. To this date, comparatively little work has been done on this level of game theory.[4]

The next level has had a much more extensive theoretical development. The transition from extensive form to normal form is provided by the concept of *strategy*, suggested by the extensive form of the game. A strategy is a program given by a player which describes what the player will do in all conceivable circumstances in which he may find himself. Since the game, as defined in game theory, is usually assumed to have a finite number of moves, there are only a finite number of strategies open to each player.[5] In this way a game with N players may be represented by an N-dimensional matrix in which each "box" represents an N-tuple of strategy choices by the respective players. The entry in each box is the outcome of the game taken to be an N-tuple of numbers, representing the respective pay-offs (in utiles) to each player.

Let there be no mistake about the "practicality" of this description. The number of strategies even in the simplest games is enormous. In tic-tac-toe, for example, taking the number of moves the least possible (i.e., five) the number of strategies open to the first player is about a trillion.[6] In chess, the number of strategies open to Black *on his first move alone* is approximately a hundred trillion trillion. But game theory is not concerned with practicality at this level. It is conceptually possible to represent each game by a matrix, as long as there are a finite number of possible moves, and that is enough. The theorems resulting from the analysis of such matrices are valid quite regardless of the "practicality" of the representation. This is not surprising and should not be disturbing. A geometric line or curve contains an infinity of points, so that it is out of the question to describe it by listing the points. Yet theorems can be derived, which are statements about "every point on such and such a curve." The usefulness of such theorems in no ways depends on the possibility of "naming" every point. In its matrix representation, the game is said to be in normal form. The associated theory has had an extensive development relating both to the classification of games and to far-reaching conclusions concerning the various classes.

One dichotomous classification is basic and leads naturally to the next phase of game theory. A two-person game may be "zero-sum" or not.[7] In a zero-sum game the sum of the two pay-offs is zero in each outcome (what one wins, the other loses). In this case, the interests of the two players are diametrically opposed. Otherwise, this is not necessarily so. The theory of the game in normal form is most complete for the zero-sum game. It rests on a fundamental theorem, which states that there exists either a strategy or a "probability mixture" of strategies[8] for each player, which enables him to do the "best he can" against an opponent. The long run outcome of this situation (where each plays the "best" strategy) is determined, in the sense that an expected value of the pay-off is determined. Each zero-sum game, therefore, has a "value" for each player—what he can expect as the average pay-off in the long run. The sum of the values is zero. This fundamental theorem can be taken as a prescription to the "rational" player of how to play under the natural assumption that he tries to maximize his expected average pay-off and that there is a definite upper limit to this pay-off set by the opponent's effort to do the same. Each has a maximum "security level." The associated strategy is called the "maximin."[9]

As a by-product of this result, indeed from intuitive considerations, we see that in a two-person zero-sum game cooperation between the players in the sense of agreements on joint choices of strategy, is pointless. Moreover, attempts at cooperation, as in divulging to the opponent the strategy one intends to use, can only harm a player, because such information can be used by the other to increase his pay-off beyond the security level and thus to depress the pay-off of the first player.

When the game is non-zero sum, security levels can still be defined in the sense of guaranteeing minimum expected pay-offs, but it no longer makes unquestioned sense to use them as determinants of strategy choices, especially if communication between players is allowed. For in a non-zero game, the players may by mutual agreement play in such a manner as to raise *both* pay-offs above the security levels. The same situation obtains generally in games with more than two players. *Coalitions* become feasible.

The concept of value (security level) then becomes the link between the normal form theory and the next level of game theory, dealing with the game in "characteristic function form." Here the problem of strategy choices is by-passed. Of central interest is now the security level commanded by each possible coalition, i.e., each subset of the N players, when their strategies can be pooled so as to guarantee the security level for the coalition. Again certain natural "taxonomies" of games suggest themselves. The theorems then deal with certain properties of the classes of games thus obtained.

We are now in a position to spell out more explicitly the achievement of game theory as a creation of the human mind. The major creative act is the transition from one level of abstraction to another with the result that, as in the cutting of the Gordian Knot, the rapidly multiplying difficulties are by-passed, a fresh start is made *on another level*, and the "conquest" of unexplored territory can continue.

If the history of physics were somewhat different, one could cite an analogy. The actual historical sequence in physics was mechanics—thermodynamics—statistical mechanics. But if statistical mechanics had preceded thermodynamics, we would have had a development to which that of game theory is analogous. As one passes from the two-body problem to the N-body problem, the mathematical difficulties become unmanageable. One therefore abandons the mathematical framework of the N-body problem (a system of non-linear differential equations) and constructs another based on probability distributions of the variables. This amounts to using a "lowing resolving power" to gain a larger field of vision. In the next step (thermodynamics) a still more drastic simplification occurs. The framework then becomes once again that of differential equations but this time of a different sort (partial linear of the first order) and the problems brought into focus are also of a different kind requiring a different conceptualization. The three areas are unified "philosophically," because a chain of reduction exists, leading from the level of low resolving power of thermodynamics to that of higher resolving power of statistical and finally to that of classical mechanics. But the links of the chain, once established, (to provide philosophical gratification), can be forgotten. Each level can be treated by a theoretical framework appropriate to it, and the levels involving large fields of vision (low resolving power) need not be encumbered by the concepts of the levels of greater resolving power. The situation thus remains intellectually manageable.

This is just what happened in game theory, except that the transitions between the levels were made in the proper order. The magnitude of the achievement can be judged by the fact that the grand design was presented in its entirety in a single work. The problem in the theory of games in extensive form is to describe the strategies, to classify them, and to classify the games according to the kinds of strategies available. The immense number of strategies soon makes this task unmanageable. One therefore takes the knowledge of the strategies and of their outcomes for granted and passes in the next formulation (the normal form) to the logic of strategy *choices* in generalized games. The possibility of coalitions in non-zero sum games and in games with more than two players soon makes this task unmanageable in turn; so one makes the conceptual jump once more: one takes the outcomes of the strategy choices of *coalitions* for granted and considers the logic of coalition formation. This is the level of the game in characteristic function form. In fact the characteristic function is a mapping of each possible coalition on to the security level which it can command.

The difficulties of the extensive form are more technical than conceptual. The task of this theory is frequently to describe a particular game as fully as possible. There is no question that an unambiguous description is always possible (since the rules are unambiguous), but of course for all but the most trivial games such a description is enormously difficult. The theory of zero-sum games in normal form is complete in the sense that the fundamental theorem (von Neumann's Minimax Theorem) establishes the existence of a unique "solution" and in principle indicates methods for computing it. The latter problem may again be extremely difficult, but the difficulties are technical.

From the point where game theory passes to non-constant sum games and to games with more than two players, *conceptual* difficulties begin to appear. And it is these difficulties that have led to developments of game theory which are especially important for behavioral science and, in my opinion, for a scientific approach to ethics.

Implications of Recent Developments

I view the implications of game theory as falling into three (certainly overlapping) categories, the philosophical, the behavioral-scientific, and the ethical. I would put into the first category primarily the implications for the philosophy of science, those related to questions of method of theory construction, in short questions related to

the strategy and perhaps the esthetics of general theoretical research. These matters were taken up in the foregoing discussion, as, for example, in the description of the "by-passing" technique, so effectively used in the over-all formulation of game theory with its prodigious "leaps" from one level of analysis to another.

The behavioral-scientific implications will be discussed next. These concern the emergence of specific research programs in behavioral science spurred on by questions which *inadvertently* arise in the pursuit of game-theoretical investigations. Here I can do no better than follow the superbly insightful "critical survey" of R. D. Luce and H. Raiffa. The scope and character of this book is described by the authors themselves from the start. The scheme of development follows von Neumann and Morgenstern, i.e., starting with an exposition of utility theory (a necessary adjunct of game theory, though not a part of it), they present the outlines of the "extensive form," then the "normal form," first of the zero-sum, then of the non-zero sum games, with and without communication; then the characteristic function form. But, as the authors point out, the emphasis is different. Concepts, rather than mathematical details and solutions of specific types of games, are at the center of interest. Moreover, the concepts selected for greatest emphasis are those that seem most germane for behavioral science. It can be maintained, therefore, with considerable justification, that the volume is a book on behavioral science, at least on theoretical behavioral science, particularly that aspect of it where inter-relations of distinct "interest" (their coincidence and conflict) are considered to be the core of social behavior.

It is, of course, impossible to make any serious book on game theory "non-mathematical." The omission of most proofs has served the purpose of by-passing arguments which would ordinarily be of interest only to mathematicians. This omission does not really make the book non-mathematical. For its thorough understanding, that is for the understanding of not only the problem areas defined, the examples, the wry humor, the arguments and counter-arguments for the various approaches, but also for an understanding of the results and especially of the limitations of the results, "mathematical maturity" is necessary. This is a quality as impossible to define in terms of courses one should have "had," as "emotional maturity" is impossible to define in terms of life experiences on which it presumably depends. Perhaps a few central mathematical notions can be pointed out with which an intimate familiarity is essential for the sort of maturity required. Fundamental is the notion of a function, whose domain consists of sets (rather than points or numbers, as in classical mathematics). Another is the notion of an existence proof, not only of what it is but a conviction of its importance. [Existence proofs are characteristically omitted or passed over in intermediate (under-graduate) mathematics. This de-emphasis does not detract from the acquisition of mathematical techniques but seriously retards mathematical maturity.] Finally the idea of seeking a mathematical function by listing its properties acquires an extremely prominent importance in game-theoretical investigations, especially in the efforts of those who have tried to transcend the limitations of game theory.

Those notions I believe to be the fundamental mathematical equipment necessary (and, I suppose, sufficient) to read the volume with understanding. *Techniques* are not involved. I don't remember seeing a single integral in the book; there are no involved equations to solve; and, as already mentioned, hardly any proofs. To force a metaphor, the presentation by Luce and Raiffa uses the full sophistication of mathematical grammar but does not involve an extensive mathematical vocabulary.

PROBLEMS OF THE NON-ZERO SUM TWO-PERSON GAME

From the point of view of behavioral science, the really important conceptual problems begin with the non-zero sum two-person game in normal form. Let us examine one such problem.

A behavioral theory may be descriptive or normative. A descriptive theory is one which makes generalizations of how people actually behave. A normative theory is one which tells how people should behave if certain standards of behavior are to be satisfied. If these standards can be exactly specified, so should behavior be in a normative theory. To take an example, consider a simple gambling situation, where a man is faced with two slot machines with different pay-off chances. It is known that in most cases, having established by experience that the pay-off chances are different, a man will distribute his bets between the two machines, playing the more generous machine more frequently. A theoretical model which aims to describe this behavior would belong to a descriptive theory. If, however, one demands that certain standards be satisfied, for example, that the expected pay-off should be the largest

possible, a little reflection shows that the more generous machine should be played exclusively. This is a conclusion of a normative theory. The discrepancy between the results of normative and predictive theories need not be discouraging to the behavioral scientist, for it provides him with a measure of how people depart from certain *a priori* established standards. Of course, if one chooses to identify these standards with criteria of "rationality," one will only too often be forced to conclude that people are not rational. One should keep in mind, however, that rationality is largely a matter of definition of standards.

On the level of the two-person zero-sum game, "rationality" can certainly be defined in a way that will enjoy wide-spread acceptance. In such a game, each player can guarantee for himself a certain minimum pay-off, and his opponent can effectively prevent him from getting more than this minimum (while guaranteeing the minimum for himself). The pair of strategies (pure or mixed) so determined are in "equilibrium" in the sense that it does not pay either player to depart from his strategy so chosen, if the other does not. Moreover, the pay-off to each is the same at all such equilibrium pairs, and the coordinates of the equilibrium pairs are interchangeable (hence the strategies can be independently chosen). Finally, there is no way that the players can *jointly* better their pay-offs (what one will gain. the other must lose), so that agreements between them are pointless. Under these circumstances, it certainly makes sense to say that game theory can assume the prerogative of a normative theory and prescribe to each player a "best" strategy.[10]

The situation is very different in non-zero-sum games. We will take as our first illustration the much used example, called the Prisoner's Dilemma, represented in normal form by a matrix of the following type:

$$
\begin{array}{cc}
 & \beta_1 \qquad \beta_2 \\
\begin{array}{c} \alpha_1 \\ \alpha_2 \end{array} & \left[\begin{array}{cc} (9, 9) & (-10, 10) \\ (10, -10) & (-9, -9) \end{array} \right]
\end{array}
$$

The rows of the matrix are the first player's strategy choices; the columns the second player's. The entries are the pay-offs (in utiles) to the players respectively. The "dilemma" comes from the discrepancy between what "rationality" dictates and what is desirable and (presumably) achievable. "Rationality," as always in game theory, means doing the best one can for oneself, where "self" is an entity with a well-defined interest. It may be an individual or a cooperating group. We see that regardless of the second player's choice, the first player does better by choosing α_2: if the second player chooses β_1, then 10 is better than 9; if β_2, then -9 is better than -10. Similar reasoning shows that regardless of the first player's choices, the second player does better by choosing β_2. The result of these "sure thing" choices is the entry $(-9, -9)$. This is what "rationality" prescribes. But the choice (α_1, β_1) with its pay-off $(9, 9)$ is clearly more desirable for *both* players!

If agreement is possible, then certainly the players should agree on (α_1, β_1). The temptation to double cross rears its ugly head, since it is of advantage to each player to break the agreement (provided the other does *not* break it!). However, if respect for the rules of the game is assumed in game theory, we may as well suppose that the rules of agreement are respected. So let us by-pass the "ethical" problem for the time being. An interesting observation and an interesting question nevertheless emerge immediately. The observation is that "group rationality" is different from "individual rationality," and not only in the well-known sense that the welfare of the group may sometimes demand "sacrifices" from the individuals in it, but in a stronger sense that departures from "selfish" decisions on the part of each individual actually result in greater *individual* rewards for both. This is often realized as part of "social wisdom," but here we have a rigorous, specific demonstration of this principle. The question which emerges, however, is more intriguing. Suppose the players cannot communicate (and so cannot explicitly agree on a joint policy). How *should* they act?

The question, "How *will* they act?" pertains to descriptive theory. Such a question can be answered empirically. It can lead to interesting investigations, for example, having very many non-communicating pairs of players play games of this type, noting whether there are statistical regularities in the distribution of outcomes, correlating these results with the characteristics of the population sample (age, cultural background), varying the pay-off entries to see if shifts occur in the distribution of outcomes, etc., etc.

But, I repeat, the intriguing question is that of the normative theory. How *should* the players act? The answer depends, of course, on the definition of "rationality." But this *is* the point. How should rationality be defined in this case? Can we accept the definition of rationality (based on "doing the best for one self") which leads to a result which definitely is not the best that each player can do for himself?

The contention of Luce and Raiffa that game theory does not prescribe behavior is now understandable. In their estimation, game theory evades the issue. It only draws logical conclusions. *If* one extends the reasoning pertinent to the zero-sum game, namely the reasoning which leads formally to maximizing one's "security level," *then* the choice (α_2, β_2) is unavoidable. This is a sad conclusion, but some investigators accept it, as harsh "reality," perhaps. Thus Nash, in his treatment of non-zero sum games in effect defines "solution" in such a way that (α_2, β_2) appears as the "solution" of the Prisoner's Dilemma. Its unsatisfactory outcome is due to the impossibility of communication (hence of agreement) between the players. For some, this settles the issue, but others have an uncomfortable feeling that "rational" players should, in contemplating the pay-off matrix, come to a tacit agreement to cooperate even in the absence of communication.

Luce and Raiffa, also unwilling to let it go at that, persist in asking further questions. Suppose, they continue, the game is played not once but 100 times. Now each play of the game can be considered as a pair of moves in a super-game with 100 such pairs of moves. How shall the strategies of the super-game now be chosen? The distressing conclusion to which one is led in the analysis of the strategies is that the "Nash solution" of the super-game prescribes the choice (α_2, β_2) in all 100 plays! Here "common sense" is violated indeed. No provision is made for a *tacit* agreement between the players not even when many plays are available, where such an agreement could be enforced by "punishing" breaches in successive plays. Luce and Raiffa admit, however, that if they were to play Prisoner's Dilemma 100 times, they would certainly not play as the "Nash solution" prescribes. In the role of player 1, they would ". . . punish 2 for each choice of β_2 early in the game by using a series of α_2's, but at the same time we want to give him another chance to get into the (α_1, β_1) routine. . . ." The "human" aspect emerges: teachability and charitable understanding ("another chance") enter strategic considerations. The game is seen not as played by two programmed robots, in which the outcomes are tautologically predictable, but between beings "rational" *in a sense other than defined in game theory.*

However, the ruthlessness of the game-theoretical orientation is again apparent in the end of the sentence just quoted: ". . . until we are ready to defect to α_2." In other words, in Luce and Raiffa's estimation, the cooperative advantage of the game, the pay-off at (α_1, β_1) should be milked for all it is worth, until it is time to double cross the partner (who had been induced to cooperate!) in order to squeeze the last bit of advantage from the situation at the other's expense![11]

One could argue that game theory "rationality" is not to be confused with morality. Well and good. Let us try to leave morality out of it. However, in game theory, "rationality" is always equi-distributed. What one can figure out, the other can. If player 1 decides to cooperate to the very last, in order to "defect" on the last play, so can player 2. This makes the last outcome (α_2, β_2), while it could have been (α_1, β_1). But let us accept it. This is "harsh reality." But then if both know that "realism" will inadvertently take over on the last play, does this not reduce the strategic considerations to the first 99 plays? In these "realism" will again prevail on the last play, and so on, until the whole cooperative house of cards collapses. "Realism," considered symmetrically, again leads to the unsanity of the Nash solution.

There is a logical flaw in this reasoning, of course, reminiscent of the famous paradox of the condemned man and his thirty days.[12] We will not pursue these questions further except to point out the moral. The results of game theory in certain contexts are intuitively unsatisfactory. One can, of course, accept the conclusions as tautologically derived from formal assumptions. But if a link is contemplated with behavioral science *either on a meaningful normative level or on a descriptive level*, concepts outside of game theory must enter the picture, for example, certain assumptions concerning the *assumptions* which people either make or should make about each other's behavior or likelihood of behavior. This brings in psychological notions and, one suspects, sociological and ethical ones (expectations). On the basis of such notions, it is easy to prescribe, in the sense of a normative theory, as follows for the players of Prisoner's Dilemma: "Assume that the other wants to cooperate and will assume that you will cooperate. Remember that any conclusion you may make about the advantage of 'defection' can also be made by your opponent."

The problem is, of course, to formalize these assumptions as strictly as the assumptions of game theory are formalized. This does not seem possible with the present conceptual repertoire of game theory. This is what is meant by the statement that the value of game theory is in its failure. The failure pinpoints the shortcomings. It reveals at what points the concepts must be modified and

extended, and it does so *compellingly* not on the basis of *a priori* convictions (prejudices) of what a given conceptual framework can or cannot do.

ETHICS IN GAME THEORY

The theme of the "ethical inadequacy" of game theory is picked up by T. C. Schelling. Schelling attributes the inadequacy to the basic orientation of game theory toward the zero-sum game, in which cooperation is pointless. He illustrates his spirited critique with examples of non-zero sum games in which the ambiguous nature of "rationality" as (implicitly) defined in game theory, especially in the zero-sum game, becomes apparent. Schelling uses the geometric representations of two-person games throughout, a useful visual aid as well as a heuristic one. In these geometric representations, every pair of pay-offs represents a point in a two-dimensional space. The usual procedure in game theory is to include the pay-off pairs resulting from mixed strategies, so that the geometric representation of a two-person game becomes a continuous region in the plane. But it suits Schelling's purposes to confine the discussion to pure strategies, and so his games are represented by sets of isolated points. He then distinguishes three classes of games: those in which the interests of the two players are always opposed; those in which they are always coincident; and those in which the interests are partially coincident and partially opposed. These he calls the mixed motive games. In their geometric representations, the opposed interests games appear as sets of points lying on curves always having a negative slope; the common interest games are portrayed by curves with positive slopes; the mixed motive games appear as sets of "scattered" points.

The first two types of games offer no conceptual difficulties: in the first kind, cooperation is impossible, and so the "classical" theory of two-person zero-sum games with its security level solution applies.[13] In the second kind, the advantage of cooperation is obvious. Even in the absence of communication, the players will choose the strategy pair which will give maximum pay-offs to both. In the mixed motive game, which Schelling maintains, represents the majority of real life situations, including social strife, war, etc., the zero-sum oriented concepts of game theory fail not merely from the moral point of view but also from the point of view of a reasonable normative theory of any kind. The fact remains that in a game of the Prisoner's Dilemma

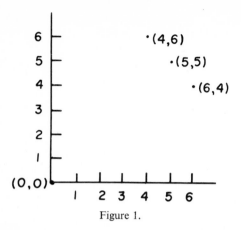

Figure 1.

type two "irrational" players do better than two "rational" ones.

Yet the situation is not hopeless, Schelling implies, if some aspects of real human behavior are taken into account. As an example he considers the game represented in Figure 1. The rules are as follows: Player 1 is to choose a horizontal coordinate; Player 2 a vertical co-ordinate, each choice to be made in ignorance of the other's. If the pair of coordinates so chosen is represented by a point in the diagram, the pay-offs are the respective coordinates. If they are not, the pay-off is (0, 0). Now player 1 would like his pay-off to be the greatest possible; but he knows that player 2 would like the same for himself. Thus player 1 dares not choose 6, because if player 2 should also choose 6, the pair (6, 6), not being in the diagram, would throw the pay-off to (0, 0). Player 1 knows this and he knows that player 2 knows this, etc. There is no way out but to trust implicitly to the good judgment of the other and to choose 5, hoping that the other will do the same. Is this choice (5, 5) not somehow an "equilibrium solution," something like the security level solution of the zero-sum game? Yet it is arrived at from considerations of co-operation rather than competition; or rather by considering the interests of the other in a realistic sense. Moreover, this "considering the interest of the other" occurs without explicit communication or agreement. It is dictated by the nature of the situation.[14]

Most significantly, communication between the players might actually lead to the dis-advantage of one or of both. To see this, suppose communication is allowed. Player 1 might then inform player 2 that he will definitely choose 6. He had a "rational" reason for doing so. If player 2 does not "go along" and choose 4 for the pay-off (6, 4), he will punish himself as well as his brow-beating partner. Not only that, but he will take a worse beating than he would if he "submitted." But suppose player 2 thinks that player 1 is bluffing. So he also announces that he will choose 6. If both are stubborn, disaster will result. If one submits, he will be worse off than if communication were impossible. If they com-promise, they will achieve the result that in all likelihood would have been achieved anyway without communication. Here is the situation where "communication" can actually throw a wrench into the machinery of tacit cooperation, *by opening up the opportunities for making threats.* (There are no such opportunities where com-munication is impossible!)

On the other hand, communication is some-times desperately needed, both if the parties have "good" intentions *and* if they have "bad" inten-tions. And communication here means not simply talking but the ability to talk so as to be believed. How can we convince the Russians, Schelling asks, that we don't want to attack them? (Read "double cross" in the context of the Prisoner's Dilemma game, where (α_1, β_1) represents armed truce, (α_2, β_2) mutual attack, and the other two pairs repre-sent unilateral attack.) On the other hand, if we have a "bad" intention (as seen from the opposing side), how can we be sure that the opponent believes our threat? If we are committed to "massive retaliation," this commitment is disas-trous to us as well as to them, if it is not believed.

Here, then, are genuine problems. They are not new. They have been raised in psychology and in social science, and even the great works of fiction show insights which are akin to the recogni-tion of these problems. But in game theory, the problems come especially sharply into focus. The consideration of these problems becomes com-pelling as game theory reaches the end of its rope.[15]

To Schelling, these dilemmas appear as failings in the over-all philosophy of game theory, and these, in turn, he attributes to the zero-sum game orientation. To this one might make two points in reply. First, while it may be true that in its original formulation these problems were virtu-ally by-passed, this was done because the non-zero-sum game appeared to von Neumann and Morgenstern largely as a situation which made for coalitions. By and large, they assumed the possibility of cooperation through communica-tion, where such cooperation made any sense. Thus the Prisoner's Dilemma is not at all a dilemma in this context. Collusion being possible, the players agree on (α_1, β_1), and that's the end of it. As for the situation depicted in Figure 1, von Neumann and Morgenstern's answer is that the solution will be one of the points in the upper right hand part of the diagram. They call these points the "negotiation set." This is the best the coalition can do. How the members will divide the spoils is *their* business. The actual result, say von Neumann and Morgenstern, will depend on the "bargaining abilities" of the individual members of the coalition.

This, of course, is again "by-passing." Actually, it is hardly justifiable to accuse an investigator for not solving a problem he has not undertaken. Not that there is any insinuation of such an accusation on the part of Schelling. But Schelling does seem to criticize game theory as an approach to behavioral science either for by-passing these problems or for illuminating them with a wrong kind of light.

The by-passing accusation (if it is actually implied, which is not entirely clear) is not justified. One needs only to peruse the literature of game theory since the publication of the principal volume to see that the workers in that field did pursue the intricacies of the non-zero-sum game with spirit, ingenuity, imagination, and with a certain degree of success. This is evident in the approach to the "bargaining situation" of Nash, Shapley, Raiffa, and others. Very considerable effort has gone into defining and clarifying the problems so as to offer reasonable solutions. The solutions are "reasonable" always according to certain criteria, of course, but in what other sense can we speak of reasonable solutions?

Whether or not the accusation of by-passing important problems was meant, the other accusation against game theory is made explicitly, namely that game theory has been a prisoner in the zero-sum game framework of thought. There is some evidence that to some extent some workers have remained within this limitation though whether this is a methodological shortcoming is open to question. Nash, for example, has felt that the non-cooperative game (i.e., where collusion is impossible) is of more fundamental importance than the cooperative one. Accordingly he concentrated much effort on the theory of non-cooperative non-zero sum games and also has tried to reduce cooperative games (where collusion is allowed) to non-cooperative ones.

In the work of some other investigators, e.g., Raiffa, the mixed motivation game is treated essentially as consisting of two components, a zero-sum game (therefore, fully competitive and non-cooperative) and a fully cooperative game. In this formulation, the opponents first jockey for an outcome in the sense of a zero-sum game. This outcome is unique, and therefore offers no ambiguities of the kind described above. Having done that, they cooperate fully to reach the final outcome, which is the best they can jointly do in the context of the preliminary zero-sum game solution. To take an example, suppose it is necessary to cooperate to get the largest joint pay-off but the division of this pay-off is something to fight about. Raiffa has shown how under certain conditions, the fight for the division can be cast into a zero-sum game. This being settled in the usual way, the erstwhile opponents can proceed to cooperate fully to get the largest rewards within the constraint imposed. This appears as an advantage from the point of view of a scientific investigation, because it removes ambiguities. But it brings up difficulties of another sort to be presently examined.

Before we take up these difficulties, we may point out that this method of using the results of the zero-sum game even in mixed motive games with opportunities for collusion (to obtain a part of the solution) does not appear as reprehensible to some authors as it does to Schelling. Braithwaite, for example, makes an eloquent defense of this technique and offers the resulting solution as an example of arriving at a "fair play" compromise by methods of game theory. In Braithwaite's setting the pay-off matrix, taken as an example, is as follows:

	Play	Not Play
Play	(1, 2)	(7, 3)
Not Play	(4, 10)	(2, 1)

The story behind it explains the labels of the strategies, "play" and "not play." Player 1 spends his evenings at home playing classical music on the piano; Player 2 plays jazz on a trumpet. The two live in the same house and are not acoustically insulated from each other. The pay-offs represent the respective measure of their preference for the four possible situations. The pianist would most prefer to play alone (7); next listen to the jazz trumpet (4); next he prefers silence (2); and what he likes least is to play the piano while his neighbor is tooting his horn (1). The trumpeter's preferences are somewhat different. He too would prefer to play alone (10); next to listen to the pianist (3); next to play while the pianist is playing (2)—that is to say, he does not mind cacophony as much as the lover of classical music; while a silent evening is the worst for him (1).

The question before us is what is a "fair" distribution of evenings so as to give as much satisfaction as possible to both, taking into account not only the preferences but also the "realities" of the situation, i.e., the "bargaining advantages" of the parties as reflected in the damage they can do to each other by violating any compromise they don't like. Note that "the greatest good for the greatest number" is easier written into a political program than achieved in practice.

There is a way of coming to a "solution" of sorts in the sense of a zero-sum game. Each has a certain "security level." By randomizing the evenings so as to play on a certain proportion of them, each can be sure of this "security level," i.e., a certain amount of satisfaction which he will get no matter what the other does. But these security levels are far below what can be attained by cooperation and a give-and-take compromise. Essentially it is a matter of deciding what is "equitable" and then getting as much as possible within the constraint imposed by this "equity."

The second part of the problem is easy. It is the first which raises the difficulties mentioned above. The choice of the "equity" solution involves a "comparison of inter-personal utilities." Such a comparison is implied in the hackneyed parental justification of physical punishment: "It hurts me more than it does you." In game theory, pay-offs are measured in "utiles," and these are in principle operationally determinable to the extent that not only the order of preferences of an individual is defined but also the relative differences of the preferences can be measured *for a given individual.* But (and this is crucial) the zero point and the unit of this measurement are not fixed by the operational procedure. Therefore these are arbitrary. If so, the utility measures of different individuals cannot be compared. In the original formulation of game theory, this limitation has always been kept in mind, and so the theory has developed in such a way that its conclusions should not depend on comparisons of inter-personal utilities. Some authors extend this limitation to bargaining problems, but others, like Raiffa and Braithwaite, do not. They introduce rules for comparing interpersonal utilities. Different rules lead to different solutions for the equity problem. For example, Raiffa proposes to assign the value 0 to each player's worst choice and the value 1 to his best choice and to distribute the other choices proportionately on this scale. This

fixes a common scale. Braithwaite, however, proposes another scheme. He notes that each player has a "security level" which he can assure by a "maximin solution." He can get more by playing a "minimax" solution,[16] which makes sure that the other does not get more than his security level, if he sticks to his maximin solution. The same is true for the other player. Braithwaite's "normalization" of the utility scales, as Luce and Raiffa point out, reduces to the following principle: the two gains thus obtained should be equal. This condition, as does Raiffa's described above, fixes a common scale to the two utility measures but this scale is different from Raiffa's. In each case, then, after the common utility scale has been established, the game can be treated as two games, namely a zero-sum game for determining equity, and a fully cooperative game to determine maximum pay-off to each within this equity.

Raiffa's arbitration scheme leads to a solution that allots 7 out of 23 nights to the pianist and 16 to the trumpeter; Braithwaite's scheme gives the pianist somewhat more, 17 out of 43 nights (26 to the trumpeter). The trumpeter is favored in both cases, presumably because his greater tolerance for noise gives him a greater bargaining advantage. His "threat power" has been taken into account.

We will have some general remarks to make about the treatment of bargaining situations by game theoretical tools in our conclusions. At this time, we will pass to the third and so far the last level of game theory, centered on the problems of coalition formation.

PROBLEMS OF COALITION FORMATION

We recall that after the difficulties of describing a real game in terms of its strategies (the extensive form) and those of finding equilibrium points of a game already represented by its strategy repertoire (normal form) have been by-passed, game theory concentrates on the coalition problem. Here the game theorist begins by considering every N-person game as a two-person game between two coalitions, indeed a zero-sum game, because a fictitious $(N+1)$th player can always be added, to whom a pay-off accrues so as to make the total pay-off to the $N+1$ players always zero. The theory of the two-person zero-sum game being complete, it remains only to examine all the possible zero-sum games which can result from all the possible partitionings of the N players into two coalitions. Since every two-person zero-sum game can be characterized by its "value" (i.e., the security level of one of the players), it follows that the N person zero-sum game can be characterized by assigning a value to each possible coalition. This assignation (of a number to each subset of the N players) is called the "characteristic function" of the game, and the game so described is said to be in the characteristic function form.

The characteristic function being given, two questions can be asked immediately: (1) Assuming "rationality" in some sense, what coalitions are likely to form? (2) How will the coalitions apportion their joint pay-offs among the participants? (Note that it is assumed here that utility is a transferable, conservative, and infinitely divisible commodity, something like money.)

One is tempted to answer the first question by saying that the coalition commanding the largest joint pay-off will form. This answer immediately leads to ambiguities. Suppose three people are to decide by majority vote how to divide a dollar. Any two can form a coalition and get the dollar between them. But since the division of the spoils is the ultimate inducement, cannot the third, outvoted player offer to take the short end of a 60–40 split to one of the others, if he will join with him? And if the other consents, cannot the deserted player offer 70 cents to the one who has agreed to take 60 cents or, better still, 50 cents to the one who has agreed to take 40 cents? No matter how the split occurs, *always* some better proposition can be made by somebody to somebody, so as to induce another arrangement.

The sets of numbers which represent the final apportionment of the utiles (and satisfy certain reasonable conditions) are called imputations. One imputation is said to *dominate* another if some subset of the players can jointly gain in passing from the second to the first and can enforce the change by forming a coalition. "Dominance" sounds like an ordering relation, that is, an anti-symmetric, and a transitive one. But the dominance relation just defined is neither anti-symmetric nor transitive. Two imputations may dominate each other, and there may be cycles of dominance.

In spite of these logical difficulties, von Neumann and Morgenstern approach the problems of the N-person game with the central concepts of the imputation and of the dominance relation among imputations. They define as a "solution" of an N-person game the *set* of imputations, such that for every other imputation there is at least one in the set that dominates it, and no imputation in the set dominates any other one in the set.

Ideally the concept of "solution" in a class of mathematical problems should satisfy an existence theorem and, preferably, a uniqueness theorem. For example, in the theory of differential equations, it can be proved that if certain conditions are satisfied, every differential equation of a certain class has a solution, and, moreover, if certain other conditions are imposed (boundary values), the solution is unique.

The trouble with the von Neumann and Morgenstern "solution" (from the point of view of the mathematician) is that an existence theorem for the general N-person game has not yet been proved, and moreover, when solutions are found, it is extremely difficult to impose "reasonable" conditions which would single out a unique solution. When one recalls that even a "unique" solution is not *an* imputation but a whole set, which may be immense (or actually infinite), one is tempted to conclude that this aspect of game theory is powerless to predict anything about human behavior, even in the normative sense (by the criterion of "rationality"). Again it appears that extra-game-theoretical concepts must be invoked to extract more definitive conclusions either in the normative or in the descriptive sense about behavior in conflict situations.

The subsequent theoretical development of the N-person game testifies to the acceptance of this challenge. We will mention here only one such development, due to Luce. Luce begins by imposing restrictions on the kind of coalition *changes* which may occur. In the original formulation, any reshuffling of the players was admissible. This lack of restraint leads to the essential "instability" of any coalition. If, however, restraints are imposed, under which one partition can pass only into a limited number of others, say those which are "not too different" from it, conditions of stability may be derived, and this stability applies both to a coalition partition and to an associated imputation. That is, the end result of an N-person game may be stabilized if there is no *permissible* coalition which can effect a change in the imputation so as to offer an inducement for its occurrence.

Luce calls such restraints sociological assumptions. And so they are or, perhaps, socio-psychological ones. They call to mind a viscosity-like or an inertia-like or friction-like quality in social behavior, which we intuitively attribute to tradition, habit, limited scope of perceived alternatives, or ethical considerations. This theory of ψ-stability, as Luce has called his approach, has led to much more definitive "solutions" of the N-person games examined than the von Neumann and Morgenstern concept of "solution," and besides, it has led to experimental and field-observational investigations (the latter in politics!) of most intriguing and promising nature. But of course all these developments could proceed only at the cost of introducing special assumptions and parameters into the theory. As attempts to make the theory empirically testable continue, these assumptions must proliferate. The logical "skeleton" of the theory will disappear under a variety of "bodies," the special theories, whose mortality, one suspects, will be very high. But the skeleton must always survive, for it is made of pure logic. Thus the inevitable failures of the special theories will not matter. As long as the general theory exists, new attempts guided by it will continue and the promise of ultimate success will remain. The review here presented barely scratches the surface of the immensely rich developments instigated by game theory since its sudden appearance as a major area of scientific investigation. It is, however, impossible to delve any deeper within the scope of a review article.

Conclusions

1. Game theory is an attempt to bring within the fold of rigorous deductive method those aspects of human behavior in which conflict and cooperation are conducted in the context of choices among alternatives whose range of outcomes is known to the fullest extent to the participants.

2. The logical analysis of such situations has led to the creation of a conceptual apparatus whose tools are derived from branches of mathematics which hitherto had not been central in applications. It is to be expected, therefore, that an impact will be given to the development of those branches under the plausible assumption that possibility of applications plays an important part in the development of the relevant branches of pure mathematics.

3. Certain problems appear to be incapable of "solution" by game theoretical tools, except tautologically, i.e., where the criteria of what constitutes a solution are dictated by the possibility of obtaining an unambiguous solution. If these criteria do not jibe with what we feel to be relevant to the behavioral situation, this points up certain inadequacies of game theory.

4. In attempting to rectify these inadequacies, workers in game theory had to introduce concepts which seem to derive from psychology, sociology, and even ethics. They have always attempted, however, because of the standards imposed by mathematical discipline, to axiomatize these concepts, so that they could be integrated into the formal method or rigorous deduction.

5. The process of axiomatization, therefore, tends to extend the scope of application of rigorous deductions to areas hitherto treated more or less intuitively. Examples, not described in our presentation for lack of space, are concepts of "power," of "social welfare," of "degree of belief," of an "index of optimism," etc. At times, this axiomatization seems destructive, as where it is shown that hardly any meaning can be assigned to certain venerable notions (the greatest good for the greatest number). But rigorization need not be destructive. It may lead to a re-thinking, to the imposition of certain conditions of validity, etc. In this *compelling* re-examination of social problems in a new light rests the greatest conceptual value of game theory and its derived disciplines.

6. In my opinion, the findings of game theory are relevant to ethics. This is not meant, of course, in the vulgar sense, to say that "science can tell us what is right and wrong." Rather game theoretical analysis helps uncover the tacit assumptions we make in our various conceptions of fairness, justice, etc. To take an example, Braithwaite's arbitration of the problem of the two musicians *exposes* the basic assumption which defines fairness, namely a particular method of normalizing the utility scales. Nash's, Shapley's, and Raiffa's arbitration schemes likewise lay bare the fundamental assumptions on which they are based. Traditional ethical systems are, of course, also often stated in terms of a few "principles," but their terminology is usually so ambiguous that it is impossible to even agree on what are the controversial points which distinguish one ethical system from another. This has been amply recognized in our legal systems, where rules and procedures are spelled out so as to give to the extent that it is possible *extensional* definitions of justice. One may agree or disagree with these, but in attacking or defending specific laws, at least it is possible to delineate the areas of agreement and disagreement. I believe game theory does the same for problems involving both parallel and opposed interests. In offering solutions, it *spells out* assumed principles of fairness as points of departure for disentangling complicated problems of settlement. Thus controversy can concentrate on these rather than on the particular solutions, where attitudes will be necessarily colored by the

results. The apparatus of reasoning seems involved, but it is the lot of civilized man to become accustomed to involved reasoning. Today we think nothing of the complications which arise in computing insurance rates, compensation laws, taxes, etc., all based on mathematical assumptions, which some centuries ago were beyond the intellectual range of almost everybody.

7. The applications of game theory are still few, certainly incommensurate with the extent of the theoretical development. But some are promising, and all are interesting. It is noteworthy that practically all applications are on the second and third levels of abstraction. Situations where opponents have a choice of a limited number of specifiable alternatives are cast into a game in normal form (as in applications to military strategy and economics); those involving formation of coalitions are cast into a game in characteristic function form. As expected, the latter have inspired several "political" models, allowing, for example, the calculation of power indices for individuals, parties, caucuses, branches of multichanger legislative bodies, etc.

8. Although the results of these applications have been anything but dramatic, one feels that conceptually all branches of behavioral science have been greatly enriched. Certain developments in psychology are also worth mentioning. Experimental determination of personal utility scales, for example, has blossomed out into a vital branch of experimental psychology. This development clearly owes its momentum to the impetus provided by von Neumann's theory of utility, which he needed as an underpinning for the theory of games. One suspects that more significant applications await the influx into the field of more behavioral scientists who are imaginative, rigorous, and technically proficient. Imagination involves insight into what is relevant and fruitful. Rigor involves the ability to translate intuitive notions into formulations which fulfill the requirements of a deductive science. Technical proficiency is know-how in handling the tools.

Notes

1. I am counting the Second and Third Laws of Motion and the Law of Gravitation.
2. For example, the integral calculus was almost invented by Archimedes during his investigations in pure geometry.
3. The usual meaning (among mathematicians) of "applied mathematics" is a formalism abstracted from some problem in the real world, regardless of whether this formalism is actually applied in the solution of a specific problem.
4. However the investigation of a simplified "poker" by von Neumann and Morgenstern is on this level.
5. There are also "infinite" games, i.e., games with an infinite number of available strategies, but we will not discuss them here.
6. This is so if the symmetries of the tic-tac-toe grid are not taken into account. Taking them into account drastically reduces the number of strategies, but raises the difficulty of describing them.
7. Practically all the results of zero-sum games hold also for constant-sum games. Therefore this class may be referred to as "zero-sum." Whenever we say non-zero-sum, we will mean non-constant sum.
8. A mixed strategy results if, in repeated plays, the choice of strategies is randomized with a certain probability assigned to each strategy.
9. This modification of the original term, "minimax," consistently used by Luce and Raiffa, is useful where maximin and minimax strategies are not identical, as in non-zero-sum games.
10. Luce and Raiffa do not think that game theory ever assumes this prerogative. I think this disclaimer derives from the ambiguity of "rationality" which emerges as soon as one passes to certain classes of non-zero-sum games.
11. It is really unfair to cast moral aspersion on the authors. It should be always borne in mind that the pay-offs in game theory are given in "utiles." This means that all aspects of the situation, including the moral ones of gaining at the other's expense in terms of whatever the *physical* pay-offs may be, have been taken into account. And so the arguments, pertaining to the ethics, of "gaining a little while the other loses much" do not apply. If one is still not satisfied with the ethical implications of the game, Luce and Raiffa point out, one can conclude that "there ought to be a law" against such games. There frequently is! One should also not forget that the two players of Prisoner's Dilemma may be anti-social characters, whose successful collusion hurts others. In this sense, we ought to be thankful for the Nash solution which doesn't let them get together. Perhaps this disposes of the "moral" argument. But the difficulty of defining "rationality," as seen from what follows, remains.
12. The man condemned to die on the gallows within thirty days requests that he be spared the expectation of being hanged "on the morrow." The request is granted. The condemned man now argues that he cannot be hanged on the 30th day, because if he is alive on the 29th, he will be sure to be hanged "on the morrow," contrary to agreement. Having established that he cannot be hanged on the 30th day, he argues that he cannot be hanged on the 29th, since being alive on the 28th and knowing that he cannot be hanged on the 30th will make him certain that he will be hanged on the 29th, again contrary to agreement. Repeating the argument, he claims that he cannot be hanged at all.
13. Strictly speaking, even here cooperation is sometimes possible, since not all such games are constant-sum. If utility is transferable, one player can agree to take less, if the gain of the other is greater than his loss, and if the other agrees to compensate him later.

14. Some would argue that mutual consideration is irrelevant here. The point (5, 5) is chosen simply because it is the only "unique" point among the desirable ones and so offers the only hope of being chosen by both players. It would have been chosen even if the pay-offs were equal at all three. Yet the mutual-consideration argument, I feel, has something to recommend it.

15. Schelling offers a formal treatment of the "balance of terror" in a paper unpublished at this writing.

16. Here is the distinction between the "maximin" and the "minimax" mentioned above (cf. Footnote 9).

59.
Society as a Complex Adaptive System

WALTER BUCKLEY

WE HAVE ARGUED at some length in another place[1] that the mechanical equilibrium model and the organismic homeostasis models of society that have underlain most modern sociological theory have outlived their usefulness. A more viable model, one much more faithful to the *kind* of system that society is more and more recognized to be, is in process of developing out of, or is in keeping with, the modern systems perspective (which we use loosely here to refer to general systems research, cybernetics, information and communication theory, and related fields). Society, or the sociocultural system, is not, then, principally an equilibrium system or a homeostatic system, but what we shall simply refer to as a complex adaptive system.

To summarize the argument in overly simplified form: Equilibrial systems are relatively *closed* and *entropic*. In going to equilibrium they typically *lose structure* and have a *minimum of free energy*; they are affected only by external "disturbances" and have *no internal or endogenous sources of change*; their component elements are *relatively simple* and *linked directly via energy exchange* (rather than information interchange); and since they are relatively closed they have no feedback or other systematic self-regulating or adaptive capabilities. The homeostatic system (for example, the organism, apart from higher cortical functioning) is open and negentropic, maintaining a moderate energy level within controlled limits. But for our purposes here, the system's main character-

istic is its functioning to *maintain the given structure of the system* within pre-established limits. It involves feedback loops with its environment, and possibly information as well as pure energy interchanges, but these are geared principally to *self-regulation* (structure maintenance) rather than adaptation (*change* of system structure). The complex adaptive systems (species, psychological and sociocultural systems) are also open and negentropic. But they are open "*internally*" *as well as externally* in that the interchanges among their components may result in *significant changes in the nature of the components themselves* with important consequences for the system as a whole. And the energy level that may be mobilized by the system is subject to relatively wide fluctuation. Internal as well as external interchanges are mediated characteristically by *information flows* (via chemical, cortical, or cultural encoding and decoding), although pure energy interchange occurs also. True feedback control loops make possible not only self-regulation, but self-direction or at least adaptation to a changing environment, such that the system may *change or elaborate its structure* as a condition of survival or viability.

We argue, then, that the sociocultural system is fundamentally of the latter type, and requires for analysis a theoretical model or perspective built on the kinds of characteristics mentioned. In what follows we draw on many of the concepts and principles presented throughout this sourcebook to sketch out aspects of a complex adaptive system model or analytical framework for the sociocultural system. It is further argued that a number of recent sociological and social psychological theories and theoretical orientations articulate well with this modern systems perspective,

---

Many of the ideas expressed here appear in more extended form in the author's *Sociology and Modern Systems Theory* (Englewood Cliffs, N.J.: Prentice-Hall, 1967).

and we outline some of these to suggest in addition that modern systems research is not as remote from the social scientists' interests and endeavors as many appear to believe.

Complex Adaptive Systems: A Paradigm

A feature of current general systems research is the gradual development of a general paradigm of the basic mechanisms underlying the evolution of complex adaptive systems. The terminology of this paradigm derives particularly from information theory and cybernetics. We shall review these concepts briefly. The *environment*, however else it may be characterized, can be seen at bottom as a set or ensemble of more or less distinguishable elements, states, or events, whether the discriminations are made in terms of spatial or temporal relations, or properties. Such distinguishable differences in an ensemble may be most generally referred to as "*variety*." The relatively stable "causal," spatial and/or temporal relations between these distinguishable elements or events may be generally referred to as "*constraint*." If the elements are so "loosely" related that there is equal probability of any element or state being associated with any other, we speak of "chaos" or complete randomness, and hence, lack of "constraint." But our more typical natural environment is characterized by a relatively high degree of constraint, without which the development and elaboration of adaptive systems (as well as "science") would not have been possible. When the internal organization of an adaptive system acquires features that permit it to discriminate, act upon, and respond to aspects of the environmental variety and its constraints, we might generally say that the system has "*mapped*" parts of the environmental variety and constraints into its organization as structure and/or "information." Thus, a subset of the ensemble of constrained variety in the environment is coded and transmitted in some way via various channels to result in a change in the structure of the receiving system which is isomorphic in certain respects to the original variety. The system thus becomes selectively matched to its environment both physiologically and psychologically. It should be added that two or more adaptive systems, as well as an adaptive system and its natural environment, may be said to be selectively interrelated by a mapping process in the same terms. This becomes especially important for the evolution of social systems.

In these terms, then, the paradigm underlying the evolution of more and more complex adaptive systems begins with the fact of a potentially changing environment characterized by variety with constraints, and an existing adaptive system or organization whose persistence and elaboration to higher levels depends upon a successful mapping of some of the environmental variety and constraints into its own organization on at least a semi-permanent basis. This means that our adaptive system—whether on the biological, psychological, or sociocultural level—must manifest (1) some degree of "*plasticity*" and "*irritability*" vis-à-vis its environment such that it carries on a constant interchange with environmental events, acting on and reacting to it; (2) some source or mechanism for *variety*, to act as a potential pool of adaptive variability to meet the problem of mapping new or more detailed variety and constraints in a changeable environment; (3) a set of *selective* criteria or mechanisms against which the "variety pool" may be sifted into those variations in the organization or system that more closely map the environment and those that do not; and (4) an arrangement for *preserving and/or propagating* these "successful" mappings.[2]

[handwritten margin note: 3 criteria for evolution]

It should be noted, as suggested above, that this is a *relational* perspective, and the question of "substance" is quite secondary here. (We might also note that it is this kind of thinking that gives such great significance to the rapidly developing relational logic that is becoming more and more important as a technical tool of analysis.) Also, as suggested, this formulation corresponds closely with the current conception of "information" viewed as the process of selection—from an ensemble of variety—of a subset which, to have "meaning," must match another subset taken from a similar ensemble.[3] Communication is the process by which this constrained variety is transmitted in one form or another between such ensembles, and involves coding and decoding such that the original variety and its constraints remains relatively invariant at the receiving end. If the source of the "communication" is the causally constrained variety of the natural environment, and the destination is the biological adaptive system, we refer to the Darwinian process of natural selection whereby the information encoded in the chromosomal material (for example the DNA) reflects or is a mapping of the environmental variety, and makes possible a continuous and more or less successful adaptation of the former system to the latter. If the adaptive system in question is a (relatively high-level) psychological or cortical system, we refer to "learning," whereby the significant environmental variety is transmitted via sensory and perceptual channels and decodings to the cortical centers where, by selective criteria (for example, "reward" and "punishment") related to physiological and/or other "needs" or "drives," relevant parts of it are encoded and preserved as "experience" for varying periods of time and may promote adaptation. Or, on the level of the symbol-based sociocultural adaptive system, where the more or less patterned actions of persons and groups are as crucial a part of the environment of other persons and groups as the non-social environment, the gestural variety and its more or less normatively defined constraints is encoded, transmitted, and decoded at the receiving end by way of the various familiar channels with varying degrees of fidelity. Over time, and again by a selective process—now much more complex, tentative, and less easily specified—there is a selective elaboration and more or less temporary preservation of some of this complex social as well as non-social constrained variety in the form of "culture," "social organization," and "personality structure."

On the basis of such a continuum of evolving, elaborating levels of adaptive system (and we have only pointed to three points along this continuum), we could add to and refine our typology of systems. Thus, we note that as adaptive systems develop from the lower biological levels through the higher psychological and sociocultural levels we can distinguish: (1) the *varying time span* required for exemplars of the adaptive system to map or encode within themselves changes in the variety and constraints of the environment; phylogenetic time scales for organic systems and for tropistic or instinctual neural systems; ontogenetic time scales for higher psychological or cortical systems; and, in the sociocultural case, the time span may be very short—days—or very long, but complicated by the fact that the relevant environment includes both intra- and inter-societal variety and constraints as well as natural environment variety (the latter becoming progressively less determinant); (2) the greatly *varying degrees of fidelity of mapping* of the environment into the adaptive system, from the lower unicellular organisms with a very simple repertoire of actions on and reactions to the environment, through the complex of instinctual and learned repertoire, to the ever-proliferating more refined and veridical accumulations of a sociocultural system; (3) the progressively greater separation and independence of the more refined "stored information" from purely

biological processes as genetic information is gradually augmented by cortically imprinted information, and finally by entirely extrasomatic cultural depositories. The implications of these shifts, and others that could be included, are obviously far-reaching.

One point that will require more discussion may be briefly mentioned here. This is the *relative* discontinuity we note in the transition from the non-human adaptive system to the sociocultural system. (The insect society and the rudimentary higher animal society make for much less than a complete discontinuity). As we progress from lower to higher biological adaptive systems we note, as a general rule, the gradually increasing role of other biological units of the same as well as different species making up part of the significant environment. The variety and constraints represented by the behavior of these units must be mapped along with that of the physical environment. With the transition represented by the higher primate social organization through to full-blown human, symbolically mediated, sociocultural adaptive systems, the mapping of the variety and constraints characterizing the subtle behaviors, gestures and intentions of the individuals and groups making up the effective social organization become increasingly central, and eventually equal if not overshadow the requirements for mapping the physical environment.[4]

It was these newly demanding requirements of coordination, anticipation, expectation and the like within a more and more complex *social* environment of interacting and interdependent others—where genetic mappings were absent or inadequate—that prompted the fairly rapid elaboration of relatively new system features. These included, of course: the ever-greater conventionalizing of gestures into true symbols; the resulting development of a "self," self-awareness, or self-consciousness out of the symbolically mediated, continuous mirroring and mapping of each unit's behaviors and gesturings in those of ever-present others (a process well described by Dewey, Mead, Cooley, and others); and the resulting ability to deal in the present with future as well as past mappings and hence to manifest goal-seeking, evaluating, self-other relating, norm-referring behavior. In cybernetic terminology, this higher level sociocultural system became possible through the development of higher order feedbacks such that the component individual subsystems became able to map, store, and selectively act toward, not only the external variety and constraints of the social and non-social environment, but also their own internal states. To speak of self-consciousness, internalization, expectations, choice, certainty and uncertainty, and the like, is to elaborate this basic point. This transition, then, gave rise to the newest adaptive system level we refer to as sociocultural. As we argued earlier, this higher level adaptive organization thus manifests features that warrant its scientific study in terms as distinct from a purely biological system as the analytical terms of the latter are from physical systems.

The Sociocultural Adaptive System

From the perspective sketched above, the following principles underlying the sociocultural adaptive system can be derived:

1) The principle of the "irritability of protoplasm" carries through to all the higher level adaptive systems. "Tension" in the broad sense— in which stress and strain are manifestations under conditions of felt blockage—is ever-present in one form or another throughout the sociocultural system—now as diffuse, socially unstructured strivings, frustrations, enthusiasms, aggressions, neurotic or psychotic or normative deviation; sometimes as clustered and minimally structured crowd or quasi-group processes, normatively supportive as well as destructive; and now as socioculturally structured creativity and production, conflict and competition, or upheaval and destruction. As Thelen and colleagues put it:

1. Man is always trying to live beyond his means. Life is a sequence of reactions to stress: Man is continually meeting situations with which he cannot quite cope.
2. In stress situations, energy is mobilized and a state of tension is produced.
3. The state of tensions tends to be disturbing, and Man seeks to reduce the tension.
4. He has direct impulses to take action. . . .[5]

2) Only closed systems running down to their most probable states, that is, losing organization and available energy, can be profitably treated in equilibrium terms. Outside this context the concept of equilibrium would seem quite inappropriate and only deceptively helpful. On the other side, only open, tensionful, adaptive systems can elaborate and proliferate organization. Cannon coined the term "homeostasis" for biological systems to avoid the connotations of equilibrium, and to bring out the dynamic, processual, potential-maintaining properties of *basically unstable* physiological systems.[6] In dealing with the sociocultural system, however, we need yet a new concept to express not only the *structure-maintaining* feature, but also the *structure-elaborating and changing* feature of the inherently unstable system. The notion of "steady state," now often used, approaches the meaning we seek if it is understood that the "state" that tends to remain "steady" is *not to be identified with the particular structure* of the system. That is, as we shall argue in a moment, in order to maintain a steady state the system may change its particular structure. For this reason, the term "morphogenesis" is more descriptive.[7] C. A. Mace recognizes this distinction in arguing for an extension of the concept of homeostasis:

The first extension would cover the case in which what is maintained or restored is not so much an internal state of the organism as some relation of the organism to its environment. This would take care of the facts of adaptation and adjustment, including adjustment to the social environment . . . the second extension would cover the case in which the goal and/or norm is some state or relation which has never previously been experienced. There is clearly no reason to suppose that every process of the homeostatic type consists in the maintenance or restoration of a norm.[8]

3) We define a system in general as a complex of elements or components directly or indirectly related in a causal network, such that at least some of the components are related to some others in a more or less stable way *at any one time*. The interrelations may be mutual or unidirectional, linear, non-linear or intermittent, and varying in degrees of causal efficacy or priority. The particular kinds of more or less stable interrelationships of components that become established at any time constitute the particular *structure* of the system at that time.

Thus, the complex, adaptive system as a continuing entity is not to be confused with the structure which that system may manifest at any time. Making this distinction allows us to state a fundamental principle of open, adaptive systems: *Persistence or continuity of an adaptive system may require, as a necessary condition, change in its structure*, the degree of change being a complex function of the internal state of the system, the state of its relevant environment, and the nature of the interchange between the two. Thus, animal species develop and persist or are continuously transformed (or become extinct) in terms of a change (or failure of change) of structure—sometimes extremely slow, sometimes very rapid. The higher individual organism capable of learning by experience maintains itself as a viable system vis-à-vis its environment by a change of structure—in this case the neural structure of the cortex. It is through this principle that we can say that the "higher" organism represents a "higher" level of adaptive system capable, ontogenetically, of *mapping the environment more rapidly and extensively* and with *greater refinement and fidelity*, as compared to the tropistic or instinct-based adaptive system which can change its structure only phylogenetically. The highest level adaptive system— the sociocultural—is capable of an even more rapid and refined mapping of the environment (including the social and non-social environment, as well as at least some aspects of its own internal state) since sociocultural structures are partially independent of both ontogenetic and phylogenetic structures, and the mappings of many individuals are selectively pooled and stored extrasomatically and made available to the system units as they enter and develop within the system.

Such a perspective suggests that, instead of saying, as some do, that a prime requisite for persistence of a social system is "pattern maintenance," we can say, after Sommerhof and Ashby,[9] that persistence of an adaptive system requires as a necessary condition the maintenance of the system's "essential variables" within certain limits. Such essential variables and their limits may perhaps be specified in terms of what some have referred to as the "functional prerequisites" of any social system (for example, a minimal level of organismal sustenance, of reproduction, of patterned interactive relations, etc.). But the maintenance of the system's essential variables, we are emphasizing, may hinge on (as history and ethnography clearly show) *pattern reorganization or change*. It is true, but hardly helpful, to say that *some* minimal patterning or stability of relations, or integration of components, is necessary—by the very definition of "system" or adaptive organization. Nor can we be satisfied with the statement that persistence, continuity, or social "order" is promoted by the "institutionalization" of interactive relations via norms and values, simply because we can say with equal validity that discontinuity or social "disorder" is *also* promoted by certain kinds of "institutionalization."

To avoid the many difficulties of a one-sided perspective it would seem essential to keep before us as a basic principle that the persistence and/or development of the complex sociocultural system depends upon structuring, destructuring, and restructuring—processes occurring at widely varying rates and degrees as a function of the external social and non-social environment. Jules Henry, among others, has made this point:

. . . the lack of specificity of man's genetic mechanisms has placed him in the situation of constantly having to revise his social structures because of their frequent failure to guide inter-personal relations without tensions felt as burdensome even in the society in which they originate . . . thus man has been presented with a unique evolutionary task: because his mechanisms for determining inter-personal relations lack specificity, he must attempt to maximize social adaptation through constant conscious and unconscious revision and experimentation, searching constantly for social structures, patterns of inter-personal relations, that will be more adaptive, as he feels them. Man's evolutionary path is thus set for him by his constant tendency to alter his modes of social adaptation.[10]

More generally, we recall from Chapter 46 that Karl W. Deutsch has seen restructuring as a basic feature distinguishing society from an organism or machine. Speaking of "the critical property which makes a given learning net into a *society*," he says:

A learning net functions as a society, in this view, *to the extent* that its constituent physical parts are capable of regrouping themselves into new patterns of activity in response to changes in the net's surroundings, or in response to the internally accumulating results of their own or the net's past.

The twin tests by which we can tell a society from an organism or a machine, on this showing, would be the freedom of its parts to regroup themselves; and the nature of the regroupings which must imply new coherent patterns of activity—in contrast to the mere wearing out of a machine or the aging of an organism, which are marked by relatively few degrees of freedom and by the gradual disappearance of coherent patterns of activity. . . .

This in turn may rest on specific properties of their members: *their capacity for readjustment to new configurations, with renewed complementarity and sustained or renewed communication.*[11]

4) The cybernetic perspective of control or self-regulation of adaptive systems emphasizes the crucial role of "deviation," seen in both negative and positive aspects. On the negative side, certain kinds of deviations of aspects of the system from its given structural state may be seen as "mismatch" or "negative feedback" signals *interpreted by certain organizing centers* as a failure of the system's operating processes or structures relative to a goal state sought, permitting—under certain conditions of adaptive structuring—a change of those operating processes or structures toward goal-optimization. (Thus, one facet of the "political" process of sociocultural systems may be interpreted in this light, with the more "democratic" type of social organization providing the more extended and accurate assessment of the mismatch between goal-attainment

on the one hand, and current policy and existing social structuring on the other.)

On the positive side, the cybernetic perspective brings out the absolute necessity of deviation—or, more generally, "variety"—in providing a pool of potential new transformations of process or structure that the adaptive systems might adopt in responding to goal-mismatch. On the lower, biological levels we recognize here the principle of genetic variety and the role of gene pools in the process of adaptive response to organismic mismatch with a changed environment. (And in regard to the other major facet of the "political" process, the more democratic type of social organization makes available a broader range of variety, or "deviation," from which to select new orientations.) Ashby, in developing his very general theory of the adaptive or self-regulating system, suggests (Chapter 15) the "law of requisite variety," which states that the variety within a system must be at least as great as the environmental variety against which it is attempting to regulate itself. Put more succinctly, only variety can regulate variety. Although such a general principle is a long way from informing more concrete analysis of particular cases, it should help provide a needed corrective to balance (not replace) the current emphasis in social science on conformity, the "control" (as against the cultivation) of "deviants," and "re-equilibration" of a given structure. (Recall also Roger Nett's argument in Chapter 48).

Thus, the concept of requisite deviation needs to be proffered as a high-level principle that can lead us to theorize: A requisite of sociocultural systems is the development and maintenance of a significant level of non-pathological deviance manifest as a pool of alternate ideas and behaviors with respect to the traditional, institutionalized ideologies and role behaviors. Rigidification of any given institutional structure must eventually lead to disruption or dissolution of the society by way of internal upheaval or ineffectiveness against external challenge. The student of society must thus pose the question—What "mechanisms" of non-pathological deviance production and maintenance can be found in any society, and what "mechanisms" of conformity operate to counteract these and possibly lessen the viability of the system?

Attempts to analyze a society from such a perspective make possible a more balanced analysis of such processes as socialization, education, mass communication, and economic and political conflict and debate. We are then encouraged to build squarely into our theory and research designs the full sociological significance of such informally well-recognized conceptions as socialization for "self-reliance" and relative "autonomy," education for "creativity," ideational flexibility and the "open mind," communications presenting the "full spectrum" of viewpoints, etc., instead of smuggling them in unsystematically as if they were only residual considerations or ill-concealed value judgments.

5) Given the necessary presence of variety or deviance in an adaptive system, the general systems model then poses the problem of the *selection* and more or less permanent *preservation* or systemic structuring of some of this variety. On the biological level, we have the process of "natural selection" of some of the genetic variety existing within the interfertile species and subspecies gene pool, and the preservation for various lengths of time of this variety through the reproductive process. On the level of higher order psychological adaptive systems, we have trial-and-error selection, by way of the so-called "law of effect," from the variety of environmental events and the potential behavioral repertoire to form learned and remembered experience and motor skills more or less permanently preserved by way of cortical structuring.[12] As symbolic mapping or decoding and encoding of the environment and one's self becomes possible,[13] the selection criteria lean less heavily on direct and simple physiological reward and more heavily on "meanings" or "significance" as manifested in existing self-group structural relations. In the process, selection from the full range of available variety becomes more and more refined and often more restricted, and emerges as one or another kind of "personality" system or "group character" structure. On the sociocultural level, social selection and relative stabilization or institutionalization of normatively interpreted role relations and value patterns occurs through the variety of processes usually studied under the headings of conflict, competition, accommodation, and such; power, authority and compliance; and "collective behavior," from mob behavior through opinion formation processes and social movements to organized war. More strictly "rational" processes are of course involved, but often seem to play a relatively minor role as far as larger total outcomes are concerned.

It is clearly in the area of "social selection" that we meet the knottiest problems. For the sociocultural system, as for the biological adaptive system, analysis must focus on both the potentialities of the system's structure at a given time, and the environmental changes that might occur and put particular demands on whatever structure has evolved. In both areas the complexities are compounded for the sociocultural system. In developing a typology of systems and their internal linkages we have noted that, as we proceed from the mechanical or physical through the biological, psychic and sociocultural, the system becomes "looser," the interrelations among parts more tenuous, less rigid, and especially less directly tied to physical events as energy relations and transformations are overshadowed by symbolic relations and information transfers. Feedback loops between operating sociocultural structures and the surrounding reality are often long and tortuous, so much so that knowledge of results or goal-mismatch, when forthcoming at all, may easily be interpreted in non-veridical ways (as the history of magic, superstition, and ideologies from primitive to present amply indicate). The higher adaptive systems have not been attained without paying their price, as the widespread existence of illusion and delusions on the personality and cultural levels attest. On the biological level, the component parts have relatively few degrees of freedom, and changes in the environment are relatively directly and inexorably reacted to by selective structural changes in the species.

Sociocultural systems are capable of persisting within a wide range of degrees of freedom of the components, and are often able to "muddle through" environmental changes that are not too demanding. But of course this is part of the genius of this level of adaptive system: it is capable of temporary shifts in structure to meet exigencies. The matter is greatly complicated for the social scientist, however, by this system's outstanding ability to act on and partially control the environment of which a major determining part is made up of other equally loose-knit, more or less flexible, illusion-ridden, sociocultural adaptive systems. Thus, although the minimal integration required for a viable system does set limits on the kinds of structures that can persist, these limits seem relatively broad compared to a biological system.[14] And given the relatively greater degrees of freedom of internal structuring (structural alternatives, as some call them) and the *potentially* great speed with which restructuring may occur under certain conditions, it becomes difficult to predict the reactions of such a system to environmental changes or internal elaboration. Considering the full complexities of the problem we must wonder at the facility with which the functionalist sociologist has pronounced upon the ultimate

functions of social structures, especially when—as seems so often the case—very little consideration is given either to the often feedback-starved social selective processes that have led to the given structures, or to the environmental conditions under which they may be presumed to be functional.

Although the problem is difficult, something can be said about more ultimate adaptive criteria against which sociocultural structures can be assessed. Consideration of the grand trends of evolution provides clues to very general criteria. These trends point in the direction of: (1) greater and greater flexibility of structure, as error-controlled mechanisms (cybernetic processes of control) replace more rigid, traditionalistic means of meeting problems and seeking goals; (2) ever more refined, accurate, and systematic mapping, decoding and encoding of the external environment and the system's own internal milieu (via science), along with greater independence from the physical environment; (3) and thereby a greater elaboration of self-regulating substructures in order—not merely to restore a given equilibrium or homeostatic level—but to purposefully restructure the system without tearing up the lawn in the process.[15]

With these and perhaps other general criteria, we might then drop to lower levels of generality by asking what restrictions these place on a sociocultural adaptive system if it is to remain optimally viable in these terms. It is possible that this might provide a value-free basis for discussing the important roles, for example, of a vigorous and independent science in all fields; the broad and deep dissemination of its codified findings; the absence of significant or long-lasting subcultural cleavages, power centers and vested interests, whether on a class or ethnic basis, to break or hinder the flow of information or feedback concerning the internal states of the system; and the promotion of a large "variety pool" by maintaining a certain number of degrees of freedom in the relations of the component parts—for example, providing a number of real choices of behaviors and goals. Thus we can at least entertain the feasibility of developing an objective rationale for the sociocultural "democracy" we shy from discussing in value terms.

6) Further discussion of the intricacies of the problem of *sociocultural selection processes* leading to more or less stable system *structures* may best be incorporated into the frame of discussion of the problem of "*structure versus process*." This is another of those perennial issues of the social

(and other) sciences, which the modern systems perspective may illuminate.

Our argument may be outlined as follows:

—Much of modern sociology has analyzed society in terms of largely structural concepts: institutions, culture, norms, roles, groups, etc. These are often reified, and make for a rather static, overly deterministic, and elliptical view of societal workings.

—But for the sociocultural system, "structure" is only a relative stability of underlying, ongoing micro-processes. Only when we focus on these can we begin to get at the selection process whereby certain interactive relationships become relatively and temporarily stabilized into social and cultural structures.

—The unit of dynamic analysis thus becomes the systemic *matrix* of interacting, goal-seeking, deciding individuals and subgroups—whether this matrix is part of a formal organization or only a loose collectivity. Seen in this light, society becomes a continuous morphogenic process, through which we may come to understand in a unified conceptual manner the development of structures, their maintenance, and their change. And it is important to recognize that out of this matrix is generated, not only *social* structure, but also *personality* structure, and *meaning* structure. All, of course, are intimately interrelated in the morphogenic process, and are only analytically separable.

Structure, Process, and Decision Theory

Though the problem calls for a lengthy methodological discussion, we shall here simply recall the viewpoint that sees the sociocultural system in comparative perspective against lower–level mechanical, organic and other types of systems. As we proceed upward along such a typology we noted that the ties linking components become less and less rigid and concrete, less direct, simple and stable within themselves. Translation of energy along unchanging and physically continuous links gives way in importance to transmission of information via internally varying, discontinuous components with many more degrees of freedom. Thus for mechanical systems, and parts of organic systems, the "structure" has a representation that is concrete and directly observable—such that when the system ceases to operate much of the structure remains directly observable for a time. For the sociocultural system, "structure" becomes a theoretical con-

struct whose referent is only indirectly observable (or only inferable) by way of series of events along a time dimension; when the system ceases to operate, the links maintaining the sociocultural structure are no longer observable.[16] "Process," then, points to the actions and interactions of the components of an ongoing system, in which varying degrees of structuring arise, persist, dissolve, or change. (Thus "process" should not be made synonymous simply with "change," as it tended to be for many earlier sociologists.)

More than a half century ago, Albion W. Small argued that, "The central line in the path of methodological progress in sociology is marked by the gradual shifting of effort from analogical representation of social structures to real analysis of social processes."[17] This was an important viewpoint for many social thinkers earlier in this century, possibly as part of the trend in physical science and philosophy toward a process view of reality developing from the work of such people as Whitehead, Einstein, Dewey and Bentley. Such views have apparently had little impact on those of recent decades who had developed the more dominant structure-oriented models of current sociology, but it seems clear that—with or without the aid of the essentially process-conscious general systems approach—a more even balance of process with structure in the analysis of sociocultural systems is gradually regaining lost ground.

C. H. Cooley, in his *Social Process*, focused on the "tentative process," involving inherent energy and growth as the dynamic agents, with ongoing "selective development" set in motion by the interaction of "active tendencies" and surrounding "conditions." He argued that for the social process, "that grows which works" is a better phrase than "natural selection" or "survival of the fittest," since "it is not so likely to let us rest in mechanical or biological conceptions."[18] R. E. Park, with his recognition of the central importance of communication, kept the notion of process in the foreground whether developing the forms of interaction or the fundations of social ecology. We should also recall the leaders of the so-called "formal" school: Whereas Simmel focused on "forms of interaction," the emphasis was always on the "interaction" as process rather than simply on the "forms"; and though the Wiese–Becker systematics developed in great detail a classification of action *patterns*, it gave equal attention to *action* patterns. For W. I. Thomas, all social becoming is viewed as a product of continual interaction of individual consciousness and objective social reality. (F. Znaniecki more recently reinforced this point of view'.[19]) And at least one unbroken thread in this vein continuing from the early part of the century is the Dewey–Mead perspective referred to as social inter-actionism, (which, we have noted, has established a strong base especially congenial to the modern cybernetic approach).[20] A reviewer of a recent collection of social interactionist essays was "reminded throughout of the continuous character of socialization, of the complexity and fluidity of interaction when it is viewed as a process rather than as the mere enactment of social forms. . . ."[21]

We can take only brief note of a few of the more recent arguments for the process viewpoint. The anthropologists, for example, have become acutely concerned in the last few years with this issue. G. P. Murdock seems to be echoing Small when he says, "All in all, the static view of social structure which seeks explanations exclusively within the existing framework of a social system on the highly dubious assumption of cultural stability and nearly perfect functional integration seems clearly to be giving way, in this country at least, to a dynamic orientation which focuses attention on the processes by which such systems come into being and succeed one another over time."[22] At about the same time, Raymond Firth was stating: "The air of enchantment which for the last two decades has surrounded the 'struc-turalist' point of view has now begun to be dis-pelled. Now that this is so, the basic value of the concept of social structure as an heuristic tool rather than a substantial social entity has come to be more clearly recognized."[23]

Soon after appeared the late S. F. Nadel's penetrating work, *The Theory of Social Structure*, which was preceded by his article on "Social Control and Self Regulation" (reprinted here as Chapter 47). This perspective is used effectively in *The Theory of Social Structures* as a basis for a critique of the current rather one-sided equilibrium model emphasizing the "complementarity of expec-tations" to the relative neglect of the several other crucial types of associative *and* dissociative social interrelationships considered equally important in earlier sociology.

Parsons' model has to do with "the conditions of relatively stable interaction in social systems," implying defined value "standards" and "institutionalized role expectations": any willful disagreement with them simply falls outside the stipulated stability and the model based on it.

I would argue that this is not necessarily so and that our model must allow for such disagreements. Even "relatively stable" social systems do not exclude them, or include them only in the form of purely fortuitous contingencies. Far from being fortuitous or idiosyncratic, the rejection of the sanctioning potential-ities of other roles may itself be anchored in the existing institutions, reflecting the presence of diverse but equally legitimate "value patterns," ideologies or schools of thought, that is, that plurality of norms we spoke of before.[24]

Nadel's book as a whole explores the thesis that structural analysis is not, and should not be treated as, static analysis: "Social structure as Fortes once put it, must be 'visualized' as 'a sum of processes in time.' As I would phrase it, social structure is implicitly an event-structure. . . ."[25] And in concluding he reiterates his argument that

. . . it seems impossible to speak of social structure in the singular. Analysis in terms of structure is incapable of presenting whole societies; nor, which means the same, can any society be said to exhibit an embracing, coherent structure as we understand the term. There are always cleavages, dissociations, enclaves, so that any description alleged to present a single structure will in fact present only a fragmentary or one-sided picture.[26]

As a final example in anthropology, we should mention the cogent argument of Evon Z. Vogt that the two concepts of structure and process must be integrated into a general theoretical model. As with Nadel, structure is seen as falsely conceived as static, with change pathological. Rather, Vogt feels, must we pose the primacy of change, considering structure the way in which moving reality is translated, for the observer, into an instantaneous and artificial observation: social and cultural structures are only the inter-sections in time and space of process in course of change and development.[27]

Among sociologists, a perennial critic of the overly-structural conception of the group is Herbert Blumer. Blumer has argued that it is from the process of ongoing interaction itself that group life gets its main features, which can-not be adequately analyzed in terms of fixed attitudes, "culture," or social structure—nor can it be conceptualized in terms of mechanical structure, the functioning of an organism, or a system seeking equilibrium, " . in view of the formative and explorative character of interaction as the participants *judge* each other and *guide* their own acts by that judgment."

The human being is not swept along as a neutral and indifferent unit by the operation of a system. As an organism capable of self-interaction he forges his actions out of a process of definition involving *choice*, *appraisal*, and *decision*. . . . Cultural norms, status positions and role relationships are only *frameworks* inside of which that process [of formative transaction] goes on.[28]

Highly structured human association is relatively infrequent and cannot be taken as a prototype of a human group life. In sum, institutionalized patterns constitute only one conceptual aspect of society, and they point to only a part of the ongoing process (and, we might add, they must be seen to include deviant and disfunctional patterns: for conceptual clarity and empirical relevance, "institutionalization" cannot be taken to imply only "legitimacy," "consent," and ultimately adaptive values).

Finally, it should be noted that Gordon Allport, viewing personality as an open-system, stresses a very similar point concerning the organization of personality:

> ... the best hope for discovering coherence would seem to lie in approaching personality as a total functioning structure, i.e., as a *system*. To be sure, it is an incomplete system, manifesting varying degrees of order and disorder. It has structure but also unstructure, function but also malfunction. As Murphy says, "all normal people have many loose ends." And yet personality is well enough knit to qualify as a system—which is defined merely as *a complex of elements in mutual interaction*.[29]

In the light of such views, we need only recall the many critiques pointing to the incapacity or awkwardness of the conventional type of framework before the facts of process, "becoming," and the great range of "collective behavior."[30]

Statements such as Blumer's, a continuation of the perspective of many neglected earlier sociologists and social psychologists, would seem to constitute a perspective that is now pursued by many under new rubrics such as "decision theory." For earlier antecedents it should be enough to mention W. I. Thomas's "definition of the situation," Znaniecki's "humanistic coefficient," Weber's "verstehen," Becker's "interpretation," and MacIver's "dynamic assessment."[31] Much of current structural, consensus theory represents a break from this focus. As Philip Selznick has argued,

> A true theory of social action would say something about goal-oriented or problem-solving behavior, isolating some of its distinctive attributes, stating the likely outcomes of determinate transformations. . . . In Parsons' writing there is no true embrace of the idea that structure is being continuously opened up and reconstructed by the problem-solving behavior of individuals responding to concrete situations. This is a point of view we associate with John Dewey and

G. H. Mead, for whom, indeed, it had significant intellectual consequences. For them and for their intellectual heirs, social structure is something to be taken account of in action; cognition is not merely an empty category but a natural process involving dynamic assessments of the self and the other.[32]

It can be argued, then, that a refocusing is occurring via "decision theory," whether elaborated in terms of "role-strain" theory; theories of cognitive dissonance, congruence, balance, or concept formation; exchange, bargaining, or conflict theories, or the mathematical theory of games. The basic problem is the same: How do interacting personalities and groups define, assess, interpret, "verstehen," and act on the situation? Or, from the broader perspective of our earlier discussion, how do the processes of "social selection" operate in the "struggle" for sociocultural structure? Instead of asking how structure affects, determines, channels actions and interactions, we ask how structure is created, maintained and recreated.

Thus we move down from structure to social interrelations and from social relations to social actions and interaction processes—to a matrix of "dynamic assessments" and intercommunication of meanings, to evaluating, emoting, deciding and choosing. To avoid anthropomorphism and gain the advantages of a broader and more rigorously specified conceptual system, we arrive at the language of modern systems theory.

Basic ingredients of the decision-making focus include, then: (1) a *process* approach; (2) a conception of *tension* as inherent in the process; and (3) a renewed concern with the role and workings of man's enlarged cortex seen as a complex adaptive subsystem operating within an *interaction matrix* characterized by *uncertainty*, *conflict*, and other dissociative (as well as associative) processes *underlying the structuring and restructuring of the larger psycho-social system*.

PROCESS FOCUS

The process focus points to information-processing individuals and groups linked by different types of communication nets to form varying types of interaction matrices that may be characterized by "competition," "cooperation," "conflict," and the like. Newer analytical tools being explored to handle such processes include treatment of the interaction matrix over time as a succession of states described in terms of transition probabilities, Markoff chains, or stochastic processes in general. The Dewey–Mead "transactions" are now discussed in terms of information and codings and decodings, with the essential "reflexivity" of behavior now treated in terms of negative and positive feedback loops linking via the communication process the intrapersonal, interpersonal and intergroup subsystems and making possible varying degrees of matching and mismatching of Mead's "self and others," the elaboration of Boulding's "Image,"[33] and the execution of Miller's "Plans" (Chapter 45). And herein we find the great significance for sociology of many of the conceptual tools (though not, at least as yet, the mathematics) of information and communication theory, cybernetics, or general systems research, along with the rapidly developing techniques of *relational* mathematics such as the several branches of set theory—topology, group theory, graphy theory, symbolic logic, etc.

CONCEPTION OF TENSION

Tension is seen as an inherent and essential feature of complex adaptive systems; it provides the "go" of the system, the "force" behind the elaboration and maintenance of structure. There is no "law of social inertia" operating here, nor can we count on "automatic" reequilibrating forces counteracting system "disturbances" or "deviance," for, whereas we do find deviance-reducing negative feedback loops in operation we *also* find deviance-maintaining and deviance-amplifying *positive* feedback processes often referred to as the vicious circle or spiral, or "escalation."[34] It is not at all certain whether the resultant will maintain, change, or destroy the given system or its particular structure. The concepts of "stress" or "strain" we take to refer only to the greater mobilization of normal tension under conditions of more than usual blockage. And instead of a system's seeking to manage *tension*, it would seem more apt to speak of a system's seeking to manage *situations* interpreted as responsible for the production of greater than normal tension.

The "role strain" theory of William J. Goode is an illustrative attack on assumptions of the widely current structural approach, using a process and tension emphasis and contributing to the decision-theory approach. Goode analyzes social structure or institutions into role relations, and role relations into role transactions. "Role relations are seen as a sequence of 'role bargains' and as a continuing process of selection among alternative role behaviors, in which each individual seeks to reduce his role strain."[35] Contrary to the current stability view, which sees social system continuity as based primarily on normative consensus and normative integration, Goode thus sees "dissensus, nonconformity, and conflicts among norms and roles as the usual state of affairs. . . . The individual cannot satisfy fully all demands, and must move through a continuous sequence of role decision and bargains . . . in which he seeks to reduce his role strain, his felt difficulty in carrying out his obligations."[36] Goode also recognizes that there is no "law of social inertia" automatically maintaining a given structure.

Like any structure or organized pattern, the role pattern is held in place by both internal and external forces—in this case, the role pressures from other individuals. Therefore, not only is role strain a normal experience for the individual, but since the individual processes of reducing role strain determine the total allocation of role performances to the social institutions, the total balances and imbalances of role strains create whatever stability the social structure possesses.[37]

It should be noted, however, that Goode accepts unnecessarily a vestige of the equilibrium or stability model when he states, "The total role structure functions so as to reduce role strain."[38] He is thus led to reiterate a proposition that—when matched against our knowledge of the empirical world—is patently false. Or, more precisely, not false, but a half-truth: it recognizes deviance-reducing negative feedback processes, but not deviance-amplifying positive feedback processes. Such a proposition appears reasonable only if we "hold everything else constant," that is, take it as a closed system. However, the proposition is unnecessary to his argument and, in fact, clashes with the rest of his formulation: ". . . though the sum of role performances ordinarily maintains a society it may also change the society or fail to keep it going. There is no necessary harmony among all role performances. . . . But whether the resulting societal pattern is 'harmonious' or integrated or whether it is even effective in maintaining that society, are separate empirical questions."[39]

STUDY OF COGNITIVE PROCESSES

A more concerted study of cognitive processes, especially under conditions of *uncertainty* and *conflict*, goes hand in hand, of course, with a focus on decision-making and role transactions. Despite the evolutionary implications of man's enlarged cortex, much social (and psychological) theory seems predicated on the assumption that men are decorticated. Cognitive processes, as they are coming to be viewed today, are not to be simply equated with the traditional, ill-defined, concept of the "rational." That the data-processing system—whether socio-psychological or electro-mechanical—is seen as inherently "rational" tells us little about its outputs in concrete cases. Depending on the adequacy and accuracy of the effectively available information, the total internal organization or "Image," the character of the "Plans" or program, and the nature of the significant environment, the output of either "machine" may be sense or nonsense, symbolic logic or psychologic, goal-attainment or oscillation.

Beyond giving us a deeper perspective on the concept of the "rational," current theories of cognitive processes give promise of transcending the hoary trichotomy of the cognitive, the conative, and the moral as analytical tools. Whether this amounts to a rejection of the distinction, or simply an insistence that what was analytically rent asunder must now be reunified, the ferment appears significant. We refer here, not only to the many neurological and schematic studies of the brain, or the processes by which it solves problems and attains concepts, but especially to the several theories of cognitive "dissonance" or "congruence" or "balance" represented in the works of Heider, Cartwright and Harary, Osgood and Tannenbaum, Festinger, and others, as well as the symbol-processing and interpersonal communication perspectives represented by the "psycholinguistics" of Osgood, the "communicative acts" of Newcomb, and the "two factor" theory of Mowrer.

The intricate meeting of the cognitive, the affective and evaluative (or attitudinal), and the semantic or symbolic in such theories is well illustrated in Osgood's treatment of "cognitive dynamics." Equating "cognitive elements" with the *meanings* of signs, Osgood proposes that "congruity exists when the evaluative meanings of interacting signs are equally polarized or intense—either in the same or opposite evaluative directions. . . ."[40] In contrast to the theories of Heider and Festinger, this theory "*assigns affective or attitudinal values to the cognitive elements themselves*, and not to their relations. . . ."[41] And in discussing the "process of inference through psycho-logic," Osgood says:

Much of what is communicated attitudinally by messages and by behavior is based on such inferences; . . . The syntax of language and of behavior provides a structural framework within which meaningful contents are put; the structure indicates what is related to what, and how, but only when the meaningful values are added does the combination of structure and content determine psycho-logical congruence or incongruence.[42]

Despite the incorporation of aspects of these several elements into their theories, however, the psychologically oriented theorist usually leaves the sociologist something to be desired, namely, something that transcends "the individual" and "his" attempts to minimize inconsistency or dissonance and maintain stability, and which views the group situation as inadequately characterized in terms of "myriad decisions in individual nervous systems." Thus Osgood hypothesizes that

laws governing the thinking and behaving of individuals also govern the "thinking" and "behaving" of groups . . . with nothing but communication to bind us together, it is clear that "decisions" and "behaviors" of nations must come down to myriad decisions in individual nervous systems and myriad behaviors of individual human organisms.[43]

We are reminded here of Robert R. Sears' complaint that "psychologists think monadically. That is, they choose the behavior of one person as their scientific subject matter. For them, the universe is composed of individuals . . . the universal laws sought by the psychologist almost always relate to a single body."[44] Arguing for the desirability of combining individual and social behavior into a common framework, Sears noted that, "Whether the group's behavior is dealt with as antecedent and the individual's as consequent, or vice versa, the two kinds of event are so commonly mixed in causal relationships that it is impractical to conceptualize them separately."[45]

Fortunately, however, there are recent statements that rally to the side of the sociological interactionist theorists, whose perspective continues to be ignored or little understood by so many personality theorists who are nevertheless gradually rediscovering and duplicating its basic principles. A good beginning to a truly interpersonal approach to personality theory and the problem of stability and change in behavior is the statement of Paul F. Secord and Carl W. Backman, which remarkably parallels Goode's theory of stability and change in social systems discussed earlier. Pointing to the assumptions of several personality theorists that when stability of behavior occurs it is solely a function of stability in personality structure, and that this latter structure has, inherently, a strong resistance to change except when special change-inducing forces occur, Secord and Backman see as consequences the same kinds of theoretical inadequacies we found for the stability view of social systems:

The first is that continuity in individual behavior is not a problem to be solved; it is simply a natural outcome of the formation of stable structure. The second is that either behavioral change is not given systematic attention, or change is explained independently of stability. Whereas behavioral stability is explained by constancy of structure, change tends to be explained by environmental forces and fortuitous circumstances.[46]

Their own theoretical view abandons these assumptions and "places the locus of stability and change in the interaction process rather than in intrapersonal structures." Recognizing the traditional two classes of behavioral determinants, the cultural-normative and the intrapersonal, their conceptualization

attempts to identify a third class of determinants, which have their locus neither in the individual nor the culture, but in the interaction process itself. In a general sense this third class may be characterized as the tendencies of the individual and the persons with whom he interacts to shape the interaction process according to certain requirements, i.e., they strive to produce certain patterned relations. As will be seen, the principles governing this activity are truly interpersonal; they require as much attention to the behavior of the other as they do to the behavior of the individual, and it cannot be said that one or the other is the sole locus of cause.[47]

They go on to analyze the "interpersonal matrix" into three components: an aspect of the self-concept of a person, his interpretation of those elements of his behavior related to that aspect, and his perception of related aspects of the other with whom he is interacting. "An interpersonal matrix is a recurring functional relation between these three components."

In these terms, Secord and Backman attempt to specify the conditions and forces leading to or threatening congruency or incongruency, and hence stability or change, in the matrix. Thus, four types of incongruency, and two general classes of resolution of incongruency, are discussed. One of these latter classes

results in restoration of the original matrix, leaving self and behavior unchanged (although cognitive distortions may occur), and the other leads to a new matrix in which self or behavior are changed.[48]

In sum, contrary to previous approaches, theirs emphasizes that "the individual strives to maintain interpersonal relations characterized by congruent matrices, rather than to maintain a self, habits, or traits."

Maintenance of intrapersonal structure occurs only when such maintenance is consistent with an ongoing interaction process which is in a state of congruency. That most individuals do maintain intrapersonal structure is a function of the fact that the behavior of others toward the individuals in question is normally overwhelmingly consistent with such maintenance.[49]

And this conception also, as most approaches do not (or do inadequately), predicts or accounts for the fact that, should the interpersonal environment cease to be stable and familiar, undergoing great change such that others behave uniformly toward the individual in new ways, the individual "would rapidly modify his own behavior and internal structure to produce a new set of congruent matrices. As a result, he would be a radically changed person."[50]

As we have said, the Secord and Backman theory and Goode's role-strain theory may be seen as closely complementary views. The former argues that *personality* structure is generated in, and continues to have its seat in, the social interactive matrix; the latter argues that *social* structure is generated in, and continues to have its seat in, the social interactive matrix. Since it is the latter that we are focusing on here, we shall conclude with additional examples of current theory and research that explore further the mechanisms underlying the genesis or elaboration of social structure out of the dynamics, especially the role dynamics, of the symbolic interaction process.

Further Examples

Ralph Turner has addressed himself to the elaboration of this perspective in that conceptual area fundamental to the analysis of institutions—roles and role-taking.[51] The many valid criticisms of the more static and overdetermining conception of roles is due, he believes, to the dominance of the Linton view of role and the use of an oversimplified model of role functioning. Viewing role-playing and role-taking, however, as a process (as implied in Meadian theory), Turner shows that there is more to it than just "an extension of normative or cultural deterministic theory" and that a process view of role adds novel elements to the notion of social interaction.

The morphogenic nature of role behavior is emphasized at the start in the concept of "*role-making*." Instead of postulating the initial existence of distinct, identifiable roles, Turner posits "a tendency to create and modify conceptions of self- and other-roles" as the interactive orienting process. Since actors behave *as if* there were roles, although the latter actually exist only in varying degrees of definitiveness and consistency, the actors attempt to define them and make them explicit—thereby in effect creating and modifying them as they proceed. The key to role-taking, then, is the morphogenic propensity "to shape the phenomenal world into roles"; formal organizational regulation restricting this process is not to be taken as the prototype, but rather as a "distorted instance" of the wider class of role-taking phenomena. To the extent that the bureaucratic setting blocks the role-making process, organization is maximal, "variety" or alternatives of action minimal, actors are cogs in a rigid machine, and the morphogenic process underlying the viability of complex adaptive systems is frustrated.

Role interaction is a tentative process of reciprocal responding of self and other, challenging or reinforcing one's conception of the role of the other, and consequently stabilizing or modifying one's own role as a product of this essentially feedback-testing transaction. The conventional view of role emphasizing a prescribed complementarity of expectations thus gives way to a view of role-taking as a process of "devising a performance on the basis of an imputed other-role," with an important part being played by cognitive processes of inference testing. In a manner consistent with models of the basic interaction process suggested by Goode and by Secord and Backman, Turner views as a central feature of role-taking "the process of discovering and creating 'consistent' wholes out of behavior," of "devising a pattern" that will both cope effectively with various types of relevant others and meet some recognizable criteria of consistency. Such a conception generates empirically testable hypotheses of relevance to our concern here with institutional morphogenesis, such as: "Whenever the social structure is such that many individuals characteristically act from the perspective of two given roles simultaneously, there tends to emerge a single role which encompasses the action."[52]

Turning directly to the implications for formal, institutional role-playing, Turner argues that the formal role is primarily a "skeleton" of rules which evoke and set into motion the fuller roles built-up and more or less consensually validated in the above ways. Role behavior becomes relatively fixed only while it provides a perceived consistency and stable framework for interaction, but it undergoes cumulative revision in the role-taking process of accommodation and compromise with the simple conformity demanded by formal prescriptions.

The purposes and sentiments of actors constitute a unifying element in role genesis and maintenance, and hence role-taking must be seen to involve a great deal of selective perception of other-behavior and relative emphasis in the elaboration of the role pattern. This selection process operates on the great variety of elements in the situation of relevant objects and other-behaviors which could become recognized components in a consistent role pattern. Not all combinations of behavior and object relations can be classed into a single role; there must be criteria by which actors come to "verify" or "validate" the construction of a number of elements into a consistent role. This verification stems from two sources: "internal validation" of the interaction itself, and "external validation" deriving from "the generalized other" of Mead. The former hinges on successful prediction or anticipation of relevant other-behavior in the total role-set, and hence on the existence of role patterns whereby coherent selection of behaviors judged to constitute a consistent role can be made. But the notion of fixed role prescriptions is not thereby implied, since, first, roles—like norms--often or usually provide a *range of alternative* ways of dealing with any other-role, or, as is most common, the small segment of it activated at any one time, and secondly, the coherence and predictability of a role must be assessed and seen as "validated," not in terms of any one other-role, but in terms of the Gestalt of all the accommodative and adjusted requirements set by the number of other-roles in the actor's role-set and generated in the ongoing role-making process.

An example is provided by the study by Gross *et al.* of the school superintendent role. It is found that incumbency in this role (1) actually involved a great deal of choice behavior in selecting among the alternative interpretations and behaviors deemed possible and appropriate, and that (2) consistency and coherence of an incumbent's behavior could be seen only in terms of the total role as an accommodation with correlative other-roles of school board member, teacher, and parent, with which the superintendent was required to interact simultaneously. As Gross puts it, a "system model" as against a "position-centric" model involves an important addition by including the interrelations among the counter positions. "A position can be completely described only by describing the total system of positions and relationships of which it is a part. In other words, in a system of interdependent parts, a change in any relationship will have an effect on all other relationships, and the positions can be described only by the relationships."[53]

Thus Turner sees the internal criterion of role validation as insuring a constant modification, creation, or rejection of the content of specific roles occurring in the interplay between the always somewhat vague and incomplete ideal role conceptions and the experience of their concrete implications by the interpreting, purposive, selectively evaluating and testing self and others.

The basis of "external validation" of a role is the judgment of the behavior to constitute a role by others felt to have a claim to correctness or legitimacy. Criteria here include: discovery of a name in common use for the role, support of major norms or values, anchorage in the membership of recognized groups, occupancy of formalized positions, and experience of key individuals as role models acting out customary attitudes, goals and specific actions.

Under the "normal loose operation of society" these various internal and external criteria of validation are at best only partially conveyant and consistent in identifying the same units and content as roles. The resulting inevitable discrepancies between formal, institutional rules and roles, and the goals, sentiments and selective interpretations arising from the experience of actually trying to play them out, make role conceptions "creative compromises," and insure "that the framework of roles will operate as a hazily conceived ideal framework for behavior rather than as an unequivocal set of formulas."[54]

In sum, "institutions" may provide a normative framework prescribing roles to be played and thus assuring the required division of labor and minimizing the costs of general exploratory role-setting behavior, but the actual role transactions that occur generate a more or less coherent and stable working compromise between ideal set prescriptions and a flexible role-making process, between the structured demands of others and the requirements of one's own purposes and sentiments. This conception of role relations as "fully interactive," rather than merely conforming, contributes to the recent trends "to subordinate normative to functional processes in accounting for societal integration"[55] by emphasizing the complex adaptive interdependence of actors and actions in what we see as an essentially morphogenic process—as against a merely equilibrial or homeostatic process.

ORGANIZATION AS A NEGOTIATED ORDER

Next we shall look at a recently reported empirical study of a formal organization that concretely illustrates many facets of the above conceptualization of Turner and contributes further to our thesis. In their study of the hospital and its interactive order, Anselm Strauss and colleagues develop a model of organizational process that bears directly on the basic sociological problem of "how a measure of order is maintained in the face of inevitable changes (derivable from sources both external and internal to the organization)."[56] Rejecting an overly structural view, it is assumed that social order is not simply normatively specified and automatically maintained but is something that must be "worked at," continually reconstituted. Shared agreements, underlying orderliness, are not binding and shared indefinitely but involve a temporal dimension implying eventual review, and consequent renewal or rejection. On the basis of such considerations, Strauss and colleagues develop their conception of organizational order as a "negotiated order."

The hospital, like any organization, can be visualized as a hierarchy of status and power, of rules, roles and organizational goals. But it is also a locale for an ongoing complex of transactions among differentiated types of actors: professionals such as psychiatrists, residents, nurses and nursing students, psychologists, occupational therapists and social workers; and nonprofessionals such as various levels of staff, the patients themselves, and their families. The individuals involved are at various stages in their careers, have their own particular goals, sentiments, reference groups, and ideologies, command various degrees of prestige, esteem and power, and invest the hospital situation with differential significance.

The rules supposed to govern the actions of the professionals were found to be far from extensive, clearly stated, or binding; hardly anyone knew all the extant rules or the applicable situations and sanctions. Some rules previously administered would fall into disuse, receive administrative reiteration, or be created anew in a crisis situation. As in any organization, rules were selectively evoked, broken, and/or ignored to suit the defined needs of personnel. Upper administrative levels especially avoided periodic attempts to have the rules codified and formalized, for fear of restricting the innovation and improvisation believed necessary to the care of patients. Also, the multiplicity of professional ideologies, theories and purposes would never tolerate such rigidification.

In sum, the area of action covered by clearly defined rules was very small, constituting a few general "house rules" based on long-standing shared understandings. The basis of organizational order was the generalized mandate, the single ambiguous goal, of returning patients to the outside world in better condition. Beyond this, the rules ordering actions to this end were the subject of continual negotiations—being argued, stretched, ignored, or lowered as the occasion seemed to demand. As elsewhere, rules failed to act as universal prescriptions, but required judgment as to their applicability to the specific case.

The ambiguities and disagreements necessitating negotiation are seen by the researchers to be patterned. The various grounds leading to negotiation include: disagreement and tension over the proper ward placement of a patient to maximize his chances of improvement; the mode of treatment selected by the physician, which is closely related to his own psychiatric ideology and training; the multiplicity of purposes and temporal ends of each of the professional groups as they maneuver to elicit the required cooperation of their fellow workers; the element of medical uncertainty involved in treating the patient as a unique, "individual case," and the consequent large area of contingency lying of necessity beyond specific role prescription; and, finally, the inevitable changes forced upon the hospital and its staff by external forces and the unforeseen consequences of internal policies and the round of negotiations themselves. What is concretely observed, then, in researching the organizational order of the hospital, is negotiation between the neurologically trained and the psychotherapeutically oriented physician, between the nurses and the administrative staff, between the nonprofessional floor staff and the physician, between the patient and each of the others.

The negotiation process itself was found to have patterned and temporal features. Thus, different physicians institute their own particular programs of treatment and patient care and in the process develop fairly stable understandings with certain nurses or other institutional gatekeepers such as to effectuate an efficient order of behaviors with a minimum of communication and special instructions. Such arrangements are not called for by any organizational role prescriptions; nevertheless, they represent a concrete part of the actual organization generated in the morphogenic process of negotiation (or role-making and -taking, in Turner's terms). Thus, agreements do not occur by chance but are patterned in terms of "who contracts with whom, about what, as well as when. . . ."[57] There is an important temporal aspect, also, such as the specification of a termination period often written into an agreement—as when a physician bargains with a head nurse to leave his patient in the specific ward for "two more days" to see if things will work themselves out satisfactorily.

In a final section of their paper, Strauss and his colleagues bring out the full implications of their negotiation model in dealing with genuine organizational change. The model presents a picture of the hospital—and perhaps most other institutionalized spheres of social life—as a transactional milieu where numerous agreements are "continually being established, renewed, reviewed, revoked, revised." But this raises the question of the relation between this process and the more stable structure of norms, statuses, and the like. The authors see a close systemic relation between the two. The daily negotiations periodically call for a reappraisal and reconstitution of the organizational order into a "new order, not the reestablishment of an old, as reinstituting of a previous equilibrium." And, we would add, it contributes nothing to refer to this as a "moving equilibrium" in the scientifically established sense of the term. The daily negotiative process not only allows the day-by-day work to get done, but feeds back upon the more formalized, stable structure of rules and policies by way of "a periodic appraisal process" to modify it—sometimes slowly and crescively, sometimes rapidly and convulsively. And, as a reading of history suggests, virtually every formal structure extant can be traced, at least in principle, from its beginnings to its present apparently timeless state through just such a morphogenic process—a process characteristic of what we have called the complex adaptive system.

THE SCHOOL SUPERINTENDENT AND HIS ROLE

We turn to the study by Gross and his associates of the role system of the school superintendent and his counter-role partners, the school board member, the teacher, and the parent. A major burden of this empirical study is to demonstrate the research sterility of the Lintonian conception of role, and structural theories built on it, due principally to the postulate of consensus on role definition. The study showed a majority of significant differences in the definitions of their own roles by a sample of incumbents of the same social position and by incumbents of different but interrelated counter positions. This fact led Gross and his associates to the demonstration of a number of important theoretical consequences derived from rejection of the postulate of role consensus. It is often assumed, for example, that the socialization process by which roles are "acquired" provides for a set of clearly defined and agreed-upon expectations associated with any particular position. But the empirically discovered fact of differential *degrees of consensus* seriously challenged this assumption. From our systems model viewpoint, recognition of degrees of consensus is tantamount to the recognition of a continuous source of "variety" in the role system, as defined earlier, which leads us to seek the various *selective*, choice processes occurring in the role transactions. At least for the occupational positions studied, it was found that the assumption of socialization on the basis of prior consensus on role definitions was untenable, and deserved "to be challenged in most formulations of role acquisition, including even those concerned with the socialization of the child."[58]

Secondly, the research showed that, instead of assuming role consensus and explaining variations of behavior of incumbents of the same position in terms of personality variables, one would better explain them in terms of the varying role expectations and definitions—which may be unrelated to psychological differences.

The implications are also great for a theory of social control. Instead of a model assuming that the application or threat of negative sanctions leads to conformity to agreed-upon norms, the research pointed to the numerous situations in which, due to variant or ambiguous role definitions, the same behavior resulted in negative sanctions by some role partners and positive sanctions by others, or failure to apply sanctions because of perceived ambiguity—or nonconformity to perceived expectations of another despite negative sanctions because other expectations were defined as more legitimate.

Another Lintonian postulate challenged by this research is that though an actor may occupy many positions, even simultaneously, he activates each role singly with the others remaining "latent." It is found, however, that individuals often perceive and act toward role partners as if simultaneous multiple roles were being activated. For example, one may hold different expectations regarding a teacher who is male, young and unmarried as against one who is female, older and married. In other words, standards and expectations are applied to the whole person as a result, in part, of the complex of positions the person is perceived as occupying at that time. A related consideration involves the time dimension over which two or more individuals interact; other positions they occupy enter progressively into their perception of each other and consequently modify evaluations and expectations. Thus the authors generalize their point to a broader theory of social interaction by suggesting that evaluative standards shift over time from those applied as appropriate to the incumbent of a particular position to those applied to a total person with particular personality features and capacities as the incumbent of multiple positions.

Finally, their rejection of the consensus model led these researchers to find a process of role-strain or role-conflict generation and resolution similar in principle to that conceptualized by others discussed above. Having defined the role set they were studying as a true *complex system* of interrelated components, and having then uncovered and analyzed the *variety* continuously introduced into the system by way of variant, ambiguous or changing role definitions, they then focused on the *selection process* whereby this variety was sifted and sorted in the give and take of role transactions. Thus, given the situation in which a role incumbent was faced with incompatible expectations on the part of two of his counter-role partners, a theory was constructed to answer the question of how the actor may choose from among four alternatives in resolving the role conflict. From our present perspective, the theoretical scheme suggested constitutes another important contribution to the forging of a conceptual link between the dynamics of the role transaction and the more stable surrounding social structure—a link that is too often skipped over by the consensus theorist's identification of social structure and consensual role playing.

This linkage is made in terms of the concepts of perceived *legitimacy* of the conflicting expectations, an assessment of the *sanctions* that might be applied, and predispositions to give primacy to a *moral* orientation, an *expedient* orientation, or a balance of the two. We face once again the reciprocal question of how role transactions are conditioned by the surrounding social structure and how that structure is generated and regenerated as a product of the complex of role transactions.

The four alternatives that Gross and colleagues see open to an actor to choose in attempting to resolve a role conflict between incompatible expectations A and B are: (1) conformity to expectation A; (2) conformity to expectation B; (3) compromise in an attempt to conform in part to both expectations; or (4) attempt to avoid conforming to either expectation. The first criterion that the theory postulates to underlie the particular alternative chosen is the actor's definition of the legitimacy of the expectations. Thus the prediction of behavior on this criterion is that, when only one expectation is perceived as legitimate the actor will conform to that one; when both are defined as legitimate he will compromise; and when neither is seen as legitimate he will engage in avoidance behavior. The second criterion is the actor's perception of the sanctions that would be applied for nonconformity, which would create pressures to conform if strong negative sanctions are foreseen otherwise. This predicts for three of the four combinations of two sets of expectations, but not for the case of both expectations being perceived as leading to weak or no negative sanctions.

It is assumed that for any role conflict situation an actor would perceive both of these dimensions and make his decision accordingly. Predictions on the basis of the theory so far provide for determinate resolutions of conflict in seven of the sixteen combinations of the four types of legitimacy and the four types of sanctions situations, but the other nine are left indeterminate with only the two criteria. This is because the criteria predispose in different directions, and at least a third criterion is needed to determine the outcome. The authors thus appeal to the actor's predisposition to give primacy to either the legitimacy or to the sanctions dimension, or to balance the two, thus leading to the postulation of three types of predisposing orientations to expectations as listed above—the *moral*, the *expedient*, and the balanced *moral-expedient*. All the combinations of situations now become predictive.

The accuracy of the predictions was tested empirically with the data from the superintendent-role study for four "incompatible expectation situations," and the evidence supported the theory, though with some incorrect predictions.

The implications of this conceptualization and empirical analysis are far-reaching, as already

suggested, for general sociological theory. The study is concerned with what must be considered "institutional" organization and process, and supports a model of that structure and process that is quite different from the more traditional models. As the authors point out, one strong advantage of the theory is its conceptualization of institutional role behavior in terms of "expectations," whether legitimate or illegitimate, rather than in terms of "obligations" (legitimate expectations) as is assumed in consensus theory. The theory thus allows for the possibility that illegitimate expectations constitute a significant part of institutional role behavior, and underlie much of the conflict occurring—as we feel intuitively to be the case—within the institutional process. It follows, further, that deviance—nonconformity to expectations—is a more intimate and normal element in institutional behavior than conformity theory would permit. And it also permits theoretical recognition of the possibility that, as Etzioni has suggested,[59] a great deal of organizational behavior is based, not on internalized norms and values, but on an expedient calculation of self-interests and of possible rewards and punishments. This, in turn, leaves open the theoretical possibility that non-legitimized power, as well as legitimized authority, may often be a controlling factor in institutional behavior.

ROLE CONFLICT AND CHANGE
AMONG THE KANURI

The final empirical study we shall sketch is explicitly based on an understanding of the modern systems approach, focusing as it does on a theory of "self-generating internal change." Ronald Cohen, an anthropologist, reports a theoretically well-organized analysis of his field study of role conflict and change among the Kanuri of Nigeria.[60] The study focuses on "goal ambiguity" and "conflicting standards" within a facet of the joint native-colonial political administrative hierarchy, particularly on the pivotal position of native "district head" which had come to combine the quite diverse cultural orientations of the colonial British and the Kanuri. This diversity between, as well as within, the two cultures made for inconsistencies, ambiguity, and conflict in political goals and in role standards and performances, which were continuously exacerbated by the variety of pressures put on district heads by the central native administration, the colonial administration, and the colonial technical departments.

The consequences of this situation for the political system are analyzed in terms of A. G. Frank's theory of organizational process and change.[61] Given the conditions of ambiguity and conflict of standards and goals, it is postulated that a process of *selective performance* and *selective enforcement* of standards will occur, with subordinates being forced to decide on which expectations to meet, and superiors required to selectively evaluate performances and hence selectively enforce some standards over others. This postulate leads to a number of predictions that Cohen proceeded to test. In essence, a continuous process is set up that appears, though in more exaggerated form, much like the "role strain," "role-making," "negotiated order" situations we met earlier. Role players fail to meet, or feign meeting some standards, and differentially select those they will meet. As a result, the role system is postulated to exhibit a strain toward substantive rationality (in Weber's sense), shifting standards for members, widespread role innovation or "deviance," ready adaptation to environmental changes, and an active and widespread circulation of information about standards and goals by "intermediary dealers in information" and by members seeking to reduce the ambiguity and conflict concerning these standards and goals.

The process is thus a circular, feedback loop whereby superiors continuously modify their standards or expectations as definitions of political objectives change, and subordinates adapt their decisions and performances to these changing expectations and surrounding circumstances, which in turn changes the states of the situation toward which superiors are acting. The role system, then, is seen as continuously receptive and responsive to external and internal pressures which demand some kind of workable "mapping" of the abundantly available situational "variety," which in turn makes possible—though does not guarantee—the evolution of more or less adaptive, institutionalized internal system procedures.

Applying this theory to the Kanuri, Cohen found the predictions to be borne out to a substantial degree. We leave the detailed description of these phenomena to the original study, which drew the general practical conclusion that—in spite of its apparent conservative, anti-progressive traditionalism—the Kanuri political role system showed greater compliance to the varied pressures of superiors and situational exigencies than to the tenets of tradition and thereby proved to be a self-generating system containing mechanisms for its own transformation. The implications of this

for policy relating to "developing countries" are of obvious importance.

On the theoretical side, Cohen clearly recognizes the implications of his mode of analysis for a genetic model of sociocultural evolution.

This model depends basically on two conditions. First, the evolving phenomenon must be shown to be *variable* in terms of its constituent units, and second, there must be analytically distinct *selective factors* which operate on the variation within the phenomenon to produce a constantly adapting and thus an evolving history of development. Although there are more or less stable orientations of tradition present in Bornu, conflicts in the political organization produce a variability of response by the actors upon which selective pressures exerted by superiors in the political hierarchy may operate to bring about innovations and changes that are incremental in their nature, i.e., evolutionary rather than revolutionary.[62]

We opened our discussion of the decision-making, process approach to complex adaptive systems with a turn-of-the-century prognosis of Albion Small. We might remind ourselves further of important ties with the past by closing with the early fundamental insight of Edward Sapir:

While we often speak of society as though it were a static structure defined by tradition, it is, in the more intimate sense, nothing of the kind, but a highly intricate network of partial or complete understandings between the members of organizational units of every degree of size and complexity. . . . It is only apparently a static sum of social institutions; actually it is being reanimated or creatively reaffirmed from day to day by particular acts of a communicative nature which obtain among individuals participating in it.[63]

Conclusion

We have suggested that much current thinking represents the coming to fruition of earlier conceptions of which Sapir's and Small's statements are harbingers. Although a science should not hesitate to forget its founders, it would do well to remain aware of their basic thought.

We have argued that a promising general framework for organizing these valuable insights of the past and present may be derived from the recent general systems perspective, embracing a holistic conception of complex adaptive systems viewed centrally in terms of information and communication process and the significance of the way these are structured for self-regulation and self-direction. We have clearly arrived at a point in the development of the "behavioral" sciences at which synthesis or conceptual unification of subdisciplines concerned with social life is challenging simple analysis or categorization. Not only is there growing demand that the "cognitive," "affective" and "evaluative" be conceptually integrated, but that the free-handed parceling out of aspects of the sociocultural adaptive system among the various disciplines (e.g., "culture" to anthropology, the "social system" to sociology, and "personality" to psychology) be reneged, or at least ignored. The potential of the newer system theory is especially strong in this regard.[64] By way of conclusion we recapitulate the main arguments.

1) The advance of science has driven it away from concern with "substance" and toward a focus on *relations* between components of whatever kind. Hence the concern with complex organization or systems, generally defined in terms of the transactions, often mutual and usually intricate, among a number of components such that some kind of more or less stable structure —often tenuous and only statistically delineated— arises (that is, *some* of the relations between components show *some* degree of stability or repetitiveness *some* of the time). Extremely fruitful advances have been taking place, especially since the rapid scientific progress made during World War II, in specifying basic features common to substantively different kinds of complex adaptive systems, as well as delineating their differences. In contrast to some of the general systems theorists themselves as well as their critics, we have argued that this is not simply analogizing, but generalizing or abstracting as well (although the former is important, and scientifically legitimate also, when performed with due caution). To say that physiological, psychological, and sociocultural processes of control all involve the basic cybernetic principles of information flow along feedback loops is no more a mere analogy than to say that the trajectories of a falling apple, an artificial satellite, or a planet all involve the basic principle of gravitational attraction.

2) Complex adaptive systems are open systems in intimate interchange with an environment characterized by a great deal of shifting variety ("booming, buzzing confusion") and its constraints (its structure of causal interrelations). The concept of equilibrium developed for closed physical systems is quite inappropriate and usually inapplicable to such a dynamic situation. Rather, a characteristic resultant is the elaboration of organization in the direction of the less probable and the less inherently stable.

Features common to substantively different complex adaptive systems can be conceptualized in terms of the perspective of information and control theory. "Information" in its most general sense is seen, not as a thing that can be transported, but as a selective interrelation or mapping between two or more subsets of constrained variety selected from larger ensembles. Information is thus transmitted or communicated as invariant constraint or structure in some kind of variety, such that subsystems with the appropriate matched internal ensembles, reacting to and acting upon the information, do so in a situation of decreased uncertainty and potentially more effective adaptation to the variety that is mapped. Unless mapping (encoding, decoding, correlating, understanding, etc.) occurs between two or more ensembles we do not have "information," only raw variety or noise.

In these terms, adaptive systems, by a continuous selective feedback interchange with the variety of the environment, come to make and preserve mappings on various substantive bases, which may be transmitted generationally or contemporaneously to other similar units. By means of such mappings (for example, via genes, instincts, learned events, culture patterns) the adaptive system may, if the mappings are adequate, continue to remain viable before a shifting environment. The transmission and accumulation of such information among contemporaneous adaptive systems (individuals) becomes more and more important at higher levels until it becomes the prime basis of linkage of components for the highest level sociocultural system.

Some of the more important differences between complex adaptive systems include the substantive nature of the components, the types of linkage of the components, the kinds and levels of feedback between system and environment, the degree of internal feedback of a system's own state (for example, "self-awareness"), the methods of transmission of information between subsystems and along generations, the degree of refinement and fidelity of mapping and information transfer, the degree and rapidity with which the system can restructure itself or the environmental variety, etc.

3) Such a perspective provides a general framework which meets the major criticisms leveled against much of current sociological theory: lack of time and process perspective, overemphasis on stability and maintenance of given structure, and on consensus and cooperative relations, to the relative neglect—or unsystematic

treatment—of deviance, conflict and other dissociative relations underlying system destructuring and restructuring.

4) Thus, the concept of the system itself cannot be identified with the more or less stable structure it may take on at any particular time. As a fundamental principle, it can be stated that a condition for maintenance of a viable adaptive system may be a change in its particular structure. Both stability and change are a function of the same set of variables, which must include both the internal state of the system and the state of its significant environment, along with the nature of the interchange between the two.

5) A time perspective is inherent in this kind of analysis—not merely historical but evolutionary. (It can probably be said that the time was ripe by 1959 for a Darwinian centennial ramifying well beyond the purely biological.) This perspective calls for a balance and integration of structural and processual analysis. As others have pointed out, the Linnean system of classification of structures became alive only after Darwin and others discovered the processes of variation, selection and recombination that gave them theoretical significance, though these discoveries leaned heavily in turn on the classification of systematically varying structures.

And among the important processes for the sociocultural system are not only cooperation and conformity to norms, but conflict, competition and deviation which may help create (or destroy) the essential variety pool, and which constitute part of the process of selection from it, such that a more or less viable system structure may be created and maintained (or destroyed).

6) In sociological terms, the "complementarity of expectations" model is an ideal type constituting only one pole of a continuum of equally basic associative and dissociative processes characterizing real societies—although the particular "mix" and intensities of the various types may differ widely with different structural arrangements. Further, the systemic analysis of a sociocultural system is not exhausted by analysis of its institutionalized patterns. By focusing on process, we are more prepared to include all facets of system operation—from the minimally structured end of the collective behavior continuum through the various degrees and kinds of structuring to the institutional pole. The particular characteristics of the process, especially the degrees and kinds of mappings and mismatchings of the interacting units, tell us whether we are in fact dealing with certain degrees of structuring and the dynamics

underlying this structuring: de facto patterning may be anchored in coercive, normative, or utilitarian compliance, making for very different kinds of system.

7) "Institutionalized" patterns are not to be construed as thereby "legitimized" or as embracing only "conformity" patterns—at least for the sake of conceptual clarity and empirical adequacy. Processes of all degrees and kinds of structuring may be seen in terms of deviant as well as conformity patterns—relative to the point of reference selected by the observer. One may select certain institutional patterns and values (to be clearly specified) as an arbitrary reference point to match against *other* institutional patterns and values, along with less structured behaviors. The concept of *the* institutionalized common value system smuggles in an empirically dubious, or unverified, proposition—at least for complex modern societies.

8) The complex adaptive system's organization *is* the "control," the characteristics of which will change as the organization changes. The problem is complicated by the fact that we are dealing directly with two levels of adaptive system and thus two levels of structure, the higher level (sociocultural) structure being largely a shifting statistical or probability structure (or ensemble of constraints) expressing over time the transactional processes occurring among the lower level (personality) structures. We do not have a sociocultural system *and* personality systems, but only a sociocultural system *of* constrained interactions among personality systems.

We can only speak elliptically of "ideas" or "information" or "meanings" in the head of a particular individual: all we have is an ensemble of constrained variety embodied in a neurological net. "Meaning" or "information" is generated only in the process of interaction with other ensembles of similarly mapped or constrained variety (whether embodied in other neurological nets or as the ensemble of causally constrained variety of the physical environment), whereby ensemble is mapped or matched against ensemble via communication links, and action is carried out, the patterning of which is a resultant of the degree of successful mapping that occurred. (Of course, "meaning" on the symbolic level can be regenerated over a long period by the isolated individual through an internal interchange or "conversation" of the person with his "self," made possible by previous socially induced mappings of one's own internal state that we call "self-awareness." But in some respects, part of the world literally loses its meaning for such a person.)

If the ensembles of variety of two interacting units, or one unit and its physical environment, have no or little isomorphic structuring, little or no meaning can be generated to channel ongoing mutual activity; or in more common terms, there is no "common ground," no "meeting of minds" and thus no meaning or information exchange— only raw variety, uncertainty, lack of "order" or organization.

Unless "social control" is taken as simply the more or less intentional techniques for maintaining a given institutional structure by groupings with vested interests, it must refer to the above transactional processes as they operate— now to develop new sociocultural structures, now to reinforce existing ones, now to destructure or restructure older ones. Thus, we cannot hope to develop our understanding much further by speaking of one "structure" determining, "affecting," or acting upon another "structure." We shall have to get down to the difficult but essential task of (a) specifying much more adequately the distribution of essential features of the component subsystems' internal mappings, including both self-mappings and their mappings of their effective environment, (b) specifying more extensively the structure of the transactions among these units at a given time, the degree and stability of the given structuring seen as varying with the degree and depth of common meanings that are generated in the transaction process, and (c) assessing, with the help of techniques now developing, the ongoing process of transitions from a given state of the system to the next in terms of the deviation-reducing and deviation-generating feedback loops relating the tensionful, goal-seeking, decision-making subunits via the communication nets partly specified by (b). Some behavior patterns will be found to be anchored in a close matching of component psychic structures (for example, legitimized authority or normative compliance); others, in threats of goal-blockage, where there is minimal matching (for example, power or coercive compliance); still others, anchored in a partial matching, primarily in terms of environmental mappings of autonomous subunits and minimally in terms of collective mappings (for example, opportunism or utilitarian compliance). As the distribution of mappings shifts in the system (which normally occurs for a number of reasons), so will the transaction processes and communication nets, and thus will the sociocultural structure tend to shift as gradients of misunderstanding, goal-blockage, and tensions develop.

9) Finally, we have tried to show how this perspective bears on, and may help to integrate conceptually, the currently developing area of "decision theory" which recognizes individual components as creative nodes in an interactive matrix. In the complex process of transactions occurring within a matrix of information flows, the resulting cognitive mappings and mismappings undergo various stresses and strains as component units assess and reassess with varying degrees of fidelity and refinement their internal states and the shifting and partially uncertain, and often goal-blocking environment. Out of this process, as more or less temporary adjustments, arises the more certain, more expected, more codified sequences of events that we call sociocultural structure. In the words of Norbert Wiener, "By its ability to make decisions" the system "can produce around it a local zone of organization in a world whose general tendency is to run down."[65] Whether that structure proves viable or adaptive for the total system is the kind of question that cannot be reliably answered in the present state of our discipline. It most certainly demands the kind of predictive power that comes with the later rather than the earlier stages of development of a science. And later stages can arrive only at some sacrifice of ideas of earlier stages.

Notes

1. *Sociology and Modern Systems Theory* (Englewood Cliffs, N.J.: Prentice-Hall, 1967).

2. See Pringle, Chapter 33; and Donald T. Campbell, "Methodological Suggestions from a Comparative Psychology of Knowledge Processes," *Inquiry*, 2 (1959), 152–67.

3. See, for example, Rapoport and MacKay selections, Chapters 16 and 24.

4. For an excellent recent overview of this transition, see A. Irving Hallowell, "Personality, Culture, and Society in Behavioral Evolution," in Sigmund Koch (Ed.), *Psychology: A Study of a Science*, Volume 6: Investigations of Man as Socius (New York: McGraw-Hill, 1963), 429–509.

5. Herbert A. Thelen, "Emotionality and Work in Groups," in Leonard D. White (Ed.), *The State of the Social Sciences* (Chicago: University of Chicago Press, 1956), pp. 184–86.

6. See Cannon selection, Chapter 32.

7. Or perhaps we might take Cadwallader's suggestion (Chapter 52) and use Ashby's term "ultrastability." I dislike, however, the connotative overemphasis on "stability," which is sure to be misunderstood by many. I prefer the term "morphogenesis" as best expressing the characteristic feature of the adaptive system. (See, for one, Maruyama's usage in Chapter 36.) Thus, we might say that physical systems are typically equilibrial, physiological systems are typically homeostatic,

and psychological, sociocultural, or ecological systems are typically morphogenic. From this view, our paradigm of the mechanisms underlying the complex system becomes a basic paradigm of the morphogenic process, perhaps embracing as special cases even the structuring process below the complex adaptive system level.

8. C. A. Mace, "Homeostasis, Needs and Values," *British Journal of Psychology*, 44 (1953), 204–205. Gordon Allport reinforces this view for personality (but note his terminology): "Some theories correctly emphasize the tendency of human personality to go beyond steady states and to elaborate their internal order, even at the cost of disequilibrium. Theories of changing energies . . . and of functional autonomy . . . do so. These conceptions allow for a continual increase of men's purposes in life and for their morphogenic effect upon the system as a whole. Although homeostasis is a useful conception for short-run 'target orientation,' it is totally inadequate to account for the integrating tonus involved in 'goal orientation.' . . . Although these formulations differ among themselves, they all find the 'go' of personality in some dynamic thrust that exceeds the pale function of homeostatic balance. They recognize increasing order over time, and view change within personality as a recentering, but not as abatement, of tension."—Gordon W. Allport, *Pattern and Growth in Personality* (New York: Holt, Rinehart & Winston, 1961), p. 569.

9. See Chapters 34 and 35.

10. Jules Henry, "Culture, Personality, and Evolution," *American Anthropologist*, 61 (1959) 221–22.

11. Karl W. Deutsch, "Some Notes on Research on The Role of Models in Natural and Social Science," *Synthese*, (1948–49), 532–33.

12. See Campbell, *op. cit.*, and Pringle, Chapter 33.

13. Recall Osgood's interpretation in Chapter 23. Also, see Charles E. Osgood, "Psycholinguistics," in S. Koch (Ed.), *Psychology, loc. cit.*, pp. 244–316; O. Hobart Mowrer, *Learning Theory and the Symbolic Processes* (New York: John Wiley, 1960), esp. Chapter 7: "Learning Theory, Cybernetics, and the Concept of Consciousness." For less behavioristic and more genetic and emergent views see, for example, George H. Mead, *Mind, Self and Society* (Chicago: University of Chicago Press, 1934), and more recently, Heinz Werner and Bernard Kaplan, *Symbol Formation* (New York: John Wiley, 1963).

14. See, for example, Marshall D. Sahlins, "Culture and Environment: The Study of Cultural Ecology," in Sol Tax (Ed.), *Horizons of Anthropology* (Chicago: Aldine, 1964), pp. 132–47.

15. See especially the selections above from Nett (Chapter 48), Deutsch (Chapter 46), Hardin (Chapter 55) and Vickers (Chapter 56).

16. However, we should not deemphasize the important structuring role of concrete artifacts, for example, the structure of physical communication nets, road nets, cities, interior layouts of buildings, etc., as limiting and channeling factors for sociocultural action and interaction.

17. Albion W. Small, *General Sociology* (Chicago: University of Chicago Press, 1905), p. ix.

18. Charles H. Cooley, *Social Process* (New York: Scribner's, 1918).

19. Florian Znaniecki, *Cultural Sciences* (Urbana: University of Illinois Press, 1952).

20. Consider the explicit "feedback" and "self-regulation" conceptions in the following statements of G. H. Mead in *Mind, Self and Society*: ". . . the central nervous system has an almost infinite number of elements in it, and they can be organized not only in spatial connection with each other, but also from a temporal standpoint. In virtue of this last fact, our conduct is made up of a series of steps which follow each other, and the later steps may be already started and influence the earlier ones. The thing we are going to do is playing back on what we are doing now" (p. 71). "As we advance from one set of responses to another we find ourselves picking out the environment which answers to this next set of responses. To finish one response is to put ourselves in a position where we see other things. . . . Our world is definitely mapped out for us by the responses which are going to take place. . . . The structure of the environment is a mapping out of organic responses to nature; any environment, whether social or individual, is a mapping out of the logical structure of the act to which it answers, an act seeking overt expression" (pp. 128–29, and footnote 32, p. 129). "It is through taking this role of the other that [the person] is able to come back on himself and so direct his own process of communication. This taking the role of the other, an expression I have so often used, is not simply of passing importance. It is not something that just happens as an incidental result of the gesture, but it is of importance in the development of cooperative activity. The immediate effect of such role-taking lies in the control which the individual is able to exercise over his own response. . . . From the standpoint of social evolution, it is this bringing of any given social act, or of the total social process in which that act is a constituent, directly and as an organized whole into the experience of each of the individual organisms implicated in that act, with reference to which he may consequently regulate and govern his individual conduct, that constitutes the peculiar value and significance of self-consciousness in those individual organisms" (p. 254, including part of footnote 7).

See also the extended discussion based on Mead's essentially cybernetic perspective in Shibutani, Chapter 39.

21. Melvin Seeman, review of Arnold Rose (Ed.), *Human Behavior and Social Processes*, in *American Sociological Review*, 27 (August 1962), 557.

22. George P. Murdock, "Changing Emphasis in Social Structure," *Southwestern Journal of Anthropology*, 11 (1955), 366.

23. Raymond Firth, "Some Principles of Social Organization," *Journal of the Royal Anthropological Institute*, 85 (1955), 1.

24. S. F. Nadel, *The Theory of Social Structure* (New York: The Free Press of Glencoe, 1957), pp. 54–55.

25. *Ibid.*, p. 128.

26. *Ibid.*, p. 153.

27. Evon Z. Vogt, "On the Concept of Structure and Process in Cultural Anthropology," *American Anthropologist*, 62 (1960), 18–33.

28. Herbert Blumer, "Psychological Import of the Human Group," in Muzafer Sherif and M. O. Wilson (Eds.), *Group Relations at the Crossroads* (New York: Harper, 1953), pp. 199–201. Emphasis added.

29. Gordon W. Allport, *Pattern and Growth in Personality* (New York: Holt, Rinehart, & Winston, 1961), p. 567. Also recall his discussion in Chapter 41.

30. For example, see Alvin W. Gouldner, "Some Observations on Systematic Theory, 1945–55," in Hans L. Zetterberg (Ed.), *Sociology in the United States of America* (Paris: UNESCO, 1956), pp. 39–40. See also Barrington Moore, Jr., "Sociological Theory and Contemporary Politics," *American Journal of Sociology*, 61 (September, 1955), 111–15.

31. Recall, for example, the excellent treatment of "decision theory" in Robert M. MacIver, *Social Causation* (New York: Ginn & Co., 1942), esp. pp. 291 ff.

32. Philip Selznick, "Review Article: The Social Theories of Talcott Parsons," *American Sociological Review*, 26 (December, 1961), 934.

33. Kenneth E. Boulding, *The Image* (Ann Arbor: University of Michigan Press, 1956).

34. Recall Maruyama's discussion in Chapter 36.

35. William J. Goode, "A Theory of Role Strain," *American Sociological Review*, 25 (August, 1960), 483.

36. *Ibid.*, 495.

37. *Ibid.*

38. *Ibid.*, 487.

39. *Ibid.*, 494.

40. Charles E. Osgood, "Cognitive Dynamics in the Conduct of Human Affairs," *Public Opinion Quarterly*, 24 (1960), 347.

41. *Ibid.*, 347–48.

42. *Ibid.*, 351.

43. *Ibid.*, 363.

44. Robert R. Sears, "A Theoretical Framework for Personality and Social Behavior," *American Psychologist*, 6 (1951), 478–79.

45. *Ibid.*, 478.

46. Paul F. Secord and Carl W. Backman, "Personality Theory and the Problem of Stability and Change in Individual Behavior: An Interpersonal Approach," *Psychological Review*, 68 (1961), 22.

47. *Ibid.*

48. *Ibid.*, 26.

49. *Ibid.*, 28.

50. *Ibid.*

51. Ralph H. Turner, "Role-Taking: Process Versus Conformity," in Arnold M. Rose (Ed.), *Human Behavior and Social Processes* (Boston: Houghton Mifflin Co., 1962), Chapter 2.

52. *Ibid.*, 26.

53. Neal Gross *et al.*, *Explorations in Role Analysis* (New York: John Wiley, 1958), p. 53.

54. Turner, *loc. cit.*, p. 32.

55. *Ibid.*, p. 38.

56. Anselm Strauss *et al.*, "The Hospital and Its Negotiated Order," in Eliot Freidson (Ed.), *The Hospital in Modern Society* (New York: The Free Press of Glencoe, 1963), p. 148.

57. *Ibid.*, p. 162.

58. Neal Gross *et al.*, *op. cit.*, p. 321. Also see Robert L. Kahn *et al.*, *Organizational Stress: Studies in Role Conflict and Ambiguity* (New York: John Wiley, 1964).

59. Amitai Etzioni, *A Comparative Analysis of Complex Organizations* (New York: The Free Press of Glencoe, 1961).

60. Ronald Cohen, "Conflict and Change in a Northern Nigerian Emirate," in George K. Zollschan and Walter Hirsch (Ed.), *Explorations in Social Change* (Boston: Houghton Mifflin Co., 1964), Chapter 19.

61. A. G. Frank, "Goal Ambiguity and Conflicting Standards: An Approach to the Study of Organization," *Human Organization*, 17 (1959), 8–13.

62. *Op. cit.*, 519. Emphasis supplied.

63. Edward Sapir, "Social Communication," *Encyclopedia of the Social Sciences* (New York: Macmillan, 1931), Vol. 4, p. 78.

64. This still remains primarily a potential, however; perusal of the general systems literature shows treatment of the sociocultural level systems to be sparse compared to that of biological, psychological and other systems. Part of the reason for this is the failure of sociologists to participate and to make what could be significant contributions to a field rapidly leaving us behind.

65. Norbert Wiener, *The Human Use of Human Beings* (Garden City, N.Y.: Doubleday Anchor, 2d ed., rev., 1954), p. 34.

Selected References

ACKOFF, RUSSELL L., 1957–58, Towards a Behavioral Theory of Communication, *Management Science*, 4: 218–34.

—— (Ed.), 1961, *Progress in Operations Research*, I. New York: John Wiley.

ALLPORT, GORDON W., 1960, The Open System in Personality Theory, *Journal of Abnormal and Social Psychology*, 61: 301–311.

ANGYAL, ANDRAS, 1939, The Structure of Wholes, *Philosophy of Science*, 6: 25–37.

ARBIB, MICHAEL A., 1964, *Brains, Machines, and Mathematics*. New York: McGraw-Hill.

——, 1966, A Partial Survey of Cybernetics in Eastern Europe and the Soviet Union, *Behavioral Science*, 11: 193–216.

ASHBY, W. ROSS, 1954, The Application of Cybernetics to Psychiatry, *Journal of Mental Science*, 100: 114–24.

——, 1956, *An Introduction to Cybernetics*. New York: John Wiley.

——, 1960, *Design for a Brain*. New York: John Wiley.

——, 1963, Induction, Prediction, and Decision-Making in Cybernetic Systems, in Henry Ely Kyburg, Jr., and Ernest Nagel (Eds.), *Induction: Some Current Issues*. Middletown, Conn.: Wesleyan University Press, 55–73.

——, 1964, The Set Theory of Mechanism and Homeostasis, *General Systems*, 9: 83–97.

ATTNEAVE, FRED, 1959, *Applications of Information Theory to Psychology*. New York: Holt, Rinehart & Winston.

BARDIS, P. D., 1965, Cybernetics: Definition, History, Etymology, *Social Science*, 226–28.

BAR-HILLEL, YEHOSHUA, 1964, *Language and Information*. Reading, Mass.: Addison-Wesley.

BARRETT, F. D., and H. A. SHEPARD, 1953, A Bibliography of Cybernetics, *Proceedings of the American Academy of Arts and Sciences*, 80: 204–222.

BEAUREGARD, O. C. de, 1961, Sur l'Equivalence Entre Information et Entropie, *Sciences*, 11: 51–58.

BECKNER, MORTON, 1959, *The Biological Way of Thought*. New York: Columbia University Press.

BEER, STAFFORD, 1959, *Cybernetics and Management*. London: English Universities Press.

BENTLEY, A. F., 1950, Kennetic Inquiry, *Science*, 112: 775–83.

BERKELEY, E. C., 1949, *Giant Brains*. New York: John Wiley.

BERLO, DAVID K., 1960, *The Process of Communication*. New York: Holt, Rinehart & Winston.

BERTALANFFY, L. von, 1950, The Theory of Open Systems in Physics and Biology, *Science*, 3: 23–29.

——, 1951, Problems of General System Theory, *Human Biology*, 23: 302–312.

——, 1952a, *Problems of Life: An Evaluation of Modern Biological Thought*. New York: John Wiley.

——, 1952b, Theoretical Models in Biology and Psychology, in David Krech and George S. Klein (Eds.), *Theoretical Models and Personality Theory*. Durham, N.C.: Duke University Press.

——, 1962, General System Theory—A Critical Review, *General Systems*, Yearbook of the Society for General Systems Research, 7: 1–20.

BLAKE, D. V., and A. M. UTTLEY (Eds.), 1959, *Proceedings of a Symposium on Mechanization of Thought Processes*, 2 vols. London: Her Majesty's Stationery Office.

BLUMER, HERBERT, 1966, Sociological Implications of the Thought of George Herbert Mead, *American Journal of Sociology*, 71: 535–44.

BOCK, KENNETH E., 1963, Evolution, Function, and Change, *American Sociological Review*, 28: 229–37.

BOULDING, KENNETH, 1956a, General Systems Theory—The Skeleton of Science, *Management Science*, 2: 197–208.

——, 1956b, *The Image*. Ann Arbor: University of Michigan Press.

BRILLOUIN, LEON, 1949, Life, Thermodynamics, and Cybernetics, *American Scientist*, 37: 554–67, 568.

——, 1950, Thermodynamics and Information Theory, *American Scientist*, 38: 594–99.

——, 1956, *Science and Information Theory*. New York: Academic Press.

BRITISH ASSOCIATION, *Symposium on Cybernetics*: E. C. Cherry, Organisms and Mechanisms; W. E. Hick, The Impact of Information Theory on Psychology; and Donald M. MacKay, On Comparing the Brain with Machines, *Advancement of Science*, 40 (March, 1954).

BUCKLEY, WALTER, 1967, *Sociology and Modern Systems Theory*. Englewood Cliffs, N.J.: Prentice-Hall.

BUKHARIN, NIKOLAI, 1925, *Historical Materialism: A System of Sociology*. New York: International Publishers.

BURGERS, J. M., 1963, On the Emergence of Patterns of Order, *Bulletin of the American Mathematics Society*, 69: 1–25.

CADWALLADER, MERVIN, 1959, The Cybernetic Analysis of Change in Complex Social Organizations, *American Journal of Sociology*, 65: 154–57.

CAMPBELL, DONALD T., 1956, Adaptive Behavior from Random Response, *Behavioral Science*, 1: 105–110.

——, 1958a, Common Fate, Similarity, and Other Indices of the Status of Aggregates of Persons as Social Entities, *Behavioral Science*, 3: 14–25.

——, 1958b, Systematic Error on the Part of Human Links in Communication Systems, *Information and Control*, 1: 334–69.

——, 1959, Methodological Suggestions from a Comparative Psychology of Knowledge Processes, *Inquiry*, 2: 152–67.

——, 1960, Blind Variation and Selective Retention in Creative Thought as in Other Knowledge Processes, *Psychological Review*, 67: 380–400.

——, 1965, Variation and Selective Retention in Socio-Cultural Evolution, in Herbert R. Barringer, George I. Blanksten, and Raymond Mack (Eds.), *Social Change in Developing Areas*. Cambridge, Mass.: Schenkman Publishing Co., 19–49.

CANNON, WALTER B., 1939, *The Wisdom of the Body*. New York: W. W. Norton.

CATTELL, JAQUES (Ed.), 1942, *Biologica Symposia*, VIII: Levels of Integration in Biological and Social Systems, Robert Redfield (Ed.). Lancaster, Pa.: Jaques Cattell Press.

CHAPANIS, ALPHONSE, 1961, Men, Machines, and Models, in Donald P. Eckman (Ed.), *Systems Philosophy*. New York: John Wiley.

CHERRY, COLIN, 1961, *On Human Communication*. New York: Science Editions.

CHIN, ROBERT, 1962, The Utility of System Models and Developmental Models for Practitioners, in Warren Bennis *et al.* (Eds.), *The Planning of Change*. New York: Holt, Rinehart & Winston, 201–214.

CHURCHMAN, C. W., and R. L. ACKOFF, 1950, Purposive Behavior and Cybernetics, *Social Forces*, 29: 32–39.

CLARK, JOSEPH T., 1962, Remarks on the Role of Quantity, Quality, and Relations in the History of Logic, Methodology, and Philosophy of Science, in Ernest Nagel *et al.* (Eds.), *Logic Methodology, and Philosophy of Science*. Stanford, Calif.: Stanford University Press.

COHN, STANTON H., and SYLVA M. COHN, 1953, The Role of Cybernetics in Physiology. *The Scientific Monthly*, 79: 85–89.

COPI, I. M., C. C. ELGOT, and J. B. WRIGHT, 1958, Realization of Events by Logical Nets, *Journal for the Association of Computing Machinery*, 5: 181–96.

CRIDER, DONALD B., 1956–57, Cybernetics: A Review of What It Means and Some of Its Implications in Psychiatry, *Neuropsychiatry*, 4: 35–58.

CULBERTSON, J. T., 1951, *Consciousness and Behavior*. London: William C. Brown.

Cybernetica. Vol. I, 1958—

Cybernetics, English translation of *Kibernetica*, published in Kiev beginning January, 1965.

DEUTSCH, KARL W., 1951a, Mechanism, Organism, and Society, *Philosophy of Science*, 18: 230–52.

——, 1951b, Mechanism Teleology, and Mind, *Philosophy and Phenomenological Research*, 12: 185–223.

——, 1952a, Communication Theory and Social Science, *American Journal of Orthopsychiatry*, 22: 469–83.

——, 1952b, On Communication Models in the Social Sciences, *Public Opinion Quarterly*, 16: 356–80.

——, 1955, Some Notes on Research on the Role of Models in Natural and Social Sciences, *Synthese*, 7: 506–533.

——, 1961, On Social Communication and the Metropolis, *Daedalus*, Vol. 90, *The Future Metropolis*, 99–110.

——, 1963, *The Nerves of Government*. New York: Free Press of Glencoe.

DEWEY, JOHN, and ARTHUR F. BENTLEY, 1949, *Knowing and the Known*. Boston: Beacon Press.

DORE, RONALD P., 1961, Function and Cause, *American Sociological Review*, 26: 843–53.

EASTON, DAVID, 1956, Limits of the Equilibrium Model in Social Research, *Behavioral Science*, 1: 96–104.

——, 1965, *A Systems Analysis of Political Life*. New York: John Wiley.

ECKMAN, DONALD P. (Ed.), 1961, *Systems: Research and Design*. New York: John Wiley.

FAIRBANKS, G., 1954, A Theory of the Speech Organism as a Servosystem, *Journal of Speech and Hearing Disorders*, 19: 133–39.

FEIBLEMAN, JAMES, and JULIUS WEIS FRIEND, 1945, The Structure and Function of Organization, *The Philosophical Review*, 54: 19–44.

——, 1954, Theory of Integrative Levels, *British Journal for the Philosophy of Science*, 5: 59–66.

FENDER, DEREK H., 1964, Control Mechanisms of the Eye, *Scientific American*, 211: 24–34.

FIELDS, WILLIAM S., and WALTER ABBOTT (Eds.), 1963, *Information Storage and Neural Control*, Tenth Annual Scientific Meeting of the Houston Neurological Society. Springfield, Ill.: Charles C Thomas.

FOERSTER, H. von (Ed.), 1953, *Cybernetics*. New York: Charles Macy.

——, MARGARET MEAD, and H. L. TEUBER, 1949–1957, *Transactions of Conference on Cybernetics*, 5 vols. New York: Josiah Macy, Jr., Foundation.

——, and GEORGE W. ZOPF, Jr. (Eds.), 1962, *Principles of Self-Organization*. New York: Pergamon Press.

FRICK, FREDERICK C., 1959, Information Theory, in Sigmund Koch (Ed.), *Psychology: A Study of a Science*, Vol. 2. New York: McGraw-Hill.

FÜRTH, R., 1952, Physics of Social Equilibrium, *Advancement of Science*, London, 8: 429–34.

GABOR, D., 1951, *Lectures on Communication Theory*. Cambridge, Mass.: M.I.T. Press.

——, 1954, Communication Theory and Cybernetics (Milan Symposium paper), in *Trans. I. R. E. Prof. Group on Non-Linear Circuits*.

GARNER, W. R., 1962, *Uncertainty and Structure as Psychological Concepts.* New York: John Wiley.

General Systems: Yearbook of the Society for General Systems Research, Ludwig von Bertalanffy and Anatol Rapoport (Eds.), Vol. 1, 1956 —.

GEORGE, F. H., and J. H. HANDLON, 1955, Towards a General Theory of Behavior, *Methodos*, 7: 24–44.

——, 1957, A Language for Perceptual Analysis, *Psychological Review*, 64: 14–25.

——, 1960, *Automation, Cybernation, and Society.* London: Leonard Hill.

——, 1965, *Cybernetics and Biology.* Edinburgh: Oliver & Boyd.

GERARD, RALPH W., 1957, Units and Concepts of Biology, *Science*, 125: 429–33.

——, 1960, Becoming: the Residue of Change, in S. Tax (Ed.), *Evolution after Darwin.* Vol. II, *The Evolution of Man.* Chicago: University of Chicago Press.

——, C. KLUCKHOHN, and A. RAPOPORT, 1956, Biological and Cultural Evolution, *Behavioral Science*, 1: 24–33.

GOLDMAN, STANFORD, 1953, *Information Theory.* New York: Prentice-Hall.

GOODE, WILLIAM J., 1960, A Theory of Role Strain, *American Sociological Review*, 25: 483–96.

GRINKER, ROY R. (Ed.), 1956, *Toward a Unified Theory of Human Behavior.* New York: Basic Books.

GUILAUD, G. T., 1959, *What Is Cybernetics?* New York: Grove Press.

HABERSTROH, CHADWICK J., 1960, Control as an Organizational Process, *Management Science*, 6: 165–71.

HALL, A. D., and R. E. FAGEN, 1956, Definition of Systems, revised introductory chapter of *Systems Engineering.* New York: Bell Telephone Laboratories. Reprinted from *General Systems* 1: 18–28.

HARDIN, GARRETT, 1959, *Nature and Man's Fate.* New York: Holt, Rinehart & Winston.

——, 1963, The Cybernetics of Competition: A Biologist's View of Society, *Perspectives in Biology and Medicine*, 7: 61–84.

HARTLEY, R. V. L., 1928, Transmission of Information, *Bell Systems Tech. J.*, 7: 535–63.

HEBB, D. O., 1949, *The Organization of Behaviour.* New York: John Wiley.

HELD, RICHARD, 1963a, Plasticity in Sensorimotor Systems, *Science*, 142: 455–62.

——, and ALAN HEIN, 1963b, Movement-Produced Stimulation in the Development of Visually Guided Behavior, *Journal of Comparative and Physiological Psychology*, 56: 872–76.

HENDERSON, L. J., 1935, *Pareto's General Sociology.* Cambridge, Mass.: Harvard University Press.

——, 1958, *The Fitness of the Environment: An Inquiry into the Biological Significance of the Properties of Matter.* Boston: Beacon Press.

HENRY, JULES, 1955, Homeostasis, Society, and Evolution: A Critique, *The Scientific Monthly*, 81: 300–309.

——, 1959, Culture, Personality, and Evolution, *American Anthropologist*, 61: 221–26.

HERBERT, P. G., 1957, Situation Dynamics and the Theory of Behavior Systems, *Behavioral Science*, 2: 13–29.

JACKSON, WILLIS (Ed.), 1953, *Communication Theory.* New York: Academic Press.

JEFFRESS, L. A. (Ed.), 1951, *Cerebral Mechanisms and Behaviour*, The Hixon Symposium. New York: John Wiley.

JOHNSON, F. CRAIG, and GEORGE R. KLARE, 1961, General Models of Communication Research, *Journal of Communication*, 11: 13–26.

——, ——, 1962, Feedback: Principles and Analogies, *Journal of Communication*, 12: 150–59.

KAPLAN, DAVID, 1965, The Superorganic: Science or Metaphysics?, *American Anthropologist*, 67: 958–76.

KATZ, KARL U., and MARGARET F. SMITH, 1966, *Cybernetic Principles of Learning and Educational Design.* New York: Holt, Rinehart & Winston.

KHAILOV, K. M., 1964, The Problem of Systemic Organization in Theoretical Biology, *General Systems*, 9: 151–57. Translated by A. Rapoport.

KILPATRICK, FRANKLIN P., 1961, *Explorations in Transactional Psychology.* New York: NYU Press.

KRECH, DAVID, 1950, Dynamic Systems as Open Neurological Systems, *Psychological Review*, 57: 345–61.

KREMYANSKIY, V. I., 1960, Certain Peculiarities of Organisms as a "System" from the Point of View of Physics, Cybernetics, and Biology, *General Systems*, 5: 221–24. Translated by Anatol Rapoport.

KUHN, ALFRED, 1961, Toward a Uniform Language of Information and Knowledge, *Synthese*, 13: 127–53.

——, 1963, *The Study of Society: A Unified Approach.* Homewood, Ill.: Irwin—Dorsey.

Kybernetica, published by Cybernetics Commission of the Czechoslavak Academy of Sciences, starting January, 1965.

Kybernetik. Vol. I, 1961—.

LANGE, OSKAR, 1965, *Wholes and Parts: A General Theory of System Behavior.* Oxford: Pergamon Press. Translated by Eugeniusz Lejsa.

LATIL, PIERRE de, 1956, *Thinking by Machine: A Study of Cybernetics.* London: Sedgwick and Jackson. Translated by Y. M. Golla.

LEAVITT, HAROLD J., and RONALD A. H. MUELLER, 1951, Some Effects of Feedback on Communication, *Human Relations*, 4: 401–410.

LEHRMAN, DANIEL S., 1964, The Reproductive Behavior of Ring Doves, *Scientific American*, 211: 48–54

LERNER, DANIEL (Ed.), 1963, *Parts and Wholes.* New York: Free Press of Glencoe.

LEWIN, KURT, 1947, Feedback Problems of Social Diagnosis and Action, Part II-B of Frontiers in Group Dynamics, *Human Relations*, 1: 147–53.

LONDON, IVAN D., 1946, Some Consequences for History and Psychology of Langmuir's Concept of Convergence and Divergence of Phenomena, *Psychological Review*, 53: 170–88.

LUCE, R. DUNCAN, and H. RAIFFA, 1957, *Games and Decisions.* New York: John Wiley.

McCLELLAND, CHARLES A., 1962, General Systems and the Social Sciences, *ETC*, 18: 449–68.

McCULLOCH, WARREN S., 1965, *Embodiments of Mind.* Cambridge, Mass.: M.I.T. Press.

——, 1949, The Brain as a Computing Machine, *Electronic Engineering.*

MACE, C. A., 1953, Homeostasis, Needs, and Values, *British Journal of Psychology*, 44: 200–210.

MacIver, Robert M., 1942, *Social Causation*. New York: Ginn & Co.

MacKay, Donald M., 1951, Mindlike Behaviour in Artefacts, *British Journal of Philosophy and Science*, 2: 105–121.

——, 1954, Operational Aspects of Some Fundamental Concepts of Human Communication, *Synthese*, 9: 182–98.

——, 1956a, Towards an Information-Flow Model of Human Behavior, *British Journal of Psychology*, 47: 30–43.

——, 1956b, The Place of "Meaning" in the Theory of Information, in Colin Cherry (Ed.), *Information Theory: Third London Symposium*. New York: Academic Press.

——, 1961, The Informational Analysis of Questions and Commands, in Colin Cherry (Ed.), *Information Theory: Fourth London Symposium*. London: Butterworth's.

Maclay, Howard, 1962, A Descriptive Approach to Communication, in Norman F. Washburne (Ed.), *Decisions, Values, and Groups*, Vol. II. New York: Pergamon Press.

Macrae, Donald G., 1951, Cybernetics and Social Science, *British Journal of Sociology*, 2: 135–49.

March, James G., and H. A. Simon, 1958, *Organizations*. New York: John Wiley.

Maruyama, Magoroh, 1963, The Second Cybernetics: Deviation Amplifying Mutual Causal Processes, *American Scientist*, 51: 164–79; Postscript, 51: 250A–56A.

Mead, George Herbert, 1934, *Mind, Self, and Society*. Chicago: University of Chicago Press.

——, 1938, *The Philosophy of the Act* (C. W. Morris, Ed.). Chicago: University of Chicago Press.

Meier, Richard L., 1956, Communication and Social Change, *Behavioral Science*, 1: 43–59.

Mesarović, Mihajlo D. (Ed.), 1964, *Views on General Systems Theory: Proceedings of the Second Systems Symposium at Case Institute of Technology*. New York: John Wiley.

Miller, George A., 1953, What is Information Measurement?, *American Psychologist*, 8: 3–12.

——, 1956, The Magical Number Seven, Plus or Minus Two: Some Limits on Our Capacity for Processing Information, *Psychological Review*, 63: 81–97.

——, E. Galanter, and K. H. Pribram, 1960, *Plans and the Structure of Behavior*. New York: Holt, Rinehart & Winston.

Miller, James G., 1955, Toward a General Theory for the Behavioral Sciences, *American Psychologist*, 10: 513–31.

——, 1965, Living Systems: Basic Concepts; Structure and Process; Cross-Level Hypotheses, *Behavioral Science*, 10: 193–237, 337–79, 380–411.

Moore, Omar K., and Donald J. Lewis, 1953, Purpose and Learning Theory, *The Psychological Review*, 60: 149–56.

Moulyn, Adrian C., 1957, *Structure, Function, and Purpose*. New York: Liberal Arts Press.

Mowrer, Orval H., 1954, Ego Psychology, Cybernetics, and Learning Theory, in Donald K. Adams *et al.* (Eds.), *Learning Theory and Clinical Research*. New York: John Wiley.

——, 1960, *Learning Theory and the Symbolic Processes*. New York: John Wiley.

Murphy, Gardner, 1961, Toward a Field Theory of Communication, *Journal of Communication*, 11: 196–201.

Nadel, S. F., 1953, Social Control and Self-Regulation, *Social Forces*, 31: 265–73.

——, 1957, *The Theory of Social Structure*. New York: The Free Press of Glencoe.

Nett, Roger, 1953, Conformity-Deviation and the Social Control Concept, *Ethics*, 64: 38–45.

Neumann, John von, 1952, *Probabilistic Logics*. Pasadena: California Institute of Technology.

——, 1958, *The Computer and the Brain*. New Haven, Conn.: Yale University Press.

——, and O. Morgenstern, 1947, *Theory of Games and Economic Behavior*. Princeton, N.J.: Princeton University Press.

Newell, Allen, J. C. Shaw, and Herbert A. Simon, 1958, Elements of a Theory of Human Problem Solving, *Psychological Review*, 65: 151–66.

Nokes, Peter, 1961, Feedback as an Explanatory Device in the Study of Certain Interpersonal and Institutional Processes, *Human Relations*, 14: 381–87.

Notterman, Joseph M., and Richard Trumbull, 1959, Note on Self-Regulating Systems and Stress, *Behavioral Science*, 4: 324–27.

Osgood, Charles E., 1952, On the Nature and Measurement of Meaning, *Psychological Bulletin*, 49: 197–237.

——, 1957, A Behavioristic Analysis of Perception and Language as Cognitive Phenomena, in Jerome S. Bruner *et al.* (Eds.), *Contemporary Approaches to Cognition: The Colorado Symposium*. Cambridge, Mass.: Harvard University Press, 75–118.

——, and Thomas A. Sebeck (Eds.), 1954, Psycholinguistics, a Survey of Theory and Research Problems, *International Journal of American Linguistics*, Memoir 10.

Ostow, Mortimer, 1951, The Entropy Concept and Psychic Function, *American Scientist*, 39: 140–44.

Pask, Gordon, 1961, *An Approach to Cybernetics*. New York: Harper & Bros.

Penrose, L. S., 1959, Self-Reproducing Machines, *Scientific American*, 200: 105–114.

Pierce, J. R., 1961, *Symbols, Signals, and Noise*. New York: Harper & Row.

Pitts, W., and W. S. McCulloch, 1947, How We Know Universals, The Perception of Auditory and Visual Forms, *Bulletin of Mathematical Biophysics*, 9: 127–47.

Powers, W. T., R. K. Clark, and R. I. McFarland, 1960, A General Feedback Theory of Human Behavior, *Perceptual and Motor Skills*, 11: 71–88.

Pringle, J. W. S., 1951, On the Parallel Between Learning and Evolution, *Behavior*, 3: 174–215.

Quastler, H. (Ed.), 1953, *Information Theory in Biology*. Urbana: University of Illinois Press.

——, 1955, *Information Theory in Psychology*. Glencoe, Ill.: Free Press.

Rapoport, Anatol, 1953, What is Information? *ETC*, 10: 247–60.

——, 1956, The Promise and Pitfalls of Information Theory, *Behavioral Science*, 1: 303–309.

——, 1959a, Critiques of Game Theory, *Behavioral Science*, 4: 49–66.

——, 1959b, Mathematics and Cybernetics, in

Sylvano Arieti (Ed.), *American Handbook of Psychiatry*, Vol. II. New York: Basic Books, 1743–59.

———, 1965, Game Theory and Human Conflict, in Elton B. McNeil (Ed.), *The Nature of Human Conflict*. Englewood Cliffs, N.J.: Prentice-Hall.

———, 1966, Some System Approaches to Political Theory, in David Easton (Ed.), *Varieties of Political Theory*. Englewood Cliffs, N.J.: Prentice-Hall.

———, and WILLIAM J. HORVATH, 1959, Thoughts on Organization Theory . . ., *General Systems*, 4: 87–91.

RASHEVSKY, N., 1955, Is the Concept of an Organism as a Machine a Useful One?, *Scientific Monthly*, 50: 32–35.

RAYMOND, RICHARD C., 1950, Communication, Entropy, and Life, *American Scientist*, 38: 273–78.

REDFIELD, ROBERT (Ed.), 1942, *Levels of Integration in Biological and Social Systems*. Lancaster, Pa.: Jaques Cattell Press.

RESCHER, NICHOLAS, 1963, Discrete State Systems, Markoff Chains, and Problems in the Theory of Scientific Explanation and Prediction, *Philosophy of Science*, 30: 325–45.

ROMAIN, JACQUES, 1959, Information et Cybérnetique, *Cybernetica*, 2: 23–50.

ROSENBERG, SEYMOUR, and ROBERT L. HALL, 1958, The Effects of Different Social Feedback Conditions upon Performance in Dyadic Teams, *Journal of Abnormal and Social Psychology*, 57: 271–77.

ROSENBLUETH, ARTURO, NORBERT WIENER, and JULIAN BIGELOW, 1943, Purpose and Teleology, *Philosophy of Science*, 10: 18–24.

———, ———, 1950, Purposeful and Non-Purposeful Behavior, *Philosophy of Science*, 17: 318–26.

ROTHSTEIN, JEROME, 1958, *Communication, Organization, and Science*. Indian Hills, Colo.: Falcon's Wing Press.

———, 1962, Discussion: Information and Organization as the Language of the Operational Viewpoint, *Philosophy of Science*, 29: 406–411.

RUESCH, JURGEN, 1952, The Therapeutic Process from the Point of View of Communication Theory, *American Journal of Orthopsychiatry*, 22: 690–701.

SAPIR, E., 1939. *Language*. New York: Harcourt, Brace.

SAPORTA, SOL (Ed.), 1961, *Psycholinguistics*. New York: Holt, Rinehart & Winston.

SCHEFF, THOMAS J., 1966, *Being Mentally Ill*. Chicago: Aldine.

SCHELLING, THOMAS C., 1960, *The Strategy of Conflict*. Cambridge, Mass.: Harvard University Press.

SCHRAMM, WILBUR, 1955, Information Theory and Mass Communication, *Journalism Quarterly*, 32: 141–46.

SCHRÖDINGER, ERWIN, 1945, *What Is Life?* Cambridge: Cambridge University Press.

SCHWEITZER, A. L. M., 1963, Sociologie en Cybernetica, *Mens en Maatschappij*, 38: 351–67.

Scientific American, 1955, *Automatic Control*. New York: Simon & Schuster.

SCOTT, WILLIAM ABBOTT, 1962, Cognitive Structure and Social Structure: Some Concepts and Relationships, in Norman Washburne (Ed.), *Decisions, Values, and Groups*, Vol. II. New York: Pergamon Press.

SELYE, H., 1956, *The Stress of Life*. New York: McGraw-Hill.

SEWARD, JOHN P., 1963, The Structure of Functional Autonomy, *American Psychologist*, 18: 703–710.

SHANNON, CLAUDE E., and J. McCARTHY (Eds.), 1956, *Automata Studies*. Princeton, N.J.: Princeton University Press.

———, and WARREN WEAVER, 1949, *The Mathematical Theory of Communication*. Urbana: University of Illinois Press.

SIMON, H. A., 1962, The Architecture of Complexity, *Proc. Amer. Phil. Soc.*, 106: 467–82.

SLACK, CHARLES W., 1955, Feedback Theory and the Reflex Arc Concept, *Psychological Review*, 62: 263–67.

SLUCKIN, W., 1960, *Minds and Machines*. Baltimore: Penguin Books.

SOMMERHOFF, G., 1950, *Analytical Biology*. London: Oxford University Press.

STANLEY-JONES, D., and K. STANLEY-JONES, 1960, *The Kybernetics of Natural Systems: A Study in Patterns of Control*. New York: Pergamon Press.

STRAUSS, ANSELM L. *et al.*, 1963, The Hospital and Its Negotiated Order, in Eliot Freidson (Ed.), *The Hospital in Modern Society*. New York: Free Press of Glencoe, 147–69.

SZILARD, L., 1929, Über die Entropieverminderung in einem Thermodynamischen System bei Eingriffen Intelligenter Wesen. *Zeitschr. f. Phys.*, 53: 840–56. Translated by A. Rapoport and M. Knoller as On the Increase of Entropy in a Thermodynamic System by the Intervention of Intelligent Beings, *Behavioral Science*, 1964, 9: 302–310.

TAYLOR, DONALD W., 1960. Toward an Information Processing Theory of Motivation, in Marshall R. Jones (Ed.), *Nebraska Symposium on Motivation*. Lincoln: University of Nebraska Press.

TAYLOR, JAMES G., 1962, *The Behavioral Basis of Perception*. New Haven, Conn.: Yale University Press.

TAYLOR, RICHARD, 1950a, Comments on a Mechanistic Conception of Purposefulness, *Philosophy of Science*, 17: 310–17.

———, 1950b, Purposeful and Non-Purposeful Behavior: A Rejoinder, *Philosophy of Science*, 17: 327–32.

TINBERGEN, N., 1953, *Social Behavior in Animals*. London: Methuen.

TOCH, HANS H., and ALBERT H. HASTORF, 1955, Homeostasis in Psychology, *Psychiatry*, 18: 81–91.

TODA, M., and Y. TAKADA, 1958, Studies of Information Processing Behavior, *Psychologica*, 1: 265–74.

TOU, JULIUS T., and R. H. WILCOX (Eds.), 1964, *Computer and Information Sciences: Collected Papers on Learning, Adaptation, and Control in Information Systems*. Washington, D.C.: Spartan Books.

TOULMIN, STEPHEN, and JUNE GOODFIELD, 1962, *The Architecture of Matter*. New York: Harper & Row.

TRANÖY, KNUT ERIK, 1959, *Wholes and Structures*. Copenhagen: Munksgaard.

TRINCHER, KARL SIGMUNDOVICH, 1965, *Biology and Information*. New York: Consultants Bureau. Authorized translation by Edwin S. Spiegelthal.

TURBAYNE, C. M., 1962, *The Myth of Metaphor*. New Haven, Conn.: Yale University Press.

TUSTIN, ARNOLD, 1952, Feedback, *Scientific American*, 187: 48–54.

UTTLEY, A. M., 1954, The Classification of Signals in the Nervous System, *Radar Research Establishment Memorandum*, 1047. England: Great Malvern.

VICKERS, GEOFFREY, 1959a, The Concept of Stress in Relation to the Disorganization of Human Behavior, in J. M. Tanner (Ed.), *Stress and Psychiatric Disorder*. Oxford: Blackwell Scientific Publications.

———, 1959b, Is Adaptability Enough? *Behavioral Science*, 4: 219–34.

VOGELAAR, G. A. M., 1962, *Communicatie, Kernproces van de Samenleving: Sociologie en Cybernetiek*. Haarlem: De Erven F. Bohn N. V.

VOGT, EVON Z., 1960, On the Concepts of Structure and Process in Cultural Anthropology, *American Anthropologist*, 62: 18–33.

WALTER, W. GREY, 1953, *The Living Brain*. London: Gerald Duckworth.

WERNER, HEINZ, and BERNARD KAPLAN, 1963, *Symbol Formation*. New York: John Wiley.

WHITEHEAD, ALFRED NORTH, 1929. *Process and Reality*. New York: Macmillan.

WHORF, B. L., 1956, *Language, Thought, and Reality*. Cambridge: Technology Press.

WIENER, NORBERT, 1954, *The Human Use of Human Beings: Cybernetics and Society*. Garden City, N.Y.: Doubleday Anchor.

———, 1961, *Cybernetics*. 2nd ed. Cambridge, Mass.: M.I.T. Press; New York: John Wiley.

———, and J. P. SCHADÉ (Eds.), 1963, *Nerve, Brain, and Memory Models*. New York: Elsevier.

———, ———, 1964, 1965a, *Progress in Bio-cybernetics*, 2 vols. New York: Elsevier.

———, ———, 1965b, *Cybernetics of the Nervous System*. New York: Elsevier.

WILKINS, LESLIE T., 1964, *Social Deviance*. London: Tavistock.

WISDOM, J. O., 1951, The Hypothesis of Cybernetics, *The British Journal for the Philosophy of Science*, 2: 1–24.

WOODGER, J. H., 1952, *Biology and Language*. Cambridge: Cambridge University Press.

YOVITS, M. C., G. T. JACOBI, and G. D. GOLDSTEIN (Eds.), 1962, *Self-Organizing Systems*. Washington, D.C.: Spartan Books.

ZEMAN, J., 1962, Le Sense Philosophique du Terme "L'information," *La Documentation en France*, 3: 19–29.

Index

521